Memoir 5

FLUVIAL SEDIMENTOLOGY

Edited by

Andrew D. Miall
Geological Survey of Canada,
Calgary, Alberta, Canada   T2L 2A7

EXPLORING FOR ENERGY
1928-1978

CANADIAN SOCIETY OF PETROLEUM GEOLOGISTS

Calgary, Alberta, Canada   August, 1978

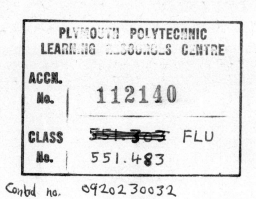
ISBN 0-920230-03-2

McAra Printing Limited
Calgary, Alberta, Canada

FOREWORD

A symposium on Fluvial Sedimentology was held on October 20-22, 1977, at The University of Calgary under the joint sponsorship of the Canadian Society of Petroleum Geologists and the Department of Geography of the university. The original concept for this symposium was proposed by Andrew D. Miall and its success was assured by the immediate and enthusiastic response of many international specialists who were invited to participate. Almost 200 scientists from around the world attended the three-day symposium and an accompanying one-day workshop. Many also participated in a two-day field trip which followed.

Miall chaired the meeting and organized the technical program; D. G. Smith (Department of Geography, The University of Calgary) co-ordinated registration and university services, and N. D. Smith (Geology Department, University of Illinois) organized the workshop discussion on bedforms.

The success of this First International Symposium on Fluvial Sedimentology is reflected not only in suggestions for similar ones in the future, but in the quality of papers contributed to this memoir. Editor Miall assembled these papers, monitored critical reading and revisions of them, and ensured their timeliness by prompt publication of this volume.

The memoir is organized to reflect several major aspects of fluvial sedimentology: examples of modern river systems, ancient fluvial systems, fluvial facies models, paleohydraulics and economic applications of these studies. It incorporates a wealth of information on the most current research relating to fluvial systems and brings together the major modern concepts. It will serve as a standard reference for many years.

The Canadian Society of Petroleum Geologists is pleased to mark its 50th Anniversary Year by the publication of this major scientific endeavour. The Society is most grateful to A. D. Miall and his colleagues for the success of both the symposium and of this memoir.

D. F. Stott
President

PREFACE

Modern fluvial sedimentology had its beginnings during the Second World War, with the work of H. N. Fisk and his colleagues on the deposits of the Mississippi River. Major advances since that time came with the careful observation and classification of cross-stratification structures and the recognition of the fining-upward cycle model in the fifties and, in the early sixties, publication of the first systematic work on the generation of bedforms in flumes. At about the same time, geomorphologists provided us with essential documentation regarding river morphology, flow characteristics and fluvial erosive and depositional behavior. Much of this work had its origins much earlier, in the classic studies of such pioneers as H. C. Sorby, G. K. Gilbert and J. Barrell, but it is only since the early nineteen sixties that the subject of fluvial sedimentology could be said to have come of age.

The last decade has seen (as with every other field of sedimentology) an explosion of research into modern sedimentary processes and ancient deposits. Facies models have become a powerful tool for interpreting the sedimentary record, and geologists are attempting paleohydraulic interpretations with increasing self-confidence. All this is considerably facilitating the task of the regional geologist in studying paleogeography and tectonic history, and that of the economic geologist in the search for coal, hydrocarbons and uranium.

In 1976 it seemed apparent to Norm Smith, Derald Smith and myself that the time had come to pull all this work together and examine the state of the art. Many of our current ideas about fluvial processes were in need of a re-examination. For example, a bewildering profusion of fluvial facies models was starting to appear, many confusions were apparent in the classification of bedforms, and research in a variety of climatic regimes was uncovering a hitherto unsuspected complexity in the sedimentary response of modern rivers to variations in discharge, sediment load and vegetation cover. An additional problem was that rivers and their deposits had been looked at from different standpoints by geomorphologists, sedimentologists and civil engineers, with consequent difficulties in communication of results and ideas. We hoped to resolve some of these problems by convening a specialized conference on the subject of fluvial sedimentology to which specialists representing a variety of disciplines would be invited. This was the first meeting of its type to be attempted.

Planning for the Calgary symposium began in September 1976 when a questionnaire inviting suggestions for papers was mailed to approximately seventy geologists in Europe, North America and elsewhere. The response was immediate. Most of the questionnaires were returned, providing numerous suggestions and enthusiastic comments, and it became clear that a specialized meeting was long overdue. A most stimulating three-day symposium ensued in Calgary in October 1977, attended by 180 delegates. It was preceeded by a one-day "Fluvial Lecture Series" at which selected papers were presented to an audience of 300 from the local petroleum geology community. Plans are currently in the making for a second international symposium to be held in England in 1981, and a fluvial newsletter has been established to keep conference attendees and other interested scientists in touch with each other.

Fifty-three formal papers and nine poster exhibits were presented at the Calgary meeting, including keynote addresses by J. R. L. Allen (Reading) and L. B. Leopold (Berkeley). Twenty-nine of those papers are included in this book, including two of the oral presentations which were each separated into two written contributions, and another ten were chosen from fourteen papers submitted after the symposium, bringing the total to forty-one.

Authors were asked to have their papers appraised by a minimum of two critical readers, preferably including at least one who attended the Calgary meeting. An independent critical reading committee was not employed by the editor except in a few doubtful cases — such a committee would, in any case, have had to consist of essentially the same people chosen by the authors. Most papers required some editorial modification before final acceptance, ranging from shortening or rewriting parts of the text, or redrafting of figures, to correcting inconsistencies in the references. Symposium papers that are not included in the book are represented by their abstract, if available, at the end of this volume.

This collection of papers represents a wide-ranging coverage of topics and areas. The work originates from a total of seven countries, including the United States (16 papers), Canada (10), United Kingdom (7), South Africa (4), Netherlands (2), Norway (1) and Australia (1). The book has been organized into eight sections:

1. Introduction: an historical review article;

2. Texture and structure of fluvial deposits: two papers dealing with sediment transport in fluvial environments;

3. Bedforms and bars: five papers describing dunes, sandwaves and bars, including point bars, transverse ribs in gravel, some comments on terminology, and a comparison between fluvial and tidal-creek point bars;

4. Modern rivers: geomorphology and sedimentation: six papers discussing channel classification, meander theory and depositional behaviour under a variety of climatic and discharge conditions;

5. Ancient fluvial systems: eleven case studies from three continents, ranging in age from Archean to Plio-Pleistocene;

6. Fluvial facies models: nine papers dealing with the development of predictive models in meandering and braided rivers and alluvial fans;

7. Paleohydraulics: an assessment of paleohydraulic methods followed by a paper dealing with mathematical modelling of flow and sediment behaviour and three analytical studies;

8. Economic applications of fluvial sedimentology: two papers describing the practical applications of the subject in the exploration and exploitation of uranium resources.

Much of this material is original; some has appeared elsewhere in different form. By bringing these papers together into a single book it is hoped that we have provided a comprehensive description of the present state of the art of fluvial sedimentology, plus much raw data and a wealth of new ideas to stimulate further study.

Andrew D. Miall
Calgary, June 1978

ACKNOWLEDGMENTS

The Fluvial Sedimentology project has been enthusiastically supported by the Executive of the Canadian Society of Petroleum Geologists from its inception in September 1976 until publication of this book. Successive C.S.P.G. presidents W. G. Ayrton, N. J. McMillan and D. F. Stott are thanked for their constant encouragement and advice. I am most grateful to my employers, the Geological Survey of Canada, for allowing me to devote so much of my time and energy to this project.

The international scope of the Calgary symposium would not have been possible without generous conference grants from the National Research Council of Canada, the Canadian Geological Foundation, and the University of Calgary. These made it possible to subsidize the travel expenses of thirteen of the speakers who would otherwise have been unable to attend. Publication grants for this volume were received from the Department of Energy, Mines and Resources, the National Research Council and the Canadian Geological Foundation, and have helped reduce the selling price of the book.

Invaluable assistance with proofreading was given by D. Armstrong. D. G. Smith and J. S. Bridge are thanked for assistance with critical reading, and I am particularly grateful to my wife, Charlene, who has remained enthusiastic and supportive throughout the two-year gestation of the project.

A. D. Miall

CONTENTS

ANCIENT FLUVIAL SYSTEMS

FLUVIAL FACIES MODELS

PALEOHYDRAULICS

ECONOMIC APPLICATIONS OF FLUVIAL SEDIMENTOLOGY

SYMPOSIUM ABSTRACTS

INTRODUCTION

FLUVIAL SEDIMENTOLOGY: AN HISTORICAL REVIEW

ANDREW D. MIALL[1]

ABSTRACT

Naturalists have been aware of the importance of rivers as depositional agents since the time of Herodotus and Aristotle although, until the mid-nineteenth century, there was a tendency to attribute "diluvial gravels" to the action of "The Deluge". Modern sedimentological concepts began with Lyell in 1830, and owe much to the subsequent work of Sorby, Walther, Gilbert, Barrell and Grabau. In the twentieth century sedimentological studies became highly specialized, with a resultant partial breakdown in communication between specialist groups. Six main themes can be traced: descriptive fluvial geomorphology, quantitative geomorphology (the study of hydraulic relationships), sediment transport and textural studies, bedforms and paleocurrents, facies studies, and paleohydraulics.

Modern concepts of fluvial sedimentary processes can be traced back to the detailed work of Fisk on the Mississippi River during and following the Second World War, the development of the flow regime concept by Simons and Richardson and of facies models by Allen. These and other developments are reviewed, and it is shown how the other papers in this book contribute to the overall development of fluvial sedimentology.

INTRODUCTION

The papers in this volume deal with many of the facets of the subject we have called fluvial sedimentology. They demonstrate the enormous blossoming of knowledge that has taken place during the last thirty-five years in the application of hydraulics to the study of bedforms, the development of the facies model concept, and the integration of purely geomorphological studies with sedimentology through the studies of modern environments being undertaken by geologists. This veritable explosion of research is, of course, not unique to fluvial, or any other branch of clastic sedimentology. It is a function of the increasing interest in sedimentology expressed by those engaged in the search for non-renewable energy resources, particularly (in the case of fluvial sediments) hydrocarbons and, more recently, coal and uranium, and it is also simply a function of the much greater number of people now engaged in sedimentological research which stemmed mainly from the post-war expansion in the universities.

Because of the enormous amount of published work now available to the research worker it was felt that an historical review of the subject of fluvial sedimentology would provide a useful introduction to the collection of papers which follows. This volume is not a texboook, in the sense that the chapters are integrated into a unified whole, and this is an additional reason for attempting to describe some of the themes which have led to the current state of knowledge. Much of this history is very poorly known, and it can generally be said of almost all our ideas about fluvial sediments that they are older than we think they are. Much of this history is also very inaccessible. Middleton (in press) provided a general discussion of the history of sedimentology; a useful review of the history of some aspects of geomorphology which bear on fluvial sedimentology has been given by Schumm (1972a); Jopling (1975) discussed early research on fluvio-glacial deposits; Middleton (1965, 1977) provided brief histories of the work carried out on bedforms and sedimentary structures; Pettijohn (1962) and Potter and Pettijohn (1963, 1977) reviewed the subject of paleocurrent analysis. Allen (1965) described the state of knowledge of alluvial sedimentology as it was in the mid-sixties, but this was not an historically oriented paper, and the main lines of development of the subject have nowhere, to the writer's knowledge, been described in detail.

It is difficult (and probably meaningless) to choose the moment at which modern studies of fluvial sedimentology could be said to have begun. Probably the monumental

---

[1]Geological Survey of Canada, 3303 - 33rd Street N.W., Calgary, Alberta, Canada, T2L 2A7.

1

work of Fisk (1944, 1947) was the most important single advance. However, of fundamental importance was work on the origin of bedforms and sedimentary structures which culminated in a major structure classification by Allen (1963a) and a statement of the flow regime concepts of bedform origin by Simons and Richardson (1961). In addition, facies model concepts, which first had a major impact on geologists following the development of the fining-upward cyclothem model by Allen (1963b, 1964) and by Bernard *et al.* (1962) were also of great importance. The development of these and other ideas is explored in the subsequent pages, and an attempt is made to show how they have led to some of the theoretical advances contained in the papers which form the remainder of this volume.

EARLY DEVELOPMENTS

From Herodotus to Playfair[1]

Rivers and their deposits did not become subjects of specialized study until the nineteenth century. However, many of the early naturalists and philosophers made notes on this topic in the context of wide-ranging discussions on the origins of the earth and its physical features. The principles of uniformitariansim were not widely accepted until the time of Lyell, but long before this it was realized that rivers could transport and deposit large quantities of sediment, so that correct geological interpretations of Recent alluvium were quickly arrived at. Application of the same ideas to consolidated and structurally deformed deposits took a little longer. The following is a brief review of some of the early writings on the physical geology of rivers.

The tendency for rivers to overflow and form thick blankets of floodplain deposits has been viewed with mixed feelings by those living in the valley of the Nile since prehistoric times. One of the first to reach correct conclusions about this process was the Greek philosopher Herodotus (born 484 B.C.) who concluded that "Egypt is the gift of the river". Aristotle (384-322 B.C.) was also aware of the depositional activity of the Nile and, in addition, he noted the rapid silting up of navigable river channels around the shores of the Black Sea and on the Bosphorous. These observations were an improvement on that of Thales of Miletus (636-546 B.C.) who thought the presence of new deltaic deposits projecting into the sea showed that water could change into earth.

Leonardo da Vinci (1452-1519) observed the gravel deposits of the River Arno and wrote in his notebooks: "which deposits are still to be seen welded together and forming one concrete mass of various kinds of stones from different localities". Elsewhere the river deposited mud containing shells, which da Vinci used as evidence to dispute the origin of fossils as products of the Flood.

Agricola, in his great treatise "De re metallica" published in 1556 made some observations on placer gold deposits. He recognized that "gold is not generated in the rivers and streams . . . but is torn away from the veins and stringers and settled in the sands of torrents and water courses . . .". He commented on a prevailing theory of his time concerning the action of the sun in drawing out the metallic material, which led to the idea that the orientation of veins or gold bearing streams was critical in their economic exploitation:

> . . .a river, they say, or a stream, is most productive of fine and coarse grains of gold when it comes from the east and flows to the west, and when it washes against the foot of mountains which are situated in the north, and when it has a level plain toward the south or west. (Hoover and Hoover, 1950, p. 75).

---

[1]Many of the early references quoted in this section are omitted from the reference list for reasons of space. They may be found in Zittel (1901), Geikie (1905), Adams (1938) or Mather and Mason (1964), from which this section is partly drawn.

Agricola devoted many pages to the extraction of gold and tin from placers, and provided us with a picture of an early fluvial sedimentologist at work with his gravels (Fig. 1).

The German physician and geographer Bernhard Varenius (1622-1650) published the first text on physical geography in 1734. He wrote "Of the Changes on the Terraqueous Globe, viz. of Water into Land or Land into Water", and listed a series of "propositions" to explain such changes. Part of Proposition V states, regarding rivers:

> If their water bring down a great deal of Earth, Sand and Gravel out of the high Places, and leave it upon the low, in process of Time these will become as high as the other, from whence the Water flows: or when they leave this Filth in a certain Place on one side of the Channel, it hems in and raises Part of the Channel which becomes dry Land.

Robert Hooke (1635-1703), the English physicist and mathematician, wrote about earthquakes changing the level of strata, and "that water counteracts these effects" by processes of erosion and sedimentation:

A—River. B—Weir. C—Gate. D—Area. E—Meadow. F—Fence. G—Ditch.

Fig. 1. Early interest in fluvial sediments: extraction of placer tin from fluvial gravels. From "De Re Metallica", by Georgius Agricola, 1556 (Hoover and Hoover, 1950).

We have multitudes of instances of the wasting of the tops of Hills, and of the filling or increasing the Plains or lower Grounds, of Rivers continually carrying along with them great quantities of Sand, Mud, or other Substances from higher to lower places . . . Egypt as lying very low and yearly overflow'd, is inlarg'd by the sediments of the Nile, especially towards that part where the Nile falls into the Mediterranean. The Gulph of Venice is almost choked with the Sand of the Po. The Mouth of the Thames is grown very shallow by the continual supply of Sand brought down with the Stream.

Nikolaus Steno (1638-1687) recognized the use of certain fossils as environmental indicators, proposing, for example, that the presence of a terrestrial fauna, rushes, grasses and tree stems indicated deposition in fresh water.

In 1770 the French naturalist Jeanne Etienne Guettard (1715-1786) published a memoir "On the degradation of mountains effected in our time by heavy rains, rivers and the sea" in which he noted the petrographic variation in the detritus of different river basins. He was aware of the way in which a river basin acts as a long-term, but nevertheless temporary, site of sediment storage in its journey to the sea, and observed how the thickness of alluvium increases toward the river mouth. Guettard interpreted the "poudingues" (English: puddingstone, an early term for conglomerate) of the basin as fluviatile in origin, comparing them to the modern gravels carried by the River Seine.

The origin of valleys by the erosive action of streams was first clearly taught from actual examples by Desmarest in a memoir published in 1774, even though others before him, including Targioni-Tozzetti, De Saussure and Guettard, had put forward similar conclusions in earlier publications.

The theory of uniformitarianism is attributed to James Hutton (1726-1797), although many of the ideas that contributed to this theory had appeared before his time. Hutton's first paper on this subject appeared as an extended abstract in 1785 (Bailey, 1967), but his principal work was published ten years later as a four-part book (Hutton, 1795). His main contribution to sedimentary geology was the principle that most of the things we see in ancient rocks can be explained by processes taking place at the present day, and that apparently imperceptible processes can have profound effects if they continue for long enough. Hutton was also, of course, the chief protagonist of the Plutonist theories of earth history, in opposition to Abraham Gottlab Werner (1749-1817) the principal Neptunist. According to Werner, alluvial deposits were part of the youngest of his four layers of water-laid rocks, immediately below the volcanics, and were always to be found in the same stratigraphic position.

Hutton's language is complex, and his uniformitarianist theories were better explained by his friend and colleague Playfair (1802), from whom the following observations on rivers and their deposits are culled: He suggested that the thickness of alluvial gravels may be used to estimate the depth of erosion in the source area, but concluded: "whether data precise enough could be found to give any weight to such a computation, must be left for future enquiry to determine". We still have much the same problem 176 years later.

Fluvial facies and textural studies could be said to have begun with Playfair's following statement:

It is a fact very generally observed, that where the valleys among primitive mountains open into large plains, the gravel of those plains consists of stones, evidently derived from the mountains. The nearer that any spot is to the mountains, the larger are the gravel stones, and the less rounded is their figure; and, as the distance increases, this gravel, which often forms a stratum nearly level, is covered with a thicker bed of earth or vegetable soil . . . The reason of this gradation is evident; the farther the stones have travelled, and the more rubbing they have endured, the smaller they grow, the more regular is the figure they assume, and the greater quantity of that finer detritus which constitutes the soil. The washing of the rains and rivers is here obvious.

These observations state in a more succinct form ideas expressed at length by Hutton (1795, Part II, Chap. IV), although Hutton doubted that a high degree of clast roundness could be attained except in the sea. All these conclusions contrast strongly with those of Finch (1823), Conybeare and Philips (1824), Hayden (1821) and others, all of whom attributed "diluvial gravels" to the Deluge (*see* Schumm, 1972a, p. 16), and ideas about the fluvial origin of valleys and alluvial deposits continued to be resisted for many years, because they conflicted with these Biblical teachings.

From Lyell To Davis

Charles Lyell's "Principles of Geology ", the first volume of which was published in 1830, is undoubtedly amongst the two or three most important books on geology ever published. The intent of his book was to establish uniformitarianism as the guiding principal for geological interpretation following Hutton's teachings, and to discredit catastrophism and diluvialism. The immediate success and wide-ranging influence of the book was due in large part to the wealth of field observation that Lyell was able to bring to bear on his thesis, based on his travels throughout Britain, France and Italy. Von Hoff (1822-1824) had published a treatise earlier on the theme "investigation of the changes that have taken place in the earth's surface conformation since historic times, and the application which can be made of such knowledge in investigating revolutions beyond the domain of history". This work provided copious documentation of recent geological activity around the shores of the Mediterranean Sea and the Black Sea, such as alluvial deposition and changes in sea level, but the work was based mainly on literature research, von Hoff not having the means for travel available to Lyell, and his treatise did not receive the recognition it deserved (Zittel, 1901, p. 188). Lyell acknowledged that much of the information he used in his Principles came from von Hoff, "but he helped me not to my scientific view of causes" (letter quoted *in* Bailey, 1962, p. 78; Wilson, 1972, p. 276).

Amongst his more important pieces of field work Lyell observed Tertiary fluvial deposits interbedded with lacustrine sediments and lava flows in the Auvergne, France, and Tertiary fluvial gravels containing beds of clay with marine fossils near Nice (Wilson, 1972, Chap. 7). These clearly demonstrated to Lyell the long and complex geological histories of these regions, explainable in terms of processes acting at the present day over long periods of time, but quite incompatible with catastrophist hypotheses.

Five chapters in the first volume of the "Principles" (1st edition) are devoted to rivers and deltas. Lyell described the way in which river meanders gradually enlarge themselves and develop neck cut-offs by wearing away at their banks. "When the tortuous flexures of a river are extremely great, the aberration from the direct line of descent is often restored by the river cutting through the isthmus which separates two neighbouring curves" (Fig. 2). Lyell thought that meanders were initiated as deflections caused by the walls of the valley or by variations in the thickness or hardness of the alluvial valley deposits. He quoted earlier work of Flint on the Mississippi River to the effect that opposite each curve "there is always a sand-bar, answering, in the convexity of its form, to the concavity of the bend", and described the development of levees, which slope back and become finer grained away from the channel.

Observations of many floods are quoted as testament to the power of rivers to erode and transport large quantities of coarse debris (Playfair was aware of this 30 years earlier), and historical evidence pertaining to the burial of Roman buildings and the

Fig. 2. A river meander, showing a potential neck cut-off (a), from Lyell (1830).

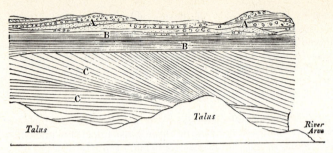

Fig. 3. Cross-stratified sand (C) overlain by laminated sand (B) and interbedded sand and gravel (A), from Lyell (1830).

No. 36.

Fig. 4. Development of an asymmetric ripple (*see* text for discussion), from Lyell (1830).

isolation of ancient ports far inland is used to demonstrate the fact that rivers can build up thick and extensive alluvial tracts in a few hundred years. The velocity gradient of river channels (slowest near the bed) was already known, and some basic data on sediment entrainment had been obtained: "A velocity of three inches per second is ascertained to be sufficient to tear up fine clay, - six inches per second, fine sand, - twelve inches per second, fine gravel, and three feet per second, stones of the size of an egg".

Lyell provided what is probably the earliest illustration of fluvial crossbedding, or "false" bedding (Fig. 3), observed in a cut-bank in the River Rhone at the confluence with the Arve. He explained the origin of the varying crossbedding angle as follows:

> Those layers must have accumulated one on the other by lateral apposition [deposition], probably when one of the rivers was very gradually increasing or diminishing in velocity, so that the point of greatest retardation caused by their conflicting currents shifted slowly, allowing the sediment to be thrown down in successive layers on a sloping bank. The same phenomenon is exhibited in older strata of all ages.

Nowadays we can recognize a reactivation surface in Lyell's illustration.

Volume 2 contains little of relevance to this paper. However, Volume 3 of the "Principles", published in 1833 after Lyell's further travels to Spain, Germany and Switzerland, deals primarily with stratigraphic geology, in particular the Tertiary, for which Lyell proposed a detailed subdivision including the terms Eocene, Miocene and Pliocene. Amongst the nonmarine deposits described are the alluvium, or "Loess" of the Rhine, and the Swiss "molasse". He observed eolian sand dunes evolving near Calais and used his observations in interpreting ripple marks and other crossbedding structures (Fig. 4):

> If a bank [ripple] has a steep side, it may grow by the successive apposition of thin strata thrown down upon its slanting side, and the removal of matter from the top may proceed simultaneously with its lateral extension. The same current may borrow from the top what it gives to the sides . . . Each ridge [ripple] had one side slightly inclined, and the other steep, the lee side being always steep, as bc, de [Fig. 4], the windward side a gentle slope, as ab, cd . . . We think we shall not strain analogy too far if we suppose the same laws to govern the subaqueous and subaerial phenomena . . . We may refer to a drawing given in the first volume [Fig. 3], to show the analogy of the arrangement of the . . . strata just considered, to that exhibited by deposits formed in the channels of rivers.

Lyell was thus aware of the hydraulic significance of "false" stratification. For example, he illustrates what we would call herringbone crossbedding and attributes the opposing dips to "the set of tides and currents in opposite directions". However, the first clear enunciation of the principles of paleocurrent analysis was left to Sorby.

In 1841 and again in 1845 Lyell visited North America; the second visit included some time spent examining the Mississippi River and delta. Both these trips provided him with much new information for revisions of his "Principles" which subsequently ran to eleven editions (the last in 1872).

Lyell's contributions to geology have been assessed at length elsewhere (Bailey, 1962; Wilson, 1972), but it is useful here to repeat Middleton's (in press) comment that although Lyell made many observations that we would regard as being sedimentological in nature, his primary concern was with establishing the working methods of historical geology, and it would not be correct to regard him as a founder of sedimentology.

The next important figure in the study of fluvial sediments is Sorby, whose earliest contribution on this topic constitutes the first paper on paleocurrent analysis (Sorby, 1852; available, along with other papers by this great innovator, in a collection edited by Summerson, 1976). The principles were stated as follows:

> . . . By observing the direction of these ripple marks the line in which the current moved may be known . . . [when ripple drift or false bedding] is observed in progress in modern sand drifts . . . it will be perceived that the line of the dip of the talus [foreset] is not constantly in the true direction of the current on account of the formation of complicated deltoid deposits, in which the line of the dip of their sloping termination varies very considerably on each side of that of the current. Its mean direction, however, coincides with it; and, hence, if a number of properly placed observations be made, their mean gives a result very closely agreeing with the true line of the current.

Sorby first studied paleocurrent patterns in the Carboniferous coal measures near Edinburgh. He classified the sandstones into four types: "level bedded", "ripple laminated" (single ripple train), "ripple drifted" and "drift bedded" (large-scale cross bedding) and referred to them using the first known set of facies symbols. He recognized that structure size is partly dependent on flow velocity.

In a paper published only seven years later, Sorby (1859) was able to boast "I must now have, in my notebooks, not less than twenty thousand recorded observations" of paleocurrent directions, a record that has probably never been equalled. He went on to say: "In various papers . . . I have explained many of my deductions, and I have shown that many peculiarities of physical geography at former epochs may be learned from a knowledge of the directions of the currents in various localities". He lamented that others had not taken up his methods of facies and paleocurrent analysis, but was not to know that such methods would not be widely used again until the nineteen fifties, over ninety years later.

Sorby, in his 1859 paper, also made the first tentative attempts at understanding bedform hydraulics. He stated: ". . . when strata are deposited under the influence of a current, the character of the resulting structure must depend upon the depth of the water, the velocity of the current, the nature of the deposits and the rate of deposition". He thought that it should be possible to deduce the rate of deposition of ripple-drift crossbedding from lamina thickness which "indicates the excess of material deposited on the sheltered side of ripples over that washed up again from the exposed side, during the time required for each ripple to advance a distance equal to its own length", and he carried out hydraulic experiments in a stream passing through the grounds of his house (*see* later section on bedforms and paleocurrents).

Sorby's contributions to the study of sedimentary structures remained unappreciated and largely unknown until exhumed by Pettijohn (1962) and reinterpreted by Allen

(1963b). They are barely mentioned in Higham's (1963) biography, and Sorby has generally been much better known for his work in establishing the use of the microscope in petrographic studies, and in metallurgy.

Several advances in the study of sedimentary structures were made by other naturalists in this period. For example Jamieson (1860) first recognized clast imbrication, while studying fluvio-glacial gravels in Scotland, and Reade (1884) described examples of ripple lamination in British fluvio-glacial deposits. Gilbert (1884, 1899) described large-scale symmetrical ripples from some of the Paleozoic marine sandstones of New York State and stated (1899) "previous to 1882, ideas as to the origin of water-made sand-ripples were crude and unsatisfactory", but quoted several papers that had appeared since that date that had improved theoretical knowledge. He proceeded to supply an explanation of oscillation ripples and current ripples which, for the latter, included Darwin's (1883) discovery of the separation eddy which forms over ripple crests "an eddy or vortex is created in the lee of the prominence, and the return current of this vortex checks travelling particles, causing a growth of the prominence on its downstream side" (Fig. 8). Gilbert (1899) provided a classification of ripple-drift crossbedding (Fig. 5) which has, in effect, only been replaced during the last ten years.

The first investigations into the nature and quantity of sediment carried by rivers were those of Everest (1832), who measured the suspended load of the River Ganges under a variety of flow conditions. Humphreys and Abbott (1861), working for the U.S. Army Corps of Topographic Engineers, carried out detailed studies of the sediment load in the Mississippi River, and their work was widely quoted by later writers (e.g. Dana, Geikie) as testament to the enormous quantities of detritus that can be moved by fluvial transport. Here is a brief extract from their report:

> . . . the sediment of the Mississippi is to the water, by weight, nearly as 1 to 1500, and, by bulk, nearly as 1 to 2900; provided long periods of time be considered . . . 812,500,000,000 pounds of sedimentary matter, constituting one square mile of deposit 241 feet in depth, are yearly transported in a state of suspension into the gulf . . . No exact measurement of the amount of the annual contributions to the gulf from [bedload transport] can be made, but from the yearly rate of progress of the bars into the gulf, it appears to be about 750,000,000 cubic feet, which would cover a square mile about 27 feet deep.

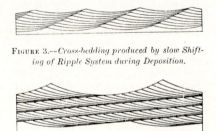

FIGURE 3.--*Cross-bedding produced by slow Shifting of Ripple System during Deposition.*

FIGURE 4.--*Compound Cross-bedding produced by Deposition and partial Erosion associated with shifting Sand-ripples.*

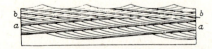

FIGURE 5.--*Complex Cross-bedding and Unconformities associated with Ripples.*

Fig. 5.   Three types of ripple marks, from Gilbert (1899).

Drew (1873) provided the first complete description of alluvial fans and their deposits, based on extensive field observations in the upper Indus Basin. A wealth of topographic details is given, and the process of fan sedimentation by distributaries wandering laterally over their own deposits is well described: "the lateral changes of position of the depositing stream, and the partial growth of each layer, are denoted by false bedding". Drew described fan-head trenching and discussed the concept of the graded stream. From the point of view of this historical review his most interesting contribution is as follows:

> We should now consider the relationship of the fans to the alluvium proper of the main valleys.
>
> If the level of this latter alluvium is remaining, on the whole, stationary (the main river neither deepening its bed nor raising it), and if the fan is undergoing increase, then the fan-stuff will just extend over the alluvium, gradually encroach on it in area; and, stratigraphically, rest upon it. But I think it is a more usual case that the river alluvium has been increasing contemporaneously with the increase of fan-stuff, that the river and the ravine-stream both were raising their beds; then there will have been an interstratification of the two deposits at the fan edges as they were at successive epochs, a lapping for a short distance of one set of alluvial beds over the other.
>
> I have sometimes observed in section an interstratification such as to suggest the above origin - beds of well-rounded materials and of sand among the less-worn fan-stuff; for indeed the latter, being nearer its source, is seldom thoroughly rounded.

Gilbert (1880) published what rapidly became a classic of geomorphological literatue in his description of the Henry Mountains, Utah. He dealt with fluvial processes extensively, including the growth of alluvial fans and the development of meanders and cut-offs and provided a detailed discussion of the balance between discharge, slope and load in a river, showing how they tended towards an equilibrium, or graded, condition. However, from the point of view of fluvial sedimentology as distinct from geomorphology little of what he said was an advance on Drew's (1873) work.

Summaries of the state of the art at this time are provided in two classic text books by Dana (1862) and Geikie (1882). Much is written in each about the erosive power of rivers, and detailed data are provided regarding the slope, discharge and sediment load of selected rivers around the world; the work of Humphreys and Abbot on the Mississippi is quoted extensively. More precise information than before regarding sediment entrainment is given and the meandering and avulsive habits of river channels are described. However, even as late as the third editions of these books, published in 1880 and 1893, respectively, little attention is paid to fluvial deposits themselves, and no mention at all is made of Sorby's paleocurrent and paleogeographic analyses. In dealing with ancient deposits the author's preoccupations are mainly with stratigraphic documentation. However, some glimmerings of fluvial sedimentology do creep in. Dana wrote "in the alluvial deposits from a flood the little layer is begun with a relatively rapid rate of deposition and finished with a slower, as the flood declines, and hence its upper and lower portions will differ as to coarseness and density". Here we have the first recognition of the cyclic nature of flood deposits. Dana also provided a good description of the shape and development of transverse sand bars, based on the work of General G. K. Warren, another of the U.S. Army topographers, on the Mississippi River. Geikie wrote:

> the deposit of alluvium on river beds is characteristically shown by the accumulation of sand or shingle at the concave side of each sharp bend of a river-course. While the main upper current is making a more rapid sweep around the opposite bank, undercurrents pass across to the inner side of the curve and drop their freight of loose detritus which, when laid bare in dry weather, forms the familiar sand-bank or shingle-beach.

This is an eminently clear description of the helical flow patterns that occur on river meanders, and of the point bar deposits that result. It remains only to provide the idea of decreasing competency and sediment sorting up the point bar slope, to arrive at the fining-upward model. But this was not to come for many years yet.

In spite of the general acceptance of the principle of uniformitarianism, these excellent observations on modern fluvial environments were not at this time being applied to interpretations of ancient rock sequences. Copious descriptions of the lithology, fauna and flora of such nonmarine units as the Old Red Sandstone, were given by Dana and Geikie, but environmental and paleogeographic interpretations were limited to general statements concerning the deposition of such rocks in lakes or "inland seas". Their general nonmarine nature was recognized, but further interpretations did not proceed beyond such statements as this, from Dana, concerning the Triassic of the Appalachians:

> The mud-cracks, raindrop-impressions and footprints - these show, wherever they occur, that the layer was for the time a half-emerged mud-flat or sand-flat; and, as they extend through much of the rock, there is evidence that the layers in general were not formed in deep water . . . The occurrence . . . of coarse conglomerate, some of the stones of which are very large and of a coarse kind of oblique lamination [crossbedding?] in much of the rock, is evidence that some of the beds were deposited by a flood of waters pouring violently down this valley.

Dana also thought that floating ice was "concerned in part of the deposition".

A detailed report on these same rocks by the geographer and geomorphologist W. M. Davis (1898a) included a detailed discussion of the various possible environments in which they could have formed, including tidal, lacustrine, fluvial, glacial or eolian settings. It is a fine example of sedimentological reasoning, in which modern analogues are clearly appealed to. For example:

> The Pre-Triassic peneplain might have been warped so as to alter the action of the quiescent old rivers that had before flowed across it . . . Such a change would set the streams to eroding in their steepened courses, and to depositing where their load increased above their ability of transportation . . . The heavy accumulation of river-borne waste on the broad plains of California, of the Po, or of the Indo-Gangetic depression all agree in testifying that rivers may form extensive stratified deposits, and that the deposits may be fine as well as coarse. They are characteristically crossbedded and variable, and they may frequently contain rain-pitted or sun-cracked layers.

However, Davis was a cautious individual. He concluded:

> Penck has suggested the name 'continental' for deposits formed on land areas, whether in lakes, by rivers, by winds, under the creeping action of waste slopes, or under all these conditions combined. This term seems more applicable than any other to the Triassic deposits of Connecticut. It withdraws them from necessary association with a marine origin, for which there is no sufficient evidence, and at the same time it avoids what is today an impossible task - that of assigning a particular origin to one or another member of the formation.

The last sentence is revealing, for it shows that application of the growing body of knowledge about modern environments to sedimentological problems was still very much in its infancy. He was somewhat more self-confident in his fluvial interpretations of the Tertiary strata of the Rocky Mountains, published only two years later (Davis, 1900).

At about this time Davis (1898b, 1899) also made a major contribution to fluvial geomorphology with his classification of rivers according to their stage of development in the erosion cycle as youthful, mature and old, and he recognized and described rivers of braided and meandering character.

A rather more advanced level of interpretation of fluvial sediments was given earlier by Medlicott and Blanford (1879), but their observations were not widely applied for some years. They described the Siwalik Formation of the Himalayas, compared it with the "Mollasse" of Switzerland and with modern alluvium, as follows:

> In the upper Siwaliks conglomerates prevail largely; they are often made up of coarsest shingle, precisely like that in the beds of the great Himalayan torrents. Brown clays occur often with the conglomerate, and sometimes almost entirely replace it. This clay, even when tilted to the vertical is indistinguishable in hand specimens from that of the recent

plains deposit; and no doubt it was formed in a similar manner, as alluvium. The sandstone, too, of this zone is exactly like the sand forming the banks of the great rivers, but in a more or less consolidated condition.

This appeal to a specific modern analogue is a good example of an early application of actualistic principles to a sedimentological problem.

GROWTH OF PRESENT DAY CONCEPTS

Increasing specialization of the twentieth century

Until the end of the nineteenth century it was possible to be both a general geologist and a geomorphologist. Middleton (in press) regards Walther (1893-1894) as the first "to draw together the many scattered observations on modern sediments and to document the actualistic method as the basis for a specific science of sedimentary rocks". At about this time, several separate themes gradually began to appear in the study of rivers, fluvial processes and their deposits. Each theme became a specialized subject, with its own body of workers, its own purposes and applications and, eventually, its own literature. To bridge these divergences has become an increasingly difficult objective for modern research workers. Likewise, it is a complex task to trace the various threads that lead to our present understanding of fluvial deposits. The following are the most important areas of specialized study that have been used to build the present day edifice of fluvial sedimentology:

1) Descriptive fluvial geomorphology: the physical characteristics of rivers, alluvial fans, flood plains, their evolution and their variations from source to mouth.

2) Quantitative fluvial geomorphology: the interrelationships between width, depth, slope, discharge, sinuosity and sediment load (hydraulic geometry).

3) Sediment transport studies: investigations of sediment entrainment, transport rates, and downstream textural changes.

4) Bedforms: their external form and internal structure, and their relationship to depth, velocity and sediment grain size. Orientation studies, in the form of paleocurrent analysis.

5) Facies studies: recognition of the various discrete lithologic components of a fluvial deposit, with their associated sedimentary structures, and investigation of the vertical (cyclic) and lateral relationships between them. Generalizations in the form of facies models.

6) Paleohydraulics: attempts to construct quantitative descriptions of vanished rivers from evidence contained in their deposits. This area of study is the most recent of the six, and draws on all the other facets of the science of fluviology listed above.

In the succeeding paragraphs an attempt will be made to trace the most significant developments in each of these branches of study down to the present day, and to show how the papers in this volume contribute to the overall body of knowledge.

Descriptive fluvial geomorphology

Rivers and drainage networks have been classified in four principal ways by geologists and geomorphologists:

1) The cycle-of-erosion classification (youth, maturity, old age) of Davis (1898b, 1899).

2) The structural control classification (consequent, subsequent, obsequent, antecedent, superimposed, etc.) developed by Powell (1875) and Davis (1898b, 1899; *see also* Johnston, 1932).

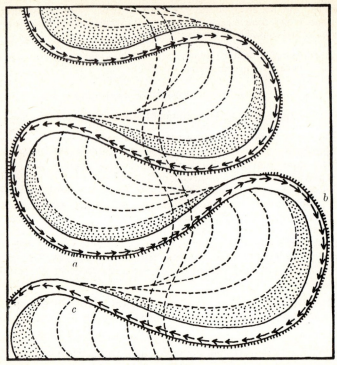

Fig. 6. Development of meanders in a high sinuosity river, showing location of cut-bank erosion (ticks on outside of curves) and of point bars (stipple); based on knowledge of the Mississippi River. From Chamberlin and Salisbury (1909).

  3)The morphological classification (braided, meandering, anastomosing, straight) which emerged slowly, mainly during this century.

  4)A process-response classification that relates channel morphology to discharge and sediment type.

  The first two classifications were the earliest, reflecting the preoccupation of nineteenth-century geologists with the cycle of erosion, landscape evolution and the power of rivers to remove and transport large quantities of debris (Dana, 1850a, b: Davis, 1899).

  The morphological study and classification of rivers does not appear to have started with the work of any single individual. The terms 'braided', 'anastomosing' and 'meandering', though poorly defined, were all in use in the nineteenth century (e.g. *see* Davis, 1898b, Chap. IX). Knowledge of the geomorphology of alluvial fans and meandering rivers was well advanced by the turn of the century, as quotes from Drew (1873), Geikie (1882) and others, earlier in this review, have indicated. The work of Gilbert (1880) on this subject has also been much quoted. The concept of the graded stream was beginning to be understood, and Powell (1875) had introduced the concept of "base level" in downward erosion.

  The Mississippi River has always loomed large in the minds of alluvial sedimentologists and geomorphologists (*see* discussion of this phenomenon at the end of the paper). Nowhere is this more obvious, in contemporary literature, than in the book by Chamberlin and Salisbury (1909), in which are provided detailed maps of the meanders, cut-offs, and bayous copied from charts of the U.S. Geological Survey and the

Mississippi River Commission. Figures 6 and 7, taken from this book, show that the surface processes of meander migration in a large, high sinuosity stream, were well understood by this time.

Davis (1898b) probably was the first to recognize clearly that there is another distinctive type of channel pattern which he called braided. He illustrated this type using the Platte River. Chamberlin and Salisbury (1909) use the same river as an example of what they term an anastomosing stream (the term had earlier been used by Jackson, 1834 and Peale, 1879). This early recognition of the distinctiveness of the Platte, and its more recent prominence in sedimentological studies as a result of Smith's (1970, 1971) work, undoubtedly qualifies the Platte as the type example of a braided river. Davis (1898b, p. 243) stated:

> If a stream has a large load of coarse rock waste, its graded flood plain must be relatively steep (a descent of from 5 to 20 ft or more a mile). In this case the stream does not turn far

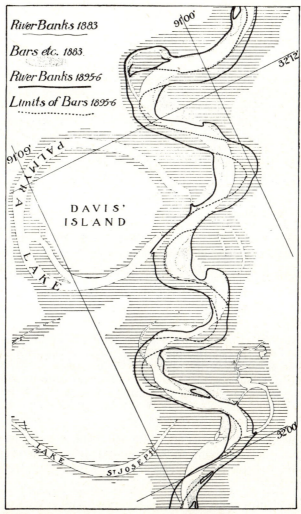

Fig. 7. "Meanders and cut-offs (ox-bow lakes) in the Mississippi Valley a little below Vicksburg. The figure also shows the migration of meanders downstream and their tendency to increase themselves". From Chamberlin and Salisbury (1909).

aside from a direct course along the flood plain; but it is constantly embarassed by the formation of bars and islands of gravel and sand, splitting its current into many shifting channels.

Chamberlin and Salisbury (1909) stated:

> In general, anything which greatly increases the load of a stream near its head is likely to cause deposition, and so the development of a flood plain, at some point farther down the valley. Streams which are actively aggrading their valleys are likely to anastomose. This results from the filling of the channels until they are too small to accommodate all the water. The latter then break out of the channel at few or many points. The new channels thus established suffer the same fate.

Nowadays it is recommended that the term anastomosing be restricted to rivers characterized by a network of stable, highly sinuous channels, and that the term braided should be used for the rivers described by Davis (1898b) and Chamberlin and Salisbury (1909), which show unstable, low sinuosity channel patterns, a wide, shallow cross-section and a coarse bedload (Miall, 1977a, b).

The fourth approach to river classification is the one of most interest to fluvial sedimentologists. Its development overlaps with that of the morphological classification. The link between channel cross-section and sediment type was first suggested by Griffith (1927), who observed that natural streams with heavy bedload tend to flow in broad, shallow channels, and concluded that a river must tend to adopt the cross-section shape which gives it the maximum sediment carrying capacity. Actual classifications began with the work of Melton (1936). This was an important geomorphological contribution, in which aerial photographs were first used extensively to illustrate fluvial landforms, but it was largely ignored by geologists (with the exception of Fisk). The reason may be that, although the paper contains numerous useful observations regarding channels, bars and floodplains, the classification proposed by Melton is complex and uses terminology that is difficult to interpret. The word 'plain' is used to refer to the bar-covered channel (as in braid-plain, meander-plain) and in a different context, its use implies the entire width of the alluvial valley. Melton's primary subdivision is into "single-crest" and "double-crest" streams. Double-crest streams are those which "possess two phases of geological activity. For example, many streams build meanders during high water and also deposit an alluvial cover on the flood-plain surface during extreme stages". Single-crest streams are those which lack this second depositional stage. Melton distinguished between vertical accretion and lateral accretion deposits; he discussed the difference between braided and meandering rivers and described some of the characteristics of intermediate morphologic types; but all this information was assigned a secondary importance in his classification scheme.

An important advance came with the extensive series of flume experiments carried out by Friedkin (1945), who investigated the effects of altering discharge, bed materials and slope on meander patterns and the stability of bars. Friedkin also produced braided channel patterns by increasing the load and slope, and concluded that an important contributary factor in the development of braided patterns was easily erodible banks. This work was much quoted by Fisk (1947) in his description of the Mississippi alluvial deposits.

Leopold and Wolman (1957) used both field and flume experiments to study bar and channel behaviour. They demonstrated that channel patterns form a continuum resulting from variations in discharge, load and slope, and showed how channel patterns change downstream in response to variations in these controls.

Schumm (1963) divided rivers into three types, bedload-, suspended load- and mixed load-streams, and showed how load characteristics are related to sinuosity and cross-section shape, and to the tendencies toward vertical or lateral erosion and sedimentation within the channel.

Galay *et al.* (1973) and Mollard (1973) carried these studies further, showing in particular the wide range of variability that can exist in channel plan and cross-section shapes and the sediment characteristics associated with them. Research is needed to relate these features in more detail to the sedimentary record, there being a particular need for quantitative studies. Rust (this volume) has developed a new braiding index and proposes coupling this with a measure of sinuosity in an inclusive channel classification. Baker (this volume) shows that many of our conceptions concerning fluvial morphology are appropriate only under certain climatic conditions, particularly temperate climates, and that rivers can show quite different patterns of sinuosity, channel shape and sediment type under tropical conditions and where rivers flow from one climatic zone to another. The implications of this work for studies of ancient fluvial sediments have yet to be explored.

Quantitative fluvial geomorphology

An understanding of the quantitative aspects of fluvial geomorphology developed simultaneously with the work carried out on descriptive fluvial morphology, that was discussed in the previoue section. The reason for separating these two aspects of the work is that they have received very different degrees of attention by clastic sedimentologists. Those studying ancient fluvial sediments have generally made some attempts (with varying degrees of sophistication) to relate the features of their deposits to the morphology and behaviour of modern rivers, but quantifications of these relationships have been attempted only recently.

In this section the development of some of the principal ideas concerning quantitative aspects of river behaviour will be described briefly, emphasizing those aspects most used by geologists. Application of this work will be considered in a later section entitled Paleohydraulics.

Early attempts to relate velocity and discharge to channel morphology were carried out by engineers for the purpose of designing irrigation canals in India, Egypt and arid parts of the United States. Kennedy (1895) published the first important work in this field, relating velocity to flow depth and width in a series of empirical graphs derived from work in India. Contributions by many later workers were summarized by Lane (1935) who showed that empirical equations were not necessarily of universal applicability, depending on a variety of factors that earlier workers, working within the confines of a limited geographic area, had not previously recognized. Lane listed the following factors that may enter into a determination of channel shapes:

> (a) Hydraulic factors (slope, roughness, hydraulic radius or depth, mean velocity, velocity distribution, and temperature); (b) channel shape (width, depth, and side slopes); (c) nature of material transported (size, shape, specific gravity, dispersion, quantity, and bank and subgrade material); and, (d) miscellaneous (alignment, uniformity of flow, and aging).

"A graded stream responds to a change in conditions in accordance with Le Chatelier's general law: - 'if a stress is brought to bear on a system in equilibrium, a reaction occurs, displacing the equilibrium in a direction that tends to absorb the effect of the stress' ". This is the primary conclusion of Mackin (1948) in an important geomorphological paper on the concept of the graded stream - the first to deal with this subject in depth since Gilbert's (1880) pioneer work. Mackin discussed the effects on a river of changes in discharge, load, base level and other parameters. His discussion is qualitative, but in setting out the logical cause-and-effect nature of river behaviour he provided an essential rationale for the detailed empirical work which was in progress at that time on rivers in the United States.

Mackin provided a warning to those engaged in the study of ancient fluvial sediments to the effect:

> (1) that deposits formed by or associated with aggrading streams differ markedly from the loads carried by them, (2) that distinguishing between channel and overbank deposits is the

first essential step in interpreting modern valley fills or ancient fluviatile sediments, and (3) that even after this distinction is made it is virtually impossible to work directly from the grade sizes represented in the channel deposits to the characteristics of the depositing streams because there is no simple relationship between the deposits of an aggrading stream and such partly independant factors as slope, discharge, channel characteristics, velocity and load. We cannot proceed directly from laboratory-determined laws relating to stream transportation processes to interpretation of ancient stream deposits. An alternative and promising route of attack on the problem is via study of deposits now being formed by natural streams of many types to determine whether the sum total of all the characteristics of given deposits is uniquely related to the particular modern streams by which they are being formed, and to proceed thence to an understanding of ancient streams by comparison of their deposits with deposits of modern streams of known characteristics.

In this statement Mackin succinctly outlined the problems we are still tackling in the area of paleohydraulics, and foreshadowed what was to become the most powerful tool in the hands of sedimentologists: the study of facies models in ancient deposits combined with the examination of modern environments.

Lane's work was continued by Leopold and Maddock (1953), who drew on "the mass of data on streamflow collected over a period of seventy years representing rivers all over the United States". This included concurrent measurements of mean velocity, width, shape and area of cross-section, discharge and suspended sediment concentration. They provided a series of graphs showing the relationships between these parameters for specific rivers, and arrived at some general relationships that theoretically could be applied to a variety of settings by adjusting the values of various exponents and coefficients. This work was continued by the same group of workers for a number of years (Leopold and Wolman, 1957; Leopold *et al.*, 1964) and additional data on the suspended-sediment load, the relationship of meander wavelength to channel width and radius of curvature, flood magnitude and recurrence interval and the relationship between discharge and drainage area were published. A later paper (Langbein and Leopold, 1966) showed that:

> . . . meanders are the result of erosion-deposition processes tending toward the most stable form in which the variability of certain essential properties is minimized. This minimization involves the adjustment of the planimetric geometry and the hydraulic factors of depth, velocity, and local slope. The planimetric geometry of a meander is that of a random walk whose most frequent form minimizes the sum of the squares of the changes in direction in each successive unit length. The direction angles are then sine functions of channel distance.

This theoretical conclusion is of great utility in attempts to model river meanders and sedimentation processes, and has also been used in attempts to reconstruct the sinuosity of a paleostream from the record of its preserved bedforms (*see* later section). The history of meander theory is considered in more detail in a paper by Allen (this volume).

From the geological viewpoint quantitative geomorphology has been brought to a relatively advanced state by the work of Schumm, in a variety of papers culminating in a recent review article (Schumm, 1972b). Schumm provided a family of empirical equations relating various hydraulic parameters to features that could be studied and measured in ancient fluvial deposits. Leeder (1973) provided a useful supplement to this, dealing with channel dimensions, and Ethridge and Schumm (this volume) provide a review of the latest developments in the methodology. There remains the problem, mentioned earlier, regarding the applicability of these relationships to rivers flowing under different climatic regimes. Even in that classic river, the Mississippi, the concept of the graded river and all its quantitative ramifications were not fully appreciated for many years, as shown by the history of the many man-made neck cut-offs made in the river since 1931, the success of which appears to have ranged from the dubious to the disastrous (Winkley, 1977).

Sediment transport and textural studies

The physics of particle movement is of importance to fluvial sedimentologists for a variety of reasons: (1) for an understanding of the gross variation in grain-size and other textural parameters in fluvial deposits (2) as a basis to the understanding of bedform generation and flow regime concepts and (3) in providing an explanation for density sorting of clastic grains, of importance in petrological studies, particularly those dealing with economic placer deposits.

Early studies were concerned with empirical measures of the current velocities required to move material of different sizes and with the total volume of sediment transported under various flow conditions. Grabau (1913a) summarized earlier work on this topic, which included the use of hazelnuts, walnuts, peas, pigeons' eggs and hens' eggs as comparative grain size standards. However, this relaxed approach to quantification was quickly lost by later workers. Grabau (1913a) also summarized early work dealing with particle sorting and rounding, it being understood that the latter depended on the size, specific gravity and hardness of the grains, their distance of transport and the agent by which they had been transported. Sorby (1908) in a classic paper on quantification in geological studies, reported on his investigations into particle settling velocities and angle of respose. He was the first to attempt to use experimental laboratory studies in investigations of geological problems (this paper is discussed again in the section on bedforms), but he appears to have been ignorant of the many advances made by this time in fluvial hydraulics.

An important, widely quoted paper by Gilbert (1914) reported an extensive series of flume investigations into the transport of sediment by streams, the stimulus for which arose from "an investigation of problems occasioned by the overloading of certain California rivers with waste from hydraulic mines". Gilbert was concerned mainly with investigating stream capacity, or the quantity of debris transported. He examined modes of transport and concluded:

> Some particles of the bed load slide; many roll; the multitude make short skips or leaps, the process being called saltation. Saltation grades into suspension.

Gilbert varied the grade and quantity of debris, width and discharge, and measured the "slope to which the stream automatically adjusted its bed so as to enable the current to transport the load". He also investigated the velocity required to move particles of different sizes and concluded that competence varied at between the 3.2 and 4.0 power of mean velocity, depending on which experimental conditions, discharge, slope and depth were varied. Gilbert's (1914) paper was an important bridge between the disciplines in that he brought many of the techniques of fluid hydraulics to bear on a subject of considerable geological significance.

Several theoretical developments followed during the nineteen twenties, including the boundary layer theory, developed mainly by aeronautical engineers, recognition of the importance of bed roughness, and the suggestion by Jeffreys (1929) that particle movement depends not upon velocity or drag force but lift induced by the velocity gradient or the rate of shear between adjacent fluid laminae. Turbulent flow in open channels, pipes and in air was investigated, and the change, from laminar to turbulent flow with increased velocity was documented with the use of the Reynolds number (Reynold's classic work in this field dates back to 1883). In 1936 Shields provided the first satisfactory theory regarding the initiation of bed load movement on a flat bed, in which the bottom shear stress or shear velocity was shown to be the critical parameter. Hjulstrom (1935) used average velocity to predict the critical conditions for the beginning of sediment movement. This approach has been used recently in paleohydraulic reconstructions, but is less accurate. Rubey (1938) re-examined Gilbert's (1914) flume data in the light of these theoretical developments, although he does not seem to have been aware of Shield's (1936) contributions. Rubey was able to explain reasonably well the movement of material of medium sand to pebble grade, but stated that "a satisfactory theory of the force required to start movement of fine sand and silt must await additional observations".

Theoretical developments on the calculation of total load are incomplete. Rouse (1939), and later Vanoni (1946), used the concept of momentum diffusion to predict concentrations of suspended sediment. The earliest successful work on bed load transportation was that of du Boys (1879), who deduced that the rate of movement of bed load is proportional to the product of the shear stress and the difference between the shear stress and the critical shear stress for the initiation of movement. This physical model has been widely used up to the present day. However, it fails to take into account bed roughness (grain roughness and form roughness - that induced by the presence of bedforms), a problem first tackled by Einstein (1950).

Developments in fluvial hydraulics up to the mid-nineteen fifties were summarized by Leliavsky (1955). Geologists made little use of this field until Sundborg's (1956) major paper on the River Klarälvan was published. In this publication Sundborg applied the available theory regarding flow dynamics, erosion and entrainment of sediment to a study of the bed load and suspended load of a specific river. The geomorphology of the river, and the texture and structure of its deposits were also described in great deal. The paper thus represented a major bridge between the disciplines of hydraulics, geomorphology and geology, and it has been widely quoted.

This brief summary has attempted to cover most of the more fundamental developments in the theory of sediment transport. A more complete discussion is beyond the scope of this paper, and the reader is referred to Brush and to Briggs and Middleton (in Middleton, 1965) for further historical details. Recent reviews of the topic have been given by Church and Gilbert (1975), Middleton and Southard (1977) and Shen (this volume).

Subaerial debris flows contribute much coarse, poorly sorted sediment to alluvial fan surfaces, as first documented in detail by Blackwelder (1928). He referred to these catastrophic events as mudflows, but data compiled by Bull (1964) and Lustig (1965) showed that their actual mud content may be less than 10%. Observations on modern floods by Chawner (1935) and Sharp and Nobles (1953) demonstrated the transporting power of debris flows and the conditions under which such events occur, notably infrequent but torrential rainfall on unvegetated upland regions where sufficient time has been allowed for the generation of a mass of loose debris. Recent studies of the mechanics of debris flows have been published by Johnson and Hampton (1969), Statham (1976) and Rodine and Johnson (1976). Other special processes operating on alluvial fans, including the tendency for flows to loose competence by infiltration leaving "sieve deposits", were described by Hooke (1967).

Textural studies, that is, investigations of grain size parameters (such as mean, sorting, skewness and kurtosis) and other grain descriptions including roundness, shape and sphericity, have long fascinated clastic sedimentologists. Since the work of Udden (1914), the hope has been (and still is) that simple laboratory measurements on a suite of samples (better still, a single sample) will reveal their depositional environment, and the literature on grain size analysis is now vast. Modern work on the subject began with the establishment of the grain size scale for pebble- to clay-sized material erected by Wentworth (1922; based on Udden, 1914) and the invention of the logarithmic phi scale for grain-size description by Krumbein (1934). Attempts to use sorting parameters as indications of mode of deposition began with the work of Inman (1949). Inman (1952) and Folk and Ward (1957) proposed methods for describing the sorting of samples of clastic rocks using percentile values of various grain-size intervals derived from graphs of the cumulative grain size distribution. These methods were reviewed by Folk (1966). Many attempts have been made to define the sorting characteristics of sand deposits formed in various environments using graphical (Passega, 1957; Friedman, 1961, 1967; Spencer,

1963; Visher, 1965a, 1969) and statistical (Sahu, 1964; Klovan, 1966) methods. None of these methods appears to be completely satisfactory for discriminating the various depositional environments, although the method of factor analysis (Klovan, 1966) and the system of graphical analysis proposed by Visher (1965a, 1969) appear to be the most consistent. Most of this work has been strictly empirical in nature.

Attempts to relate grain-size distribution to sediment transport mechanics have recently been made by Middleton (1976) and Sagoe and Visher (1977). Middleton (1976) showed that the size break between the traction and intermittent suspension subpopulations in a river bed depends on the dominant shear velocity of the flow - a fact of some potential paleohydraulic significance.

Recent work includes that of Steidtmann (this volume), who carried out experimental work into the rate of transport of grains of different size and density in which rates were shown to depend on the grain- and form-roughness of the bed. Schumm (this volume) investigated the behaviour of heavy mineral grains on alluvial fans subjected to fan-head trenching and reworking. Davies *et al.* (this volume) measured the sorting and rate of transport of fresh pyroclastic debris in an active volcanic region subject to violent flash floods. Shaw and Kellerhals (this volume) measured the downstream grain size changes in some Albertan rivers and separated the effects of grain size reduction into those caused by abrasion during transport, abrasion in place, and differential transport.

Methods for the study of particle roundness, sphericity and shape (a more comprehensive description than sphericity) have been provided by Krumbein (1941) and Sneed and Folk (1958). More recent developments have been summarized by Pettijohn *et al.* (1972). Roundness increases downstream due to abrasion; mean and maximum grain size decrease in the same direction. These parameters are therefore useful indicators of paleoslope and have been used in many studies of fluvial sediments too numerous to mention (*see* Pettijohn, 1962, for an earlier review). However, their use demands much field measurement and data processing, and as paleoslope indicators they are less flexible tools than paleocurrent analysis. An additional problem is that of inherited textural characteristics - sand which has been recycled from earlier deposits may retain size and shape characteristics which obscure those imprinted by the later environment. Other complications are described in the four papers on textural topics included in this volume (referred to above).

Bedforms and paleocurrents

It was realized early in sedimentological studies that the morphology of individual hydrodynamic sedimentary structures, or their assemblage and relative arrangement, might yield clues as to the depositional environment under which they were formed, and that their scale was somehow related to the energy level during deposition. Some of the conclusions of Lyell, Sorby and others along these lines have been quoted above. Other early work included that of Spurr (1894a, b), who discussed the strength and constancy (directional consistency) of currents required to produce ripple marks, and Darwin (1883) who demonstrated the existence of separation eddies in the lee of ripple crests (Fig. 8) using a rotating flume device. Cornish (1899) observed the formation of antidunes in natural systems and Hunt (1882) studied the growth of ripples on tidal flats.

Fig. 8. Flow lines over an asymmetric ripple, showing separation eddies. From Darwin (1883).

Hobbs (1906) observed crossbedding with a consistent dip orientation and used this as part of his evidence for interpreting the enclosing strata as "torrential" in origin (*see* under Fluvial facies models).

Sorby carried out experimental work on the "effects of current on sand". He explained (in his major 1908 paper, published shortly after his death):

> Fifty-nine years ago, when I was living at Woodbourne, a country-house on the east side of Sheffield, there was at the bottom of the small park a brook entirely under my control. In order to investigate a number of questions, I constructed a place for experiment with some self-registering appliances. I could easily regulate and measure the depth and velocity of the current within certain limits.

The work carried out here and elsewhere, plus careful observations of sedimentary structures in ancient rocks, led Sorby to divide sandstones into four types:

> 1. Thinly or thickly bedded rock, without ripples or drift bedding, and showing little or no graining of the surface in the line of the current . . . This could be explained by supposing that the water was at considerable depth, and the material mainly deposited from above, not drifted along the bottom where the velocity of the current was much less than 6 inches per second.

> 2. Thinly bedded rock, with well marked graining in the surface in the line of the current, indicating a mean velocity up to about 6 inches per second, but showing few or no ripple marks.

> 3. More or less thick masses of rock almost entirely made up of ripple drift. This must have been when the velocity of the current was something like a foot per second . . .

> 4. What I have called drift bedding in numerous published papers . . . The velocity of the current is indicated by the nature of the sand; and probably further experiments would enable us to learn the approximate depth, which probably was small, since an increase of a very few feet made so great a difference in the strength of the current.

As pointed out by Allen (1963b), sedimentary structures in types 2 to 4 would now be termed primary current lineation, small-scale cross-stratification and large-scale cross-stratification. The order in which Sorby described the sandstone types was one of increasing flow velocity - which was the beginning of the flow regime idea, although, as Allen (1963b) stated, Sorby was mistaken in that type 2 is now known to be a higher energy structure than any of the other three. The sequence dune, smooth, antidune, with increased flow velocity was documented by Owens (1908), although the term antidune was first coined by Gilbert (1914). The term dune was in use by Swiss workers before the turn of the century, and had been proposed for water-formed bedforms on the basis of their similarity to eolian dunes. The term was introduced to American workers by Gilbert (1914) in preference to the term sand wave, also in current use (e.g. Hider, 1882), and this started a nomenclature confusion that is still in the process of being resolved (Smith, this volume).

Gilbert's (1914) detailed flume experiments confirmed the results of Sorby's pioneer work and also reproduced the bedform sequence described by Owens (1908). The mechanism of grain and water movement was observed in detail. It was recognized that dune spacing depends on depth and velocity (Gilbert quotes earlier workers who had made similar observations) and the significance of grain size was realized (although not explored in detail). Similar experimental work was carried out by Hahmann (1912) but has not been widely quoted by English speaking workers.

Several detailed descriptive papers on crossbedding structures appeared at about this time, most of which attempted to develop criteria by which such structures could be used as environmental indicators (Grabau, 1907; King, 1916; Kindle, 1917; Bucher, 1919). A few papers used orientation data in paleogeographic reconstructions but, it appears, no systematic measurements were made in alluvial deposits (*see* Pettijohn, 1962). Epry (1913) stated of ripple marks that "no one seems until now, to have definitely ascertained their cause" and "ripple marks are the work of the tide and of that alone" but he obviously had not been keeping up with the literature. Bucher provided data concerning wavelength, amplitude, "horizontal form index" ("the degree of asymmetry may be expressed by the ratio of the horizontal length of the lee-side to that of the stoss-side") and "vertical form index" (wavelength/amplitude). He proposed the terms rhomboid and linguoid for the principal three-dimensional ripple forms, and illustrated their morphology and the flow lines passing over them (Fig. 9). Oscillation and current ripples were clearly distinguished on the basis of morphology, origin and environmental significance.

Bucher emphasized the usefulness of ripples as current direction indicators:

> While in marine sediments the trend of ripples seems of little value to the paleogeographer, its study may yield important results in fluviatile deposits, the true nature and wide distribution of which among the sediments of the past we are just beginning to realize . . . In the deposits of a river obviously a great majority of all ripples should be found facing approximately in the direction of flow of the river. Locally, of course, they can be found facing even an opposite direction, as along parts of meanders or under the influence of local eddies. Such cases can not, however, seriously affect the average direction obtained from numerous determinations.

This was contrasted with the more varied ripple patterns to be expected in marine environments, for which Bucher provided what are probably the first published current rose diagrams. Bucher discussed the initiation of ripples on a flat bed, a process that still has not been satisfactorily explained. Darwin (1883) thought that discontinuous water

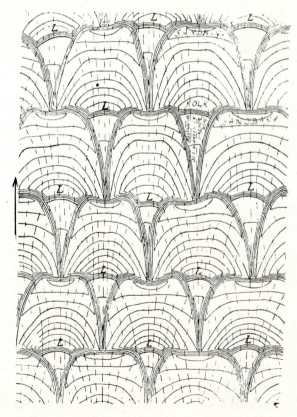

Fig. 9. "Contour map representing the ideal form of linguoid current-ripples. L=lowest point; dotted lines=lines of flow of water". From Bucher (1919), copied from earlier work by Blasius.

motion was necessary, whereas Gilbert (1914) thought that a pre-existent water rhythm was the cause, and Bucher assumed the need for a small obstacle to flow. Modern thinking on this subject is summarized by Brush (*in* Middleton, 1965), and Middleton (1977).

A useful review of crossbedding structures was provided by Twenhofel (1932; parts of this section of Twenhofel's book were contributed by Kindle and Bucher). A description of the development of sand bars (probably what are now termed linguoid or transverse bars) was provided:

> Sandbars move over bottoms of water as plateau-like areas with steep slopes on the advancing sides. They may range in height from less than an inch to 10 or more feet, most of them being a foot or less. The sand is rolled over the top, which usually undergoes some erosion. Reaching the edge of the plateau, the sands roll down the slope at the front. The direction of the front slope is usually not constant and the inclinations have a similar range in direction . . . Each bed thus formed has its cross-lamination in one direction, which may or may not be the same as that of the beds below and above.

Apart from Sorby's work, the first systematic mapping of cross-bedding orientation in fluvial deposits appears to have been that of Rubey and Bass (1925). According to Pettijohn (1962) "Brinkmann (1933) was, perhaps, the first of the 'modern' workers to have a clear concept of the objectives of paleocurrent research and to develop methods to fulfill these aims".

> . . . through measured observations and statistical analysis a simple facies characteristic - crossbedding - might allow a number of basic, reliable paleogeographic conclusions to be drawn.

Important advances in the statistical treatment of paleocurrent measurements were made by Reiche (1938), including the use of stereonets for plotting poles to crossbed foresets, and vector methods for data reduction. McKee (1938, 1939) described the sedimentary structures of the Colorado River in the Grand Canyon, and proposed a classification of ripple cross-lamination; Knight (1929) described festoon crossbedding.

Except for the German work, led by Brinkman (1933), the pace of research in this area was slow, especially compared to the explosion of paleocurrent studies which took place in the nineteen fifties. Before this time the attention of most sedimentologists was held by petrographic studies, as revealed by the focus of the first edition of Pettijohn's (1949) textbook, and by the title of the first professional sedimentological journal, the Journal of Sedimentary Petrology, founded in 1931. As Pettijohn (1962) has pointed out, there was much misunderstanding as to the origins of many sedimentary structures which, in the case of crossbedding, was not fully resolved until the various structures were reproduced in exhaustive flume experiments, beginning in the nineteen fifties. The work of Fisk (1944, 1947) stimulated a renewed interest in alluvial sedimentation after World War II, particularly amongst oil companies, who increasingly realized the value of studies of modern environments, and the importance of statistics in geological applications also gained general recognition. These developments all came together in the early fifties to start the surge in knowledge of fluvial processes that has occurred during the last twenty-five years.

A contribution by McKee and Weir (1953) focussed attention on the internal arrangement and contact relations of hydrodynamic sedimentary structures and provided a structure classification that became widely used. Statistical techniques began to be applied to paleocurrent measurements on a systematic basis, including the use of perfected vector methods of calculating mean and dispersion (Curray, 1956), moving average methods for clarifying regional trends (Potter, 1955; Pelletier, 1958), and the use of analysis of variance methods to unravel the sources of dispersion in samples of different outcrop scale (Olson and Potter, 1954). Tanner (1955) discussed, with a variety of examples based on his own field work, the use of paleocurrents in basin analysis. A much expanded classification of hydrodynamic sedimentary structures was proposed by

Allen (1963a), and this remains the most comprehensive treatment of the subject, although other classifications have been proposed since. The use of paleocurrent measurements in basin analysis and regional paleogeographic reconstructions was demonstrated in an innovative case study by Pryor (1960) and reviewed in two important state-of-the-art publications by Pettijohn (1962) and Potter and Pettijohn (1963; revised edition, 1977), and the latter authors produced an illustrated manual of sedimentary structures for use in field work, a year later (Pettijohn and Potter, 1964). Another useful book on this topic was published later by Conybeare and Crook (1968).

A renewed interest in flume work produced some results of fundamental importance in the early sixties. Various workers experimented with the production of crossbedding (McKee, 1957; Brush, 1958; Jopling, 1963, 1965), but the most important developments were those of Simons and Richardson (1961) in which the flow regime concept was established. This work later appeared (with C. F. Nordin as an additional co-author) in a collection of papers on the hydrodynamics of primary sedimentary structures edited by Middleton (1965). Much of the work of earlier years on fluid mechanics and sediment transport was brought to bear on sedimentological problems in this book. The statement of the flow regime concept and the demonstrations in the field and flume of the relationship between bedforms and sedimentary structures were the two principal features of the book, and it has had a major influence on sedimentologists over the last thirteen years.

The development of ideas on the hydraulic interpretation of sedimentary structures in the last dozen years is described in detail by Middleton (1977). The remainder of this section deals briefly with those aspects of the subject of most relevance to fluvial sedimentology.

Parallel to the North American work, mentioned above, was that of Allen, who reviewed much engineering work on bedforms and carried out his own flume experiments on the flow patterns associated with the various ripple types. Most of this work is contained in his two books (Allen, 1968, 1970a). The most widely used part of this work has been Allen's diagram showing the control of flow velocity and sediment grain size on the stability of the principal bedforms.

Recent descriptive papers on bedforms and bars in modern sandy rivers include those of Carey and Keller (1957), Harms *et al.* (1963), Harms and Fahnestock (1965), Coleman (1969), Collinson (1970), Smith (1970, 1971), Williams (1971), and Jackson (1976a). All provide much useful detail regarding morphology and internal structures, orientation measurements are provided by some, and most of these papers have been widely used in interpretations of the ancient record. Further details are provided by Ashley (this volume) and Levy (this volume).

Paleocurrent variance in alluvial systems has been discussed by a few workers. Potter and Pettijohn (1977, p. 108-111) summarized the early research in this area. An important concept was introduced by Allen (1966, 1967), who classified sedimentary structures into a hierarchy of size and attempted to relate this to the variability shown by paleocurrent measurements. Miall (1974, 1976) developed these ideas further. Collinson (1971) and Schwartz (this volume) showed how paleocurrent dispersion varies with discharge. Jackson (1975) showed the time-dependent nature of the elements in the bedform hierarchy, and Allen (1973, 1974a) discussed the problems associated with bedform lag - their slowness to respond to changes in flow conditions and the implications for hydraulic interpretation (the concept of lag had been introduced by Kennedy, 1963). Southard (1971) and Harms *et al.* (1975) demonstrated the importance of a third parameter, depth, in controlling bedform morphology, and the recognition of this important point has allowed a much greater degree of precision in the prediction of bedform morphology from flow conditions. Harms *et al.* (1975) also recorded the recognition of a new, ''bar'' or ''sand wave'' bed phase, generally developed between the ripple and dune phase. Such features had been first recognized in experiments by French workers in 1963, and were subsequently recorded in the natural state in intertidal environments (*see* Middleton, 1977, p. 7 for details). Southard (this volume) has expanded the range of conditions amenable to laboratory experimentation by employing hot water, a device that enables wider and deeper flows than before to be modelled without resorting to excessively large flume equipment.

Pebbly rivers have been studied by Johansson (1963), Williams and Rust (1969), Rust (1975), Smith (1974), Koster (this volume) and Gustavson (this volume). Measurement and interpretation of clast imbrication has been an important part of this work. Hein and Walker (1977) proposed a model for the evolution of bars in gravel rivers (*see* also Rust, this volume), and Shaw and Kellerhals (1977) discussed the interpretation of antidunes in fluvial gravels.

Middleton (1977) reviewed developments in the hydraulic interpretation of sedimentary structures. The recent descriptive literature on bedforms and bars in sandy and pebbly braided rivers was summarized by Miall (1977a), who discussed the problems that have arisen with nomenclature and classification. Smith (this volume) summarizes the latest thinking on this subject, including results of some of the discussion at the Fluvial Symposium.

Fluvial facies models

From Hobbs to Fisk

In 1917 Grabau wrote:

> From the very first, students of sediments have been in the main, marinists, though when the organic evidence pointed unmistakeably to the presence of fresh waters they resorted to a lacustrine variant. Even today the interpretation of sediments as of other than marine or lacustrine origin meets with a good deal of skepticism in some circles, and the fluviatilists are not much in favour, especially when they intrude upon territory hitherto preempted by the marinists . . . Fortunately however, American geologists are the most tolerant of men, and even the most startling ideas get a hearing if they are supported by facts and are the result of logical reasoning.

Some of the early attempts at fluvial interpretations have been quoted earlier in this paper. That by Medlicott and Blanford (1879) was well ahead of its time, probably because exposures of the Siwalik sediments are located close to the great rivers of the Indo-Gangetic plain - an inescapable modern analogue. A similar spatial proximity led to an interpretation by Hobbs (1906) of various Tertiary sandstone and conglomerate deposits of Spain and Italy as "torrential" in origin, by comparison with modern flood deposits he observed in the process of formation in the same areas. Hobbs quoted Italian workers who had reached the same conclusion in 1895, and concluded that such deposits probably are very common in the ancient record. Here are some of Hobbs' (1906) sedimentological criteria:

> It should be characterized by included lenticular areas of coarser or finer material and by layers of fine material in sharply defined films and plates which reveal its bedding . . . Locally, at least, it may show a type of crossbedding . . . which, though often observed in ancient sandstones is not adequately explained by the changing currents along a marine shore . . . The dominance of ripple marks and the paucity of marine fossils just where marine life should have been most abundant are facts difficult to explain on the theory of marine origin for the great sandstone formations.

Kindle (1911) discussed crossbedding and fossils as evidence of continental origin, and Graubau (1906) used gross-stratigraphic relationships as indicators of environment, suggesting that a "non-marine progressive overlap" (what we would term progradation, with interfingering of marine and non-marine facies) or an unconformable overlap "away from the sources of supply of the material" (as in an alluvial fan) were good evidence of continental origin.

Virtually nobody was carrying out detailed sedimentological work in alluvial environments. A major exception was Trowbridge (1911) who made a careful study of the geomorphology and deposits of some modern alluvial fans in California, "to discuss the causes and processes involved in their deposition" with the objective of deducing "certain criteria whereby materials so deposited may be distinguished from other deposits such as those of lakes and seas, even after cementation has taken place". The criteria so deduced are similar to those of Hobbs (1906) to whom Trowbridge does not refer. Fenneman (1906) used the terms lateral accretion and vertical accretion (in reference to what we now term point-bar and overbank deposits, respectively) in a discussion of the origin of floodplains in modern streams. Melton (1936) made use of this sedimentological concept but the first geologists to fully exploit it were Happ *et al.* (1940) and Fisk (1944, 1947).

Barrell's (1912) classic paper on the "criteria for the recognition of ancient delta deposits" provided a detailed discussion of the differences in bed geometry, sedimentary structures, colour, texture and fossils between marine, fluvial and eolian deposits and is, in fact, a sophisticated approach to integrated facies studies following Walther's (1893-1894) teachings, which belies Grabau's pessimistic statement published five years later (quoted above). The paper had a profound impact on the development of sedimentology. As regards fluvial sedimentation Barrell stated (in part):

> In rivers where sands are being deposited the channel is subject to meandering. It cuts laterally into the banks and scours down into older floodplain deposits. The sands of the abandoned channels cut across the bedding of the floodplain deposits on the convex sides of the meanders and on the concave side are interlaminated with them.

This last statement is, of course, incorrect. Barrell continued:

> The river works across the floodplain and buries channel structures widely in the fluviatile deposits. The river bars work regularly downstream, being continually cut out above and deposited below . . . Gravel tends to be concentrated along such channel bottoms. Lateral discontinuity of the sandstone lenses is also a feature, the ancient channel deposits forming a meshwork.

Barrell was also aware of another style of fluvial sedimentation:

> Effects of sheet-flood deposition: - Many aggrading streams overloaded with sand exhibit at low stage shallow braided channels within the main channel. At higher water the main channels may likewise form a braided system, and at highest water the whole floodplain may be covered by a shallow moving water body . . . It is such conditions which seem to be required to produce the great depths of regularly bedded and widely extended sandstones which mark certain continental deposits . . . They succeed each other without inter-lamination of clays and commonly show neither structure nor fossils. False bedding oblique to the even regular bedding is occasionally observed, but the homogeneous material conceals its frequent presence. Ripple and current marks, however, are rare or absent.

This reads like a description of the Bijou Creek style of braided river sedimentation (McKee *et al.*, 1967; Miall, 1977a).

Detailed criteria such as those described by Barrell (1912) were used by Grabau (1913b) and Barrell (1913, 1914) in their lengthy descriptions of the Paleozoic deltas of the Appalachians. The alternation of sandstone and shale units in these deposits was noted, and Barrell (1913, p. 458) stated of the alternations in the Catskill Formation:

> A gray or olive sandstone member is commonly sharply delimited at bottom, but at the top grades first into maroon argillaceous sandstone, and this in turn into red sandy shale.

Elsewhere (p. 466) Barrell attributed the alternation of fine and coarse units to "the lateral shifting of distributaries, the red shales representing flood plain areas temporarily removed from the presence of currents". Here, then, is the first inkling of the fining-upward cycle and its origin, although Barrell was not aware of the general significance of this observed sequence. This seems to have first occurred to Dixon (1921, p. 32) in his studies of the South Wales coalfield:

> The relations of each sandstone-band (b) to the marls (c and a) above and below are, typically, as follows:—
>
> (c) Red marls; passing down into
>
> (b) Flaggy sandstones, the upper red, the lower green; at the base of the whole band one or more conglomeratic cornstones; eroding
>
> (a) Marl, green immediately below (b) to a depth which varies between a mere skin and several feet, but, in each case, is fairly uniform; often a purple band below; lower still the marl is red and passes down into another sandstone band, which repeats the features of (b).
>
> This sequence is repeated interminably and in all parts of the series.

Other contributions made at this time include a study of the fine banding in Tertiary fluvial sediments in Burma by Stamp (1925), who attributed it to the effects of seasonal floods, and a study of the origins and transportation processes of gravel, including fluvial gravel, by Barrell (1925).

A good summary of the state of knowledge of fluvial sedimentology is included in Twenhofel's (1932) great treatise. Alluvial fan, valley flat and fluvio-deltaic environments are distinguished, and the general facies descriptions are much as we would use today, except Twenhofel thought that "channel deposits are entirely ephemeral" and that the distribution of the various lithologies is "extremely erratic".

The next important contributions were those of Happ *et al.* (1940) and Fisk (1944, 1947). Based on their work in modern rivers, Happ *et al.* divided fluvial sediments into six types: channel-fills, vertical accretion deposits, flood plain splays, colluvial deposits (hill wash), lateral accretion deposits and channel lag deposits. They thought that "the last two types are of little importance among the modern deposits, but they may have been more important under pre-modern conditions". Good descriptions of the facies were given, and four facies associations were proposed: normal flood-plain or valley-flat, alluvial-fan, valley-plug and delta. Their description of the valley-flat association includes the first block diagram of a fluvial model (Fig. 10) which clearly shows the fining-upward nature of meandering stream deposits, but this is nowhere explicitly stated. For example:

> The modern deposits of vertical accretion usually lie upon older deposits of similar origin, which in turn are typically underlain by sandy channel deposits that were accumulated when the channel occupied other positions than at present. These older channel sands may represent lateral accretion on the inside of gradually shifting channel bends, or they may represent filling of former channels abandoned by avulsions. More recent deposits of lateral accretion, in generally crescentic shapes along the inside of stream bends, may still be exposed at the surface but are typically lower than the surface of the adjacent older flood-plain which has been covered by deposits of vertical accretion.

Happ *et al.*'s (1940) work was descended from that of Melton (1936; quoted earlier) and Mackin (1937), who discussed lateral accretion in pebbly rivers and illustrated diagrammatically the development of what we now call epsilon crossbedding (Fig. 11). It was incorporated into the geological literature on facies models by Allen (1964) but, apart from this it has rarely been referred to. This is in marked contrast with Fisk's (1944, 1947) work, which undoubtedly constitutes the most widely cited source in alluvial sedimentology. Some Russian work (e.g. Shantzer, 1951; Botvinkina *et al.*, 1954) rivals that of Fisk in its scope but, being in Russian, it has had little influence on the Western "school" of sedimentology (again, except for Allen's work).

Fisk's two classic reports (1944, 1947) contributed a wealth of detail on the Mississippi fluvial and deltaic plain. The post-glacial history of the Mississippi was documented in

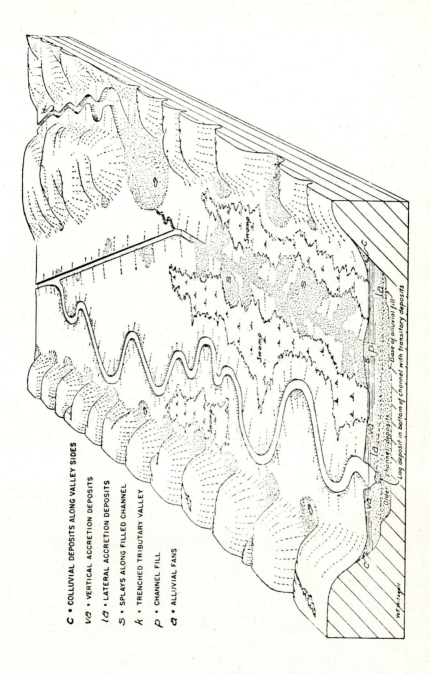

Fig. 10. A fluvial block diagram model ("diagram illustrating typical relations of the various types of deposits in the valley accumulation"). From Happ *et al*. (1940).

ANDREW D. MIALL

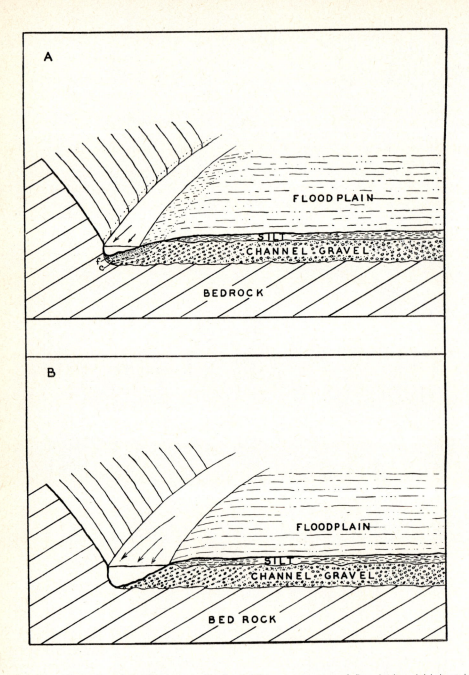

Fig. 11. "Formation of the Shoshone valley floor: (A) Low water stage f, fine detritus, laid down in the stream channel during periods of normal flow; c, coarse detritus, laid down in the channel at the end of a period of high water. (B) High water stage." From Mackin (1937). This figure illustrates the principal of lateral accretion, and shows the accumulation of heterogenous lithologic units in the point bar, preserving large-scale, low-angle crossbedding surfaces - the structure later termed epsilon crossbedding by Allen (1963a).

detail, paleo-channels were mapped, the sediments were described and the processes of fluvial sedimentation including meander migration and cut off, and the formation of clay plugs, levee and splay deposits were all discussed in detail. The reports were based on detailed field work, 16,000 boreholes and examination of air photos. They remain as one of the most detailed studies of an alluvial system ever undertaken, and this is the most valuable aspect of Fisk's contribution. Most of the main fluvial sedimentary processes and products were fairly well understood in a qualitative way by the nineteen forties, and Fisk filled out this knowledge with the necessary detailed documentation. Fisk's main sedimentological contribution was to demonstrate the magnitude and complexity of the deposits of a major river and to demonstrate the response of that river to fluctuations in base level, in the form of variations in channel morphology and sedimentary facies. Fisk's contributions to deltaic sedimentology were probably more fundamental than this, in that they represented the first major advance in the study of modern deltas since Gilbert's work in lacustrine deltas in the latter part of the nineteenth century. In terms of the development of scientific thought Fisk provided an authoritative demonstration of the value of combining detailed sedimentological analysis with a close examination of sedimentary processes in a modern environment. This was essential groundwork for the development of actualistic facies models, which became a major trend of sedimentological progress in the nineteen sixties following an innovative conference on the facies models concept in 1958 (Potter, 1959). Fisk's reports are still being mined for information at the present day, for example *see* Jackson (this volume).

Meandering-river deposits: development of modern facies model concepts

A complete description of the characteristics of fining-upward cycles was provided by Bersier (1948) based on his work in the Alpine molasse (Fig. 12). The following is a free translation of his description of the typical cycle:

> The most marked characteristic is the base of the sandstone beds, which overlie the shale with a sharp contact, without transition. This is where the grain size is coarsest, with

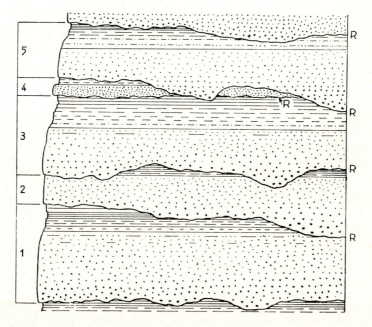

Fig. 12. A suite of typical fining-upward fluvial cyclothems in the Alpine molasse. R = erosion surface. From Bersier (1948).

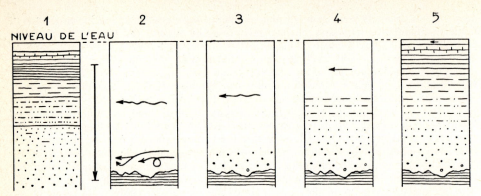

Fig. 13. Development of a molasse cyclothem. From Bersier (1948). *See* text for explanation.

minor gravel and reworked shale pebbles. The hard sandy bed sits on the surface of the shale with its impressions (sole structures?) borings and tracks. The basal surface is not planar, but undulatory and irregular. The deposit is preceded by an erosion surface in the subjacent shale by which the beds are cut out.

   Higher in the sandstone bed the grain size decreases. At the same time fine argillaceous beds appear mixed with the sand. One passes progressively from the sandstone-molasse to argillaceous, immature sandstone, then to fine-sandy shale, with intermediate gradations. The cycle is always terminated at the top by fine-grained units, coloured shale and calcareous mud or lithified fresh water limestone. A new erosion surface, with a recurrence of coarse material marks the end of one cycle and the beginning of the next.

Bersier thought that these cycles were tectonic in origin, in keeping with prevailing theories of the time that had been developed to explain, among other things, the Carboniferous coal-bearing cycles of the United States. Figure 13 is his diagram showing the development of the cycles, and the following is a translation of his figure caption:

   Development of a molasse cycle by subsidence. The length of the horizontal arrows is proportional to the strength of the currents.

   1. Completion of underlying cycle; 2. Subsidence (vertical arrow) with simultaneous or subsequent erosion by turbulent currents developed in the deeper channel; 3. Sedimentation of coarse detritus commences; 4. Deposition of fine sand, filling the channel, and slowing of currents by diminution of water depth; 5. Deposition of fine mud at top. In beds of Chattian age fresh water limestone and coal.

   Later Bersier (1958) provided further details on the molasse cyclothems, including their variability and thickness, and proposed a different explanation of their origin. He ascribed them to vertical aggradation in channels which shift in position on a delta plain, using the modern Mississippi delta as his analogue. Allen (1962) proposed a similar mechanism for some Old Red Sandstone (Devonian) cycles in England, suggesting that the upward-fining was due to a decrease in slope and current strength as the channel became clogged with sediment. These proposals dispensed with the need for rhythmic subsidence as a cause, but there was still no general comprehension by geologists of the sedimentary mechanism of lateral accretion. This idea first appeared in the geological literature as a result of Wright's (1959) attempts to explain laterally persistent crossbedding in fluvial deposits. Wright quoted Van Straaten (1954) who had "described deposition of material on the inside curves of beach gullies", and Dunbar and Rodgers (1957) who "suggested that some wide lenses of cross-stratified sediments may have been deposited as point bars on the inside curves of a meandering stream." Wright does not seem to have been aware of the work on alluvial sediments by Fenneman (1906), Melton (1936), Mackin (1937) or Happ *et al.* (1940) and he confused point bar accretion surfaces with crossbedding. Sundborg (1956) described growth of point bars in the River Klarälven and illustrated what we now realise to be lateral accretion surfaces, but his description focuses on texture, and on smaller scale bedforms. The lateral accretion concept, as currently conceived, was first proposed for modern estuarine channel-fill sediments by Oomkens and Terwindt (1960), and later was combined with the fluvial fining-upward cyclothem into a complete process-response facies model by Allen (1963b, 1964) working on the Old Red Sandstone of southern Britain and, more or less simultaneously, by Bernard *et al.* (1962) and Bernard and Major (1963) who studied Recent and Pleistocene fluvial deposits. The latter papers reflected part of a major sedimentological research effort by petroleum geologists in the United States, particularly the Shell Development group based at Houston. They had identified upward-fining cyclic fluvial deposits in the subsurface some years previously (Nanz, 1954) but the origin of such sequences was not discussed in detail (in print) at that time.

Allen's (1963b) depositional model is quite different from his earlier (1962) interpretation. He discussed the origins of the various facies present in a fluvial cyclothem and showed that their vertical arrangement is

> . . . consistent with offlap deposition on the channel bar (? point bar) of a laterally migrating stream. As can be observed on stream channel and point bars today . . . both the intensity and speed of flow tend to decline as the bar is ascended from the deepest part of the river bed to the top of the channel.

In a later paper Allen (1964) documented the lateral accretion model in detail. This paper was an important bridge between the disciplines in that it drew on an extensive review of geological and geomorphological literature including European, Russian and North American work. Elsewhere Allen (1963a) described a type of large-scale, low-angle crossbedding, termed "epsilon cross-stratification", and stated that:

> Wright's (1959) interesting suggestion that point bar construction generally may explain cross-stratified units where the cross-strata are laid down parallel to the surface of the bar, has not been borne out by a study of the literature on point-bar deposits.

Mackin (1937, Fig. 2) had, in fact, illustrated what we now term epsilon crossbedding based on his work in Big Horn Basin (Fig. 11) as had Sundborg (1956, Fig. 47) but its significance for interpreting the ancient record was missed. But with the new fluvial facies model available students of ancient rocks rapidly began to find examples of point bar deposits containing large-scale, low-angle epsilon crossbedding, thereby providing abundant evidence of lateral accretion as the primary depositional mechanism (Allen, 1965b; Moody-Stuart, 1966).

Fluvial and deltaic sedimentary processes and their deposits were described in a detailed review paper by Allen (1965a), which summed up the state of knowledge as it existed in the mid-sixties. The influence of this paper has been far reaching, and next to Fisk's work it is probably the most widely quoted work yet published on fluvial sedimentology. In it Allen provided, among other things, a series of block diagrams showing various theoretical facies associations (facies models) which have done much to stimulate the generation of detailed regional paleogeographic reconstructions by later workers.

A useful paper by Beerbower (1964) divided cycles into those of autocyclic origin, generated by sedimentary processes such as channel migration, and those generated by allocyclic processes such as tectonic movement.

The importance of Allen's work on the fining-upward cycle was emphasized by the work of Visher (1965a) and it was also seen that this cycle was one of a range of possible stratigraphic variations that could be developed in marine and continental sedimentary environments, so that Visher's (1965b) other, more generalized paper (which actually appeared first), firmly established vertical profile models as one of the most useful tools available for studying ancient clastic deposits. In subsurface studies it remains the most widely used technique, aided by the fact that cyclic character can almost invariably be guessed at from geophysical logs alone, without detailed core or sample studies (Pirson, 1970).

The most definitive paper to date on the fining-upward model is that of Allen (1970b) in which statistical studies of a suite of Devonian cyclothems in Britain and North America were described, and hydraulic interpretations were provided. Little work on the study of modern meandering river analogues was carried out until McGowen and Garner (1970) described some point bars in the southern United States, Bluck (1971) studied the Endrick River in Scotland and Jackson (1976b) analysed a reach of the Wabash River (Indiana-Illinois). All of these workers documented the variability that can occur in point bar deposits and showed that the point bar cycle may be much more complex than is commonly envisaged. Excellent examples of ancient point bar deposits were described by Puigdefabregas (1973) and are included in this volume in papers by Mossop, Puigdefabregas and van Vliet, Nijman and Puigdefabregas, and Leeder and Nami. Collinson (this volume) and Jackson (this volume) both provide critiques of current meandering-river models. Jackson (this volume) relates cycle characteristics to sediment load and discharge variations and proposes several new facies associations.

Braided rivers

The distinctiveness of the braided channel pattern was recognized in the nineteenth century but not until 1962 was an attempt made to relate this pattern to depositional behaviour and to recognize a distinctive sedimentary style, different from that of the meandering river. In that year Doeglas published his analysis of some braided rivers in southern France. Ore (1964) discussed criteria for distinguishing the deposits of meandering and braided streams, and Moody-Stuart (1966) suggested differences in sedimentary facies that might be generated by low, as distinct from high, sinuosity rivers (not necessarily the same distinction as that of Ore, as some of Moody-Stuart's criteria may apply to single-channel straight streams).

The widespread distribution of the braided environment was made apparent by the work of McKee et al. (1967), Coleman (1969), Williams and Rust (1969) and Smith (1970, 1971) who described, respectively, flash flood deposits of a small ephemeral stream; the sedimentology of the Brahmaputra River - the first work of its kind on one of the world's major rivers; a gravel-dominated mountain river; and a sandy river of the American Great Plains. (Glacial outwash rivers are also almost invariably braided and have received much sedimentological study). All these papers, except the first, have been widely quoted in later studies of ancient braided stream deposits.

By contrast with meandering river deposits, those of braided rivers have generally been thought of as somewhat random in character, and lacking recognizable cyclicity. Application of the statistical technique of Markov Chain Analysis to bedding sequences showed that this is not always the case (Miall, 1973; Cant and Walker, 1976) and, in a review of the braided river environment, Miall (1977a) developed four facies models for braided rivers, one of which was strongly cyclic. Generation of these models was aided by the development of a simple lithofacies coding system, which permitted a unified description of braided river deposits, including those already described in the literature.

Cyclic braided deposits were assigned by Miall (1977a), to his Donjek model, but subsequent work has shown that this name would be best confined to gravel dominated cyclic deposits, as in the original description of the Donjek sediments (Williams and Rust, 1969; Rust, 1972). Cant and Walker (1976) and Cant (this volume) described a sand dominated cycle which has been erected as a fifth model and termed the South Saskatchewan type (Miall, this volume). Rust (this volume) discusses a slightly different grouping of braided facies associations using his own field examples.

Alluvial fans

Because of their distinctive morphology, alluvial fans have always been recognizeable as discrete components of any alluvial system, and sedimentologists have attempted to carry this distinction into their interpretation of the ancient record. Several authors have distinguished an alluvial fan facies from a braided river facies, which is misleading, because much of the sedimentation that takes place on alluvial fans is brought about by the activity of braided streams. The most distinctive components of alluvial fan environments are debris flow deposits, although not all fans necessarily contain this facies (*see* Miall, this volume; Rust, this volume).

Modern alluvial fans were studied by Tolman (1909), who introduced the term "bajada", and Trowbridge (1911), whose contribution has been referred to earlier. Lawson (1913) coined the word "fanglomerate" and Blackwelder (1928) first appreciated the importance of mudflows in alluvial fan deposits. More recent, detailed work on the geology of alluvial fan deposits by Blissenbach (1954) has been much quoted. Other recent descriptions have been given by Bull (1972) and Boothroyd and Ashley (1975). Ancient fan deposits were described, amongst others, by Nilsen (1968) and Miall (1970) who used textural parameters and clast composition to map fan outlines and transport directions. Heward (this volume) provides a review of depositional processes on alluvial fans and describes some Carboniferous examples in Spain. Boothroyd and Nummedal (this volume) document the depositional facies occurring on humid alluvial fans.

Other facies models

Recent work on fluvial geomorphology (eg. Galay *et al.*, 1973; Mollard, 1973; Rust, this volume) has emphasized that there is great variability in fluvial morphology, and that our conceptions regarding the two so-called principal channel patterns, braided and meandering, are simplistic. Miall (1977a) illustrated four principal morphological patterns, the other two being straight and anastomosing but, in fact, there is a continuum between all these four "end members".

The sedimentological ramifications of this variability are far from being understood. Moody-Stuart (1966) discussed possible depositional products of a straight river, D. G. Smith (1973 and unpublished data) has examined the sedimentology of a modern anastomosing system, and Rust (this volume: channel classification paper) examines the general problem of interpreting channel sinuosity and channel multiplicity in the ancient record. It has been shown that under certain circumstances sequences of apparently braided origin can be formed in a high-sinuosity, single-channel environment (Jackson; Nijman and Puigdefabregas; both this volume) and Collinson (this volume) argues that some sand members in fluvial sequences should not be interpreted as channel deposits at all, but as the product of overbank sheet floods. Baker (this volume) and Taylor and Woodyer (this volume) describe modern fluvial systems in, respectively, the Amazon Basin and New South Wales (Australia) that obey very few of our 'norms' of depositional behaviour.

To conclude, it is apparent that the area of fluvial facies models still requires a great deal of work, both in modern rivers and in ancient deposits. It is hoped that a careful reading of the papers in this volume will provoke a re-examination of existing data and will, perhaps, help stimulate some of the necessary research.

Fluvial architecture

Fluvial architecture is a term used by J. R. L. Allen (during the Fluvial Symposium) to describe the geometry and internal arrangement of channel and overbank deposits in a fluvial sequence. Most work by fluvial sedimentologists up to the present day has concentrated on vertical profiles, and less attention has been paid to the way in which the various elements in a fluvial sequence stack on one another, although this is of considerable importance with regard to reservoir or ore-body continuity and the analysis of present-day or ancient ground water systems. Allen's (1965a) facies model block diagrams, referred to earlier, were a theoretical approach to this problem, following the preliminary work of Happ *et al.* (1940). The geometry of sand units on a delta plain was considered by Ferm and Cavaroc (1968). In a later paper, Allen (1974b) emphasized the importance of pedogenic carbonate units as indicators of stability and non-sedimentation on a fluvial plain, and used them in building a variety of hypothetical models showing the variations to be expected in the deposits of a meandering stream system as a result of the interplay of various autocyclic and allocyclic mechanisms. Knowledge of the three-dimensional (or even two-dimensional) architecture of ancient fluvial deposits is at present limited. Campbell (1976) provided a detailed cross-section through a Jurassic fluvial deposit of possible braided origin; Horne and Ferm (1976) described the architecture of some Carboniferous coal-bearing fluvio-deltaic rocks in the central Appalachians based on large highway and railroad exposures. Cant (this volume) erected a hypothetical model for the two dimensional geometry of a braided system of South Saskatchewan type; Nami and Leeder (this volume) describe the geometry of a well-exposed meandering river deposit; Friend (this volume) comments on gross architectural features of several ancient fluvial sequences; and Leeder (this volume) has developed a theoretical model for channel superimposition and interconnectedness in relation to the controls discussed above.

Another important control on fluvial architecture is vegetation, which affects channel morphology and shifting behaviour through increasing bank stability and, by its presence or absence in headwater catchment areas, plays a large part in determining the nature of the river hydrograph. The geomorphic effects of vegetation were recognized many years ago (eg. Surell, 1870) but Schumm (1968a) was the first to clearly state its paleoclimatic implications, and he pointed out that river systems of pre- and post-Devonian age must have been fundamentally different because of the first appearance at that time of an extensive cover of land vegetation. Long (this volume) discusses the implications of this difference for studies of the Proterozoic fluvial record, and Cotter (this volume) examines the Phanerozoic record.

On all scales ranging from the local to the continental, a knowledge of the geometry and dispersal patterns of alluvial deposits can lead to a better understanding of contemporary tectonics, as several papers in this volume (Steel and Aasheim, McLean and Jerzykiewicz, Friend) and elsewhere (*see* Potter, 1978 and his references) amply demonstrate. Conversely, an appreciation of tectonics may help us better understand local fluvial architecture. In our enthusiasm for studying fluvial processes we have, perhaps, tended to forget about allocyclic controls on the development of fluvial sequences. Tectonic activity is the most important of these, and should receive greater attention in studying basin-wide alluvial deposits.

Paleohydraulics

The study of fluvial facies models in vertical profiles or in two or three dimensions is an essentially qualitative approach to the reconstruction of past river systems, even where statistical techniques are used to provide quantitative information regarding facies assemblages. Paleohydraulics is the study of the quantitative relationships between the hydraulic parameters of a river (depth, width, slope, discharge, sediment type etc.) and its preserved deposits.

The study of paleohydraulics is fraught with difficulties, and some fluviologists feel that our knowledge of the depositional and preservational processes of rivers is still so inadequate that numerical estimates of past hydraulic parameters verge on the fictional. However, some progress has undoubtedly been made, and continued attempts at paleohydraulic reconstruction can only serve to encourage fluvial sedimentologists to structure their future research into modern rivers so as to ask the right questions, in the hope that some of the difficulties can gradually be overcome.

There are essentially two approaches to paleohydraulics, the engineering approach, which uses theoretical relationships and empirical data regarding sediment transport mechanics and bedform generation to reconstruct the depositional conditions for individual beds or sedimentary structures, and the geomorphological approach, which relates empirical data about the morphology of modern rivers to some of the gross features of their deposits. Each method has its limitations - the engineering approach concerns itself primarily with instantaneous depositional conditions of specific bedding units, the overall, long term significance of which may be hard to judge because of our lack of knowledge of facies preservability in fluvial environments. And, as Allen (1973, 1974a) has shown, bedforms are slow to react to changes in flow conditions and are therefore inaccurate hydraulic indicators. The geomorphological approach is potentially more useful, in that it deals with long-term, statistical averages, but suffers, at present, from a gross lack of data from all but a limited range of modern climatic, discharge and sediment load conditions. As several papers in this volume (those by Baker, Jackson, and Friend, in particular) and elsewhere (Miall, 1977a) demonstrate, current sedimentological concepts regarding rivers are derived from a very few, possibly atypical modern examples, most of which are located in populated or otherwise accessible, relatively temperate climates. The range of different conditions represented by ancient fluvial deposits of all ages has barely been touched, and research which integrates sedimentology and hydraulics is sparse. We are hampered, also, by the influence on modern rivers of the rapid sea level changes during and following the Quaternary glaciation, and an uncertainty as to how this distorts our mental projections back into times of greater geological stability.

Developments in sediment transport mechanics, bedform classification, flow regime concepts and quantitative geomorphology on which the study of paleohydraulics rests have been summarized earlier in this review. The remainder of this section contains a brief discussion of some of the published applications of such knowledge to geological problems.

The fining-upward point-bar cycle has been the most widely used paleohydraulic indicator, almost since the genesis of the cycle as the product of lateral accretion was first deduced in the early sixties. The presence of epsilon crossbedding and its interpretation as superimposed point bar accretion slopes, helped confirm this interpretation. Allen (1965b) was the first to realize the paleohydraulic significance of these cycles, in his study of the Old Red Sandstone of Anglesey, North Wales:

> . . . it appears that the streams which deposited the alluvial facies were already substantial rivers on reaching Anglesey. Each epsilon cross-stratified unit has been interpreted as a point-bar deposit. The thickness of such units therefore corresponds to the stream channel depth at the bankfull state (Wolman and Leopold, 1957, pp. 92, 95), and the mean unit thickness is six feet three inches. Leopold and Maddock (1953, app. A) gave data leading to relationships between stream width, mean depth and drainage basin area, and Hack (1957, p. 63) later presented an equation connecting stream drainage area and stream length. When the mean unit thickness is substituted for the mean channel depth in these relationships, it becomes statistically very unlikely that the streams were less than forty miles long, including meanders, and seventy-feet wide. Statistically the streams were most likely to have been four hundred miles long and three hundred feet wide. Needless to say, these figures do not represent predictions, for very many factors combine to determine stream geometry, but are intended merely to convey the probable order of importance of the streams which deposited the Old Red Sandstone of Anglesey.

Schumm(1968b) studied the present and prior courses of the Murrumbidgee River, Australia, and used the empirical geomorphological relationships he had been developing for American and Australian rivers to determine their various paleohydraulic parameters. Sedimentological information such as sedimentary structures and facies relationships, were not used in this work.

Eicher (1969) attempted to estimate river size, discharge and slope for a Cretaceous fluvial system in Colorado. His only input datum consisted of stream length, derived from regional paleogeographic reconstruction, whence the other hydraulic parameters were derived using empirical geomorphological relationships. This building of an edifice of interpretation from one or two items of data by using one estimated parameter as input for the next estimate, is one of the principal weaknesses of the paleohydraulic method.

Allen (1970b) used knowledge of bedform hydraulics and flow patterns in meander bends to deduce a generalized paleohydraulic model for a suite of Devonian fining-upward cycles. No attempt was made to proceed from this model to generalizations regarding discharge or drainage area.

Cotter (1971) combined Allen's (1965b, 1970b) observations (in part quoted above) with Schumm's family of equations to produce a paleohydraulic interpretation of a Cretaceous fluvial unit in Utah, including estimates of river length, drainage area, discharge, sinuosity and slope. Cotter was aware of the possible sources of error in his estimates deriving from differences in climate and vegetation in the past, and emphasized that "the results obtained are only reasonable estimates that must fit the nature of the environment interpreted in other ways".

Similar exercises to those quoted above were used by Friend and Moody-Stuart (1972), Padgett and Ehrlich (1976), Cant and Walker (1976), and Miall (1976) in paleohydraulic reconstructions performed on fluvial units in Spitsbergen, Morocco, Atlantic and Arctic Canada. The routes through the various equations differ in each case; for example Friend and Moody-Stuart (1972) and Cant and Walker (1976) base much of their work on flow velocity and power deductions derived from sedimentary structures and grain size. Miall (1976) introduced a sinuosity estimation based on paleocurrent measurements.

Baker (1973) studied the catastrophic Lake Missoula flood and provided a paleohydraulic reconstruction of some giant bedforms. Leeder (1973) compiled data concerning epsilon crossbedding and suggested some refinements of the methods for using this structure and cycle thickness as paleohydraulic indicators. Bridge (1975) used much of the earlier work on bedform hydraulics and flow patterns in a computer simulation of sedimentation in meandering streams. Allen (1977) and Shaw and Kellerhals (1977) examined ripples and antidunes, respectively, as paleohydraulic indicators.

Several papers on fluvial paleohydraulics are included in this volume. Ethridge and Schumm assess the empirical geomorphological approach, with examples; Bridge provides a more elaborate computer model of sedimentation in meander bends, and individual examples of paleohydraulic reconstruction are given by Church, Morton and Donaldson, and Padgett and Ehrlich. Church emphasizes the utility of carrying out such work in Quaternary fluvial or fluvioglacial deposits for which many of the estimates can be checked against existing landform elements.

CONCLUSIONS

Two opposed thoughts come to mind at the conclusion of this review. Firstly, it is surprising how many good ideas were thought of too early, and lay around for so many years until the level of sophistication advanced enough for them to be properly exploited. The concepts of vertical and lateral accretion, and the use of sedimentary structures in paleocurrent analysis, are good examples. Secondly, it is astonishing how far we have come in recent years in so short a time. For example, the fining-upward cyclothem process-response model is only fifteen years old, yet not only is it an entrenched part of every clastic sedimentologist's memory bank but it is already being viewed by some as simplistic and about to be replaced by a whole assemblage of facies models which better reflect our improved level of observation. The value of the facies model concept as a distillation of observations and a prediction tool for future work (Walker, 1976) is in danger of being overwhelmed in the process.

The dialogue between sedimentologists, geomorphologists and hydraulic engineers is increasing, but there remain several conceptual differences between the various specialists. For example the time frame of concern to geologists is quite different from that of other fluviologists; secondly, geologists are interested in buried fluvial deposits beneath an alluvial plain, as well as with surface deposition and the processes of sediment preservation, whereas geomorphologists and engineers are concerned primarily with surface processes and with events that may appear to the sedimentologist to be of ephemeral importance.

Our knowledge of modern fluvial environments reflects a research effort carried out in both large and small rivers, located in a variety of climatic and tectonic settings. However, this work is very incomplete; for example, most studies have concentrated on one or two facets of the work such as bedform description or textural analysis. Therefore it is of more than historical interest to comment on the geography of fluvial research, as this has influenced the development of ideas as much as the personality of the researchers, and it may assist us to detect biases in our knowledge that are affecting our interpretation of the ancient record (Friend, in this volume, also comments on this point). Firstly, it is important to emphasize the profound influence that the Mississippi River has had on geomorphologists and sedimentologists, far exceeding that of any of the other great rivers of the Western world. This undoubtedly reflects the ready accessibility of the river and the numerous geologists who are familiar with it but, more importantly, it stems from the great research effort expended on the Mississippi and its deposits, primarily for the purposes of river engineering (navigation, flood control), by various arms and agencies of the U.S. Army extending over a period of more than one hundred years. The work of Humphreys and Abbot, and Warren, of the Corps of Topographic Engineers (referred to earlier) had an important influence on Dana, Geikie, Chamberlin and Salisbury, and Grabau. More recently the work of the Corps of Engineers through their research establishment at Vicksburg, Mississippi (Waterways Experiment Station; Mississippi River Commission) lead to Fisk's (1944, 1947) monumental reports, which firmly established the Mississippi as the best known river in the Western World. A recent report by Winkley (1977) shows that the work of this group is still continuing.

Important contributions to the study of river mechanics originated with canal engineers in the Indian subcontinent during the days of the British Empire (Kennedy, 1895; Lane, 1935), but the two great rivers of the area, the Indus and Ganges and their tributaries, have been little studied by sedimentologists. The work of Everest (1832) and Medlicott and Blanford (1879) has been referred to earlier. That of Coleman (1969) is the most important recent contribution dealing with this region, and papers by Geddes (1960) and Gole and Chitale (1966), which are more geomorphological in emphasis, have been used by some geologists.

Shantzer (1951) and Botvinkina *et al.* (1954) discussed some Russian rivers and their deposits, Chien (1961) studied the Yellow River, and Belcher (1975) documented the history of the Rio Grande; but none of the other major rivers of the world have received the attention from fluvial sedimentologists that they deserve. The mouth's of the Rhine, Rhone, Niger, Klang-Langat, Mekong and various other rivers, have been much studied by those interested in estuarine and deltaic processes, but the deposits of inland alluvial basins of such rivers as the Amazon, Orinoco, Tigris, Euphrates and the Nile are little known (Russian geologists probably could tell us much about the Nile deposits beneath Aswan). Much of our knowledge of fluvial processes comes from relatively small rivers such as the Platte, Donjek, Wabash, Brazos, Endrick, Tana and South Saskatchewan. None of these are actively aggrading on a large scale, and future research may show that this has had a detrimental effect on the development of ideas about fluvial architecture and facies models. Large fluvial systems also have much to tell us about the sedimentary and tectonic history of major sedimentary basins (Potter, 1978).

One area that has not been touched on in this paper is the history of practical or economic applications of fluvial sedimentology. This would justify a paper of its own. In terms of resource extraction fluvial deposits act as hosts for a variety of non-renewable resources, including coal, hydrocarbons, and many placer deposits, including uranium, diamonds and gold. Uranium can also occur as a post-depositional accumulation. Techniques required for finding and exploiting such deposits include a knowledge of depositional systems, porosity and grain size trends, for which the papers in this book should provide ample documentation. Examples of the practical application of sedimentological techniques are given in the three papers at the end of this book, by Horne and Saxena, Minter, and Turner.

As with every other branch of sedimentology the number of skills required to be a useful practitioner increases year by year. The ideal fluvial research worker would be a Quaternary sedimentologist with experience in petroleum geology and river engineering, a passion for hydraulics, statistics and scuba diving, a more than passing interest in tectonics and a lot of money for coring equipment. In this paper an attempt has been made to show how the development of such a list of requirements came about, but even for the majority of us who can claim only a few of these skills, there is still much work to be done.

ACKNOWLEDGEMENTS

J. R. McLean, J. R. L. Allen and G. V. Middleton read and commented on an earlier draft this paper. I am particularly grateful to G. V. Middleton for a preprint of his article on the history of Sedimentology and for many additional comments on this topic that have, I hope, helped improve the completeness and balance of this paper.

Preparation of this review could not have been achieved without constant access to the excellent library facilities of the Geological Survey of Canada (Calgary and Ottawa) and the University of Calgary. J. Graff (G. S. C. librarian, Calgary) was instrumental in unearthing many of the earlier references.

REFERENCES

Adams, F. D., 1938, The birth and development of the geological sciences (republished by Dover Pub. Inc., New York, 1954, 506 p.).

Allen, J. R. L., 1962, Petrology, origin and deposition of the highest Lower Old Red Sandstone of Shropshire, England: J. Sediment. Petrol., v. 32, p. 657-697.

———, 1963a, The classification of cross-stratified units, with notes on their origin: Sedimentology, v. 2, p. 93-114.

———, 1963b, Henry Clifton Sorby and the sedimentary structures of sands and sandstones in relation to flow conditions: Geol. Mijnbouw, v. 42, p. 223-228.

———, 1964, Studies in fluviatile sedimentation: six cyclothems from the Lower Old Red Sandstone, Anglo-Welsh Basin: Sedimentology, v. 3, p. 163-198.

———, 1965a, A review of the origin and characteristics of recent alluvial sediments: Sedimentology, v. 5, p. 89-191.

———, 1965b, The sedimentation and palaeogeography of the Old Red Sandstone of Anglesey, North Wales: Yorks. Geol. Soc. Proc., v. 35, p. 139-185.

———, 1966, On bedforms and paleocurrents: Sedimentology, v. 6, p. 153-190.

———, 1967, Notes on some fundamentals of paleocurrent analysis, with reference to preservation potential and sources of variance: Sedimentology, v. 9, p. 75-88.

———, 1968, Current ripples, their relation to patterns of water and sediment motion: North Holland Pub. Co., Amsterdam, 433 p.

———, 1970a, Physical processes of sedimentation: George Allen and Unwin, London, 248 p.

———, 1970b, Studies in fluviatile sedimentation: a comparison of fining-upward cyclothems, with special reference to coarse member composition and growth: J. Sediment. Petrol., v. 40, p. 298-323.

————, 1973, Phase differences between bed configurations and flow in natural environments, and their geological relevance: Sedimentology, v. 20, p. 323-329.

————, 1974a, Reaction, relaxation and lag in natural sedimentary systems: general principles, examples and lessons: Earth Sci. Revs., v. 10, p. 263-342.

————, 1974b, Studies in fluviatile sedimentation: implications of pedogenic carbonate units, Lower Old Red Sandstone, Anglo-Welsh outcrops: Geol. J., v. 9, p. 181-208.

————, 1977, The plan shape of current ripples in relation to flow conditions: Sedimentology, v. 24, p. 53-62.

Bailey, E. B., 1962, Charles Lyell: Thomas Nelson and Sons, London, 214 p.

————, 1967, James Hutton - the founder of modern geology: Elsevier, Amsterdam, 161 p.

Baker, V. R., 1973, Paleohydrology and sedimentology of Lake Missoula flooding in eastern Washington: Geol. Soc. Am. Spec. Paper 144.

Barrell, J., 1912, Criteria for the recognition of ancient delta deposits, Geol. Soc. Am. Bull., v. 23, p. 377-446.

————, 1913, 1914, The Upper Devonian delta of the Appalachian geosyncline: Am. J. Sci., v. 36, p. 429-472; v. 37, p. 87-109, 229-253.

————, 1925, Marine and terrestrial conglomerates; Geol. Soc. Am. Bull. v. 36, p. 279-342.

Beerbower, J. R., 1964, Cyclothems and cyclic depositional mechanisms in alluvial plain sedimentation: Geol. Surv. Kansas Bull. 169, v. 1, p. 31-42.

Belcher, R. C., 1975, The geomorphic evolution of the Rio Grande: Baylor Geol. Studies, Bull. 29.

Bernard, H. A., and Major, C. J., 1963, Recent meander belt deposits of the Brazos River: an alluvial "sand" model (abs.): Am. Assoc. Petrol. Geol. Bull., v. 47, p. 350.

————, Leblanc, R. J., and Major, C. J., 1962, Recent and Pleistocene geology of southeast Texas:*in* E. H. Rainwater and R. P. Zingula, *eds.*, Geology of the Gulf Coast and central Texas; Geol. Soc. Am. Guidebook for 1962 Ann. Mtg., p. 175-224.

Bersier, A., 1948, Les sédimentations rythmiques synorogéniques dans l'avant-fosse molassique alpine: 18th Int. Geol. Congr., Part IV, p. 83-93.

————, 1958, Séquences détritiques et divagations fluviales: Eclog. Geol. Helv., v. 51, p. 854-893.

Blackwelder, E., 1928, Mudflow as a geologic agent in semiarid mountains: Geol. Soc. Am. Bull., v. 39, p. 465-484.

Blissenbach, E., 1954, Geology of alluvial fans in semi-arid regions: Geol. Soc. Am. Bull., v. 65, p. 175-190.

Bluck, B. J., 1971, Sedimentation in the meandering River Endrick; Scott. J. Geol., v. 7, p. 93-138.

Boothroyd, J. C., and Ashley, G. M., 1975, Process, bar morphology, and sedimentary structures on braided outwash fans, northeastern Gulf of Alaska: *in* A. V. Jopling and B. C. McDonald, *eds.*, Glacio-fluvial and glaciolacustrine sedimentation; Soc. Econ. Paleont. Mineral. Spec. Pub. 23, p. 193-222.

Botvinkina, L. N., *et al.*, 1954, A study of the textures and conditions of deposition of the recent alluvium and certain deposits of the lower reaches of the Don Valley and the shore of the Sea of Azov; Trans. Inst. Geol., Nauk, Akad. Nauk. S.S.S.R., Geol. Ser. 151, p. 30-89.

Bridge, J. S., 1975, Computer simulation of sedimentation in meandering streams: Sedimentology, v. 22, p. 3-44.

Brinkman, R., 1933, Ueber Kreuzschichtung im deutschen Buntsandsteinbecken: Nachr. von der Gesselschaft der Wissensch. zu Gottingen, Math. Phys. Kl. Fachtgruppe IV, Nr. 32.

Brush, L. M., jr., 1958, Study of stratification in a large laboratory flume (abs.): Geol. Soc. Am. Bull., v. 69, p. 1542.

Bucher, W. H., 1919, On ripples and related sedimentary surface forms and their paleogeographic interpretation: Am. J. Sci., v. 47, p. 149-210, 241-269.

Bull, W. B., 1964, Alluvial fans and near surface subsidence in western Fresno County, California: U.S. Geol. Survey Prof. Paper 437-A.

————, 1972, Recognition of alluvial fan deposits in the stratigraphic record; *in* J. K. Rigby and W. K. Hamblin, *eds.*, Recognition of ancient sedimentary environments; Soc. Econ. Paleont. Mineral. Spec. Pub. 16, p. 63-83.

Campbell, C. V., 1976, Reservoir geometry of a fluvial sheet sandstone: Am. Assoc. Petrol. Geol. Bull., v. 60, p. 1009-1020.

Cant, D. J., and Walker, R. G., 1976, Development of a braided-fluvial facies model for the Devonian Battery Point sandstone, Quebec: Can. J. Earth Sci., v. 13, p. 102-119.

Carey, W. C., and Keller, M. D., 1957, Systematic changes in the beds of alluvial rivers: Am. Soc. Civil Eng., Proc., v. 83, HY4, p. 1-24.

Chamberlin, T. C., and Salisbury, R. D., 1909, Geology: processes and their results: John Murray, London, 2nd ed., 684 p.

Chawner, W. D., 1935, Alluvial fan flooding, the Montrose, California, flood of 1934: Geogr. Rev., v. 25, p. 225-263.

Church, M., and Gilbert, R., 1975, Proglacial fluvial and lacustrine environments: *in* A. V. Jopling and B. C. McDonald, *eds.*, Glacio-fluvial and glaciolacustrine sedimentation; Soc. Econ. Paleont. Mineral. Spec. Pub. 23, p. 22-100.

Coleman, J. M., 1969, Brahmaputra River: channel processes and sedimentation: Sediment. Geol., v. 3, p. 129-239.

Collinson, J. D., 1970, Bedforms of the Tana River, Norway: Geogr. Ann., v. 52A, p. 31-55.

————, 1971, Current vector dispersion in a river of fluctuating discharge: Geol. Mijnbouw, v. 50, p. 671-678.

Conybeare, C. E. B., and Crook, K. A. W., 1968, Manual of sedimentary structures: Australian Bur. Min. Resources, Geol. and Geophys. Bull. 102, 327 p.

Conybeare, W. D., and Phillips, W., 1824, Outline of the geology of England and Wales: Am. J. Sci., v. 7, p. 203.

Cornish, V., 1899, On kinematology. The study of waves and wave structures of the atmosphere, hydrosphere and lithosphere: Geogr. J., v. 13, p. 624-626.

Cotter, E., 1971, Sedimentary structures and the interpretation of paleoflow characteristics of the Ferron Sandstone (Upper Cretaceous), Utah: J. Sediment. Petrol., v. 41, p. 129-138 (*see also* discussion and reply, v. 43, p. 1176-1180).

Curray, J. R., 1956, The analysis of two-dimensional orientation data: J. Geol., v. 64, p. 117-131.

Dana, J. D., 1850a, On denudation in the Pacific; Am. J. Sci., v. 9, p. 48-62.

————, 1850b, On the degradation of the rocks of New South Wales and formation of valleys: Am. J. Sci., v. 9, p. 289-294.

————, 1862, Manual of Geology: Ivison, Blakeman, Taylor & Co., New York.

Darwin, G. H., 1883, On the formation of ripple marks: Roy. Soc. London, Proc., v. 36, p. 18-43.

Davis, W. M., 1898a, The Triassic Formation of Connecticut: U.S. Geol. Survey Ann. Rept. 18, pt. 2, p. 1-192.

————, 1898b, Physical Geography: Ginn and Co., Boston, 432 p.

————, 1899, The geographical cycle: Geogr. J., v. 14, p. 481-504.

————, 1900, The Fresh-water Tertiary formations of the Rocky Mountain region: Proc. Am. Acad. Arts and Sciences, v. 35, p. 345-373.

Dixon, E. E. L., 1921, The geology of the South Wales Coalfield; Part XIII: The country around Tenby: Geol. Survey Great Britain, Memoir.

Doeglas, D. J., 1962, The structure of sedimentary deposits of braided rivers: Sedimentology, v. 1, p. 167-190.

Drew, F., 1873, Alluvial and lacustrine deposits and glacial records of the Upper-Indus Basin: Quart. J. Geol. Soc. London, v. 29, p. 441-471.

Du Boys, P. F. D., 1879, Le Rhone et le rivier a lit affouillable: Annales des Ponts et Chaussees, v. 18, ser. 5, p. 141-195.

Dunbar, C. O., and Rodgers, J., 1957, Principles of stratigraphy: John Wiley and Sons Inc., New York, 356 p.

Eicher, D. L., 1969, Paleobathymetry of Cretaceous Greenhorn sea in eastern Colorado: Am. Assoc. Petrol. Geol. Bull., v. 53, p. 1075-1090.

Einstein, H. A., 1950, The bed-load function for sediment transportation in open channel flows: U.S. Dept. Agriculture, Soil Cons. Serv. Tech. Bull. 1026.

Epry, C., 1913, Ripple marks: Ann. Rept. Smithsonian Inst., p. 307-318.

Everest, R., 1832, A quantitative study of stream transportation: J. Asiatic Soc. Bengal, v. 1, p. 238-240.

Fenneman, N. M., 1906, Floodplains produced without floods: Am. Geogr. Soc. Bull., v. 38, p. 89-91.

Ferm, J. C., and Cavaroc, V. V., jr., 1968, A nonmarine sedimentary model for the Allegheny Rocks of West Virginia: in G. deV. Klein ed., Late Paleozoic and Mesozoic continental sedimentation, northeastern North America; Geol. Soc. Am. Spec. Paper 106, p. 1-20.

Finch, J., 1823, Geological essay on the Tertiary formations in America: Am. J. Sci., v. 7, p. 31-43.

Fisk, H. N., 1944, Geological investigation of the alluvial valley of the Lower Mississippi River: Mississippi River Commission, 78 p.

———, 1947, Fine-grained alluvial deposits and their effect on Mississippi River activity: Mississippi River Commission, 82 p.

Folk, R. L., 1966, A review of grain-size parameters: Sedimentology, v. 6, p. 73-93.

———, and Ward, W. C., 1957, Brazos River bar: a study in the significance of grain size parameters: J. Sediment. Petrol., v. 27, p. 3-26.

Friedkin, J. F., 1945, A laboratory study of the meandering of alluvial rivers: Mississippi River Commission, Vicksburg.

Friedman, G. M., 1961, Distinction between dune, beach and river sands from their textural characteristics; J. Sediment Petrol., v. 31, p. 514-529.

———, 1967, Dynamic processes and statistical parameters compared for size frequency distribution of beach and river sands; J. Sediment. Petrol., v. 37, p. 327-354.

Friend, P. F., and Moody-Stuart, M., 1972, Sedimentation of the Wood Bay Formation (Devonian) of Spitsbergen: Regional analysis of a late orogenic basin: Norsk Polarinstitutt Skrifter 157.

Galay, V. J., Kellerhals, R., and Bray, D. I., 1973, Diversity of river types in Canada: in Fluvial Processes and Sedimentation, Proc. Hydrology Symp., Edmonton, National Research Council, Canada, p. 217-250.

Geddes, A., 1960, The alluvial morphology of the Indo-Gangetic Plains: Trans. Inst. Brit. Geogr., no. 28, p. 253-276.

Geikie, A., 1882, Text book of geology: Macmillan and Co., London.

———, 1905, The Founders of Geology (republished in 1962 by Dover Pub. Inc., New York, 486 p.).

Gilbert, G. K., 1880, Land, sculpture, geology of the Henry Mountains: U.S. Geogr. and Geological Survey of the Rocky Mountain Region, 2nd ed., 160 p.

———, 1884, Ripple marks: Science, v. 3, p. 375-376.

———, 1899, Ripple marks and cross-bedding: Geol. Soc. Am. Bull., v. 10, p. 135-140.

———, 1914, The transportation of debris by running water; U.S. Geol. Survey Prof. Paper 86.

Gole, C. V., and Chitale, S. V., 1966, Inland delta building activity of Kosi River: J. Hydraulics Div., Am. Soc. Civil Eng., v. 92, HY2, p. 111-126.

Grabau, A. W., 1906, Types of sedimentary overlap: Geol. Soc. Am. Bull., v. 17, p. 567-636.

———, 1907, Types of cross-bedding and their stratigraphic significance: Science, v. 25, p. 295-296.

———, 1913a, Principles of Stratigraphy: New York, A. G. Seiler and Co., 1185 p.

———, 1913b, Early Paleozoic delta deposits of North America: Geol. Soc. Am. Bull., v. 24, p. 399-528.

———, 1917, Problems of the interpretation of sedimentary rocks: Geol. Soc. Am. Bull., v. 28, p. 735-744.

Griffith, W. M., 1927, A theory of silt and scour: Inst. Civil Engr. Proc., v. 223-314.

Hack, J. T., 1957, Studies of longitudinal stream profiles in Virginia and Maryland: U.S. Geol. Survey Prof. Paper 294-B.

Hahmann, P., 1912, Die bildung von sandduenen bei gleichmaessiger stroemung: Ann. Phys., p. 637-676.

Happ, S. C., Rittenhouse, G., and Dobson, G. C., 1940, Some principles of accelerated stream and valley sedimentation: U.S. Dept. Agriculture Tech. Bull., 695.

Harms, J. C., and Fahnestock, R. K., 1965, Stratification, bed forms, and flow phenomena (with an example from the Rio Grande): *in* G. V. Middleton, *ed.*, Primary sedimentary structures and their hydrodynamic interpretation; Soc. Econ. Paleont. Mineral. Spec. Pub. 12, p. 84-115.

——, Mackenzie, D. B., and McCubbin, D. G., 1963, Stratification in modern sands of the Red River, Louisiana: J. Geol., v. 71, p. 566-580.

——, Southard, J. B., Spearing, D. R., and Walker, R. G., 1975, Depositional environments as interpreted from primary sedimentary structures and stratification sequences: Soc. Econ. Paleont. Mineral. Short Course 2, Dallas, 161 p.

Hayden, H. H., 1821, Geological essays, or an enquiry into some of the geological phenomena to be found in various parts of America and elsewhere: Am. J. Sci., v. 3, p. 47-57.

Hein, F. J., and Walker, R. G., 1977, Bar evolution and development of stratification in the gravelly, braided Kicking Horse River, British Columbia: Can. J. Earth Sci., v. 14, p. 562-570.

Hider, A., 1882, Appendix D, Report of assistant engineer Arthur Hider upon observations at Lake Providence, November, 1879 to November, 1880: *in* Rept. Mississippi River Comm., p. 80-98.

Higham, N., 1963, A very scientific gentleman: the major achievements of Henry Clifton Sorby: Pergamon Press, Oxford, 160 p.

Hjulström, F., 1935, Studies in the morphological activity of rivers as illustrated by the River Fyris: Geol. Inst. Univ. Uppsala, Bull., v. 25, p. 221-528.

Hobbs, W. H., 1906, Gaudix Formation of Granada, Spain: Geol. Soc. Am. Bull., v. 17, p. 285-294.

Hoff, K. E. A. von, 1822-1824, Geschichte der durch Ueberlieferung nachgewiesen naturlichen Veranderungen der erdoberflache: 2 vols, J. Perthes, Gotha.

Hooke, R. LeB., 1967, Processes on arid region alluvial fans: J. Geol., v. 75, p. 438-460.

Hoover, H. C., and Hoover, L. H., 1950, Georgius Agricola: De Re Metallica: translated from the first Latin edition of 1556; Dover Pub. Inc., New York, 638 p.

Horne, J. C., and Ferm, J. C., 1976, Carboniferous depositional environments in the Pocahontas Basin, eastern Kentucky and southern West Virginia: Guidebook, Dept. Geology, Univ. South Carolina.

Humphreys, A. A., and Abbot, H. L., 1861, Report upon the physics and hydraulics of the Mississippi: U.S. Army Corps Topographic Eng., Prof. Paper 4.

Hunt, A. R., 1882, On the formation of ripple marks; Roy. Soc. London Proc., v. 34, p. 1-19.

Hutton, J., 1795, Theory of the Earth (2 vols., reprinted, 1959, by Engelmann, Wheldon and Wesley Ltd., Weinheim, Bergstr.).

Inman, D. L., 1949, Sorting of sediments in the light of fluid mechanics: J. Sediment. Petrol., v. 19, p. 51-70.

——, 1952, Measures for describing size distribution of sediments: J. Sediment. Petrol., v. 22, p. 125-145.

Jackson, J. R., 1834, Hints on the subject of geographical arrangement and nomenclature: Roy. Geogr. Soc. J., v. 4, p. 72-88.

Jackson, R. G., 1975, Hierarchical attributes and a unifying model of bed forms composed of cohesionless material and produced by shearing flow: Geol. Soc. Am. Bull., v. 86, p. 1523-1533.

——, 1976a, Largescale ripples of the lower Wabash River: Sedimentology, v. 23, p. 593-624.

——, 1976b, Depositional model of point bars in the lower Wabash River: J. Sediment. Petrol., v. 46, p. 579-594.

Jamieson, T. F., 1860, On the drift and rolled gravel of the North of Scotland: Quart. J. Geol. Soc. London, v. 16, p. 347-371.

Jeffreys, H., 1929, On the transport of sediment by streams: Cambridge Philos. Soc. Proc., v. 25, p. 272-276.

Johansson, C. E., 1963, Orientation of pebbles in running water. A laboratory study: Geogr. Ann., v. 45, p. 85-112.

Johnson, A. M., and Hampton, M. A., 1969, Subaerial and subaqueous flow of slurries: unpub. final report to U.S. Geol. Survey, Branner Library, Stanford, California.

Johnson, D., 1932, Streams and their significance: J. Geol., v. 40, p. 481-496.

Jopling, A. V., 1963, Hydraulic studies on the origin of bedding: Sedimentology, v. 2, p. 115-121.

———, 1965, Hydraulic factors controlling the shape of laminae in laboratory deltas: J. Sediment. Petrol., v. 35, p. 777-791.

———, 1975, Early studies on stratified drift: *in* A. V. Jopling and B. C. McDonald, *eds.*, glaciofluvial and glaciolacustrine sedimentation; Soc. Econ. Paleont. Mineral Spec. Pub. 23, p. 4-21.

Kennedy, J. F., 1963, The mechanics of dunes and antidunes in erodible-bed channels: J. Fluid Mech., v. 16, p. 521-544.

Kennedy, R. G., 1895, The prevention of sitting in irrigation canals: Inst. Civil Eng., v. 119, p. 281-290.

Kindle, E. M., 1911, Cross-bedding and absence of fossils considered as criteria of continental deposits: Am. J. Sci., v. 32, p. 225-230.

———, 1917, Recent and fossil ripple marks: Geol. Survey Canada, Mus. Bull. 25.

King, W. S. H., 1916, The nature and formation of sand ripples and dunes: Geogr. J., v. 46, p. 189-209.

Klovan, J. E., 1966, The use of factor analysis in determing depositional environments from grain-size distributions; J. Sediment. Petrol., v. 36, p. 115-125.

Knight, S. H., 1929, The Fountain and Casper Formations of the Laramie Basin: Univ. Wyoming Pub. Sci., Geology, no. 1, p. 1-82.

Krumbein, W. C., 1934, Size frequency distribution of sediments: J. Sediment. Petrol., v. 4, p. 65-77.

———, 1941, Measurement and geologic significance of shape and roundness of sedimentary particles: J. Sediment. Petrol., v. 11, p. 64-72.

Lane, E. W., 1935, Stable channels in erodible material: Trans. Am. Soc. Civil Eng., v. 63, p. 123-142.

Langbein, W. B., and Leopold, L. B., 1966, River meanders - theory of minimum variance: U.S. Geol. Survey Prof. Paper 422-H.

Lawson, A. C., 1913, The petrographic designation of alluvial fan formations: Univ. California Pub., Dept. G, v. 7, p. 325-334.

Leeder, M. R., 1973, Fluviatile fining-upward cycles and the magnitude of paleochannels: Geol. Mag., v. 110, p. 265-276.

Leliavsky, S., 1955, An introduction to fluvial hydraulics: Constable and Co., London (reprinted, 1966, by Dover Pub. Inc., New York, 257 p.).

Leopold, L. B., and Maddock, T., jr., 1953, The hydraulic geometry of stream channels and some physiographic implications: U.S. Geol. Survey Prof. Paper 252.

———, and Wolman, M.G., 1957, River channel patterns; braided, meandering, and straight: U.S. Geol. Survey Prof. Paper 282-B.

———, ———, and Miller, J. P., 1964, Fluvial processes in geomorphology: W. H. Freeman and Co., San Francisco, 522 p.

Lustig, L. K., 1965, Clastic sedimentation in Deep Springs Valley, California: U.S. Geol. Survey Prof. Paper 352-F.

Lyell, C., 1830-1833, Principles of Geology: John Murray, London, 3 vols. (reprinted by Johnson Reprint Corp., New York, 1969).

Mackin, J. H., 1937, Erosional history of the Big Horn Basin, Wyoming: Geol. Soc. Am. Bull., v. 48, p. 813-894.

———, 1948, Concept of the graded river: Geol. Soc. Am. Bull., v. 59, p. 463-512.

Mather, K. F., and Mason, S. L., 1964, A source book in geology: Hafner Pub. Co., New York, 702 p.

McGowen, J. H., and Garner, L. E., 1970, Physiographic features and stratification types of coarse-grained point bars; modern and ancient examples: Sedimentology, v. 14, p. 77-112.

McKee, E. D., 1938, Original structures in Colorado River flood deposits of Grand Canyon: J. Sediment. Petrol., v. 8, p. 77-83.

———, 1939, Some types of bedding in the Colorado River delta: J. Geol., v. 47, p. 64-81.

———, 1957, Flume experiments on the production of stratification and cross-stratification: J. Sediment. Petrol., v. 27, p. 129-134.

———, Crosby, E. J., and Berryhill, H. L., 1967, Flood deposits, Bijou Creek, Colorado: J. Sediment. Petrol., v. 37, p. 829-851.

———, and Weir, G. W., 1953, Terminology for stratification and cross-stratification in sedimentary rocks: Geol. Soc. Am. Bull., v. 64, p. 381-390.

Medlicott, H. B., and Blanford, W. T., 1879, Manual of the Geology of India, pt. II.

Melton, F. A., 1936, An empirical classification of flood-plain streams: Geogr. Rev., v. 26, p. 593-609.

Miall, A. D., 1970, Devonian alluvial fans, Prince of Wales Island, Arctic Canada; J. Sediment. Petrol., v. 40, p. 556-571.

———, 1973, Markov chain analysis applied to an ancient alluvial plain succession: Sedimentology, v. 20, p. 347-364.

———, 1974, Paleocurrent analysis of alluvial sediments - a discussion of directional variance and vector magnitude: J. Sediment. Petrol., v. 44, p. 1174-1185.

———, 1976, Paleocurrent and paleohydrologic analysis of some vertical profiles through a Cretaceous braided stream deposit: Sedimentology, v. 23, p. 459-484.

———, 1977a, A review of the braided river depositional environment: Earth Sci. Revs., v. 13, p. 1-62.

———, 1977b, Fluvial sedimentology: Fluvial lecture series notes, Can. Soc. Petrol. Geol., 111 p.

Middleton, G. V., ed., 1965, Primary sedimentary structures and their hydrodynamic interpretation: Soc. Econ. Paleont. Mineral., Spec. Pub. 12.

———, 1976, Hydraulic interpretation of sand size distributions: J. Geol., v. 84, p. 405-426.

———, 1977, Introduction - progress in hydraulic interpretation of sedimentary structures: in G. V. Middleton, ed., Sedimentary process: hydraulic interpretation of primary sedimentary structures; Soc. Econ. Paleont. Mineral., Reprint Series no. 3, p. 1-15.

———, in press, Sedimentology - history: in R. W. Fairbridge, ed., Encyclopedia of Sedimentology.

———, and Southard, J. B., 1977, Mechanics of Sediment movement: Soc. Econ. Paleont. Mineral., Short Course no. 3.

Mollard, J. D., 1973, Airphoto interpretation of fluvial features: in Fluvial Processes and Sedimentation, Proc. Hydrology Symp., Edmonton, National Research Council, Canada, p. 341-380.

Moody-Stuart, M., 1966, High and low sinuosity stream deposits, with examples from the Devonian of Spitzbergen: J. Sediment. Petrol., v. 36, p. 1102-1117.

Nanz, R. H., jr., 1954, Genesis of Oligocene sandstone reservoir, Seeligson field, Jim Wells and Kleberg Counties, Texas: Am. Assoc. Petrol. Geol. Bull., v. 38, p. 96-117.

Nilsen, T. H., 1968, The relationship of sedimentation to tectonics in the Solund Devonian district of southwestern Norway: Universitetsforlaget, Oslo, Norges Geologiske Undersökelse, No. 259, 108 p.

Olson, J. S., and Potter, P. E., 1954, Variance components of cross-bedding direction in some basal Pennsylvanian sandstones of the eastern Interior Basin: statistical methods: J. Geol., v. 62, p. 26-49.

Oomkens, E., and Terwindt, J. H. J., 1960, Inshore estuarine sediments in the Haringvliet (The Netherlands): Geol. Mijnbouw, v. 39, p. 701-710.

Ore, H. T., 1964, Some criteria for recognition of braided stream deposits: Wyoming Contr. Geol., v. 3, p. 1-14.

Owens, J. S. 1908, Experiments on the transporting power of sea currents: Geogr. J., v. 31, p. 415-420.

Padgett, G. V., and Ehrlich, R., 1976, Paleohydrologic analysis of a Late Carboniferous fluvial system, southern Morocco: Geol. Soc. Am. Bull., v. 87, p. 1101-1104.

Passega, R., 1957, Texture as characteristic of clastic deposition: Am. Assoc. Petrol. Geol. Bull., v. 41, p. 1952-1984.

Peale, A. C., 1879, Report on the geology of the Green River District: *in* F. V. Hayden, U.S. Geol. and Geogr. Survey Terr., 9th Ann. Rept., 720 p.

Pelletier, B. R., 1958, Pocono paleocurrents in Pennsylvania and Maryland: Geol. Soc. Am. Bull., v. 69, p. 1033-1064.

Pettijohn, F. J., 1949, Sedimentary rocks: Harper and Row, New York, 526 p.

———, 1962, Paleocurrents and paleogeography: Am. Assoc. Petrol. Geol. Bull., v. 46, p. 1468-1493.

———, and Potter, P. E., 1964, Atlas and glossary of primary sedimentary structures: Springer-Verlag, New York, 370 p.

———, ———, and Siever, R., 1972, Sand and sandstone: Springer-Verlag, New York, 618 p.

Pirson, S. J., 1970, Geologic well log analysis: Gulf Pub. Co., Houston, 370 p.

Playfair, J., 1802, Illustrations of the Huttonian theory of the Earth: Cadell and Davies, London, 528 p. (Reprinted by Dover Pub. Inc., New York, 1964).

Potter, P. E., 1955, The petrology and origin of the Lafayette Gravel, Part I: mineralogy and petrology: J. Geol., v. 63, p. 1-38.

Potter, P. E., 1959, Facies models conference: Science, v. 129, p. 1292-1294.

———, 1978, Significance and origin of big rivers: J. Geol., v. 86, p. 13-33.

———, and Pettijohn, F. J., 1963, 1977, Paleocurrents and basin analysis: Academic Press Inc., New York, 296 p., 2nd ed. 1977, Springer-Verlag, Berlin, 425 p.

Powell, J. W., 1875, Exploration of the Colorado River of the West: Washington, 291 p. (*see* also U.S. 43rd Congr. 1st session, H. Misc. Doc. 265, 29 p., 1874).

Pryor, W. A., 1960, Cretaceous sedimentation in upper Mississippi Embayment: Am. Assoc. Petrol. Geol. Bull., v. 44, p. 1473-1504.

Puigdefabregas, C., 1973, Miocene point-bar deposits in the Ebro Basin, northern Spain: Sedimentology, v. 20, p. 133-144.

Reade, T. M., 1884, Ripple marks in drift in Shropshire and Cheshire: Quart. J. Geol. Soc. London, v. 40, p. 267-269.

Reiche, P., 1938, An analysis of cross-lamination: the Coconino Sandstone: J. Geol., v. 46, p. 905-932.

Rodine, J. D., and Johnson, A. M., 1976, The ability of debris, heavily freighted with coarse clastic materials to flow on gentle slopes: Sedimentology, v. 23, p. 213-234.

Rouse, H., 1939, Experiments on the mechanics of sediment suspension: 5th Internat. Congr. Applied Mech., Cambridge, Mass., p. 550-554.

Rubey, W. W., 1938, The force required to move particles on a stream Bed: U.S. Geol. Survey Prof. Paper 189-E, p. 121-141.

———, and Bass, N. W., 1925, The geology of Russell County, Kansas: Kansas State Geol. Survey Bull., v. 10, p. 1-86.

Rust, B. R., 1972, Structure and process in a braided river: Sedimentology, v. 18, p. 221-246.

———, 1975, Fabric and structure in glaciofluvial gravels: *in* A. V. Jopling and B. C. McDonald, *eds.*, Glaciofluvial and glaciolacustrine sedimentation; Soc. Econ. Paleont. Mineral. Spec. Pub. 23, p. 238-248.

Sagoe, K-M. O., and Visher, G. S., 1977, Population breaks in grain-size distributions of sand - a theoretical model: J. Sediment. Petrol., v. 47, p. 285-310.

Sahu, B. K., 1964, Depositional mechanisms from the size analysis of clastic sediments: J. Sediment. Petrol., v. 34, p. 73-83.

Schumm, S. A., 1963, A tentative classification of alluvial river channels: U.S. Geol. Survey Circ. 477.

———, 1968a, Speculations concerning paleohydrologic controls of terrestrial sedimentation: Geol. Soc. Am. Bull., v. 79, p. 1573-1588.

———, 1968b, River adjustment to altered hydrologic regimen - Murrumbidgee River and paleochannels, Australia: U.S. Geol. Survey Prof. Paper 598.

———, *ed.*, 1972a, River morphology: Benchmark papers in geology, Dowden, Hutchinson and Ross, Stroudsburg, Pennsylvania, 429 p.

———, 1972b, Fluvial paleochannels: *in* J. K. Rigby and W. K. Hamblin, *eds.*, recognition of ancient sedimentary environments; Soc. Econ. Paleont. Mineral. Spec. Pub. 16, p. 98-107.

Shantzer, E. V., 1951, Alluvium of plains rivers in a temperate zone and its significance for understanding the laws governing the structure and formation of alluvial suites. Tr. Inst. Geol. Nauk, Akad. Nauk S.S.S.R., Geol. Ser. 135, 271 p.

Sharp, R. P., and Nobles, L. H., 1953, Mudflow in 1941 at Wrightwood, southern California: Geol. Soc. Am. Bull., v. 64, p. 547-560.

Shaw, J., and Kellerhals, R., 1977, Paleohydraulic interpretation of antidune bedforms with applications to antidunes in gravel: J. Sediment. Petrol., v. 47, p. 257-266.

Shields, A., 1936, Anwendung der Ahnlichkeitsmechanik und der Turbulenzforschung auf die Geschiebebewegung: Mitteilungen der Preuss. Versuch aust. f. Wasserbau u Schiffbau, Berlin, Heft 26, 26 p.

Simons, D. B., and Richardson, E. V., 1961, Forms of bed roughness in alluvial channels: Am. Soc. Civil Eng. Proc., v. 87, HY3, p. 87-105.

———, ———, and Nordin, C. F., jr., 1965, Sedimentary structures generated by flow in alluvial channels: *in* G. V. Middleton, *ed.*, Primary sedimentary structures and their hydrodynamic interpretation; Soc. Econ. Paleont. Mineral. Spec. Pub. 12, p. 34-52.

Smith, D. G., 1973, Aggradation of the Alexandra-North Saskatchewan River, Banff Park, Alberta: *in* M. Morisawa, *ed.*, Fluvial geomorphology; Proc. 4th Ann. Geomorph. Symp., Pub. in Geomorph., SUNY-Binghamton, New York, p. 201-219.

Smith, N. D., 1970, The braided stream depositional environment: comparison of the Platte River with some Silurian clastic rocks, north central Appalachians: Geol. Soc. Am. Bull., v. 81, p. 2993-3014.

———, 1971, Transverse bars and braiding in the Lower Platte River, Nebraska: Geol. Soc. Am. Bull., v. 82, p. 3407-3420.

———, 1974, Sedimentology and bar formation in the upper Kicking Horse River, a braided outwash stream: J. Geol., v. 82, p. 205-224.

Sneed, E. D., and Folk, R. L., 1958, Pebbles in the lower Colorado River, Texas, a study in particle morphogenesis: J. Geol., v. 66, p. 114-150.

Sorby, H. C., 1852, On the oscillation of the currents drifting the sandstone beds of the southeast of Northumberland, and on their general direction in the coalfield in the neighbourhood of Edinburgh: Proc. W. Yorks. Geol. Soc., v. 3, p. 232-240.

———, 1859, On the structures produced by the currents present during the deposition of stratified rocks: The Geologist, v. 2, p. 137-147.

———, 1908, On the application of quantitative methods to the study of the structure and history of rocks: Quart. J. Geol. Soc. London, v. 64, p. 171-233.

Southard, J. B., 1971, Representation of bed configurations in depth-velocity-size diagrams: J. Sediment Petrol., v. 41, p. 903-915.

Spencer, D. W., 1963, The interpretation of grain size distribution curves of clastic sediments: J. Sediment. Petrol., v. 33, p. 180-190.

Spurr, J. E., 1894a, False bedding in stratified drift deposits: Am. Geologist, v. 13, p. 43-47.

———, 1894b, Oscillation and single current ripple marks: Am. Geologist, v. 13, p. 201-206.

Stamp, L. D., 1925, Seasonal rhythms in the Tertiary sediments of Burma: Geol. Mag., v. 62, p. 515-528.

Statham, I., 1976, Debris flows on vegetated screes in the Black Mountain, Carmarthenshire: Earth Surf. Proc., v. 1, p. 173-180.

Summerson, C. H., *ed.*, 1976, Sorby on sedimentology: A collection of papers from 1851 to 1908 by Henry Clifton Sorby: Geological Milestones I, Comparative Sedimentology Laboratory, Univ. Miami, 225 p.

Sundborg, A., 1956, The River Klarälven, a study of fluvial processes: Geogr. Ann., v. 38, p. 217-316.

Surell, A., 1870, Etude sur les torrents des Hautes-Alpes: Paris, 2nd ed.

Tanner, W. F., 1955, Paleogeographic reconstructions from cross-bedding studies: Am. Assoc. Petrol. Geol. Bull., v. 39, p. 2471-2483.

Tolman, C. F., 1909, Erosion and deposition in the southern Arizona bolson region: J. Geol., v. 17, p. 136-163.

Trowbridge, A. C., 1911, The terrestrial deposits of Owens Valley, California: J. Geol., v. 19, p. 706-747.

Twenhofel, W. H., 1932, Treatise on sedimentation: 2nd ed., Williams and Wilkins, New York, 926 p.

Udden, J. A., 1914, Mechanical composition of clastic sediments: Geol. Soc. Am. Bull., v. 25, p. 655-744.

Vanoni, V. A., 1946, Transportation of suspended sediment by water: Am. Assoc. Civil Eng. Trans., v. 111, p. 67-102.

Van Straaten, L. M. J. U., 1954, Sedimentology of Recent tidal flat deposits and the Psammites du Condroz (Devonian): Geol. Mijnbouw, v. 16, p. 25-47.

Visher, G. S., 1965a, Fluvial processes as interpreted from ancient and recent fluvial deposits: in G. V. Middleton, ed., Primary sedimentary structures and their hydrodynamic interpretation; Soc. Econ. Paleont. Mineral. Spec. Pub. 12, p. 84-115.

———, 1965b, Use of vertical profile in environmental reconstruction: Am. Assoc. Petrol. Geol. Bull., v. 49, p. 41-61.

———, 1969, Grain size distributions and depositional processes: J. Sediment. Petrol., v. 39, p. 1074-1106.

Walker, R. G., 1976, Facies models 3: sandy fluvial systems: Geoscience Canada, v. 3, p. 101-109.

Walther, J., 1893-1894, Einleitung in die geologie als historische wissenschaft: Gustav Fischer, Jena, 3 vols., 1055 p.

Wentworth, C. K., 1922, A scale of grade and class terms for clastic sediments: J. Geol., v. 30, p. 377-392.

Williams, G. E., 1971, Flood deposits of the sandbed ephemeral streams of central Australia: Sedimentology, v. 17, p. 1-40.

Williams, P. F., and Rust, B. R., 1969, The sedimentology of a braided river: J. Sediment. Petrol., v. 39, p. 649-679.

Wilson, L. G., 1972, Charles Lyell, the years to 1841: the revolution in geology: Yale Univ. Press, New Haven, Conn., 553 p.

Winkley, B. R., 1977, Manmade cutoffs on the lower Mississippi River; conception, construction and river response: U.S. Corp. of Engineers, Vicksburg, Miss., 213 p.

Wolman, M. G., and Leopold, L. B., 1957, River flood plains: some observations on their formation: U.S. Geol. Survey Prof. Paper 282-C.

Wright, M D., 1959, The formation of cross-bedding by a meandering or braided stream: J. Sediment. Petrol, v. 29, p. 610-615.

Zittel, K. A., von, 1901, History of geology and palaeontology to the end of the nineteenth century (translated by M. M. Ogilvie-Gordon): Walter Scott, London, 562 p.

TEXTURE AND STRUCTURE OF FLUVIAL DEPOSITS

SEDIMENT TRANSPORT MODELS

H. W. Shen[1]

Abstract

Sediment transport models are classified into the following categories: 1) wash load, 2) bed load, 3) total bed material load. For each category, the more popular models are presented in similar format as a basis for comparison. Selected current research efforts by various investigators are also discussed.

Introduction

Both fluvial sedimentologists and river engineers are interested in the behavior of alluvial rivers. However, these two groups have traditionally approached the problem from different viewpoints.

Fluvial sedimentologists have attempted to use field evidence to trace previous developments of rivers, and they are particularly interested in long-term effects. Engineers, on the other hand, have started from theory to investigate what would happen to rivers under certain conditions. From these conceptual models they would like to investigate what would be the ultimate results and they are usually interested in short-term effects say within one hundred years. Sedimentologists are more interested in finding *why* from *how*, and engineers are trying to predict *how* from *why*. There is an apparent need for these two groups of people to get together and share their knowledge.

The purpose of this paper is to briefly review the current state of knowledge on sediment transports. An attempt is being made to convert many different transport models to a similar basis for comparison.

Classification of Sediment Transport

The total sediment transport rates can be divided according to the following three different ways: 1) *By the mechanics of movements:* The total sediment transport rate can be divided into bed load plus suspended load. Bed load is the rate of the movement of the sediment particles very close to the beds. Most of the time these particles are supported by the sediment bed. The suspended load is the rate of that part of the sediment particles which are supported by the moving fluid. 2) *By methods of measurement*: In this case, it consists of measured load and unmeasured load. Since bed load is very difficult to measure, the measured load consists entirely of suspended load. It does not include the entire suspended load because we can only measure from the water surface down to about 10 centimeters from the top of the bed surface. Therefore, the measured load is the part of suspended load which is at least 10 centimeters from the top of the bed to the water surface. The unmeasured load consists of the entire bed load plus suspended load within 10 centimeters from the top of the bed surface. 3) *According to the method of calculations*: In this case, it is equal to wash loads plus bed material load. The concept of the wash load and bed material load may not be very clear to many geologists and the following explanation may be necessary. In general, the wash load is limited by the supply of that part of particles from the watersheds to the stream. The bed material load is entirely determined by the capability of the flow to transport sediments.

The best that a sediment transport equation based on river flow condition can do is to predict the sediment transport capability of a given flow for a certain sediment mixture.

---

[1]Engineering Research Center, Colorado State University, Fort Collins, Colorado 80523

For instance, one may hope to obtain a relationship between transport capability and sediment size for a flow discharge q on a particular river as shown by curve COD in Figure 1.

One may plot the available supply of various sediment sizes from the upslope area for the same river discharge q, as indicated by AOB on the same figure. The intersection of these two curves may not be as distinct as shown; however, let us assume that they intersect at point O for illustration. The sediment size is d_s^* at point O. For sediment sizes equal to and greater than d_s^*, the available supply rate from the upslope area is equal to, or greater than, the river can carry and therefore deposition of these sizes will occur at the upstream reach; the river is transporting these sediment sizes at its transport capability at the lower reach. One may hope to establish a sediment transport equation based on the sediment transport capability of the river to agree with the actual amount transported in the lower reach. However, for sediment sizes less than d_s^*, the transport capability of this river exceeds the amount of supply from upslope, the actual amounts of sediment transport for these sizes are determined by the rates of the upslope supply or production and not by the capability of the river to transport. Obviously, in order to predict the actual amounts transported for these sizes, one must study the upslope erosion rates and not the sediment transport capabilities of the river. Knowledge of d_s^* is rather important because for a sediment size equal to or greater than that size, the actual transport rate equals the sediment capability and may be determined by a sediment transport equation (based on river flow condition), and for sediment below that size, the actual transport rate equals the upslope sediment supply rate, and therefore must be estimated at that point. Consequently, the best that a sediment transport equation (based on river flow) can do is to predict sediment transport rates of sizes equal to and greater than d_s^*.

Wash load is defined as the portion of sediment load governed by the upslope supply rate and is considerably less than the sediment transport capability of a river. The sizes of

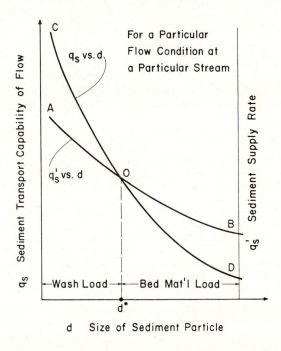

Fig. 1.   Division between wash load and bed material load.

wash load are all less than d_s^* as defined in Figure 1. Bed material load is defined as the portion of sediment load that is governed by the sediment transport capability of a river and is less or equal to its upslope supply rate. The sizes of bed material load are all equal to or greater than d_s^*. The sum of the wash load and the bed material load equals the total sediment load.

Since d_s^* is determined by the intersection of the sediment transport capability curve and the sediment supply available from upslope curve, the value of d_s^* will certainly be changed if any one or both of the curves change.

The remainder of this paper is devoted to discussing wash load, bed load and bed material load.

WASH LOAD

By definition, wash load is that portion of the sediment transport rate which is supplied to the streams from the watershed. Soil erosion formulas must be used to estimate what would be erosion rate and thus the supply rate to the streams. Almost all these equations are obtained by years of experiments collected from different watersheds.

The U. S. Department of Agriculture began to study soil loss about 1930 when the first ten Federal-State Cooperative Stations began operation. Thirty-two additional stations were established in the next 25 years. Measurements of precipitation, runoff and soil loss at these 42 stations in 23 states (east of the Rocky Mountains) were collected continuously for periods of from 5 to 30 or more years. The field plots were rectangular to facilitate normal flow row spacing for cultivated units. A "unit plot" of 22.1 meters long on a nine percent uniform slope, continuously in bare fallow soil and tilled to break surface crusts was arbitrarily selected to serve as a common reference point for evaluation. These dimensions were used because most of the plots in U. S. erosion studies by the Department of Agriculture from 1930 to 1960 were 22.1 meters long and nearly nine percent slope.

Six major factors to describe rainfall (R), soil erodibility (K), slope length (L), slope steepness (S), cropping and management (C), and supplemental erosion-control practices such as contouring, terracing, etc. (P) were used to develop the following Universal Soil Loss Equation based on the field data described above and rainfall simulation data by a multiplicative model:

$$\underline{E} = R\,K\,L\,S\,C\,P \tag{1}$$

where L, S, C and P are all dimensionless. Thus, the computed soil loss $\underline{E}$ has the time period of R and soil loss dimension of K. Smith and Wischmeier (1957), Meyer (1971), and Wischmeier (1973) gave detailed descriptions of this equation and provide methods to estimate these factors.

The Universal Soil Loss Equation is designed to predict average annual soil losses on sheet and rill erosion on upslope areas such as farmland and construction sites. It can be helpful for prediction of contributions from these sources to downstream sediment loads, but its capabilities and limitations for this use must be recognized. It is an erosion equation and is not designed to predict deposition. Its predictions do not include sediment contributions from gully erosion, and it does not include factors to account for sediment losses or gains between the field and the stream or reservoir. These items must be evaluated separately.

Many other investigators also formulate different equations to estimate soil erosion from experimental data. Table 1 gives a brief summary of some available soil loss equations, where L_o in the plot length in feet, S_o is the bed slope, and S_c and L_c are values of S_o and L_o at which erosion begins.

Table 1..   Soil Loss Equations

Investigator	Relationship Presented
Zingg (1940)	$\underline{E} \alpha L_0^{0.66} S_0^{1.37}$
Musgrave (1947)	$\underline{E} \alpha R K L_0^{0.37} S_0^{1.35} C$
Wischmeier and Smith (1965)	$\underline{E} \alpha R K L_0^{0.5}(0.00076 S_0^2 + 0.0053 S_0 + 0.0076) C P$
Meyer and Monke (1965)	$\underline{E} \alpha L^{0.9} S_0^{3.5}$
Meyer (1965)	$\underline{E} \alpha (L_0 - L_c)^{1.\sim 1.5}, \ \underline{E} \alpha (S_0 - S_c)^{1.5}$
Meyer and Kramer (1968)	$\underline{E} \alpha L^{0.5}(S_0 - S_c)^{1.4}$
Young and Mutchler (1969)	$\underline{E} \alpha L^{1.24} S_0^{0.74}$
Kilinc (1972)	$\underline{E} \alpha L^{1.035} S_0^{1.664}$

Shen and Li (1976) presented a theoretical analysis of these soil loss regression equations. Out of the six factors affecting soil erosion, it is impossible at this stage to make a theoretical analysis of the cropping management factor C and the erosion-control factor P. It is also difficult to analyze the soil erodibility factor K, because the relationship between soil characteristics and soil erosion loss was not given in a closed form.

Assume that: (1) the overland-flow reach is wide and one-dimensional; (2) hydrostatic pressure distribution is valid across the flow depth; (3) the variation of the momentum coefficient β along the direction of flow is negligible; (4) the soil layer is loose and uniformly consists of soil with fine-sediment size; (5) the sediment concentration is small so that the equation of motion for sediment-laden water can be approximated by the equation of motion for water only; (6) the variations of bottom slope S_0 is negligible; (7) the rainfall intensity and the infiltration rate are constant; the approximate solutions to the mechanics of steady sheet flow can be developed; (8) the continuity equation for sediment can be expressed as

$$\frac{dq_s}{dx} = p_s$$

(2)

where q_s is the sediment discharge per unit width of channel, and P_s is the fine-sediment pickup rate per unit area; (9) the rate of fine-sediment pickup is assumed to be a power function of τ, the boundary shear stress of surface runoff. The relation is

$$p_s = a \ \tau_0^b$$

(3)

where a is some constant describing the erodibility of a specific soil, and b is some exponent. Judging from the existing sediment transport equations, the exponent b is assumed to be 2.0. However, a and b can be determined by an optimization scheme. Shen and Li found that for small Reynolds number of the flow

$$\underline{E} = R_k L_o^{0.67} S_o^{1.33}$$

(4)

and for large Reynolds number of the flow

$$\underline{E} = R_k L_o^{1.17} S_o^{1.33}$$

(5)

where R_k is a factor describing the rainfall characteristics, soil erodibility, and fluid properties.

The exponents of slope length and percent slope in the above equations are consistent with the regression equations given in Table 1. The analytical results indicate the exponents of L_o and S_o for most of these empirical equations are reasonable and these exponents should be different for different Reynolds numbers.

Although these equations seem to be reasonable, one must be very cautious in using any of these. Erosion of soil in watersheds is an extremely complex matter and there are so many factors involved. These equations can only at best provide us with qualitative answers.

THE BED LOAD

Generally, the amount of bed load that includes surface creep, saltation, and part of the suspended load near the bed is small compared with suspended load in the main flow. In most of the canals and rivers in West Pakistan, the bed load is only about five percent of the total load and may even be neglected for computing the total sediment load. If this should be the case why should one even consider bed load? The answer is that although the amount of bed load may be small as compared with the total sediment load, it is nevertheless important because it shapes the bed and is a major factor in determing the stability of the channel, the form of the sediment bed surface, etc.

The bed load equation can be classified into the following four categories. From a great deal of practical experience, engineers have found that bed load is a function of the flow velocity, V, and the characteristics of sediment. Thus, the following type of equation is developed:

$$q_B = \text{bed load (rate)} = k_1 V^{k_2}$$

(6)

They have found the velocity is much more important than flow depths or slopes or other parameters in the determination of the bed load. However, it was found later that this equation does not satisfy the condition that there is a critical velocity below which no bed load movement would occur. In order to satisfy that condition, engineers have proposed a second equation

$$q_B = k_1 (V - V_c)^{k_2}$$

(7)

when V_c is defined as the critical velocity below which no sediment transport would occur. This equation does satisfy the initial condition that when $V \leqslant V_c$, q_B is zero or even negative, which is impossible. Although this equation does satisfy this initial condition for incipient motion, there is no physical reason to believe that if $V > V_c$, the q_B should be a function of $V - V_c$ rather than as a function of V as indicated in Eq. (6). In the last decades it was found that shear stress on the bed may be a better parameter to indicate the bed load rate rather than the flow velocity. Thus, engineers proposing Eq. (8) that

$$q_B = k_1 \tau^{k_2}$$

(8)

In order to satisfy the initial condition for incipient motion, engineers propose the Eq. (9) where

$$q_B = k_1 (\tau - \tau_c)^{k_2}$$

(9)

In the above equation τ is the shear stress on the bed and τ is the shear stress on the bed for incipient motion. Similar to Eq. (7) this Eq. (9) also satisfies the initial condition that when $\tau \leq \tau_c$, q_B is zero or negative which is impossible. Laursen (1956) reviews several well-known bed load equations into the following forms:

	Original form	Reduced Form
DuBoys (Straub, 1954)	$q_B = A_1 (\tau - \tau_c)$	$= B_1 n^4 \dfrac{v^4}{d^{2/3}}$
Schoklitsch (Shulits, 1935)	$q_B = \dfrac{A_2}{d_s^{1/2}} S^{3/2} (q - q_c)$	$= B_2 \dfrac{n^3}{d_s^{1/2}} \dfrac{v^4}{d}$
Meyer-Peter et al., (1934)	$q_B = (A_3 q^{2/3} S - A_4 D)^{3/2}$	$= B_3 \, n^3 \dfrac{v^4}{d}$
Wes (1935)	$q_B = \dfrac{A_5}{n} (\tau - \tau_c)^m$	$= B_5 n^{2m-1} \dfrac{v^{2m}}{d^{m/3}}$
Shields (1936)	$q_B = \dfrac{A_6}{d_s} \dot{q}\, S (\tau - \tau_c)$	$= B_6 \dfrac{n^4}{d_s} \dfrac{v^5}{d^{m/3}}$
Brown-Einstein (1950)	$q_B = \dfrac{A_7}{d_s^{3/2}} \tau^3$	$= B_7 \dfrac{n^3}{d_s^{3/2}} \dfrac{v^6}{d}$
Brown-Kalinske (1950)	$q_B = \dfrac{A_8}{d_s} \tau^{5/2}$	$= B_8 \dfrac{n^5}{d_s} \dfrac{v^5}{d^{5/6}}$

where m, A_1 through A_8 (inclusive) and B_1 through B_8 (inclusive) are constants, n is Manning's roughness, d is the flow depth, and d_s is the sediment size.

Among all these equations the one developed by Einstein (1950) probably is the most reliabel because it is based on a very comprehensive study and has been tested the most. It should also be pointed out that all these equations are applicable only the cohesionless material in steady and uniform flow. Another limitation could probably be added; that is that these equations are not useful if the ration between the flow depth and the sediment particles is too small. A possible improvement over these equations is through the formulation of stochastic bed load. For a general discussion of the stochastic bed load models the reader is referred to Shen (1976).

Sediment Bed Material Load

Rational Approaches. Lane and Kalinske (1941) and Kalinske (1947) probably made the first successful attempt to determine the suspended sediment discharge by integrating Eq. (10). They found first the bed load sediment concentration and used that as the reference sediment concentration at a certain level to integrate Eq. (10) to give the total bed material load.

Einstein (1950) also integrates Eq. (10) to obtain suspended load from bed load.

$$\frac{C_y}{C_a} = \left(\frac{d-y}{y} \frac{a}{d-a} \right)^Z$$

(10)

where C_y and C_a are suspended sediment concentration at distances y and a from the top of the bed surface. Z is a function of shear velocity and sediment size. He first determines the bed load for a particular particle size and assumes from experimental evidence that the bed load concentration occurs at two grain diameters from the bed in order to integrate Eq. (10) to obtain the suspended load for that size. Einstein gives an example to illustrate the application of his procedure in calculating the total sediment transport rate for Big Sand Creek, Mississippi.

Einstein's procedure, although theoretically sound, does involve many assumptions. Some of these assumptions are necessary if no other data are available. However, if sediment concentration and stream flow can be measured in a river, one may find the total load according to a modified procedure.

Colby and Hembree (1955) propose a modified Einstein procedure to obtain the total sediment transport rate in a river. The term "modified Einstein procedure" usually gives the false impression that it serves the same purpose as Einstein's procedure. Actually, these two procedures, although based on similar principles, serve entirely different purposes. As stated in the previous paragraph, Einstein's procedure estimates total sediment bed material load for different river discharges based on channel cross section and sediment bed sample in a selected river reach with uniform flow. This procedure is mainly for design purposes. The modified Einstein procedure as developed by Colby and Hembree only estimates the total sediment load (including wash load) for a given discharge from the measured depth integrated suspended sediment load, the stream flow measurements, the bed material samples, and the water temperature for this discharge at a given cross section.

The major modifications used in the modified Einstein procedure from Einstein's original procedure are:

1. The calculation is based on a measured mean velocity rather than on the slope and the depth is observed for each velocity.

2. The friction velocity and the corresponding suspended-load exponent Z in Eq. (10) are determined from the observed Z value for a dominant grain size. Values of Z for other grain sizes are derived from that of the dominant size and are assumed to change with the 0.7 power of the settling velocity.

3. A slight change in the hiding factor is introduced.

4. Flow depth is used to replace the hydraulic radius.

5. The value of Einstein's intensity of bed load transport is arbitrarily divided by a factor of two to fit the data more closely. Since the modified Einstein procedure is essentially to estimate the total sediment load from the measured sediment load, it no doubt can give better agreement with field data than the Einstein procedure which is based on more assumptions and much less data. Since the modified Einstein procedure relies on depth integrated suspended sediment samples, it should be more reliable with shallow streams where the sediment concentration variation is less than that in deep rivers.

Bishop *et al.*, (1965) present other modifications of the procedure presented by Einstein (1950).

Toffaleti (1968, 1969) presents a procedure for the analytical determination of sediment transport based on the concepts of Einstein (1950) and Einstein and Chien (1953). According to Toffaleti (1969) ''. . . the comparison of approximately 600 cases (339 river cases and 282 laboratory cases) of computed versus measured loads that covered a very wide range of conditions shows that his proposed procedure to be consistently satisfactory for all conditions tested. There are a number of cases showing wide discrepancy in this comparison, but consideration must be given to the degree of accuracy of the reported datad . . .''. He also provided a computer program to simplify the computation procedure.

Laursen (1958) developed empirical relationships for evaluating the composition and the rate of transport of the total load, the suspened load, and the bed load from the hydraulic characteristics of the stream. He stated that these relationships were basically empirical and, thus, can be used with confidence only within the range of conditions for which they were originally tested.

Bagnold (1966) introduced the concept that sediment transport mechanism is related to the availability and the efficiency of stream power to transport sediments.

Empirical Approaches. Because most of the rational approaches described previously involve rather complex procedures and many questionable assumptions, there is definitely a need to develop a simple relationship between the sediment transport rate and the flow condition, based entirely on available data, to provide the design engineer with an estimation that is in the correct order of magnitude.

In order to achieve a successful empirical approach, one must first select all the important factors involved. It is generally agreed that the flow velocity, the flow depth (or hydraulic radius), the energy slope, the characteristics of the sediment, and the temperature of the fluid are important factors; whether or not all these factors must be included in analysis is, of course, subject to interpretation. In making the analysis, one must recognize that the sediment concentration increases much faster than the increase of flow discharge. There is a generalized rule that sediment transport rate increases with flow velocity to the fourth power at low flow discharges and increases with flow velocity to the eighth power at high flow discharges. In other words, there is not much hope in finding a single combination of flow and sediment characteristics to describe sediment concentration for all flow conditions. Recognizing this fact, Colby (1964) developed different sediment discharge relationships with flow fro 0.1-foot, 1-foot, 10-foot, and 100-foot flow depths; Maddock (1969) provided different sediment discharge relationships with flow for low-, mid-, and high-velocity ranges, and Shen and Hung (1972) constructed a sediment-transport parameter and determined the concentration as the function of this parameter.

Colby (1964) investigate the effect of mean flow velocity, shear, shear velocity computed from mean velocity, stream power of flow, flow depth, viscosity, water temperature, and concentration of fine sediment on the discharge of sand per foot of channel width. He recommended three diagrams. In spite of many inaccuracies in the available data and uncertianties in the graphs, Colby found that ". . . about 75 percent of the sand discharges that were used to define the relationships were less than twice or more than half of the discharges that were computed from the graphs of average relationship. The agreement of computed and observed discharges of sands for sediment stations whose records were not used to define the graphs seemed to be about as good as that for stations whose recored were used." Note that all curves of 100-ft depth, most curves of 10-ft depth and part of the curves of 1.0 ft and 0.1 ft for Colby's diagrams are not based on available data and are extrapolated.

Shen and Hung (1971) began with the assumption that sediment transport is such a complex phenomenon that no single Reynolds number, Froude number, or combination of them can be found to describe sediment motion under all conditions. They recommended a regression method be used to develop a formula based on all available data for immediate engineering purposes. The disadvantage of this approach is, of course, the final flow parameter will probably be dimensional. But, the approach has merit in that if all previous data are found to correlate well, it is likely that other data (within the same range) will follow the same trend. They selected the sediment concentration (bed material load) as the dependent variable and the fall velocity of the median sediment particle of the bed sample as well as the flow velocity and flow depth as the independent variables.

Based on available reliable flume data and a few river data, they found that the sediment concentration is a function of $V^{0.57} S/w^{0.32}$ when V is the flow velocity in feet per second, S is the energy slope and w is the fall velocity in feet per second of the median sediment size. Since this is an empirical dimensional equation it is not easy to convert it to cgs. units.

Yang (1972) presented another regression equation based on his stream power concept. He found that the sediment concentration C is a function of stream power (product of flow velocity and energy slope), sediment particle size, shear velocity, fall velocity of sediment particle, etc.

Table 2 gives a comparison of many equations by Yang (1977) based on all data available to him.

Table 2. Summary of Accuracies of Different Equations

Equations	Date	Data with discrepancy ration between ½ and 2
Yang	1973	91 percent
Shen and Hung	1972	85
Ackers and White	1973	68
Engelund and Hansen	1967	63
Rottner	1959	56
Einstein	1950	46
Bishop *et al.*,	1965	39
Toffaleti	1969	37
Bagnold	1966	22
Meyer-Peter and Muller	1948	10

Should not be applied to large rivers.

CONCLUSIONS

Because of the tremendous uncertainties, it is difficult to make any final recommendations on the estimation of sediment transport which is, unfortunately, an extremely important problem. However, the following procedures are suggested for analyzing field data:

1. Use the modified Einstein's method (Colby and Hembree, 1955) to estimate the unmeasured suspended load and bed load based on measured data. There is a question of whether Einstein's intensity of bed load transport should be arbitrarily divided by a factor of two.

2. Separate bed material load from wash load and analyze them separately.

3. Decide which available sediment transport equations best agree with the measured data and use it to estimate the sediment transport load for the design flow, where actual measurement is not available.

When no measured data is available, the writer is inclined to:

1. Use Einstein's (1950) procedure if bed load is a significant portion of the total bed material load. Otherwise see 4 below.

2. Use Colby's (1964) method for rivers with flow depth less than or about 10 feet; also see 4 below.

3. Use Toffaleti's (1969) method for large rivers.

4. Use methods by either Shen and Hung (1971) or Yang (1972) for reference purposes.

ACKNOWLEDGMENT

Financial support for this study was provided by the U. S. National Science Foundation through Grant ENG 76-05773.

REFERENCES

Ackers, P. and White, W. R., 1973, Sediment Transport: New Approach and Analysis; J. Hydraulics Div., AM. Soc. Civil Eng., v. 99, p. 2041.

Bagnold, R. A., 1966, An Approach to the Sediment Transport Problem from General Physics; U.S. Geol. Survey Prof. Paper 422-J.

Bishop, A. A., Simons, D. B., and Richardson, E. V., 1965, Total Bed-Material Transport: J. Hydraulics Div., Am. Soc. Civil Eng., v. 91, p. 175.

Brown, C. B., 1950, Sediment Transportation: *in* Hunter Rouse, *ed.,* Engineering Hydraulics, John Wiley and Sons, p. 796-99.

Colby, B. R., 1964, Discharge of Snads and Mean-velocity Relationships in Sand-bed Streams: U. S. Geol. Survey Prof. Paper 462-A.

———, and Hembree, C. H., 1955, Computations of Total Sediment Discharge Niobrare River near Cody, Nebraska: U.S. Geol. Survey Water Supply Paper 1357.

Einstein, H. A., 1950, The Bedload Function for Sediment Transport in Open Channel Flows: U. S. Department of Agric. Soil Conservation Service, Technical Report No. 1026.

———, and Chien, N., 1953, Transport of sediment mixtures with large ranges of grain size: Missouri River Div. Sediment Series No. 3, U. S. Army Engr. Div., Missouri River, January.

Engelund, F. and Hansen, E., 1967, A Monograph on Sediment Transport in Alluvial Streams: Danish Technical Press, Copenhagen, Denmark, Revised edition 1972.

Kalinske, A. A., 1947, The movement of sediment as bed load in rivers: Trans. Am. Geophys. Union, v. 28, p. 615-620.

Kilinc, M. Y., 1972, Mechanics of Soil Erosion from Overland Flow Generated by Simulated Rainfall: Unpub. Ph. D. thesis, Dept. Civil Eng., Colorado State University, Fort Collins, Colorado.

Lane, E. W., and Kalinske, A. A., 1941, Engineering Calculations of Suspended Sediment: Trans. Am. Geophys. Union, v. 22, p. 603-607.

Laursen, E. M., 1956, The Application of Sediment-Transport Mechanics to Stable-Channel Design: J. Hydraulics Div., Am. Soc. Civil Eng., v. 82, No. HY4, Proc. Paper 1034.

———, 1958, The Total Sediment Load of Streams: J. Hydraulics Div., Am. Soc. Civil Eng., v. 84, No. HY1, Proc. Paper 1530.

Maddock, T., Jr., 1969, The Behavior of Straight Open Channels with Movable Beds: U. S. Geol. Survey Prof. Paper 622-A.

Meyer, L. D., 1965, Mathematical Relationships Governing Soil Erosion by Water: J. Soil Water Cons., v. 20, No. 4.

———, 1971, Soil Erosion by Water on Upland Areas: in H. W. Shen, ed., River Mechanics, v. II, Chap. 27, Colorado State Universiy, Fort Collins, Colorado.

———, and Kramer, L. A., 1968, Relation between Land-slope Shape and Soil Erosion: paper presented at 1968 winter meeting, ASAE, Chicago, Illinois.

———, and Monke, E. J., 1965, Mechanics of Soil Erosion by Rainfall and Overland Flow: Transactions, ASAE, v. 8, p. 572-580.

Meyer-Peter, E., Favre, H., and Einstein, A., 1934, Neuere Versuchsresultate uber den Geschiebetrieb; Schweizerische Bauzeitung 103.

———, and Muller, R., 1948, Formula for Bedload Transport; Internat. Assoc. Hydraulic Res., 2nd Meeting, Stockholm, p. 39.

Musgrave, G. W., 1947, The Quantitative Evaluation of Factors in Water Erosion, a First Approximation: J. Soil Water Cons., v. 2, p. 133-138.

Rottner, J., 1959, A Formula for Bed Material Transport: Houille Blanche 4.

Schoklitsch, A., 1934, Geschiebetrieb und die Geschiebefracht: Wasserkraft Wasserwirtsch, 39.

Shen, H. W., 1976, Sediment Transport Models: in H. W. Shen, ed., Stochastic Approaches to Water Resources, v. II.

———, and Hung, C. S., 1972, An Engineering Approach to Total Bed Material Load by Regression Analysis: in H. W. Shen, ed., Proceedings, Sedimentation Symposium, Berkeley, California, p. 14.

———, and Li, R. M., 1976, Watershed Sediment Yield: in H. W. Shen, ed., Stochastic Approaches to Water Resources, Vol. II.

Shields, A., 1936, Application of Similarity Principles and Turbulence Research to Bedload Movement: transl. into English by W. P. Ott and J. C. Van Uchelen at California Institute of Technology, Pasadena, California.

Shulits, S., 1935, The Schoklitsch Bed Formula: Engineering, v. 139, p. 644-646, 687.

Smith, D. D., and Wischmeier, W. H., 1957, Factors Affecting Sheet and Rill Erosion: Trans. Am. Geophys. Union, v. 38, p. 889-896.

Straub, L. G., 1954, Terminal Report on Transportation Characteristics — Missouri River Sediment: University of Minnesota, St. Anthony Falls Hydraulics Lab., Sediment Series No. 4.

Toffaleti, F. B., 1968, A procedure for computation of the total river sand discharge and detailed distribution, bed to surface: Tech. Report No. 5, Committee on Channel Stabilization, Corps of Engineers, U. S. Army.

———, 1969, Definitive Computations of Sand Discharge in Rivers, "Journal of the Hydraulic Div., ASCE, Vol. 95, No. HY1, January, Proc. Paper 6350.

U.S. Waterways Experiment Station, 1935, Studies of river bed materials and their movement with special reference to the lower Mississippi River: Paper 17.

Wischmeier, W. H., 1973, Upslope Erosion Analysis; in H. W. Shen, ed., Environmental Impact of Rivers, Chap. 15, Colorado State University, For Collins, Colorado.

———, and Smith, D. D., 1965, Predicting Rainfall-erosion Losses from Cropland East of the Rocky Mountains: Agriculture Handbook, No. 282, USDA.

Yang, C. T., 1972, Unit Stream Power and Sediment Transport: J. Hydraulics Div., Am. Soc. Civil Eng., v. 98, No. HY10, Proc. Paper 9295, p. 1805-1826.

————, 1973, Incipient Motion and Sediment Transport: J. Hydraulics Div., Am. Soc. Civil Eng., v. 99, p. 1679.

————, 1977, The Movement of Sediment in Rivers: Geophysical Surveys, v. 3, D. Reidel Publishing Company, Dordrecht-Holland, p. 39-68.

Young, R. A., and Mutchler, C. K., 1969, Soil Movement on Irregular Slopes, Water Resources Research, v. 5, p. 1084-1089.

Zingg, A. N., 1940, Degree and Length of Land Slope as it Affects Soil Loss in Runoff, Agricultural Engineering, p. 59-64.

FLUVIAL TRANSPORT AND DOWNSTREAM SEDIMENT MODIFICATIONS IN AN ACTIVE VOLCANIC REGION

David K. Davies[1], Richard K. Vessell[1], Robert C. Miles[2],
Michael G. Foley[3], and Samuel B. Bonis[4]

Abstract

River systems draining the active volcano Fuego are dominantly braided, and characterized by a high sediment transport rate. In a single year (1976) approximately 6 million tonnes of sediment are estimated to have been eroded from the cone by two river systems, with some 4 million tonnes reaching the sea.

Downstream modifications in sediment texture and composition are closely interrelated with fluvial transport mechanics. Grain size decreases exponentially downstream from 398 mm to 1.0 mm during some 90 km of fluvial transport. This results from decreasing flood-flow competence, and does not reflect in situ abrasion or abrasion or abrasion during transport. Abrasion results only in a 6% reduction of the volume of individual boulders, and causes surface pitting and loss of sharp edges. Concomitantly, boulder roundness increases downstream from 0.4 to 0.8. Sediment composition is also modified during fluvial transport. Grains in transport are derived largely from the products of the 1974 eruption of Fuego. They consist of fresh, unweathered feldspar rock fragments, pyroxene rock fragments, and olivine rock fragments together with free crystals of feldspar, pyroxene, and olivine. Rock fragment abundance decreases some 45% during 90 km of transport, while free crystal abundance increases 48%. The rate of rock fragment breakdown is different for different species of fragment, feldspathic rock fragments being apparently the most susceptible to physical breakdown. The physical breakdown of rock fragments results from fracturing in the glass groundmass which binds component crystals.

The most rapid rates of breakdown of sand-size rock fragments occurs in areas of high slope (the Volcanic Highlands), during the first 30 km of transport. Physical breakdown of rock fragments is probably directly related to the mode of sand grain transport, and results from impact shattering during transportation in intermittent suspension (saltation).

Introduction

Two facors are of significance in the control of downstream modifications of sediment composition and texture: 1) the relative resistance of the individual grains in transport, and 2) the sediment transport mechanics of the flow itself. Relative resistance of grains involves both the physical durability and chemical stability of individual mineral and rock fragment species. As Blatt (1967) pointed out, geologists intuitively "know" that certain grain species break down rapidly during transport, while other species appear to be more resistant, but hard data substituting their intuitions are not readily available. In one sense this is surprising, since studies of the effects of fluvial transport on sediments have been undertaken since the latter part of the nineteenth century (Mackie, 1896). In another sense, however, the lack of hard data is not surprising since most of these studies have encountered several common problems which have rendered their conclusions either conflicting or confusing.

Principal amongst these problems is the effect of tributary dilution. Sediment input from tributaries often masks the effects of abrasion and selective sorting. This is particularly true when tributaries add sediment of similar composition to that being transported in the trunk stream. Another problem involves the nature of the source-terrane. Several investigators have studied modifications in sediments of highly

[1]Department of Geosciences, Texas Tech University, Lubbock, Texas 79409.
[2]Amoco Production Company, P.O. Box 3092, Houston, Texas 77001.
[3]Department of Geology, University of Missouri at Columbia, Columbia, Missouri 65201.
[4]División de Geológica, Instituto Geográfico Nacional, Guatemala City, Guatemala.

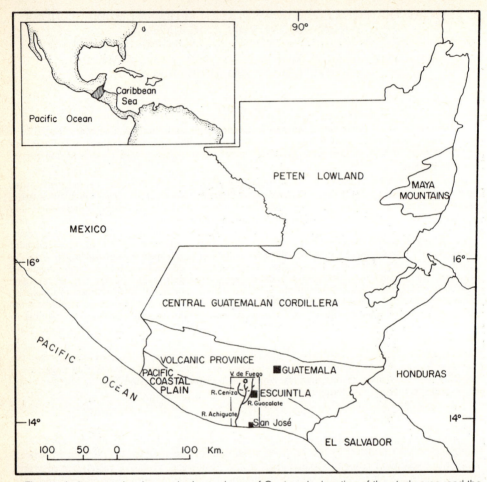

Fig. 1. Index map showing geologic provinces of Guatemala, location of the study area, and the volcano Fuego.

varied composition which were eroded from ancient consolidated deposits. Thus individual grains in transport have been subjected to long periods of non-fluvial weathering, diagenesis, and perhaps even several previous cycles of erosion. A third common problem in many of these studies is that the interrelationship between sediment transport mechanics and downstream sediment modifications is generally either ignored, or only paid lip-service. Downstream changes in sediment transport mechanics may play a significant role in the way in which individual grains are modified during fluvial transport. Thus it is important to have some knowledge of the hydraulics of the streams under investigation.

The present study was designed in an attempt to eliminate or reduce these and related problems. The area selected for study is an area of active volcanism in Guatemala (Fig. 1). The sediments in transport are fresh olivine basalts produced by major eruptions of Fuego. Between 87 and 99% of the sediment load in transport is derived from loose, easily erodible glowing avalanche deposits produced by the 1974 eruption (Figs. 2 and 3). Thus the great majority of the sediments consist predominantly of fresh, single-cycle grains of feldspar, pyroxene, olivine, feldspathic rock fragments, pyroxene rock

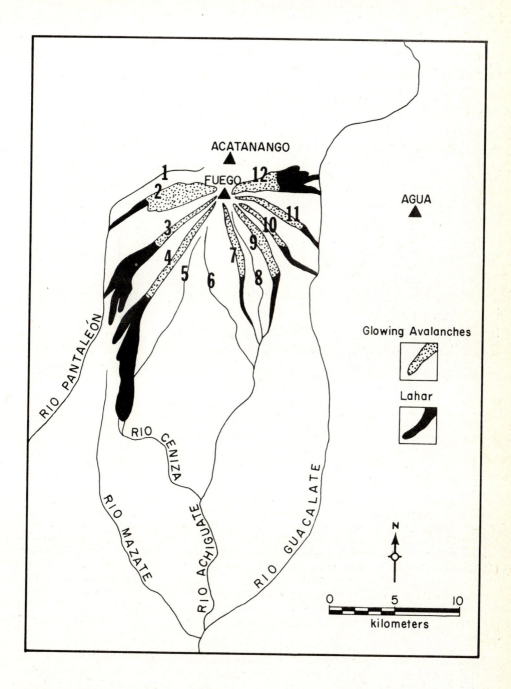

Fig. 2. Map displaying the distribution of glowing avalanche and lahar deposits in canyons on the flanks of Fuego. The glowing avalanche deposits were the product of the 1974 eruption. The relationship of these deposits to the major streams is as follows: 1, La Playa; 2, Seca; 3, Taniluya; 4, Ceniza; 5, Platanares; 6, Achiguate; 7, Trinidad; 8, Las Cañas; 9, El Jute; 10, Las Lajas, 11,

Fig. 3. An example of extremely rapid rate of denudation on the volcano flanks. Photograph shows a canyon which was completely filled with glowing avalanche deposits from the 1974 eruption, together with minor airfall. The 40 meter deep channel was cut through these deposits in two wet season, predominantly by flood surges. (El Jute canyon — for location see figure 2.)

fragments, and olivine rock fragments. The sediment supply is not limited by erosion rates within the drainage basin (Vessell, 1977).

Downstream sediment modifications were studied for all streams carrying significant sediment loads (Figs. 2 and 4). Dilution by bed or bank erosion is negligible due to the high sediment transport rate and rapidly aggrading nature of these wide, shallow streams. The presence of high and low gradient stream sections within the study area provides an opportunity to study and contrast the relative transport mechanics and sediment breakdown under different slope conditions. The hydraulics of these streams have been studied in detail (Vessell, 1977), enabling us to integrate both sediment transport mechanics and relative resistance of individual grain species in this analysis of downstream sediment modifications.

STUDY AREA

The volcano Fuego, with over sixty eruptive events documented since the Spanish conquest, is one of the most active members of an imposing chain of stratovolcanoes occurring along the Pacific coast of Central America. The cone, 3763 m in height, is composed predominantly of pyroxene andesite although recent eruptions have produced high alumina basaltic tephra (Davies *et al.*, 1978).

The climate of the area is well known, and extensive records are kept at coffee plantations on the flanks of Fuego. The geology of the area has been field-mapped on a

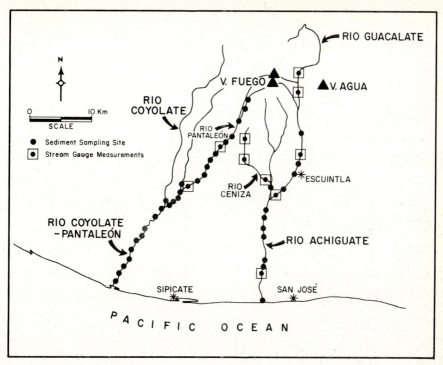

Fig. 4. Major drainage systems of the study area, showing location of sediment sampling and stream gauging sites.

scale of 1:50,000 (Boothby, 1978; Greer, 1978; Hebberger, 1977; Hunter, 1976; Tharpe, 1977; Quearry, 1975; Vessell, 1977). Sediment supply and transport rates have been evaluated for all rivers (Vessell, 1977).

Two major river systems head in tephra-filled canyons on Fuego's flanks (Figs. 2 and 4). The Achiguate, draining 1321.5 km² of the volcanic flanks and adjacent highlands, flows due south 60 km, entering the Pacific Ocean near the city of San José. The Achiguate system forms from the confluence of the Guacalate, El Jute, Las Cañas, Trinidad, Ceniza, and Mazate. The Pantaleón, draining 1131.6 km² of Fuego's western flank and adjacent highlands, flows a distance of 88 km southwest and enters the Pacific to the west of the town of Sipicate. The Pantaleón system forms from the Taniluyá, Seca, and La Playa, joined later by the Cristobal and Coyolate which drain the highlands west of the cone.

Average annual rainfall along the Pacific coast of Guatemala ranges from 450 cm on the volcanic cones to 230 cm over the coastal plain. In an exceptionally wet year, annual precipitation on the cone may exceed 600 cm (Fig. 5) (Anon., 1974). Nearly all rain falls during intense storms between May and early October. During the wet season, one thunderstorm may produce as much as 37 cm of rainfall in a single evening (Neira, 1970).

METHODS

A. Fluvial Hydraulics

Ten stream gauging stations were established on the two river systems draining the volcano Fuego (Fig. 4). In the course of two consecutive wet seasons the following

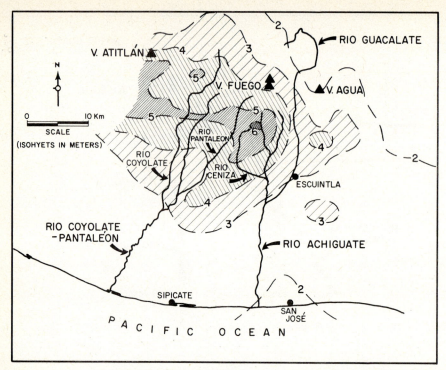

Fig. 5. Annual rainfall distribution in the study area, and its relationship to the major drainage systems.

parameters were measured — velocity and velocity distribution, flow resistance, bed and suspended sediment discharge, water discharge, stream competance, and flood flow. Instrumentation included a Helley-Smith bedload sampler (with 3 inch orifice), DH48 suspended load sampler, and U.S.G.S. type 622AA and pygmy current meters. Indirect calculations were obtained as a result of actual observations, and subsequent application of the slope-area method (Benson and Dalrymple, 1967). The Meyer-Peter and Muller (1948) formula was used for flood sediment discharge calculations. The precision of indirect calculation procedures was enhanced through integration with measurements, whenever possible.

B.   Sediment Sampling

Sediment sample sites were established at intervals along the Rio Pantaleón and the Rio Achiguate systems (Fig. 4). At each site the following field measurements were made: a) roundness and sphericity of particles coarser than 2.5 cm, using Krumbein's (1941) and Zingg's (1935) techniques; b) mean size (b axis) of randomly selected boulders; c) percentage of meterial coarser and finer than 2.5 cm was visually estimated. At each site several samples of sediment finer than 2.5 cm (herein termed matrix) were collected for laboratory analysis. Individual matrix samples were sieved, and grain size distributions plotted. Statistical moments were calculated using the graphic statistics of Inman (1949) and Folk (1974). A bulk matrix sample and a sieved sample were selected from each sample locality, impregnated and thin sectioned. Petrographic analyses, using grain counts of more than 200 grains per thin section, were used to obtain compositional data for analyses of downstream sediment modification. A total grain size distribution was calculated for each sample site using the technique described by Hunter (1976). Further details of sampling procedure and analytical design may be found in Miles (1977).

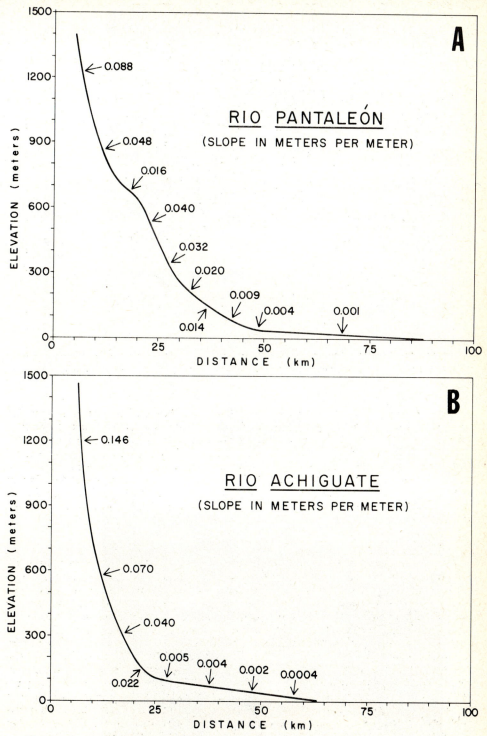

Fig. 6. Longitudinal profiles of two major streams: A. The Rio Pantaleón; B. The Rio Achiguate.

Fig. 7. Streams in the volcanic highlands, A. Rio Guacalate, 8 km from crater; B. Rio Guacalate, 9 km from crater, just below confluence with barranca Honda — a major source of sediment. Note the difference in sediment load and width of the floodplain between D and E; C. Braided channel, 15 km from crater — Rio Guacalate; D. Numerous braid bars, strewn with boulders, 15 km from crater — Rio Guacalate; E. Typical braided stream, 15 km from crater — Rio Guacalate.

FLUVIAL TRANSPORT

A. Channels

Channel gradients range from 0.146 on the volcano flanks to 0.0004 at the coast (Fig. 6). Channel profiles can be divided into three segments, 1) the Volcanic Highlands, 0 to 30 km from the crater; 2) the Transition Region, some 30 to 40 km from the crater, and 3) the Coastal Plain, 40 to 90 km from the crater. Characteristics of the fluvial sediments, channel form, and flow conditions for each segment are summarized in Table 1. Photographs of streams and the sediments in each segment are presented in Figures 7 through 9.

B. Rates of Erosion and Deposition

Sediment discharge measurements throughout the study area indicate rapid rates of erosion and transport of the fresh ejecta from the 1974 eruption. The Achiguate system is currently eroding approximately 4.1 million metric tons (tonnes) annually. Some 1 million tonnes are deposited in the Transition Region, and 100,000 tonnes are deposited on the Pacific Coastal Plain, and 3 million tonnes reach the sea. The Pantaleón system is eroding 1,726,000 tonnes annually from the cone. Some 691,000 tonnes of this material are deposited at the coastal plain transition while 1,035,000 tonnes reach the junction with the Rio Coyolate and the lower coastal plain. No data are presently available to give the annual sediment discharge of the Pantaleón system to the sea. Tributary input to the Pantaleón system is more important than in the case of the Rio Achiguate. Measurements indicate that the Rio Coyolate, which drains Pleistocene and Tertiary sediments to the west of the cone, annually transports some 255,000 tonnes of sediment to the Pantaleón system (approximately 13% of the sediment load of the lower Pantaleón system). These sediments, however, contain large quantities of pumice and acidic, weathered Pleistocene ash which may be readily distinguished from sediments eroded from the flanks of Fuego.

Floods are the dominant process controlling the transportation of coarse debris. In one such flood, May 19, 1976, the Guacalate (Fig. 4) grew from its nominal discharge of 5.5 m³/sec to a calculated peak discharge of 2200 m³/sec in a period of approximately 20 minutes. During peak flow this river transported boulders up to 2.3 m in diameter.

Table 1. Characteristics of fluvial sediments, channel form, and flow conditions for each segment of fluvial systems draining the active volcano, Fuego.

	Volcanic Highlands	Transition	Coastal Plain
Slope	0.146 — 0.004	0.004 — 0.001	0.001 — 0.0004
Grain Size Range	10 m — clay	0.6 m — clay	0.2 m — clay
Mean Grain Size	209 mm	12 mm	2 mm
Percent Sediment			
Coarser than 2.5 m	24 — 66	15 — 24	1 — 15
Sorting (ϕ)	4 — 6	3.3 — 4	0.9 — 3.3
Floodplain Width (m)	70 — 350	100 — 410	—
Channel Width (m)	6 — 25	35 — 60	43 — 100
Channel Depth (m)	0.3 — 1.2	0.2 — 0.4	0.8 — 3.0
W/D	10 — 40	200 — 400	86 — 406
Channel Form	Sinuous — Braided	Braided	Sinuous — Straight
Bedform	Antidunes — Flat Bed	Antidunes	Antidunes — Flat
Regime	Upper — Transition	Upper	Upper — Transition
Velocity (mean) (m/s)	1.25 — 2.00	2.00 — 1.85	1.85 — 1.60
Shear Stress (lb/ft²)	2.00 — 0.85	0.85 — 0.80	0.80 — 0.35

Fig. 8. Streams at the transition between the volcanic highlands and the Pacific Coastal Plain. A. Confluence of the Rios Achiguate and Ceniza. Bars are composed of coarse sand and gravel. In the foreground remnants of a major highway bridge may be seen. This was destroyed by a single flood in 1969. In the background are the volcanoes Fuego (left) and Agua (right). B. The Rio Pantaleón.

Fig. 9. The Rio Achiguate in the Pacific Coastal Plain, 50 km from the crater. Note stable banks and longitudinal bars.

Meyer-Peter and Muller (1948) calculations indicate that the two hour flood event was capable of transporting 390,000 tonnes of sediment (Vessell, 1977; Vessell *et al.*, 1977).

DOWNSTREAM SEDIMENT MODIFICATIONS

A. Texture

Sediments exhibit a pronounced decrease in mean size with increasing transport distance (Fig. 10). This rapid reduction is a reflection of the steady decrease in both abundance and size of boulders deposited on the stream floodplains. Deposition of boulders is primarily related to a rapid downstream loss of stream flood competence. In the volcanic highlands, flood competence (a function of channel slope and flow depth) is controlled by steadily decreasing channel slope and the presence of terraces. Terraces constrict flood flows, and thus increase flow depth and shear stress. Terraces terminate at the coastal plain, and this results in a drastic increase in floodplain width and a concurrent decrease in flow depth and shear stress.

Plots of mean boulder diameter and flood flow competence (as a function of distance) display a strong interdependence (Fig. 11). This suggests that flow competence rather than abrasion is primarily responsible for the observed decrease in mean grain size with increasing transport distance. Actual field measurements (in which painted boulders were observed after transport in flood flows) further indicate that abrasion is insignificant during transport. Textural modification of the boulders involved only the production of

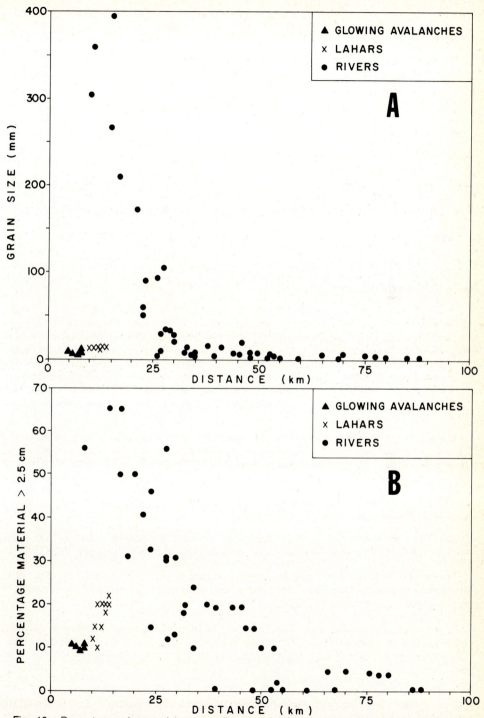

Fig. 10. Downstream changes in mean sediment size (A) and concentration of material coarser than 2.5 cm (B). In this and subsequent figures, glowing avalanche data are from Quearry (1975); laharic data are from Hebberger (1977).

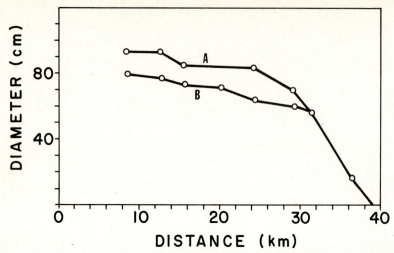

Fig. 11. Competence of flood flows on the Rio Guacalate. A. Competence predicted by the extrapolated Shield's curve and flood flow data from the May 19, 1976 flood. B. Mean boulder diameter (b-axis) measured at sample sites.

percussion marks on the surface and the abrasion of sharp edges. (Boulders too large to be moved during observed floods, but which have been subjected to in-place erosion by sand-blasting, have scarcely lost their coat of paint after two years in the stream beds). Abrasion accounts for a 6% reduction in the volume of individual boulders during transport from the cone to the coastal plain — far less than the observed reduction in boulder size (Fig. 10). Hence it is the decrease in the abundance of sediment coarser than 2.5 cm (Fig. 10B) which controls the reduction of mean size with increasing distance (Fig. 10A).

Sorting improves with increasing transport distance (Fig. 12A). This reflects the reduction in extreme grain size variation resulting from gradual drop-out of coarse sediments. Values range from 5.9 ϕ (extremely poorly sorted) on the volcano flanks to 0.84 ϕ (moderately poorly sorted) at the coast. The improvement of sorting with increasing transport distance is exponential. Sorting improves at a more rapid rate in upper stream reaches than on the coastal plain. This is a reflection of the rapid drop-out of coarse-grained material in these upper reaches.

As expected, sediments in these high slope streams are coarsely skewed wih some tendency displayed toward becoming less coarsely skewed downstream (Fig. 12B).

Grain roundness increases with increasing distance of transport (0.41%/km) (Fig. 13A). Increasing roundness in pebble and cobble size sediment is a result of the abrasion of rough edges of particles during transport.

Grain sphericity values range from 0.55 to 0.74 for various sampling sites, but display no downstream change (Fig. 13B). Zingg sphericity data show no downstream trends although there appears to be a slight decrease in the percentage of spheres with transport distance.

B. Composition

Feldspathic rock fragments (feldspar crystals set in a groundmass of vesicular glass) and free crystals of plagioclase feldspar are the dominant components of fluvial sediments in this area. Crystals of olivine and pyroxene (present in near equal abundance) comprise less than 30% of the sediment. Rock fragments composed dominantly of olivine and

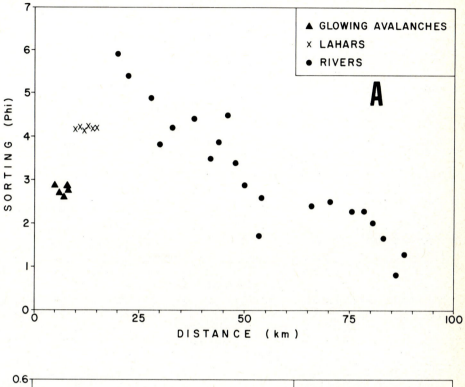

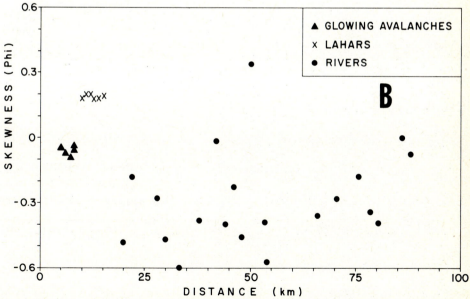

Fig. 12.   Downstream variation in A, sorting; and B, skewness.

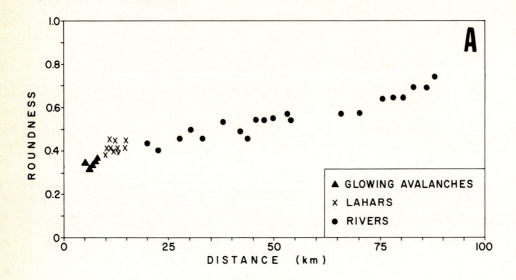

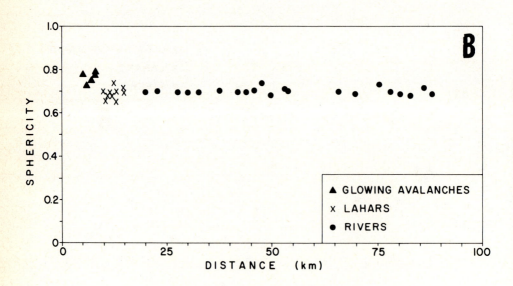

Fig. 13.　Downstream variation in A, roundness; and B, sphericity.

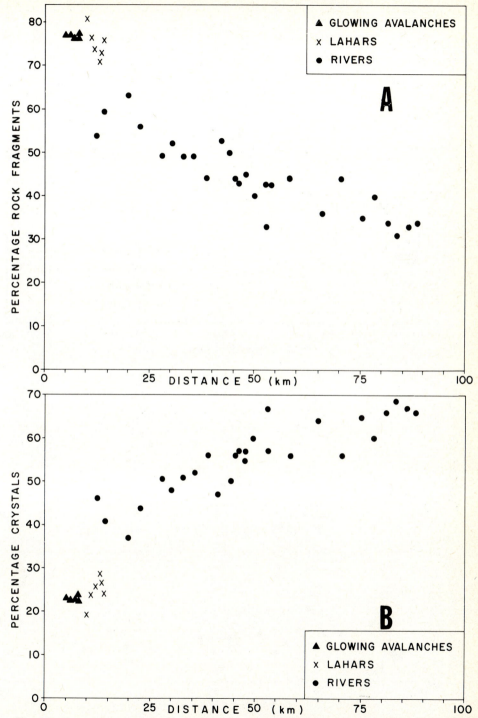

Fig. 14.   Downstream variation in abundance of rock fragments (A); and abundance of free crystals (B). (In this and subsequent diagrams, all compositional data relate only to the 1 to 3 ϕ size range, 0.5 — 0.125 mm.)

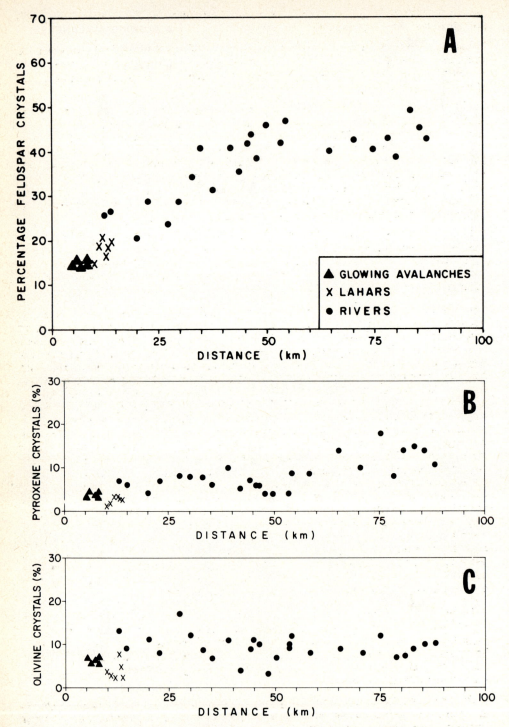

Fig. 15. Downstream variation in abundance of free crystals. A. Feldspar crystals, B. Pyroxene crystals, C. Olivine crystals.

pyroxene set in a glassy groundmass are minor components, accounting for less than 10 and 5% of the sediment, respectively.

The coarse sand-size fraction is dominated by rock fragments, while the medium and fine-sand fractions are dominated by free crystals. Rock fragment abundance decreases exponentially downstream while the abundance of free crystals increases exponentially (Fig. 14).

The plot for rock fragment modification (Fig. 14A) is a total plot which includes feldspathic rock fragments, pyroxene rock fragments, and olivine rock fragments. Since most rock fragments are feldspathic, this plot is dominated by the influence of feldspathic rock fragments (compare Figs. 14A and 16A). Rock fragments consisting dominantly of pyroxene show no observable downstream variation in abundance (Fig. 16B). Olivine rock fragments show only a small downstream decrease in abundance (Fig. 16C). The rate of downstream decrease in abundance of the various types of rock fragments thus is not uniform. Feldspathic rock fragments break down more rapidly than do pyroxene and olivine rock fragments (compare Figs. 16A, B and C).

Feldspar crystals increase exponentially in abundance downstream (Fig. 15A) [This plot is similar to the plot for total crystals (Fig. 14B) reflecting the fact that virtually all free crystals in transport are feldspathic]. Free crystals of pyroxene show a downstream increase in abundance (Fig. 15B) while free olivine crystals show no observable variation in abundance over the 90 km of fluvial transport (Fig. 15C).

Explanations for downstream modifications of sediment composition must take into account both the rate of the modification and changes in the sediment transport mechanics. Total rock fragments decrease 45% in abundance from the cone to the sea, while free crystals increase 48% in abundance (Fig. 14). This indicates that the rock fragments break down into their component crystals on a one to one basis. One rock fragment dissintegrates into one free crystal. Petrographic evidence (and common sense) suggest that this correlation is artificial. Rock fragments are generally composed of several crystals held together by a groundmass of vesicular glass (Fig. 17A). If the rock fragment is subjected to shattering during transport, it follows that the fractures will occur in the weakest component — the glass groundmass. If a single rock fragment is broken into several components, it will produce several free crystals together with some volcanic glass fragments. Thus, the rate of increase of free crystals should be more rapid than the rate of decrease in rock fragments. However, it would appear from Figure 14 that this is not the case. The explanation probably reflects the size distribution of crystals within the rock fragments. Virtually all rock fragments analyzed consist generally of one large crystal together with one or more small crystals (Fig. 17A). Shattering of the rock fragment releases the large crystal, which is large enough to be incorporated in the 0.5 to 0.125 mm (1 to 3 ϕ) size range used in compositional analyses in this study. The small crystals and fragments of glass are smaller than 0.125 mm, and they are added to the sediment load in the finest grained fraction. The one to one relationship between rock fragment decrease and free crystal increase (Fig. 14) only reflects the fact that there is one crystal available to be added to the 0.5 to 0.125 mm size range.

There is a 9% increase in pyroxene crystals downstream (Fig. 15B), but no corresponding downstream decrease in pyroxene rock fragments (Fig. 16B). It may be suggested that this small increase in abundance of pyroxene crystals results from liberation of a few large pyroxene crystals in the feldspathic rock fragments. Not all feldspathic rock fragments consist solely of feldspar and glass. They occasionally contain a significant pyroxene component.

The exponential decrease in rock fragments with increasing transport distance suggests that the physical breakdown of rock fragments occurs at a more rapid rate in upper stream reaches than on the coastal plain. The most rapid rock fragment decrease occurs within the first 30 km of stream transport. The inflection point coincides with the

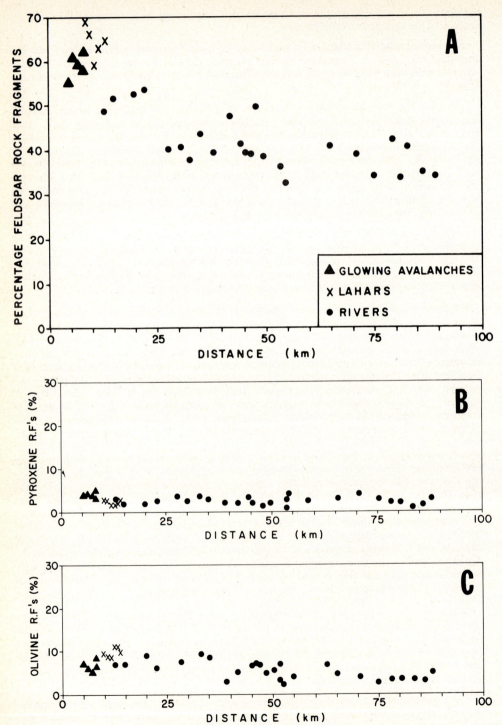

Fig. 16.   Downstream variation in rock fragments of the 1 to 3 φ size range. A. Feldspar Rock Fragments, B. Pyroxene Rock Fragments, C. Olivine Rock Fragments.

break-in-slope between the Volcanic Highlands and the Coastal Plain (i.e. the Transition Region). We suggest that the differential rate of rock fragment breakdown may be related to a downstream change in sediment transport mechanics for particles in the 0.5 to 0.125 mm size range. In particular, it may be related to the mode of transport of these grains — whether they are transported as bedload or in intermittent suspension (saltation). In the Volcanic Highlands (0 to 30 km from the cone) most grains in this size range are transported in intermittent suspension and are collected in both bedload and suspended load samples. In the Coastal Plain, these sediments are collected entirely in the bedload of the streams — suspended load samples contain few or no grains of this size range.

These field observations are substantiated by theoretical calculations of the ability of the streams to hold grains in suspension. The theoretical calculations used were Rouse Numbers generated from stream gauge measurements. The Rouse Number (Rouse, 1937; Vanoni, 1975) is the ratio of the fall velocity of a particle (of a certain size) to the turbulent energy of the flow. Thus the Rouse Number (Z) enables estimations to be made of the transport mechanics of grains in fluvial systems.

Rouse Number calculations (Table 2) indicate that in the Volcanic Highlands, grains in the 0.5 to 0.125 mm size-range are transported in intermittent suspension. They comprise 25% of the sediment in transport. Coarser grains comprise 70% of the total sediment in transport, and are generally transported as bedload. In the Transition Region grains in the 0.5 to 0.25 mm size range are transported as bedload. Grains between 0.25 and 0.125 mm are transported in intermittent suspension, and comprise 20% of the total sediment load. On the Coastal Plain only grains of 0.125 mm may be transported in intermittent suspension. (They comprise 10% of the total load.) Coarser grains are transported as bedload.

The greatest rates of rock fragment breakdown occur in areas where the 0.5 to 0.125 mm sediment is transported in intermittent suspension. We suggest that the rapid rate of breakdown in these areas is a result of impact shattering during transport. Impact of saltating rock fragments results in the shattering of the rock fragments into component crystals and glass (Fig. 17B).

The different rates of reduction in abundance of feldspar, pyroxene, and olivine rock fragments (Fig. 16) may reflect two facts: 1) Olivine and pyroxene rock fragments compose only a small portion of the sediment in the fluvial systems, and 2) Olivine and pyroxene rock fragments, with their high specific gravity, are not transported in intermittent suspension in the upper stream reaches. Under conditions of bedload transport they would be less susceptible to breakdown resulting from impact shattering. The feldspathic rock fragments, being less dense, would be transported in intermittent suspension, and be more susceptible to impact shattering.

Table 2. Rouse numbes (Z) computed for various sediment sizes for measured flow conditions at 4 gauging stations on the Achiguate System ($\kappa = 0.3$).

DISTANCE FROM CONE	SEDIMENT SIZE					
	10 mm	1 mm	0.5 mm	0.25 mm	0.10 mm	0.05 mm
12 km	5.34	1.89	0.72	0.43	0.11	0.04
15 km	6.19	2.19	0.84	0.50	0.13	0.05
33 km	9.57	3.39	1.30	0.77	0.19	0.70
51 km	14.34	5.08	1.94	1.16	0.30	0.104

(**Note:** In these computations, von Karman's constant κ, and bed shear velocity U^* are measured data.)

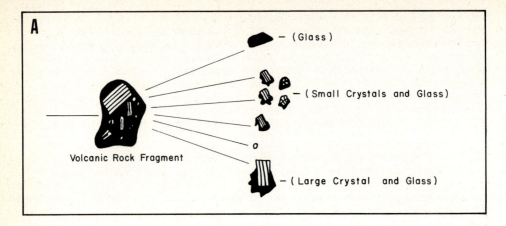

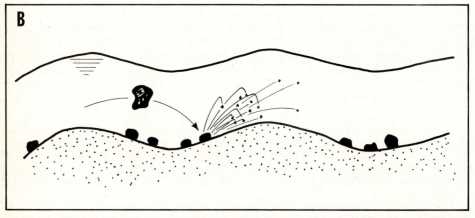

Fig. 17. Proposed model for physical breakdown of rock fragments under fluvial transport in upper stream reaches. A. Impact shattering of a rock fragment releases the major component (a single, large free crystal), several smaller crystals, and some glassy matrix. Note that most free crystals observed generally have small amounts of glassy matrix (less than 10%) attached to their edges. B. Rock fragment in 1 to 3 ϕ size range is transported in intermittent suspension ("saltation"). Impact on other grains causes shattering.

There is no petrographic evidence of chemical weathering of crystals or rock fragments at any sampling site in any size range. Sediments are exceptionally fresh, angular, and lacking evidence of chemical attack. With the extreme range of grain sizes present in the system, abrasion or physical processes may be extremely rapid while chemical processes do not have time to occur until sediment is buried offshore or beneath successive flood deposits. Our studies of rates of erosion and transport indicate that sand size sediment may have an extremely short residence time in the fluvial systems on the flanks of Fuego.

Conclusions

Channel patterns, morphologies, and flow conditions in streams draining the active volcano Fuego alter in response to rapid downstream gradient change. Fluvial sediments in transport exhibit downstream changes in texture and composition which are closely related to fluvial transport mechanics. Decreasing flood-flow competence (caused by decreasing slope and flood-flow depth) results in major downstream modification in grain size, sorting, and skewness. Sediment composition is also modified. There is an exponential downstream decrease in the abundance of volcanic rock fragments, and a concomitant exponential increase in free crystals. These changes are apparently related to the mechanics of grain transport. Most rock fragments probably break as a result of impact shattering during transportation in intermittent suspension ("saltation"). Greatest rates of intermittent suspension occur in areas of high slope, i.e. the Volcanic Highlands. The highest rates of compositional modifications occur during the first 30 km of fluvial transport.

ACKNOWLEDGEMENTS

This study was made possible through the financial support of the National Science Foundation. Office of International Programs, under grant INT78-12365, and the División Geológica, Instituto Geográfico Nacional de Guatemala. Oscar Salazar, Chief of the División Geológica, Instituto Geográfico Nacional, Guatemala, was responsible for the well-organized logistical support provided during the field seasons. James Berkeley, Eugene Schweig, and Neil Skilton assisted in the field studies. Chiqui Bonis and Fraterno Vila Betoret provided generous hospitality in Guatemala. The administrators of Fincas San Diego and Pantaleón kindly allowed free access to their properties.

The manuscript was reviewed by Dr. Frank G. Ethridge and Dr. Stanley Schumm, and we are grateful to them for their constructive criticisms. The manuscript was typed by Mrs. Madonna Norris.

REFERENCES

Anon, 1974, Estudio Integral de los recursos hidraulicas del departmento de Escuintla: Instituto Geográfico Nacional, Guatemala, 166 pp.

Benson, M. A., and Dalrymple, T., 1968, General field and office procedures for indirect discharge measurements: U.S. Geol. Survey Techniques Water Resources Inv. book 3, chap. A1, 30 pp.

Blatt, H., 1967, Provenance determinations and recycling of sediments: J. Sediment. Petrol., v. 37, no. 4, p. 1031-1044.

Boothby, D. R., 1978, Coastal sedimentation — Guatemala: Unpub. M.S. thesis, Texas Tech University, (in preparation).

Davies, D. K., Quearry, M. W., and Bonis, S. B., 1978, Glowing avalanches from the 1974 eruption of the volcano Fuego, Guatemala: Geol. Soc. Am. Bull., v. 89, p. 369-384.

Folk, R. L., 1974, Petrology of Sedimentary Rocks: Hemphill, Austin, Texas, 182 p.

Greer, E. W., 1978, Sedimentation patterns of an estuarine delta system — Guatemala: Unpub. M.S. thesis, Texas Tech University, (in preparation).

Hebberger, J. J., Recent laharic and glowing avalanche sediments — Guatemala: Unpub. M.A. thesis, University of Missouri, Columbia, 117 p.

Hunter, B. E., 1976, Fluvial sedimentation on an active volcanic continental margin: Rio Guacalate, Guatemala: Unpub. M.A. thesis, University of Missouri, Columbia, 135 p.

Inman, D. L., 1949, Sorting of sediments in the light of fluid mechanics: J. Sediment. Petrol., v. 19, p. 51-70.

Krumbein, W. C., 1941, Measurement and geological significance of shape and roundness of sedimentary particles: J. Sediment. Petrology, v. 11, p. 64-72.

Mackie, W., 1896, The sands and sandstones of eastern Moray: Edinburgh Geol. Soc. Trans., v. 7, p. 148-172.

Meyer-Peter, E., and Muller, R., 1948, Formulas for bed-load transport: Report on second meeting of International Association for Hydraulic Research, Stockholm, Sweden, p. 39-64.

Miles, R. C., 1977, Modifications of sediment composition and texture in two fluvial systems, Guatemala: Unpub. M.A. thesis, University of Missouri, Columbia, 91 p.

Neira, H. C., 1970, Las Crecidas de los Rios Achiguate y Guacalate en Septiembre de 1969: Publicación No. 49 del PHCA, San José, Costa Rica, Julio de 1970, 63 p.

Quearry, M. W., 1975, Continental volcanic sediments in the region of Volcan de Fuego, Guatemala: Unpub. M.A. thesis, University of Missouri, Columbia, 105 p.

Rouse, H., 1937, Nomogram for the settling velocity of spheres: Div. Geol. Geo. Exhibit D, Rept., Comm. Sedimentation (1936-1937) Natl. Res. Council, Washington, D.C., p 57-64.

Tharpe, L. W., 1976, Fluvial sediments of the Rio Achiguate and its tributaries, Guatemala: Unpub. M. A. thesis, University of Missouri, Columbia, 127 pp.

Vanoni, V. A., 1975, Sedimentation Engineering: Am. Soc. Civil Eng. Manual No. 54, 745 pp.

Vessell, R. K., 1977, Morphology, hydrology, and sedimentology of the Rio Guacalate — Volcanic Highlands — Guatemala: Unpub. M.A. thesis, University of Missouri, Columbia, 144 pp.

Vessell, R. K., Davies, D. K., Foley, M. G., and Bonis, S. B., 1977, Sedimentology and hydrology of flood flow on the active volcano Fuego, Guatemala: (abst.), Geol. Soc. Am., Abstracts with Programs, v. 9, p. 1210.

Zingg, T. H., 1935, Beitrag Zur Schotteranalyse: Schewiz, Min. v. Pet. Mitt. vol. 15, p. 39-140.

BEDFORMS AND BARS

SOME COMMENTS ON TERMINOLOGY FOR BARS IN SHALLOW RIVERS

NORMAN D. SMITH[1]

INTRODUCTION

Fluvial sedimentologists, geomorphologists, and engineers are becoming increasingly aware of the confusion that exists in present terminology dealing with large scale fluvial bedforms. This is especially true for non-periodic forms occurring in shallow depths where small changes of flow may cause considerable variation in bedform morphology. These forms, most frequently referred to as "bars," have long been noted by fluvial workers but remain poorly described and understood when compared to the smaller scale periodic forms (ripples, dunes, antidunes), of which an abundant and varied literature exists. Nomenclatural deficiences are no better manifested than in the recent surge of sedimentological literature on braided streams (Miall, 1977), in which bars comprise the characteristic depositional forms. It is quite clear to workers of modern streams as well as even casual surveyors of the literature that, for example, the commonly used words "longitudinal bar" and "transverse bar" have evolved into virtual garbage terms, especially when applied to ancient sediments where the two terms are often regarded as essentially synonomous with, respectively, gravelly and sandy braided stream deposits. Whether the popularity of these two terms (and several others) among sedimentologists arises from sloppiness or from creditable attempts at generalization is a moot point but perhaps an unimportant one. What is interesting however, is that in recent years a large number of bar terms has been introduced from river studies, but that only a very few have caught the fancy of geologically-oriented sedimentologists, for whatever reasons. In a non-exhaustive survey of recent English-language fluvial literature, the writer found no less than 32 specific terms preceding the word "bar" that identified particular bedforms in modern streams, most of which could be called braided (Table 1). A perusal of the definitions and/or contexts of these bars leaves the reader with at least two impressions: (1) there is considerable lack of clarity embodied in many of the terms, and (2) there is much overlapping in their intended or implied definitions. Furthermore, that only a handful of these terms has come into common use appears to be more a case of fortune than design.

SOME PROBLEMS

Several problems obstruct the way of an inclusive and effective nomenclature for bars. Three of the more significant ones include the following: (1) The word "bar" itself is too vague; the term has been used to designate a great variety of unrelated bedforms in river channels. (2) Hydraulic origins of bars have been little investigated and remain, as a whole, poorly understood. No classification scheme can yet have a genetic basis comparable to the common small-scale bedforms (ripples, dunes, etc.) whose relationships between sediment size and flow variables are fairly well known. (3) Bar morphology, upon which present terminology is principally based, may develop from either depositional or erosional processes, and they are often difficult to distinguish in the field or on aerial photographs. Moreover, bars range from relatively simple depositional features to complex forms resulting from multiple depositional and erosional events.

DISCUSSION

Whatever its shortcomings may be, the term "bar" is probably with us forever. Before coming under the scrutiny of modern fluvial workers, the term has historically served well

---

[1]Department of Geological Sciences, University of Illinois at Chicago Circle, Chicago, Illinois 60680.

for any exposed or slightly submerged major positive element of the river bed. As descriptions of fluvial bedforms have proliferated, however, so too has the size of the umbrella covered by the word "bar". The term now includes features that range from very small to enormous, solitary to periodic, mid-channel to bank-attached, quasi-stable to freely migrating, and simple depositional forms to highly complex ones. It would seem desirable to restrict usage of the term in future classifications. One possibility is to eliminate the word "bar" from the freely-migrating periodic forms so common in shallow sand-bed streams and substituting one of the common terms for large periodic forms such as "dune", "megaripple", or "sand wave". Under this guideline, the transverse bars of Sundborg (1956) and many of the linguoid bars of Collinson (1970) and Boothroyd and Ashley (1975) would no longer be termed "bar". The same consideration could be given for repetitive forms attached alternately to opposite banks but which migrate independently of the banks, for example, the alternating bars of Task Force (1966) or the migrating forms of Kellerhals *et al.'s* (1976) side bars. Another possible way of restricting the use of "bar" is to eliminate it from certain complex depositional areas known to have formed by overriding or lateral coalescence of smaller bedforms. Cant (1976, also Walker, 1976) has done this with his "side flats" and "sand flats" for, respectively, bank-attached and mid-channel positive areas having compound depositional and erosional histories.

Our poor understanding of the hydraulic conditions which bring about the origin and development of various bar forms has unquestionably hampered the development of a satisfactory classification. Because of scaling problems, bars have attracted scant attention from experimentalists, and normal field conditions make it difficult to obtain appropriate hydraulic data in natural channels at the times when bars are forming, especially in gravel-bed streams. Some workers indeed believe that an effective bar classification cannot be developed until a firmer genetic basis is established. While there is no doubt some validity to this view, the present problem in not entirely one of trying to put the cart before the horse. Effective terminology should be able to evolve at the same pace as our understanding of the features being considered, and there is no question that at least certain aspects of our understanding of bars in now ahead of present terminology. Much useful sedimentological and morphological data have been accumulated without a thorough comprehension of the processes involved, and that we are still in a primitive stage of understanding these processes should not discourage continuous efforts to more effectively and precisely categorize those observations we now have and continue to make. Poor terminology works against these efforts.

Bar classifications are at present, and probably will continue to be, mainly based upon morphologic criteria. Although probably necessary, this is also unfortunate because of the vast spectrum of morphologic variation that exists in river bars. Only those whose existence depends more on the long-term position and shape of the channel (e.g., point

Table 1..   Bar terms from recent fluvial literature

Alternating	Lateral	Sheet
Braid	Linguoid	Side
Channel	Lobate	Spool
Channel-Junction	Longitudinal	Transverse
Chute	Meander	Transverse-Lobate
Compound	Medial	Transverse-Lunate
Cross-Channel	Mid-Channel	Transverse-Riffle
Diagonal	Point	Triangular
Diamond	Remnant	Tributary
Foreset	Riffle	Unit
Horseshoe	Scroll	

bars in bends; Bluck's, 1976, lateral bars) than on short-term variations in flow and sediment supply are likely to maintain a more-or-less stable form over a long period. The more transient bars, however, including most of those in braided channels, are usually out of equilibrium with the inherently unsteady fluvial flows; resulting lag effects, including late-stage deposition and erosional modifications, combine to produce a bewildering variety of forms. Even in early stages of development during rising or steady flows, one bar form may evolve into another (e.g., Hein and Walker, 1977). Thus, two bars having different origins and initial forms may end up appearing superficially similar; or conversely, two initially identical bars may, after waning-flow modifications, lose all resemblance to each other by the time they are exposed and no longer active. Rate of discharge fall appears to be an important factor in determining the extent of bar modification (Jones, 1977) and, ultimately, bar form. Rapidly falling flows, for example, have the least time to effect late-stage modification, and active bars may emerge with depositional morphologies relatively intact. Such considerations, although appreciated by most fluvial workers, are generally not reflected in present classification schemes. For example the rhomboidal or petal shapes that are so typical of bars in gravel-bed braided streams can be either unmodified depositional forms, erosional remnants, or products of multiple erosional and depositional events. A number of terms have been used to identify these forms, with perhaps "longitudinal bar" the most frequently used among sedimentologists, but only rarely are distinctions between erosional and depositional morphologies attempted. Such distinctions may seem unimportant to some workers, but to sedimentologists wishing to understand sedimentation patterns or trying to interpret ancient bar deposits, the difference is hardly trivial. Sequences of sedimentary textures and structures could be expected to follow some orderly internal patterns in bars whose form is depositional; on the other hand, a similar form produced by dissection may be a mere remnant of compound depositional events, and internal sedimentary arrangements will be commensurately complex. To give both bars the same name because of superficial similarities is to obscure these essential (to sedimentologists) relationships. Authors, therefore, should always attempt to clarify such relationships in their observations of both modern and ancient bar deposits rather than hide behind jargon that, because of its imprecision, tends to obfuscate such details.

CONCLUSIONS

The time appears to be ripe for a thorough revision of present classifications for bars in stream channels. Present schemes contain too much imprecise and overlapping terminology that is difficult to apply in many natural situations, either in modern streams or in ancient fluvial deposits. Unqualified use of poor terminology inevitably results in loss of information. Until a more satisfactory nomenclature is introduced, fluvial workers should shun the use of popular but imprecise terms; at the very least, such terminology should not be substituted for observational precision or description clarity. While awaiting the development of a more precise terminology, the following suggestions are offered:

10. Bars formed by complex successions of depositional and/or erosional events should be regarded as separate from those with simple depositional histories and unmodified forms, the "unit bars" of Smith (1974). Although in some cases the distinction may not be operationally easy, the conceptual difference is important enough to justify more attention than has been given up to now.

11. Insofar as it is possible, attempts should be made to distinguish bars that are relatively stable and permanent parts of the channel system from those that are mobile and unstable. Jackson (1975) discusses a conceptual basis for this.

12. Distinction between periodic, quasi-periodic, and solitary bar forms should be clearly made. This can be readily done in modern streams and could probably be

interpreted in many ancient deposits as well. Periodic and non-periodic forms should be classified separately, and future classifications should consider dropping the term "bar" from periodic forms.

13. When describing bar morphology, attempts should be made to determine whether the form in predominantly depositional or a product of erosional modification. This simple but important distinction is often overlooked in descriptions of modern river bars.

ACKNOWLEDGEMENTS

The foregoing comments were largely inspired by a workshop entitled "Large-Scale Bedforms in Shallow Rivers" held during the evenings of October 20 and 21, 1977, at the First International Research Symposium on Fluvial Sedimentology. In addition to the writer, who organized and co-ordinated the workshop, participants included G. M. Ashley, V. R. Baker, J. C. Boothroyd, B. J. Bluck, D. J. Cant, M. Church, J. D. Collinson, T. C. Gustavson, F. J. Hein, R. G. Jackson, I. P. Martini, and J. B. Southard. Numerous others attended the sessions and contributed to discussions, of which J. Shaw and I. Karcz were among the most active. In brief commentaries such as this, it is impossible to properly credit the origins of viewpoints. These comments reflect some of the writer's opinions on the subject, whatever their origin, and should not be construed as a consensus or statement representing the views of the entire workshop group.

REFERENCES

Bluck, B. J., 1976, Sedimentation in some Scottish rivers of low sinuosity: Roy. Soc. Edinburgh Trans., v. 69, p. 425-456.

Boothroyd, J. C. and Ashley, G. M., 1975, Process, bar morphology, and sedimentary structures on braided outwash fans, northeastern Gulf of Alaska: *in* A. V. Jopling and B.C. McDonald, *eds.,* Glaciofluvial and Glaciolacustrine Sedimentation, Soc. Econ. Paleont. Mineral. Spec. Pub. 23, p. 193-222.

Cant, D. J., 1976, Braided stream sedimentation in the South Saskatchewan River: Unpub. Ph.D. thesis, McMaster Univ., 248 p.

Collinson, J. D., 1970, Bedforms of the Tana River, Norway: Geogr. Ann., v. 52A, p. 31-56.

Hein, F. J. and Walker, R. G., 1977, Bar evolution and development of stratification in the gravelly braided Kicking Horse River, B.C.: Can. J. Earth Sci., v. 14, p. 562-570.

Jackson, R. G. II, 1975, Hierarchial attributes and a unifying model of bedforms composed of cohesionless material and produced by shearing flow: Geol. Soc. Am. Bull., v. 86, p. 1523-1533.

Jones, C. M., 1977, Effects of varying discharge regimes on bedform sedimentary structures in modern rivers: Geology, v. 5, p. 567-570.

Kellerhals, R., Church, M. and Bray, D. I., 1976, Classification and analysis of river processes: J. Hydraulics Div., Am. Soc. Civil Eng., v. 102, HY7 , p. 813-829.

Miall, A. D., 1977, A review of the braided-river depositional environment: Earth Sci. Revs., v. 13, p. 1-62.

Smith, N. D., 1974, Sedimentology and bar formation in the upper Kicking Horse River, a braided outwash stream: J. Geol., v. 82, p. 205-224.

Sundborg, A., 1956, The River Klarälven, a study of fluvial processes: Geogr. Ann., v. 38, p. 125-316.

Task Force on Bedforms in Alluvial Channels, 1966, Nomenclature for bedforms in alluvial channels: J. Hydraulics Div., Am. Soc. Civil Eng., v. 92, HY3, p. 51-64.

Walker, R. G., 1976, Facies models 3: Sandy fluvial systems: Geoscience Canada, v. 3, p. 101-109.

BEDFORMS IN THE PITT RIVER, BRITISH COLUMBIA

GAIL M. ASHLEY[1]

ABSTRACT

The Pitt River is a meandering river channel linking the Fraser River estuary and Pitt Lake, which acts as a temporary reservoir for tidally diverted Fraser River flow. Large-scale bedforms can be divided into dunes and sand waves on the basis of three dimensional geometry. Sand waves have linear crests and height/spacing ratio of <1:20; dunes (2-3 m spacing) have sinuous crests and height/spacing ration of >1:10 and are found mainly on the stoss side of sand waves. Sand waves do not occur in a continuous range of sizes but rather form discreet height/spacing groups: .8 m/10-15 m; 1.5 m/25-30 m; 3 m/50-60 m. Most (70%) are of the intermediate size. These groups fall on a straight line on a log height-log spacing plot implying a common hydraulic control. Repeated depth soundings revealed reorganization of bedforms from one scale range to another, by halving or doubling. The size and morphology of bedforms appear to be independent of depth. Small forms are characteristic of areas with flat channel topography whereas large forms predominate in reaches of shallowing depths (slope ≥ 1°). Bedforms are flood-oriented, reflecting dominant flow conditions and direction of net sediment transport, and appear to be in quasi-equilibrium with tidal and seasonal flow variations. Both bedforms and channel geometry appear to be scaled solely to the winter flood peak discharge (2400 m^3 .sec^{-1}) which is thus considered the effective discharge, Q_e.

INTRODUCTION

Pitt River — Pitt Lake (Fig. 1) is a freshwater system, yet strongly influenced by the tides. The single meandering channel of the Pitt River acts as a conduit for water passing in (flood) and out (ebb) of Pitt Lake during tidally induced flow reversals. Flow in Pitt River is not only bidirectional but varies in magnitude due to seasonal changes in basin runoff and differences in tidal range. Studies of channel geometry (Fig. 2) revealed that the river has regular meanders (λ_m = 6100 m) and evenly spaced pools and riffles which are scaled to the strongest flow (winter flood current, Q_e). Meander point bars are accreting on the upstream side indicating deposition by the flood-oriented current.

A predominance of flood-oriented bedforms were found to cover the sandy meandering thalweg, but were not found in the finer grained portions of the channel adjacent to the bank.

The purpose of the study was to examine the dimensions and morphology of the bedforms as well as their distribution and to attempt to relate these characteristics to flow conditions.

GEOLOGIC SETTING

The Pitt River — Pitt Lake system is situated in a glacially scoured valley within the Coast Mountains, British Columbia, approximately 30 km inland from the port of Vancouver. Both the river and lake are tidal and are connected to the Pacific Ocean (Strait of Georgia) by the Fraser River. During flood tide, Fraser River water is diverted into the Pitt River; and conversely, water drains from the Pitt system during the ebb. Although flow is reversing, sediment transport appears to be completely dominated by flood-oriented flow (Ashley, 1977). Flow pattern that occurs during flood is shown in Figure 3A.

---

[1]Department of Geological Sciences, University of British Columbia, Vancouver, British Columbia; Present Address: Department of Geological Sciences, Rutgers, The State University of New Jersey, New Brunswick, New Jersey 08903.

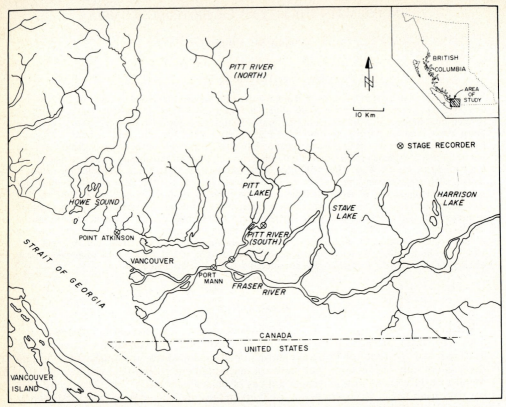

Fig. 1.   Location map of Pitt River — Pitt Lake freshwater tidal system.

Sediment is supplied to the system by the Fraser River, with insignificant amounts being added by smaller streams. Mean grain size of sediments decreases from 1ϕ at the Fraser-Pitt confluence to 9ϕ at mid lake. A bedrock sill near the lake's northern end prevents all but small amounts of clay-sized material (contributed by rivers draining into the northern end of Pitt Lake) from reaching the southern end. An estimated total of $150\pm20 \times 10^3$ tonnes are deposited annually in Pitt Lake.

STREAM FLOW

Velocity is one of the more important parameters used to characterize flow. Unfortunately, velocity in the Pitt River is variable in direction and magnitude, both daily and seasonally. It was considered important to determine the range of velocities that could be expected to occur during any year, in particular, the maximum velocities and their duration. Two different methods of current measurements were used: (1) readings taken at 7.5-minute intervals, one meter from bottom, for 36 days; (2) current profiles taken at 30-minute intervals for flood or ebb cycles. Portions of 50 days of velocity data in both river and lake channel were taken in hopes of obtaining a representative sampling of the broad spectrum of flow conditions existing in the Pitt system.

Peak mean velocities measured at four sites (Fig. 3B) are summarized in Table I and show that flood flows are, in general, stronger than ebb. This is true, in particular, when the hydrodynamic control conditions (tidal range in the Strait and Fraser discharge) are the same for the flood and ebb flows being compared. Although flood currents have a

A = 3250 sq. m	Q WINTER-FLOOD = 2400 cu m/sec (85,000 cu.ft./sec.)
W = 610 m	Q WINTER-EBB = 2080 cu.m/sec.(73,500 cu.ft./sec.)
d = R = 12.1 m	Q FRESHET-FLOOD=1800cu.m/sec (64,500 cu.ft./sec.)
λ= 6100 m	Q FRESHET-EBB = 950cu.m/sec.(33,500 cu.ft./sec.)
W/d= 50	S WINTER FLOOD = + .000053
CHANNEL LENGTH – 19.8 Km	S WINTER EBB = – .000032
	S FRESHET FLOOD = +.000018
SINUOSITY 1.20	S FRESHET EBB = – .000016

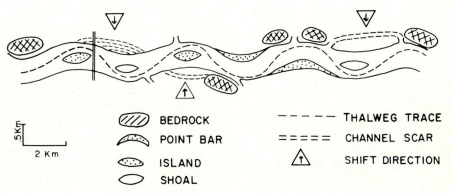

⬭(hatched) BEDROCK	– – – – – THALWEG TRACE
⬭ POINT BAR	===== CHANNEL SCAR
⬭ ISLAND	⬭↑ SHIFT DIRECTION
⬭ SHOAL	

.5 Km
2 Km

Fig. 2.   Diagrammatic sketch of hydraulic geometry of the Pitt River.

higher velocity, they flow for a shorter time period than the corresponding ebb (Fig. 4). Total discharge is greater on the ebb because of the volume contributed by basin drainage into the lake.

Current profiles[1] were made from a boat anchored at fixed positions in the thalweg during ebb and flood flows and under freshet (May-July) and "winter" (August-April) conditions. Each profile consisted of 8 points (10 cm from bottom, 30 cm from bottom, one meter from bottom, 0.2d, 0.4d (mean), 0.6d, 0.8d, and surface). The measurements (both magnitude and direction) at each depth were based on readings averaged over a two-minute period, thus each profile spans 15 to 20 minutes. A digital counter integrating electrical pulses over a 10-second period was used to average velocity fluctuations caused by micro- and macroturbulence (Matthes, 1947).

Semi-log plots of velocity data for entire flow depth (boundary layer) reveal that most profiles are composed of two distinct zones. The break between the zones occurs between 2 m and 4 m above the bottom (Fig. 5). Within each zone the profile generally shows a logarithmic variation of velocity with depth (d, distance above the bed), however, the slope ($d\mathrm{V}/d\log d$) is higher in upper zone. The slope steepness changes during acceleration and deceleration of both flood and ebb oriented flows. As large scale bedforms (1 - 3 m in height) cover the channel bottom, it is possible that their presence is instrumental in the development of the two zones. Considerably more data is needed to determine how flow structure changes during a tidal cycle and to ascertain the relationship between bedforms and the presence of flow zones.

---

[1]Hydro Products, Savonius Rotor with a direct readout for current speed (model #460A) and direction (model #465A)

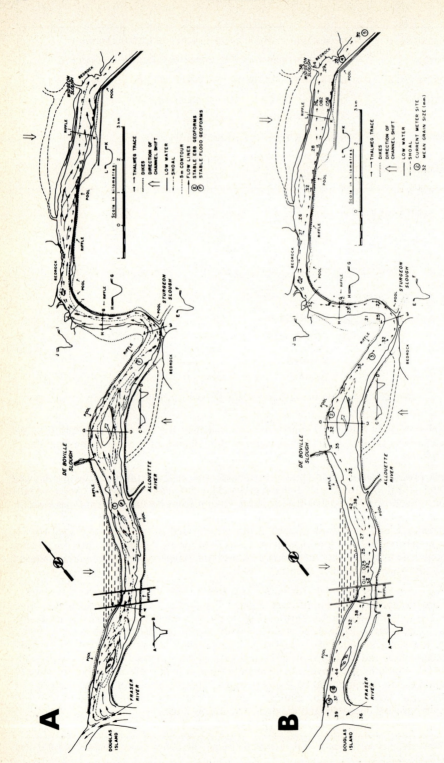

Fig. 3. A. Flow pattern of flood-oriented currents. Flow lines drawn perpendicular to crest of large scale two-dimensional bed configurations (15-60 m in spacing, 1-3 m in height). B. Mean grain size (in mm) distribution map and location of current measurement sites.

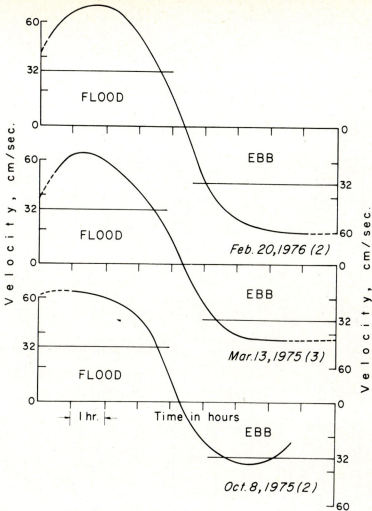

Fig. 4. Comparison of time duration of flood and ebb currents above the predicted critical velocity (32 cm/sec) for Pitt River (Ashley, 1977).

Table 1. Summary of peak mean velocities (cm/sec [1]) determined from profile measurements; mean is at .4d, measured from bed.

Date	Site	Flood	Ebb	Date	Site	Flood	Ebb
March 11, 1975	(3)	52	40	August 6, 1975	(4)		57
March 13, 1975	(3)	64	42	August 11, 1975	(1B)	59	44
May 9, 1975	(2)		56	August 13, 1975	(4)	34	
May 21, 1975	(2)	40		Sept. 4, 1975	(1B)		62
June 12, 1975	(4)	50		Oct. 8, 1975	(2)	64	38
June 24, 1975	(4)	47		Feb. 20, 1976	(2)	70	
July 9, 1975	(4)		33				

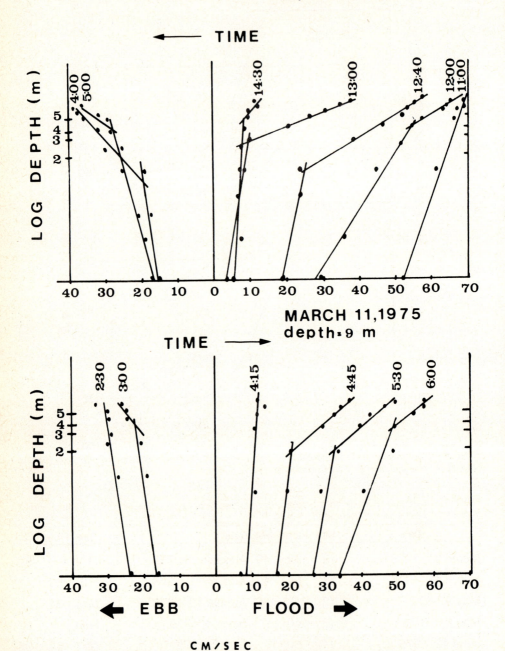

Fig. 5. Velocity profiles taken at A. site 2 B. site 3 (see Fig. 3B). Flow is divided into two distinct zones. Velocity varies logarithmically with depth, but at different rates in each zone. Division between the zones is at 2-4 m from bed. Flow structure may be related to bedforms (1-3 m in height) present on channel bottom.

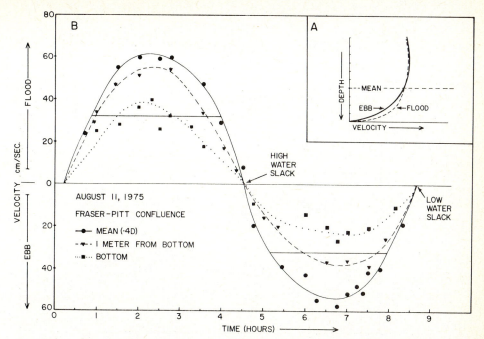

Fig. 6. A. Diagrammatic comparison of "typical" flood and ebb current profiles. B. Velocity vs. time plot of data taken August 11, 1975 at current measurement site 1B (Fig. 3B). Line drawn at 32 cm/sec illustrates amount of flood time above critical velocity is greater than that of ebb (Ashley, 1977). Bottom actually is 10 cm from bottom.

Arithmetic depth-velocity plots indicate that the shape of "typical" flood and ebb curves are distinctly different for a given mean velocity (Fig. 6A). The shape of a velocity-depth plot depends mainly on extent of drag imposed on the flow. For a given depth, the greater the drag the greater the turbulence, which in turn produces a more gradual velocity profile toward the bed. Bedforms on channel bottom are predominantly flood oriented and the form resistnace (drag coefficient) would be expected to change between flood and ebb. The slope angle of the exposed bedform surface (stoss side) presented to ebb flow is greater than slope angle of the surface (lee side) opposing flood-oriented flows (Fig. 7). Thus, it is interpreted that more drag occurs on the ebb resulting in a more gradual velocity profile toward the bed. Although Znamenskaya (1967) has attempted to relate bedform geometry and flow resistance, few quantitative data are available for the types of large-scale bedforms found in the Pitt River.

Velocity measurements taken near the Pitt-Fraser confluence (Fig. 3, site 1B; Fig. 6B) are considered typical of most profiles measured. These time-velocity curves were drawn by eye to average scatter which presumably is due to low-frequency velocity fluctuations. Plots of mean velocity, velocity one meter from bottom and 10 cm from bottom all exhibit similar shapes. The difference between flood and ebb mean velocity peaks is 7 cm/sec, however, the difference at one meter from bottom is substantially greater (17 cm/sec.). Visher and Howard (1974) noted a 20 cm/sec difference at one meter in Altahama Estuary, whereas, Klein (1970) found only 6 cm/sec difference at one meter from bottom in the portion of the Midas Basin characterized by flood-oriented sand waves. It should be noted that, in the Pitt, if an ebb or flood flow were of equal strength (equal mean velocities) the velocity near the base and thus basal shear stress and sediment entrainment potential would be greater for the flood. This is a consequence of the basic difference between the flood and ebb profiles (Fig. 6A). At first glance this interpretation

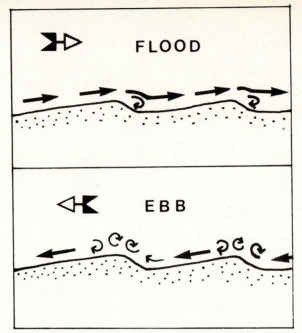

Fig. 7. Diagrammatic comparison between flood and ebb flows of turbulence created by the flood-oriented flows.

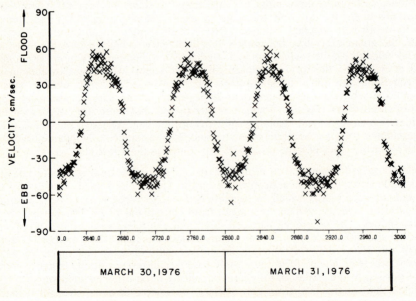

Fig. 8. Computer plot of "continuously" recorded velocity data. Each X represents an instantaneous velocity (magnitude and direction) measurement at 7.5-minute intervals. Measurements are one meter from bottom. Note flood velocities are higher than ebb.

may appear to contradict theory which assumes that turbulence intensity and shear increase together. However, the portion of total flow resistance represented by form resistance or that utilized on the grains differs between ebb and flood. On ebb, a greater portion of the resistance is borne by form resistance (Einstein and Barbarossa, 1952) because of increased drag (Fig. 7). The opposite is true of the flood where a greater portion of resistance is available for shear at grain level. It is only that portion of total resistance imparted on the grain which may lead to sediment entrainment.

The continuously recorded velocity measurements were taken by a positively bouyant meter. [General Oceanics, Inc. film recording current meter (model #2010)], which was anchored to the channel bottom (Fig. 3A, site 5) but free to sway with changing currents. The meter recorded on movie film instantaneous readings of magnitude and direction of flow (one meter off bottom) at 7.5-minute intervals. Figure 8 illustrates the rhythmical change in flow magnitude and direction which occurs in response to the tides.

In summary, velocity profile data indicate the flow structure changes at a distance 2-3 m above bottom and that flood flows generally exhibit higher mean velocities, but for shorter durations than corresponding ebb flows.

Bed Configurations

Observations

All observations of the configuration of the channel bottom were made remotely by depth sounders and a side-scan sonar. The precision of the depth sounding records is within 30 cm (vertical) and approximately 15 m (horizontal). The records have a vertical exaggeration of between 1:10 and 1:15. Side-scan sonar records can be read to within 1 m vertically and 15 m horizontally. Vertical exaggeration of side-scan records is 1:3. A distortion of the true bedform shape occurs on some depth sounding records due to orientation of slopes of varying steepness relative to direction of boat motion. However, as only the gross form (length and height) and proportion of length of stoss and lee sides of the bedforms were being examined from the depth soundings the distortion was not considered important.

Repeated depth soundings were taken along all reaches of the river, concentrating on the thalweg. Most runs were carried out between the months of May and September with a lesser number in the winter to detect seasonal changes. Because a bewildering variety of bedform shapes and sizes was revealed during the 18-month survey, the side-scan sonar was used for a two-day period (June 1, 2, 1975) to aid in the interpretation of the forms by determining their 3-dimensional geometry.

The large range of sizes and shapes of bedforms found in the channel presented a problem of bedform terminology. The classification of large-scale alluvial bedforms is in a state of confusion reflecting a general lack of understanding of their genesis. The most recent review of bedform terminology (Task Force, 1966) is based on an attempt to extrapolate bedform-hydraulic relationships developed in flume studies such as that of Simons *et al.*, (1965) to the natural environment. Unfortunately, the range of conditions found in rivers is not easily reproduced in flumes. In addition, the large-scale, low-amplitude bedforms found in shallow marine, estuarine, and riverine environments have no equivalent forms in the flume results of Simons *et al.*, (1965). These large bedforms have been termed giant ripples, sand ridges, dunes, sand waves, super ripples, transverse bars, sand dunes, and large scale ripples by various authors. Clearly, for the present, a viable bedform classification should be independent of genetic assumptions and based only on descriptive morphology.

A detailed study of the 3-dimensional geometry of bedforms in Pitt channel was made from the sounding records. Using plan geometry, form height and spacing, and form orientation with respect to flood or ebb flow direction, a classification of bedform shapes

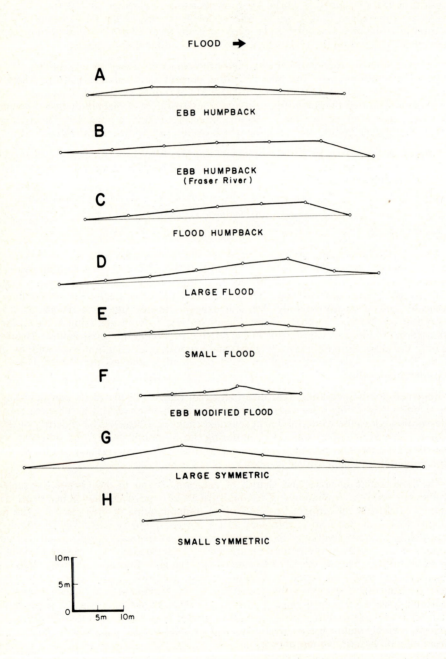

Fig. 9. Bedform shapes found in Pitt River drawn to true scale.

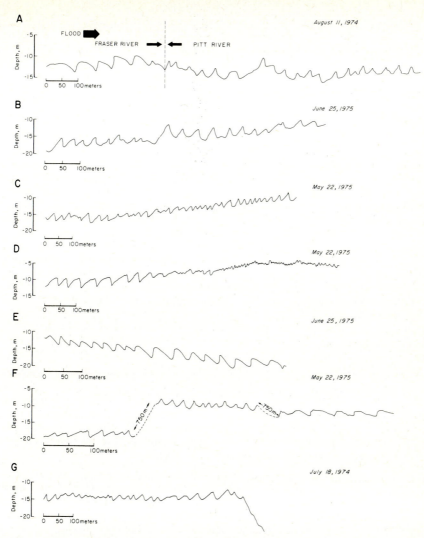

Fig. 10. Depth-sounding profiles: A. Fraser-Pitt confluence; Fraser has ebb forms, Pitt floodforms. B. Flood-oriented sand waves with some ebb modified crests (km 3). C. Flood-oriented small scale sand waves (10-15 m spacing (km 8). D. Flood "humpback" forms and dunes on flat topography (km 9). E. Ebb "humpback" forms south of Point Addington (E on Fig. 3A). F. Flood-oriented, symmetric and ebb-modified, and ebb-oriented sand waves that occur on large mounds in channel. G. Bedforms at Pitt Lake outlet.

found in Pitt River was developed (Fig. 9). On the basis of height/spacing proportions two major groups of forms can be discerned: small forms (spacing 5 m, ht/spacing ratio = 1/10) equivalent to "dunes" of Simons *et al.*, (1965), and large forms (spacing 10 - 60 m, ht/spacing ratio = 1/20) equivalent to "sand waves" of Harms *et al.*, (1975). Spacings in the large forms do not represent a continuum but occur in discrete groups (10 - 15 m, 25 - 30 m, 50 - 60 m) with few bedforms of 15 - 25 m and 30 - 50 m observed. The dominant form is 25 - 30 m in length, composing 70% of the total. The smaller forms (10-15 m) make up 25% and the large forms (50 - 60 m) about 5%.

Dunes (height/spacing = 0.3 m/2 - 3 m) are 3-dimensional with sinuous crests and are found on the backs of the large 2-dimensional straight-crested sand waves (Fig. 11C).

Sand waves occur in two basic shapes. Type one, characterized by a rounded stoss side (usually covered with dunes) and fairly steep lee slope (Fig. 9A, C) occurs in trains at only a few sites (E and F of Fig. 3A). This form referred to as a "humpback" sand wave (Fig. 10D, 11A, flood forms; Fig. 10F, ebb forms) is identical to ones found in Fraser River (Fig. 9B; Fig. 10A) developed under unidirectional flow. The second type (Fig. 9D-H) has uniformly sloping stoss and lee sides. However, the angle of slope and proportion of length of sides varies continuously from flood-oriented forms (60% of total) through ebb-modified flood forms and fairly symmetrical shapes (25%) to ebb-oriented types (15%).

Bedforms occur along the entire sandy thalweg from Fraser River (Fig. 10A) to the lake outlet (Fig. 10G). The thalweg is approximately 100 -150 m wide and bedforms usually cover the entire area. Sand wave crests are perpendicular to the orientation of the thalweg and the superimposed dunes have crests parallel to the sand wave crests. Dunes also occur in sandy channel areas off the thalweg and on sandy shoals surrounding the islands. In contrasts, side-scan records and visual observation indicate that ripples are typical of finer grained (silty) areas.

A general relationship between both bedform size and type and topography was found at the base of ramps (areas of rapid shoaling, with slopes $\leqslant 1°$) and the smallest forms are found on relatively flat topography. For example, sand waves (4.5 m/60 m) occur on the ramp at the constricted area near a bedrock outcrop at km 15 (Fig. 3A). The channel shallows from 21 m toward the shoal to 10 m at the north end of the wave train. The reach at cross section G-H (Fig. 3A) is deep (24 m) but relatively flat. Sand waves here are 1 m/30 m), identical with the average size found in other parts of the river at only 5 m depth. Several trains of small sand waves (.8 m/10-15 m) occur on flat shoals surrounding the islands.

In most reaches large longitudinal "mounds" occur with a wavelength of 3 km (one half the meander wavelength, λ_m) and are thus presumably related to the meander sequence. Elevation difference from crest to trough of the mounds is 7 - 8 m. These major channel features are located on inside bends of midchannel islands or at riffles and present a ramp of shallowing channel to both flood and ebb oriented currents. Flood-oriented sand waves are found on the downstream side of these ramps, symmetrical forms on the shallow top, and ebb-oriented forms on the upstream side (Fig. 10F).

Repeated soundings over a period of months indicated a reorganization of different scales of sand waves. For example, it appeared that several (10 - 15 m) forms merged to create a 30 m or 60 m bedform or, inversely, a large form would be replaced by smaller ones. Similar jumps in scale of bedforms have been noted in other rivers by Pretious and Blench (1951) in Fraser River, Znamenskaya (1963) in Polometi River (U.S.S.R.) and Neill (1969) in Red Deer River (Alberta, Canada). The nature of the transformation is not clear; however, intermediate size forms are transient if they exist at all. Although the exact flow conditions which might have caused "regrouping" in the Pitt could not be determined, the changes occurred mainly within the flood-oriented forms during winter flows and on the downstream side of the "mounds" (2 km, 6 km and 12 km: Fig. 3).

Repeated depth sounds over a tidal cycle provided evidence of bedform modification. Flood-oriented forms developed ebb-modified crests (Fig. 9F; Fig. 10F) during strong ebb flows (mean velocity of 50 - 60 cm/sec). A complete change from flood-oriented to ebb-oriented form or from ebb-to-flood was not observed.

Interpretation

Because Pitt River has bidirectional flow, the state of equilibrium of the bedforms becomes an important factor. Unidirectional rivers commonly experience an annual high discharge event (spring flood) lasting several days or weeks. Bedforms have been observed to have a delayed response to the increasing or decreasing discharge, and this

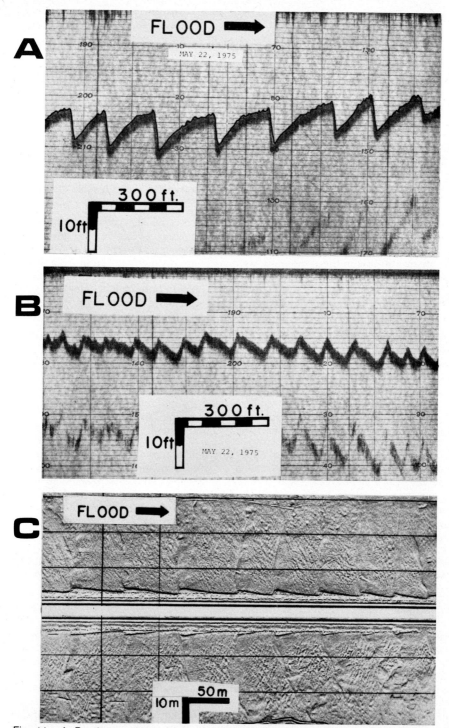

Fig. 11. A. Depth soundings of flood-oriented sand waves. B. Depth sounding of ebb-oriented sand waves. C. Side-scan sonar record. Note mottled pattern of dunes on backs of sand waves.

delay has been termed "lag". In tidal flows the discharge fluctuations have a time scale of hours, apparently insufficient for one event to have a significant effect on large bedforms (Ludwick, 1974). Thus a lag phenomenon would be difficult to measure in the tidal environment. In the Pitt, geometry of each bedform represents the summation of the modifications of both flow directions. Since sounding over an 18-month period determined that the majority of bedform types (Fig. 9) remained constant throughout the year, the forms are interpreted to be a quasi-equilibrium with the bidirectional flow. Thus, bedform shape maintains the imprint of the dominant flow conditions at that site. The humpback forms (Fig. 9A, C) are found in fairly protected areas of the channel. As they appear to be affected by only one current direction humpback forms are considered to be an equilibrium form. In general, the size and shape of the sand waves appears to be related to channel geometry and not to depth as was suggested by Allen (1968), Yalin (1974), and Jackson (1976b). Similarly Coleman (1969, Brahmaputra River) and Whetten and Fullam (1967, Columbia River) both found no correlation between scale of bedforms and depth, in agreement with the data from the Pitt.

Bed configurations in the Pitt are difficult to interpret because of the multiplicity of forms and the fact that they appear to reflect the average hydraulic conditions. Most research on sand waves has been done in shallow marine and estuarine environments. Characteristics of flow in these infinitely wide areas are distinctly different from those in channelized flow where lateral-boundary conditions are important. Although the Pitt River is tidal, it is clearly similar to unidirectional rivers in both channel morphology and bed configuration. Significantly, analysis reveals systematic relationships between the various bedform groups and channel parameters.

Plotting the average height and spacing of sand waves on a logarithmic scale results in a linear relationship. As these sand waves form at least 3 distinct groups, this linear relationship appears significant and suggests a common mode of genesis. A similar plot of estuarine sand waves (Boothroyd and Hubbard, 1975; Fig. 3) shows a scatter of values but does suggest a distinction between subtidal, or deeper water forms, and intertidal, or more shallow water forms. The Pitt values fall within the field of the deep water sand waves on Boothroyd and Hubbard's plot. Dunes from the Pitt plot in the middle of the megaripple field on the same diagram and are assumed to be the same form.

Theoretical models for the genesis of dunes and sand waves are in a state of flux. Dune formation is thought by some to be related to large scale turbulence (Velikonov and Mikhailova, 1950; Znameskaya, 1963). The regular spacing of dunes appears to be a direct function of the scale of the largest turbulent eddies (Znamenskaya, 1963; Grishanin, 1972; Jackson, 1975). Flow separation and associated slipface development provide alternating areas of erosion and deposition critical to sediment entrainment and transport. A detailed model of dune genesis has been presented by Costello (1974).

Although considerable progess has been made in understanding the mechanics of dune formation, none of the theoretical models proposed (Kennedy, 1969; Smith, 1970) predict the existence of sand waves.

Recent flume studies (Pratt and Smith, 1972; Pratt, 1973; Costello, 1974) reveal bedforms intermediate between ripples and dunes. The flume bedforms were called "intermediate flattened dunes" by Pratt and "bars" by Costello, who equates them with sand waves. Costello adapted kinematic wave theory to explain the genesis of these forms. Although shallow water (≤3 m) transverse bars (Smith, 1971; Jackson, 1976a) may be adequately explained by "shock wave" aggradation of sediment, several characteristics of deep water sand waves in river channels are not consistent with Costello's model. For example, the regular geometry and spacing of sand wave trains in Pitt and other rivers (Pretious and Blench, 1951; Whetten and Fullam, 1967; Carey and Keller, 1957) conflict with Costello's conclusion that sand waves "are randomly generated and (that) this randomness carries over into their spacing and height". The occurrence of dunes in apparent equilibrium with sand waves at velocities considerably lower than predicted by Costello's depth-velocity diagram (his Fig. 26) has been documented in a number of studies (Pretious and Blench, 1951; Coleman, 1969; Neill, 1969; Singh and Kumar, 1974; and many others). Jackson (1976a) has presented evidence that dunes and sand waves not only occur together but also migrate under essentially steady flow conditions. It is unfortunate that these large-scale forms, characteristic of sandy rivers, estuaries, and marine shoals, are so poorly understood.

CONCLUSIONS

The Pitt River is a flood dominated system; although flow duration is shorter, flood flows have slightly higher peak velocities than the ebb. The area adjacent to thalweg is covered with large-scale, low amplitude, and straight-crested bedforms herein termed sand waves. These bedforms appear to be in quasi-equilibrium with bidirectional and seasonal changes in discharge. 60% are flood oriented, 25% symmetric or ebb-modified flood forms, and 15% are ebb oriented. The majority have smaller bedforms (dunes) on their stoss sides. Three distinct sizes (height/spacing = .8 m/10-15 m; 1.5 m/25-30 m; 3 m/50-60 m) of sand waves were found in the river and the linear relationship between log height and log spacing suggests a common hydraulic control. The position of the various sand wave types and sizes appears to be related to channel geometry and not to depth of flow.

ACKNOWLEDGEMENTS

The guidance and encouragement of W. H. Mathews during all phases of the Ph. D. thesis which produced this study is greatly appreciated. I am grateful to W. C. Barnes, M. Church and J. D. Milliman for reading earlier versions of the manuscript and making many helpful suggestions. I. J. Duncan and J. C. Boothroyd kindly reviewed the final version.

I am thankful to Jon Jolley, Inc., Seattle, Wash. for providing the Klein side-scan sonar; H. Van der Meulen who aided in the field; and to R. Macdonald for providing technical assistance.

National Research Council of Canada Grant A-1107 (Prof. W. H. Mathews, U.B.C.) provided support for both field work and preparation of the manuscript.

REFERENCES

Allen, J. R. L., 1968, The nature and origin of bedform hierarchies: Sedimentology, v. 10, p. 161-182.

Ashley, G. M., 1977, Sedimentology of a freshwater tidal system, Pitt River-Pitt Lake, British Columbia: Unpub. Ph. D. thesis, Univ. British Columbia, 404p.

Boothroyd, J. C., and Hubbard, D. K., 1975, Genesis of bedforms in mesotidal estuaries: *in* L. E. Cronin, *ed.*, Estuarine Research; Academic Press, N.Y., v. 2, p. 217-235.

Carey, W. C., and Keller, M. D., 1957, Systematic changes in the beds of alluvial rivers: J. Hydraulics Div., Am. Soc. Civil Eng., v. 83, Proc. Paper 1331, 24p.

Coleman, J. M., 1969, Brahmaputra River; channel processes and sedimentation: Sediment. Geol., v. 3, p. 129-239.

Costello, W. R., 1974, Development of bed configurations in coarse sands: Rept. 74-1, Earth and Planetary Sciences Dept., M.I.T., Cambridge, Mass, 120p.

Einstein, H. A., and Barbarossa, N. L., 1952, River channel roughness: Am. Soc. Civil Eng. Trans., v. 117, p. 1121-1146.

Grishanin, K. V., 1972, Stability of river channels and kinematic waves: Trans. State Hydrol. Inst., no. 190, p. 37-47.

Harms, J. C., Southard, J. B., Spearing, D. R., and Walker, R. G., 1975, Depositional environments as interpreted from primary sedimentary structures and stratification sequences: Soc. Econ. Paleont. Mineral. Short Course 2, 161p.

Jackson, R. G., 1975, Velocity-bedform-texture pattern of meander bends in the lower Wabash River of Illinois and Indiana: Geol. Soc. Am. Bull., v. 86, p. 1511-1522.

———, 1976a, Large scale ripples of the lower Wabash River: Sedimentology, v. 23, p. 593-623.

———, 1976b, Sedimentological and fluid-dynamic implications of the turbulent bursting phenomenon in geophysical flows: J. Fluid Mech., v. 77, p. 531-560.

Kennedy, J. F., 1969, The formation of sediment ripples, dunes and antidunes: *in* Annual Review of Fluid Mechanics, 1969: p. 147-168.

Klein, G. de V., 1970, Depositional and dispersal dynamics of intertidal sand bars: J. Sediment. Petro., v. 40, p. 1095-1127.

Ludwick, J. C., 1974, Tidal currents and zig-zag sand shoals in a wide estuary entrance: Geol. Soc. Am. Bull., v. 85, p. 717-726.

Matthes, G. H., 1947, Macroturbulence in natural stream flow: Am. Geophys. Union Trans., v. 28, p. 255-262.

Neill, C. R., 1969, Bedforms in the lower Red Deer River, Alberta: J. Hydrol., v. 7, p. 58-85.

Pratt, C. J., 1973, Bagnold approach and bedform development: J. Hydraulics Div., Am. Soc. Civil Eng., v. 99, p. 121-137.

———, and Smith, K. V. H., 1972, Ripple and dune phases in a narrowly graded sand: J. Hydraulics Div., Am. Soc. Civil Eng., v. 98, p. 859-874.

Pretious, E. S., and Blench, T., 1951, Final report on special observations of bed movement in lower Fraser River at Badner Reach during 1950 freshet: Nat. Res. Council, Canada; Vancouver, British Columbia, 12p.

Simons, D. B., Richardson, E. V., and Nordin, D. F., Jr., 1965, Sedimentary structures generated by flow in alluvial channels: *in* G. V. Middleton, *ed.,* Primary sedimentary structures and their hydrodynamic interpretation: Soc. Econ. Paleont. Mineral. Spec. Pub. 12, p. 34-52.

Singh, I. B., and Kumar, S., 1974, Mega-and giant ripples in the Ganga, Yamuna, and Son Rivers, Uttar Pradesh, India: Sediment. Geol., v. 12, p. 53-66.

Smith, J. D., 1970, Stability of a sand bed subjected to a shear flow of low froude no.: J. Geophys. Res., v. 75, p. 5928-5939.

Smith, N. D., 1971, Transverse bars and braiding in the lower Platte River, Nebraska: Geol. Soc. Am. Bull., v. 82, p. 3407-3420.

Task Force on Bedforms in Alluvial Channels, 1966, Nomenclature for bedforms in alluvial channels: J. Hydraulics Div., Am. Soc. Civil Eng., v. 92, (HY3), p. 51-64.

Velikonov, M. A., and Mikhailova, N. A., 1950, The effect of large-scale turbulence on pulsations of suspended sediment concentration: Izv., Akad., Navk. SSSR. Ser. Geogr. Geofiz., v. 4, p. 421-424.

Visher, G. S. and Howard, J. D., 1974, Dynamic relationship between hydraulics and sedimentation in the Altahama Estuary: J. Sediment. Petrol., v. 44, p. 502-521.

Whetten, J. T., and Fullam, T. J., 1967, Columbia River bedforms: Proc. Int. Assoc. Hydraul. Res., 12th Cong., Fort Collins, p. 107-114.

Yalin, M. S., 1974, On the formation of dunes and meanders: Int. Assoc. Hydraulic Res., c. 13, p. 1-8.

Znamenskaya, N. S., 1963, Experimental study of the dune movement of sediment: Trans. State Hydrologic Inst., no. 108, p. 89-114.

———, 1967, The analyses and estimation of energy loss: Proc. Int. Assoc. Hydraulic Research, 12th Cong. Proc., Fort Collins.

BED-FORM DISTRIBUTION AND INTERNAL STRATIFICATION
OF COARSE-GRAINED POINT BARS
UPPER CONGAREE RIVER, S.C.

RAYMOND A. LEVEY[1]

ABSTRACT

Ground and aerial studies demonstrate that point bars along the upper Congaree River show considerable variation in size, topography, morphologic features, facies and grain size patterns. A generalized depositional model for these coarse grained meander deposits consists of bar-apex, midbar, distal-bar, and the scour-pool and crossover facies. Point-bar facies are distinguished by their surficial morphologic form, internal stratification, and grain-size characteristics.

Well-developed coarse-grained point bars contain sand and gravel near the bar-apex. The mid-bar facies is covered by sandy megaripples that commonly migrate over the stoss side of large transverse bars attached to the point bar margin closest to the channel thalweg. Distal portions of the bar may develop both chute channels and chute bars resulting from high-discharge events. The scour-pool and crossover facies are characterized by transverse bars, sandwaves, and megaripples in between consecutive meander bends and deep scour troughs within the meander bend.

Observations of sediments at the surface and in shallow trenches indicate that the apex gravel material represents a lag deposit. Large-scale tabular cross beds of transverse bars and small trough and tabular bedding of megaripples characterize the mid-bar facies. Chute bars contain complex bedding that indicates their deposition during multiple high-stage flow events.

Grain-size characteristics, stratification and physiographic features that occur in these coarse-grained meander deposits establish criteria for the recognition of point bar sequences in ancient rocks.

INTRODUCTION

In this study, the term point bar refers to sediment deposited along the inside of meander bends occupying that area of channel between the thalweg and the lower limit of the flood plain. Coarse-grained point bar deposits are important constituents of many ancient sandstone and conglomerate units (Klein, 1975). However, only a few studies have been carried out on modern coarse-grained point bars, the studies by McGowen and Garner (1970), Bluck (1971) and Jackson (1976a) being notable exceptions.

The three main objectives for this study of a coarse-grained point-bar system are (1) to develop a depositional model for a Coastal Plain-Piedmont boundary river system; (2) to relate the geomorphic expression of point bar facies and bed forms to their internal stratification; (3) to compare the results of this study to the classical point bar sequence model.

A meander train along the upper Congaree River near Columbia, South Carolina, contains a sequence of coarse-grained (mixed sand and gravel) point bars. In the present study, the seven point bars which form this meander train were monitored during a one-year period for changes in bed form pattern and grain size. Velocity distribution of the flow around the meander bends was also measured.

Table 1 outlines the bed form terminology used in this study. Terminology describing stratification produced by migrating bed forms follows the description outlined by Harms et al. (1975; Figs. 3-1 to 3-4) of structures typical of unidirectional flow over sand size material.

STUDY AREA

The Congaree River is formed by the confluence of the Broad and Saluda Rivers at Columbia, South Carolina (Fig. 1). This confluence is located on the fall line between the

---

[1]U.S. Geological Survey and Coastal Research Division, Department of Geology, University of South Carolina, Columbia, S.C. 29208

Table 1.. CLASSIFICATION OF BED FORMS WITHIN THE CONGAREE RIVER

	Ripples	Megaripples	Sandwaves	Transverse Bars
Spacing (trough to trough)	Less than 60 cm	60 cm to 6 m	6 m to 60 m	Greater than 25 m
Height (crest to trough)	Less than 5 cm	10 to 30 cm	30 to 130 cm	50 cm to 2 m
Geometry	Slightly asymmetric.	Asymmetric form often highly three-dimensional, straight to cuspate crests, scour in troughs.	Both asymmetric and symmetric, straight to sinuous crests, presence of scour troughs variable.	Triangular form highly asymmetric, straight crest.
Bed form Superimposed on stoss side	None	Ripples rare, stoss side usually planed off.	Megaripples usually superimposed.	Megaripples always superimposed on stoss side.

Piedmont and the Coastal Plain. Approximately 80 km below Columbia, the Congaree joins the Wateree River to form the Santee River. The Congaree River basin drains approximately 22,000 km².

The study area is located from 11 to 22 km below Columbia and forms the first meander train on the Coastal Plain province. The gradient along this portion of the river is approximately 0.5 m/km, and the sinuosity index is 1.75, using the method of Brice (1964) (channel length divided by the meander-belt axis length). Within the meander train, the radius of curvature varies from 1.7 to 6.0 (Fig. 1), where Rc (radius of curvature) equals the radius of the best inscribed circle through the middle of the channel divided by the average width of the bankfull channel through the meander bend. Maps from 1820, 1889, 1972 and airphotos from 1938 indicate that the channel pattern of the first meander train has not changed drastically since 1820 (Fig. 2).

During periods of low water, the subaerial exposure, topography, and facies characteristics differ among the point bars studied. The largest point bar is over thirty times greater in area and shows considerably greater topographic variability than the smallest point bar within the first meander train.

HYDROLOGY

Mean annual discharge of the Congaree River at Columbia, S.C. was 264 m³/s during 1939-76 (U.S. Geological Survey, 1976). Figure 3 shows the discharge hydrograph for the Congaree River gage at Columbia during the study period from October 1975 through October 1976. Examination of the flow-duration records for the Congaree River at Columbia indicate a flashy rather than steady flow pattern.

METHODS

Fence posts were hammered into sites selected for both topographic and hydraulic surveys. Hydraulic data were collected during both seasonal-low-flow and bankfull conditions. Within the meander wavelength containing point bars 2 and 3, twelve cross-sectional traverses were marked to obtain detailed hydraulic and bed form data. Current velocity measurements were collected by either anchoring or maneuvering a small outboard craft into a stationary position aligned parallel with a cross-sectional traverse. A remote reading current meter (Endeco Type 110) was used to obtain the

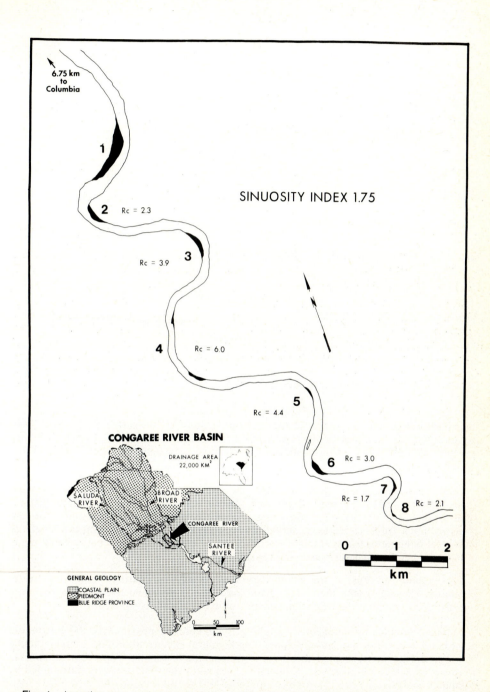

Fig. 1. Location map and general geology of the Congaree River basin. Point bars, of the first meander train below Columbia, S.C. are consecutively numbered. Radius of curvature (Rc) varies from 1.7 to 6.0.

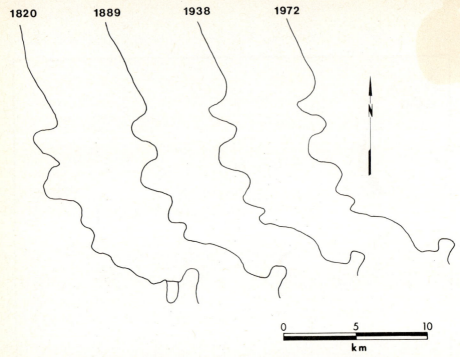

Fig. 2. Comparison of channel centerline positions in the study area. 1820 centerline taken from the Atlas of the State of South Carolina (1825). 1889 centerline taken from the Annual Report of the Chief of Engineers, U.S. Army part 2 (1889). 1939 centerline taken from the U.S. Department of Agriculture, aerial photographs. 1972 centerline taken from 7½ minute U.S. Geological Survey topographic maps.

magnitude, direction and depth of the current velocity at the surface, mid-depth, and bottom at three sample localities along each cross-sectional traverse. Current velocity was recorded at various points within a full meander wavelength for both seasonal low flow and bankfull discharges.

Continuous fathometer traces of the channel were obtained from a portable echo-sounding survey recorder (Bludworth Marine Div. Model ES130) mounted from a small outboard craft. Ranges, consisting of painted plywood sheets mounted on the cross-sectional traverse markers, served as fix indicators for determining the length of individual fathometer traces. Fathometer traces were used to construct bathymetric maps of meander bends and determine bed form spacing.

Topographic surveys of the point bars were conducted during similar low water conditions, using both theodolite and auto-level. Pace-and-Brunton maps were constructed to monitor the migration distance of individual transverse bars.

Sediment samples were collected by taking 15 cm cores (approximately 200 g) from the point bar surface. For the samples containing high gravel concentrations, a larger sample of 0.5 to 2.0 kg was collected to obtain a representative sample. All sediment samples were sieved at 0.25 phi intervals. Cumulative frequency distributions were determined by computer analysis.

Shallow trenches were dug by shovel, and the sidewalls were photographed. A Brunton compass was used to measure the dips of individual laminae.

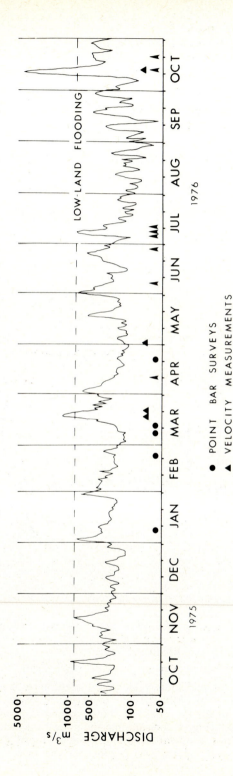

Fig. 3. Discharge hydrograph of the Congaree River at Columbia, S.C. for the study period from October 1975 through October 1976. Data from the Water Resources Division of the U.S. Geological Survey.

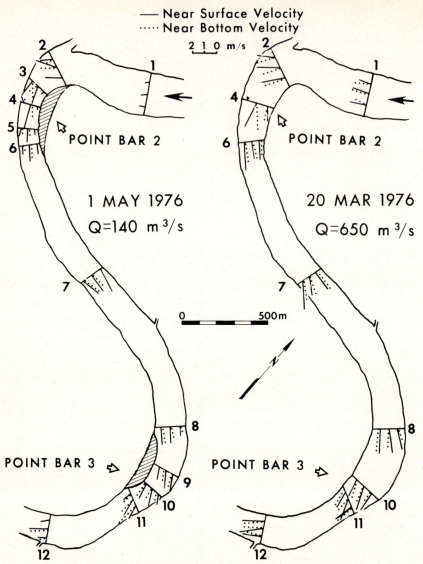

Fig. 4. Current velocity comparison of seasonal low flow and bankfull discharge for the meander containing point bars 2 and 3. All near-bottom velocities taken within 1 m of the bed.

RESULTS

Velocity distribution through a meander wavelength

During a rise in water level, the maximum velocity at the entrance of a meander bend shifts from the outer to the inner bank of the meander bend (Fig. 4). Measurements during bankfull conditions at the upstream ends of both meander bends, shown by cross-sectional traverses 2 and 8 (Fig. 4) indicate that the inner bank velocities near the surface and bottom are at least as great as the velocities toward the outer bank. Measurements in the middle and downstream portions of the meanders, shown by cross-sectional traverses 4 and 11 (Fig. 4) indicate that the locations of maximum velocities shifted toward the cutbank side of the channel during high flow.

In this study, areas of both no flow and reversed flow were measured on the inner and outer banks of a meander bend. This phenomenon of stagnant zones and reversed flow occurs during both seasonal low flow and bankfull discharges near the outer bank where slump blocks cause a zone of flow separation between the main channel and the outer bank. Along the inner bank of the Congaree River point bars, areas of reversed flow are caused by low river stages in which the crest of a transverse bar is exposed and the trough submerged. These areas of reversed flow adjacent to point bars are often covered by organic-rich silt deposits several centimeters thick deposited downstream from the bar crest. Similar patterns of flow separation and stagnation along the outer bank at high water and the inner bank at low water have been observed in meander bends along intertidal mudflat channels (Bridges and Leeder, 1976).

The radius of curvature is a critical determinant of the velocity distribution through a meander bend. Similar velocity patterns along the lower Wabash River by Jackson (1975) and along the Congaree River (this study) support Jackson's hypothesis that both very tightly and gently curved meander bends are incapable of developing a strongly asymmetrical flow pattern in which the maximum current velocities are confined to the outer or cutbank portions of the bend. Meanders of intermediate curvature are capable of developing the classical asymmetrical flow patterns described by Leopold and Wolman (1960, Fig. 5).

Grain size distribution along point bars

A point bar in the study area may contain as many as three sections or facies: a bar-apex or upstream facies, a mid-bar facies, and a distal-bar or downstream facies. There is also a scour-pool and crossover facies, consisting of the thalweg and cutbank adjacent to the point bars and the crossover section that occupies the channel between meander bends.

Figure 5 shows representative grain-size distributions for the three point-bar facies. A composite size distribution curve for each facies was calculated by taking the cumulative weight percent at one phi interval and averaging these values based on the number of samples classified as that facies type.

Grain size on the point bars ranges from silt to gravel with coarse to medium sand being the most common. A trend of decreasing grain size in a downstream direction along the point bar surface is apparent only when the bar-apex facies is present in conjunction with the mid- and distal-bar facies. There is no major difference in grain size between the mid-bar and distal-bar facies.

The meandering stream models of Visher (1965) and Allen (1970) postulate a decrease in mean particle size and in bed form height from the thalweg upward to the flood plain. Jackson (1975) reported that along the Wabash River, these conditions are satisfied only where an asymmetrical flow pattern is developed, a conclusion supported by this study. This in turn is related to the radius of curvature of a meander bend as mentioned above.

Facies characteristics within the meander train

Bar-apex facies

The bar-apex facies is either a pure gravel subfacies (a deposit with gravel-size material over at least 90% of its surface) or a mixture of coarse sand and gravel which results in a bimodal sediment-size distribution. The gravel size material is usually found supported by a sand matrix. Shallow trenches reveal poorly developed cross-stratification.

Gravel accumulation at the apices of point bars 1, 2, and 3 may represent a lag accumulation that results from a winnowing of sand-size material between flood

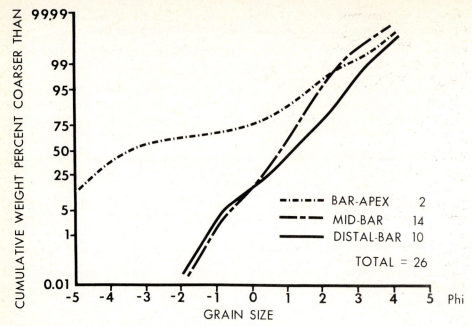

Fig. 5. Grain-size comparison of the apex-, mid-, and distal-bar facies. Size distribution curve for each facies is calculated by averaging the cumulative weight percent at one phi intervals by the number of samples classified as that facies.

discharges capable of gravel transport. A similar hypothesis to explain high gravel concentrations on Mississippi point bars was proposed by Fisk (1944).

Reconnaissance of the river channel upstream from the study area during annual low water indicated no immediate source of gravel. The greater concentration of gravel on point bars 1, 2, and 3 in comparison to those point bars further downstream may result from the higher local river slope and proximity to the Piedmont provinces. No other large, pure gravel deposits exist along the other point bars studied, but there are isolated pockets of gravel in the megaripple scours on all the point bars.

The mean vector and mean angular deviation for the gravel subfacies along the apices of point bars 2 and 3 were calculated using a circular normal or Von Mises distribution. Azimuth readings perpendicular to the long axis of the gravel clasts, spaced every 5 m apart, were used to determine the scope of long-axis orientation. The gravel subfacies of point bars 2 and 3 show an approximate 90° transverse orientation of the long axes of clasts to the general channel trend.

Gravel imbrication (dip of the plane containing the long and intermediate axes) appears to be poorly developed. Poorly developed imbrication may result from the mode of transport of gravel-size material within this meandering system. Flume studies by Johansson (1963) showed that material moving as contact load acquires a more transverse orientation than saltating pebbles. It seems possible that a high percentage of the gravel-size material within this system is transported by saltation as a result of the high velocities close to the inner bank of the meander bend during bankfull flow conditions (see Fig. 4).

In October 1976, a flood with a recurrence interval of approximately 16 years occurred along the Congaree River. Field reconnaissance at point bar 2 after the flood indicated

Fig. 6. Oblique downchannel view of subaerial portion along bar 6 shows transverse bars (crests are marked T) and megaripples over the point bar surface. Photo taken 13 February 1977. Point bar 6 is approximately 500 m long.

vertical accretion and bed form migration over the entire point bar. The most drastic change was the disappearance of the apex facies, which had contained a high gravel concentration. Trenches dug during low water following the flood revealed that the gravel apex material was not buried by the migration of bed forms but transported along the point bar surface. A small concentration of gravel-size material was lodged within a log protruding 50 cm above the post-flood point bar surface. By measuring the intermediate axes lengths and referring to the critical erosion velocity curves of Sundborg (1956), a velocity of 200 cm/s is estimated for this portion of the point bar surface during the October 1976 flood. Gravel-size material was deposited on top of the highest chute bar within point bar 2. Preservation of this sedimentological pattern would not exhibit the upward decrease in grain size on a point bar as proposed by Allen (1970) and Visher (1965). The observation of gravel movement indicates that gravel size material is capable of being transported by an event of this frequency along the Congaree River.

Mid-bar facies

The middle portion of well-developed point bars consists of coarse to medium sand. Megaripples are the dominant bed form of the mid-bar facies. Megaripple spacing (37 measured) ranged from 1.3 to 5.7 with an average of 2.8 m. Heights (45 measured) ranged from 9 to 28 cm and averaged 18 cm.

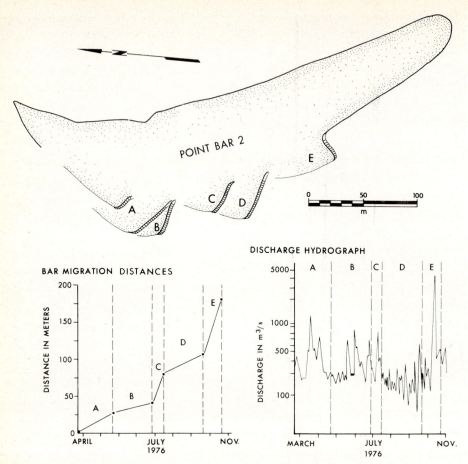

Fig. 7. Map showing sequential migration of a single transverse bar along the margin of point bar 2. Transverse bar migration distances and the hydrograph are shown.

Internal stratification consists of small-scale tubular cross beds with set thickness less than 50 cm. Commonly, set thickness is 10 to 20 cm, and laminae dip averages 32°. Reactivation surfaces are common and probably represent an interruption of steady, continuous bed form movement.

Megaripple crests varied from straight to sinuous. The stoss sides of megaripples rarely are covered by ripples and are generally planar. The plane beds probably result from upper-flow regime transport over the stoss sides of megaripples and represent a plane-bed condition in shallow water described by Harms and Fahnestock (1965). Horizontal stratification results from the plane-bed condition.

The low number of ripples relative to the other types of bed forms present on the Congaree River point bars may be related to the coarse sediment size. Experimental work, using flumes to model bed form phases, has shown that in coarse sands (greater than 0.6 mm), ripples are uncommon regardless of the velocity (Harms *et al.*, 1975; Fig. 2-5).

The mid-bar facies also contains transverse bars which are defined as large asymmetrically-shaped bed forms with a wavelength greater than 25 m and a maximum

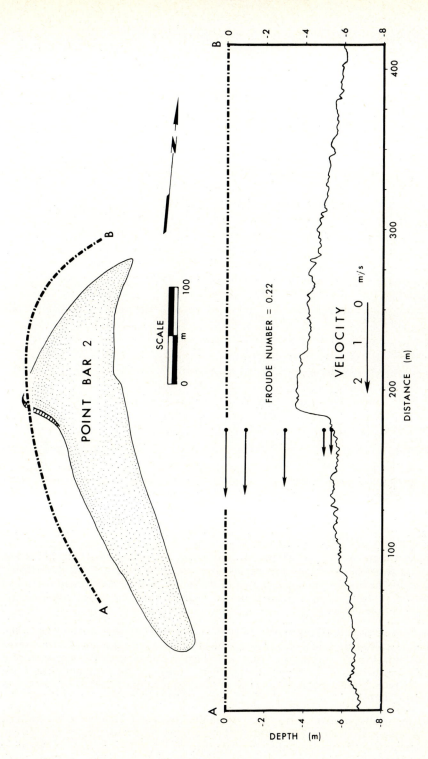

Fig. 8. Fathometer trace over a single transverse bar along the margin of point bar 2. Downstream is to left. 6 July 1976 2045 EST. Approximate discharge 860 m³/s.

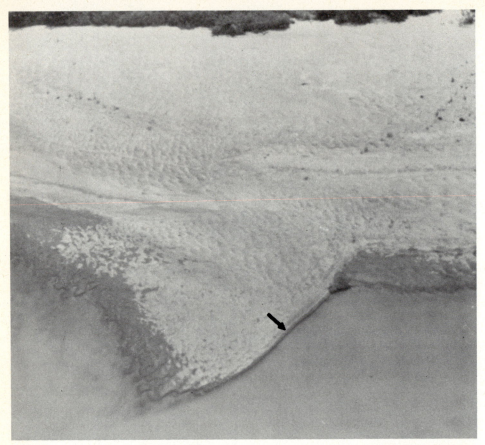

Fig. 9. Oblique aerial photograph of transverse bar along point bar 2. Megaripples present on the stoss side of the transverse bar and along the point bar surface toward the vegetation. Arrow marks the location of trench through the transverse bar slipface in figure 11. Downstream is to right. 9 June 1976.

slipface height of 2 m. Large transverse bars are found along the point bar margin closest to the channel thalweg (Fig. 6). Transverse bar crests are straight and oriented either perpendicular or at oblique angles with the local channel flow direction. Jackson (1976a) observed similar bed forms along the margin of the lower Wabash River point bars. The megaripples described above characteristically migrate over the stoss sides of the Congaree River transverse bars.

Transverse bars migrate during periods of lowland flooding or higher water stages when most of the point bars are submerged. Measurements during the migration of a transverse bar indicated the Froude number was only 0.22. Figure 7 documents the migration of a single transverse bar along the margin of point bar 2. Figure 8 shows a fathometer trace and the velocity measurements over the crest of the migrating transverse bar documented in time phase C of Figure 7.

Figure 9 shows the transverse bar that was studied in detail. Transverse-bar bedding consists of large-scale tabular cross beds, often with set thickness greater than 50 cm (Fig. 10). The cross beds are alternating thin laminae of silt and thick laminae of coarse to medium sand. The individual sand laminae may represent the migration of megaripples

Fig. 10. Trench through slipface of transverse bar shown in Figure 9. Downstream is to right. Photograph shows zone of soft-sediment deformation separating tabular cross-beds. Small-scale trough cross-bedding with reactivation surfaces resulting from megaripple migration forms upper cross bedded unit. Scale is 20 cm.

Fig. 11. Downchannel view of terminus of chute bar within the distal-bar facies of point bar 2.

over the crests of the transverse bars. Both flume and field studies (Jain and Kennedy, 1974; Boothroyd and Hubbard, 1974) have found that smaller bed forms migrate at a faster pace than larger forms in the same system. Therefore, differential rates of bedform migration may be indicated by different thicknesses of cross beds.

Soft-sediment folds of cross bedding are sometimes observed within the tabular bedding of transverse bars. These folds may result from rapid fluctuations in water stages which would upset grain-to-grain contact pressure. Similar features of deformed cross bedding have been noted within the cross beds of point bars along the Red River, Louisiana (Harms *et al.*, 1963).

Distal-bar facies

The distal, or downstream, portion of well-developed, coarse-grained point bars is characterized by chute channels and chute bars similar to features described by McGowen and Garner (1970). Chute channels originate as areas of scour, usually forming partially confined, straight-sided channels from a few tenths to one meter deep and several meters in width. Chute bars, located at the termini of chute channels, are characteristically narrow (2 - 8 m) and long (>10 m) and lobate on their downstream end (Fig. 11).

Chute channel and chute bar features occur toward the highest and downstream parts of point bars 2, 3, and 6. Chute bars and chute channels are active only during above average water stages when part of the flow shifts toward the inner bank, allowing active and rapid migration of bedload along distal portions of the point bars. Along the Congaree River, the chute channels are usually located between rows of trees paralleling point-bar margins. During high discharge, chute channels are often covered by ripples and megaripples. Subsequent deposition of the suspended load occurs during falling-stage conditions, covering these areas with a mud drape. Mud cracks develop within several hours of subaerial exposure of the mud drape.

Fig. 12.   Closeup view of trench in terminus of a chute bar within point bar 2. The thickness of tabular cross bedding is approximately 1 m. Downstream is to right. Scale is 20 cm.

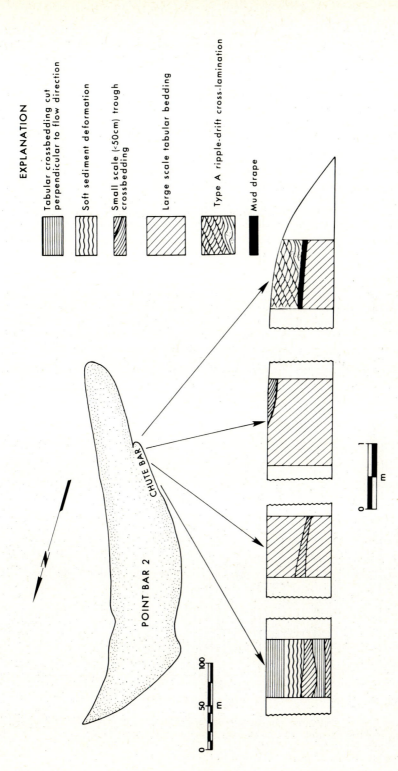

EXPLANATION

Tabular crossbedding cut perpendicular to flow direction

Soft sediment deformation

Small scale (<50cm) trough crossbedding

Large scale tabular bedding

Type A ripple-drift cross-lamination

Mud drape

Fig. 13.   Schematic diagram and map of point bar 2 showing the variations in cross stratifiction observed in a chute bar of the distal facies.

Trenches within chute channels show abundant cut-and-fill stuctures when viewed perpendicular to the flow direction. Variable cross bed directions of the distal-bar facies probably results from divergent current patterns which occur during flood conditions. Delta-like lobes composed of tabular cross beds may result from divergent flow patterns. Tabular beds forming oblique or perpendicular to the main channel direction were explained by Singh and Kumar (1974) as small delta lobes prograding toward the main river channel.

Along point bar 2, chute-bar heights vary between 1 and 3 m. Their elevation may be lower than or equal to the local flood plain elevation. Figure 12 shows large-scale tabular beds consisting of thin alternating sand and silt laminae, similar to structures observed in the transverse bar bedding. Trenches in chute bars reveal that these features are complex products of multiple high-stage events. Figure 13 illustrates the variations of cross bedding observed within the lower chute bar of point bar 2. Cosets of cross-bedded sands are sometimes separated by mud drapes that serve as time markers between high discharge events when the chute bars are activated. The presence of type A ripple-drift cross-lamination (as defined by Jopling and Walker, 1968) indicates that deposition was not rapid enough to prevent cannibalism of the stoss sides of migrating ripples or megaripples. Bedding of this type indicates a low suspension/traction ratio (Jopling and Walker, 1968).

McGowen and Garner (1970), in their study of coarse-grained point bars along the Colorado River, Texas and Amite River, Louisiana, listed low angle foresets and regressive ripples as evidence that chute bars are the result of rapid flow conditions. The presence of similar structures, along with high-angle tabular bedding indicates that both rapid and slow transport are viable mechanisms for deposition of chute bars.

A review of braided-river deposits by Miall (1977) indicates that chute bars of coarse-grained point bar deposits are similar to the linguoid bar type of braided rivers

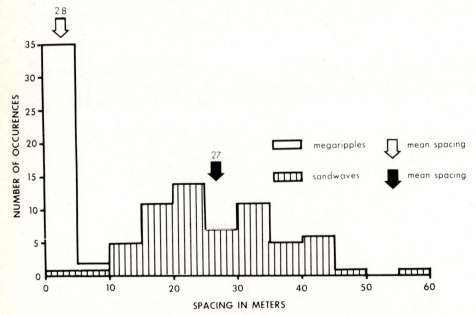

Fig. 14. Frequency histogram comparison of megaripples and sandwaves. 63 sandwave spacings are determined from fathometer trace through the first meander train. 10 June 1976. Approximate discharge 200 m³/s. 37 megaripple spacings are determined from measurements along point bars 4, 5 and 7. 100 bed form measurements.

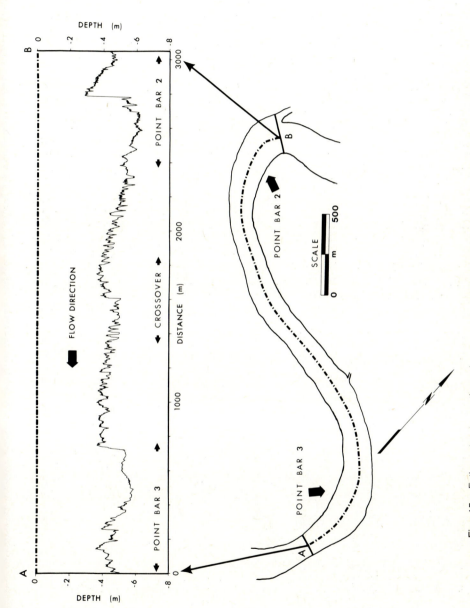

Fig. 15. Fathometer trace through one meander wavelength containing point bars 2 and 3. Sandwave spacing is about 27 m, megaripple spacing 3 m. Downstream is to left. 6 July 1976 1130 EST. Approximate discharge 710 m³/s.

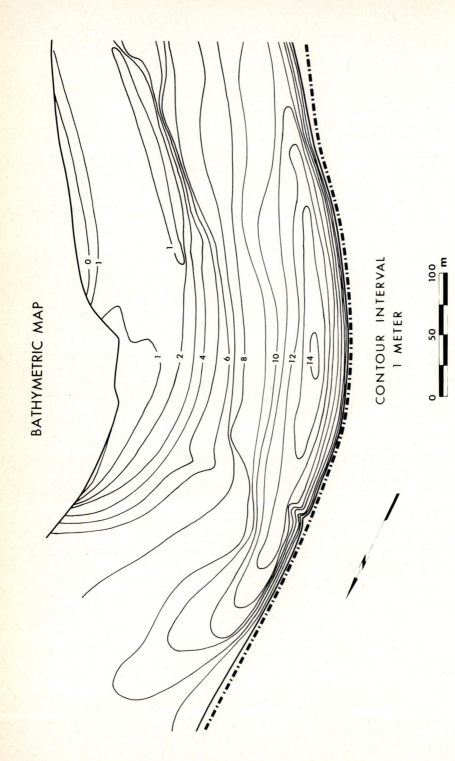

BATHYMETRIC MAP

CONTOUR INTERVAL
1 METER

Fig. 16. Bathymetric map of meander bend containing point bar 2. Note the deep pool area adjacent to the cutbank side of channel. Based on theodolite survey of point bar 2 on 21 February 1976 and fathometer traces recorded 11 April 1976.

(Collinson, 1970; Smith, 1971, 1974). The resemblance of braided-river linguoid bars and meandering-river chute bars may be related to the similar flashy discharge pattern which characterize river systems that exhibit these features. The dominant internal structure of chute bars within the Congaree River is tabular cross-beds, similar to those described for linguoid bar structures.

Scour-pool and crossover facies

Transverse bars, sandwaves and megaripples characterize the crossover section between meander bends. During average discharge conditions, crossover sections are distinguished by megaripples superimposed on sandwaves. Sandwaves are the dominant bed form in the relatively straight reaches, or crossovers, between meander bends. Sixty-three sandwaves were measured from a continuous fathometer trace through the entire meander train. Sandwave spacing ranged from 5 to 59 meters, with a mean spacing of 27 meters. Heights ranged from 50 to 130 cm and averaged 85 cm. Sandwaves along the Congaree form both symmetric and asymmetric bed forms similar to those measured by Jackson (1976b) along the Wabash River.

Figure 14 shows a frequency histogram comparison of megaripple and sandwave spacing measured within the Congaree River. Studies of bed form populations of both fluvial (Jackson, 1976b; Coleman, 1969) and intertidal environments (Allen and Friend, 1976; Boothroyd and Hubbard, 1974) have recognized a break in bed form spacings. Flume and field studies (Costello and Southard, 1974; Boothroyd and Hubbard, 1974) indicated that sandwaves are lower order bed forms with respect to flow strength than megaripples. Analysis of bed form data from several different systems by Allen and Collinson (1974) showed that superimposition of such bed forms occurs primarily where hydrological variables change rapidly through time. Jackson (1976b) found the migration of megaripples over both sandwaves and transverse bars to occur simultaneously even over long periods of relatively constant discharge. Fathometer traces of the Congaree River during various flow conditions indicate that superimposition of megaripples over sandwaves may cease during periods of high discharge. The bankfull discharge along the Upper Congaree River is approximately 650 m³/s and is characterized by substantial bed form migration and reactivation of point bar surfaces. During bankfull discharge conditions, sandwaves are the most abundant bed form within the crossover sections. Figure 15 illustrates a fathometer trace through one meander wave length during a bankfull discharge.

The scour-pool within the meander bends is generally the deepest part of the river channel. Figure 16 shows a bathymetric map of point bar 2 and the adjacent scour-pool that features the deepest pool along the upper Congaree River. Cross-sectional fathometer profiles between the point bars and the opposite cutbanks indicate that these pools can obtain a maximum depth of 14 m below the adjacent flood plain elevation. The channel floor within the scour-pools contains uprooted trees and slump blocks eroded from the cutbank side. Additional accumulations of gravel-size material sometimes occur in these pools within the channel thalweg. These isolated deposits probably represent the channel lag material commonly preserved in fluvial deposits.

DEPOSITIONAL MODEL

Using morphologic features, bed-form stratification and grain size from point bars, a model was developed to illustrate the development of coarse-grained point-bar deposits along the upper Congaree River.

This model (Fig. 17) shows the three separate facies types that occur. Well-developed point bars in the study reach consist of: (A) an apex facies of coarse sand and gravel which represents a lag deposit characterized by poorly developed cross stratification; (B) a mid-bar facies composed of megaripples migrating over transverse bars with internal

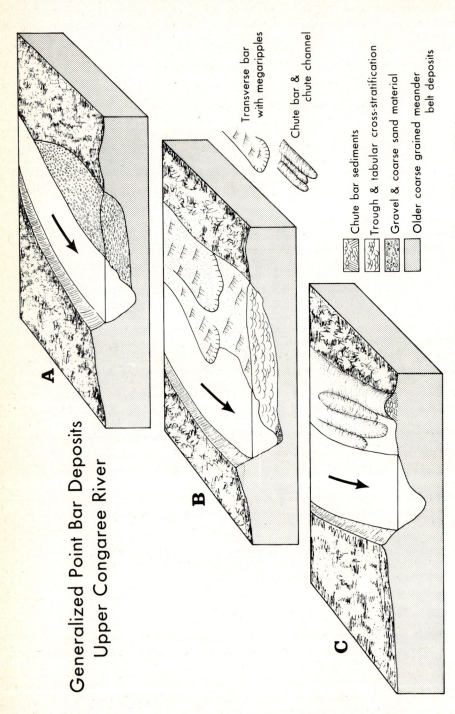

Generalized Point Bar Deposits
Upper Congaree River

Transverse bar
with megaripples

Chute bar &
chute channel

Chute bar sediments

Trough & tabular cross-stratification

Gravel & coarse sand material

Older coarse grained meander
belt deposits

Fig. 17. Generalized depositional model of a well-developed coarse-grained point bar within the
upper Congaree River. Block A represents the bar-apex facies of gravel and coarse sand. Block B
shows the mid-bar facies characterized by migrating transverse bars and megaripples. Block C
displays the distal-bar facies with chute bars and chute channels.

stratification that includes both large-scale tabular and small-scale trough and tabular crossbedding, and (C) a distal-bar facies, composed of chute bars and chute channels that result from high discharge events. Distal-bar stratification is highly variable in both bedding type and the orientation of local dip with respect to the direction of flow in the main channel.

A hypothetical vertical sequence for a well-developed coarse-grained point bar within a meander bend of intermediate curvature ($2.3 < Rc < 3.9$) may consist of a coarse channel lag at the base. This material would be overlain by coarse to medium sand consisting of both small scale (< 50 cm set thickness) trough and tabular beds from megaripples and large scale (>50 cm thickness) tabular beds resulting from transverse bar bedding. The chute channel and chute bars of the distal-bar facies occur along the highest portion of the point bar surface. Internal stratification of chute bars often contains large tabular crossbedding similar in grain size and scale to that resulting from transverse bars. Mud drapes deposited during the falling stage may separate the beds formed during one depositional event from the other sand beds. Bedding plane orientation is highly variable in the distal-bar facies.

The preservation of any vertical sequence in this system depends on the complex interaction among sediment size and availability, direction of meander bend migration, radius of curvature, and discharge pattern.

CONCLUSIONS

(1) Within the first meander train along the upper Congaree River, point bars show wide variation in size, topography, facies and morphologic features.

(2) Well-developed point bars along the upper Congaree River are distinguished by bar-apex, midbar and distal-bar facies. A scour-pool and crossover facies characterizes the deep areas adjacent to point bars and the relatively straight sections between meander bends.

(3) Point bars 2, 3, and 6, where the flow displays the classical asymmetrical pattern, contain the apex, mid and distal-bar facies. In comparison, the other point bars which occupy either very tight or gentle meander bends ($Rc < 2.3$ or > 3.9) lack either the apex and/or distal-point bar facies.

(4) Crossover sections may contain a wide range of bed-form sizes that depend on various discharges for their formation.

(5) Most of the point bars studied do not show the pattern of decreasing grain size upward from the thalweg that is postulated in the facies models of Visher (1965) and Allen (1970).

(6) The response of facies type, morphologic features and bed forms is complex and depends on local channel geometry, velocity distribution, sediment size and availability and discharge pattern.

ACKNOWLEDGMENTS

This research was part of a masters thesis at the Geology Department, University of South Carolina. Support was provided by the Coastal Research Division (Miles O. Hayes, Director) and the Water Resources Division of the U.S. Geological Survey (Graduate Thesis Support Program).

A critical review of the manuscipt by R. T. Getzen (U.S.G.S., Reston, Va), R. G. Jackson (Northwestern Univ., Evanston, IL), G. P. Williams (U.S.G.S., Denver, CO), and J. C. Boothroyd (Univ. of Rhode Island, Kingston, RI) is gratefully acknowledged.

Linda McKenna, Jeff Knoth, and Susan Fakkema provided valuable assistance in the field and laboratory. Burk Scheper and Nannette Muzzy helped prepare the figures.

REFERENCES

Allen, J. R. L., 1970, Studies in fluviatile sedimentation: A comparison of fining-upwards cyclothems, with special reference to coarse-member composition and interpretation: J. Sediment. Petrol., v. 40, p. 298-323.

——, and Collinson, J. D., 1974, the superimposition and classification of dunes formed by unidirectional aqueous flows: Sediment. Geol., v. 12, p. 169-178.

——, and Friend, P. F., 1976, Changes in intertidal dunes during two spring-neap cycles, Lifeboat Station Bank, Wells-next-the-Sea, Norfolk (England): Sedimentology, v. 23, p. 329-346.

Bluck, B. J., 1971, Sedimentation in the meandering River Endrick: Scott. J. Geol., v. 7, p. 93-138.

Boothroyd, J. C., and Hubbard, D. K., 1974, Bed form development and distribution pattern, Parker and Essex Estuaries, Massachusetts: U.S. Army Corps of Engineers, Coastal Eng. Research Center, Misc. Paper 1-74, 45p.

Brice, J. C., 1964, Channel patterns and terraces of the Loup Rivers in Nebraska: U.S. Geol. Survey Prof. Paper, 422-D.

Bridges, P. H., and Leeder, M. R., 1976, Sedimentary model for intertidal mudflat channels, with examples from the Solway Firth, Scotland: Sedimentology, v. 23, p. 533-552.

Coleman, J. M., 1969, Brahmaputra River: Channel processes and sedimentation: Sediment. Geol., v. 3, p. 129-239.

Collinson, J. D., 1970, Bedforms of the Tana River, Norway: Geogr. Ann., v. 52A, p. 31-56.

Costello, W. R., and Southard, J. B., 1974, Development of sand bed configurations in coarse sands: Am. Assoc. Petrol. Geol. — Soc. Econ. Paleont. Mineral. Ann. Mtg., Abstracts, v. 1, p. 20-21.

Fisk, H. N., 1944, Geological investigations of the alluvial valley of the Lower Mississippi River: Mississippi River Comm., Vicksburg, Mississippi, 78p.

Harms, J. C., and Fahnestock, R. K., 1965, Stratification bedforms and flow phenomena (with an example from the Rio Grande): in G. V. Middleton, ed., Primary sedimentary structures and their hydrodynamic interpretation; Soc. Econ. Paleont. Mineral. Spec. Pub. 12, p. 84-115.

——, MacKenzie, D. B., and McCubbin, D. G., 1963, Stratification in modern sands of the Red River, Louisiana: J. Geol., v. 71, p. 566-580.

——, Southard, J. B., Spearing, D. R., and Walker, R. G., 1975, Depositional environments as interpreted from primary sedimentary structures and stratification sequences: Soc. Econ. Paleont. Mineral. Short Course 2, 161p.

Jackson, R. G., II, 1975, Velocity-bedform-texture patterns of meander bends in the lower Wabash River of Illinois and Indiana: Geol. Soc. Am. Bull., v. 86, p. 1511-1522.

——, 1976a, Depositional model of point bars in the lower Wabash River: J. Sediment. Petrol., v. 46, p. 579-594.

——, 1976b, Large scale ripples of the lower Wabash River: Sedimentology, v. 23, p. 593-623.

Jain, S. C., and Kennedy, J. F., 1974, The spectral evolution of sedimentary bedforms: J. Fluid Mech., v. 63, p. 301-314.

Johansson, C. E., 1963, Orientation of pebbles in running water: a laboratory study: Geogr. Ann., v. 45, p. 85-111.

Jopling, A. V., and Walker, R. G., 1969, Morphology and origin of ripple-drift cross-lamination with examples from Pleistocene of Massachusetts: J. Sediment. Petrol., v. 38, p. 971-984.

Klein, G. deV., 1975, Sandstone depositional models for exploration for fossil fuels: Continuing Education Publishing Co., Champain, Ill., 109p.

Leopold, L. B., and Wolman, M. G., 1960, River meanders: Geol. Soc. Am. Bull., v. 71, p. 769-794.

McGowen, J. H., and Garner, L. E., 1970, Physiographic features and stratification types of coarse-grained point bars: Modern and ancient examples: Sedimentology, v. 14, p. 77-111.

Miall, A. D., 1977, A review of the braided-river depositional environment: Earth-Sci. Revs., v. 13, p. 1-62.

Singh, I. B., and Kumar, S., 1974, Mega-and giant ripples in the Ganga, Yamuna, and Son Rivers, Uttar Pradesh, India: Sediment. Geology, v. 12, p. 53-66.

Smith, N. D., 1971, Transverse bars and braiding in the lower Platte River, Nebraska; Geol. Soc. Am. Bull., v. 82, p. 3407-3420.

————, 1974, Sedimentology and bar formation in the upper Kicking Horse River, a braided outwash stream: J. Geol., v. 82, p. 205-224.

Sundborg, A., 1956, The river Klaralven, a study in fluvial processes; Geogr. Ann., v. 38, p. 125-316.

U.S. Geol. Survey, 1976, Water Resources data for South Carolina: Water-data report, S.C. 75-1, 224p.

Visher, G. S., 1965, Use of the vertical profile in environmental reconstruction: Am. Assoc. Petrol. Geol. Bull., v. 49, p. 41-61.

SEDIMENTOLOGY OF SOME SOUTH CAROLINA TIDAL-CREEK POINT BARS, AND A COMPARISON WITH THEIR FLUVIAL COUNTERPARTS

JOHN H. BARWIS[1]

ABSTRACT

The geometry of intertidal point bars in South Carolina's predominantly mesotidal (tidal range 1.4 m to 2.6 m) marsh creek system is determined by tidal current ebb-dominance and meander geometry, primarily radius of curvature. The distribution of sedimentary facies is controlled by lateral segregation of flood and ebb currents set up by bar geometry and is further modified by tidal current time-velocity asymmetry. Ebb current dominance causes all point bars to be skewed toward, to trail away from the ebb side of a meander bend. This ebb dominance is purely a function of the entire inlet-marsh system and is present even in systems where freshwater influx is absent.

On tight meanders (r/w <2.5) point bars are relatively small with steep flanks and are fully welded to the bank. Chutes are absent. Semidiurnal current reversals generate no significant differences between flood and ebb current patterns, with the exception of turbulence associated with occasional ebb-oriented transverse bars. Bar surfaces are predominantly rippled fine sand with minor amounts of mud, and are bioturbated on their crests.

Intermediate meanders (2.3< r/w <3) produce complex multilobed bars, with flood-dominant chutes and interlobe ramps. The chute and inner bar may be exposed to no ebb discharge and are composed of fine-grained sediments. The interlobe ramp displays a wide variety of structures and textures, ranging from flood-oriented sand waves (clean fine sand) to bidirectionally oriented ripples (muddy sand), depending on the bar orientation with respect to adjacent meanders. The outer bar comprises structures and textures similar to those on gentle meander point bars.

Gentle meanders (r/w >3) produce long narrow bars with relatively deep, flood-dominant chutes. Chute bars occur infrequently, but where present are ebb-oriented. Bar extremities are sand wave and megaripple fields. Their ebb- and flood-orientations (on the floodward and seaward ends, respectively) are preserved throughout the tidal cycle by shielding of the higher mid-bar area, where smaller bedforms and flat beds predominate. Mud occurs as drapes in larger bedform troughs and as thin beds on mid-bar crests. Chutes are silty muds, usually devoid of bedforms.

Point bars on very gentle meanders (r/w >>3) are fully welded to the bank, and resemble the "skew shoals" of fluvial systems. There is little or no current segregation on this type of bar. Gentle swales behind bar crests are filled with bioturbated muddy sand that is gradational into the adjacent marsh surface.

The fundamental contrast between fluvial and tidal channels is the difference in the temporal relationship between water surface elevation and the concommitant flow velocity. In fluvial systems highest velocities accompany the highest river stages. In most tidal channels, however, highest velocities occur at an intermediate stage. This contrast results in vastly different modes of floodplain development. Because tidal channels migrate more slowly than river channels, and because in tidal systems floods are far more frequent, a relatively larger proportion of tidal creek sediments occurs as overbank deposits.

Tidal creek and fluvial floodplain lithosomes are distinguishable not only on this basis of the relative importance of channel versus overbank deposits, but on the bases of faunal diversity, bioturbate structures, orientations of cross-stratification, and characteristics of vertical point bar sequences that reflect tidal current segregation.

INTRODUCTION

Geologists investigating ancient sediments have at their disposal detailed studies of fluvial facies and their associated hydrodynamics (this volume; Reineck and Singh, 1975). Primarily because of the difficulties involved in three-dimensional observations of modern depositional environments, the classic fining-upward point bar sequence (Allen, 1970)

---

[1]Coastal Research Division, Department of Geology, University of South Carolina, Columbia, South Carolina 29208, U.S.A.

Table 1

CHARACTERISTIC	ENVIRONMENT	
	FLUVIAL (1, 2)	TIDAL (3)
GEOMORPHIC FEATURE		
Channel		
Slope	Highly variable.	Very low
Depth	Variable, up to 70 m.	Varies with tidal range: mesotidal, 1-15 m. macrotidal, 1-30 m.
Network	Varies with structure, tectonics, climate.	Dendritic, meandering.
Point Bar	Crest inactive during normal flow.	Crest active every tidal cycle.
Chute	Subaqueous only during floods.	Subaqueous through entire tidal cycle.
Chute Bar	May be coarser than main bar.	Rare; may be ebb- or flood-oriented.
Sediment	Grain size varies, fine sand to pebbles.	Grain size generally fine to medium sand.
Bank		
Levee	Often well-developed.	Poorly developed to absent.
Crevasse-splay	Often well-developed.	Absent or rare.
Sediment	Sand, silt.	Mud.
Flood Basin		
Deposits	point bar 60-90% overbank 10-40%	point bar <30% overbank >70%
Preservation	Flood plain elevation modified by lateral erosion.	Elevation of marsh tends towards MSL.
FLOW CONDITIONS		
Flood Freq.	<1/yr	>6/mo
Velocity Channel Maximum Velocity	Occurs near bankfull stages; usually >100 cm/sec.	Flood: occurs before bankfull. Ebb: occurs at lower stages than maximum flood velocity. usually <100 cm/sec.
Overbank Average Velocity	Often competent to move medium sand.	Competent to transport mud.
Discharge	Highest at maximum river stage.	Low near tidal extremes; maximum at intermediate stages.
Suspended Sediment Conc.	Less than capacity at maximum discharge.	Highest during maximum discharge (spring ebb).

1 Leopold and Wolman (1957, 1960).
2 Leopold, et. al. (1964).
3 This study.

owes its interpretive utility as much to studies of ancient sediments as to their modern analogs. Similarly, sedimentological studies of tidal environments are today more commonly involving this comparative approach to the study of present and past environments in the syntheses of facies models (Hobday and Eriksson, 1977). However, although there is a voluminous literature on various aspects of both modern and ancient tidal hydraulics and sedimentology (Barwis, 1976), very few papers have dealt with the point bars that are so common in meandering tidal creeks of modern mesotidal and macrotidal salt marsh environments. The purposes of this study are to describe observed variations in the planform geometries of tidal creek point bars, to relate both large- and small-scale sediment distributions to tidal creek hydrodynamics, and to compare point bars of tidal creek and fluvial environments, particularly with regard to their recognition in the rock record.

Flow in Fluvial vs. Tidal Channels

Sediment transport processes in both fluvial and tidal channels are governed by similar principles of open-channel flow. However, the actual processes produce very dissimilar sedimentary results. This is attributed not only to regular flow reversals in tidal channels, but also to differences in depth-discharge-velocity relationships, and the texture of the sediment itself (Table 1).

The fundamental hydrodynamic contrast between fluvial and tidal channels is the difference in the temporal relationship between water surface elevation and the concommitant flow velocity. In fluvial systems highest velocities accompany the highest river stages. In most tidal channels however, highest velocities occur at an intermediate stage. This contrast is enhanced by time-velocity asymmetry and ebb-dominance.

Time-Velocity Asymmetry

Maximum tidal current velocities do not occur at midtide, when the rate of change of tide stage is at a maximum. Rather, the maximum velocities occur later, at times nearer to, but not co-incident with, times of high and low water (Fig. 1). This delay is caused by inertial effects, by frictional effects of the channels and flats, and by changes in the effective "flood-basin" area as the marsh system floods or drains (Keulegan, 1967; King, 1974). As a result of time-velocity asymmetry, velocity maxima in a channel occur at different stages and shallow areas are often emergent before maximum ebb occurs (Van Veen, 1950; Price, 1963). The effects of this phenomenon on fine-grained sedimentation were discussed by Postma (1967). Field studies have shown this phenomenon to be the primary control on the morphologies of estuarine sand bodies and tidal deltas (Daboll, 1969; Hayes *et al.*, 1973). The most important result of time-velocity asymmetry is the development of segregated flood and ebb channels (Van Veen, 1950; Oomkens and Terwindt, 1960; Price, 1963; Schou, 1967; Daboll, 1969; Hayes *et al.*, 1973).

Depth vs. velocity differences are largely responsible for the striking contrast between fluvial and tidal flood-basin lithosomes. In meandering stream environments, as much as 60 — 90% of a normal flood plain may be composed of lateral accretion (point bar) deposits, while 10 — 40 percent is composed of over-bank sediments (Wolman and Leopold, 1957; Lattmann, 1960; Ferm and Cavaroc, 1968; Ferm, 1974). Conversely, less than 30 percent of marsh sediments may be deposited as tidal creek point bars. This is due largely to differences in flood frequency in marshes and rivers; floods occur hundreds of times more often in the marsh with conditions exceeding bankfull during every spring tide. The highest tidal discharges, during spring tides, carry the highest concentrations of suspended sediment (Boon, 1974; Ward, 1978). Maximum discharges of rivers, however, carry suspended sediment loads that are less than capacity (Einstein *et al.*, 1940). These factors combined with very low tidal current velocities at high water cause a higher proportion of sediments in the marsh and tidal flat environments to be deposited as overbank material. Irrespective of stage, tidal channels are less likely to rework their

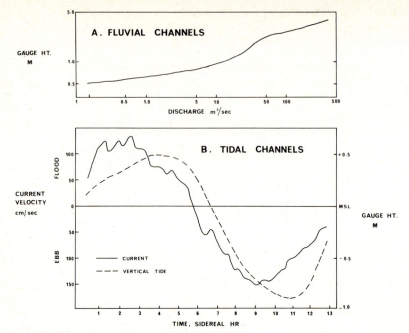

Fig. 1. Comparison of fluvial and tidal stage-discharge relationships. a) Rating curve for Seneca Creek near Dawsonville, Maryland (modified after Leopold *et al*., 1964, Fig. 6-10). b) Vertical and horizontal tides in the Texel area, Wadden Zee (modified after Postma, 1967, Fig. 5). Note that peak velocities occur at different stages, and that at highest stages flood velocities range from an average value to zero.

associated overbank deposits by meandering, as maximum velocities in most tidal creeks are lower than those in streams. This reworking is lowest in the salt marsh, where the channel banks are composed of coherent muds stabilized by grasses, particularly *Spartina alterniflora*. Aerial photo studies of South Carolina marshes show channel migration rates of less than 2 m/yr. On lower, non-vegetated, more open tidal flats however, meandering is more easily accomplished. There, channel migration rates vary from less than 25 m/yr in muddy sediments (Trusheim, 1929) to 25 — 100 m/yr in sandy sediments (Luders, 1934; Homeier, 1969). On subtidal flats, Reineck (1958) found 58 percent reworking by tidal channel migration in only 58 years.

Time-velocity asymmetry not only preferentially apportions fine-grained sediments to overbank deposition (Postma, 1967), but exerts a strong geomorphic control on levees and point bars. Levees are often well-developed in fluvial systems, where sudden dumping of sediments occurs as current velocities rapidly attenuate on leaving the main flow. In marsh creeks, levees are poorly developed (ht. <20 cm) or absent because current velocities are already low when bankful stages are attained. When present, marsh levees display less striking grain-size contrasts with adjacent overbank surfaces. They are composed primarily of mud trapped by the baffling effect of *Spartina* which grows more luxuriantly along the bank crests. River point bars have inactive crests and chutes at low discharges, when the meandering channel completely contains the flow. Chutes and crests of tidal creek point bars, on the other hand, are active to some degree during every tidal cycle.

Ebb-Dominance

As the tidal wave propagates into a mesotidal salt marsh system, the storage characteristics of that system may generate shallow-water tidal components (overtides)

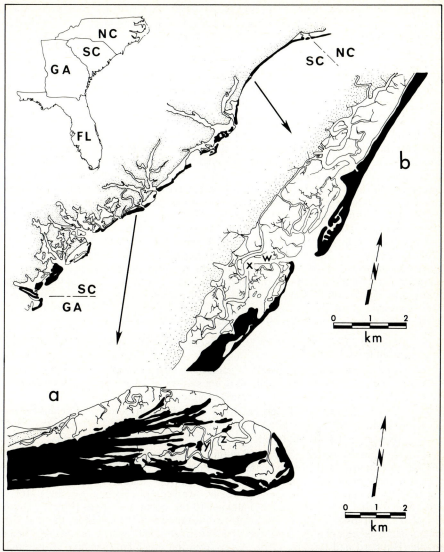

Fig. 2. Generalized coastal geomorphology in South Carolina. Black: mainland- and barrier-beaches; stippled: pre-Holocene coastal plain sediments; white: salt-marsh. a) Back-barrier marsh system at Murrells Inlet. w, x: locations of point bars shown in Figs. 8 and 12, respectively. b) Bass Creek marsh system on Kiawah Island, occupying swales between Holocene beach ridges. y, z: locations of point bar shown in Figs. 7f and 9 respectively.

that lengthen the duration of the flood current (Byrne *et al.*, 1975). For equal flood and ebb tidal prisms (which assumes no net water exchange with other subsystems), mean ebb discharges must be higher to accomodate the relatively shorter ebb current duration. This regular occurrence of higher ebb velocities is termed ''ebb dominance'' and was predicted theoretically by Mota Oliviera (1970). Ebb-dominance is important in determining not only the ability of tidal channels to remain open, but in controlling the position and shape of point bars. Thus, as the net sediment transport direction below certain depths in the channels is seaward, point bars occupy positions seaward of their associated meander bend apexes (Van Veen, 1950; Land and Hoyt, 1966). This asymmetry is very similar to that observed in fluvial meanders, where both deposition on the point bar and erosion on the cutbank tend to be greatest downstream from the point of maximum curvature (Mackin, 1937; Eardley, 1938; Fisk, 1947; Dietz, 1952). The combination of ebb-dominance and time-velocity asymmetry produces ebb-preferred bedform orientations. Ebb-dominance occurs in the deepest parts of channels even in tidal basins where the overall storage characteristics produce flood-dominance (Wright *et al.*, 1975).

South Carolina Marsh System

The largest salt marsh system in North America lies behind the 700 km long chain of barrier islands in the Georgia Bight. As in the German North Sea Bight and other major shoreline re-entrants, hydrographic regime and resultant coastal geomorphology vary systematically alongshore (Hayes, 1975). Tidal range is a function of shelf width (Redfield, 1958), and is the primary control on barrier geometry, as well as the size and spacing of tidal inlets (Nummedal *et al.*, 1977). In the outer, more exposed areas (North Carolina and Florida) where tidal range is often less than 1 m, marshes are less well-developed, and occur predominantly as fringes to large shallow lagoons or as narrow strips between Holocene barriers and the Pleistocene mainland. Spring tidal range exceeds 2 m in the central portion of the bight near the Savannah river mouth, and the resultant broad expanses of marsh often extend for tens of kilometers. Although tidal creek point bars are more common in these larger marsh systems, many of those areas are influenced by significant freshwater discharges. The point bars described in this paper lie in the mesotidal (spring tidal range 1.4 — 2.6 m) marsh system of South Carolina (Fig. 2). The areas described were chosen as examples for three reasons:

1)   there is negligible freshwater influx, which allows tidal effects to be isolated from fluvial effects;

2)   each marsh drainage system is geographically isolated, so that total flood and ebb discharges are identical, allowing studies of coarse-grained sediments to be related to quantitative fine-grained sediment budgets (Ward, 1978);

3)   by eliminating fluvial sources from consideration, studies of relatively small, isolated systems allow qualitative estimations of coarse fraction provenance-dispersal trends.

The two tidal creek systems discussed in this study are in the vicinities of Murrells Inlet and Kiawah Island, South Carolina (Fig. 2). Point bars in each of these semi-enclosed marsh drainage systems display a wide range of planform, hydrodynamic and sedimentological variation. The area drained by Murrells Inlet is enclosed by a Holocene barrier that welds to the Pleistocene mainland to the north and south (Fig. 2), (Johnson and Dubar, 1964; Brown, 1976). As local freshwater runoff is negligible, the pod-shaped marsh system is completely controlled by the inlet itself, which is sand-choked and unstable (Hubbard and Barwis, 1976). Away from the immediate vicinity of the inlet, all channels display time-velocity asymmetry and ebb-dominance, with ebb velocities exceeding flood velocities by about 20 percent. Spring tidal range averages 1.8 m (U.S. Dept. of Commerce, 1976). In contrast to the back-barrier Murrells Inlet marsh system, the Bass Creek marsh of Kiawah Island lies completely within swales of Holocene beach ridges.

Tidal channel sand bodies display a gentle transition from flood-tidal delta forms near inlets (Hubbard, 1977) to point bars in the more distal portions. Several facies characteristics reflect this trend, including sediment texture, changes in the degree of wave influence, current segregation, and biota.

Textural Trends

In the Murrells marsh, an overall decrease in grain size away from the inlet occurs for several reasons. Firstly, mean diameter of the quartz fraction is greatest near the inlet, the sole source of sand. Modal grain-size analyses show means in the fine to medium sand range ($\mu = 2.54\phi$, $\sigma = 0.34\phi$; to $\mu = 1.21\phi$, $\sigma = 0.62\ \phi$) within a kilometer of the inlet. However, in the smaller, lower energy tidal creeks, quartz is relatively finer ($\mu = 2.78\ \phi$, $\sigma = 0.72\phi$; to $\mu = 1.86\ \phi$, $\sigma = 0.51\ \phi$). Secondly, mud content increases away from the inlet. Whereas point bars near the inlet contain mud only as occasional drapes in ripple or megaripple troughs, or as intraclastic and fecal pellet debris on slipfaces, some of the lower energy bars away from the inlet are composed almost entirely of thixotropic mud. Thirdly, mean grain sizes of bulk samples are highest near the inlet because of higher percentages of carbonate bioclastic debris (chiefly strongly-abraded bivalve fragments) which is much coarser than the hydraulically equivalent quartz fraction.

In Bass Creek the primary sand source is the beach ridges that semi-enclose the 6 km² marsh. In several places (Fig. 2) meander cutbanks have intersected these ridges, creating steep banks from 1 to 8 m high. Occasional mass-wasting combined with continuous erosion during spring high tides provides a long-term source not only of fine, well-sorted quartz sand but also of organic debris such as trees and forest litter. Point bars occur adjacent to these cutbanks, but because of the well-sorted nature of the source sands, display no quartz-fraction grain-size trends. An overall decrease in grain-size away from the mouth of Bass Creek in the flood direction is due to an increase in mud content. Other authors have commented on beach ridges or buried Pleistocene ridges as a source of point bar sand (Land and Hoyt, 1966; Mayou and Howard, 1969; Andrews *et al.*, 1973).

Grain-size trends in tidal creeks offer an interesting contrast to fluvial systems where velocities and grain-size both generally decrease in a downstream direction (see Pettijohn, 1975, pp. 512-515, for a literature review). There is no "downstream" direction in a purely tidal marsh-creek, which is subject to current reversals. Furthermore, there is no net water flux, asuming an isolated system with no fresh water inflow. Yet, based on thalweg slope and ebb-dominance, the term "downstream" may be synonymous with "seaward." In both fluvial and tidal streams grain-size decreases with bed shear stress, which makes the fluvial-tidal disparity purely an apparent one.

Wave Influence

Point bars proximal to the inlet reflect a higher degree of wave influence, both from chop generated locally during bar emergence and from swell passing through the inlet at high tide. The crests of these bars display a higher proportion of upper flow-regime flat beds and wavy beds developed from washed-out bedforms. Oscillation ripples occur on both sandy and muddy beds, and display no preferred orientation, but frequently form interference features in combination with current ripples on the crests of bars and sand waves. Wave activity enhances mud winnowing, as also noted by Luneberg (1961), Daboll (1969) and Gadow (1970). To a limited extent wave activity is also significant in river systems. The author has observed wind waves 40 cm high that winnowed fine sand on large Mississippi River point bars near Vicksburg, Mississippi (during the period 1972 - 1975).

Bank Slumping

In rivers bank retreat is enhanced by slumping during falling stages due to excess pore pressures in bank materials; return seepage then causes the bank to lose competence, or even become quick (Inglis, 1949, pt. 1, p. 152). Leopold and Miller (1956) note that in arroyos most bank erosion occurs *after* flow. River bank slump blocks may retain their internal fabric, and are often preserved in ancient fluvial sequences (Horne and Ferm, 1976, p. 99). They contribute to unstable roof conditions in mines of alluvial valley coal deposits (Horne, *et al.*, 1978; Horne and Saxena, this volume).

In the South Carolina marshes, slumping due to pore pressure effects is largely limited to bank sediments below the densely rooted *Spartina* mat which may be 30 - 80 cm thick. Muds below this horizon are less stable, and lie at angles of repose from 15 to 45 degrees.

Rooted muds of the marsh surface are much more competent and form banks with slopes of 45 to 90 degrees. Weakening and slumping of grassy blocks is aided by extensive burrowing of the mud crab *Panopeus herbstii*. Subsequent incorporation of these blocks in the lower bank produces chaotic bedding (Howard *et al.*, 1973) which has been preserved in ancient tidal-channel sequences (Horne and Ferm, 1976, p. 121). Diving observations reveal occasional undeformed, rooted blocks up to 1 m in diameter along the thalweg. Small rooted mudballs (up to 20 cm) are much more common, and often litter the channel floor, especially near meander apexes. These mudballs are usually partially armored with shell fragments and fine sand but, because of lower energy conditions, are less well-armored and more angular than those described by Kugler and Saunders (1959) from beaches, or those reported from fluvial environments (Bell, 1940; Pettijohn, 1975).

Fauna

Much of the shell debris on point bars and in channels is derived from open marine fauna which has often been swept several kilometers landward. Large shells of the whelks *Busycon carica* and *B. canaliculatum,* remains of the echinoderm *Mellita quinquiesperforata*, echinoid tests, and byrozoan-encrusted corals are frequently found as allochthonous debris on bars up to 5 km from an inlet. Similar occurrences have been reported from a Dutch estuary by Noorthoorn van der Kruijff and Lagaaij (1960). Nearly all of this material has been severely broken and abraded and would be easily recognizable in the rock record as a death assemblage. Nevertheless, deceptively high species diversities could be attached to this environment if based on fossil counts alone. Infaunal species diversity increases from bar to bar in a seaward direction (Howard, 1971) and is higher than diversities on fluvial point bars (Pryor, 1967; Mayou and Howard, 1969). Diving and surface observations in the Carolina marshes show bioturbation to be much more

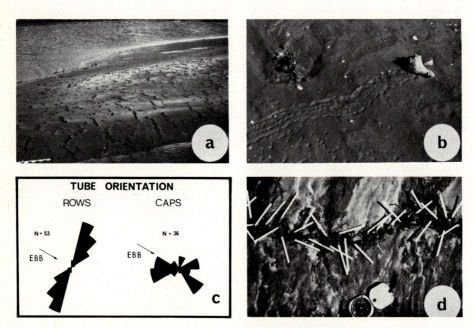

Fig. 3. Burrows of the polychaete *Diopatra cuprae* on the point bar shown in Fig. 7f, Bass Creek, Kiawah Island. a) Bar apex area, ebb direction to the right. Note tube rows. Scale is one meter. b) Solitary tube cap ornamented with shell fragments and pine needles. Note scour pit around tube and *Busycon*-dwelling hermit crab trail. c) Summary of *Diopatra* preferred-orientations. Cap orientation becomes less pronounced in muddier, lower velocity area. d) Single row of Diopatra tubes, with soda straws in apertures to indicate orientations. Note current lineations at a slight angle to row because of late-stage ebb directional changes.

significant along channel margins and on point bar crests, whereas physical structures are dominant in channels, as noted by Howard and Frey (1973).

Point bar crests display numerous biogenic structures. In addition to feeding tracks of birds, feeding pits of rays, and an occasional alligator wallow, surface markings include trails of the snails *Nassarius obsoletus*, *N. vibex* and *Littorina irrorata*. High intertidal and supratidal portions of bars have especially large snail populations. These higher areas are also densely populated by the hermit crabs *Uca pugilator* and *U. pugnax*, often with over 150 burrow entrances per square meter. Frey and Howard (1969) provide an excellent guide to related zonations of fauna and their biogenic structures, which closely reflect conditions observed in the Bass Creek and Murrells marshes. Additional helpful details have been presented by Land and Hoyt (1966), Howard and Dorjes (1972), Howard and Frey (1973), and Howard *et al.*, (1973).

Polychate burrows were found to provide good indicators of current direction in several different ways. As noted by Frey (1970), exposed mucilagenous burrow linings of *Streplospio benedicti* tend to lie recumbent on the sediment surface and align parallel to current flow. More rigid burrows like *Onuphis microcephala* and *Diopatra cuprae* provide obstructions to flow, which form aligned scour pits and current lineations (Fig. 3). Of greatest interest is the frequent alignment of *Diopatra* tubes into rows (Fig. 3) ranging from 15 to 150 cm in length. These rows are linear to gently undulatory, and are aligned perpendicular to the ebb current direction. Similar alignments are described by Frey and Howard (1969), and Myers (1972). Population densities are much higher on the floodward ends of point bars, probably because that position offers more exposure to ebb currents, which carry higher concentrations of organic debris (Ward, 1978). In addition, large-scale bedforms do not form there, so the sedimentation rate is sufficiently low to allow habitation (Howard, 1975).

Tube-caps have the shape of a periscope or an inverted "J", which helps provide feeding efficiency (Mangum *et al.*, 1968). The caps protrude from 1 to 6 cm above the sediment surface, and are armored with shell fragments and sand, which provides tube strength and aids in predator detection (Brenchley, 1976). Myers (1972) found that the plane of the tube-cap aperture was parallel to flow. He suggests that this orientation allows the maximum development of laminar flow past the opening, which aids flushing. As the animal detects food by chemo-reception (Mangum and Cox, 1971), laminar flow conditions would also aid in feeding. The tubes observed in Bass Creek show tube-cap orientations at angles oblique to flow (Fig. 3c, d). This may be a compromise between the maintenance of optimum hydraulic conditions and the avoidance of facing into another tube. Such an avoidance would maximize the area of substrate on which the animal could feed, a strategy which is supported by the alternating orientations of adjacent individuals (note bimodalism in Fig. 3d).

MEANDER GEOMETRY

Flow in curved channels is helicoidal and therefore involves inward-directed transverse components (NEDECO, 1959; Bagnold, 1960; Leopold and Wolman, 1960). Point bars accrete at meander bends because of both the inward-directed flow of these macroturbulent eddies, and the shifting toward the cutbank of the maximum velocity line (Prandtl, 1952).

Yalin (1972) has shown that theoretically a large macroturbulent eddy is systematically affected by a discontinuity such that identical and reverse perturbations are repeated downstream at equal intervals. He demonstrates that these intervals in fully developed turbulent flows are equivalent to meander wavelength λ, which is related to width ω, by:

$$\lambda = 2\pi\omega \qquad\qquad (1)$$

Further, he states that if turbulence is low, wavelength depends on Reynolds number and
relative bed roughness, so that:

$$\lambda = 20\,\omega \qquad\qquad\qquad (2)$$

if the ratio of channel width to median bed grain-size is greater than or equal to 1000.

Hey (1976) shows a discrepancy between Yalin's (1972) theory and the field studies of
Leopold and Wolman (1960) who relate wavelength to width by:

$$\lambda = 10.9\omega^{1.01} \qquad\qquad (ft) \qquad\qquad (3)$$

Hey suggests that as turbulence is fully developed at bankfull stage, and as the
channel-forming discharge corresponds to that stage, Leopold and Wolman's coefficient
of 10.9 should be closer to 2. He attributes this disparity to the repeating distance, which
should be measured along the channel as meander arc length, z, and not in a straight line.
He then solves Leopold and Wolman's equations relating radius of curvature, r, to
channel width, and arrives at:

$$r = 2.37\omega^{1.03} \qquad\qquad (ft) \qquad\qquad (4)$$

Linearity was assumed and the equation modified to:

$$r = 2.40\omega \qquad\qquad (ft.\ or\ m) \qquad\qquad (5)$$

for convenience. All of Hey's data plot above the line of equation (5), however (Fig. 4),
which he explains as a function of meander arc angle, θ. This assumes that equation (5)
corresponds to "well-developed" meanders (150°). Hey's subsequent family of curves

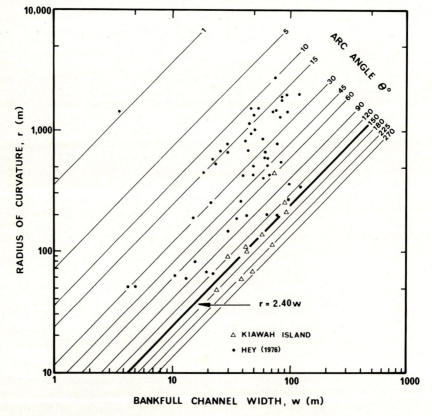

Fig. 4. Meander geometry of Bass Creek in contrast to the fluvial data of Hey (1976). Heavy line
is predicted radius/width ratio of Leopold and Wolman (1960) (after Hey, 1976).

demonstrates that the relationships between wavelengths, radius of curvature and channel width vary with arc angle, and that the meander arc length, z, is a function of channel width.

$$z = 4\pi\omega \tag{6}$$

Hey argues that as it is more logical to take z as the repeating distance rather than λ, a comparison of equations (1) and (6) shows that Yalin's (1972) model underestimates that distance by a factor of two. This discrepancy arises because in Yalin's analysis he assumes the width of the helicoidal flow cell to be equal to the channel width. At bankfull stage in natural streams however, as many as three macroturbulent eddies may exist (Eakin, 1935), and at least two have been observed (Hey and Thorne, 1975). Thus, if the cell diameter is half the channel width, equation (6) is a better approximation for natural meanders (Hey, 1976)

Bass Creek meander geometry data more closely approximate Leopold and Wolman's (1960) meander relationships (eq. 5), as shown by Fig. 4. To examine this apparent discrepancy between Yalin's (1972) and Hey's (1976) hypotheses on meander repeating distance, surface drogues were released during peak ebb velocities, on calm days, at each of ten Bass Creek meanders. No surface patterns of flow convergence or divergence were observed at either pools or riffles, indicating the presence of a single helicoidal cell. Therefore, assuming cell and channel widths to be equivalent, the macroturbulent eddy repeating distance (z, not λ) supports Yalin's (1972) model (Fig. 5). Although the data given in Fig. 5 are sparse and scattered, a regression line constrained through the origin plots nearly directly on Yalin's curve.

How are the data from this study to be reconciled with Hey's results? A possible explanation might be the basic differences between tidal and fluvial channels (Table 1).

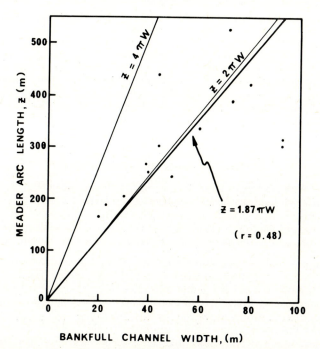

Fig. 5. Repeating distance of helicoidal cells. Yalin's (1972) curve (z = 2πω) assumes cell width equal to channel width. Hey's (1976) curve (z = 4πω) assumes twin cells. Heavy line is regression constrained through origin, assuming cell and channel width equal.

More specifically, bankfull discharge in tidal creeks is not the channel-forming discharge, as at bankfull stage the velocity is decreasing to zero. The lower average and maximum velocities of tidal creeks, and corresponding lower Reynolds numbers, may be insufficient to develop two or more helicoidal cells. But available information argues against this interpretation. Firstly the banks of most marsh creeks are very steep (often vertical) so that meander geometry parameters are essentially equal for maximum discharge and bankfull stages. Secondly, even in small tidal channels average peak velocities are high enough to fully develop turbulence, with Reynolds numbers usually of at least 10,000. A better explanation involves the constraints of channel size. In streams where width/depth ratio is high, a single cell is likely to be unstable and multiple cells develop (Nemenyi, 1946; Leopold and Wolman, 1960). Because of bank coherence and time-velocity asymmetry, smaller tidal creeks are relatively deep for their width; on tight meanders width/depth ratios are often as low as 4 or 5 (see Fig. 6).

Although the geometry of individual meanders is similar for tidal and fluvial channels, tidal creeks appear more tortuous in plan. These higher sinuosities are no doubt the result of very low slopes (Schumm, 1963; Leopold et al., 1964; Schafer, 1973) but may also be influenced by the way in which the meander form migrates. In fluvial channels bank erosion is greatest downstream from the meander apex, and the whole meander thus moves downstream (Fisk, 1944; Sundborg, 1956; Bluck, 1971). This mode of meander translation may be due to constrictions caused by point bar growth (Leopold and Wolman, 1960) or to lag in development of the helicoidal cell until half-way around the meander (Jackson, 1975, 1976a). Regardless of mechanism, laboratory studies show that in uniformly eroding material this migration takes place without an increase in meander amplitude, and with no tendency for meander loops to cut-off (Friedkin, 1945). In tidal creeks, on the other hand, although ebb-dominance places point bars in the same relative "downstream" position, reversing flow directions often provide for preferential cutbank erosion on both sides of the meander apex. This process is believed to cause corresponding increases in arc angle and sinuosity. The relative paucity of abandoned channels in marsh systems in comparison to meandering rivers could be the result of several causes. Although relatively less homogenous, tidal creek banks are more coherent than the sand banks in Friedkin's (1945) experiments. Secondly, a contributing factor to the lower migration rates must be the more weakly erosive nature of the lower velocity tidal currents, as discussed previously (see Time-Velocity Asymmetry). Finally, tidal creek meanders may display a relative inability to cut off because maximum velocities occur at low stages (Van Veen, 1950) as shown by Fig. 1.

POINT BAR GEOMETRY AND SEDIMENTS

Leopold and Wolman (1960) have shown that in fluvial meanders the ratio radius of curvature/width (r/w) tends toward a constant value between 2 and 3 (Fig. 4), which they suggest is the shape offering minimum resistance to flow. Extensive aerial photographic and ground studies of the South Carolina marshes show tidal meanders to be comparable (Fig. 4), with a range of r/w between 1 and 7 (Barwis and Ward, in preparation).

Point bars display basic shape differences that depend primarily on the ratio r/w (Fig. 6), inasmuch as time-velocity asymmetry and ebb-dominance operate approximately equally on all meanders. On very gentle meanders or nearly straight reaches (r/w >>3) bars are long and narrow, fully welded to the bank, and resemble the "skew shoals" of Quraishy (1944) in both form and origin (Fig. 6a). Inner bank slopes are relatively gentle. Depth varies inversely with radius of curvature (Leliavsky, 1955, p. 118), so that on very tight meanders (r/w <2.5) both inner and outer banks are steep (Fig. 6d). Due to the resultant space limitations, point bars are relatively small and welded to the bank. Both of these bar types (Fig. 6a, d) are skewed toward the ebb direction. Time-velocity asymmetry generates little to no flow segregation.

Meanders of intermediate radius develop more complex point bars. Gentle meanders (r/w >3) produce long, narrow bars (Fig. 6b) with length to width ratios often greater than 10. These bars are detached from the inner bank at all but their floodward extremities, and are sometimes completely detached, occupying a mid-channel position. They separate the ebb-dominant main channel from a smaller flood-dominant channel that lies against the inner bank. The most complex bars occur near flood-tidal deltas, at stream junctions, or on meanders where 2.3< r/w <3. These bars comprise two or more lobes which are joined at their floodward extremities and may or may not be detached from the inner bank (Fig. 6c). Current segregation is well-developed on these bars, as it is on the flood-tidal deltas to which they are geomorphically similar (Daboll, 1969). Current segregation and bedform distribution on these bars is controlled by topographic shielding caused by bar geometry, by shifting of streamlines on flood vs. ebb, caused by the geometries of adjacent meanders, and by time-velocity asymmetry.

The distribution of sedimentary textures and structures is distinct on each type of bar, and is distinctive enough in general for all bars that their tidal origin would be easily recognizable in the rock record. About 50 point bars were visited at several localities in the South Carolina marsh during 1975 - 1978. Four bar types are recognized: linear welded bars; steep apical bars; linear mid-channel bars; and multilobed bars (Figs. 6, 7).

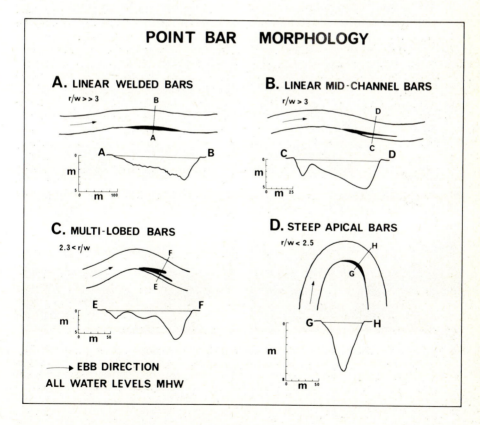

POINT BAR MORPHOLOGY

A. LINEAR WELDED BARS
r/w >> 3

B. LINEAR MID-CHANNEL BARS
r/w > 3

C. MULTI-LOBED BARS
2.3 < r/w

D. STEEP APICAL BARS
r/w < 2.5

→ EBB DIRECTION
ALL WATER LEVELS MHW

Fig. 6. Point bar geometry as a function of the ratio radius of curvature/width. Note that all bars are skewed in the ebb direction. Profiles are fathometer traces from bars in the Murrells and Bass Creek systems, and show inverse proportionality between depth and radius of curvature. Profiles are of bars shown in the following figures: A, Fig. 8; B, Fig. 10; C, Figs. 7f, 14.

Bedform terminology is that used by Boothroyd and Hubbard (1975). Crest to crest spacings are: ripples, <60 cm; megaripples, 60 cm - 6 m; sand waves, >6 m.

Linear, Welded Bars

The long, narrow point bars on very gentle meanders (r/w >3) are fully welded to the bank over most of their length. Channel profiles are the most asymmetrical for this type of bar (Fig. 6a). Subtidal bar flanks vary in slope between 5 and 10 degrees, as is the case for many fluvial bars (Allen, 1963). Behind the bar crest is a very shallow, gently ebb-sloping high intertidal to supratidal swale that extends from mid-bar to tail. Apart

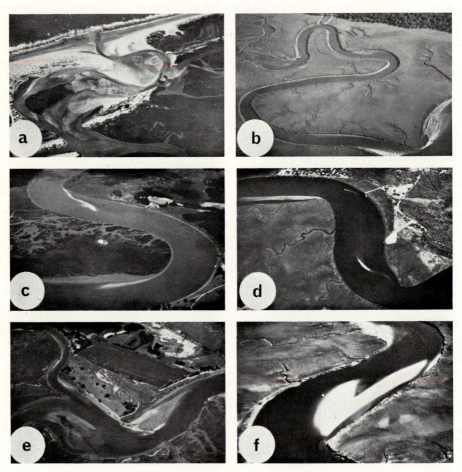

Fig. 7. Examples of tidal creek point bar geomorphology. All photographs are low-obliques. a) Mad Inlet, showing gradational nature between tidal point bar, flood-tidal delta, and spit. b) Series of 8 meanders showing bar variation as a function of r/w. Two tight bends in background with steep apical bars, separated by a broader, lobate bar on the intervening meander. Note complex multilobed bars in left background, and bank-welded linear bar ("skew-shoal") in foreground. c) As r/w decreases bars lie closer to inner bank. Sand waves on bar in foreground are flood oriented. d) Linear, mid-channel bars in a very large creek. Bar in foreground is actually closer to the cutbank than the inner bank. Small wisp in background is a boat wake. e) Multi-lobed bars on meanders of intermediate radius. Top of bar on left has become *Spartina*-stabilized. f) Multi-lobed bar in Bass Creek. Boat is 5 m long. Note two flood channels, with flood-oriented sand stringer originating at bar apex. Despite high albedo (infrared film) interior lobe is entirely mud.

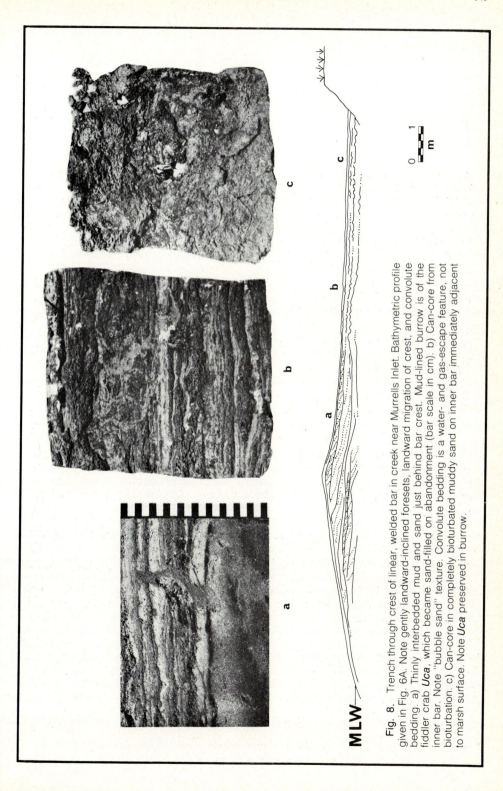

Fig. 8. Trench through crest of linear, welded bar in creek near Murrells Inlet. Bathymetric profile given in Fig. 6A. Note gently landward-inclined foresets, landward migration of crest, and convolute bedding. a) Thinly interbedded mud and sand just behind bar crest. Mud-lined burrow is of the fiddler crab *Uca*, which became sand-filled on abandonment (bar scale in cm). b) Can-core from inner bar. Note "bubble sand" texture. Convolute bedding is a water- and gas-escape feature, not bioturbation. c) Can-core in completely bioturbated muddy sand on inner bar immediately adjacent to marsh surface. Note *Uca* preserved in burrow.

from weak flood currents in these swales during spring tides (<15 cm/sec), no other current segregation is evident on or near the bar.

In small tidal creeks (w<100 m), large-scale bedforms were completely confined to subtidal bar flanks, these usually consisting of ebb-oriented sand waves and megaripples. Current and oscillation ripples dominate the intertidal zone, although small, weakly developed fields of lunate megaripples are present at both bar extremities. Even where these fields comprise little more than a few bedforms, diving observations reveal that they do not reverse orientations during the tidal cycle, an effect of current shielding by the bar crest. Because of higher velocities, bars in larger creeks (w>200 m) frequently display well-developed sand wave and megaripple fields on intertidal flanks and extremities.

Internal structures of a linear, welded bar near Murrels Inlet are illustrated in Fig. 8. The outer flank consists of thin flat beds with obvious beach-face similarities. Laminae are often accentuated by heavy minerals and very fine shell hash. These beds pass laterally into 5 to 40 cm thick internally cross-stratified units that dip gently bankward. Foresets in these units vary from steep tabular laminae in the thinner sets to very low angle tangential surfaces in the thicker sets. The lower angle foresets usually contain intrasets of ripple trough cross-laminae, the orientations of which are bipolar in directions parallel to the bar crest.

Mud is present as fecal pellets, as 1 cm rip-up clasts, and as finely disseminated suspension sediment. Mud content increases bankward, occurring both as thin (1 - 2 cm) continuous beds and in flaser- and linsen-structures (Figs. 8a and b). Organic content of the mud is high, with strong release of H_2S on trenching. This depth to black-stained sediments (hydrotroilite) lies close to mean low water under the outer, more permeable portions of the bar but climbs to within 20 cm of the surface on the muddier bankward portions.

The degree of bioturbation increases dramatically over very short distances in a bankward direction. *Mercenaria* is very common, but most of the bioturbate structures are attributable to burrowing activities of the fiddler crab *Uca* (Fig. 8a). Burrowing on the innermost portions of the bar is intense, and, enhanced by the relatively slow sedimentation rate, may completely disrupt bedding and result in homogenized muddy sand (Fig. 8c).

Interstitial gasses are an important modifying agent of primary sedimentary structures on the bar crest (cf. Cloud, 1960). Cleaner sands often display a soft, spongelike "bubble-sand" fabric, an effect of entrapped air being pumped through dry sand by the tidally rising water table (Emery, 1945; Hoyt and Henry, 1964). In muddier, interbedded sediments where sands cannot dry sufficiently to entrap air, gasses from decaying organics generate both bubble sand fabric and convolute bedding (Fig. 8b), the latter being dominant (cf. Lowe, 1975; Johnson, 1977). Small sand volcanoes up to 5 cm in diameter were also observed during high stages.

In addition to displaying abundant faunal tracks and trails as previously described, bar crests are littered with occasional tree debris, and fragments of the marsh grasses *Spartina alterniflora* and *Borrichia frutescens*. *Spartina* is established on the bar crest only at the higher, floodward extremity. This elevation control of marsh flora with respect to mean high water (Chapman, 1960) highlights another important difference between fluvial and tidal point bars. As very little of a tidal creek point bar is free to become vegetated, the bar form cannot be stabilized by plant growth. Thus, ridge-and-swale topography, the development and preservation of which in fluvial systems is partially dependent on the binding and baffling effects of vegetation, cannot develop. This biological control is probably secondary to basic differences in flood duration and strength (Table 1).

Steep, Apical Bars

The relatively small bars on tight meander bends (r/w<2.5, as shown by Figs. 6d, 7b and 9) are completely attached to the bank, with no swales on the inner bar. Bar flanks at the minimum width cross-section are steep (15 — 20 degrees), and rise to the elevation of the marsh itself. This is the only type of bar for which bar accretion would produce the epsilon cross-bedding form of Allen (1963) in set thickness equal to the channel depth. Ebb-oriented features resembling the transverse bars of Jackson (1976a) are often present (Fig. 9), but these are stationary features. Scour pools behind these bars are probably the result of the flow separation which occurs when r/w is less than 2, and the inner zone of relatively stagnant water becomes unstable and eddies (Bagnold, 1960, p. 139). During spring ebb conditions on the bar shown in Fig. 9, a 5 cm hydraulic jump was set up across

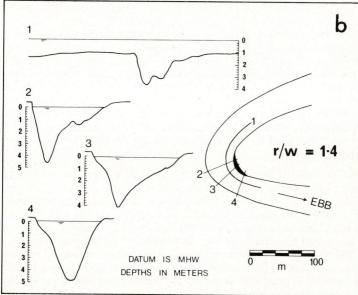

Fig. 9. Tight meander bend in Bass Creek. a) View from cutbank looking in flood direction at neap low tide. Note "transverse bar" in foreground, a relatively stationary feature on many tight meanders. b) Bathymetric profiles of bar in a). Eddy scour-pit on profile 1 is a function of low r/w ratio (Bagnold, 1960).

the transverse bar, and small (λ = 30 cm) standing waves developed for a few minutes as the bar crest became emergent. No other large-scale bedforms were observed during diving observations on any tight bends, either in the channel or on the intertidal bar. A few, isolated, ebb-oriented megaripples (spacings up to 1 m) ornament the floodward ends of the intertidal bar areas. The development of all other bedforms is restricted to linear and cuspate current ripples which, on the intertidal bar, become washed-out during emergence.

These bars are generally cleaner than linear, welded bars, although intraclasts and fecal pellets are sparsely present on the bar crests. Most mud occurs on the highest portions of the bar as thin, patchy drapes that when present are continuous with the adjacent marsh surface. Convolute bedding and bubble-sand are present, but bioturbate structures are subordinate. This is attributed to the absence of a swale area protected from current reworking, so the relative sedimentation/bioturbation rate is higher. Additionally, the area with elevations appropriate for *Uca* habitats is minimal, as the bar profile is steep.

Linear, Mid-Channel Bars

Gentle meanders (r/w >3) produce long, curvilinear bars that taper out in the ebb direction at low angles to the bank. Bar position is related to, and also reinforces flood- and ebb-current segregation (the "drempel" bars of Van Veen, 1950). Bars may be attached to the inner bank at their floodward extremities, as is the case in most small channels, or completely detached in larger channels where flood currents are strong (Van Veen, 1950, Figs. 16 and 24, respectively). Where bars are fully detached, their crests segregate ebb and flood currents (op. cit.) as do the "sand tongues" described by Gohren (1969) in the Nordergrunde, where velocities on the bar crests are fifty percent of those in the adjacent channels. Bars vary in position, close to the bank on tighter bends, and more toward mid-channel on more gentle bends (Fig. 7). Regardless of bar shape or position, flood channels are always shallower than their associated ebb channels (Fig. 6b).

Attached Bar

The attached bar studied in detail is located in Bass Creek (Figs. 2, 10). The main channel is ebb-dominant and of relatively low-energy; ebb velocities seldom exceed 50 cm/sec. The inside channel carries only flood currents which very rarely exceed 15 cm/sec. In the main channel, the thalweg and lower bar flank consist of current-rippled muddy sand. Megaripples are linear to undulatory (spacings 1 — 3 m), ebb-oriented, and largely restricted to the upper flank of the subtidal bar. A few megaripple crests extend as high as 30 - 40 cm onto the intertidal bar, including some that are flood-oriented on the seaward bar extremity, but these become progressively washed-out with elevation above mean low water. Bedforms in the flood channel are confined to flood-oriented current ripples in muddy sand.

The intertidal portion of the bar displays a marked fining-upward texture from fine sand into thinly inter-bedded mud and sand (Figs. 10b, c). On the bar crest, beds are horizontal and continuous with the gently dipping faces of the flanks, and are occasionally truncated there by spring tide and storm erosion. Mud on the bar crest has settled from suspension during neap tides, and has accumulated more rapidly on the ebb side of the oyster mound shown in Fig. 10a due to current-baffling. The degree of desiccation increases with height on the bar, and because the resultant mudcrack polygons are separated by thin sand beds, they are easily ripped up to form intraclasts (Fig. 10b). Erosion of these mudflakes occurs only during the peak velocities of spring tides. Higher spring tide velocities also introduce more sand, which fills the pits left by rip-up clasts, as well as the remaining polygon cracks (cf. Tankard and Hobday, 1977, Fig. 15), as shown by Fig. 10d. All forms of flaser and lenticular bedding (Reineck and Wunderlich, 1968) are present in the upper one meter of the intertidal bar. These grade from flaser types in the lower 40 cm to wavy and lenticular forms in the upper 60 cm as mud content increases upward.

Fig. 10.   Linear, mid-channel bar in Bass Creek. a) View toward bar apex, along flood channel. Note oyster (**Crassostrea**) bank on bar crest. b) Inner flank of bar, showing both vertical and lateral accretion. Note surface littered with mudcrack rip-up clasts (scale is 1 m). c) Closeup of trench in b; thin interbedding, flaser and linsen structures, and a paucity of bioturbation (scale is 1 m). d) Genesis of sand-filled mudcracks on bar crest.

Bioturbation is less evident than on the two previously mentioned bars, as *Uca* is less abundant. Surface markings include occasional hermit crab crawling traces and the ubiquitous bird tracks and feeding pits.

Detached Bar

The detached bar observed near Murrells Inlet (Fig. 2) resembles the lower energy, attached bar just described, but lies in a channel where maximum ebb velocities are 60 - 80 cm/sec. The smaller, inside channel is open to ebb- as well as flood-currents. As a result, a small (1.5 m high) ebb-oriented chute bar occurs at the channels' seaward end. In form it resembles the fluvial chute bars described by McGowen and Garner (1970) and Levey (1976), but is smaller and subtidal, and does not occur on the upper bar.

Bedforms in the main channel are bimodal current ripples superimposed on ebb-oriented megaripples and sandwaves with spacing varying between 4 and 10 m (average 7 m). On this bar and all others visited, no discreet natural grouping of bedform spacings was observed, as suggested by Boothroyd and Hubbard (1975).

Sand wave crests are continuous from the channel onto the crest of the bar, with lower height/spacing ratios on the intertidal portions. This is probably due not only to the depth restrictions placed on bedform height, but to crest erosion associated with higher Froude numbers during emergence.

Except for the extreme floodward end of the bar, where intertidal sand waves remain ebb-oriented throughout the tidal cycle, the highest portions of the bar are covered by flood-oriented sandwaves and megaripples. Along their crests and channelward down the outer bar flank, these bedforms grade into ebb-oriented features with roughly the same spacings. The transition between the two orientations is a low (10 — 30 cm) rounded form with no preferred orientation, and occurs just above mean low water. These forms were not observed during maximum velocity conditions, although they were observed through a range of spring to neap conditions. Inasmuch as they changed position and developed weak asymmetries with varying water depth and velocity conditions, they probably represent a zone of multiple, bidirectional reactivation surfaces. Ebb-oriented megaripples reflect a transport component toward the inside of the meander bend (5 — 10°), a result of helicoidal flow (Allen, 1970; Puigdefabregas, 1973; Jackson, 1976b).

Sedimentary structures comprise mostly planar-tabular and planar-tangential cross-bed sets up to 20 cm thick, with rippled toesets and topsets often displaying washed-out geometries. Although as a whole slipface orientations are bimodal, flood- and ebb-oriented bedforms are sufficiently segregated that no herringbone features larger than ripple-size were observed. The only large-scale trough cross-stratification present is that associated with ebb-oriented cuspate megaripples on the floodward bar extremity and along the outer bar flank just below mean low water.

The bar contains very little mud; where present, it occurs as simple to bifurcating flasers, and thin (1 — 5 mm) meter-wide drapes in sand wave and megaripple troughs. Weakly graded foresets are observable at the bar extremities, where sorting of the quartz fraction is poorer than on the bar crests. Bubble-sand is present on the highest portions of the bar, in beds up to 20 cm thick. Bioturbation rates are low; *Mercenaria* is present only in the lower intertidal bar, and *Uca* only sparsely populates the bar crest.

Rainstorms have a pronounced effect on the bar surface if they coincide with low tide. Within minutes after the onset of heavy rain, ripples, tracks, trails and *Uca* burrow entrances are quickly erased. Bubble sand layers are deflated as air escapes to the surface in small boils. As runoff increases, the crests of sand waves and megaripples become

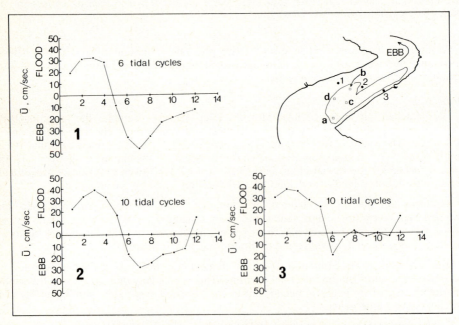

Fig. 11. Time-averaged current velocity distributions, multilobed bar in Bass Creek (Fig. 7f). Times are lunar hours. Each data point is the vector mean of 16 velocities for each hour; e.g. station 1, hour 1 represents the vector mean of 36 velocities. Note well-developed current segregation, and ebb-dominance of main channel. At station 3, low velocity fluctuations during late ebb are wind effects. a-d: locations of cores shown in Fig. 14.

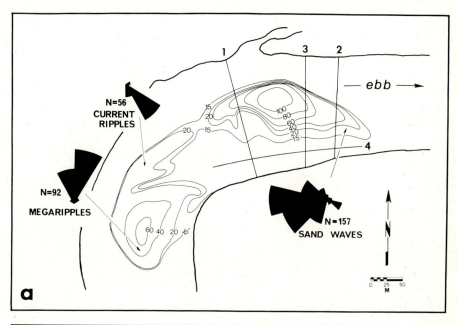

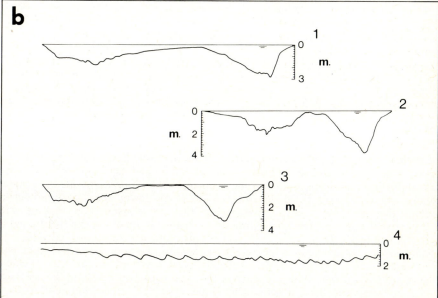

Fig. 12.   Multilobed bar near Murrells Inlet. a) Bedform distribution, with locations of bathymetric profiles given in b). Nearly all allocthonous marine faunal remains lie in megaripple troughs at west end of bar. Contours in cm above mean low water. b) Bathymetric profiles. Note flood-oriented sand waves of interior channel, which become progressively flattened in shallower water.

rounded by sheetflow erosion and slumping. Troughs of these bedforms become small streams that erode gullies up to 25 cm deep in the bar flanks. Small (1 m) Gilbert-type deltas are built at the mouths of these gullies on both bar flanks. All surface mud drapes are removed in suspension. Within an hour, all that remains on the bar crest are shallow, nondirectional vestiges of sand waves and megaripples, with rill marks abundant on their gentle slopes. No raindrop impressions were observed. Subsequent preservation of the rain-influenced surface would offer many possibilities for incorrect interpretations of the rock record, including current-generated reactivation surfaces, the influence of upper flow regime tidal currents, or even tectonic deformation of cross-stratification.

Multilobed Bars

Meander bends of intermediate radius of curvature/width ratios ($2.3 < r/w < 3$) produce the most complex bar geometries. These bars comprise two or more long, curvilinear lobes that trail away from the meander apex in the ebb direction (Figs. 6c, 7f). Well-developed current segregation results from the combined effects of ebb-dominance, time-velocity asymmetry and bar/meander geometry. The distribution of sedimentary textures and structures reflects these current patterns. Examples from both the Bass Creek and Murrells systems were examined in detail.

The outermost channel is ebb-dominant, the middle channel is flood-dominant, and the innermost channel carries flood discharges almost exclusively. These patterns were observed through one tidal cycle on three separate bars by half-hourly spot checks with a Price current meter. The resultant data were then used as a guide to site recording current meters on one of the bars.

Fig. 11 (inset) shows the locations of recording current meters stationed on the Bass Creek bar illustrated in Fig. 7f. These meters measured, at a height 50 cm from the bed, the instantaneous horizontal components of speed and direction 17 times per lunar hour over a ten day period encompassing the spring- to neap-tide transition. As bedform and textural distributions are more closely related to longer-term hydraulics than the unique conditions of a single tidal cycle, velocities are presented as an ''average'' tidal cycle for the observation period considered. Each point on the velocity curves represents the vector mean (Potter and Pettijohn, 1963) of all velocities taken during the nth hour of the 12 lunar hour cycle, calculated as a resultant for all tidal cycles. It should be noted that as vector mean data, actual times of peak velocities, slack water, and even sense of direction cannot be read directly, e.g. the vector means of low velocities comprise both flood and ebb readings.

The current segregation patterns of Van Veen (1950) are clearly shown by Fig. 11. The most obvious result of this current pattern is the control of sediment textures in the meander bend. Very coarse material, such as shells and larger mudclasts, are not confined solely to the deepest parts of the main channel as is primarily the case in fluvial systems. Although peak flood-channel velocities are lower than peak ebb-channel velocities, they are competent to move much of this coarse material up onto the surface of the bar itself. Functionally, this central flood channel is similar to the flood-ramp on a flood-tidal delta (Hayes et al., 1973). The flood channel serves to introduce local coarsening-upward sequences to the overall fining-upward nature of the bar, as do chutes on fluvial point bars. On many higher energy bars, ebb currents in the interior channels are more pronounced due to the presence of ebb spillover lobes. These spillover lobes are identical in function to these described by Hayes et al. (1973) on flood-tidal deltas, and those described from estuarine shoals by Van Veen (1950). These features are less ephemeral on tidal creek point bars, probably due to space limitations imposed on current patterns by the presence of constricting channel banks.

Bedform and Sediment Distribution

Because of the current segregation shown by Fig. 11, bedforms are also segregated with respect to both type and orientation. The bar crest provides a shielding effect so that bar

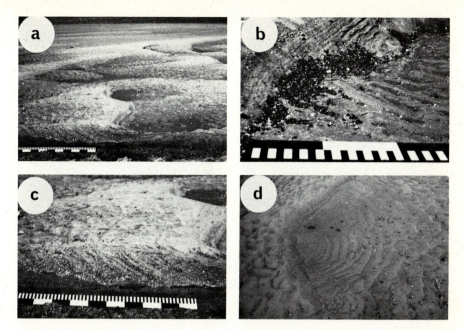

Fig. 13. Typical multilobed bar bedforms, Bass Creek. a) Ebb-oriented megaripples. Note trough cross-stratification set thickness and clay-draped troughs. b) Incorporation of mud both as clasts (left) and as ripple drapes (right). c) Closeup of a, with abundant shell material and mud clasts on foresets. Note rippled toesets and topsets, and darker, lower Eh conditions at base of trench. d) Local scour in lee of flood-oriented sand-wave. Note flow divergence, and small-scale bedform transition from linear ripples to cuspate forms (small units on all scales are cm).

extremities have bedforms oriented only with the current to which they are exposed, with forms larger than ripples seldom reversing orientation. Sand waves are usually flood-oriented, as ebb velocities exceed those for sand wave stability in areas that are ebb-dominant (Boothroyd and Hubbard, 1975). This typical pattern is demonstrated particularly well by a larger, higher-energy bar near Murrells Inlet (Fig. 12, profile 4).

The floodward end of this bar is covered by ebb-oriented cuspate megaripples which pass upward into flat beds and washed-out ripples. Troughs of these megaripples contain extensive shell material, much of it allochthonous. The area just below mean low water is stabilized by a dense population of the polychaete *Streplospio benedicti*. The seaward extremity displays bedforms with spacings varying from 4 — 8 m (average 7 m). On the outer flank of this portion of the bar these bedforms grade into ebb-oriented cuspate megaripples that lie just above mlw. Mud units are absent from both bar extremities, although mud does occur as intraclasts (up to 2 cm) on megaripple foresets. Crests and troughs of all larger scale bedforms are to some degree ornamented with ripples, the orientations of which are widely divergent from the main flow direction (Fig. 13d).

The bar crest comprises mainly flat beds and washed-out ripples superimposed on low, rounded forms that look like non-directional, washed-out megaripples. These features failed to exhibit any directionality when observed at both water-level extremes. Bubble-sand is common on the bar crest. The flood-channel is floored by flood-oriented sand waves with spacings up to 20 m (Fig. 12b). These sand waves grade laterally up the inner bar flank into smaller, flood-oriented forms with spacings of 4 — 6 m. Between the two bar extremities is an ebb-spillover lobe covered with linear to rhomboid current ripples which display all degrees of being washed-out on emergence.

Fig. 14. Latex peels of can-cores from multilobed bar, Bass Creek. a) Bar apex, with in-situ *Diopatra cuprae* tube, which has slightly warped the adjacent laminae. b) Tail of outer lobe, with laminated sand and mud overlying planar-tabular megaripple set composed largely of shell material. c,d) Bar crest interbedded sediments. Note bubble-sand in c, and low bioturbation levels everywhere but bar apex. Core locations given in Fig. 11.

The low energy Bass Creek bar (Figs. 7f, 11) has no well-developed megaripple or sand wave fields, as velocities are too low. The only large-scale bedforms are flood-oriented megaripples (spacings 3 to 7 m) on the seaward end of the outer lobe. Higher velocities associated with spring tides reverse the orientations of the topographically highest megaripples. Despite the general paucity of large-scale bedforms, current patterns are markedly reflected on the bar.

The inner lobe is completely mud, and is thixotropic in many places. The crest is deeply cracked into polygons averaging 20 cm in diameter and 30 cm deep. Although subaerially exposed at low tide, the surface is never completely dry, and does not provide a source for rip-up clasts.

The outer lobe is predominantly sand, with sediments and bedforms displaying a zonation along the long axis. The floodward end comprises bioturbated muddy sand with ebb-oriented current ripples, and current lineations (Fig. 3). Flaser and lenticular bedding is common, with mud layers usually being reworked as rip-up clasts. Bedding is distinct, as bioturbation is almost entirely due to *Diopatra cuprae* (Fig. 14a). The seaward end is much cleaner, and displays a small flood-oriented megaripple field. The orientations of these megaripples reverse only during a four to five day period around spring tide when ebb velocities are higher. They produce planar-tangential to trough cross-stratification in sets 20 to 30 cm thick (Fig. 13a). Mud is present both as thin drapes in bedform troughs (Fig. 13b) and as intraclasts. As the amount of mudclast and shell debris is quite high, foresets often display an obvious grading (Fig. 13c). Cross-bed sets produced on this area of the bar are intercalated with flaser- and lenticular-beds which also have a high shell debris content, and with horizontally laminated very fine sand and mud (Fig. 14b). The crest of the outer lobe comprises flaser-bedded muddy sands in 10 to 30 cm interbeds with much cleaner, ripple cross-laminated fine sand (Figs. 14c, d).

The two flood channels are floored by finer-grained sediments than is the main ebb channel. The central channel contains rippled muddy sand that grades laterally into the coarser and finer sediments of the outer and inner lobe, respectively. Large, allochthonous shell material is common. The innermost channel is composed of mud, with thin sand streaks at the seaward end, brought in by flood currents. Small oyster mounds (to 1 m wide, 30 cm high) along the intertidal inner bank act as traps for fine sediment.

VERTICAL SEQUENCES

Although no complete sequences were observed in any of the bars studied, a combination of short cores (up to 3 m), can-cores, trenches and SCUBA observations permits hypothetical sequences to be constructed (Fig. 15). The general fining-upward nature of each sequence obviously resembles the now classic fluvial sequence (c.f. Allen, 1962, 1964, 1965, 1970; Bernard and Major, 1963; Visher, 1965, 1972). Bedform scale decreases upward, as in fluvial point bars (Harms *et al.*, 1963). These general trends have been briefly described in modern channel-fill sequences by Van Straaten (1952), Wunderlich (1970), and Reineck (1972). Van Straaten (1957, 1959), De Raaf and Boersma (1971), and Oomkens (1974) provide more detail in terms of actual textures and structures, but do not relate these vertical sequences to variations in lateral facies distributions set up by bar geometry. The most detailed descriptions of tidal channel sequences have been provided by Oomkens and Terwindt (1960), Terwindt (1971), and Van Beek and Koster (1972), who illustrate deposits grading upward from a shell-lag through cross-bedded sand, into horizontally laminated sand and clay. Sequences very similar to these have been described from ancient sediments by Selley (1968), Campbell and Oaks (1973) and Horne and Ferm (1976).

Several features of tidal creek point bars clearly distinguish them from fluvial point bars:

1. Channel-lag deposits are predominantly shell material and mud balls, often in beds up to 30 cm thick.

2.   Large-scale cross-stratification (associated with megaripples and sand waves) does not immediately overlie the lag deposit.

3.   The orientations of both small- and large-scale bedforms display bimodal to bipolar distributions.

4.   Small-scale bedforms tend to generate flaser and lenticular bedding on the bar crests, in preference to the climbing ripple cross-stratification prevalent in Allen's (1970) bar crests. This is because the sedimentation rate on bar crests is relatively low compared to fluvial systems, and the higher "flood" frequency causes constant reworking (Table 1).

5.   Bioturbation is far more intense on tidal point bars and should occur to some degree in any sequence, increasing in intensity inversely with grain size.

6.   Organic material capping the point bar sequence is unique to the tidal environment. *Spartina*-rooted muds form a thin, dense mat that contains little else in the way of plant matter but the *Spartina* roots themselves. The rooted zones of fluvial bars on the other hand contain a wide variety of plant material, including deep tap roots.

When considered in their stratigraphic context, tidal creek point bars are strikingly different from their fluvial analogs. In their studies of the Appalachian Carboniferous sediments, Horne *et al.* (1978) have shown that lower delta plain coals, many of which ride on tidal channel sequences, are thin and relatively continuous; coals associated with fluvial sequences on the other hand are thicker, and relatively discontinuous.

Tidal creek point bar sequences are less easily distinguishable from tidal inlet channel-fill sequences, in view of the continuum between inlet-related sand bodies and tidal creek sand bodies (Fig. 7a). Even though both types of sequences display upward decreases in grain size and bedform scale, point bars are distinct in that they are capped by muddy sediments instead of the wave-dominated deposits associated with spits or tidal deltas (Kumar and Sanders, 1974; Barwis and Makurath, 1978).

Two features of cross-bed orientations have important implications for paleocurrent analyses. First, as a result of flood- and ebb-current segregation and time-velocity asymmetry, flood- and ebb-oriented cross-beds will occur at different elevations in a vertical section, and herringbone structures are rare on a scale larger than ripple-size. This characteristic is shared by flood tidal deltas and other estuarine sand bodies as well, and was noted by Terwindt (1971). Second, ebb-dominance of the tidal creek system generates a corresponding preferred orientation of bedforms. As flood-oriented bedforms are more common toward the top of a sequence, truncation by channel migration could yield multistoried sequences that reflect only unimodal ebb-current directions. Such is the case in the black-barrier tidal flat sediments that Tankard and Hobday (1977) describe from the Ordovician of South Africa.

Reactiviation surfaces were observed during formation in channels and on point bar crests. Very seldom were they generated by reversing flow, although such a mode of origin did occur. More often than not, they formed by rounding of brinkpoints during higher Froude number conditions at lower tide stages, and by local reorientation of crests on undulatory forms.

Although soft sediment deformation features have been attributed by some workers to water escape following earthquakes, these studies indicate such an explanation to be unnecessarily elegant in the vast majority of cases. All of the features observed by Johnson (1977) for example, are common aseismic occurrences in modern estuaries, and can be related to gas escape or tidally induced pore pressure changes. No recumbently overturned cross-beds were observed.

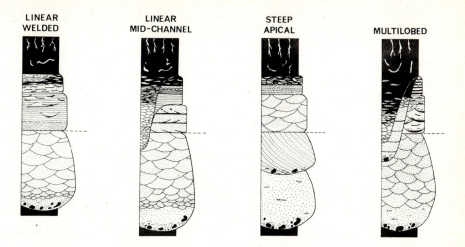

Fig. 15. Tidal creek point bar sequences constructed from pooled data, including can-cores, vibro-cores, shallow trenches and diving observations. Dashed line is mean low water.

DISCUSSION

In addition to fluvial point bar affinities, tidal creek point bar sequences display similar vertical trends in textures and structures to those produced by prograding tidal flats as first described in detail by Evans (1965). As in rivers, the sequence thickness is equivalent to channel depth, which is proportional to discharge, and on a regional scale therefore varies with tidal range (Nummedal *et al.*, 1977). In very large outcrops the channel profile encompassing these sequences should be reflected by the large, low-angle accretion deposits of Wright (1959), which is the epsilon form of Allen (1963). Unfortunately, these outcrops are the exception. Great care must therefore be exercised in making inferences about the origins of these sequences as channel deposits. Such care is particularly warranted for paleotidal range estimates. Klein (1971, 1972) for example has suggested that the thickness of a fining-upward tidal flat sequence provides a quantitative measure of paleotidal range, as this thickness represents the vertical distance between the low tide stage and the crests of higher mud flats. Crucial to such interpretations is the preservation of sedimentary structures produced by emergence runoff, without which the level of mean low water cannot be established. Use of an entire point bar sequence would yield anomalously high tidal ranges, a problem which is compounded (as Klein notes) by basin subsidence.

Point bar characteristics described here are purely observational; very little is known about the hydrodynamics of tidal flow in natural meandering channels with cohesive banks. No detailed hydrography has been published, nor are any data available on meander geometry/discharge relationships. It has been shown that even though tidal and fluvial meander geometries are similar, associated point bars are readily distinguishable both in planform and in vertical section.

ACKNOWLEDGEMENTS

This work was part of a Ph.D. thesis at the University of South Carolina, with Miles O. Hayes as supervisor. I am grateful to many students who assisted in the field, particularly Dan Domeracki, Ian Fischer, Dennis Hubbard, Judy Jackson, Helen Johnson, Frank Lesesne, Raymond Levey, Chris Ruby, Mike Waddell and Steve Zenger. Miles O. Hayes, John C. Horne, John C. Kraft, Raymond Levey and Dag Nummedal provided ideas in their helpful discussions. I appreciate the efforts of Bruce Baganz who showed me ancient analogs in the Carboniferous, and Dennis Hubbard who provided helpful comments on coastal geomorphology. I also benefitted from informative talks on polychaetes with Gail Brenchley, John Day, James Howard, and Alan Myers. Thanks go to Frank Stapor and the South Carolina Marine Resources Division for providing current meters and analysis software. H. Edward Clifton, David Hobday, Raymond Levey, and Anthony Tankard reviewed the manuscript. Finally, I am indebted to Larry G. Ward without whose generous help, ideas, companionship, and good humor this study would have been quickly overrun by mud, mosquitoes, and broken equipment.

REFERENCES

Allen, J. R. L., 1962, Petrology, origin and deposition of highest Lower Old Red Sandstone of Shropshire: J. Sediment. Petrol., v. 32, p. 657-697.

———, 1963, The classification of cross-stratified units, with notes on their origin: Sedimentology, v. 2, p. 93-114.

———, 1964, Studies in fluviatile sedimentation: six cyclothems from the lower Old Red Sandstone, Anglo-Welsh Basin: Sedimentology, v. 3, p. 163-198.

———, 1965, A review of the origin and characteristics of recent alluvial sediments: Sedimentology, v. 5, p. 89-191.

———, 1970, Studies in fluviatile sedimentation: a comparison of fining-upwards cyclothems, with special reference to coarse-member composition and interpretation: J. Sediment. Petrol., v. 40, p. 298-323.

Andrews, E., Stephens, D., and Colquhoun, D. J., 1973, Scouring of buried Pleistocene barrier complexes as a source of channel sand in tidal creeks, North Island quadrangle, South Carolina: Geol. Soc. Am. Bull., v. 84, p. 3659-3662.

Bagnold, R. A., 1960, Some aspects of the shape of river meanders: U.S. Geol. Survey, Prof. Paper 282-E, p. 135-144.

Barwis, J. H., 1976, Annotated bibliography on the geologic, hydraulic, and engineering aspects of tidal inlets: U.S. Waterways Exp. Sta., and Coastal Engrg. Res. Cen., GITI Rept. 4, 333p.

——— and Makurath, J. H., 1978, Recognition of ancient tidal inlet sequences; an example from the Upper Silurian Keyser Limestone in Virginia: Sedimentology, v. 25, in press.

Bell, H. S., 1940, Armored mud balls — their origin, properties and role in sedimentation: J. Geol., v. 48, p. 1-31.

Bernard, H. A. and Major, C. F. Jr., 1963, Recent meander belt deposits of the Brazos River: an alluvial "sand" model (abs.): Am. Assoc. Petrol. Geol. Bull., v. 47, p. 350.

Bluck, B. J., 1971, Sedimentation in the meandering River Endrick: Scott. Jour. Geol., v. 7, p. 93-138.

Boon, J. D. III, 1974, Sediment transport processes in a salt marsh drainage system: Unpub. Ph.D. thesis, Sch. of Mar. Sci., Coll. of Wm. and Mary, Williamsburg, Va., 238p.

Boothroyd, J. C., and Hubbard, D. K., 1975, Genesis of bedforms in mesotidal estuaries: in L. E. Cronin, ed., Estuarine Research; Academic Press, New York, v. II, p. 217-234.

Brenchley, G. A., 1976, Predator detection and avoidance: ornamentation of tube-caps of Diopatra spp. (Polychaeta-Onuphidae): Mar. Biol., v. 38, p. 179-188.

Brown, P. J., 1976, Variations in South Carolina coastal geomorphology: in M. O. Hayes and T. W. Kana, eds., Terrigenous Clastic Depositional Environments; Univ. South Carolina, Dept. Geology, Tech. Rept. No. 11-CRD, II, p. 2-15.

Byrne, R. J., Bullock, P., and Tyler, D. G., 1975, Response characteristics of a tidal inlet: A case study: in L. E. Cronin, ed., Estuarine Research; Academic Press, New York, v. II, p. 201-216.

Campbell, C. V., and Oaks, R. Q., 1973, Estuarine sandstone filling tidal scours, Lower Cretaceous Fall River Formation, Wyoming: J. Sediment. Petrol., v. 43, p. 765-778.

Chapman, V. J., 1960, Salt marshes and salt deserts of the world: Interscience Publishers, London, 392p.

Cloud, P. E. Jr., 1960, Gas as a sedimentary and diagenetic agent: Am. J. Sci., v. 258-A, p. 35-45.

Daboll, J. M., 1969, Holocene sediments of the Parker River Estuary, Massachusetts: Contr. No. 3-CRG, Dept. of Geol., U. Mass., 138p.

De Raaf, J. F. M., and Boersma, J. R., 1971, Tidal deposits and their sedimentary structures: Geol. Mijnbouw, v. 50, p. 479-504.

Dietz, R. A., 1952, The evolution of a gravel bar: Mo. Bot. Garden Ann., v. 39, p. 249-254.

Eakin, H. M., 1935, Diversity of current-direction and load-distribution on stream beds: Trans. Am. Geophys. Union, v. 2, p. 467-472.

Eardley, A. J., 1938, Yukon channel shifting: Geol. Soc. Am. Bull., v. 49, p. 343-358.

Einstein, H. A., Anderson, A. G., and Johnson, J. W., 1940, A distinction between bed load and suspended load: Trans. Am. Geophys. Union, v. 21, p. 628-633.

Emery, K. O., 1945, Entrapment of air in beach sand: J. Sediment. Petrol., v. 15, p. 39-49.

Evans, G., 1965, Intertidal flat sediments and their environments of deposition in the Wash: Quart. J. Geol. Soc. London, v. 121, p. 209-245.

Ferm, J. C., 1974, Carboniferous environmental models in the eastern United States and their significance: *in* G. Briggs, *ed.*, Carboniferous of the Southeastern United States; Geol. Soc. Am. Spec. Paper 148, p. 79-96.

———, and Cavaroc, V. V. Jr., 1968, A nonmarine sedimentary model for the Allegheny rocks of West Virginia: *in* G. de V. Klein, *ed.*, Late Paleozoic and Mesozoic continental sedimentation, northeastern North America: Geol. Soc. Am. Spec. Paper 106, p. 1-20.

Fisk, H. N., 1944, Geological investigation of the alluvial valley of the Lower Mississippi River: U.S. Waterways Exp. Sta., Vicksburg, Mississippi, 78p.

———, 1947, Fine-grained alluvial deposits and their effects on Mississippi River activity: U.S. Waterways Exp. Sta., Vicksburg, Mississippi, 2 vol., 82p.

Frey, R. W., 1970, Environmental significance of Recent marine lebensspuren near Beaufort, North Carolina: J. Paleont., v. 44, p. 507-519.

———, and Howard, J. D., 1969, A profile of biogenic sedimentary structures in a Holocene barrier island-salt marsh complex, Georgia: Gulf Coast Assoc. Geol. Socs. Trans., v. 19, p. 427-444.

Friedkin, J. F., 1945, A laboratory study of the meandering of alluvial rivers: U.S. Waterways Experiment Station, Vicksburg, Mississippi, 40p.

Gadow, S., 1970, Sedimentologie und Makrobenthos der Nordergrunde und der Aussjade (Nordsee): *in* J. Dorjes, S. Gadow, H. E. Reineck, and I. B. Singh; Senckenbergiana Marit., v. 2, p. 31-59.

Gohren, H., 1969, Die Stromungsverhaltnisse im Elbmundungsgebiet: Hamburger Kustenforsch, v. 6, p. 1-83.

Harms, J. C., Mackenzie, D. B., and McCubbin, D. G., 1963, Stratification in modern sands of the Red River, Louisiana: J. Geol., v. 71, p. 566-580.

Hayes, M. O., 1975, Morphology of sand accumulation in estuaries: an introduction to the symposium: *in* L. E. Cronin, *ed.*, Estuarine Research; Academic Press, New York, v. II, p. 3-21.

———, Owens, E. H., Hubbard, D. K., and Abele, R. W., 1973, The investigation of form and process in the coastal zone: *in* D. R. Coates, *ed.*, Coastal Geomorphology; Proc. 3d. Ann. Geomorph. Sym., Binghamton-Suny, p. 11-41.

Hey, R. D., 1976, Geometry of river meanders: Nature, v. 262, p. 482-484.

———, and Thorne, C. R., 1975, Secondary flows in river channels: Area, v. 7, p. 191-195.

Hobday, D. K., and Eriksson, K. A., *eds.*, 1977, Tidal Sedimentation, with particular reference to South African examples: Sediment. Geol., v. 18, No. 1/3, 287p.

Homeier, H., 1969, Das Wurster Watt — Eine historisch-mor-morphologische Untersuchung des Kusten — und Wattgebiets von der Weser — bis zur Elbmundung: Forsch.-Stelle Norderney, Jahrb. 1967, v. 19, p. 31-120.

Horne, J. C., and Ferm, J. C., 1976, Carboniferous depositional environments in the Pochahontas Basin, Eastern Kentucky and Southern West Virginia: Am. Assoc. Petrol. Geol. Field Course, Dept. Geol. Univ. South Car., 129p.

————, ————, Caruccio, F. T., and Baganz, B.P., 1978, Depositional models in coal exploration and mine planning: Am. Assoc. Petrol. Geol. Bull., in press.

Howard, J. D., 1971, Trace fossils as paleoecological tools: *in* J. D. Howard, J. W. Valentine, and J. E. Warme, *eds.*, Recent Advances in Paleoecology and Ichnology; Am. Geol. Inst. Washington, Short Course Lecture Notes, p. 184-212.

————, 1975, The sedimentological significance of trace fossils: *in* R. W. Frey, *ed.*, The Study of Trace Fossils; Springer Verlag, New York, p. 131-146.

Howard, J. D., and Dorjes, J., 1972, Animal-sediment relationships in two beach-related tidal flats; Sapelo Island, Georgia: J. Sediment. Petrol., v. 42, p. 608-623.

————, and Frey, R. W., 1973, Characteristic physical and biogenic sedimentary structures in Georgia estuaries: Am. Assoc. Petrol. Geol. Bull., v. 57, p. 1169-1184.

————, ————, and Reineck, H. E., 1973, Holocene sediments of the Georgia coastal area: *in* R. W. Frey, *ed.*, The Neogene of the Georgia coast, Ga. Geol. Soc., 8th Ann. Field Trip pub. by Dept. Geol. Univ. Ga., p. 1-58.

Hoyt, J. H., and Henry, V. J., 1964, Development and geologic significance of soft beach sand: Sedimentology, v. 3, p. 44-51.

Hubbard, D. K., 1977, Variations in tidal inlet processes and morphology in the Georgia Embayment: Univ. South Carolina, Coastal Research Div., Tech. Rept. No. 14-CRD, 79p.

————, and Barwis, J. H., 1976, Discussion of tidal inlet sand deposits: examples from the South Carolina coast: *in* M. O. Hayes, and T. E. Kana, *eds.*, Terrigenous Clastic Depositional Environments: Univ. South Carolina, Dept. Geol. Tech. Rept. No. 11-CRD, II, p. 128-142.

Inglis, C. C., 1949, The behaviour and control of rivers and canals: Central Waterpower Irrigation and Nav. Res. Sta., Poona, Res. Pub. 13, vols. 1, 2, 486p.

Jackson, R. G., 1975, Velocity-bedform-texture patterns of meander bends in the lower Wabash River of Illinois and Indiana: Geol. Soc. Am. Bull., v. 86, p. 1511-1522.

————, 1976a, Depositional model of point bars in the Lower Wabash River: J. Sediment. Petrol., v. 46, p. 579-594.

————, 1976b, Large-scale ripples of the Lower Wabash River: Sedimentology, v. 23, p. 593-624.

Johnson, H. D., 1977, Sedimentation and water escape structures in some Late Precambrian shallow marine sandstones from Finmark, North Norway: Sedimentology, v. 24, p. 389-412.

Johnson, H. S. Jr., and Dubar, J. R., 1976, Geomorphic elements of the area between the Cape Fear and Peedee Rivers, North and South Carolina, Southeastern Geology, v. 6, p. 37-48.

Keulegan, G. H., 1967, Tidal flow in entrances; water-level fluctuations of basins in communication with seas: U.S. Army Corps of Engineers, Com. on Tidal Hydraulics, Tech. Bull. 14, Washington.

King, D. B., 1974, The dynamics of inlets and bays: Tech. Rept. 22, Coastal and Oceanographic Engrg. Lab., Univ. Fla., Gainesville, 82p.

Klein, G. de V., 1971, A sedimentary model for determining paleotidal range: Geol. Soc. Am. Bull., v. 82, p. 2585-2592.

————, 1972, Determination of paleotidal range in clastic sedimentary rocks: Proc. Int. Geol. Congr., v. 24(6), p. 397-405.

Kugler, H., and Saunders, D., 1959, Occurrence of armored mud balls in Trinidad, West Indies: J. Geol., v. 67, p. 563-565.

Kumar, N., and Sanders, J. E., 1974, Inlet sequence: a vertical succession of sedimentary structures and textures created by the lateral migration of tidal inlets: Sedimentology, v. 21, p. 491-532.

Land, L. S., and Hoyt, J. H., 1966, Sedimentation in a meandering estuary: Sedimentology, v. 6, p. 191-207.

Lattmann, L. H., 1960, Cross section of a floodplain in a moist region of moderate relief, J. Sediment. Petrol., v. 30, p. 275-282.

Leliavsky, S., 1955, An introduction to fluvial hydraulics: Constable and Co., London, 257p.

Leopold, L. B., and Miller, J. P., 1956, Ephemeral streams, hydraulic factors and their relation to the drainage net: U.S. Geol. Survey Prof. Paper 282-B, p. 1-37.

————, and Wolman, M. G., 1957, River channel patterns: Braided, meandering and straight: U.S. Geol. Survey Prof. Paper 282-B, p. 39-85.

————, and ————, 1960, River meanders: Geol. Soc. Am. Bull., v. 71, p. 769-794.

————, ————, and Miller, J. P., 1964, Fluvial processes in geomorphology: Freeman, San Francisco, 522p.

Levey, R. A., 1976, Characteristics of coarse-grained point bars, Upper Congaree River, South Carolina: *in* M. O. Hayes, and T. W. Kana, *eds.*, Terrigenous Clastic Depositional Environments; Am. Assoc. Petrol. Geol. Field Course, Univ. South Car., Tech. Rept. No. 11-CRD, II, p. 38-51.

Lowe, D. R., 1975, Water escape structures in coarse-grained sediments: Sedimentology, v. 22, p. 157-204.

Lüders, K., 1934, Uber das Wandern der Priele: Abhandl. Naturwiss. Ver., Bremen, v. 29, p. 19-32.

Lüneburg, H., 1961, Zur Sedimentverteilung in der Aussenweser Zwischen Hoheweg und Rotersand: Veroeffentl. Inst. Meeresforsch. Bremerhaven, v. 7, p. 1-14.

Mackin, J. H., 1937, Erosional history of the Big Horn Basin, Wyoming: Geol. Soc. Am. Bull., v. 48, p. 813-894.

Mangum, C. P., and Cox, C. D., 1971, Analysis of the feeding response in the Onuphid Polychaete *Diopatra Cuprae* (Bosc.): Biol. Bull., v. 140, p. 215-229.

————, Santos, S. L., and Rhodes, W. R., 1968, Distribution and feeding in the Onuphid Polychaete, *Diopatra Cuprae* (Bosc.): Mar. Biol., v. 2, p. 33-40.

Mayou, T. V., and Howard, J. D., 1969, Recognizing estuarine and tidal creek sandbars by biogenic sedimentary structures (abs.): Am. Assoc. Petrol. Geol. Bull., v. 53, p.731.

McGowen, J. H., and Garner, L. E., 1970, Physiographic features and stratification types of coarse-grained point bars; modern and ancient examples: Sedimentology, v. 14, p. 77-112.

Mota Oliviera, I. B., 1970, Natural flushing ability of tidal inlets: Proc. 12th Conf. Coastal Engineering, p. 1827-1845.

Myers, A. C., 1972, Tube-worm sediment relationships of *Diopatra Cuprae* (Polychaeta: Onuphidae): Mar. Biol., v. 17, p. 350-356.

NEDECO (Netherlands Engineering Consultants), 1952, River studies and recommendations on improvement of Niger and Benue: North Holland, Amsterdam, 1000p.

Nemenyi, P. F., 1946, Discussion of Vanoni, V., Transportation of suspended sediment by water: Trans. Am. Soc. Civ. Engrs., v. 111, p. 116-125.

Noorthoorn van der Kruijff, L. J., and Lagaaij, R., 1960, Displaced faunas from inshore estuarine sediments in the Haringvleit: Geol. Mijnbouw, v. 39, p. 711-723.

Nummedal, D., Oertel, G., Hubbard, D. K., and Hine, A. C., III, 1977, Tidal inlet invariability — Cape Hatteras to Cape Canaveral: *in* Coastal Sediments '77, Am. Soc. Civ. Engrs., in press.

Oomkens, E., 1974, Lithofacies relations in the late Quaternary Niger Delta Complex: Sedimentology, v. 21, p. 195-222.

————, and Terwindt, J. H. J., 1960, Inshore estuarine sediments in the Haringvliet (Netherlands): Geol. Mijnbouw, v. 39, p. 701-710.

Pettijohn, F. J., 1975, Sedimentary Rocks, 3rd Ed.: Harper and Row, New York, 628p.

Postma, H., 1967, Sediment transport and sedimentation in the estuarine environment: *in* G. H. Lauff, *ed.*, Estuaries: Am. Assoc. Adv. Sci., Wash., D.C., p. 158-179.

Potter, P. E., and Pettijohn, F. J., 1963, Paleocurrents and basin analysis: Springer-Verlag, Berlin, 296p.

Prandtl, L., 1952, Essentials of fluid dynamics: Blackie, London-Glasgow, 452p.

Price, W. A., 1963, Patterns of flow and channeling in tidal inlets: J. Sediment. Petrol., v. 33, p. 279-290.

Pryor, W. A., 1967, Biogenic directional features on several recent point bars: Sediment. Geol., v. 1, p. 235-245.

Puigdefabregas, C., 1973, Miocene point-bar deposits in the Ebro Basin, northern Spain: Sedimentology, v. 20, p. 133-140.

Quraishy, M. S., 1944, The origin of curves in rivers: Current Sci., v. 13, p. 36-39.

Redfield, A. C., 1958, The influence of the continental shelf on the tides of the Atlantic coast of the United States: J. Mar. Res., v. 17, p. 432-448.

Reineck, H. E., 1958, Longitudinale Schragschisten im Watt: Geol. Rundschau, v. 47, p. 73-82.

————, 1972, Tidal Flats: *in* J. K. Rigby and W. K. Hamblin, *eds.*, Recognition of Ancient Sedimentary Environments; Soc. Econ. Paleont. Mineral. Spec. Pub. 16, p. 146-159.

————, and Singh, I. B., 1975, Depositional Sedimentary Environments: Springer-Verlag, New York, 439p.

————, and Wunderlich, F., 1968, Classification and origin of flaser and lenticular bedding: Sedimentology, v. 11, p. 99-104.

Schafer, W., 1973, Der Oberrhein, sterbende Landschaft?: Natur. Mus., v. 103, p. 1-29.

Schou, A., 1967, Estuarine research in the Danish Moraine Archipelago: *in* G. H. Lauff, *ed.*, Estuaries; Am. Assoc. Adv. Sci., Wash., D.C., p. 129-145.

Schumm, S. A., 1963, Sinuosity of alluvial rivers n the Great Plains: Geol. Soc. Am. Bull., v. 74, p. 1089-1100.

Selley, R. C., 1968, Nearshore marine and continental sediments of the Sirte Basin, Libya: Quart. J. Geol. Soc. London, v. 124, p. 419-460.

Sundborg, A., 1956, The River Klaralven, a study of fluvial processes: Geogr. Ann., v. 38, p. 125-316.

Tankard, A. J., and Hobday, D. K., 1977, Tide-dominated back-barrier sedimentation, early Ordovician Cape Basin, Cape Peninsula, South Africa: Sediment. Geol., v. 18, p. 135-159.

Terwindt, J. H. J., 1971, Lithofacies of inshore estuarine and tidal-inlet deposits: Geol. Mijnbouw, v. 50, p. 515-526.

Trusheim, F., 1929, Zur Bildungsgeschwindigkeit geschichteter Sedimente im Wattenmeer, besonders solcher mit schrager Parallelschictung: Senckenbergiana, v. 11, p. 47-55.

U.S. Dept. of Commerce, 1977, Tide Tables, East coast of North and South America: Nat. Oceanic and Atmos. Admin., National Ocean Survey, Rockville, Md., 288p.

Van Beek, J. L., and Koster, E. A., 1972, Fluvial and estuarine sediments exposed along the Oude Maas (The Netherlands): Sedimentology, v. 19, p. 237-256.

Van Straaten, L. M. J. U., 1952, Biogene textures and the formation of shell beds in the Dutch Wadden Sea: Koninkl. Ned. Akad. Wetenschap., Proc. Ser. B., v. 55, p. 500-516.

————, 1957, The Holocene Deposits: *in* L. M. J. U. Van Straaten, and J. M. de Jong, *eds.*, The Excavation at Velsen, Konikl. Ned. Geol. Mijnbouw, Genoot., Verhandel. Geol. Ser., v. 17, p. 158-173.

————, 1959, Minor structures of some recent littoral and neritic sediments: Geol. Mijnbouw, v. 21, p. 197-216.

Van Veen, J., 1950, Ebb- and flood-channel systems in the Dutch tidal waters: Tijdschr. Koninkl. Ned. Aardrijkskundig Genootschap (2), v. 67, No. 3, p. 303-325.

Visher, G. S., 1965, Use of vertical profile in environmental reconstruction: Am. Assoc. Petrol. Geol. Bull., v. 49, p. 41-61.

————, 1972, Physical characteristics of fluvial deposits: *in* J. K. Rigby and W. K. Hamblin, *eds.*, Recognition of Ancient Sedimentary Environments: Soc. Econ. Paleont. Mineral. Spec. Publ. 16, p. 84-97.

Ward, L. G., 1978, Hydrodynamics and sedimentological processes in salt marsh systems: Unpub. Ph.D. thesis, Univ. South Carolina, Columbia, S.C.

Wolman, M. G., and Leopold, L. B., 1957, River flood plains. Some observations on their formation: U.S. Geol. Survey Prof. Paper 282-C, p. 87-109.

Wright, M. D., 1959, The formation of crossbedding by a meandering or braided stream: J. Sediment. Petrol., v. 29, p. 610-615.

Wright, L. D., Coleman, J. M., and Thom, B. G., 1975, Sediment transport and deposition in a macrotidal River Channel: Ord. River, Western Australia: *in* L. E. Cronin, *ed.*, Estuarine Research, Academic Press, New York, v. II, p. 309-322.

Wunderlich, F., 1970, Schichbanke: *in* H. E. Reineck, *ed.*, Das Watt, Ablagerungs- und Lebensraum, Frankfurt, Kramer, p. 48-55.

Yalin, M. S., 1972, Mechanics of sediment transport: Oxford, Pergamon Press, 290p.

TRANSVERSE RIBS: THEIR CHARACTERISTICS, ORIGIN AND PALEOHYDRAULIC SIGNIFICANCE

EMLYN H. KOSTER[1]

ABSTRACT

Transverse ribs — a gravel bedform in shallow, high-energy fluvial systems — are a series of narrow, current-normally orientated accumulations of large clasts. On Spring Creek alluvial fan, Yukon, they are common and occur in braided channels and on longitudinal bars. Aspects of their occurrence indicate that rib sequences form very rapidly (via abrupt stages) upon waning flow over surfaces that are often immobile.

Previous flume experiments showed that ribs can form beneath an upstream-migrating hydraulic jump. On the grounds that super-critical flow is rare in Spring Creek, and prompted by observations in a large flume, this study supports and extends the viewpoint that transverse ribs are relict antidune bedforms. This alternative hypothesis receives support from the nature of bivariate relationships of rib wavelength versus rib width and coarseness of rib material, as well as from a successful prediction of paleoflow parameters, using data from Spring Creek and another study. Paleovelocity is calculated on the basis of antidune wavelength and an equation for competence. Two methods, based on clast size versus shear stress relations, are selected for estimating paleodepth: the results indicate that large error arises if reconstruction is attempted for flow over abnormally low-sloped sections of the alluvial surface. The analysis concludes that a large majority of rib-forming flows are sub-critical, and it is suggested that the development of kinematic waves assists rib formation.

The preservation potential and/or ease of recognition of rib sequences in ancient alluvium appears to be low. However, because direct hydrologic investigation in rib-forming environments during high-stage flow is hampered by practical difficulties, ribs assume considerable importance by supplying this information indirectly, with reasonable accuracy.

INTRODUCTION

The considerable recent rise in interest in proximal braided river sedimentation has been largely directed to an increasing knowledge of facies (McDonald and Banerjee, 1971; Rust, 1972a; Gustavson, 1974), bar evolution (Church, 1972; Fahnestock, 1973; Bluck, 1974; Smith, 1974; Hein and Walker, 1977) and clast fabric (Rust, 1972b, 1975; Koster, 1977a). As a result of this kind of work, a variety of minor (i.e. within-channel and -bar) structures have been recognized. These include 'transverse ribs' (McDonald and Banerjee, 1971; Koster, 1977b) also called 'clast stripes' by Boothroyd (1972) 'Ostler lenses' (Martini and Ostler, 1973), 'stone cells' (Gustavson, 1974), 'gravel mounds' (Shaw and Kellerhals, 1977) and 'gravel lenses' (Foley, 1977). Preserved equivalents of these presumably high stage bedforms in stratigraphic successions have useful potential as indicators of the paleoflow regime, although such reconstructions do not apply to the entire flow for braided systems.

McDonald and Banerjee (1971, p. 1290) defined transverse ribs as "a series of regularly spaced pebble, cobble or boulder ridges orientated transverse to flow." This paper describes ribs from the alluvial fan of Spring Creek, Yukon (Fig. 1), considers their origin in the light of field characteristics and some flume observations, and outlines methods for their paleohydraulic reconstruction.

Since their initial description by McDonald and Banerjee (1971) from proglacial outwash in Alberta, transverse ribs have also been reported from fans on the Malaspina Glacier foreland (Boothroyd, 1972; Boothroyd and Ashley, 1975; Gustavson, 1974) and from two small streams in south-eastern Canada (Laronne and Carson, 1976; Martini, 1977). McDonald and Day (1978, p. 441) report that they have also been found in Himalayan streams, on alluvial fans in Nevada and in numerous locations in the Canadian

---

[1]Department of Geological Sciences, University of Saskatchewan, Saskatoon, Sask. S7N 0W0

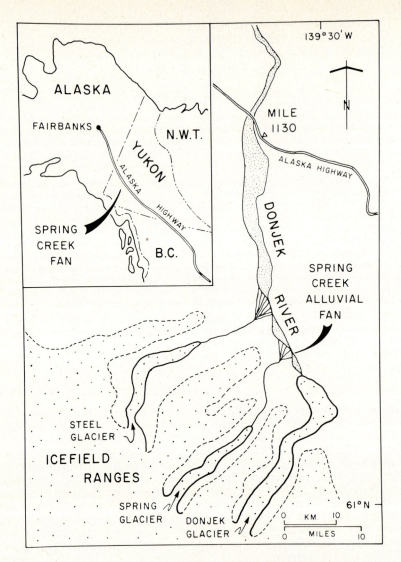

Fig. 1. Location of study area in south-west Yukon. Shaded area at lower left represents icefields and glaciers; stippled zone outlines the proximal valley sandur of the Donjek River (Rust, 1972a; Williams and Rust, 1969).

and Colorado Rockies: the present author has also observed them in a tributary of the Susquehanna River, New York State. Ribs are therefore widely distributed but were not formally recognized prior to the work of McDonald and Banerjee (1971). Gustavson (1974, p. 379) considers that ribs "occur on the gravel reaches of many clear-water streams, but simply have not been recognized because there is no sand in the inter-rib areas to emphasize the ribs."

Table 1 lists the features which rib-forming environments appear to have in common. Table 2 indicates that the majority of these features are displayed by the alluvial fan of Spring Creek.

Table 1. Features common to rib-forming environments.

Environment	Mountainous relief with moderate-steep channel slopes. Climate may be paraglacial (Church and Ryder, 1972), humid-temperature or hot-arid.
Channel characteristics	Either braided with longitudinal bars or low sinuosity, single channels: high width-depth ratios.
Flow characteristics	Abrupt attainment of high energy conditions during which flows have high velocity (ie. 1 - 2 m sec[-1]) but fairly shallow depth (ie. approx. 1 m): high competence and capacity for sediment transport.
Bedload	Dominated by poorly-sorted pebble-, cobble-, or boulder-gravel, giving rise to high clast size to depth ratios.

Table 2. Data for Spring Creek alluvial fan.

Fan geomorphology	Paraglacial; glacierized basin of 470 km², second-order stream, 37 km long. Maximum radius 3.7 km; average slope 0.0019.
Flow characteristics	Diurnal and seasonal discharge variation. At fanhead (based on 1972 gauging): — bankfull discharge 30 m³ sec[-1] ; max. surface velocity 3.6 m sec[-1] , max. deduced bankfull depth 1.6 m.
Sedimentary features	Distributary system of braided channels with longitudinal bars. Dominated by poorly-sorted, clast-supported bouldery-gravel with maximum sizes grading downfan from -10.5ϕ to -7.5ϕ.

DESCRIPTIVE PARAMETERS

McDonald and Banerjee (1971, p. 1290) proposed four properties for the description of transverse ribs; namely, local bed-slope, "average spacing between ribs . . . average widths of ribs . . . and average b-axis of the ten largest stones in ribs." On Spring Creek, the ten largest clasts in each rib were measured, rather than for the entire sequence. The term 'rib spacing' as defined by McDonald and Banerjee (1971, Figs. 10b and 12) is equivalent to what is conventionally termed the wavelength of a bedform, since it includes rib width as well as the space between ribs. On Spring Creek, inter-rib spacings were measured, so that mean wavelength equals mean spacing plus mean width. Gustavson (1974, p. 379) also measured "mean inter-rib widths" as spacing, while Church and Gilbert (1975, p. 57) have interpreted Boothroyd and Ashley's (1975) use of the term spacing as synonymous with wavelength. To avoid this confusion, it is recommended that future studies of transverse ribs adopt the parameters illustrated in Figure 2.

FIELD CHARACTERISTICS

Ribs are common on Spring Creek fan, both in abandoned channels (Fig. 3) and on inclined lateral surfaces of longitudinal bars (Fig. 4). Usually they contain the coarsest grades of the locally available bed material. Ribs on Spring Creek display some characteristics which have not been reported by the previous studies: Table 3 catalogues the field characteristics of ribs, and constitutes a supplement to the list of features provided by McDonald and Day (1978, p. 442). There appears to be general agreement that a majority of the listed properties hold genetic significance.

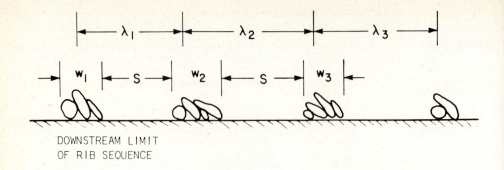

DOWNSTREAM LIMIT
OF RIB SEQUENCE

NOTATION	DEFINITION	CALCULATION
$\overline{\lambda}_b$	AVERAGE DOWNSLOPE DISTANCE BETWEEN ADJACENT RIB CRESTS	$\dfrac{\Sigma \lambda_i}{j}$
$\overline{b}_{max}$	AVERAGE b-AXIS LENGTH OF TEN LARGEST CLASTS COMPRISING EACH RIB	$\dfrac{\Sigma \left(\dfrac{\Sigma b_{max_i}}{10} \right)}{j}$
$\overline{W}$	AVERAGE WIDTH OF RIBS	$\dfrac{\Sigma W_i}{j}$
L	TOTAL LENGTH OF RIB SEQUENCE	$\Sigma \lambda_i$
S_b	AVERAGE SLOPE OF RIB SEQUENCE	–

Fig. 2. Parameters for description of transverse rib sequences.

Table 4 summarizes the Spring Creek rib data, and distinguishes between ribs in major, 'first-order' channels and those on large, 'first-order' bars (Williams and Rust, 1969, Fig. 2b; Rust, this volume). Topographically lower, within-channel rib sequences are significantly longer, coarser-grained and of larger scale than those occurring on bars. Average bed-slope is marginally steeper for the bar category.

IMPLICATIONS FOR THE NATURE OF BEDLOAD TRANSPORT

Analysis-of-variance tests performed on the size measurements of the rib sequences on Spring Creek indicate that, with two exceptions, there is no significant between-rib difference in the coarseness of successive ribs. This is interpreted as indicating that the competence level was essentially constant during the formation of a sequence.

Fig. 3. Rib sequence #1 (see Table 4). Upstream view of a channel reach containing 8 fairly well-defined ribs. The rib sequence is partly concealed along the right-bank by wind-blown sediment and material collapsed from the bank. Evidence is present to suggest that this rib sequence is being preserved in the stratigraphic succession of the alluvial fan deposits: immediately to the right of the shovel (70 m long) less coarse bar sediment has (?)prograded laterally towards the right-bank over part of the rib sequence.

Fig. 4. Rib sequence #5 (see Table 4). Series of poorly-defined pebbly ribs on an inclined bar surface. Flow was from right to left. Laterally, away from the staff (4.5 ft [1.37 m] long) the ribs bifurcate and become indistinct, and eventually display transition to an unribbed surface.

Table 3.   Field characteristics of transverse ribs.

Scale of field observation	Features
A) Individual rib	**a)** Current-normal orientation.
	b) Clasts display prominent imbrication and current-normal orientation (Fig. 6; McDonald and Banerjee, 1971, Fig. 10a).
	c) Larger clasts tend to be located on the downstream rib margin*.
B) Individual rib sequence	**a)** Length from 2 to 19 m*, with number of ribs ranging from 2 to 12*, although bar surface sequences may contain considerably more (Boothroyd, 1972, Fig. 29).
	b) Wavelengths generally uniform (upstream decrease, if any, is accompanied by a slight reduction in clast size*).
	c) Rib widths fairly uniform.
	d) Occasional transition laterally to a flat bed (Boothroyd and Ashley, 1975, Fig. 15b) and/or to a 'cellular structure' (Boothroyd, 1972), and/or to 'stone cells' (Gustavson, 1974).
	e) Occasional bifurcation of ribs onto sloping channel margins, and/or vague continuation onto adjacent bar surfaces*.
	f) Tendency for gentle convexity downstream at points of channel widening (McDonald, pers. comm.).
	g) Original inter-rib, and sub-rib, material either slightly cohesive (McDonald and Banerjee, 1971, Fig. 8), or clast-supported, generally cobble-gravel*.
C) Alluvial surface	Positive linear interrelationships between clast size and rib wavelength, and rib wavelength and width, as first shown by McDonald and Banerjee (1971).

* denotes evidence from Spring Creek fan

Although rib formation was not observed in Spring Creek, timing of rib growth relative to the daily meltwater hydrograph can be deduced. Because within-channel ribs contain the coarsest percentiles of the bedload and ribs on bars occupy the topographically higher parts of the alluvial surface, near peak flow conditions are obviously required for their formation. It is therefore considered that rib growth commences at or shortly after the hydrograph peak and is terminated when the flow wanes to a level that is incompetent for bedload transport, and/or by upstream migration of the rib sequence into an unfavourable environment. However, two lines of evidence presented below suggest that rib growth occupies a short interval of the recession limb on the hydrograph.

Firstly, stream gauging in June 1972 revealed that during falling stages, which typically last 14 hours, discharge decreased four-fold. According to at-a-station hydraulic geometry determined at the fanhead, this involves a two-fold decrease in velocity. Using the so-called "2.6 power law" (Fig. 18 of Church and Gilbert, 1975) this corresponds to a 40% decrease in the size of bed material capable of being entrained. Thus a time-span for rib formation much less that of the recession limb is strongly implied.

Fig. 5. A well-defined rib in sequence #1 (see Fig. 3). Flow was from the upper left, parallel to the staff. Note the current-normal alignment of the rib, as well as of the component clasts, and the imbrication of oblate and bladed cobbles on its upstream side. Later deposition of sand with current crescents in the inter-rib areas serves to emphasize the bedform.

Fig. 6. Upstream view of an abandoned channel (containing rib sequence #4 — see Table 4) on Spring Creek fan which is compared to a reach of the Middle River, Baffin Island (Church, 1972, Fig. 38 [Section M2]) in the text. Note the good definition and coarseness (b-axis of the largest boulder is 47 cm) of the ribs. As the channel shallows toward the left-bank, the ribs show transition to a cellular pattern. Recent high-stage flow has partly dissected the inter-rib sediment. Shovel is 70 cm long.

Table 4.   Data of transverse rib sequences on Spring Creek fan.

Sub-Environment	Sequence	S_b	L (m)	λ_b (cm)	W (cm)	b_{max} (cm)	Range of b_{max} (cm)
Within channels	1	.0320	18.5	104	46.5	18.8	11 - 30
	2	.0203	17.0	190	90.5	22.3	14 - 36
	3	.0349	5.4	116	41.0	19.3	13 - 32
	4	.0233	18.8	249	110.5	30.9	17 - 46
	7	.0116	9.5	158	83.5	24.4	17 - 49
	8	.0204	4.1	166	77.0	26.9	16 - 39
Averages		.0238	12.2	147	75.0	23.8	15 - 39
On bars	5	.0291	4.5	113	67.5	14.3	9 - 20
	6	.0087	6.0	169	80.5	22.6	14 - 40
	9	.0291	3.1	132	49.0	17.8	12 - 26
	10	.0495	1.9	64	33.0	12.8	9 - 18
	11	.0146	5.1	150	57.0	19.8	14 - 35
	12	.0408	1.6	36	15.5	2.6	1.7 - 4.2
Averages		.0286	3.7	111	50.5	15.0	10 - 27

Note: Underlining of averages denotes the sub-environment with the greater value of the parameter.

Secondly, the channel containing sequence #4 (Fig. 6) has similar cross-sectional size and shape to that of the Middle River at Ekalugad Fjord (Fig. 38 of Church, 1972). Using the Meyer-Peter and Müller (1948) formula, Church computed a bedload discharge of 9.7 m^3 sec^{-1} across a bed width of 20 m. If this transport rate is applied to the 6 m wide ribbed reach containing sequence #4, the calculations (based on the 10 largest clasts in six well-defined ribs: $\bar{b}_{max}$ = 30.9 cm) suggest that the rib sequence formed in a matter of a few seconds. Although this estimate undoubtedly involves some error, the result suggests that rib sequences form very rapidly.

Inter-rib areas are frequently mantled with planar and/or cross-laminated sand during low-stage flow following rib information (Fig. 6). The nature of the 'sub-rib' material is highly relevant to the hydraulic and sedimentologic significance of transverse ribs. McDonald and Banerjee (1971, p. 1290, Fig. 8) noted that pebbly ribs lay on weakly cohesive silty sand. Ribs on Spring Creek rest ubiquitously on imbricated, clast-supported gravel. Inter-rib areas of coarse within-channel ribs generally lack clasts of similar size to those comprising the ribs. This indicates that rib formation proceeds in abrupt stages with almost instantaneous shifts in loci of deposition. Conversely, pebbly ribs on bar surfaces have sub- and inter-rib material which are lacking in grain size contrast. Transverse ribs of this sort may represent the redistribution of gravel from a flat bed condition: this was the sequence of events in the flume experiments of Shaw and Kellerhals (1977).

Evidence for two causes of transverse ribs is discussed below.

Hydraulic Jump

Flume experiments reported by McDonald and Day (1978) showed that transverse ribs can be formed by upstream migration of a hydraulic jump. They observed clasts being deposited in rib-like fashion in the zone where super-critical flow changes abruptly to sub-critical flow. Individual ribs formed in seconds, after which ribs further downstream become "fossilized" in the less competent sub-critical flow. Formed in this way, the morphology of a rib sequence reflects the dimensions of a hydraulic jump, while downstream a train of standing waves is stabilized over the inactive part of the sequence.

On Spring Creek fan, rib sequences were located in inactive vegetated channels, as well as in abandoned channels and bar surfaces exposed as recently as the day before. Prolonged observation of water surfaces during the 1972 meltwater season failed to locate hydraulic jumps. This may be due to spill resistance (Leopold *et al.*, 1964), created by chaotic deformation of flow around clasts in transport which involves sufficient energy loss to inhibit major increases in Froude number. Church (1972, p. 72) found that super-critical flow was approached at peak runoff at only two sections in Baffin Island sandurs. On the White River, Fahnestock (1963, p. 18) found that "Froude numbers for single channels containing the entire flow of the river, and with few exceptions for smaller anabranches with discharges larger than 10 ft^3. sec^{-1} [0.3 m^3 sec^{-1}], ranged from 0.5 - 0.9". However, from neither of these outwash environments were rib-like bedforms described. Simons and Richardson (1960, p. 45) consider that "rarely does a Froude number, based on average velocity and depth, exceed unity for any extended time period in a stream with erodible banks."

Antidunes

The rarity in rib-forming environments, such as Spring Creek, of super-critical flow with upstream-migrating hydraulic jumps points to an alternative cause for the formation of transverse ribs.

Sub-critical antidune flow has been observed to produce rib-like bedforms in the flume described by McDonald (1972). T. J. Day (pers. comm.), Judd and Peterson (1969), Shaw and Kellerhals (1977) and Foley (1977) have since reported similar findings. In the author's (Koster, 1977a) experiments, mean depth and velocity were 10.3 cm and 1.99 m sec^{-1}, over a flume floor roughened by a mat with a coarse granular (-1 to +1ϕ) surface, which effectively retarded the flow near the bed. 122 inequidimensional clasts, with b-axes between 3.3 and 13.4 cm, were introduced singly at 5 second intervals at the flume's head in random orientation. Deposition at the downstream end was induced by a low, coarse-mesh barrier which occupied 0.8 of the flow depth. Upstream of this barrier, the bed developed a short sequence of imperfect ribs, which were observed to lie beneath the crests of quasi-stable antidune waves. Abrupt upstream migration of the surface waveform occurred when the width of the rib undergoing formation became too great. A new antidune was then stabilized as less transportable clasts became stationary in the lower velocity crest zone, which then acted as a locus for deposition of ensuing clasts. Clasts comprising each rib showed 'contact imbrication' (Laming, 1966) and current-normal a-axis orientation. Because the clasts were numbered and entered in numerical order, it was possible to compare the entry order with the order of deposition in the rib sequence. The majority of clasts were initial rib material, but some smaller clasts were subsequent additions to rib material after passing the rib(s) already formed upstream.

This study therefore supports and extends the viewpoints of Boothroyd (1972) and Gustavson (1974) that transverse ribs are relict antidune bedforms. The development of standing and upstream-breaking waves is common in proximal reaches of braided systems; most are located in riffles, but at high stages they are also fairly abundant in pools and across lateral surfaces of longitudinal bars. McDonald and Banerjee (1971, p. 1288) found that the flow condition of the "largest active channels in the late afternoon" on Peyto outwash was standing waves.

Gustavson (1974) distinguished between three morphological types of upstream-breaking antidunes in streams draining Malaspina Glacier: conical 'rooster tails' (Kennedy, 1961), ones with considerable lateral extent approaching channel width, and a reticulate or 'checkerboard' pattern. Gustavson postulated that transverse ribs and stone cells are developed by linear and reticulate antidunes, respectively. Shaw and Kellerhals (1977) envisaged that the narrow, rib-like gravel mounds they observed in overbank areas of the North Saskatchewan river were formed by the 'rooster-tail' variety.

Downstream from an obstruction, like a large boulder, undulatory water waves are commonly present; wave amplitude declines uniformly downstream from the point of propagation. Elsewhere in shallow, braided-river environments undulatory water waves are presumably associated with waves (or 'ribs') on the bed surface, thereby making them antidunes in the sense proposed by Kennedy (1961), ie. bedforms that are in-phase with surface waveforms.

Coarse-grained antidune deposits may be composed of: —

  (i) Sand or pebbly sand over which an isolate bedload of cobbles and/or boulders is transported and temporarily deposited beneath antidune crests (Fahnestock and Haushild, 1962). Reversion to a plane bed during falling stage leaves no evidence of former antidune flows.

  (ii) Cobble or boulder bedloads in which individual clasts arrive singly at the site of potential rib formation. Because the largest percentiles of the available bedload are involved, lower stage flows are incapable of modification, and the evidence of former antidune flows remains.

Discussion

The similarities between the hydraulic jump and antidune hypotheses for the formation of transverse ribs are as follows: —

  (i) growth of the ribbed sequence proceeds upstream in response to a water disturbance migrating upstream in discrete steps, and

  (ii) once a rib has formed, those 'fossilized' downstream stabilize a standing wave train with wavelength and amplitude governed by the spacing and height of the in-phase ribs.

The differences are: —

  (i) flow is not necessarily super-critical in the antidune hypothesis, and

  (ii) the two models differ in the mechanism which initiates the formation of the first, downstream rib.

The latter difference warrants attention and two possible triggering mechanisms are proposed. In the first, chance deposition of a larger, less transportable clast causes ensuing clasts in a similar trajectory to be deposited with contact imbrication. Such a combination of features was observed in the laboratory and inferred from the coarser, within-channel ribs of Spring Creek. Initially, the waveform would probably be of the 'rooster tail' variety, but lateral propagation subsequently took place over flat-bottomed channels as more clasts became stationary in this lower velocity portion of the flow.

Langbein and Leopold (1968) showed that incohesive bedloads (without size limitation) subject to unidirectional flow frequently develop 'kinematic waves' (Lighthill and Whitham, 1955), and proposed that they are manifested by bedforms on a wide variety of scales. In an experiment on sediment transport, they observed (p. 15) that "as the rate of feed was increased, the size of groups increased, and the average speed decreased so that ultimately a point was reached where a group became so large that a jam formed and all motion halted." Bowman (1977) postulated that the cyclic distribution of boulder accumulations observed on ephemeral stream beds in Israel resulted from the imposition of kinematic waves on bedload transport, and likened his observations to those of Scott and Gravlee (1968) who described 'boulder fronts' resulting from a catastrophic flood. It is possible that the development of kinematic waves during bedload transport could abruptly trigger formation of the initial transverse rib. Thereafter, a standing wave would be anchored over the first-formed rib and ensuing clasts would be influenced by dynamic waves — in this case, antidunes — which are principally governed in their wavelength by mean flow velocity (Kennedy, 1961).

INTER-RELATIONSHIPS OF RIB PARAMETERS

On the assumption that ribs are formed in association with antidunes, use of least-squares regression for relating rib parameters is invalid, since coarseness of rib material and rib wavelength are independent functions of flow properties, principally mean velocity. The absence of clearly defined dependent and independent variables causes rib data to be best related by means of reduced major axes (Till, 1974). Data collected by McDonald and Banerjee (1971, Table 3) were re-calculated using this alternative method.

Rib width versus wavelength

Figure 7 shows a strong tendency for average width of ribs to be slightly less than half of the mean wavelength. Intercept values of reduced major axes relating these parameters cluster about zero. Data from Spring Creek and the North Saskatchewan River both generate gradient values of 0.48 and very high correlation coefficients. The data from Petyo outwash are only moderately correlated and have a somewhat lower slope. The combined data yield a reduced major axis having the form[1]

$$\overline{W} \simeq 0.47\overline{\lambda}, \qquad r = 0.966.$$

A rib width to wavelength ratio of about 0.5 is predictable assuming an antidune origin. On sandbeds, antidunes characteristically produce a sinusoidal profile (Allen, 1970; p. 81), and trough regions are therefore smoothly transitional to the crest regions. Transverse ribs, however, are generally conspicuous for the sharp definition of rib (ie. crest) and inter-rib (ie. trough) regions, and rib width is a practical measure of the bedform. If antidune flow over coarse gravelly beds approaches a sinusoidal waveform, then the high relief crest region would be, on average, slightly less than one half the wavelength.

In a particular sequence of ribs, values of width deviate about the mean value, but in each wavelength a relatively large width is compensated by a relatively narrow inter-rib area (see Fig. 10b, McDonald and Banerjee, 1971).

Clast size versus wavelength

McDonald and Banerjee (1971, p. 1291) noted that bivariate relationships amongst rib parameters "are better defined where data from different localities are considered separately." A summary (Table 5) of their data and that from Spring Creek shows that confidence intervals for slopes and intercepts of the best-fit equations are non-overlapping (Fig. 8). A method of statistical discrimination (Imbrie, 1956) substantiates that, with the exception of lines representing Spring Creek and Peyto outwash, locations generate significantly different relationships. Correlations are highly significant in spite of small sample sizes. Intercept values are evidently distributed about zero, giving a simple equation of the form

$$\overline{b}_{max} = m\,\overline{\lambda}$$

where, based on four locations thus far studied, the coefficient m varies between 0.075 and 0.165.

---

[1]See Appendix for notations

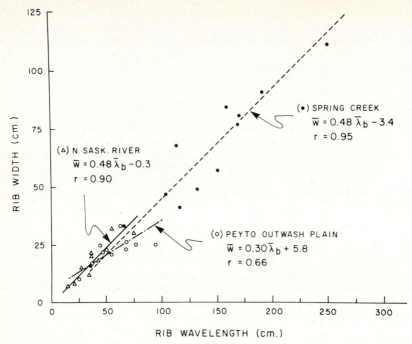

Fig. 7. Rib width versus wavelength relations for Spring Creek and two proglacial outwash environments in Alberta (data from McDonald and Banerjee, 1971, Table 3).

Church and Gilbert (1975, p. 57) analysed the combined data of McDonald and Banerjee (1971) and Boothroyd and Ashley (1975) and found a m-value of 0.15. Using this value in Kennedy's (1961) relation

$$\bar{u}^2 = \frac{g\,\lambda}{2\,\pi}, \tag{1}$$

and assuming ribs to have an antidune origin under critical flow, they showed that maximum clast sizes approximate flow depth in ribbed reaches. In the field, McDonald and Banerjee (1971) observed this to be the case, but it is possible that ribs may have already formed prior to their observation.

RECONSTRUCTION OF PALEOFLOWS

For paleohydraulic determinations, ribs are more effectively analysed using average values of wavelength and clast size for the sequence, rather than individual rib data. This assumes that flow conditions were essentially constant during the period of rib formation; the earlier discussion indicates that little error arises from this assumption.

Paleovelocity

Accepting the antidune origin of transverse ribs, there are two independent methods for estimating mean paleovelocity for flow over ribs.

The first utilizes equation 1 (Kennedy, 1961) relating antidune wavelength to mean flow velocity. Student's t-tests on the experimental data of Shaw and Kellerhals (1977, Table 1) confirm the equivalence of wavelengths on the bed and water surfaces. Kennedy's

TABLE 5. Bivariate relationships between rib wavelength and clast size for the various locations studied.

Location	Reference	Sample size	Reduced major axis equation	*Correlation coefficient	95% confidence intervals intercept	coefficient
Spring Creek	This study	12	$\overline{b}_{max} = 0.129\overline{\lambda}_b + 1.73$	0.921	-2.49 to 5.95	0.10 to 0.16
Peyto outwash		11	$\overline{b}_{max} = 0.165\overline{\lambda}_b - 0.90$	0.866	-3.82 to 2.02	0.12 to 0.21
North Sask River	McDonald and Banerjee (1971, Table 3)	9	$\overline{b}_{max} = 0.075\overline{\lambda}_b + 1.20$	0.952	0.57 to 1.83	0.06 to 0.09
No-See-Um Creek		2	$\overline{b}_{max} = 0.150\overline{\lambda}_b - 1.20$	—	—	—
Malaspina foreland	Gustavson (1974, Fig. 10)	10	$**\overline{a}_{max} = 0.130\overline{\lambda}_b$	—	—	—
Malaspina foreland	Boothroyd and Ashley (1975, Fig. 16)	11	$**\overline{a}_{max} = 0.224\overline{\lambda}_b + 3.80$	—	—	—

* Using the method of Till (1974, p. 87), the correlation coefficients are highly significant with the probability that the samples stem from populations with zero correlation being <0.005.

** Relations deduced from authors' figures.

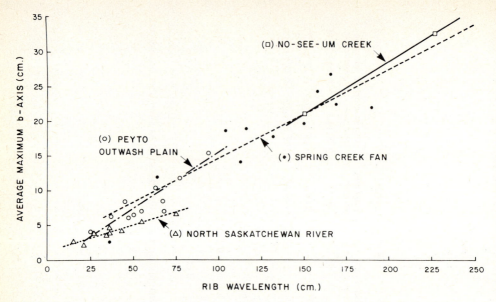

Fig. 8. Rib wavelength versus clast size: see upper part of Table 5 for statistics of the bivariate relationships.

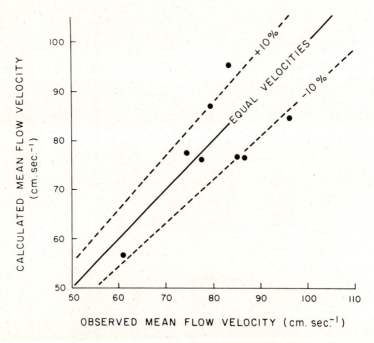

Fig. 9. A comparison of measured velocities with those calculated on the basis of antidune wavelength (equation 1; Kennedy, 1961), using the experimental data of Shaw and Kellerhals (1977, Table 1).

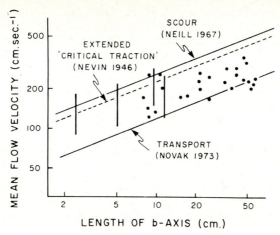

Fig. 10. Results of various studies to predict movement of coarse gravel by shallow riverflows. Dots show transport data of Fahnestock (1961, Fig. 87.1); vertical lines represent the range of entrainment velocities calculated by Boothroyd (1972) using Helley's (1969) method.

equation assumes that the antidune wave train is stationary, two-dimensional and perfectly sinusoidal. Parker (1975) developed equations for antidune flow which have the form $Fr = f(k\overline{Y})$, and Shaw and Kellerhals (1977) concluded that their laboratory antidune flows were most accurately described by an equation of this type. However, the maximum error is $\pm14\%$ when the same data are expressed by Kennedy's equation (Fig. 9). Parker's (1975) equations appear to be of less practical value in paleohydraulic analysis because depth and wavelength must be known independently; alternatively, paleovelocity can be estimated by assuming a Froude number for antidune formation (Shaw and Kellerhals, 1977; p. 264).

A second independent method for estimating paleovelocity is based on a competence relation. As concluded earlier, ribs are formed when waning flow becomes marginally competent to maintain transport of the coarse fraction. Estimating velocity from maximum bedload sizes should therefore use an equation for depositional, rather than erosional velocities. Diagrams by Hjulstrom (1935) and Sundborg (1967) suggest that movement of gravel-sized material can be maintained by velocities below those required for entrainment. There is growing evidence that coarse-particle transport follows the so-called "2.6 power law" (Nevin, 1946; Fahnestock, 1961; and Church and Gilbert, 1975); however the range of values for the coefficient of the power relation is large (Fig. 10). The specific equation (in cm) used for the present study is that calculated from metric equivalents of the data compiled by Novak (1973, Table 1),

$$\overline{U} = 49 \ d^{0.381}, \ r = 0.825 \ (n = 112).\tag{2}$$

In comparison with other studies, it appears to predict the minimum velocities necessary for transport. In general, the data compiled by Novak describe particle size in terms of b-axes, so that $\overline{b}_{max}$ can be substituted in rib studies. Use of equation 2 to estimate paleovelocity assumes that transport has occurred in fluvial systems which are 'graded' (as defined by Mackin, 1948), ie. that the largest available grades of bedload are in equilibrium with the prevailing peak flow velocities. This is not always the case; for example, the 1965 Bijou Creek flood transported bridge girders over a sandbed (McKee *et al.*, 1967). Application of equation 2 to ungraded paleochannels will result in under-estimation of paleovelocity. A second assumption is that velocity is a major variable affecting competence. If this were not so, poorly-defined relationships would prevail between entraining or transporting velocities and particle size: however, Fahnestock's (1961) study for example shows high correlation. Also, Keller (1970) concluded that about 70% of the variability in transport behaviour could be explained in terms of paticle diameters and effective flow velocities. This finding raises another important point in competence studies. Briggs and Middleton (1965; p. 14) stressed that transport studies should take into account particle size in relation to the vertical velocity profile. In proximal outwash systems, size to depth ratios are typically large and flows are extremely turbulent: these factors probably result in flow velocities ambient to clast motion being close to the mean value of the profile.

   Paleovelocity estimates based on equations 1 and 2 correspond surprisingly well (Table 6). Student's t-tests confirm that for Spring Creek and the North Saskatchewan River, and for No-See-um Creek by visual inspection, there are no significant differences at the 95% level between the two sets of independent velocity determinations. For Peyto outwash, equation 2 yields paleovelocity values which, on average, are 17% greater than those calculated using equation 1; reasons for this discrepancy are unclear.

   In figure 11, equation 1 has been used to place a broadly equivalent scale of mean flow velocity beneath that of wavelength. Using the velocity axis, the line representing equation 2 is parallel to, and lies in the vicinity of, the data scatter. M. R. Stauffer (pers. comm., 1977) has pointed out the effect of different measures of clast size in determining the relative positions of the competence relationship and the data scatter.

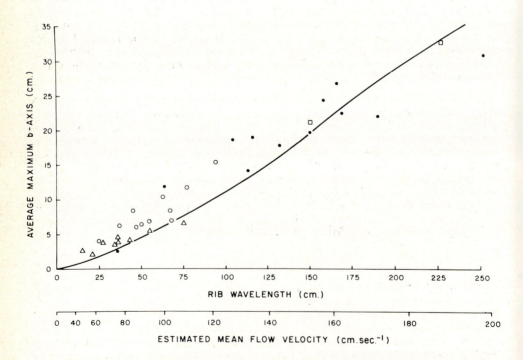

Fig. 11. Same scatter diagram as in Fig. 8, but with a relation for competence superimposed, as explained in the text. The line represents transport velocities, and is slightly sigmoidal on account of the modified abscissal scale, the exponential nature of the clast size versus velocity relation and the fact that the graph coordinates are linear.

Paleodepth

The literature contains various methods for estimating depth, but all are based on DuBoys' (1879) expression for boundary shear stress: —

$$\tau = \rho_f g R S_e.$$ (3)

In paleohydraulic work, R and S_e are normally exchanged for $\overline{Y}$ and S_b on the assumption that the paleochannels were wide and shallow and that the paleoflows were uniform. Width-depth ratios of braided channels are highly variable (see Fig. 42 in Church and Gilbert, 1975). For rectangular channels with width-depth ratios between 20 and 40 (20% on Baffin Island sandurs; 60% on White River), mean depth may exceed the hydraulic radius by a factor of 10-15%. Thus, under-estimates of depth can result from this assumption. The effect of non-uniformity of flow on paleodepth estimates is discussed later. The common practice has been to insert the standard value for ρ_f (ie. 1.00 g. cm⁻³) in equation 3. Ostrem (1975) found suspended sediment concentrations of up to 12 g. l⁻¹ in Norwegian proglacial streams, and Bryan (1972) found densities of Slims River (Yukon) water approaching 1.0035 g. cm⁻³. Although the differences in computed shear stress are slight, it seems appropriate to recognize that proglacial rivers contain water with specific weights of about 1.005 g. cm⁻³.

Five methods for depth estimation (equations 4 to 8) are described below and subsequently assessed for their relevance to shallow, gravelly, braided rivers.

Clague (1975) considered that the particle size - shear stress relation determined by Church (1970) from Baffin Island Sandurs, ie.

$$\tau_c = 1800 \ d \ \text{(mks units)},$$

was applicable to some Late Wisconsian outwash, and combining this relation with equation 3, proposed the following equation for estimating depth: —

$$\overline{Y}_{est} = \frac{0.184 \ d}{S_e} \quad \text{(mks units)}$$ (4)

Baker (1974) used a similar equation, ie.

$$\overline{Y}_{est} = \frac{0.099 \ d}{S_e} \quad \text{(mks units)},$$ (5)

for his paleohydraulic interpretation of Quaternary alluvium. This was derived by combining equation 3 with Shields' (1936) dimensionless shear stress parameter, β, which for uniform flow acting on homogenous bed material coarser than 7 mm is generally regarded to equal 0.06.

Novak (1973, Fig. 2) showed that tractive force and mean flow velocity are related with high correlation by the following equation: —

$$\overline{U} = 0.4\tau^{0.333} \quad \text{(mks units)}$$

Combining this with equation 3 gives

$$\overline{Y}_{est} = [\frac{\overline{U}}{8.32 \ S_e^{0.333}}]^3 \quad \text{(mks units)}$$ (6)

The fourth method is based on a modified version of equation 5. The data of Lane (1955) and Fahnestock (1963) for shallow gravel-bed channels plotted on a modified (as proposed by Bagnold, 1966) Shields' diagram, as done by Baker (1973, Fig. 16), are fitted by a negatively-sloped line with the equation

$$\beta = 0.085d^{-0.333}.$$

When this is substituted into the expression for β, ie.

$$\beta = \frac{\tau_c}{(\gamma_s - \gamma_f)d}$$

and equation 3 is inserted, the following expression is obtained: —

$$\overline{Y}_{est} \; (cm) \; = \; \frac{0.15 \; d^{0.67}}{S_e} \; , \tag{7}$$

where ρ_s has been assigned the value of 2.75 g. cm^{-3}.

The final method is based on the work of Baker and Ritter (1975). They compiled the available data on tractive force for particles coarser than 2 cm, and derived a best-fit equation, which in mks units (wherein τ is in newtons m^{-2}) takes the form

$$d = 0.007\tau^{0.54}, \; r = 0.92.$$

This, combined with equation 3 leads to the following expression: —

$$\overline{Y}_{est} \; = \; \frac{1.1 \; d^{1.852}}{S_e} \; (mks \; units). \tag{8}$$

It is probable that the various estimates of depth, for a given set of conditions, are normally-distributed. Thus it is better to take an average of theoretically acceptable methods. Equations 4, 5 and 8 are considered unsuitable for the present purposes for the following reasons. Figure 12 shows that, for a slope of 0.025, equations 4 and 5 indicate clast size to depth ratios of 0.25, and 0.14, respectively: the evidence (eg. Church and Gilbert, 1975; p. 57) suggests that these are too small for rib-forming flows. Equation 8, indicating that at a given slope value, depth varies as almost the square of clast dimensions (Fig. 12), is derived from data including catastrophic deep floods which are exceptional in the general fluvial context. Paleodepth estimates based on equations 6 and 7 — which are considered preferable because they require different combinations of rib measurements — are given in Table 6. Over the range of rib-forming conditions, the latter equation yields somewhat greater paleodepths. Average values of paleodepth, included in Table 6, are discussed below.

Discussion

Froude numbers in the range 0.7 - 1.3 are associated with natural antidune flow. Using these values, two related methods are available for estimating the minimum and maximum possible depths over ribbed sections of the bed. The first,

$$\frac{\overline{U}_{est}^{2}}{1.69g} \; \leqslant \; \overline{Y} \; \leqslant \; \frac{\overline{U}_{est}^{2}}{0.49g} \; , \tag{9}$$

utilizes the previous estimates of paleovelocity (average of the results obtained using equations 1 and 2) in the formula for the Froude number. The second,

$$(1.69k)^{-1} \; \leqslant \; \overline{Y} \; \leqslant \; (0.49k)^{-1}, \tag{10}$$

where k is as previously defined, used a modified version of equation 1 (Shaw and Kellerhals, 1977; p. 261). Depth ranges calculated with these methods are not exact since the first is based on velocity estimates and the second assumes validity of Kennedy's equation. Because both methods relate to the Froude number formula, only the ranges based on equation 9 are shown in Tables 6a and 6b. Specific values of paleodepth calculated by the above-described techniques should fall within these ranges. However, for certain ribbed reaches in each location the averaged depth estimate is greater than the value calculated for a Froude number of 0.7.

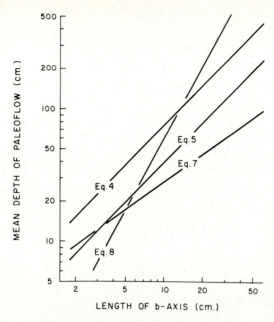

Fig. 12. Prediction of mean depth on the basis of clast size for uniform flows over a bed-slope of 0.025 by four of the five methods described in the text. Equation 6 can not be represented because it predicts paleodepth using velocity data.

Channel reaches and bar surfaces with transverse ribs have slopes which are commonly greater than or less than the average slope of the river system. Estimates of flow depth based on equations which assume uniformity of the paleoflow involve most error when applied to sections of the alluvial surface with bed slopes considerably less than average. In braided systems thalweg slope is highly variable and it is probable that flow decelerates over low-sloped sections, causing deepening in the downstream direction. Shaw and Kellerhals (1977, Fig. 4) showed that antidunes can form over horizontal, or even negatively-sloped, beds.

Average slopes are 0.019 for Peyto outwash (McDonald and Banerjee, 1971; Fig. 3a) and Spring Creek (Table 2), while for the North Saskatchewan River the average slope for the headwater reaches where McDonald and Banerjee observed ribs is 0.009 (Dept. Mines and Tech. Survs. 1957; Plate 34). Significantly, it is only those ribbed reaches with bed slopes less than average for which estimated paleodepths were found to exceed the calculated maximum values. It is concluded that methods for estimating paleodepth which are based on DuBoys' boundary shear stress relation should only be applied to reaches that have a bedslope similar to, or greater than, the average value for the system. For those ribbed reaches with estimates of paleodepth falling within the range stipulated by equation 9 (ie. those not in parentheses), a Froude number for the paleoflow has been calculated (Table 6).

CONCLUSIONS

Transverse ribs have been studied because their explanation by an upstream-migrating hydraulic jump did not appear to be a viable mechanism for their widespread development on Spring Creek fan, where super-critical flows are rare. Experimental observations prompted a detailed enquiry of the possibility that antidunes of sub-critical

TABLE 6a. Estimated parameters of rib-forming antidune paleoflows on Spring Creek: for an explanation of the underlining of certain average values, see Table 4.

Sub-Environment	#	ANTIDUNE GEOMETRY (cm)		PALEOVELOCITY (cm sec⁻¹)			PALEODEPTH (cm)					***Local bedslope in relation to average slope	FROUDE NUMBER
		$\bar{\lambda}_w$	*H_w	$\bar{U}$ by eq.1	$\bar{U}$ by eq.2	**$\bar{U}_{est}$	Min. $\bar{Y}$ by eq.9	Max. $\bar{Y}$ by eq.9	$\bar{Y}$ by eq.6	$\bar{Y}$ by eq.7	***$\bar{Y}_{est}$		Fr_{est}
Within-channels	1	104	15	128	150	139	12	40	13	34	24	+	.91
	2	190	27	172	160	166	17	57	37	59	48	≈	.77
	3	116	17	135	151	143	12	43	14	31	23	+	.95
	4	249	35	196	181	189	22	74	47	64	56	+	.81
	7	158	22	157	166	162	16	55	59	110	(85)	−	
	8	166	24	161	172	167	17	58	37	67	52	≈	.74
Averages		164 cm	23 cm			161 cm/sec					41 cm		.84
On bars	5	113	16	133	135	134	11	37	13	31	22	+	.91
	6	169	24	163	161	162	16	55	79	139	(109)	−	
	9	132	19	143	147	145	13	44	17	36	27	+	.89
	10	64	9	100	129	115	8	27	5	17	11	+	1.40
	11	150	21	153	153	153	14	49	40	76	(58)	−	1.04
	12	36	5	75	71	73	3	11	2	7	5	+	1.04
Averages		111 cm	16 cm			130 cm/sec					16 cm		.99

*Maximum amplitude of water waves given by 0.142λ (Kennedy, 1961).

**Average of values in two preceding columns.

***Symbols +, ≈ and − denote local bed-slopes that are higher, similar and lower than the average slope, respectively.

TABLE 6b. Estimated parameters of rib-forming antidune paleoflows on Peyto outwash, North Saskatchewan River and No-See-Um Creek, Alberta (field measurements of ribs from Table 3 of McDonald and Banerjee, 1971). For explanation of *, ** and *** see footnotes under Table 6a.

Location	#	ANTIDUNE GEOMETRY (cm)		PALEOVELOCITY (cm sec^{-1})			PALEODEPTH (cm)					***Local bedslope in relation to average slope	FROUDE NUMBER
		$\bar{\lambda}_w$	*$\bar{H}_w$	$\bar{U}$ by eq.1	$\bar{U}$ by eq.2	**$\bar{U}_{est}$	Min. $\bar{Y}$ by eq.9	Max. $\bar{Y}$ by eq.9	$\bar{Y}$ by eq.6	$\bar{Y}$ by eq.7	**$\bar{Y}_{est}$		F_{est}
Peyto Outwash	1	25	4	62	85	74	3	11	5	28	(17)	−	
	2	45	6	84	110	97	6	20	5	23	14	+	.83
	3	50	7	88	100	94	5	18	22	87	(55)	−	
	4	63	9	99	120	110	7	25	31	103	(67)	−	
	5	77	11	110	126	118	8	29	267	788	(528)	≈	.71
	6	47	7	86	97	92	5	18	7	26	17	−	
	7	55	8	93	103	98	6	20	14	50	(32)	−	.80
	8	67	10	102	111	107	7	24	9	27	18	+	
	9	68	10	103	103	103	6	22	18	55	(37)	−	.84
	10	37	5	76	105	91	5	17	4	20	12	+	
	11	94	13	121	140	131	10	36	5	14	10	+	1.32
Averages		57 cm	8 cm			101 cm/sec					14 cm		.90
North Saskatchewan River	1	15	2	48	71	60	2	8	1	7	4	+	.96
	2	27	4	65	81	73	3	11	3	19	11	+	.70
	3	36	5	75	81	78	4	13	3	14	9	+	.83
	4	21	3	57	66	62	2	8	1	7	4	+	.99
	5	36	5	75	88	82	4	14	4	18	11	+	.79
	6	55	8	93	94	94	5	19	5	19	12	+	.87
	7	75	10	108	101	105	7	23	8	24	16	+	.84
	8	43	6	82	84	83	4	15	87	386	(237)	−	
	9	34	5	73	78	76	3	12	4	18	11	+	.73
Averages		38 cm	5 cm			79 cm/sec					10 cm		.84
No-See-Um Creek	1	226	32	188	185	187	21	72	18	27	23	?	1.25
	2	150	21	153	157	155	14	50	8	16	12	?	1.43
Averages		188 cm	27 cm			171 cm/sec					18 cm		1.34

flow could form the ribs observed in the field. The antidune hypothesis receives support from an analysis of paleoflow on the following accounts:

1) Independent velocity estimates made using Kennedy's (1961) simple antidune formula and a competence relationship deduced from data compiled by Novak (1973) are in excellent agreement.

2) For all four locations studied, there is an exact correspondence between those rib sequences which are developed on relatively low-sloped sections of the alluvial surface with those for which estimated depths [a mean value using equations derived from the work of Baker (1973) and Novak (1973)] indicate a paleoflow that is too tranquil (ie. $Fr < 0.7$) for antidune formation. This result implies that DuBoys' (1879) shear stress relation should only be applied to reaches in which flow is fairly steady or mildly accelerating.

3) Estimated values of flow parameters are realistic for the fluvial environments studied. Rib-forming flows are generally sub-critical, relatively shallow ($Y < 1$ m) and highly competent because of their high velocity (U typically 1 - 2 m sec⁻¹). Rib-forming flows in Spring Fan channels have greater depths and higher velocities than those over bars (Table 6a). Accordingly, channels possess larger antidune waves. Because flow over bars tends to be shallower and on steeper slopes (Table 4), the Froude numbers of rib-forming flows over bars are somewhat greater than those within channels. Rib-forming flows on the Peyto outwash plain and in the North Saskatchewan River appear to be quite similar. However, in the former location, complete paleoflow analysis is hampered by the development of rib sequences on very low or even negative slopes (see Table 3 and Fig. 7a, respectively, McDonald and Banerjee, 1971). No-See-Um Creek appears to generate the highest flows of the four locations studied, with rib-forming flows being super-critical.

Where present, transverse ribs are conspicuous surface features of gravelly alluvium, both from the standpoint that they are often the only small-scale bedform and because they involve the largest sizes of the available bedload. Therefore, they reflect the highest flow events, and this makes the set of equations for paleoflow reconstruction a potentially useful addition to sedimentology.

Transverse ribs have yet to be recognized in ancient or Recent successions and it may be assumed that ribs have a low preservation potential; alternatively, they might be widely preserved but difficult to recognize. McDonald and Banerjee (1971, p. 1294) speculated that an upstream-dipping 'pseudo-bedding' would evolve if ribs migrated downstream and were aggrading. Since these features have not been confirmed, it appears that a buried rib sequence could be recognized only on the basis of evenly-spaced clusters of large clasts along a bedding surface. Perhaps paleoflow reconstruction based on rib sequences will find most application in modern environments, such as Spring Creek. Proglacial braided rivers are not generally amenable to direct hydrologic or hydraulic studies and so ribs assume considerable importance by supplying this information indirectly.

More field measurements of rib sequences are required to clarify the reasons for variable inter-relationships between the coarseness of ribs and their wavelength in different fluvial environments. Also, the merit of the antidune origin should be further investigated. This could perhaps be achieved by detailed monitoring of channel and over-bar flow which is observed during high stage to have undulatory surface wave forms. The equations for calculation of paleoflow parameters could then be tested using rib sequences exposed as a result of avulsion or abandonment at lower stages.

ACKNOWLEDGEMENTS

The paper stems from the author's Ph.D. thesis written under the direction of Dr. Brian R. Rust at the University of Ottawa. The many suggestions for improvement of the manuscript by Dr. Rust and Dr. G. V. Middleton, as well as those of Dr. T. J. Day on an earlier draft, are gratefully acknowledged. I also thank Dr. B. C. McDonald for arranging the use of the flume, the National Research Council of Canada for financial support, the Arctic Institute of North America for logistic support in Yukon, A. B. Ross for field assistance, W. McMillan for drafting, and S. Babcock for typing.

REFERENCES

Allen, J. R. L., 1970, Physical processes of sedimentation: George Allen & Unwin Ltd., 248p.

Bagnold, R. A., 1966, An approach to the sediment transport problem from general physics: U.S. Geol. Survey prof. Paper 422-I, 37p.

Baker, V. R., 1973, Paleohydrology and sedimentology of Lake Missoula flooding in eastern Washington: Geol. Soc. Am. Spec. Paper 144, 79p.

————, 1974, Paleohydraulic interpretation of Quaternary alluvium near Golden, Colorado: Quaternary Res., v. 4, p. 94-112.

———— and Ritter, D. F., 1975, Competence of rivers to transport coarse bedload material: Geol. Soc. Am. Bull., v. 86, p. 975-978.

Bluck, B. J., 1974, Structure and directional properties of some valley sandur deposits in southern Iceland: Sedimentology, v. 21, p. 533-554.

Boothroyd, J., 1972, Coarse-grained sedimentation on a braided outwash fan, northeast Gulf of Alaska: Office of Naval Research Tech. Rep. No. 6-CRD, Univ. South Carolina, 127p.

———— and Ashley, G. M., 1975, Processes, bar morphology and sedimentary structures on braided outwash fans, northeastern Gulf of Alaska: *in* A. V. Jopling and B. C. McDonald, *eds.*, Glaciofluvial and glaciolacustrine sedimentation, Soc. Econ. Paleont. Mineral. Spec. Pub. 23, p. 193-222.

Bowman, D., 1977, Stepped-bed morphology in arid gravelly channels: Geol. Soc. Am. Bull., v. 88, p. 291-298.

Briggs, L. I., and Middleton, G. V., 1965, Hydromechanical principles of sediment structure formation: *in* G. V. Middleton, *ed.*, Primary sedimentary structures and their hydrodynamic interpretation, Soc. Econ. Paleont. Mineral. Spec. Pub. 12, p. 5-16.

Bryan, M. L., 1972, Variations in quality and quantity of Slims River water, Yukon Territory: Can. J. Earth Sci., v. 9, p. 1469-1478.

Church, M., 1970, Baffin Island sandur, a study in Arctic fluvial environments: Unpub. Ph.D. thesis, Univ. British Columbia, 680p.

————, 1972, Baffin Island sandurs: Geol. Surv. Can. Bull. 216, 205p.

———— and Gilbert, G., 1975, Proglacial fluvial and lacustrine environments: *in* A. V. Jopling and B. C. McDonald, *eds.*, Glaciofluvial and glaciolacustrine sedimentation, Soc. Econ. Paleont. Mineral. Spec. Pub. 23, p. 22-100.

———— and Ryder, J. M., 1972, Paraglacial sedimentation: consideration of fluvial processes conditioned by glaciation: Geol. Soc. Am. Bull., v. 83, p. 3059-3072.

Clague, J. J., 1975, Sedimentology and paleohydrology of Late Wisconsian outwash, Rocky Mountain Trench, south-eastern British Columbia: *in* A. V. Jopling and B. C. McDonald, *eds.*, Glaciofluvial and glaciolacustrine sedimentation, Soc. Econ. Paleont. Mineral. Spec. Pub. 23, p. 223-237.

Department of Mines and Technical Surveys, 1957, Atlas of Canada: Geogr. Branch, Ottawa.

DuBoys, P. D. F., 1879, Le Rhône et le riviére afouillable: Annales des ponts et chaussées, v. 18, series 5, p. 141-195.

Fahnestock, R. K., 1961, Competence of a glacial stream: U.S. Geol. Survey Prof. Paper 424-B, p. 211-213.

————, 1963, Morphology and hydrology of a glacial stream — White River, Mount Ranier, Washington: U.S. Geol Survey Prof. Paper 422-A, 70p.

————, 1973, Knik and Matanuska Rivers, Alaska — a contrast in braiding: *in* M. Morisawa, *ed.*, Fluvial Geomorphology, Pub. Geomorph., S.U.N.Y.-Binghamton, p. 220-250.

———— and Haushild, W. L., 1962, Flume studies of the transport of pebbles and cobbles on a sandbed: Geol. Soc. Am. Bull., v. 73, p. 1431-1436.

Foley, M. G., 1977, Gravel-lens formation in antidune-regime flow — a quantitative hydrodynamic indicator: J. Sediment. Petrol., v. 47, p. 738-746.

Gustavson, T. C., 1974, Sedimentation on gravel outwash fans, Malaspina Glacier foreland, Alaska: J. Sediment. Petrol., v. 44, p. 374-389.

Hein, F. J., and Walker, R. G., 1977, Bar evolution and development of stratification in the gravelly, braided, Kicking Horse River, British Columbia: Can. J. Earth Sci., v. 14, p. 562-570.

Helley, E. J., 1969, Field measurement of the initiation of large particle motion in Blue Creek, near Klamath, California: U.S. Geol. Survey Prof. Paper 562-G, 19p.

Hjulstrom, F., 1935, Studies of the morphological activity of rivers as illustrated by the River Fyris: Geol. Inst., Univ. Uppsala Bull., v. 25, p. 221-527.

Imbrie, J., 1956, Biometrical methods in the study of invertebrate fossils: Am. Museum Nat. Hist. Bull., v. 108, p. 211-252.

Judd, H. E., and Peterson, D. F., 1969, Hydraulics of large bed element channels: Utah Water Res. Lab., Rept. PRWG 17-6, Utah State Univ., Logan, 115p.

Keller, E. A., 1970, Bed-load movement experiments: J. Sediment. Petrol., v. 40, p. 1339-1344.

Kennedy, J. F., 1961, Stationary waves and antidunes in alluvial channels: W. M. Keck Lab. Hydraulics and Water Resources, California Inst. Tech., Pasadena, Rept. No. KH-R-2, 146p.

Koster, E. H., 1977a, Experimental studies of coarse-grained sedimentation: Unpub. Ph.D. thesis, Univ. Ottawa, 221p.

————, 1977b, Paleohydraulic interpretation of coarse-grained alluvium using transverse ribs (abstract): Geol. Soc. Am., Abstracts with Programs, v. 9, no. 3, p. 287-288.

Laming, D. J. C., 1966, Imbrication, paleocurrents and other sedimentary features of Lower Old Red Sandstone, Devonshire, England: J. Sediment. Petrol., v. 36, p. 940-959.

Lane, E. W., 1955, Design of stable channels: Am. Soc. Civil Eng. Trans., v. 120, p. 1234-1260.

Langbein, W. B., and Leopold, L. B., 1968, River channel bars and dunes — theory of kinematic waves: U.S. Geol. Survey Prof. paper 422-L, 20p.

Laronne, J. B., and Carson, M. A., 1976, Interrelationships between bed morphology and bed-material transport for a small, gravel-bed channel: Sedimentology, v. 23, p. 67-86.

Leopold, L. B., Wolman, M. G., and Miller, J. P., 1964, Fluvial processes in geomorphology: W. H. Freeman, 522p.

Lighthill, M. J., and Whitham, G. B., 1955, On kinematic waves — 1. Flood movement in long rivers: Proc. Royal Soc. London, v. 229, series A, p. 281-316.

Mackin, J. H., 1948, Concept of a graded river: Geol. Soc. Am. Bull., v. 59, p. 463-512.

Martini, I. P., 1977, Gravelly flood deposits of Irvine Creek, Ontario, Canada: Sedimentology, v. 24, p. 603-622.

———— and Ostler, J., 1973, Ostler lenses — possible environmental indicators in fluvial gravels and conglomerates: J. Sediment. Petrol., v. 43, p. 418-422, and v. 44, p. 274.

McDonald, B. C., 1972, The Geological Survey of Canada sedimentation flume: Geol. Surv. Can. Paper 71-46, 12p.

———— and Banerjee, I., 1971, Sediments and bedforms on a braided outwash plain: Can. J. Earth Sci., v. 8, p. 1282-1301.

———— and Day, T. J., 1978, An experimental flume study on the formation of transverse ribs: Geol. Surv. Can. paper 78-1A, p. 441-451.

McKee, E. D. Crosby, E. J., and Berryhill, H. L., 1967, Flood deposits, Bijou Creek, Colorado: J. Sediment. Petrol., v. 37, p. 829-851.

Meyer-Peter, E., and Muller, R., 1948, Formula for bedload transport: Int. Assoc. Hydraulic Structures Research, Stockholm, v. 3, p. 39-65.

Neill, C. R., 1967, Mean-velocity criterion for scour of coarse uniform bed-material: Proc. 12th. Congr. Internat. Assoc. Hydraulic Res., Colorado State Univ., v. 3, p. 46-54.

Nevin, C., 1946, Competency of moving water to transport debris: Geol. Soc. Am. Bull., v. 57, p. 561-674.

Novak, I. D., 1973, Predicting coarse sediment transport: the Hjulstrom curve revisited, *in* M. Morisawa, *ed.,* Fluvial Geomorphology: Pub. Geomorph., S.U.N.Y.-Binghampton, p. 13-25.

Østrem, G., 1975, Sediment transport in glacial meltwater streams: *in* A. V. Jopling and B. C. McDonald, *eds.,* Glaciofluvial and glaciolacustrine sedimentation, Soc. Econ. Paleont. Mineral. Spec. Pub. 23, p. 101-122.

Parker, G., 1975, Sediment inertia as cause of river antidunes: J. Hydraulics Div., Am. Soc. Civil Eng., v. 101, no. HY2, p. 211-221.

Rust, B. R., 1972a, Structure and process in a braided river: Sedimentology, v. 18, p. 221-245.

―――, 1972b, Pebble orientation in fluvial sediments: J. Sediment. Petrol., v. 42, p. 384-388.

―――, 1975, Fabric and structure in glaciofluvial gravels: *in* A. V. Jopling and B. C. McDonald, *eds.,* Glaciofluvial and glaciolacustrine sedimentation: Soc. Econ. Paleont. Mineral. Spec. Pub. 23, p. 238-248.

Scott, K. M., and Gravlee, G. C., 1968, Flood surge on the Rubicon River, California — hydrology, hydraulics and boulder transport: U.S. Geol. Survey Prof. Paper 422-M, 40p.

Shaw, J., and Kellerhals, R., 1977, Paleohydraulic interpretation of antidune bedforms with applications to antidunes in gravel: J. Sediment. Petrol., v. 47, p. 257-266.

Shields, A., 1936, Anwendung der Ahnlichkeitsmechanik und der Turbulenzforshung auf die Geschiebebewegung: Mitteilungen der Preuss, Versuch anst. f. Wasserbau u. Schiffbau, Berlin, Heft 26, 26p.

Simons, E. V., and Richardson, E. V., 1960, Discussion of resistance properties of sediment-laden streams: Am. Soc. Civil Eng. Trans., v. 125, p. 1170-1172.

Smith, N. D., 1974, Sedimentology and bar formation in the upper Kicking Horse River, a braided outwash stream: J. Geol., v. 82, p. 205-223.

Sundborg, A., 1967, Some aspects of fluvial sediments and fluvial morphology, I. General views and graphic methods: Geogr. Ann., v. 49A, p. 333-343.

Till, R., 1974, Statistical methods for the earth scientist — an introduction: John Wiley & Sons, 154p.

Williams, P. F., and Rust, B. R., 1969, The sedimentology of a braided river: J. Sediment. Petrol., v. 39, p. 649-679.

APPENDIX: MATHEMATICAL NOTATIONS

Overbar indicates mean value.

a	long axis dimension (cm) of clast
a_{max}	as above, with subscript denoting average size of 10 largest clasts in sampling area
b	intermediate axis dimension (cm) of clast
b_{max}	as above, with subscript denoting average size of 10 largest clasts in sampling area
d	unspecified measure of clast size, usually b-axis (mm, cm or m)
est	subscript denoting value is an estimate
Fr	Froude number
g	gravitational acceleration (cm. sec^{-2} or m. sec^{-2})
H_w	amplitude (cm) of water-surface undulations in antidune waves
i	subscript denoting particular transverse rib in sequence

j	number of transverse ribs in sequence
k	antidune 'wave number' defined as $2\pi/\lambda$
L	length (m) of rib sequence
m	coefficient in relationships of rib width versus rib wavelength
n	sample size
r	correlation coefficient
R	hydraulic radius (m)
S	spacing (cm) between two adjacent ribs
S_e	slope of energy grade line
S_b	slope of bed-surface
U	flow velocity (cm.sec^{-1} or m.sec^{-1})
W	width (cm) of transverse rib
Y	flow depth (cm or m)
ρ_f	mass density (g.cm^{-3} or kg.m^{-3}) of fluid
ρ_s	mass density (g.cm^{-3} or kg.m^{-3}) of clast
γ_f	$= \rho_f g$, specific weight (g.cm^{-2}.sec^{-2} or kg.m^{-2}.sec^{-2}) of fluid
γ_s	$= \rho_s g$, specific weight (g.cm^{-2}.sec^{-2} or kg.m^{-2}.sec^{-2}) of clast
λ	antidune wavelength (cm), with subscripts b and w referring to bed- and water-surfaces, respectively
τ	boundary shear stress (newtons.m^{-2})
τ_c	as above, with subscript denoting critical value

Footnote: B. R. Rust and V. A. Gostin (Rust, written communication, July 1978) have found a sequence of transverse ribs ($\overline{y_b}$ = 300 cm, $\overline{b}$ max = 35.7 cm) preserved in Late Holocene deposits of Depot Creek alluvial fan, 30 km north of Port Augusta, South Australia. The mode of preservation is the same as that predicted on p. 182. Based on a preliminary estimate of S_b, and using the equations outlined in this paper, the rib-forming flow was subcritical with $\overline{U}$ = 203 cm. sec.$^{-1}$ and $\overline{Y}$ = 0.75 cm.

MODERN RIVERS:
GEOMORPHOLOGY AND SEDIMENTATION

A CLASSIFICATION OF ALLUVIAL CHANNEL SYSTEMS

BRIAN R. RUST[1]

ABSTRACT

A new classification of alluvial channel systems is proposed, based on the braiding parameter: the number of braids per mean meander wavelength, braids being defined by the mid-line of the channels surrounding each braid bar. Single-channel and multi-channel systems are defined as having braiding parameters less than and more than one, respectively. Channel systems are further divided into low- and high-sinuosity categories at the boundary 1.5, giving four types, of which single-channel high-sinuosity (meandering) and multi-channel low-sinuosity (braided) are by far the most common. The other two types are much less abundant: single-channel low-sinuosity (straight), and multi-channel high-sinuosity (anastomosing).

The classification can be applied to ancient alluvial deposits through an understanding of the processes that relate channel morphologies to their resulting sedimentary suites. As the processes are imperfectly understood, so the application to paleochannels is imprecise. However, it can be made satisfactorily in most cases, provided maximum use is made of the channel-process information in alluvial sedimentary models.

INTRODUCTION

The interpretation of ancient alluvial deposits is largely dependent on our knowledge of modern alluvial systems. Formulation of a suitable classification is an important step towards better interpretation, and should be based primarily on modern systems, but be capable of application to the ancient record. Plan morphology is the most easily recognised feature of modern channel systems, and is the basis of Leopold and Wolman's (1957) classification, probably the most widely used at present. However, there are possibilities of overlap between the channel categories defined by Leopold and Wolman, and they made no quantitative distinction between single and multiple channel systems. This paper proposes a morphological channel classification which to a large extent overcomes these disadvantages. It also attempts to relate paleochannel features to the major sedimentary factors that govern alluvial deposition.

PLAN MORPHOLOGY

Introduction

Leopold and Wolman (1957) classified alluvial channel systems into straight, braided and meandering categories. Leaving aside cases of bedrock control as non-alluvial, straight channels are rare. Leopold and Wolman (p. 53) defined a braided river as "... one which flows in two or more anastomosing channels around alluvial islands". They regarded a single channel division around a bar or island as constituting a braid, and showed that hydraulic factors (notably slope) adjust to the presence of individual braids. Meandering rivers were defined as having a channel sinuosity equal to or greater than 1.5 (Leopold and Wolman, 1957, p. 60).

A disadvantage of Leopold and Wolman's classification is that different criteria are used to distinguish the two most important channel patterns: the meandering type is defined by sinuosity, whereas channel multiplicity defines braided systems. As a result, a channel may have characteristics of both types, for example if it has a sinuosity greater than 1.5, but also has alluvial islands. A second problem is that the definition of a braided reach was not quantified: How many islands per length of reach are required? In addition, a numerical definition would allow one to distinguish between moderately and highly

---

[1]Geology Department, University of Ottawa, Ottawa K1N 6N5, Canada

braided reaches. These problems can be overcome by defining channel categories on the basis of two quantifiable parameters: sinuosity, and a measure of channel multiplicity. To avoid giving undue significance to isolated bars or islands within otherwise single-channel systems, the parameters should normally be mean values for a reach of several meander wavelengths.

Sinuosity

Leopold and Wolman (1957, p. 60) defined sinuosity as the ratio of thalweg length to valley length. Brice (1964, p. D25-26) introduced a sinuosity index, defined as the ratio of channel length to the length of the meander-belt axis. This modification is useful for distinguishing between meandering systems with straight as opposed to sinuous meander belts; in the latter case the sinuosity index is less than the sinuosity. However, the concept of a meander belt is not usually applicable to braided reaches, and Leopold and Wolman's sinuosity is accepted here as a suitable parameter for distinguishing the main types of channel system. Likewise, the concept of meander symmetry (Brice, 1964, p. D25-26) is here viewed as a refinement which can be disregarded for the purposes of defining principal channel types.

On the basis of experience, Leopold and Wolman (1957) chose the arbitrary sinuosity value of 1.5 to distinguish between meandering and non-meandering channels. The same value is retained in this paper, but is proposed as the boundary between high- and low-sinuosity channels. Moody-Stuart (1966) placed this boundary at a sinuosity of 1.3, whereas Leeder's (1973a) preference was for 1.7. Amongst these arbitrary choices, 1.5 has the merit of being the median value, and is selected here as probably the most acceptable, but it is recognised that later work may necessitate revision.

Channel Multiplicity

Brice (1964) introduced the braiding index, a quantitative measure of channel multiplicity:

$$\text{Braiding index} = \frac{2 \,(\text{Sum of lengths of islands and (or) bars in a reach})}{\text{Length of reach measured midway between banks}}$$

Bars were recognised as being non-vegetated, and submerged at bankfull stage, whereas islands are vegetated and normally emergent at bankfull stage. The braiding index can be expressed as "transient" for bars, "stabilised" for islands, or "total" for the two combined. Brice recognised that the transient-braiding index varies greatly with water level, but claimed that the stabilised-braiding index remains nearly constant with changing stage. The fact is, however, that the stabilised index is of little use for describing active braided tracts dominated almost exclusively by non-vegetated bars. For example, the Donjek River, Yukon has vegetated tracts formed at earlier times and under different hydrologic regimes from the presently active tract (Williams and Rust, 1969). Any descriptor based on vegetated islands is therefore a poor indication of present-day fluvial activity. It would also be misleading in terms of ancient alluvial processes, because parts of the older, vegetated levels were eroded during formation of more recently active tracts (Williams and Rust, 1969, Fig. 27; Miall, 1977, Fig. 2), reducing the stabilised-braiding index from its original value. The disadvantages of the transient- or total-braiding indices are equally serious: because they vary with water level, they are indicative more of stage than of basic morphological characteristics of the channel system.

The problem of stage-dependence can be tackled by defining the perimeter of a *braid* as the mid-line of the channels surrounding each bar or island, whether or not these channels contain water (Fig. 1). It follows that the *braid length* is the straight line distance between the extremities of the braid, as defined above. A measure of braiding intensity can then be expressed as the number of braids per mean meander wavelength, which is here proposed as the *braiding parameter* (Fig. 2). Mean meander wavelength is easily determined for

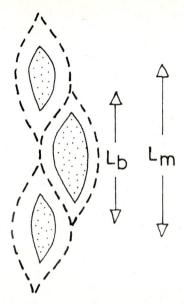

Fig. 1. Definition diagram for braid, braid length (Lb) and meander wavelength (Lm). Dots indicate exposed portions of braid bars; dashes are mid-lines of channels.

reaches with few braids, but for rivers with a high intensity of braiding it is less obvious. Inspection of several braided systems showed that the mean meander wavelength can be approximated as 1¼ to 1½ times the mean braid length, depending on the amount of overlap between the adjacent braids (Fig. 1). This approximation can be used for cases in which the mean meander wavelength is difficult to measure directly.

Compared with Brice's (1964) braiding index, the braiding parameter varies much less with river stage. Provided none of the bars are flooded, each defines a braid, and each can be counted when determining the braiding parameter. The measured value of the parameter starts to fall once bar inundation commences, but there seems to be no way of avoiding this problem, short of remote sensing to detect submerged bars. As with Brice's index, the braiding parameter can be determined separately for bars and vegetated islands, or for a combination of both. For rivers like the Donjek, however, it is more realistic to determine braiding parameters for vegetated and non-vegetated tracts separately. As a general rule, the parameter should only be obtained from tracts in which all parts are laterally continuous, without intervening erosional banks.

A further problem, which has not been tackled by previous investigators, is that of bar order, a concept introduced by Williams and Rust (1969, Fig. 2), and here illustrated in Fig. 3. Clearly, the braiding parameter for a reach can be expressed in terms of first, second, or any other recognisable order of bar and associated braid. It is important therefore, to specify which order of braid is used to determine the braiding parameter, and it may be useful to express the parameter in terms of more than one order. Illustrations of this will be given later in the paper.

An important function of the braiding parameter is to define quantitatively the difference between single-channel and multi-channel systems. After inspecting reaches of various rivers, it was concluded that a braiding parameter of one serves as a suitable boundary between these two categories (Fig. 2.). The parameter can also be used to define various intensities of braiding. These definitions are quite arbitrary, but in the writer's opinion a system with a braiding parameter between 1 and 4 would commonly be

SINGLE-   MULTI-CHANNEL
CHANNEL   Moderately   Highly
                    braided   braided

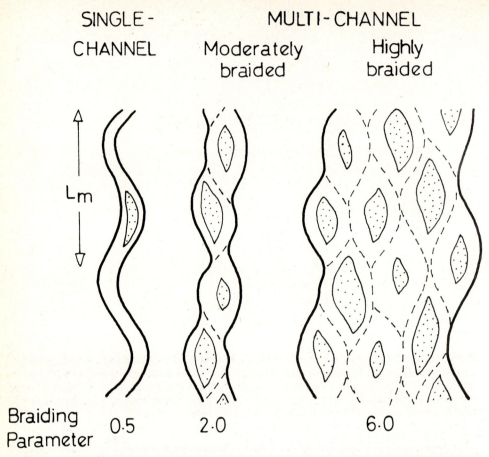

Braiding   0·5   2·0   6·0
Parameter

Fig. 2. Diagrammatic representation of single-channel and multi-channel (moderately and highly braided) alluvial systems. Ornament as in Fig. 1.

regarded as moderately braided, whereas those with a higher parameter would be assessed as highly braided (Fig. 2). Further divisions at higher values probably have little meaning, because variance between operators for judging braid order and determining mean meander wavelength increases as the braiding parameter rises.

The definition of a multi-channel system given here differs from that of Schumm (1968, p. 1579), who stated: ". . . the multiple-channel rivers considered here are those that are flowing on an alluvial surface and are distributary systems." Alluvial fan and alluvial plain systems were illustrated as purely distributary (Schumm, 1968, Fig. 3), whereas fans are invariably braided, as are many alluvial plains, for example the Skeidara (Fig. 4). These systems are distributary only in the sense that the braiding parameter increases downstream, because more braids develop as the tract expands to occupy more space. Similar effects result from constrictions in valley-confined braided rivers, for example the upper reaches of the Donjek (Rust, 1972), and the Slims River (Fahnestock, 1969). The braiding parameter decreases towards each constriction, and increases again downstream from it (Fig. 5).

In common with most authors in the geological literature (for example Miall, 1977), braided systems are defined here as multi-channeled, in contrast with Schumm (1968, p.

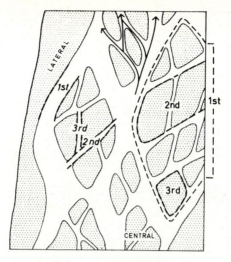

1st, 2nd, 3rd, =   Channel (Orders)
1st, 2nd, 3rd, =   Bar (Orders)

Fig. 3. Definition diagram for channel and bar order (from Williams and Rust, 1969, Fig. 2b). Braid order corresponds to the respective bar order, as shown in Fig. 1.

1579): "Braided channels are single-channel bedload rivers which at low water have islands of sediment or relatively permanent vegetated islands . . .''. Schumm's view results from classifying channels at bankfull stage, when many or all of the braid bars and channels are covered. However, for many braided systems such as the Skeidara (Fig. 4) and the Cooper (Fig. 6), bankfull stage cannot be determined by inspection. Bankfull discharge could be estimated by assuming a similar recurrence interval to that of well-defined single-channel rivers, but the assumption is of uncertain validity, as there are few good hydrological data for braided rivers. Hence, if bankfull stage were the criterion, there would always be uncertainty in assigning braided rivers to single- or multi-channel categories. There are significant differences between the bedforms and internal stratification of channels and bars, especially in distal braided systems (Rust, this volume). From the geological viewpoint the multi-channel nature of braided systems should be recognised, even when all channels are flooded. It is therefore recommended that channel types be defined at a stage approximately mid-way between bankfull and minimum discharge. This approach also avoids the problem that at very low stage some single channels diverge around larger bedforms such as dunes, but should not then be regarded as multi-channeled.

The Proposed Channel Classification

The classification proposed here has the form of a simple two-by-two matrix, which defines four morphological channel types (Table 1). The two most abundant are the single-channel, high-sinuosity (*meandering*) and the multi-channel, low-sinuosity (*braided*) systems. The others are *straight* (more accurately termed single-channel, low-sinuosity) and *anastomosing* (multi-channel, high-sinuosity) systems. Leopold and Wolman (1957) used ''anastomosing'' as a synonym for ''braided'', as did the A.G.I. glossary (1972). However, this seems to be a waste of good term, for ''braided'' is so well established that it hardly needs a synonym. Smith (1976) and Miall (1977) used the term ''anastomosing'' in the same sense as proposed here, and its general adoption, following the quantitative definition given in this paper, is recommended.

Fig. 4. Air photo mosaic of the active tract of the Skeidará outwash plain, Southeast Iceland (16°55'W, 63°50'N). Reproduced by permission of Dag Nummedal, Coastal Research Division, University of Southern Carolina.

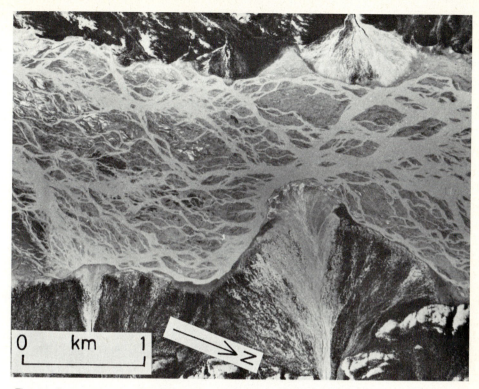

Fig. 5. Part of vertical air photo (A15517-20) of upper reach of the Slims River, Yukon (60° 52'N, 138° 35'W). Original supplied by Surveys and Mapping Branch, Department of Energy, Mines and Resources, Canada. Note the tributary alluvial fans, which constrict the river, causing variation in its braiding parameter.

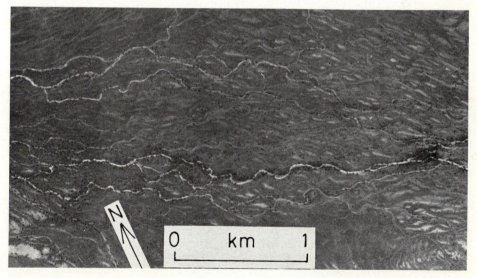

Fig. 6. Part of vertical air photo (1448-264) of Cooper Creek, South Australia, 70km west of the Queensland border (27°48'S, 140°20'E). Original supplied by Department of Lands, South Australia.

Table 1.   Channel classification

	Single-channel (Braiding parameter< 1)	Multi-channel (Braiding parameter>1)
Low-sinuosity (<1.5)	Straight	BRAIDED
High-Sinuosity (>1.5)	MEANDERING	Anastomosing

Table 2.   Data for braided reaches

			1st order braids	2nd order	3rd order
Slims River, Yukon	Mean meander wavelength ($\overline{\text{Lm}}$)		1040m	310m	
	Braiding parameter for bars (Bb)		8.7	15	
Skeidará outwash plain, Iceland	Upper reach	$\overline{\text{Lm}}$	540m	150m	
		Bb	9.5		
	Middle reach	$\overline{\text{Lm}}$	810m	330m	150m
		Bb	32	≈ 70	
	Lower reach	$\overline{\text{Lm}}$			≈75m
		Bb			≈160
Cooper Creek, S. Australia		$\overline{\text{Lm}}$	380m		
	Bb+i (bars and islands)		21		

The use of the new classification to distinguish between straight and braided systems is shown in Fig. 2, while the recognition of meandering and anastomosing patterns is not different from that proposed elsewhere (Miall, 1977, Fig. 1, for example). However, the classification of braided systems is more difficult, and requires illustration with examples. Fig. 5 shows a gravel reach of the Slims River, Yukon, 2.5 to 4 km downstream from the outermost recent moraine of the Kaskawulsh Glacier (Fahnestock, 1969). The mean meander wavelength for the first order braids in this reach is estimated to be 1040 m, while that for second order braids is 310 m. The respective braiding parameters are 8.7 and 15, both based entirely on non-vegetated bars (Table 2).

Another example with proglacial braided deposits is from the Skeidara outwash plain, Iceland (Nummedal et al., 1974). Fig. 4 is a photo mosaic of the active tract, illustrating two inter-related downslope trends: increased dominance of lower order braids, and increased braiding parameter (Table 2). These trends are related to features observed by Nummedal et al., (1974) on the alluvial surface. The upper part of the active reach is characterised by gravel in longitudinal bars, which have high relative relief, and a few well-defined, relatively deep channels, whereas the lower plain is dominated by linguoid bars of coarse sand, with low relative relief. The downslope trends towards lower order braids and increased braiding parameter are common in braided systems which change from gravel to sand distally. However, an additional factor increasing the braiding parameter on alluvial plains like the Skeidara is the space available for the active tract to spread.

The third example is an ephemeral braided channel system with some vegetation, in an arid environment: Cooper Creek, South Australia (Fig. 6). The braids are less distinct than those of the other two examples, and different braid orders could not be distinguished. Cooper Creek floods after monsoon rainfall in Queensland; the floods are very variable, sometimes reaching Lake Eyre, the centre of a large internal drainage basin (Bonython, 1963). Within a 60 km section starting 10 km west of the Queensland border, the creek has reaches exhibiting straight, meandering, anastomosing and braided channel patterns, which will be the subject of further investigation by the the author.

APPLICATION TO ANCIENT ALLUVIAL DEPOSITS

The critical problem in applying a unified classification to modern and ancient alluvial channel systems is the link between plan morphology and the sedimentary suites observable in ancient rocks. Because our knowledge of channel processes is incomplete, this link is somewhat tenuous, whether one starts with channels defined morphologically and then predicts sedimentary suites, or vice versa. The important questions are: What factors cause variation in sinuosity and channel multiplicity? Can their lithological products be recognised in paleochannels? The answers ultimately lie in the perfection of process-response models for alluvial systems. This stage has not been reached, but sufficient progress has been made to attempt classification of modern and ancient channels by the same system.

Sinuosity

Schumm *et al.*, (1972) showed that for very low slopes ($[0.4$ to $1.8] \times 10^{-6}$) sinuosity in the Mississippi River varies with slope, but it is unclear what effect, if any, holds at higher gradients. A more generally applicable relationship is the inverse correlation between the silt and clay content of channel perimeters and their width/depth ratio (Schumm, 1960). Width/depth ratio is also inversely related to sinuosity, which in turn correlates strongly with channel perimeter silt-clay content, but less so with the silt-clay content of banks alone (Schumm, 1963b). The association of relatively deep, narrow channels with a silt-clay rich perimeter can be explained by the cohesive nature of the fine sediment, which stabilises steep banks. The same phenomenon appears to influence sinuosity, although one would expect sinuosity to correlate better with the silt-clay content of banks rather than channel perimeters. Silty braided systems such as the Lower Yellow River (Chien, 1961) and lower reaches of the Slims River (Fahnestock, 1969) do not fit these relationships, for they commonly have 50% or more channel sediment finer than very fine sand, but have wide, shallow channels of low sinuosity. Rapid caving indicates low bank stability in both the Yellow River (Chien, 1961) and the lower Slims. This suggests that braided silts have low cohesion because they lack clay (unfortunately size analyses are not available), which would explain the departures from the relationships discussed above.

Except for silty braided deposits, paleochannel sinuosity can be obtained from the width/depth ratio measured in cross-sections, using an equation developed by Schumm (1963b, p. 1091). Failing this, silt-clay content can be estimated from the stratigraphic succession (again excepting silty braided deposits), and sinuosity derived from the relationship given by Schumm (1963b, Fig. 4). Miall (1976) estimated paleochannel sinuosity for a braided system as a function of between-bar directional variance, assuming planar cross-bed sets to represent individual bars. Sinuosity was determined by comparison with modern systems, or from an equation relating it to angular change of channel orientation (Langbein and Leopold, 1966).

Channel Multiplicity

Unlike sinuousity, the relationship between channel multiplicity (as expressed by the braiding parameter), channel processes and their resulting products is known only qualitatively. Most authors have discussed the causes of channel multiplicity in terms of braided versus meandering systems. According to Walker (1976, p. 105), braided systems are characterised by more rapid discharge fluctuations of greater absolute magnitude, and by higher slope, coarser load, and more easily eroded banks. Leopold and Wolman (1957, p. 53) observed that braided reaches are steeper, wider and shallower than undivided reaches with the same discharge. Combined with Schumm's (1960) correlation between channel shape and silt-clay content, this supports the general observation that braided systems deposit little silt and clay without subsequent reworking. The lack of fine, cohesive sediment results in easily-eroded banks; channels therefore divide and relocate with minimum difficulty.

Once again, silty braided systems are significant exceptions to some of the above generalisations. Their proportion of channel sediment finer than sand is similar to that of many single-channel meandering rivers. However, like other braided systems, they show little or no grain size differentiation within the active tract between channel and overbank or bar top deposits. The silt is intimately mixed with fine sand, and is abundantly cross-laminated, indicating deposition from traction currents (Rust, this volume). In contrast, meandering systems show strong differentiation into sandy channel deposits transported as bedload, and overbank silt and clay which settled from suspension. The clay content of the latter material gives rise to relatively stable banks on which plants can grow.

Silty braided systems are rare today, because most alluvial systems accumulating fine sediment have high-sinuosity channels (meandering or anastomosing), stabilised to an important degree by vegetation (Smith, 1976). Terrestrial vegetation was essentially absent before the Late Paleozoic, and hence braided (bedload) channel systems were predominant prior to that time (Schumm, 1968). It is likely that silty braided deposits were abundant on distal reaches of pre-Late Paleozoic alluvial plains, and should therefore be taken into account when classifying paleochannels.

Paleochannel Classification

Schumm (1963a, 1968) used the correlations discussed earlier to establish three channel types on the basis of sediment transport: suspended-load, mixed-load and bedload channels. In practice, suspended load refers to wash load, and bedload to bed material load, as defined by Einstein (1950). Schumm assumed that the mode of transport is reflected in the nature of the channel sediment, so for paleochannels the three types are defined on essentially the silt-clay content of their deposits (Schumm, 1968, Table 1). Because it relates directly to lithological features observable in paleochannels, Schumm's classification has been adopted widely. However, it is subject to certain limitations. The correlations were established for rivers with minor amounts of gravel, and with channels that were neither aggrading nor degrading. Some rivers are dominated by gravels, and all ancient alluvium was subject to net aggradation. These limitations are probably not significant, for gravel can be included with sand as bedload, and the effects of net aggradation on the established correlations appear to be minor. However, the relationship between load type and channel morphology has not been fully developed for this classification, and there is a difficulty with the assumption that silt-clay content reflects the proportion of suspended load. The suspended load is normally deposited only when flood waters stagnate on the floodplain, and its subsequent preservation occurs where channels and floodplain are well differentiated, that is in meandering and anastomosing systems. The banks are vegetated and stable, and only a portion of the floodplain deposits is destroyed by later channel migration. Silty braided systems have none of these characteristics, for they are not suspended-load systems, despite their abundance of silt.

Classifying paleochannels on the basis of plan morphology requires full integration of data on channel processes, as outlined in various depositional models (Walker, 1976; Miall, 1977 and this volume; Rust, this volume). The principal component of the meandering model is a cycle comprising a sandy channel unit in which grain size and the scale of sedimentary structures decrease upwards, and a fine-grained overlying floodplain unit. Variations in the cycle result from different modes of initiation of the channel unit (Walker, 1976, Fig. 4), and from variation in stream power or sinuosity (Allen, 1970). However, fining-upward cycles are not diagnostic of meandering paleochannels, for they also develop in distal braided systems, both gravelly and sandy. The former can be distinguished by the dominance of framework gravel, whereas the latter are in some cases transitional to deposits of meandering systems (Rust, this volume). Thus, although there are no empirical equations relating features of ancient successions to values of the braiding parameter, careful consideration of facies assemblages allows one to assign most successions to groups with parameters below one, approximately one, and above one.

Another consideration is the possibility of recognising straight-channel and anastomosing-channel deposits in the stratigraphic record. Moody-Stuart (1966) and Leeder (1973b) interpreted coarse units which lack upward fining and consistent upward change in type and scale of sedimentary structures as deposits of low-sinuosity, probably single paleochannels. These features could also be explained by the interruption and superposition of successive channel units in meandering or distal braided systems, but existing depositional models cannot make the distinction with certainty. Anastomosing-channel sediments should be distinctive in the ancient record on account of abundant suspended-load deposits, and evidence of intense plant activity. Many Late Paleozoic and younger delta platform and coastal plain deposits have these charcteristics, and were probably formed in anastomosing channel systems.

CONCLUSIONS

Values of sinuosity and the braiding parameter, a measure of channel multiplicity, are proposed as boundaries defining four types of channel system. The boundaries are arbitrary, and may need revision in the light of subsequent work, but it is maintained that a morphological classification with quantified boundaries is more useful than those in current use.

In the present state of knowledge, paleochannels cannot be placed in quantitatively defined categories with the same precision as modern channel systems. However, this does not mean that quantitative definitions should be abandoned for paleochannels; ambiguous definitions can only lead to greater confusion when applied to the ancient record.

The approach used here is to seek the relationships between morphological channel types and their resulting sedimentary suites through application of all available information in alluvial depositional models. An alternative method proposed by Schumm is equally valid, and recognises types of paleochannels on lithological grounds. However, it encounters some difficulty with respect to silt-dominant braided systems, and has not developed the relationship between lithologically-defined paleochannels and their morphological equivalents.

ACKNOWLEDGEMENTS

I am indebted to Prof. R. W. R. Rutland for the provision of facilities at the University of Adelaide, South Australia, and to V. A. Gostin, J. R. Hails, A. D. Miall and S. A. Schumm for comments on the manuscript.

References

Allen, J. R. L., 1970, Studies in fluviatile sedimentation: A comparison of fining-upward cyclothems, with special reference to coarse-member composition and interpretation: J. Sediment. Petrol., v. 40, p. 298-323.

Bonython, C. W., 1963, Further light on river floods reaching Lake Eyre: Proc. Roy. Geograph. Soc. Australasia, S. Australian Branch, v. 64, p. 9-22.

Brice, J. C., 1964, Channel patterns and terraces of the Loup Rivers in Nebraska: U.S. Geol. Surv. Prof. Paper 422-D.

Chien Ning, 1961, The braided stream of the lower Yellow River: Scientia Sinica, v. 10, p. 734-754.

Einstein, H. A., 1950, The bed-load function for sediment transportation in open channel flows: U.S. Dept. Agriculture Tech. Bull. 1026.

Fahnestock, R. K., 1969, Morphology of the Slims River: Icefield Ranges Research Project, Scient. Results, v. 1, p. 161-172.

Langbein, W. B. and Leopold, L. B., 1966, River meanders — theory of minimum variance: U.S. Geol. Surv. Prof. Paper 422-H.

Leeder, M. R., 1973a, Fluviatile fining-upwards cycles and the magnitudes of paleochannels: Geol. Mag., v. 110, p. 265-276.

———, 1973b, Sedimentology and paleogeography of the Upper Old Red Sandstone in the Scottish Border Basin: Scott. J. Geol., v. 9, p. 117-144.

Leopold, L. B. and Wolman, M. G., 1957, River channel patterns: braided, meandering, and straight: U.S. Geol. Survey Prof. Paper 282-B, p. 39-85.

Miall, A. D., 1976, Paleocurrent and paleohydrologic analysis of some vertical profiles through a Cretaceous braided stream deposit, Banks Island, Arctic Canada: Sedimentology, v. 23, p. 459-484.

———, 1977, A review of the braided-river depositional environment: Earth Sci. Revs., v. 13, p. 1-62.

Moody-Stuart, M., 1966, High- and low-sinuosity stream deposits, with examples from the Devonian of Spitsbergen: J. Sediment. Petrol., v. 36, p. 1102-1117.

Nummedal, D., Hine, A. C., Ward, L. G., Hayes, M. O., Boothroyd, J. C., Stephen, M. F. and Hubbard, D. K., 1974, Recent migrations of the Skeidárasandur shoreline, Southeast Iceland: Coastal Res. Div., Univ. of S. Carolina, Report N60921-73-C-0258.

Rust, B. R., 1972, Structure and process in a braided river: Sedimentology, v. 18, p. 221-246.

Schumm, S. A., 1960, The shape of alluvial channels in relation to sediment type: U.S. Geol. Survey Prof. Paper 352-B, p. 17-30.

———, 1963a, A tentative classification of alluvial river channels: U. S. Geol. Survey Circular 477.

———, 1963b, Sinuosity of alluvial rivers on the Great Plains: Geol. Soc. Am. Bull., v. 74, p. 1089-1099.

———, 1968, Speculations concerning paleohydrologic controls of terrestrial sedimentation: Geol. Soc. Am. Bull., v. 79, p. 1573-1588.

———, Khan, H. R., Winkley, B. R. and Robbins, L. G., 1972, Variability of river patterns: Nature (Phys. Sci.), v. 237, p. 75-76.

Smith, D. G., 1976, Effect of vegetation on lateral migration of anastomosed channels of a glacial meltwater river: Geol. Soc. Am. Bull., v. 87, p. 857-860.

Walker, R. G., 1976, Facies models 3: Sandy fluvial systems: Geoscience Canada, v. 3, p. 101-109.

Williams, P. F. and Rust, B. R., 1969, The sedimentology of a braided river: J. Sediment. Petrol., v. 39, p. 649-679.

L. VAN BENDEGOM: A NEGLECTED INNOVATOR
IN MEANDER STUDIES

J. R. L. ALLEN[1]

ABSTRACT

An analytical solution to the problem of the three-dimensional geometry of a river meander of specified plan shape, discharge, and calibre of sediment load was obtained more than 30 years ago by L. Van Bendegom, a Dutch hydraulic engineer. His work has since been almost totally overlooked. The measure of his achievement, however, is established by means of a review of the now substantial and rapidly growing literature concerned with this question, originating partly in engineering and partly in sedimentological circles. It is only in the past few years that the position reached by Van Bendegom has been regained.

INTRODUCTION

Science advances hesitantly, partly because of human frailty and limitations, and partly due to the barriers to communication imposed by language, geography, and educational system. Some of these points are well-illustrated in the field of fluvial sedimentology by the fate of the penetrating and comprehensive investigation made by L. Van Bendegom, an engineer of the Rijkswaterstaat in the Netherlands, into the movement of water and sediment in river meanders. Van Bendegom's paper setting out this work was published in 1947 in Dutch in the somewhat obscure journal *Die Ingenieur,* but appeared again in 1963 in an English translation by the National Research Council of Canada. It has scarcely been noticed since in either form, yet Van Bendegom succeeded in deriving a very satisfactory mathematical first approximation to the motion of water and sediment in meandering channels, at much the same time as there began in the U.S.S.R. the work that a decade later culminated in Rozovskii's (1961) outstanding and now well-known monograph. Van Bendegom's investigation has in the past 30 years been repeated in various ways, independently once in the field of fluvial sedimentology, and at least three times, in two cases apparently independently, within the field of hydraulic engineering.

It is my purpose in this paper to provide a short historical review of the development of analytical studies of meander geometry, partly because this is now appropriate in view of the growing literature on this topic, and partly to ensure that Van Bendegom's (1947) work is assigned its proper place as a major innovation. Water travelling round a river bend develops a helical spiral path, because of an imbalance between the radial pressure gradient imposed on fluid elements by the superelevation of the water surface and the centrifugal force on fluid elements, which varies in the vertical due to the action of viscosity. As a consequence of this secondary flow, which is directed toward the inner bank over the channel bed, the transported sediment tends to accumulate on the gently shelving point bar enclosed by the meander. In harmony, the zone of largest flow velocity moves towards the steep, erosive outer bank at a certain distance round the bend, and is then said to be fully developed. In order for an engineer to provide the best possible conditions for navigation on a river, and to design training works that will successfully withstand floods, it is essential that there exist a sound knowledge of the dynamics of channel bends. Fluvial sedimentologists also require this knowledge, because it underpins an understanding of processes in meanders and assists in the interpretation of textural and structural patterns in alluvial deposits.

[1]Sedimentology Research Laboratory, Department of Geology, The University, Reading, RG6 2AB, U.K.

Van Bendegom's (1947) Results

First Van Bendegom considered the forces acting on a fluid element travelling under conditions of steady flow round a channel bend. These are due to: (1) the longitudinal water-surface slope, (2) the transverse water-surface slope (superelevation), (3) the centrifugal force, and (4) the Coriolis force. Making use of conventional hydraulic relations, Prandtl's mixing-length theory, and a power-law velocity profile, he deduced that the mean bed shear-stress vector deviated from the direction of the mid-channel line (x-direction) by

$$\tan \delta = 10 \frac{h}{r} \tag{1}$$

in which δ is the deviation angle in the horizontal plane (positive when measured inward), h is the local flow depth, and r is the local radius of channel curvature (Fig. 1). The numerical constant of 10 followed from the choice of 50 m$^{\frac{1}{2}}$ s^{-1} for the Chezy bed friction coefficient (Darcy-Weisbach $f \approx 0.03$). Equation (1) states that the strength of the secondary flow is directly proportional to flow depth and inversely proportional to radius of flow curvature. It is the first major achievement of the analysis.

By adapting the relations which gave eq. (1) to the engineer's standard procedure for estimating a gradually varied flow by the method of channel segments, Van Bendegom was able to calculate in detail the flow path taken by water particles through a specified series of channel bends. Calculations of this sort were subsequently coupled to an analysis of the motion of sand particles travelling over the shelving point bar in a meander.

He considered the significant forces which determine the path of a sand particle travelling at the bed to be: (1) the component of the particle weight acting in the x-direction, due to the slope of the channel bed in the direction of the mean flow, (2) the component of the particle weight acting radially (r-direction), due to the transverse slope of the point bar, (3) the component of the fluid drag on the particle in the x-direction, and (4) the component of the fluid drag on the particle in the r-direction due to the secondary flow. The radial forces, for example, he wrote as

$$G = \frac{\pi (\sigma - \rho) gD^3 \sin \beta}{6} \tag{2}$$

$$F \propto \frac{\pi \rho D^2 U^2 h}{4r} \tag{3}$$

in which G and F are the weight and drag forces, respectively, σ the solids density, ρ the fluid density, g the acceleration due to gravity, D the particle diameter (assumed spherical), β the point-bar transverse slope, and U the local mean flow velocity. In eq. (3) a number of coefficients peculiar to Van Bendegom's analysis have for simplicity been omitted. When a channel bend of uniform width with a mobile bed has attained a stable shape (on the time-scale of the sediment motion), the particles on the bed travel on curved paths that lie parallel with the mid-channel line and with the banks, because the radial components of the particle body and drag forces are now in balance. Referring to eq. (1) and to his scheme of forces, Van Bendegom derived for this condition

$$\sin \beta = \frac{10 Sh^2}{Dr} \tag{4}$$

approximately, in which S is the local slope of the water surface. The slope combined with the square of the depth appear here because U^2 in eq. (3) is proportional to Sh in the conventional hydraulic relations. The numerical coefficient includes the sediment and fluid densities and a constant bed friction coefficient.

Equation (4) is Van Bendegom's second major achievement. According to this relationship, the local slope of the point bar is a decreasing function of the particle size and radius of curvature, an increasing function of the water-surface slope, and a steeply increasing function of the flow depth. Van Bendegom did not give in his paper an explicit solution to this equation, but it seems likely that he obtained one, as he used the formula in conjunction with his estimation of the varied flow to obtain by iteration the stable shape of a pair of bends through which a medium-coarse quartz sand was transported under conditions of steady discharge (Fig. 2). This calculation showed the deepest part of the channel to lie against the outer bank of the meander, a little downstream from the apex as in real rivers (Leopold and Wolman, 1960) and laboratory channels (Hooke, 1975). Similarly, the highest point on the point bar also lay further round the channel than the apex. Such realistic conclusions were not again achieved theoretically until Engelund (1974) and Gottlieb (1976) made their analyses almost thirty years later.

SUBSEQUENT ENGINEERING CONTRIBUTIONS

Van Bendegom's (1947) results were later used by NEDECO (1959, p. 526-533), a consortium of Dutch hydraulic engineers, in an investigation into the navigability of the Rivers Niger and Benue, West Africa. Choosing a numerical coefficient of 8 instead of 10 in eq. (1), and noting that for point bars $\sin \beta \approx \tan \beta = dh/dr$, they wrote eq. (4) as

$$\frac{dh}{dr} = \frac{12 \, \rho r_o S_o h^2}{D \, (\sigma - \rho) \, r^2} \tag{5}$$

in which σ and ρ are the sediment and fluid densities respectively, r_o is the radius of curvature of the outer bank, and S_o is the longitudinal water-surface slope measured at

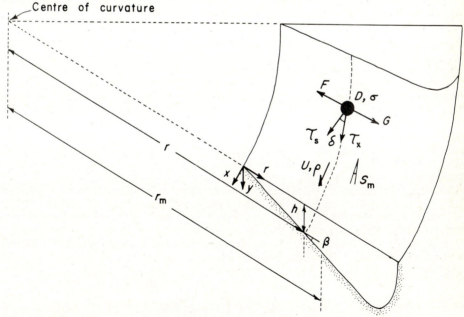

Fig. 1. General definition diagram for flow and sediment movement in a channel bend of uniform curvature and with steady fully-developed secondary flow. Note that the deviation angle is shown as lying in the plane of the bed, on the assumption that the transverse bed-slope is small. Strictly, the deviation angle is defined as lying in the *xr*-plane (Van Bendegom, 1947; Rozovskii, 1961).

this bank. The square of the local radius of curvature appeared in the denominator as the result of account being taken of the radial variation in the local longitudinal slope of the water surface

$$S = \frac{S_o \, r_o}{r} \tag{6}$$

around the bend. On substituting this equation into eq. (5), we regain Van Bendegom's eq. (4), in which the density ratio $\rho \, / \, (\sigma - \rho)$ is implicit in the numerical coefficient.

NEDECO (1959, p. 528) solved eq. (3) to obtain an expression for the radial profile of a point bar under known conditions of channel width, slope, curvature, and bed-material calibre. They obtained the flow discharge, however, only after calculating the cross-sectional profile.

Independently in the U.S.S.R., hydraulic engineers had by this time begun an intensive study of the dynamics of flow in river meanders. Rozovskii (1961) has summarized most of their work, together with his own salient contributions, in a monograph which first appeared in Russian in 1957. His approach to open-channel bends is essentially theoretical, with the emphasis falling overwhelmingly on the dynamics of the fluid flow. The only attempt to examine the behaviour of sediment is a tentative and purely kinematic study of the profile of a non-eroding bend. However, unaware of Van Bendegom's (1947) work, Rozovskii (1961, p. 194, 196) obtained for the deviation of the mean bed shear-stress vector under steady conditions in an open-channel bend where the secondary flow is fully developed the equation

$$\tan \delta = 11 \frac{h}{r} \tag{7}$$

where the symbols have their previous meanings. This relationship is identical except for the numerical constant to Van Bendegom's eq. (1). Rozovskii was able to justify it by reference to substantial field and laboratory data.

Interest amongst engineers in theoretical studies of the dynamics of meandering channels then shifted to the U.S.A. and to Japan, where considerable difficulties are encountered in designing training works on powerful, often flashy, and frequently unstable streams. Unaware of the work of Van Bendegom (1947) and NEDECO (1959), Yen (1970) wrote out the equations of motion of a near-bed sediment particle travelling under steady conditions around a stable channel bend with a fully developed secondary flow on the assumption that the particle is spherical and unaffected as regards its mean path by fluid turbulence. These equations are complex, and so will not be repeated here, but the system of forces they express differs significantly from Van Bendegom's (1947) in that the lift on the particle is included. Another difference is that Yen wrote his equations in terms of a coordinate system involving one angular and one curvilinear coordinate in addition to the vertical direction. On the basis of order-of-magnitude considerations, Yen reduced his equations to the following relationship for the local transverse slope of the point bar

$$\frac{\partial y}{\partial s} = - K \left(\frac{\rho}{\sigma - \rho} \right)^{\frac{1}{2}} \frac{U_r}{(gD)^{\frac{1}{2}}} \tag{8}$$

in which y is the elevation of the bed above an arbitrary datum, s is distance measured radially outward along the point bar surface, K is a general proportionality constant lumping all unknown coefficients, and U_r is the maximum radial component of the fluid

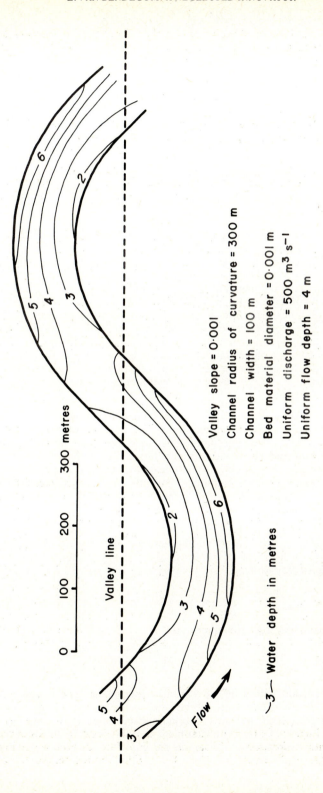

Valley slope = 0·001
Channel radius of curvature = 300 m
Channel width = 100 m
Bed material diameter = 0·001 m
Uniform discharge = 500 m³ s⁻¹
Uniform flow depth = 4 m

Fig. 2. Bottom contours for a pair of meanders in equilibrium with a secondary flow (after Van Bendegom, 1947, Fig. 7).

velocity measured near the bed. For small point bar slopes, $|\partial y/\partial s| \approx dh/dr$ in the previous coordinate system, whence

$$\frac{dh}{dr} = K \left(\frac{\rho}{\sigma - \rho} \right)^{\frac{1}{2}} \frac{U_r}{(gD)^{\frac{1}{2}}} \tag{9}$$

approximately. Yen's result has some resemblance to eq. (5). That dh/dr is an increasing function of the water surface slope is implied by the construction of K, while the presence of U_r in the numerator implies (see eqs. 1 and 7) that the bed slope is a decreasing function of the radius of curvature, but an increasing one of the flow depth. The equation explicitly shows that the point bar slope is a decreasing function of grain size, as NEDECO (1959) had found. Yen did not develop an expression for U_r, and so was able to solve his equation only on the basis of some assumed or empirical variation of U_r across the channel.

Likewise unaware of Van Bendegom's (1947) and NEDECO's (1959) results, Ikeda (1974, 1975) closely followed Yen (1970) in deriving the balance of forces acting on a bed-load particle travelling around a stable channel bend under conditions of fully developed secondary flow. He obtained an equation which can be written identically to Yen's eq. (8). However, the magnitude of the secondary flow in the channel was derived from a consideration of the equations of motion of a fluid in a curved channel as given by Rozovskii (1961), and on similar lines to him. The vertical distribution of the flow velocity in the x-direction was represented by the velocity-defect law. Ikeda then combined his equations for the fluid and sediment motion to obtain an expression for the cross-sectional shape of the point bar, the objective attained much earlier by Van Bendegom (1947) and NEDECO (1959). Ikeda later elaborated his analysis in collaboration with Kikkawa and Kitagawa (Kikkawa et al., 1976), and showed that it afforded results consistent with field and laboratory observations.

Suga (1967) made an additional but minor independent approach in Japan to the problem of the dynamics of open-channel bends. He derived two equations, one for the point bar profile during static equilibrium (no bed-load transport), and a second applicable during dynamic equilibrium (bed-load transport but no net sediment accumulation in the bend). Suga's approach, although partly analytical, does not rest on general principles and requires knowledge of numerous empirical constants appearing in equations to describe the flow and sediment transport. It has so far not been extended.

Engelund (1974, 1975) has led the second important independent attack within engineering circles on the problem of the stable cross-sectional profile of a point bar. Taking Rozovskii's (1961) analysis of the fluid motion, he obtained two equations for the deviation from the x-direction of the bed shear-stress vector under conditions of fully-developed secondary flow,

$$\tan \delta = 7 \frac{h}{r} \tag{10}$$

for an open channel under field conditions, and

$$\tan \delta = 21 \frac{h}{r} \tag{11}$$

for a rectangular conduit with a lid moving in its own plane. These equations have the same form, and numerical constants of the same order, as eq. (1) by Van Bendegom (1947) and NEDECO (1959), of whose work Engelund appears to have been ignorant. Gottlieb (1976) of Engelund's Institute later made a sophisticated analysis of the equations of motion to obtain expressions to describe the fluid flow in an open-channel bend of specified geometry.

The essential novelty of Engelund's (1974) work lies in the specification of the forces acting on a sediment particle travelling over the point bar. In the general case, the balance in the x-direction of forces acting on a particle travelling in the bend under conditions of fully developed secondary flow is written as

$$F = \left(\frac{\pi \, (\sigma - \rho) \, gD^3}{6} - L \right) \tan \alpha \cos \beta \tag{12}$$

in which F and L are the fluid drag and lift forces, respectively, and α is described as the sediment dynamic friction angle. But the forces on the particle have the component

$$\left(\frac{\pi \, (\sigma - \rho) \, gD^3}{6} - L \right) \sin \beta - F \tan \delta \tag{13}$$

in the radial direction whence, relating the transverse to the longitudinal force,

$$\tan \beta = \tan \alpha \tan \delta \tag{14}$$

in a channel of stable cross-sectional form. On substituting from eq. (10)

$$\frac{dh}{dr} = 7 \tan \alpha \; \frac{h}{r} \tag{15}$$

approximately, and then integrating, Engelund obtained

$$h = Cr^{\,7 \tan \alpha} \tag{16}$$

where C is an integration constant, to describe the cross-sectional shape of a stable point bar. It is interesting that Saplyukov (1973) has also used a power relationship to describe point bar profiles, though no derivation is given. Since Engelund took $\tan \alpha$ as a constant for the purpose of his experiments, eqs. (15) and (16) describing the form of the bar include no explicit dependence on grain size, which is not the case with Van Bendegom's (1947) eq. (1) or with eq. (9) advanced independently by Yen (1970) and Ikeda (1974, 1975). Engelund offers no discussion of this important difference, as he makes no reference to these earlier studies.

The remainder of Engelund's (1974) and Gottlieb's (1976) contribution is concerned with the calculation of the three-dimensional form of a meandering channel of specified discharge and plan geometry, on the basis of the flow-equations and eq. (15). Engelund's first approximation to the solution of this problem is a channel that is deepest along the outer bank of the meander, but a pattern of bed contours that is symmetrical about the axis of meander symmetry. His second approximation, together with Gottlieb's result, is more realistic. Like Van Bendegom (1947) before them, Engelund and Gottlieb found a phase shift between the point of maximum curvature of the meander in plan and the deepest and shoalest parts of the point-bar, in conformity with both field and laboratory evidence (Leopold and Wolman, 1960; Hooke, 1975).

An important recent contribution is from Zimmerman (1977), who showed theoretically and confirmed experimentally that the coefficient in eqs. (1) and (7) for the deviation angle is sensitive to the bed friction coefficient. Rozovskii (1961) had previously argued that the coefficient is virtually a constant.

Contributions From Sedimentologists

The quantitative study of meander dynamics by sedimentologists began much later than in the engineering field, but in harmony with an awakening of interest in fluvial processes and deposits. Not knowing of the results obtained by Van Bendegom (1947) and NEDECO (1959), Allen (1970a, 1970b, 1971) deduced that the forces acting radially on a spherical sediment particle travelling over the bed of a meander of stable cross-sectional form (Fig. 1) are the component of the particle weight acting downslope

$$G = \frac{\pi (\sigma - \rho) gD^3 \sin \beta}{6} \qquad (17)$$

and the fluid force acting upslope

$$F = \frac{\pi D^2 \tau_s \sin \delta}{4} \qquad (18)$$

in which τ_s is the mean bed shear-stress vector. In contrast to Yen (1970) and Ikeda (1974, 1975), but in harmony with Van Bendegom (1947), the lift force is omitted as likely to be negligible for moving bed-load particles (Bagnold, 1974). These equations were combined with conventional hydraulic relationships and arbitrary functions describing the point bar cross-sectional form to obtain expressions for the distribution of grain size across a stable meander bend in a zone of fully developed secondary flow. For example, the equation relating particle size to the local slope of the point bar is

$$\frac{dh}{dr} = \frac{3 \tau_x \tan \delta}{2 (\sigma - \rho) gD} \qquad (19)$$

approximately, in which $\tau_x = \tau_s \cos \delta$ is the component of the bed-shear stress vector parallel with the x-direction. This equation is almost identical with Van Bendegom's (1947) eq. (4), as τ_x is proportional to Sh. The distribution of grain size over the surface of the bar given by eq. (19) is, of course, the same as the distribution in the corresponding vertical profile. In a further extension of the analysis, Allen obtained the distribution of bed forms (internal sedimentary structures) by reference to empirically and theoretically established limits on the existence of these features. Bridge (1975a, 1975b) has considerably widened the scope of this analysis by including some of the effects of time-variation of flow discharge, flood-plain aggradation, and meander growth pattern. His results, obtained using a computer providing a partly graphical output, amount to the view of a point bar that might be obtained from a large quarry-face or cliff.

The existence of the analysis by Van Bendegom (1947) and NEDECO (1959) had become known by the time Allen (1977) published a paper in which the growth of a simple meander was considered. The transverse variation of the water surface slope is expressed as

$$S = \frac{S_m r_m}{r} \qquad (20)$$

in which S_m and r_m are the slope and radius at the channel centreline, respectively, since eq. (4) proposed by NEDECO (1959) is incorrect for a sequence of bends (see also Engelund, 1974; Bridge and Jarvis, 1976). Recognizing that Rozovskii (1961) originally defined the deviation angle δ as lying in the horizontal plane, the equation for the local transverse slope of the point bar should strictly be written from eqs. (5), (19) and (20) as

$$\sin \beta \cos \beta = \frac{33 \rho S_m r_m h^2}{2D (\sigma - \rho) r^2} \qquad (21)$$

However, provided that β is a small angle, $\sin \beta \cos \beta \approx \tan \beta = dh/dr$. On this assumption, Allen (1977) solved eq. (21) to obtain

$$h = \frac{2D (\sigma - \rho) r}{33 \rho S_m r_m [1 - 2D (\sigma - \rho) Cr]} \qquad (22)$$

where C is an integration constant. This equation resembles an expression given in a corresponding form by NEDECO (1959, p. 528), but differs in the numerical constants and the use of S_m rather than S_o. Allen calculated C in cases of known channel width and flow discharge by invoking the condition of continuity. Although there are objections to the use of eq. (22) close to channel edges, mathematically because of the assumption made with respect to β, and physically because of the vertical components of flow, it does afford cross-sections that are tolerably realistic and gives a valuable insight into the controls on some observed patterns of meander growth (Popov, 1965; Kondrat'yev, 1969; Hickin and Nanson, 1975). This work has since been considerably extended, for example, to include phase effects between channel plan and flow.

Bridge (1976, 1977) combined the Van Bendegom-Allen analysis for grain size in relation to point bar transverse slope, eq. 19, with Engelund's (1974) simple relationship, eq. 14, for the transverse profile, but using Rozovskii's (1961) numerical constant pertaining to the deviation angle for preference. In Engelund's equation, there is no explicit dependence of point bar transverse slope on grain size, because tan α is taken to be a constant for the purpose of specific experiments. Bridge has explicitly interpreted tan α as Bagnold's (1954) dynamic friction angle relating to the normal and tangential collision forces arising between sheared grains. It is then theoretically possible for point bar transverse slope to become dependent on grain size, at least under certain circumstances.

When bed-load transport is fully developed, tan α ranges between two limits (Bagnold, 1954, 1956, 1966), between which it is inversely related to grain size for a constant non-dimensional bed shear stress. The lower limit describes the development of wholly inertial conditions of grain-shearing, and the upper limit the occurrence of wholly viscous shearing conditions. Beyond these limiting conditions, tan α is independent of grain size, when point bar slope should also become independent of the calibre of bed-load sediment. In contrast, the equations developed by Van Bendegom (1947), NEDECO (1959), Allen (1970a, 1970b, 1971, 1977), Yen (1970) and Ikeda (1974, 1975) all indicate that the transverse slope is inversely proportional to grain size for all calibres of particle.

An inverse relationship between transverse slope and grain size applicable at all calibres, however, would become possible for conditions of incompletely developed bed-load transport, provided that one assumed with Bagnold (1966) that in this regime tan α varied inversely as the size. As yet there is no experimental evidence against which to test this assumption, but there are objections of a general nature. When bed-load transport is incompletely developed, the grains in motion are relatively dispersed over the bed, and so behave individually. Bagnold's concept of the dynamic friction angle pertains not to individual particles, but to a continuum composed of grains densely arrayed and sheared together in a fluid. Even the variation of tan α with grain size when bed-load transport is fully developed has not passed without question. Bagnold (1973) has latterly seemed to shift from his original position, indicating that the collision-related friction angle is a universal constant of the same order as the static friction angle. Clearly, the use of tan α in the specification of the forces acting on bed-load particles in meander bends demands closer study, together with actual variations of point-bar transverse slope under as wide a range of transport conditions as can be achieved in the field and the laboratory.

CONCLUSIONS

(1) A theoretical solution to the problem of the stable (on the time-scale of sediment motion) form of a river meander was first obtained by Van Bendegom (1947) working in the Netherlands. The question has since been studied independently at different times by several investigators in the field of hydraulic engineering and in sedimentology. But the spread of knowledge within and between these fields has been slow, with the result that perhaps progress has been delayed and effort spent unnecessarily.

(2) There is widespread agreement that the strength of the secondary flow, as measured by the deviation of the mean bed shear-stress vector from the mean flow direction, can be adequately described by an expression of the form $\tan \alpha = kh/r$, where k is a numerical constant (or a friction-related coefficient) of order 10.

(3) There is less agreement regarding the effect of grain size on point bar slope. Whereas the independent analyses of Van Bendegom (1947), Yen (1970), Ikeda (1974, 1975), and Allen (1970a, 1970b, 1971, 1977) suggest that the transverse slope bears a simple inverse relationship to grain size, Bridge (1976, 1977), who bases his approach partly on Engelund's (1974, 1975) work, implies that a more complex inverse dependence may obtain.

ACKNOWLEDGEMENTS

I am indebted to Dr. J. S. Bridge and Dr. P. F. Friend for suggestions which have led to an improvement of the manuscript.

REFERENCES

Allen, J. R. L., 1970a, Studies in fluviatile sedimentation: a comparison of fining-upwards cyclothems, with special reference to coarse-member composition and interpretation: J. Sediment. Petrol., v. 40, p. 298-323.

———, 1970b, A quantitative model of grain size and sedimentary structures in lateral deposits: Geol. J., v. 7, p. 129-146.

———, 1971, Rivers and their deposits: Science Progress, Oxford, v. 59, p. 109-122.

———, 1977, Changeable rivers: some aspects of their mechanics and sedimentation: in K. J. Gregory, ed., River Channel Changes, Wiley, London, p. 15-45.

Bagnold, R. A., 1954, Experiments on a gravity-free dispersion of large solid spheres in a Newtonian liquid under shear: Proc. Royal Soc. London, v. A225, p. 46-63.

———, 1956, The flow of cohesionless grains in fluids: Phil. Trans. Royal Soc. London, v. A249, p. 235-297.

———, 1966, An approach to the sediment transport problem from general physics: U.S. Geol. Survey Prof. Paper 422-I, 37p.

———, 1973, The nature of saltation and of 'bed-load' transport in water: Proc. Royal Soc. London, v. A332, p. 473-504.

———, 1974, Fluid forces on a body in shear-flow: experimental use of 'stationary flow': Proc. Royal Soc. London, v. A340, p. 147-171.

Bridge, J. S., 1975a, Computer simulation of sedimentation in meandering streams: Sedimentology, v. 22, p. 3-43.

———, 1975b, Mathematical model and Fortan IV program for computer simulation of sedimentation in meandering streams: Computer Applications, v. 2, p. 217-266.

———, 1976, Bed topography and grain size in open channel bends: Sedimentology, v. 23, p. 407-414.

———, 1977, Flow, bed topography, grain size and sedimentary structure in open channel bends: a three dimensional model: Earth Surf. Proc., v. 2, p. 401-416.

——— and Jarvis, J., 1976, Flow and sedimentary processes in the meandering River South Esk, Glen Clova, Scotland: Earth Surf. Proc., v. 1, p. 303-336.

Engelund, F., 1974, Flow and bed topography in channel bends: J. Hydraulics Div., Am. Soc. Civil Eng., v. 100, no. HY 11, p. 1631-1648.

———, 1975, Instability of flow in a curved channel: J. Fluid Mechanics, v. 72, p. 145-160.

Gottlieb, L., 1976, Three-dimensional flow pattern and bed topography in meandering channels: Institute of Hydrodynamics and Hydraulic Engineering, Technical University of Denmark, Series Paper no. 11, 79p.

Hickin, E. J. and Nanson, G. C., 1975, The character of channel migration on the Beatton River, northeast British Columbia, Canada: Geol. Soc. Am. Bull., v. 86, p. 487-494.

Hooke, R. leB., 1975, Distribution of sediment transport and shear stress in a meander bend. J. Geol., v. 83, p. 543-565.

Ikeda, S., 1974, On secondary flow and dynamic equilibrium of transverse bed profile in alluvial curved open channel: Proc. Japanese Soc. Civil Eng., no. 229, p. 55-65.

———, 1975, On secondary flow and dynamic equilibrium of transverse bed profile in alluvial curved open channel: Proc. 16th Conf., Internat. Assoc. Hydraulics Res., v. 2, p. 105-112.

Kikkawa, H., Ikeda, S. and Kitagawa, A., 1976, Flow and bed topography in curved open channels: J. Hydraulics Div., Am. Soc. Civil Eng., v. 102, no. HY 9, p. 1327-1342.

Kondrat'yev, N. Y., 1969, Hydromorphological principles of free meandering. 1. Sign and indexes of free meandering: Soviet Hydrology, 1968, p. 309-335.

Leopold, L. B. and Wolman, M. G., 1960, River meanders: Geol. Soc. Am. Bull., v. 71, p. 769-794.

NEDECO (Netherlands Engineering Consultants), 1959, River Studies and Recommendations on Improvement of Niger and Benue: North-Holland, Amsterdam, 1000p.

Popov, I. V., 1965, Hydromorphological principles of the theory of channel processes and their use in hydrotechnical planning: Soviet Hydrology, 1964, p. 188-195.

Rozovskii, I. L., 1961, Flow of Water in Bends of Open Channels: Israel Program for Scientific Translations, Jerusalem, 233p.

Saplyukov, F. V., 1973, Development of river meandering: Soviet Hydrology, 1971, p. 465-468.

Suga, K., 1967, The stable profiles of the curved open channels: Proc. 12th Congr. Internat. Assoc. Hydraulics Res., v. 1, p. 487-495.

Van Bendegom, L., 1947, Eenige beschouwingen over riviermorphologie en rivierbetering: De Ingenieur, v. 59(4), p. 1-11. (National Research Council of Canada, Technical Translation No. 1054, 1963).

Yen, C. L., 1970, Bed topography effect on flow in a meander: J. Hydraulics Div., Am. Soc. Civil Eng., v. 96, no. HY 1, p. 57-73.

Zimmerman, C., 1977, Roughness effects on the flow direction near curved stream beds: J. Hydraulic Res., v. 15, p. 73-85.

ADJUSTMENT OF FLUVIAL SYSTEMS TO CLIMATE AND SOURCE TERRAIN IN TROPICAL AND SUBTROPICAL ENVIRONMENTS

Victor R. Baker[1]

Abstract

Rivers in western Brazil and northeastern Argentina are used to illustrate the complex adjustment of fluvial patterns and sediments to the shifting arid-humid climatic regimens of the tropics and subtropics. Many of these rivers show pronounced deviations from the sediment size-sinuosity relationships that were developed by studies of rivers in temperate semi-arid and humid environments. In the Amazon Basin, a broad range of sinuosity (P) and floodplain depositional features occurs in rivers that lack appreciable bedload. Rivers draining the tropical lowlands, e.g. the Juruá River, have very regular streamflow, nearly all solution and suspended load, numerous meander cutoffs, poorly developed point bars and scroll topography, and single channels of high sinuosity (P = 1.9-3.0). Major trunk rivers draining Andean source terrains, e.g. the Solimões River, have pronounced anastomosing patterns of lower sinuosity (P = 1.2-1.5), prominent scroll topography on flood plains, and high suspended load. These differences result from the relative abilities of different rivers to rework coarse, relict alluvium that was deposited during the relatively arid full-glacial phases of the Pleistocene.

Rivers of the Gran Chaco of northeastern Argentina display a detailed record of altered hydrologic regimen during the Quaternary. Many streams show a late Quaternary reduction of channel-forming discharge that always produced greatly reduced channel widths but only occasionally produced reduced meander wavelengths. However, not all these changes need be attributed to temporal climatic change. Major changes in both sediment load and discharge can be achieved in this piedmont environment through stream captures that delete or add the Andean source terrain to a given river's drainage basin.

Hydrologic calculations show that empirical formulae derived from studies of humid-temperate and semi-arid streams do not necessarily explain the regimen behavior of these complex tropical rivers. The low-sinuosity Amazon Basin streams are much less dominated by relatively coarse-grained sediment than are streams of equivalent sinuosity described for semi-arid regions by S. A. Schumm. Discharge characteristics and bank stability (related to both vegetation and lithology) appear to play the dominant role in accounting for channel pattern variations in the Amazon Basin.

Introduction

Fluvial sedimentologists may be justifiably pleased that their discipline has now progressed from the incipient scientific phase of mere description to the more advanced phase of prediction. The current trend that heralds this change is the formulation of facies models. Ultimately the success of any fluvial model will depend, among other factors, on its ability to predict new geological situations and to serve as a basis for hydrodynamic interpretation (Walker, 1976). This paper will address critically only one aspect of the evolving fluvial facies models: their ability of make quantitatively precise paleohydrologic predictions.

Paleohydrology is the science of terrestrial water, its distribution, occurrence, and movements on ancient landscapes. This developing subdiscipline is an excellent example of the geological exercise that is termed a "think back" by Bayly (1968). The genesis of a suite of ancient fluvial sediments cannot be solved by even the most meticulous description of their attributes as they exist today. Questions must be asked — What were the conditions under which these sediments were deposited? What were the ultimate controls on sedimentation and erosion in the river basins? The answers to such questions can only be achieved by making certain assumptions about sedimentary environments as they existed in the past. Paleohydrology follows the uniformitarian principle in making such assumptions.

---

[1]Department of Geological Sciences, The University of Texas at Austin, Austin, Texas 78712

A schematic linkage of processes and information involved in performing a paleohydrologic think back is illustrated in Fig. 1. Only the information of level 1 ("Field Evidence") is available at the outcrop. Information for the succeedingly higher levels must be reconstructed according to models of fluvial processes. The success of the reconstruction will depend on the applicability of the chosen model to the fluvial system of interest. The reconstruction can occur by deduction, proceeding from the general rules that govern the higher levels of information to the implications of those rules for lower levels of information. Alternatively the reconstruction can be by induction, proceeding by extrapolation from the secure details of the outcrop to general predictions of information at higher levels. In practice, the more successful investigators will combine both approaches, using the most secure predictions of one method as checks on the more tenuous implications of the other method.

A relatively recent approach to the paleohydrologic think back is to evaluate certain aspects of the chain in at least a semi-quantitative fashion. The linkage of sedimentary textures to local hydraulic conditions (level 2) can be estimated by paleohydraulic analysis (e.g. Baker, 1974). A more generalized approach derives from the principle that the habits or characteristics of individual streams, termed their "regimen", can sometimes be expressed by empirical formulae (level 3). Most important are the relationships between channel morphology, sediment load, and streamflow derived from Schumm's (1960, 1963) studies of rivers in the Great Plains of the United States and of the Murrumbidgee River of the Australian Riverine Plain (Schumm, 1968a). The results of these studies provide the only rigorous model that we have for evaluating levels 3 and 4 in the paleohydrologic think back (Fig. 1). This model has been applied to certain ancient fluvial sediments (Schumm, 1968b, 1972; Cotter, 1971) and to studies of modern river metamorphosis (Schumm, 1969). In this paper Schumm's model will be considered in relation to modern and Quaternary fluvial systems in several tropical and subtropical environments. The reader should remember, however, that the model was not derived from data obtained in such environments. The following test is merely for universality of the model in fluvial paleohydrologic work. The model has already proven its applicability to temperate environments.

The most general of the known linkages in the paleohydrologic think back is that connecting the modern field evidence to the climatic and source terrain controls (Baker and Penteado-Orellana, 1978). The basic principles of this linkage (level 5) were eloquently stated by Joseph Barrell (1908). He argued in classical deductive terms that widespread sand and conglomerate formations might be interpreted as the result of climatic change. Moreover, he recommended that this hypothesis should be tested by studies of Quaternary fluvial systems because of the relative ease in interpreting climatic change where large segments of the drainage systems are still preserved. This paper will follow Barrell's suggestion while also providing an overall framework for testing the more specific fluvial regimen approach to paleohydrology.

THE GRAN CHACO

The Gran Chaco of northern Argentina and western Paraguay is a great alluvial plain about 650 km wide that extends from the Andean foothills to the Rio Paraguay. Only a few modern rivers, notably the Rio Pilcomayo and the Rio Teuco - Rio Bermejo, completely traverse this plain from sources in the relatively dry rain shadow of the eastern Andes. Gradients for these streams range from 2 m/km (10 ft/mi) in the Andean foothills to 0.2 m/km (1 ft/mi) at the Rio Paraguay.

The streams studied in this report occur in the Provincia de Formosa, between the Rio Bermejo and the Rio Pilcomayo in northern Argentina (Fig. 2). Channel morphology was mapped from color infrared photographs of the NASA Earth Resources Aircraft Program Mission 97 on July 19, 1969. The photographs were taken along a flightline that followed a railroad for about 340 km from the city of Formosa to between Juan G. Barzan and Pozo

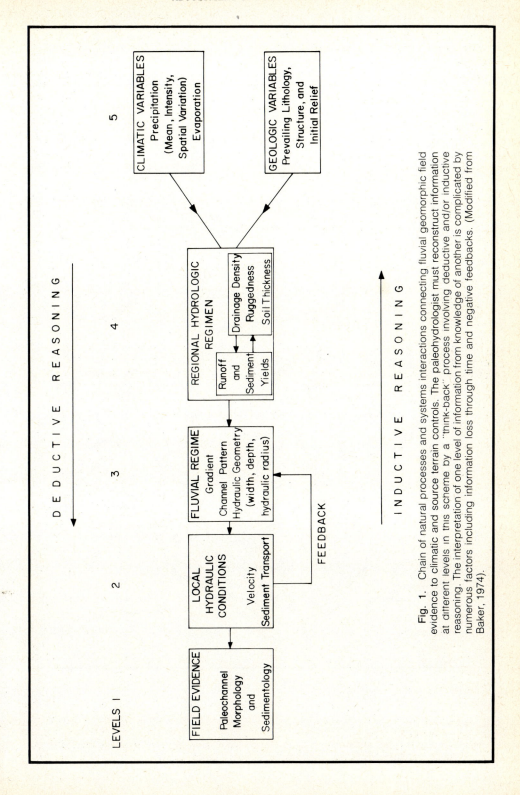

Fig. 1. Chain of natural processes and systems interactions connecting fluvial geomorphic field evidence to climatic and source terrain controls. The paleohydrologist must reconstruct information at different levels in this scheme by a "think-back" process involving deductive and/or inductive reasoning. The interpretation of one level of information from knowledge of another is complicated by numerous factors including information loss through time and negative feedbacks. (Modified from Baker, 1974).

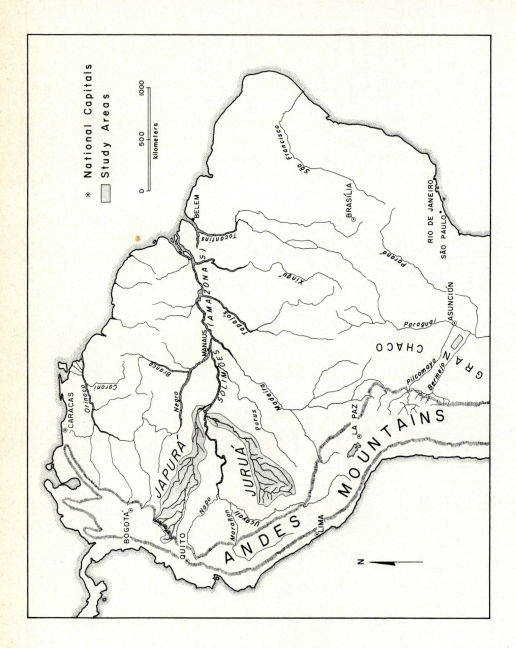

Fig. 2. Location map showing the Gran Chaco study area and the drainage basins of the Japurá, Juruá, and Solimões Rivers of the western Amazon Basin.

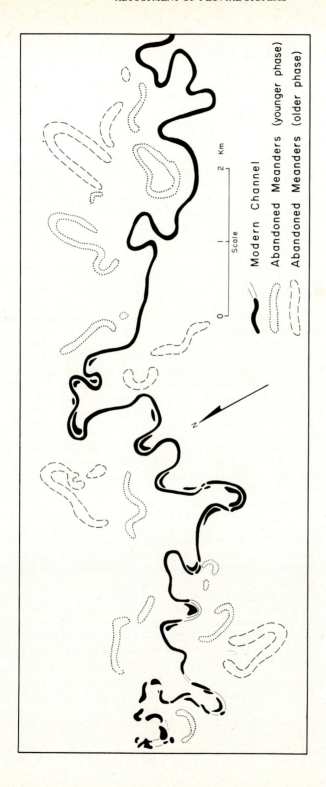

Fig. 3. Geomorphic map of a typical underfit stream on the Gran Chaco of northern Argentina. The approximate location of this map is shown in Fig. 2.

del Mortero. A complete description of the imagery and of the regional hydrologic conditions is given by Conway and Holz (1973).

The modern climate of this region is transitional from a tropical savanna climate in the northwest to a humid maritime climate farther east. The annual rainfall averages 700-1000 mm, but most of this comes in the summer months. The months of May through September are nearly rainless. Rainfall variability increases westward as mean annual rainfall decreases. The variation coefficient ranges from 30% at the study site to 50% at the headwaters of major rivers draining into Provincia de Formosa (Prohaska, 1976).

Streams in the study area display several varieties of underfitness as classified by Dury (1964). Manifest underfit streams generally meander with high sinuosity and short wavelength within much more ample valleys of relatively long wavelength. Another variety of underfit stream is one which lacks definite borders cut into the alluvial plain. Such streams have significantly smaller wavelengths than do their surrounding abandoned meanders (Cucchi, 1973). Another variant, however, has a modern channel of similar wavelength to the abandoned meanders, but it has a smaller channel width than that of its immediate predecessors. Dury (1966, 1970) has named this variety an "Osage-type" underlift stream. Such streams supposedly suffer a reduction of channel-forming discharge in such a way that channel width and pool-and-riffle spacing adapt to the change, but meander wavelength does not.

Abandoned meanders adjacent to a typical Osage-type underfit stream in the study area can be grouped into size classes based on their relict paleochannel width (Fig. 3). The oldest abandoned meanders are the widest, and the modern channel represents the narrowest width. Another abandoned channel is intermediate in width between these two extremes. Ratios of meander wavelengths to channel width (λ/W) indicate that the two abandoned channels fall within the normal range (8-11) for stable river channels. However, the ratio (λ/W) for the modern river (24) is much higher that the typical maximum value of 11. The prominance of larger prior stream landforms around the modern channel suggests that the overlarge wavelength of the relatively narrow modern channel has been inherited from the larger prior streams.

Underfit stream morphology results from a decrease in the channel-forming discharge. The remarkable preservation of Quaternary underfit streams has led to several suggestions for using empirical correlations of meander wavelength, channel width, and sediment size to discharge as an aid in paleohydrologic interpretation (Dury, 1965, 1976; Carlston, 1965; Schumm, 1969). Such interpretations use the assumption of Schumm (1968a) that any difference in paleochannel morphology should reflect changes in the sediment load and in the runoff from the catchment of the river under study. The most exhaustive approaches to this application are those of Dury (1976) and Schumm (1969). Table 1 illustrates the discharge estimates that can be made by directly applying the appropriate empirical formulae to the channels. The two estimates are not necessarily very different because the most probable flood discharge (Qp) is always considerably in excess of the mean annual discharge (Qm). Table 1 can be viewed as an exercise that quantifies the observation that channel-forming discharge has progressively declined for the evolving river shown in Figure 3. Unfortunately there are several unresolved problems that must be addressed before such a quantitative understanding can be fully accepted (Baker and Penteado-Orellana, 1977).

The first problem involves the frequencey of discharge that dominates in producing a morphological change. Dury (1965, 1968) has argued that the causative discharge for a given meander wavelength and channel width is the bankfull discharge, with an average recurrence interval of 1.58 years (i.e. the most probable annual flood). This hypothesis is also supported by Wolman and Miller (1960) who suggest that channel shape as well as dimensions of meandering rivers appear to be associated with bankfull stage flows, which recur on the average once every year or two. Most studies emphasize alluvial rivers in

TABLE 1
PALEOCHANNELS OF THE GRAN CHACO

	Modern Channel	Younger Abandoned Channel	Older Abandoned Channel
Width (W)	24m	55m	90m
Wavelength (λ)	580m	580m	630m
Sinuosity (P)	2.2	2.2	2.0+
λ/W	24	10	7
Silt-Clay % (M*)	30	30	~30
Qp* (m³/sec)	44	200	550
Qm* (m³/sec)	6	55	260

* Parameters Calculated from the following:

$P = 0.94\ M^{0.25}$ (Schumm, 1963)

$Qp = (W/2.99)^{1.81}$ (Dury, 1976)

$W = 37Q\ m^{0.38}/M^{0.39}$ (Schumm, 1969)

humid climates, where most of the work is done during relatively frequent events. However, evidence from streams in relatively arid regions (Baker, 1977), characterized by highly variable flow regimes, indicates that major fluvial channel transformations take place during relatively infrequent flows. The second problem, demonstrated by Schumm (1967, 1968a), is that meander wavelength is dependent not only on water discharge, but also on the type of sediment transported by the stream. The meander wavelength for bedload type streams will be greater than the meander wavelengths of suspended-load channels of similiar mean annual discharge. Moreover, the two problems are really interwoven, since relatively coarse sediment poses a discharge threshold for its removal that can only be accomplished by a less frequent, larger flood. In other words, a coarser load will necessitate a change in the channel-forming discharge. Until these problems are resolved, estimates such as given in Table 1 will be quantitatively suspect, although they may still provide a useful qualitative picture of river metamorphosis.

The causes of underfitness pose even more problems than its quantification. Dury (1965, 1970) has argued a world-wide cause of underfit streams induced by reduced bankfull (most probable flood) discharges that were brought about by climatic change. Cucchi (1973, 1974) followed this suggestion for the underfit streams of the Gran Chaco. However, Iriondo (1974) provided an alternative hypothesis. Rivers of the Gran Chaco, he noted, have experienced amazing changes in their courses. In 1871 the Rio Bermejo changed its course by capture through a reach of over 800 km, and the abandoned channel became an intermittant stream, now known as Antiguo Cauce del Rio Bermejo (Soldono, 1947). The Rio Pilcomayo changed course nine times between 1917 and 1970 (Cordini, 1947, and Noriegas, 1971). Flooding associated with the Rio Pilcomayo course change of 1967 created a large marsh that fed several small streams (Noriegas, 1971). It is clear that changes in river course can affect streamflow in various ways.

The juxtaposition of alluvial plains and active orogenic belts like the Andes produces an ideal situation for creating pronounced changes in river courses. As shown in studies by Mackin (1936) and Ritter (1972), rivers draining mountainous source terrains often have steeper gradients than nearby rivers that head on the piedmont. The piedmont rivers receive only the weathered sediments of the alluvial plain; their finer load caliber results in gentler gradients than the adjacent mountain-draining rivers. This disparity of gradients on adjacent rivers produces a disequilibrium that can only be resolved by capture on the alluvial plain. Streams that formerly drained the mountains are abandoned, thereby reducing their channel-forming discharge. In this way the locus of active mountain sedimentation shifts about the alluvial plain. Moreover, the paleochannels of the plain will show immense lateral and temporal variations in channel-forming discharges as a result of the captures. These discharge changes could easily be misinterpreted as climatic changes, especially in the case of ancient systems which often only preserve portions of an entire fluvial landscape.

THE AMAZON BASIN

It could be argued that the rivers of the Gran Chaco occur in a transitional semi-arid to humid environment that is similar to the one studied by Schumm (1963, 1968a, 1969, 1972). The next example will consider rivers in a decidedly different environment, the western Amazon Basin.

The Amazon is easily the world's largest river. The flood discharge recorded near Óbidos (near river mouth) on June 5, 1972 totalled 255,000 m^3/sec, the largest discharge ever measured on earth (Sternberg, 1975). The Amazon has a mean annual discharge of 160,000 m^3/sec (Sternberg, 1975), four times the mean annual flow of the second largest river, the Zaire (formerly named the "Congo"). Approximately 15 per cent of all fresh water passing from the continents to the oceans is discharged by the Amazon. Certainly the Amazon Basin constitutes one of the most important fluvial environments on earth. How do the various empirical equations apply to describing fluvial regimen in this region?

Regional Setting

The Amazon River (Fig. 2) drains an area of over 6 million km^2 (2.3 million mi^2). The lower Amazon (downstream from Manaus) flows between two stable crustal blocks, the Guiana and Central Brazil Shields. To the west the basin widens as high-discharge tributaries drain the Cis-Andean Plains, which slope eastward from the Andes of Columbia, Peru, and Bolivia. Surrounded by uplands to the north, south, and west, the Amazon Basin experienced active sedimentation throughout the late Tertiary and Quaternary (Neogene). The most rapid sedimentation occurred in the west as coarse debris was shed eastward from the Andes. This Neogene alluvium thickly mantles interfluves in the basins of the Solimões, Ucayali, Japurá, Juruá, Purus, Javari, and Madeira Rivers. East of Manaus the structural trough of the lower Amazon, only 160 km wide between the two shields, was also filled with Neogene sediments.

The gradient of the Amazon mainstem is exceptionally gentle. The average slope is less than 0. 1 m/km from the Andean foothills of Peru to the Atlantic Ocean. Downstream from Manaus, the lower 1440 km of the Amazon has a gradient of less than 0.03 m/km (2 inches/mile). Through much of this lower reach the flood plain is remarkably wide, nearly 100 km in some places. The reasons for these relationships were pointed out by Russell (1958), who described the lower Amazon as a deltaic plain. The late-glacial Amazon was deeply entrenched during the last low sea level stand. The subsequent sea level rise (Flandrian Transgression) caused the river level to rise. The rising base level transformed the lower Amazon into a vast freshwater gulf or *ria* (Sternberg, 1975). As sea level began to stabilize in the Holocene, the western Amazon rapidly built a delta into this gulf. Nurtured by the high sediment loads of its Andean tributaries, the Amazon filled its ria so rapidly from the west that its flood plain (*varzea*) now blocks the mouths of huge tributaries like the Tapajós and Xingu. Today these rivers flow into great wedgelike lakes, called *rias fluviales,* where their mouths are blocked by the Amazon *varzea.*

The entire Amazon Basin is today characterized by heavy rainfall, but the highland margins, especially on the Andean west, usually receive well over 2000 mm of precipitation yearly; some areas receive 3000 mm. The "wet", or rainy, season is from

January to June, and the "dry" season is really just a period of less rain that extends from July to December. This climate supports the world's largest single expanse of forest. The tropical rain forest is the dominant vegetation of the Basin; although on some of the highlands, especially those underlain by crystalline rocks, the forest is replaced by grassy savanna with scattered trees. The rain forest maintains and is, in turn, maintained by highly leached soils. The luxuriance of the vegetation derives from the perfect climatic conditions for vegetative growth rather than from the fertility of the soils.

River Sediments

The Neogene uplands of the Amazon basin are called "terras firmes." Active Holocene river flood plains, or *varzeas,* are inset into these uplands. The rivers flowing on these flood plains are distinguished by their sediment loads. In the local descriptive language the rivers that lack appreciable suspended load are called "green rivers". Typical of this class are the Amazon's major southern tributaries, the Tapajós and Xingu Rivers. The western Amazon tributaries are called "rios brancos" (white rivers). The color, really more a muddy-yellow than a white, is caused by the very large suspended sediment loads. The important "rios brancos" are the Solimões (western Amazon), Juruá, Purus, Madeira, Javari, and Japurá Rivers (Fig. 2). Several rivers draining from the north are colored black by organic solutions. These include the Nego, Jari, and Trombetas Rivers.

The different sediment-load characteristics of the Amazon tributaries are directly attributable to sedimentary source terrains. The "black" and "green" rivers drain crystalline shield areas and sandstone plateaus. The Negro, Branco, and other northeastern tributaries drain the Guiana Shield, while the Tapajós and Xingu Rivers drain the Central Brazil Shield. In contrast the "rios brancos" drain the Neogene sedimentary blankets that spread eastward from the Andes across the Cis-Andean Plains. Neogene sediments also fill the narrow, fault-bounded block between the two shields downstream from Manaus.

The Modern Regimen of Rivers in the Amazon Basin

The preliminary results cited here are based on examinations of color photographs generated during the Apollo-Soyuz space mission of July, 1975 (Holz *et al.,* in press). The orbital photographic study was locally supplemented with large-scale side-looking radar imagery (SLAR) generated by the Brazilian government for their project "Radar Amazon" (RADAM). Descriptions of ground conditions were supplied by Dr. M. M. Penteado-Orellana (written communication, 1977), and hydrologic information was summarized from the existing Brazilian government reports.

Three rivers are chosen to illustrate the remarkable variations in fluvial regimen and morphology that exist in the Amazon Basin. Two of these, the Japurá and Juruá, are tributary to the third, the Solimões (Brazilian name for the western Amazon River mainstem). All three join near Fonte Boa, Brazil, where their diverse character is highly evident (Fig. 4). The properties of these rivers are compared in Table 2.

The channel pattern of the Solimões River in western Brazil is both anastomosing and meandering. This poses an immediate problem for regimen analysis employing sinuosity, because the significance of sinuosity for anastomosing channels has not been adequately investigated. Nevertheless, the relevant parameters can be easily measured (Table 2). It is only their significance that remains in doubt. Whereas the empirical equations predict a width/depth ratio of 23 for the Solimões River, actual measurements downstream at Manaus (where the ratio should be even lower) indicate ratios of 40-60 (Oltman *et al.,* 1964). One would expect a relatively high bed load transport by such a wide, shallow stream. Gibbs' (1967) sediment load data, however, fail to substantiate this extrapolation of Schumm's model. The Solimões transports an immense suspended load and very little bed load (Table 2). This apparent paradox is even more pronounced for the Japurá River,

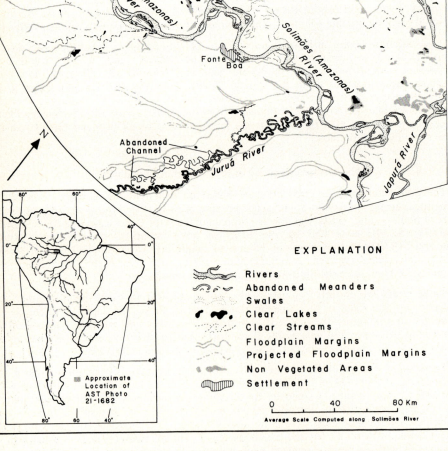

Fig. 4. Geomorphic map of the Solimões River flood plain near Fonte Boa, Brazil. Major tributaries shown include the Japurá and Juruá Rivers. The map was prepared from a color orbital photograph (AST 21-1682) taken in July 1975 during the Apollo-Soyuz space mission (Holz *et al.*, in press).

TABLE 2

PROPERTIES OF RIVERS IN THE WESTERN AMAZON BASIN

Rivers	Juruá (Jct. Solimões)	Japurá (Jct. Solimões)	Solimões (near Fonte Boa)
Environment[1]	Tropical	Mixed	Mixed
Drainage Area[1] (10^3km²)	217	289	1263
Discharge[1] (10^{12}m³/yr.)	0.197	0.351	1.161
Solid Load (% total load)	40	48	~30
Bedload[1] (% solid load)	<1	2.3	~5
Sediment Yield[1] (10^6g/km²/yr.)	49.4	120.2	not available
Pattern	Moderate to Tortuous Sinuosity	Straight, Moderately Anastomosing	Moderately Sinuous, Highly Anastomosing
Meander Form[2]	Compound Asymmetrical	Simple Symmetrical	Simple Asymmetrical
P	1.8-3.0	1.1	1.5
M[3]	13.5-100	1.1	6.5
F[4]	12.0-1.8	73	23

[1]Parameters determined by Gibbs (1967)
[2]Classified according to Brice (1974)
[3]Calculated from $P = 0.94M^{0.25}$ (Schumm, 1960, 1963)
[4]Calculated from $P = 3.5F^{-0.27}$ (Schumm, 1960, 1963)

which transports only 2% bed load, yet has a wide, shallow channel and a nearly straight, but anastomosing, pattern. Obviously the empirical equations do not adequately describe these rivers.

The channel width and wavelength of the Solimões River, combined with data on sediment load, can be used to estimate river discharge. The relationships are, of course, subject to the same problems of interpretation discussed for the Gran Chaco. However, here the exercise can be compared to actual measured discharges (Table 3). Note that the unusual sediment load-sinuosity relationships noted above are responsible for the overestimate of mean annual discharge by Schumm's (1968a) relationship.

The Juruá River contrasts sharply with both the Solimões and the Japurá (Table 2). Unlike the latter two rivers which head in the Andes, the Juruá headwaters rise to only about 600 m. The Juruá is representative of Gibbs' (1967) tropical fluvial environment. Sinuosity is quite high, ranging from 1.8 to 3.0. Meanders are typically compound and asymmetrical (according to the classification of Brice, 1974), and many channel reaches have irregular and tortuous patterns (according to the classification of Schumm, 1963). The pattern suggests that the Juruá is a relatively narrow, deep, suspended load steam.

Gibbs' (1967) analysis of another, similar tropical Amazon tributary, the Xingu, shows that the suspended load is probably made up entirely of particles finer than about 8 microns. Gibbs' (1967) data suggest that less than one percent of the solid sediment load transported by the tropical rivers is bed load. If the suspended sediment load is an indication of the total sediment load of these rivers, then even less than one percent of the total solid load may meet Schumm's (1969) criterion for bed load (coarser than 0.074 mm).

Along some reaches of the Juruá, e.g. Fig. 4, the sinuosity may vary from about 2 to 3 in a relatively short distance. According to Schumm's (1963) empirical formulae, channel width-to-depth ratios for this range of sinuosities should range from about 10 to less than 1, and the percent of silt and clay in channel-bank sediments (M) should range from about 20 to about 100. For this range of M, the proportion of bed load in the total sediment load should range from about 3 percent to virtually nothing. The suggested range of these variables is quite large, especially in light of environmental constraints on the character of sediment available to this river. It is likely that the actual variance of channel width-to-depth ratio, the amount of silt and clay in channel sediments, and the proportion of bed load in the total sediment load is actually less than is suggested by the variance of sinuosity. If so, then the wide range of sinuosities results from a relatively small range of environmental parameters.

Recent geomorphic research on tropical river systems may help explain some of these paradoxes for the Amazon River system. Savat's (1975) study of the Zaire (Congo) Basin has shown that discharge magnitude exerts an important control such that larger rivers

TABLE 3
ESTIMATING THE DISCHARGE OF THE SOLIMÕES RIVER AT FONTE BOA

River Properties			
Bank Full Width W (km)	Meander Wavelength λ (km)	Sinuosity P	% Silt Clay M (Calculated)
2.4	29	1.5	6.5

Discharge Estimates			
Equation	Reference	Discharge (m³/sec)	(cfs)
$\lambda = 30Qp^{0.5}$	Dury, 1965	2.8×10^5	1.0×10^7
$\lambda = 106Qm^{0.5}$	Carlston, 1965	2.3×10^4	8.1×10^5
$\lambda = 1890Qm^{0.34}/M^{0.74}$	Schumm, 1968	1.7×10^5	6.0×10^6
Measured at Fonte Boa	Gibbs, 1967	3.5×10^4	1.3×10^6
Measured at Manaus	Oltman et al. 1964	1.0×10^5	3.7×10^6

Symbols for Equations:
   λ   Meander Wavelength (ft.)
   Qp Most Probable Flood Discharge (cfs)
   Qm Mean Annual Flood Discharge
   M   Percent Silt-Clay in Channel Wetted Perimeter (Calculated from $P = 0.94M^{0.25}$)
   P   Sinuosity

tend to have more complex anastomosing patterns. This apparently occurs because both small and large rivers have about the same overbank flood depth of about 6 m. Large rivers are not a hydraulic unity, but rather a juxtaposition of several active thalwegs. Both the single-channel and the anastomosing streams appear to have stable banks promoted mainly by the dense forest vegetation. The stability of islands between the anastomosing channels is maintained by the rapid plant colonization of overbank sediments. Even regions of nearly pure sand sources (e.g. the "Kalihari" sands of the Kwango and Kasai Basins of Zaire) fail to develop true braiding because of the vegetation effect. That braiding could occur is documented by its development along reaches where improper agricultural practices have induced gullying of interfluves, thereby introducing abnormal quantities of sand into the river channels.

Tricart (1977) has recently described the processes along the Rio Juruá of Brazil. The phenomenal impoverishment of this river in bed load results in unusual widening of the channel at meander bends. The concave bank retreats much faster than the width can be maintained by deposition on the point bar. This disparity results in great broadening at the bends as banks retreat by undercutting of the superficially rooted rain forest vegetation and by landsliding during flood recession. The high sinuosity occurs because vegetation on the flood plain resists cutoff by chuting. Thus, sinuosity increases rapidly to extreme values, limited only by the attainment of neck cutoffs (Fig. 5).

Point bars in the Rio Juruá behave as a lag of relict coarse alluvium that is slowly being reworked by a modern stream transporting nearly 100% suspended and solution load. As in Zaire, modern phytostabilization prevents the influx of coarse sediment into the system. Nevertheless, this is a system delicately poised for future alteration by climatic change.

Climatic Change and Altered Fluvial Regimen in the Amazon Basin

Recent work by Brazilian investigators (RADAM project) has shown that the principal source of sediment in the Amazon Basin rivers is alluvium that was shed from the Andes during drier climatic episodes than prevail today. The sediment characteristics of the rivers have varied through the Quaternary depending on the rivers' abilities to rework the coarser fraction of their own relict alluvium. The fluctuating Quaternary climate of the Amazon region is the key to understanding the modern variability in fluvial processes.

A variety of studies have now converged to provide a qualitative understanding of Quaternary climates and environments in the Amazon Basin (Table 4). Evidence for full-glacial aridity comes from the dated fluctuations of Andean lake levels (Van Geel and Van der Hammen, 1973) and from the speciation patterns of Amazonian forest birds (Haffer, 1969). The available pollen data suggest that during full-glacial episodes of the Pleistocene mean annual temperatures were 4-5°C cooler in the Amazon Basin (Van der Hammen, 1972, 1974). The combination of lower temperatures and lower rainfall resulted in extensive replacement of forest by open grassland and savanna. Furthermore, Tricart (1975) has shown that the extensive Neogene uplands, now stablized by dense rainforest, have a well-preserved relict topography of intense fluvial dissection. He suggests that the relatively dry full-glacial episodes of the Pleistocene were periods of high sediment yield from these uplands contributing coarse bed load to the Amazon River system.

Full-glacial tropical aridity has now been thoroughly simulated in a general circulation model of the earth's atmosphere (Manabe and Hahn, 1977). Continental portions of the tropics, such as the western Amazon Basin, were drier during glacial periods because of stronger surface outflow from the continents in contrast to weaker inflow by maritime air masses. Damuth and Fairbridge (1970) give a local explanation for the western Amazon involving a full-glacial strengthening of westerly atmospheric flow off the cold Peru current and across the Andes. The low-pressure cell commonly situated today over the Amazon Basin was absent, and the current warm-wet air flow from the Atlantic was

Fig. 5. Photograph of the upper Juruá River between its junctions with the Liberdadade (left) and the Gregório Rivers (flowing across bottom right). The fine textured dissection of the uplands is considered to be relict from drier climatic episodes of the Pleistocene (Tricart, 1975). The Holocene flood plain of the Juruá River is inset into the upland as is a Pleistocene terrace of the river. (LANDSAT Image E-2226-14112-7, 5 Sept. 1975).

TABLE 4

GENERALIZED PALEOHYDROLOGIC REGIMENS OF THE WESTERN AMAZON BASIN

Stage	Full Glacial (18,000 yrs. B.P.)	Interglacial (Modern)
Climate	Semiarid Cooler	Humid Warmer
Vegetation	Savanna	Rain Forest
Interfluves	Dissected	Phytostabilized
Streamflow	Flashy	Uniform
Sediment	Coarser	Finer

effectively blocked. Full-glacial dry winds flowing down the eastern flanks of the Andes insured a relatively arid interior in the rain shadow of the mountains.

The channel patterns and deposits of the Amazon Basin rivers largely reflect the influences of these rather drastic changes in their hydrologic regimens. Under the modern hydrologic conditions only the major rivers are able to rework the coarser alluvium that was deposited during more arid phases. Mixed load rivers draining Andean source regions, like the Solimões and Japurá, have definite periods of high discharge that alternate with low flows. As a result, they can scour the coarser, noncohesive sediment that underlies the modern overbank sediment. Huge landslides, called *terras caidas,* occur during falling stage on the Solimões River, and aid in the erosion of relict alluvium (Fig. 6). Such rivers characteristically have anastomosing patterns of moderate to low sinuosity. The alluvial islands, in contrast to braided stream bars, are extremely stable (Sternberg, 1959). They are covered by dense forest, and their bounding levees give them a saucer-like morphology.

The actively evolving Holocene flood plains of the major Andean tributaries are characterized by accretionary ridge-and-swale topography at meander bends (Fig. 4), well-developed levees, and paleomeander scars. The contrast between these rivers and their sediment-impoverished tributaries flowing from the forested uplands is evinced by the clear-water lakes (*rias fluviales*) along the flood plain margin (Figs. 4 and 6). Each tributary has been dammed by the active flood plain growth of the trunk stream (Solimões). Field observations by Tricart (1977) show that Holocene sedimentation by these tributaries has only been sufficient to form small deltas at the heads of the lakes.

In contrast to the Andean rivers, the larger rivers draining extensive tropical lowlands, and lacking mountain sources have relatively narrow flood plains. The Juruá River is representative of this group. Discharge from the lowlands is relatively uniform, and the flow transports essentially no bed load. The amount of overbank deposition is severely limited by the extremely low solid sediment yield of this phytostabilized environment. Scour on concave banks occurs more rapidly than deposition on convex banks, resulting in tortuous sinuousity and numerous abandoned meanders (Figs. 5 and 7). Ridge-and-swale topography, however, is absent along the Juruá, showing the much less active flood-plain depositon by this river in comparison with the Solimões. The fine-grained varzea sediments are thinner along these rivers (Fig. 7) than along the Solimões. The combination of cohesive bank sediments, stabilization by vegetation, and relatively low, uniform discharges exert the major influences on channel morphology in the Juruá River.

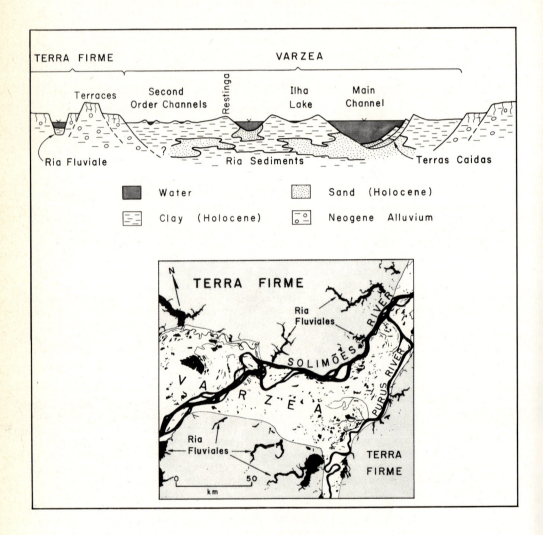

Fig. 6. Morphology and sedimentology of the Solimões River at its junction with the Purus River. The upper diagram is a schematic cross section (not to scale) showing the hypothetical distribution of sediments underlying the Holocene flood plain (*varzea*). The surficial features are as observed on SLAR imagery and orbital photographs. The lower map shows typical morphologies, including levees (*restingas*), alluvial islands (*ilhas*), second order channels (smaller channels on the *varzea*), landslides (*terras caidas*), and tributaries draining sediment-deficient uplands (*terra firme*) and blocked at their mouths by Holocene *varzea* sedimentation to form lakes (*rias fluviales*). The lower map was prepared from LANDSAT image E-1008-13481-7 (31 July 1972).

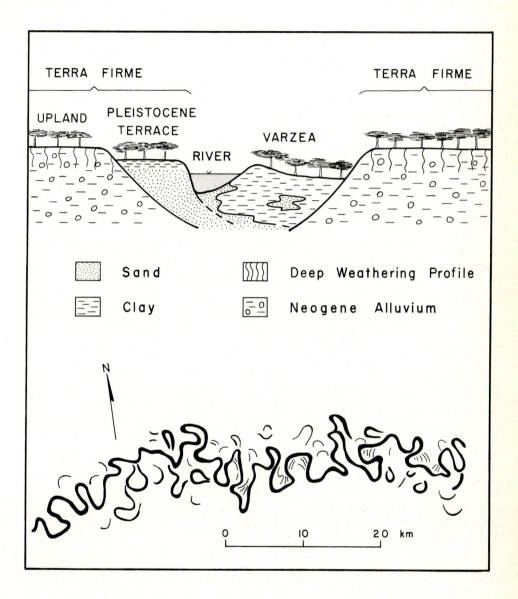

Fig. 7.   Morphology and sedimentology of the Juruá River near its junction with the Tarauacá River (Eirunepé, Brazil). The upper diagram is a schematic cross section (not to scale) showing the hypothetical distribution of sediments underlying the Holocene flood plain (*varzea*). A prominent Pleistocene terrace occurs along the upper reaches of the Juruá and is characterized by coarse alluvium deposited during the relative dry phases of the Pleistocene (Tricart, 1977). The lower map shows the modern channel pattern and recently abandoned meanders between the Gregório River junction at Canindé (Fig. 5, right) and the Tarauacá River junction at Eirunepé. The map was prepared from side-looking airborne radar (SLAR) imagery (RADAM project mosaic SB 19-Y-B).

DISCUSSION

The above examples illustrate several methodological difficulties that can arise when the modern fluvial regimen behaviour derived from one environment is used to characterize rivers in a very different environment. Although the empirical equations developed by Schumm (1963, 1968a, 1969, 1972) for temperate, semi-arid rivers may be useful in a variety of geomorphic settings, they certainly do not form a precise quantitative model for all fluvial systems. Obviously the problems of assessing the validity of predictions from the Schumm model are greatly magnified if one indiscriminantly applies the model to ancient fluvial paleochannels. Any such paleohydrologic application requires first a question: How much of the information in the paleohydrologic think-back (Fig. 1) can one adequately reconstruct from the available field evidence?

In the tropical rivers of the western Amazon Basin, fluvial regimen is influenced by relict sediments and by the ability of the interfluves to contribute those sediments to the active channelways. Even given the intensely phytostabilized environments of the modern interglacial climate, tropical rivers may, nevertheless, exhibit a broad range of channel patterns and flood-plain sedimentation. Where the depth factor is limited in hydraulic geometry scaling, the complexity of channel anastomosis will increase with discharge magnitude. Where flow variability allows bank and channel scour, coarser relict alluvium may be introduced into the modern river system. Distant mountainous sources, such as the Andes, may result in immense disparities in solid sediment loads between the mountain-draining rivers and nearby high-discharge rivers with basins confined to tropical lowlands. Indeed, the RADAM project geomorphologists have concluded that no existing theory of fluvial morphology explains the complexities of the Amazon environment (M.M. Penteado-Orellana, written communication, 1977).

Obviously an alluvial plain like the Gran Chaco offers a potential modern analogue to ancient fluvial rock systems of broad lateral extent. In fact the analogy has already been drawn between the Gran Chaco and the basal Cretaceous fluvial sequence of the western interior United States (Moberly, 1960; Eicher, 1969). Ancient alluvial plain sequences also preserve numerous opportunities to calculate paleodischarges in a manner similar to that shown in Table 1. However, opportunity should not be confused with success. The application of several empirical formulae to a preserved paleochannel should not be viewed as a precise quantitative reconstruction of an ancient hydrologic environment. At best the exercise might characterize several single-channel phases of a great alluvial complex. Moreover, that characterization will be based on modern knowledge of river regimen that has been extrapolated from data sets known to be limited to distinct climatic and physiographic settings. Finally, the paleohydrologic variation calculated for the rock system may be the product of climatic change, or, alternatively, the result of dynamic adjustments of river courses that occur along active mountain fronts. Distinguishing the two causes can be difficult even in late Quaternary examples where nearly all the channel morphology is preserved.

The Amazon Basin examples are perhaps even more illustrative of the limitations of modern regimen theory in explaining diverse fluvial phenomena. Where discharge characteristics and bank stabilization by vegetation become important, the precise scaling of form ratio (width/depth) and sinuosity to channel perimeter sediment characteristics will no longer hold. Until detailed field studies of river morphology, hydrology, and sedimentology can be made in the tropics, paleohydrologists will have to be content with the more generalized (and less quantitative) characterization of these environments in terms of climatic and source terrain controls.

ACKNOWLEDGMENTS

The research reported here was supported by the Smithsonian Institution under contract number PC6-22305. I thank R. K. Holz, M. M. Penteado-Orellana, and S. M.

Sutton, Jr. for their work with me on that project. The manuscript was read critically by R. A. Morton and C. M. Woodruff, Jr. Publication was supported by The Geology Foundation, The University of Texas at Austin.

REFERENCES

Baker, V. R., 1974, Paleohydraulic interpretation of Quaternary alluvium near Golden, Colorado: Quaternary Research, v. 4, p. 94-112.

———, 1977, Stream-channel response to floods with examples from central Texas: Geol. Soc. Am. Bull., v. 88, p. 1057-1071.

Baker, V. R., and Penteado-Orellana, M. M., 1977, Adjustment to Quaternary climatic change by the Colorado River in central Texas: J. Geol., v. 85, p. 395-422.

———, 1978, Fluvial sedimentation conditioned by Quaternary climatic change in central Texas: J. Sediment. Petrol, v. 48, p. 433-451.

Barrell, Joseph, 1908, Relations between climate and terrestial deposits: J. Geol., v. 16, p. 159-190, 255-295, 363-384.

Bayly, Brian, 1968, Introduction to petrology: Englewood Cliffs, N. J., Prentice-Hall, 371p.

Brice, J. C., 1974, Evolution of meander loops: Geol. Soc. Am. Bull., v. 85, p. 581-586.

Carlston, C. W., 1965, The relation of free meander geometry to stream discharge and its geomorphic implications: Am. J. Sci., v. 263, p. 864-885.

Conway, Dennis, and Holz, R. K., 1973, The use of near-infrared photography in the analysis of surface morphology of an Argentine alluvial floodplain: Remote Sensing of Environment, v. 2, p. 235-242.

Cordini, R., 1947, Los ríos Pilcomayo en la región del Patiño: Dirección de Minas y Geología, Anales 1, no. 22, 82p.

Cotter, Edward, 1971, Paleoflow characteristics of a late Cretaceous river in Utah from analysis of sedimentary structures in the Ferron Sandstone: J. Sediment. Petrol., v. 41, p. 129-138.

Cucchi, R. J., 1973, Aspectos geomorfologicos de la Llanura Formoseña, los ríos sin proporcion y su significado climafico: Revista de la Asociación Geológica Argentina, v. 28, p. 156-164.

———, 1974, Respuesta a una hipotesis alternativa sobre el origen del desajuste de los riós del este de Formosa: Revista de la Asociación Geológica Argentina, v. 29, p. 138-139.

Damuth, J. E., and Fairbridge, R. W., 1970, Equatorial Atlantic arkosic sands and ice-age aridity in tropical South America: Geol. Soc. Am. Bull., v. 81, p. 189-206.

Dury, G. H., 1964, Principles of underfit streams: U. S. Geol. Survey Prof. Paper 452-A, 67p.

———, 1965, Theoretical implications of underfit streams: U. S. Geol. Survey Prof. Paper 452-C, 43p.

———, 1966, Incised valley meanders on the Colo River, New South Wales: Austr. Geogr., v. 10, p. 17-25.

———, 1968, Bankfull discharge and magnitude frequency series: Austr. J. Sci., v. 30, p. 371.

———, 1970, General theory of meandering valleys and underfit streams: *in* G. H. Dury, *ed.,* Rivers and river terraces; Macmillan, London, p. 264-275.

———, 1976, Discharge prediction, present and former, from channel dimensions: J. Hydrology, v. 30, p. 219-245.

Eicher, D. L., 1969, Paleobathymetry of Cretaceous Greenhorn Sea in eastern Colorado: Am. Assoc. Petrol. Geol. Bull., v. 44, p. 156-194.

Gibbs, R. J., 1967, The geochemistry of the Amazon River System. Part 1. The factors that control the salinity and composition and concentration of the suspended solids: Geol. Soc. Am. Bull., v. 78, p. 1203-1232.

Haffer, J., 1969, Speciation in Amazonian forest birds: Science, v. 165, p. 131-137.

Holz, R. K., Baker, V. R., Sutton, S. M., Jr., and Penteado-Orellana, M. M., in press, An analysis of South American river morphology and hydrology from ASTP imagery: U. S. National Aeronautics and Space Administration Spec. Publ. 412, v. 2.

Iriondo, M. H., 1974, Los ríos desajustados de Formosa, una hipotesis alternativa: Revista de la Asociación Geológica Argentina, v. 29, no. 1, p. 136-137.

Mackin, J. H., 1936, The capture of the Greybull River: Am. J. Sci., 5th Ser., v. 31, p. 373-385.

Manabe, Syukuro, and Hahn, D. G., 1977, Simulation of the tropical climate of an ice age: J. Geophys. Res., v. 82, p. 3889-3911.

Moberly, R. M., Jr., 1960, Morrison, Coverly, and Sykes Mountain Formations, northern Bighorn basin, Wyoming and Montana: Geol. Soc. Am. Bull., v. 71, p. 1137-1176.

Noriegas, R., 1971, Comportamiento del río Pilcomayo y su influencia en al provincia de Formosa: 5th Congreso Nacional del Agua, Sante Fe, v. 2, no. 8, p. 1-12.

Oltman, R. E., Sternberg, H. O'R., Ames, F. C., and Davis, L. C., Jr., 1964, Amazon River investigations reconnaissance measurements of July 1963: U. S. Geol. Survey Circ. 486, 15p.

Prohaska, Fritz, 1976, The climate of Argentina, Paraguay, and Uruguay: in Werner Schwerdtfeger, ed., Climates of Central and South America; Elsevier, Amsterdam, p. 13-112.

Ritter, D. F., 1972, The significance of stream capture in the evolution of a piedmont region, southern Montana: Zeitschr. Geomorphologie n. f., v. 16, p. 83-92.

Russell, R. J., 1958, Geological geomorphology: Geol. Soc. Am. Bull., v. 69, p. 1-22.

Savat, Jan, 1975, Some morphological and hydraulic characteristics of river-patterns in the Zaire Basin: Catena, v. 2, p. 161-180.

Schumm, S. A., 1960, The shape of alluvial channels in relation to sediment type: U. S. Geol. Survey Prof. Paper 352-B, p. 17-30.

————, 1963, Sinuosity of alluvial channels on the Great Plains: Geol. Soc. Am. Bull., v. 74, p. 1089-1100.

————, 1967, Meander wavelength of alluvial rivers: Science, v. 157, p. 1549-1550.

————, 1968a, River adjustment to altered hydrologic regimen Murrumbidgee River and paleochannels, Australia: U. S. Geol. Survey Prof. Paper 598, 65p.

————, 1968b, Speculations concerning paleohydrologic controls of terrestrial sedimentation: Geol. Soc. Am. Bull., v. 79, p. 1573-1588.

————, 1969, River metamorphosis: J. Hydraulics Div., Am. Soc. Civil Eng., v. 95, no. HY1, p. 255-273.

————, 1972, Fluvial paleochannels: in J. K. Rigby, and W. K. Hamblin, eds., Recognition of ancient sedimentary environments; Soc. Econ. Paleont. Mineral. Spec. Pub. 16, p. 98-107.

Soldono, F., 1947, Régimen de la red fluvial argentina: Editorial Cimera, Tomo I, 277p.

Sternberg, H. O'R., 1959, Radiocarbon dating as applied to a problem of Amazonian morphology: 17th Intern. Geogr. Congr., C. R. 2, p. 399-424.

————, 1975, The Amazon River of Brazil: Franz Steiner Verlag GMBH, Wiesbaden, 74p.

Tricart, J., 1975, Influence des oscillations climatiques récentes sur le modelé en Amazonie Orientale (Region de Santarém) d'après les images radar latéral: Zeitschr. Geomorphologie n. f., v. 19, no. 2, p. 140-163.

————, 1977, Types de lits fluviaux en Amazonie brésilienne: Annales de Geographie, v. 84, no. 473, p. 1-54.

Van der Hammen, T., 1972, Changes in vegetation and climate in the Amazon basin and surrounding areas during the Pleistocene: Geol. Mijnbouw., v. 51, p. 641-643.

————, 1974, The Pleistocene changes of vegetation and climate in tropical South America: J. Biogeography, v. 1, p. 3-26.

Van Geel, B., and Van der Hammen, T., 1973, Upper Quaternary vegetational and climatic sequence of the Fuquene area (Eastern Cordillera, Columbia): Palaeogeog. Palaeoclim. Palaeoecol., v. 14, p. 9-92.

Walker, R. G., 1976, Facies models 3. Sandy fluvial systems: Geoscience Canada, v. 3, no. 2, p. 101-109.

Wolman, M. G., and Miller, J. C., 1960, Magnitude and frequency of forces in geomorphic processes: J. Geol., v. 68, p. 54-74.

HYDROLOGY AND CURRENT ORIENTATION
ANALYSIS OF A BRAIDED-TO-MEANDERING TRANSITION:
THE RED RIVER IN OKLAHOMA AND TEXAS, U.S.A.

DANIEL E. SCHWARTZ[1]

ABSTRACT

Current orientation patterns in the sand bed of the braided-to-meandering transition zone of the Red River vary with discharge and flow character, and do not correspond with overall river trend. Patterns produced by flood and falling discharge have high azimuthal dispersions in channels with wedge-shaped cross sections and areas of low topographic relief, and low dispersion in areas where the channel is constricted or where a chute is formed during the flood. Low discharge patterns have high dispersion in tabular channels where bed relief is high.

Three types of flow which combine to form a sedimentation cycle were discerned during the study. They are: (1) channel flow/low discharge; (2) sheet flow/flood discharge; and (3) post-sheet flow/falling discharge. These types of flow are differentiated on the basis of their characteristic sedimentation load, flow pattern, channel morphology, and bed form occurrence.

Overall current directions measured during low discharge, when channel flow was dominant, show a 21.78° mean deviation from the river trend, and an azimuthal spread of 285°. Current vectors measured after a flood, when sheet flow and post-sheet flow conditions prevailed, show a 25.08° mean deviation from the river trend, and a 332.5° azimuthal spread. Variations in river bed relief, flow behavior, sedimentation, and channel bank stability resulted in variable orientation patterns which do not have a downstream trend in azimuthal character.

Paleocurrent studies of deposits which formed under conditions similar to those described in this study must take into account the great variability in flow patterns as well as the local river bed topography which determine the type and extent of sediment depositon.

INTRODUCTION

Paleocurrent direction is a significant factor in surface and subsurface sand body analysis. Cross beds formed during sand transport have been analyzed to determine the local flow direction and integrated over the entire body to determine the overall trend. For example, Steinmetz (1962, 1972), Campbell (1968), McDaniel (1968), and Shelton and Noble (1974), have interpreted sand body orientation from modern and ancient deposits using cross bed direction and dip to determine paleocurrent direction. In addition, Bluck (1971) and Smith (1971, 1974)) have described active river sedimentation, and presented detailed analyses of bar migration and cross bed development in relation to flow patterns which were present in the channel at the time of bar formation. Steinmetz (1972) has pointed out that a high degree of cross bed orientation variability is associated with flood generated fluvial deposits, and Collinson (1971) determined that flood-produced scours in a pebble-sand stream have lower variance than low and falling discharge generated sand ribbons on the same bar. In order to explain this variability, an analysis of a single sedimentation cycle (low discharge, flood discharge, and falling discharge) should be carried out. This report presents data obtained from such an analysis of a modern fluvial system — the Red River.

The Red River flows eastward from the High Plains of eastern New Mexico and the Texas Panhandle to its confluence with the Atchafalaya River near Alexandria Louisiana. This report is based on a study of the transition from a braided to a meandering pattern observed on the Red River between the cities of Burkburnett, Texas and Terral, Oklahoma (Fig. 1A). The purpose of the overall study is to examine the

---

[1]Programs in Geosciences, The University of Texas at Dallas, P.O. Box 688, Richardson, Texas 75080

Contribution No. 339

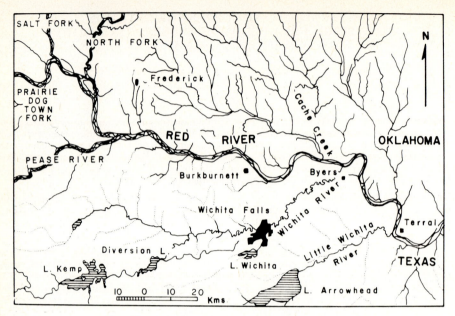

Fig. 1A.   Location map of the Red River study area.

sediment distribution patterns and processes occurring in such a transition, and relate these patterns to regional and local features (Schwartz, 1976, 1977).

Field work was carried out over a two year period between August, 1975 and August, 1977, and was augmented with two aerial reconnaissance flights in February and July of 1976. Information gathered during ground and aerial observations was combined with data collected by gaging stations maintained by the United States Geological Survey at river crossings near Burkburnett and Terral.

GENERAL

Regional Setting

The Red River drainage system crosses the High Plain, Osage Plain, and West Gulf Coastal Plain sections of the Central Lowlands Province of North America (Leifeste, *et al.*, 1971). The main tributaries of the Red River; the North Fork, the Salt Fork, and the Prairie Dog Town Fork, join to form the main channel of the Red River west of Frederick, Oklahoma (Fig. 1A), approximately 95 km upstream of the study area. The Peace River, a major southern tributary, enters the Red south of Frederick, approximately 60 km upstream of the study area. These tributaries carry intermittent-to-continuous discharge resulting from runoff from surrounding plains and agricultural lands. Tributaries are generally wide and shallow, and have braided-anastomosing channels.

The Red River gradually changes from a predominantly braided to a predominantly meandering stream over a 230 km portion of its reach, beginning 50 km upstream of Burkburnett, to 80 km downstream of Terral. The central 100 km segment of the river between these cities is the subject of this study.

The Red River study reach is located in rolling prairie underlain by Permian red beds which rarely crop out along the flanks of the flood plain. The flood plain is bordered to

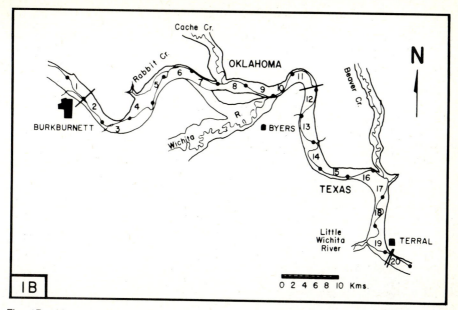

Fig. 1B. Map of the present Red River channel and flood plain within the study reach. The reach is divided into twenty 5-kilometer segments which have straight, meandering, or braided characteristics.

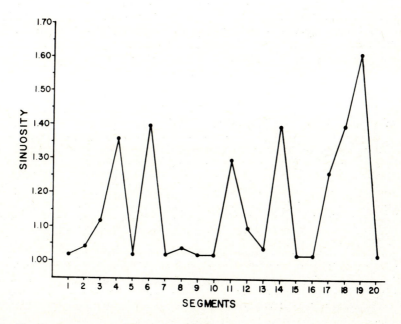

Fig. 2. The sinuosity of each of the twenty segments on Figure 1B. Sinuosity was determined using the method of Schumm (1963).

the north and south by alternating Pleistocene fluvial and eolian terraces of Early Wisconsinan, Kansan, and Nebraskan age (Frey and Leonard, 1963), which represent previous river elevations and courses (Leet and Judson, 1971).

Morphology

The Red River drainage basin above Terral has an area of 74,393 km², of which approximately 15,375 km² are considered as non-contributory to stream flow (Water Resources Data, U.S.G.S., 1977). The river has a ninth order dendritic drainage pattern (Strahler, 1952) upon entering the Atchafalaya River, with tributaries of the sixth order flowing into the main channel within the study reach.

The Red River flood plain meanders with a regular sinuosity (p = 1.32; Schumm, 1963), and has a maximum depth of 21 m and a width varying between 1.60 and 3.61 km. The river bed forms a braid-meander belt within the flood plain, composed of braided, meandering and straight segments, occasionally anastomosing, which do not occur in a sequential or a predictable pattern (Fig. 1B). The overall sinuosity of the channel in the reach is 1.60, while the sinuosities of the twenty 5-km segments of the river range between 1.02 and 1.62 (Fig. 2).

Highway bridges occur at both ends of the reach, northeast of Burkburnett and south of Terral, as well as in the center of the reach, northeast of Byers, Texas (Fig. 1B). A railroad bridge crosses the river 0.8 km upstream of the Terral highway bridge. The channel is slightly constricted at these bridges where banks are stabilized, but there is no modification of the flow pattern in these areas.

A longitudinal water-surface profile (Fig. 3) was drawn from topographic maps of the region, from upstream of Burkburnett (point "A") to downstream of Terral (point "G"). Because the river flows only within unconsolidated alluvium, the profile is gently curved, and shows a gradual flattening toward the east. The profile is divided into two sections near point "C", where there is a reduction in slope from .00076 to .00044. This point corresponds with the entrance of the Wichita River into the Red. When compared with the published values of slopes from other braided and meandering rivers (Stricklin, 1961; Bernard, et al., 1970; McGowen and Garner, 1970; Smith, 1970, 1972, 1974; and Shelton and Noble, 1974), the study reach occupies an intermediate position between these two patterns. The reason for the slope break is not fully understood, but may be related to the

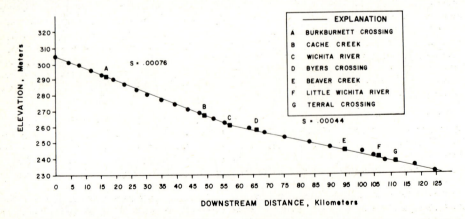

Fig. 3.   Downstream water-surface profile of the Red River from upstream of Burkburnett, Texas to downstream of Terral, Oklahoma. The solid circles represent elevation inflection points, while the squares represent highway bridge crossings or tributary entrance points. "S" is the slope of the profile which is divided into two sections.

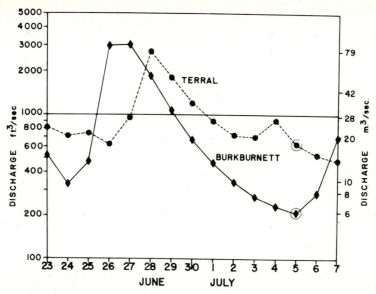

Fig. 4. Mean daily discharge measured at the Burkburnett and Terral gaging stations between June 23, 1976 and July 7, 1976. The solid round points represent discharges measured at Terral, while the solid diamonds represent discharge measured at Burkburnett. The circled points on July 5, 1976 represent the day of the aerial reconnaissance flight.

entrance of the Wichita River, which carries a fine-grained mixed load, or as discussed by McDonald and Banerjee (1971) and Smith (1974), may represent the boundary between downstream aggradation and upstream degradation. No "nick point" (the contact between the aggrading and degrading morphologies) has been observed, but there is some channel incisement in the Terral area, the upstream extent of which is not certain.

HYDROLOGY

Hydrological records for the Red River have been kept by the United States Geological Survey and the Water Resources Division of the state of Texas. Within the reach, gaging stations are maintained at the highway bridges at Burkburnett and Terral. Monthly discharge was recorded at Burkburnett from July, 1924 until August, 1925. No discharge was gauged from that time until December, 1959. Discharge has been continuously monitored since late in 1959, and water quality and temperature measuring equipment have also been in service since July, 1968. Discharge at Terral has been monitored continuously since January, 1938; while water quality and temperature have been analyzed since October, 1967.

For the sixteen years from 1960 to 1976, the mean daily discharge measured at Burkburnett has been 23.14 m³/s, with a maximum dischage of 1,780 m³/s on October 19, 1965. There are many times when there is no flow in the channel at this location. Over the thirty-eight-year gaging period at Terral, the mean daily discharge is 62.11 m³/s, with a maximum of 5,580 m³/s on June 8, 1941, and a minimum of 1.22 m³/s on March 15, 1939.

River bed features observed during the aerial reconnaissance flights reflect the discharge patterns of at least the two weeks prior to the flight. During the period from February 11, 1976 to February 25, 1976, discharge was at a low level at both gaging stations, with Burkburnett registering a more variable pattern. Both stations showed discharges well below the long-term mean values. Discharge ranged between 6 and 9 m³/s at Burkburnett, and between 11 and 14 m³/s at Terral. Both stations showed a gradually decreasing trend in discharge. Figure 4 displays the period between June 23, and July 7, 1976. Flow was at a moderate level prior to the rapid onset of flood stage. It is interesting to note that peak flood stage (85 m³/s) occurred on June 26, and was maintained through June 27 at the Burkburnett station, whereas the peak flood stage occurred on June 28 at the Terral station, reaching 79 m³/s, and being maintained only one day. Also of interest is the post-flood period. Burkburnett's gauge measured a continuously decreasing discharge until July 5, the day of the flyover. The Terral station measured a discharge decrease until July 4, when a small increase was measured. The small flood displayed on Fig. 4 is of interest for two reasons: (1) the magnitude of peak mean daily discharge measured at the Terral gauge was less than that measured at Burkburnett; and (2) the Terral gauge measured an increased discharge on July 4, which was not measured at Burkburnett. Over the entire gaging period, discharge measured at Terral has been consistently higher than that measured at Burkburnett. The reason for this is the increased drainage area which contributes added discharge. In the case of the June 26-27 flood, a different set of circumstances prevailed. The source of the water which caused the flood was localized in the upper portion of the study reach. Precipitation from an east-to-west moving storm system fell only on the Burkburnett area and did not reach the ground past Byers. Flow produced from the storm dissipated rapidly as indicated by the relatively low peak stage registered at Terral on June 28. Because there had not been a flow of this magnitude since late May, 1976, much of the water was probably taken into the river bed as ground water. The small increase measured at Terral on July 4, was probably caused by the passage of a small weather system related to the June 26-27 system, which affected only the lower portion of the reach.

With the onset of flood discharge, the hydraulic character of the river changes from channel flow to sheet flow. Small tabular channels which previously directed water movement in braided patterns quickly give way to an influx of sediment-laden water which covers much of the river bed. Low discharge water depths generally range between .033 and 0.88 m at Burkburnett, between 0.155 and 2.23 m at Byers, and between 0.256 and 3.47 m at Terral. During high discharge, water depth may increase by 1.0 to 3.0 m; depths of 2.95 m and 3.493 m correspond with discharges of 357 m³/s and 1,780 m³/s respectively from the Burkburnett station. Depths of 5.051 and 8.571 m correspond with discharges of 588 and 5.580 m³/s respectively from the Terral station.

Channel width increases dramatically during sheet flow periods. Width-to-average-depth ratios for the active channels are variable and discharge-dependent. At Burkburnett, the range is from 13.88 to 621.05, averaging 216.45; at Byers the range is between 23.41 and 471.32, and averages 163.12; and at Terral, the range is between 31.53 and 308.71, averaging 109.46. During sheet flow the entire river bed may be covered with water. Average values for river bed width are shown on Fig. 5.

Decreasing river bed width and increasing channel depth in a downstream trend combine to cause the downstream constriction of the river. Point bars tend to increase in height and decrease in width in a downstream trend (Schwartz, 1977). Because of this bed relief change, extensive effects of sheet flow are generally restricted to the upper portions of the reach, where gradient is high, point bars are low and wide, and where width-to-depth ratios are high.

Mean daily water velocities have been measured at Burkburnett and Terral. Between September 24, 1969 and August 23, 1976 at Burkburnett, the range was between 0.0853 and 1.585 m/s, the mean was 0.6097 m/s. At Terral between August 18. 1969 and July 28, 1976, the water velocity ranged from 0.2011 to 1.9476 m/s, with a mean of 0.6266.

SEDIMENT
Bed Material

Clay- through cobble-sized material was collected at Burkburnett and Byers, while clay- through coarse sand-sized material was collected at Terral. Compositionally, the

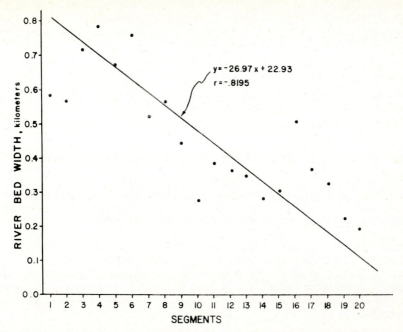

Fig. 5. Diagram showing the river bed width of each segment. Values plotted are the means of ten measurements from aerial photographs and topographic maps. Downstream of the study reach, the river bed width is generally between 0.2 and 0.3 kilometers, while upstream, the width varies between 0.5 and 1.2 kilometers.

sand-sized material would be classified as a lithic arkose (Folk, 1974), having abundant quartz, orthoclase, and igneous and sedimentary rock fragments. Mean diameter values in Phi units for the sand fraction from Burkburnett, Byers, and Terral are: 2.11 ϕ; 2.25 ϕ and 3.13 ϕ respectively, corresponding to: 0.227 mm; 0.210 mm; and 0.114 mm. The range in sand-sized material is between 1.0 ϕ (0.50 mm) and 3.75 ϕ (0.074 mm), with most sands being moderately to well sorted (Folk, 1974). Gravel-sized sediment was collected at Burkburnett and Byers. The clasts at Burkburnett ranged in size from: 2.0 to 20.2 cm (A-axis); 1.5 to 12.1 cm (B-Axis); and 0.3 to 11.2 cm (C-axis). Clasts at Byers ranged in size from: 1.6 to 13.5 cm (A-axis); 1.3 to 6.0 cm (B-Axis); and 0.4 to 4.0 cm (C-Axis). No clasts were observed at Terral. In addition to lithic material, armored and unarmored mud balls, fossils, disarticulated modern animal remains, and plant debris were also observed.

Suspended Sediment

Suspended sediment concentration is variable and directly related to discharge and temperature (Simons and Richardson, 1962; Briggs and Middleton, 1965; and Simons, *et al.*, 1965). During the February aerial reconnaissance flight, active channel bed forms were easily discernible; yet during the July flight, the suspended sediment content was so high that active channel features were almost totally obscured.

Samples collected on February 22, 1976, two days before the flyover, had suspended sediment concentrations between 63.0 and 75.0 mg/l. Flow velocity was 0.495 m/s and the water temperature was 9.0°C. Sieve analysis of the dried, weighed, and re-wetted material revealed that 70 to 90% of the suspended material from the six samples was finer than 0.062 mm. No samples were collected during or soon after the June 26-27 flood, but material collected by the U.S.G.S. at Burkburnett on July 22, 1975 had concentrations in the area of 2000 mg/l. The current velocity at the time of sampling was 1.737 m/s and the water temperature was 28.0°C. These conditions were similar to those which existed after the flood peak of June 27, 1976. Of the material collected by the U.S.G.S., 53% was finer than 0.062 mm. Suspended sediment collected at Burkburnett on May 27, 1976, one month prior to the flood of interest in this study had concentrations of 7,490 mg/l. The instantaneous discharge at the time of sampling was 96.288 m³/s (slightly higher than that for the June 26-27 flood) and the water temperature was 20.0°C. The water temperature at the time of the June 26-27, 1976 flood was between 28.0 and 29.0°C at Burkburnett and between 25.0 and 27.0°C at Terral (Water Res. Data for Texas, 1977).

CURRENT ORIENTATION

Measurements were made of 533 current orientation vectors discerned from analysis of photographs of bed forms at seven different locations within the reach. During each of the flights, aerial photographs were taken of nearly the entire reach. Photographs and some associated ground observations from the two periods were compared in order to determine the current orientation patterns which occurred during channel, sheet, and post-sheet flow. The same sections of the river bed were used for comparison in five cases, while two different areas in the same segment were compared in the sixth case. Figures 6A and 6B display rose diagrams developed from the directional analysis. Diagrams from segments 1, 4, 6, 11, and 20 are from the same portion of the river in each figure, with the diagrams from segment 17 being from different parts of the same segment, 1.8 km apart.

Current orientation was determined by analysis of photographs of large-scale bed forms which were either in formational stage (those seen during the February flight) or those formed during or soon after the flood (as seen during the July flight). Sand streaks on bed form surfaces and crestal outlines were also used in the analysis. Large-scale bed forms which were used in the study include: linguoid, longitudinal, braid, transverse, and scroll-like bars; sand waves; and caternary, sinuous, and straight dunes (terminology of Allen, 1968; Boothroyd and Ashley, 1975; Smith, 1971, 1974). Only Rank 4 bed forms (Miall, 1974) were used in the study, as they could be seen easily from the air, and because they were shown to be oriented more consistently with the river trend than smaller or larger bed forms of the hierarchy presented by Miall (1974). Bed forms of this type also form most of the stratification found within the river system.

No vector weight was assigned to the measurements, as dip was not taken into account in this strictly *two-dimensional* study. Weighting factors as described by Curray (1956) and as used by Miall (1974) were not employed in the study.

The summary of directional features is displayed in Fig. 7. The range in mean directions is 97° for channel flow bed forms and 160° for sheet and post-sheet flow deposits. The standard deviations of the means are 38° and 56° respectively, with variance being 1194.28 and 2608.62 respectively. The overall means for the two sets of measurements are 93° for the February flight and 90° for the July reconnaissance flight. These values vary from the valley trend (115°) by 22° and 25° respectively. The azimuthal spreads of all the measurements are 285° and 332.5° respectively.

The overall means of the two data sets correspond quite well, yet vary somewhat from the river trend. Individual diagrams in Figs. 6A and 6B show that in some cases the mean orientations shifted as much as 54.5° (Segment 1) and as little as 1.0° (Segment 6). In the case of Segment 1, the mean orientation vector (arrowed) shifted to the opposite side of the valley trend at that locality. Figures 8A1 through 8F2 are sketch maps drawn from aerial photographs and ground mapping of the individual study areas used in the preparation of the rose diagrams. Figures 8A1 through 8F1 are sketches of the river bed as seen during the February flight. Figures 8A2 through 8F2 are sketches of the bed as seen during the July flight. Active bed forms occupying the channel were measured for

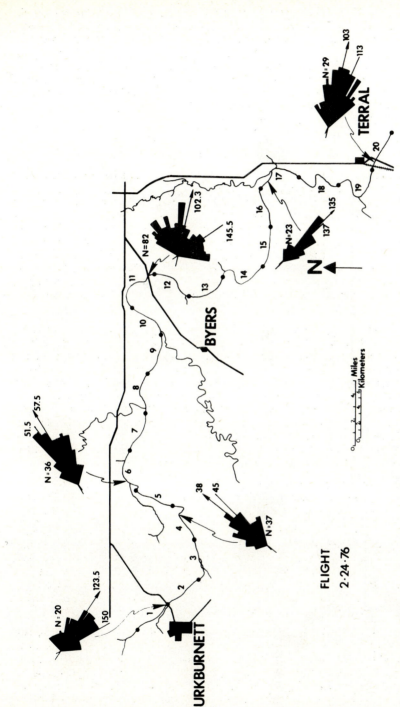

Fig. 6A. Current analysis for the Red River study reach during low discharge. Azimuths represented by non-arrowed vectors are the local channel trend. Arrowed vectors are the mean current direction of the measured bed forms. "N" represents the total number of measurements for each individual diagram. A total vector spread of ten degrees should be assumed for each measurement.

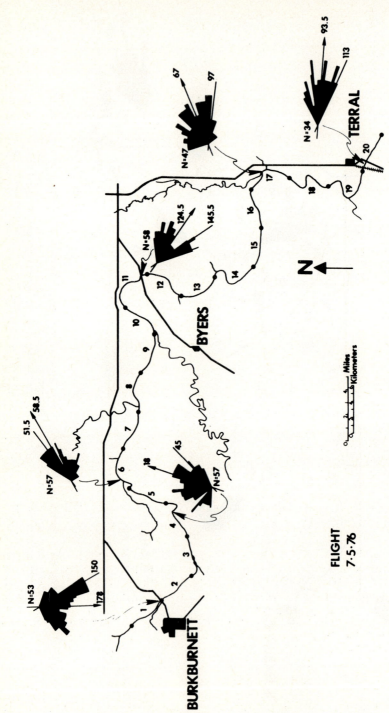

Fig. 6B. Current analysis for the Red River study reach after a period of flood and falling discharge. Azimuths represented by non-arrowed vectors are the local channel trend. Arrowed vectors are the mean current direction of the measured bed forms. "N" represents the total number of measurements for each individual diagram. A total vector spread of ten degrees should be assumed for each measurement.

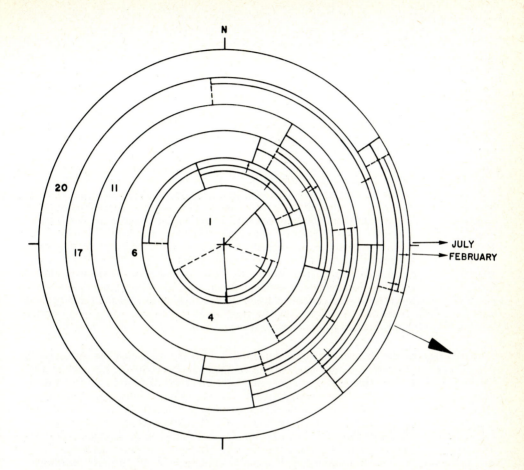

Fig. 7. Composite diagram showing the azimuthal spreads of the rose diagrams for both flights. The numbers within the concentric rings represent the segment numbers. Arcs bounded by dashed lines represent the spreads of the July 5th measurements. Arcs bounded by solid lines represent azimuthal spreads from the February 24th measurements. The mean azimuths for each segment are shown within that segment's ring, bisecting the inner arc for that specific flight. Overall means for each flight are designated by two labelled arrows to the right of the diagram. The large arrow represents the overall channel trend.

the February (1-series) diagrams; and bed forms formed by the flood waters, either during or soon after the flood peak, were measured for the July (2-series) diagrams. The diagrams drawn of the two areas in Segment 17 should not be compared directly, as Fig. 8E1 depicts a narrow channel segment in low discharge, and Fig. 8E2 shows a point bar which was extensively reworked by sheet and post-sheet flow discharge. These two areas are 1.8 km apart, and illustrate the great variability in channel morphology within the braided-to-meandering transitional sedimentary environment. In Figs. 8A1 through 8F2, stippling represents vegetation.

DISCUSSION

Current orientation studies of modern fluvial systems by Collinson (1971) and Steinmetz (1972) arrived at different conclusions with respect to the variance associated with flows of variable discharge. Collinson (1971) analyzed a large pebble- and sand-covered bank on the Tana River in Norway. This bank sloped gradually upstream and was more inclined along its flanks and downstream portion, forming a lobate, linguoid-like side channel bar. The upstream portion of the bank was pebble-covered, with the downstream end sandy. Pebble, scour, and sand ribbon orientations in the upstream portion were analyzed to compare the effect of flood and falling discharge. Scour features were formed during peak flood stage and sand depositional features formed during falling stage, when the topographic effect of the pebble surface resulted in increased dispersion of the orientation features. Pebble and scour orientations had lower dispersions than the sand ribbons.

Steinmetz (1972) analyzed flood deposits of the Arkansas River in Oklahoma. Cross-bed dips in the sandy deposits were oriented in highly variable patterns, and did not always coincide with the overall river trend. He studied several generations of flood deposits, but did not differentiate between flood and falling stage-generated forms. Primary and secondary mode vectors were determined from detailed analysis of sedimentary structures exposed in a large trench and cut bank. P-mode vectors were oriented toward the northeast, while S-mode vectors point toward the southwest. The channel trend in the area was west-northwest to southeast. Bed forms which produced the P-mode vectors were formed during the peak flood stage and reflect innerchannel flow patterns. S-mode vectors were generated by discharge flowing toward the thalweg, and may have been generated by falling flow. In most cases the P-mode vectors had higher variance than S-mode vectors from the same study section.

Comparison of the depositional environments of both the Tana and the Arkansas is needed in order to determine the cause of the variance, and to compare these areas with the Red. In both cases, the analyzed bed forms were point bar-like forms, attached to an escarpment. The Tana bank had low topographic relief in relation to the channel (between 150 and 200 cm), a bimodal sediment distribution, and was affected by a flood with rapid onset and decline (Collinson, 1971, p. 675, Fig. 3). The Arkansas River point bar had relatively high relief (between 210 and 400 cm), upward fining sandy sediment, and was affected by a series of fluctuating flood stages over a six month period (Steinmetz, 1972, p. 35, Fig. 6). It is apparent that there are major differences between these two fluvial systems. Why are there directional variance differences between the two areas? The reasons are complex, and probably reflect local flow patterns in relation to river bed topography. The pebble bed of the Tana acted to localize sand deposition by causing shear and turbulence changes as the flood waters subsided. The irregular relief of the pebble surface caused the slowing water to flow around the high pebbles and deposit the sandy load as ribbons oriented toward the thalweg. The scours were formed during the peak flood period, when water depth was increased, and relative topographic effect of the bottom was less significant.

Steinmetz analyzed cross beds on the Arkansas River formed by flood-generated bed forms, possibly large-scale lunate ripples (terminology of Allen, 1965). The form of these ripples accounts for the high P-mode variance. Complex flow patterns within the channel at the time of the flood also resulted in high variable current patterns. During falling discharge, water rushed off the point bar surface toward the thalweg, forming low variance bed forms.

The analysis of the Red River presented in this paper entails the investigation of two-dimensional features which resulted from a complete sedimentation cycle. The current pattern shows a good deal of variability in dispersion both from the same locality at different times, and from different localities at the same time. There does not appear to be a trend in the variance, nor can the channel trend always be determined from flow pattern orientation.

Figs. 8A1 and 8A2. Sketch maps of the Burkburnett locality in Segment 1, at the Highway 277-281 crossing. Fig. 8A1 was drawn from aerial photographs and ground observations made on February 22 and 24, 1976. The arrows are the current direction vectors measured from active channel-bed forms (linguoid bars) which migrated through and around exposed longitudinal and braid bars. Fig. 8A2 was drawn from aerial photographs taken July 5, 1976. The arrows represent current direction vectors measured from exposed point and side-bar surfaces. These areas were inundated by the June 26-27 flood, resulting in the formation of large-scale ripples on point and side-bars.

Figs. 8B1 and 8B2. Sketch maps of the large meander-bend in Segment 4, east of Burkburnett. Fig. 8B1 was drawn from aerial photographs taken on February 24, 1976. The arrows represent current direction vectors measured from active linguoid and braid bars which migrated around exposed longitudinal and braid bars. Figure 8B2 was drawn from aerial photographs taken July 5, 1976. Arrows represent current orientation vectors measured from sand waves, linguoid bars, and dunes formed during or soon after the June 26-27 flood crest. Note the extensive coverage of the flood waters in comparison to Fig. 8B1. The five arrows crossing the point bar surface were drawn from bed forms within a pair of chutes which were activated during the flood.

Figs. 8C1 and 8C2. Sketch maps of a portion of the Red River in Segment 6, northeast of Burkburnett. Figure 8C1 was drawn from aerial photographs taken February 24, 1976. Arrows represent orientation vectors measured from active linguoid and transverse bars. Active dissection of the downstream portion of the point bar shown on the south side of the river was taking place at the time of the reconnaissance. The longitudinal bars on the south side of the channel are erosively formed, not depositionally. Figure 8C2 was drawn from aerial photographs taken July 5, 1976. Arrows represent current direction vectors measured from large-scale ripples on exposed longitudinal bar surfaces. Catenary and sinuous dunes were exposed. Note that the point bar which was shown in Fig. 8C1 is largely removed. The longitudinal bars shown in this figure are also of erosional origin, although a large longitudinal sediment mass of depositional origin was the precursor for their occurrence.

Figs. 8D1 and 8D2. Sketch maps of the Highway 79 crossing near Byers, in Segment 11. In Fig. 8D1, arrows represent current orientation vectors measured from active linguoid and transverse bars and dunes. These bed forms were partially or completely submerged and migrated within the channel between exposed longitudinal bars. This map was drawn from aerial photographs taken February 24, 1976. Figure 8D2 was drawn from aerial photographs taken July 5, and from ground observations made on July 11, 1976. The arrows represent orientation vectors measured from dunes, washover-lobes, side bars, and chute-delta dunes exposed on the point bar surface, within the chute, and on exposed channel bars. The two barb-shaped bars on the point bar flank are scroll-like bars, with the upstream one being formed after the other, when the flood waters of the June 26-27 flood were subsiding.

Figs. 8E1 and 8E2. Sketch maps of portions of the Red River in different areas of Segment 17. Figure 8E1 shows an area upstream of the entrance of Beaver Creek into the Red. It was drawn from aerial photographs taken February 24, 1976, and arrows represent vectors measured from active linguoid bars. The large longitudinal bars is an erosional remnant of the upstream end of the point bar which is partially shown in the figure. Figure 8E2 is a sketch map of a large point bar which occurs downstream of the Beaver Creek entrance, north of Terral. The map was drawn from aerial photographs taken July 5, 1976. Arrows represent orientation vectors measured from bed forms occurring on the point bar surface. A variety of dunes were observed, as well as linguoid and side-bar remnants.

Figs. 8F1 and 8F2. Sketch maps of the point bar at the Highway 81 crossing in Segment 20, south of Terral; 8F1 was drawn from photographs taken February 24, 1976. Arrows represent current orientation vectors measured from active channel bed forms (linguoid and transverse bars). Figure 8F2 was drawn from aerial photographs taken July 5, 1976. The surface of the point bar was completely covered during the flood of June 26-27, resulting in the formation of sinuous, straight, and catenary dunes on the bar surface. Arrows represent current vectors measured from these bed forms. In comparison with Fig. 8F1, note that the two longitudinal bars in the area of the railroad bridge are the remnants of the point bar which was present in that location in February. The arrows in the channel represent measurements from the surfaces of slightly submerged bed forms formed during the post-flood period.

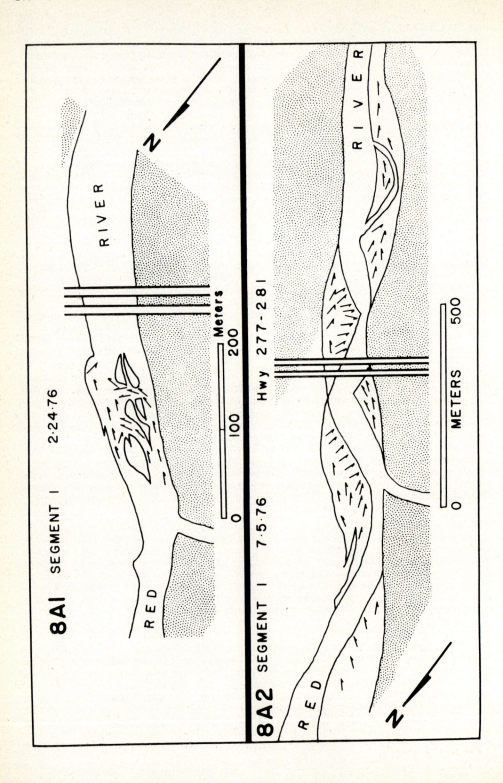

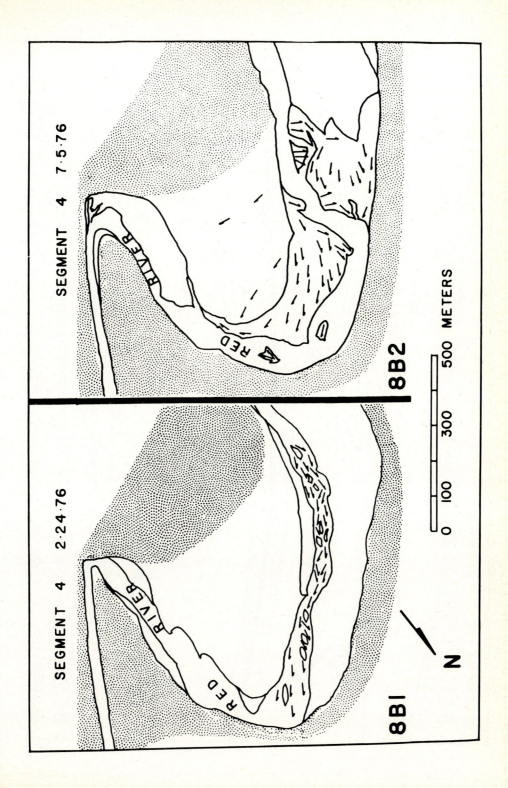

SEGMENT 4 7·5·76

SEGMENT 4 2·24·76

RED RIVER

RED RIVER

8B2

8B1

0   100   300   500 METERS

N

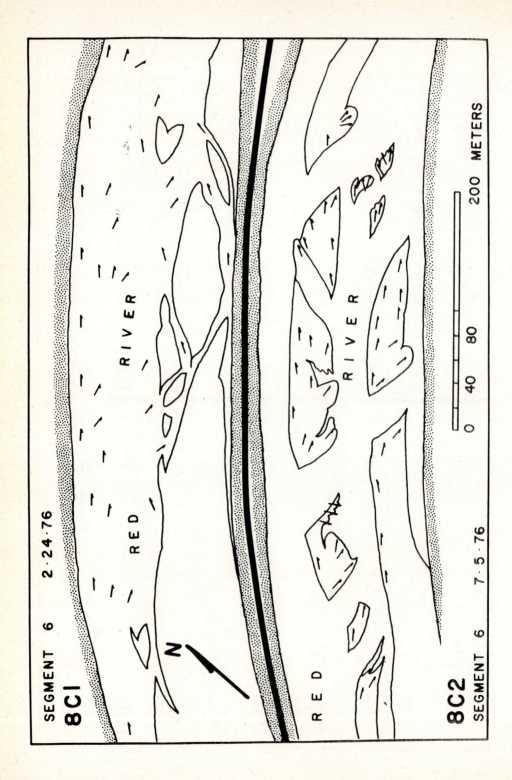

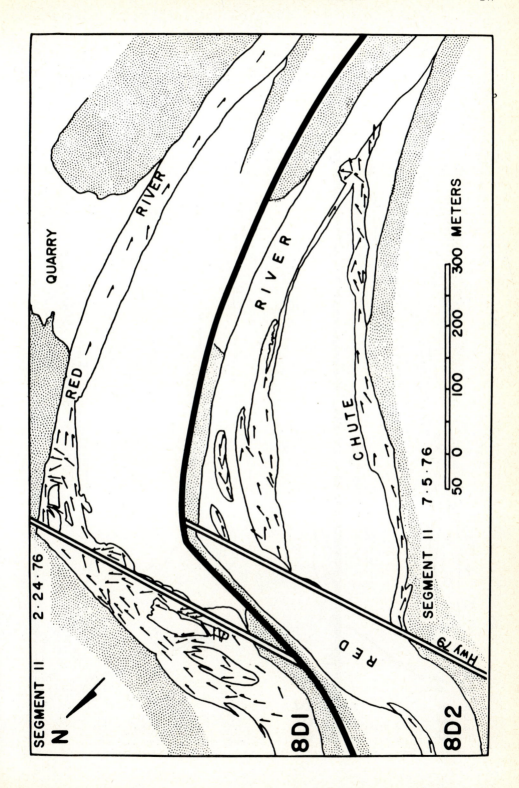

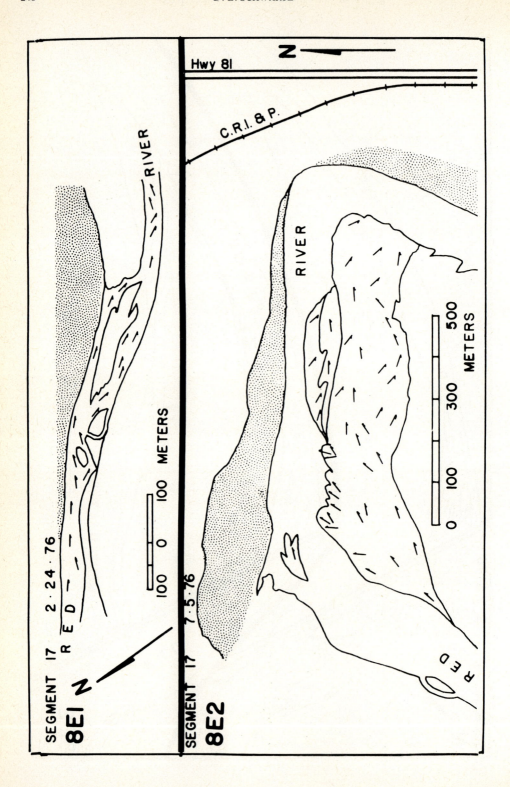

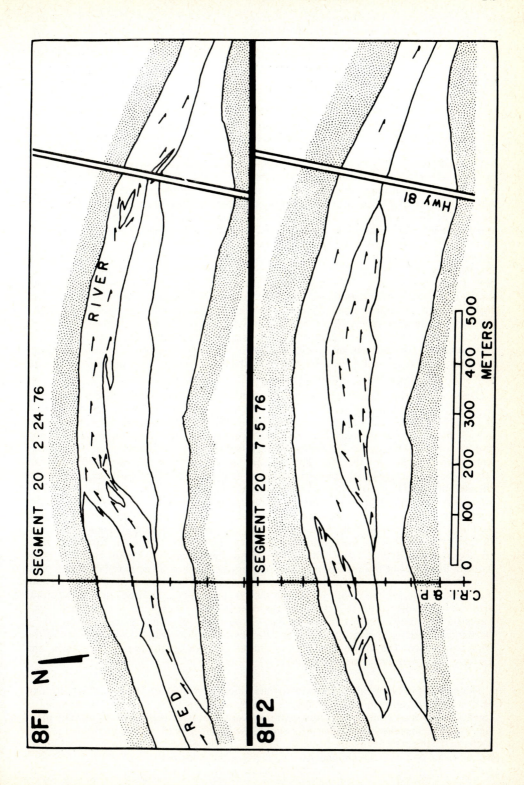

The various aspects of the three flow patterns which make up the sedimentation cycle are displayed in Table I. The vertical axis contains the three patterns resulting from varying discharges, while the horizontal axis displays the characteristic features of the area in which they occur.

Figures 9 and 10 are aerial photographs of the large point bar in segment 4 (see also Figs. 6A, 6B, 8B1, and 8B2), taken in February and July respectively. They illustrate the effect of discharge variation on river bed morphology. Figure 9 shows the area during low discharge, when channel flow conditions were acting. There are numerous small channels carrying discharge in an anastomosing pattern. Figure 10 shows the point bar after the June-July flood, displaying flood peak and falling discharge bed forms (dashed lines are sand wave crests, dotted lines are linguoid bar crests). The ability to maintain the flood-bed forms is a function of the rapidity of the flood subsidence. The peak discharge was maintained long enough to rework the previous point bar surface and to form sand waves. The flood discharge then fell rapidly, without totally removing the sand wave crests. The effect of this rapidly subsiding discharge was local breaching and dissection of sand waves and the formation of linguoid bars. The resultant flow patterns for this area show that the sheet- and falling-flow deposits have higher azimuthal spreads than the channel flow bed forms. In Segments 11 and 20, the opposite orientation pattern is seen. Low discharge channels are dominated by longitudinal and linguoid bars, and the variance in vectoral spread is high. Sheet-flow and falling-flow deposits have less dispersion and are more closely aligned with the local river trend. Segment 6 (Figs. 6A, 6B) displays another type of flow pattern, which may represent an intermediate stage between previously discussed patterns. In this segment, the channel flow and sheet flow plus post-sheet flow current orientation patterns are nearly the same. The reason behind this occurrence may be the constriction of the channel by high banks and bars, and bank stabilization by vegetation in Segment 6.

Channel character, specifically the shape of the river bed normal to the trend, appears to be a significant factor in determining the flow pattern. Channel flow-dominated channels are somewhat tabular in form with multiple thalwegs and low elongate bed forms. Sheet flow-dominated channels tend to be wedge-shaped, occupy much of the river bed, and contain large diagonally oriented bed forms. In Segments 1 and 4, the areas which were reworked by the June flood had gradual slopes (between .01 and .018) toward the thalweg. The slopes of these convex bed features directed flow back toward the channel, resulting in radially diverging flow pattern orientations (Figs. 8A2 and 8B2). This mode of point bar formation is contrary to the model of Allen (1968), in which point bars were formed by successive scroll bar accretion. Flow patterns around channel bars cause high dispersion of orientation vectors from the channel flow period at Segments 11 and 20. The irregular topography formed by the bars themselves caused the flow to deviate greatly from the overall trend. During the flood, increased flow depths in these areas reduced the effect of the topographic inconsistencies and low dispersion bed forms were generated in response to the high discharge. In the case of Segment 11 (Fig. 8D2), a chute was activated during the flood, bypassing the meander curve, and reducing the vectoral dispersion. As discharge fell after the flood peak, a high dispersion chute delta was formed at the downstream end of the bar. Scroll bars were also formed during falling discharge. Segment 20 (Fig. 8F2) was affected by sheet flow which covered much of the point bar surface. The low relief of the bar did not greatly alter the flood flow pattern, and low dispersion bed forms were generated.

The patterns in Segment 17 fit into two categories. The channel flow area (Figs. 6A and 8E1) would probably fit into the same category as Segment 6. Because the channel is constricted by high banks, sheet flow patterns would not have high dispersions. The point bar which is sketched in Figure 8E2 would be grouped with the high discharge-high dispersion category (Segments 1 and 4). The sheet flow pattern from this bar would have higher dispersion than the channel flow deposits of the same bar.

Table 1. Characteristics of the sedimentation cycle flow levels.

FLOW TYPE	CHANNEL CHARACTER	BED FORMS	FLOW BEHAVIOR	DISPERSION	SEDIMENTATION	EFFECT OF TOPOGRAPHY & VEGETATION
LOW DISCHARGE/CHANNEL FLOW	-SHALLOW -WIDE -1 OR MORE SM. CHNLS -TABULAR X-SECTION	LINGUOID, BRAID. & LONGITUD. BARS WITH DUNES	-MODIFY "STREAM-LINE" -ELONGATE EARLIER BED FORMS	MODERATE TO HIGH	PREDOMINANTLY BED LOAD	-CONTROLS FLOW PATTERN -LOCALIZES SEDIMENTATION -CONSTRICTS CHANNEL MIGRATION
HIGH DISCHARGE/SHEET FLOW	-1 LARGE CHANNEL -WEDGE-LIKE X-SECTION	SAND WAVES, TRANSV., SIDE. CHUTE. & SCROLL-LIKE BARS	-BAR FORMATION -EROSION -HIGH SED. TRANSPORT RATE	MODERATE TO HIGH	MIXED LOAD (HIGH SUSP'D) (HIGH BED)	-LOCALIZES SEDIMENTATION -RESTRICTS CHANNEL OVERFLOW -REDUCES BANK EROSION
FALLING DISCHARGE/POST-SHEET FLOW	-1 CHANNEL WHICH REDUCES IN WIDTH AS DISCHARGE FALLS	LINGUOID, CHUTE DELTA & SCROLL BARS WITH DUNES	-MODIFY FLOOD DEPOSITS (INCISE & DISSECT)	HIGH	MIXED LOAD (HIGH SUSP'D) (MOD. BED)	-FLOOD FORMED TOPOGRAPHY CONTROLS FLOW PATTERN -VEGETATION HAS LITTLE EFFECT

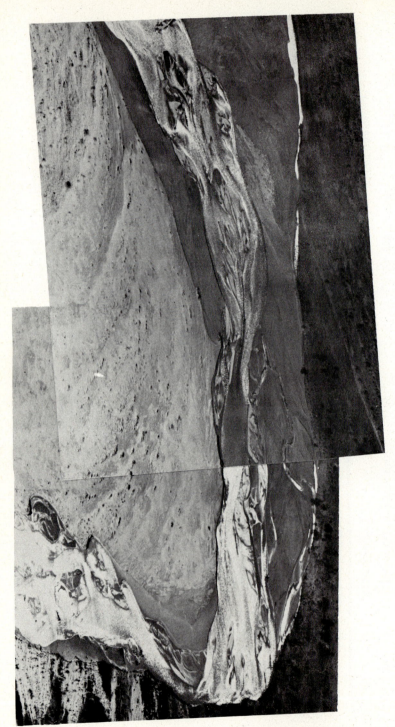

Fig. 9. Oblique aerial photograph of an active channel adjacent to the large point bar in Segment 4 (Figure 8B1). Photograph was taken on February 24, 1976, when channel flow conditions prevailed. There is a two-rut road in the lower center of the photomosaic, which indicates the scale. The overall lateral field of view is approximately 1100 meters. Flow is from right to left, and north is toward the lower left.

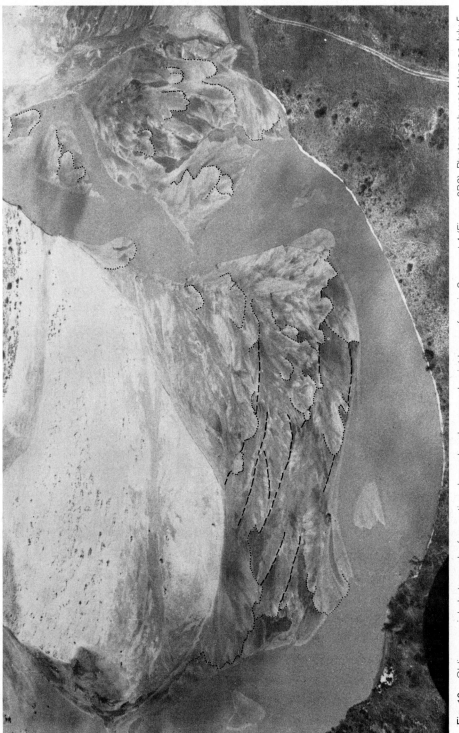

Fig. 10. Oblique aerial photograph of an active channel and exposed point bar surface in Segment 4 (Figure 8B2). Photograph was taken on July 5, 1976, after the flood of June 26-27, 1976, when sheet flow conditions were active; followed by falling (post-sheet) discharge. The bed forms exposed on the point bar surface were formed and modified by the flood and falling discharge. There is a two-rut road in the lower right-hand corner which indicates the scale. The overall lateral field of view is approximately 900 meters. Flow is from the upper right, and north is toward the lower left. Dashed lines on the point bar surface indicate sand wave crests, with dotted lines indicating linguoid bar crests.

Conclusions

The Red River of Oklahoma and Texas undergoes a downsteam change from braided to meandering over a distance of 230 km. The central 100 km of this reach, between Burkburnett, Texas and Terral, Oklahoma displays the major characteristics of this braided-to-meandering transition. The river in this reach is characterized by:

(1) a complex fluvial pattern with braided, meandering, and straight segments;

(2) a variety of large-scale bed forms which occupy inner-channel and side-channel positions;

(3) discharge which varies directly with local rainfall as well as with seasonal climatic changes;

(4) a sedimentation cycle composed of three flow levels,
a) low discharge/channel flow,

b) high (flood) discharge/sheet flow, and

c) falling discharge/post-sheet flow, which are each characterized by sediment load reflecting the flow conditions;

(5) current orientation patterns which reflect variations in,
a) discharge, and

b) local topographic relief and vegetation; and

(6) highly variable azimuthal dispersions within deposits.

The great variability in current patterns shown by the braided-to-meandering transition zone of the Red River is probably characteristic of most fluvial transitions of this type. Deposits formed by all levels of flow must be analyzed if the true current orientation pattern is to be derived. Although the preservation potential is higher for the deposits of sheet and post-sheet flow conditions, channel flow deposits have a significant effect on modifying previous bed forms and complicating the overall orientation pattern.

Acknowledgements

This study is a portion of a doctoral dissertation submitted to the Program for Geosciences at the University of Texas at Dallas. The critical review of the paper by R. J. Moiola and D. E. Eby is gratefully acknowledged. Financial support for the study was provided by a Penrose Bequest research grant from the Geological Society of America, and by a Grant-in-Aid from Sigma Xi. Moral support and inspiration were lovingly given by my wife, Ineke.

References

Allen, J. R. L., 1965, A review of the origin and characteristics of recent alluvial sediments: Sedimentology, v. 5, p. 89-191.

——, 1968, Current Ripples: Their Relation to Patterns of Water and Sediment Motion: North Holland Publ. Co., Amsterdam, 433p.

Bernard, H. A., Majors, C. F. Jr., Parrot, B. S., and LeBlanc, S. J. Sr., 1970, Recent sediments of southeast Texas: Texas Bur. Econ. Geology, Guidebook 11.

Bluck, B. J., 1971, Sedimentation in the meandering River Endrick: Scott. J. Geol., v. 7, p. 93-138.

Boothroyd, J. C. and Ashley, G. M., 1975, Process, bar morphology, and sedimentary structures on braided outwash fans, northeastern Gulf of Alaska: *in* A. V. Jopling and B. C. MacDonald, *eds.*, Glaciofluvial and glaciolacustrine sedimentation: Soc. Econ. Paleont. Mineral. Spec. Pub. 23, p. 193-222.

Briggs, L. I. and Middleton, G. V., 1965, Hydromechanical principles of sediment structure formation: *in* G. V. Middleton, *ed.*, Primary sedimentary structures and their hydrodynamic interpretation: Soc. Econ. Paleont. Mineral. Spec. Pub. 12, p. 5-16.

Campbell, R. L. Jr., 1968, Stratigraphic applications of dipmeter data in Mid-Continent: Am. Assoc. Petrol. Geol. Bull., v. 52, p. 1700-1719.

Collinson, J. D., 1971, Current vector dispersion in a river of fluctuating discharge: Geol. Mijnbouw, v. 50, p. 671-678.

Curray, J. R., 1956, The analysis of two-dimensional orientation data: J. Geol., v. 64, p. 117-131.

Folk, R. L., 1974, Petrology of Sedimentary Rocks: Hemphill Pub. Co., Drawer M, University Station Austin Texas, 182p.

Frey, J. C. and Leonard, A. B., 1963, Pleistocene geology of Red River basin in Texas: Texas Bur. Econ. Geol., Rept. Invest., No. 49.

Leet, L. D. and Judson, S., 1971, Physical Geology: 4th ed., Prentice-Hall Inc., Englewood Cliffs, N.J., 687p.

Leifeste, D. K., Blakey, J. F., and Hughes, L. S., 1971, Reconnaissance of the chemical quality of surface waters of the Red River basin, Texas: Texas Water Devel. Bd. Report No. 129.

McDaniel, G. A., 1968, Application of sedimentary directional features and scalar properties to hydrocarbon exploration: Am. Assoc. Petrol. Geol. Bull., v. 52, p. 1689-1699.

McDonald, B. C. and Banerjee, I., 1971, Sediments and bed forms on a braided outwash plain: Can. J. Earth Sci., v. 8, p. 1282-1301.

McGowan, J. H. and Garner, L. E., 1970, Physiographic features and stratification types of coarse-grained point bars: Modern and ancient examples: Sedimentology, v. 14, p. 77-111.

Miall, A. D., 1974, Paleocurent analysis of alluvial sediments: A discussion of directional variance and vector magnitude: J. Sediment. Petrol., v. 44, p. 1174-1185.

Schumm, S. A., 1963, Sinuosity of alluvial rivers on the Great Plains: Geol. Soc. Am. Bull., v. 74, p. 1089-1100.

Schwartz, D. E., 1976, Sedimentology of the braided-to-meandering transition zone of the Red River in Texas and Oklahoma: Geol. Soc. Am. Abs. with program, Denver, Colo. v. 8, p. 1092.

————, 1977, Flow patterns and bar morphology in a braided-to-meandering transition zone — The Red River in Texas and Oklahoma: Am. Assoc. Petrol. Geol.-Soc. Econ. Paleont. Mineral., Abs. with program, Washington, D.C., p. 107.

Shelton, J. W. and Noble, R. L., 1974, Depositional features of a braided-meandering stream: Am. Assoc. Petrol. Geol. Bull., v. 58, p. 742-752.

Simons, D. B. and Richardson, E. V., 1962, The effect of bed roughness on depth-discharge relations in alluvial channels: U.S. Geol. Survey Water Sup. Paper 1498-E, 26p.

———— and Nordin, C. F. Jr., 1965, Bedload equations for ripples and dunes: U.S. Geol. Survey Prof. Paper 462-H.

Smith, N. D., 1970, The braided stream depositional environment: Comparison of the Platte River with some Silurian clastic rocks, north-central Appalachians: Geol. Soc. Am. Bull., v. 81, p. 2993-3013.

————, 1971, Transverse bars and braiding in the Lower Platte River, Nebraska: Geol. Soc. Am. Bull., v. 82, p. 3407-3420.

————, 1972, Some sedimentologic aspects of planar cross-stratification in a sandy braided river: J. Sediment. Petrol., v. 42, p. 624-634.

————, 1974, Sedimentology and bar formation in the Upper Kicking Horse River: A braided outwash stream: J. Geol., v. 82, p. 205-233.

Steinmetz, R., 1962, Wasatch sandstones in the eastern Green River Basin: Unpub. Ph.D. thesis, Northwestern Univ., Evanston, Ill., 181p.

————, 1972, Sedimentation of an Arkansas River sand bar in Oklahoma: A cautionary note on dipmeter interpretation: Shale Shaker, v. 16, p. 32-37.

Strahler, A. N., 1952, Dynamic basis for geomorphology: Geol. Soc. Am. Bull., v. 63, p. 923-938.

Stricklin, F. L. Jr., 1961, Degradational stream deposits of the Brazos River, central Texas: Geol. Soc. Am. Bull., v. 72, p. 19-36.

U.S. Geol. Survey, 1977, Water Resources Data for Texas: Water Year 1976, v. 1, U.S.G.S., Water-Data Report TX-76-1.

BANK DEPOSITION IN SUSPENDED-LOAD STREAMS

GRAHAM TAYLOR[1] & K. D. WOODYER[2]

ABSTRACT

Many of the western-flowing streams of eastern Australia are suspended-load streams. Most of these streams deposit the majority of their load within the channel as bank benches. Bench is used here to describe a flat surfaced depositional body of sediment on the channel bank. It is distinct from a terrace produced by incision. These benches commonly occur at three levels and form the most notable morphologic feature of these rivers. The three levels form as a result of the interaction of the hydrologic and sedimentologic regimes within the stream.

The Barwon River in northern New South Wales is typical of these Western Rivers and is discussed in detail here. Floods along the Barwon are very slow moving (less than 0.5 ms^{-1}) and the peak sediment concentrations usually preceed the flood wave by up to 14 days along the middle reaches. Sediment concentrations are in general low, rarely exceeding 1500 ppm at peak concentration and the sediment carried is extremely fine grained, 80% being less than 2 μm. These loads are related to the low bed gradients (5×10^{-5}). The river is highly sinuous (2.3) but does not actively meander and has a width-depth ratio of about 8.

The benches along the Barwon develop along the banks in straight reaches as well as at the inside and outside of bends. They are narrow and long and have a lensoid sectional shape. The sediments composing the benches vary according to elevation and their situation along the channel. Those benches occuring low in the channel and at point situations contain more sand than those at the outside of bends and those occuring higher up the banks.

The deposits of the lowest bench are dominantly cross-bedded sands, up to 20 cm thick with thin mud interbeds, which drape the underlying sands. The muds protect the sands from erosion during the next rise, hence enabling the bench to develop. The growth of the bench often occurs with limited erosion of the outside bank of the channel, or no erosion of the bench itself.

The deposits which form the two higher benches are thinly interbedded sands and muds which drape the top and front of each deposit. The sand beds attain a maximum thickness of 16 cm and the muds 14 cm. The beds commonly contain parallel or wavy fine lamination, are often graded and rarely cross-bedded. The graded bedding is commonly normal, but may also be reversed or threefold (e.g. sand/mud/sand). No erosional contacts are evident with the exception of very localised scours. Because of the notable lack of erosion and the geometry of the deposits, the benches accrete both laterally and vertically at rates depending on their elevation, and to a lesser extent their situation in the channel.

INTRODUCTION

Suspended-load streams are not common in most countries but there are many amongst the west flowing rivers of Eastern Australia. The term suspending-load stream in used as defined by Schumm (1963). He defines it as a river which carries over 85% of its total load in suspension and has a channel perimeter weighted mean per cent silt-clay (M) value greater than 30. The term suspended-load is used in the same way as Schumm (1963), i.e.: "as synomomous with wash-load, that part of the sediment load not significantly represented in the bed of the stream (Einstein, 1950)".

Most of the west flowing streams in Eastern Australia have many of the other morphological characteristics described by Schumm (1963). They are generally highly sinuous (sinuosity greater than 2.0), width/depth ratios are less than 10 and they have

[1]Department of Geology, Australian National University, P.O. Box 4, Canberra, ACT — presently at School of Applied Science, Canberra College of Advanced Education, P.O. Box 1, Belconnen, ACT 2616, Australia.
[2]Division of Land Use Research, C.S.I.R.O., Black Mountain, A.C.T. Australia.

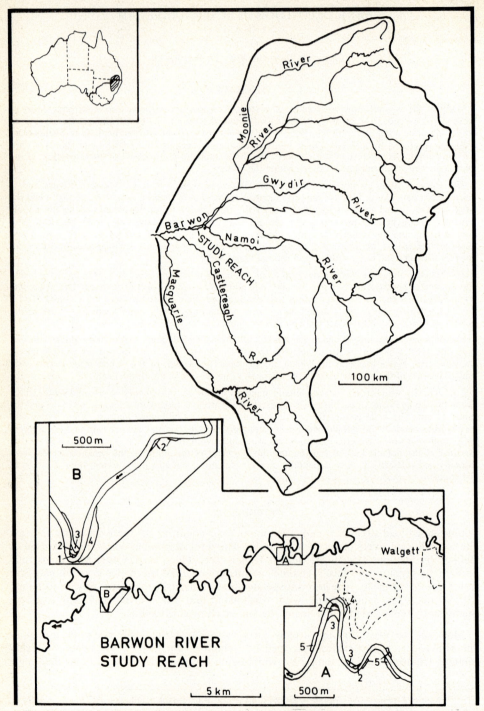

Fig. 1. Locality Map showing the Darling River, its Catchment downstream to the Macquarie confluence and a detailed map of the study reach. The insets A and B show detailed maps of the deposits discussed herein; 1: Low Bench, 2: Middle Bench, 3: High Bench, 4: Concave-Bank Bench, and 5: Ti-tree Benches on straight reaches.

extremely low gradients. Schumm (1968) describes the morphology and Quaternary history of one west-flowing Eastern Australian River, the Murrumbidgee. We have made a detailed study of the Cainozoic history, sedimentation, hydrology and morphology of a second of these rivers, the Barwon River of Northern New South Wales (Fig. 1).

THE BARWON RIVER

The Barwon rises in the eastern highlands of New South Wales at an average elevation of 1500 m and falls to about 250 m after flowing west for about 200 km. West of the highlands the river continues to travel west and south (as the Darling River) to the Southern Ocean, about 2000 km distant. Average gradients across the plain are about 5×10^{-5}. After leaving the highlands, the river develops an anastomosing pattern with many anabranches (Riley, 1973; Riley and Taylor, 1978; Woodyer et al., 1978).

The Barwon River, along the study reach (Fig. 1), has a sinuosity of 2.3 with a mean width-depth ratio of 8. Floods are very slow-moving (less than 0.5 ms^{-1}) and peak suspended sediment concentrations usually preceed the flood wave by up to 14 days (Fig. 2). Sediment concentrations are in general low, rarely exceeding 1500 ppm at peak concentration. The sediment carried is extremely fine grained, some 80% being less than $2 \mu m$. Sedimentation rates locally can reach 20 mm per year (Woodyer, 1978).

Rapid deposition occurs only along the channel banks on structures called benches (Woodyer, 1968) and is more rapid on the Low and Middle Benches. Despite this rapid deposition on the channel banks the channel is stable. Two river surveys from 1880 and 1969 show no significant shift in the channel either laterally or downstream. Further, general bank erosion is not evident; however, occasional localised erosion does occur.

These benches develop at three levels within the channel and form the most obvious morphologic feature of the channel (Figs. 1, 3). The Low Bench forms low cuspate sand bodies near the channel bed at bends in the river. These develop to a maximum height of about 2.5 m above the thalweg, their height depending on the height of bed-load movement. The Middle Bench is a depositional surface about 10 m above the thalweg located along straight reaches and at bends. The High Bench occurs about 2 m above the Middle Bench at bends and along straight reaches. Another significant in-channel deposit is the Concave-bank Bench (Woodyer, 1975). This forms at the outside of hairpin bends up-stream of the bend apex.

DEPOSITIONAL BANK BENCHES

Low Benches

These are restricted to point localities (Figs. 1, 3) and are low cuspate bodies with gentle to steep slip-off faces. They may deposit thin extensions both upstream and downstream of the point. In section these deposits are lensoid, thinning both bankwards and towards the stream.

The Low Benches are composed dominantly of very fine to medium size quartz sand with minor silt and clay (Fig. 4). The silts are dominantly quartz and the clays dominantly montmorillonite and mixed-layered montmorillonite/illite with minor kaolinite. The sand beds are commonly lenticular and reach maximum thicknesses of 20 cm, and in general thin up sequence. The sand beds are separated by thin mud beds which in general increase in frequency up sequence. They are for the most part thin (2 to 10 mm) but thicken up sequence. The mud beds drape the underlying sand beds with no evidence of erosion of the sand prior to mud deposition. The mud beds show little internal structure with the exception of occasional coarse, wavy, parallel lamination which is very rarely normally graded. The mud beds are frequently discontinuous and curled in places as a result of dessication.

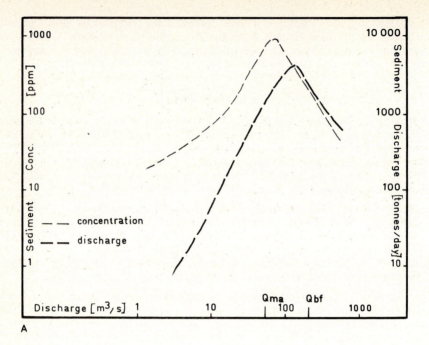

A

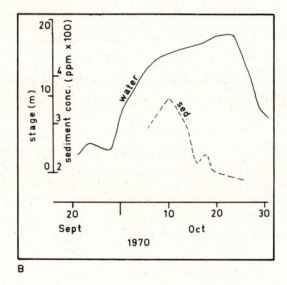

B

Fig. 2. Graph (A) showing lines of best fit through two years data of suspended-sediment concentration and suspended-sediment discharge versus water discarge at Walgett on the Barwon River. B is a sample hydrograph showing the relation between water discharge and sediment concentration.

Fig. 3. Photographs of benches at two bends on the Barwon River. 1: Low Bench, 2: Middle Bench, 3: High Bench, and T: terrace of earlier deposits.

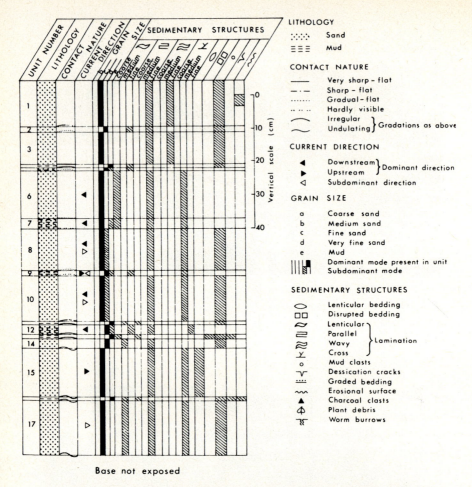

Fig. 4. Graphic log of a portion of the Low Bench deposits.

The sand beds are typically cross-bedded. The cross-bedded sets are either tabular or trough with a set height of up to 14 cm. The cross-beds indicate current directions dominantly downstream; however, occasional upstream dipping sets occur (Taylor *et al.*, 1972). The sands also commonly contain fine parallel laminae (both parallel to principal- and cross-bedding). Mud clasts occur throughout the sand beds but are most common near their bases.

When exposed, the surface of these low benches are generally mud draped, but the bedforms active at higher stages are preserved. Commonly dunes are preserved low on the bench (Fig. 5) with plane bedding or occasionally ripples higher on the bench. The Low Bench is rarely vegetated, even after extended periods of low flow.

Middle and High Benches

These benches are best developed on the inside bank of bends but also occur along straight reaches. At points where a low bench is exposed the Middle and High Benches are developed immediately above and behind them (Fig. 3). At bends these benches are cresentic in plan with a length of up to 1200 m and widths up to 30 m. The deposits forming these benches are lensoid in section and up to 6 m thick. In the straight reaches the benches and their deposits have a similar geometry with the exception that the benches are straight, paralleling the bank. They frequently support thick vegetation and often trees (*Eucalyptus camaldulensis,* Dehn.)

The deposits of these benches are made up of thin interbeds of fine sands and muds, similar in many respects to "Laminites II" of Lombard (1963), (Fig. 6). They contain very little material coarser than 200 μm. The beds are in general 5-10 cm thick and almost without exception their upper surface conforms to their depositional surface. In general they lack any indication of having been deposited by current activity, and there are virtually no erosional contacts between beds. The deposits are rather uniform, with minor variations corresponding to different localities within the channel. The overall nature of the unit is best described from the point localities; the variations at other localities are discussed later.

The sand beds vary in thickness from 0.1 to 16 cm, averaging 5 cm. The grain size is less than 300 μm. They rarely contain more than 10% mud. The mud beds varys in thickness
are rare and are usually very thin (less than 0.5 cm). Both sand and mud beds are relatively continuous, both laterally and up-dip draping the entire area covered during their disposition. However, they do wedge out up-dip. The height at which they wedge out corresponds to the height of the flood from which they were deposited. Hence there are approximately twice as many beds in the Middle Bench deposits as in the High Bench deposits. Further, although few data are available, it follows that the deposits are about twice the thickness under the Middle Bench as under the High Bench. At the 'front' of the Middle Bench the beds continue down the bank, often dipping at angles up to 40°. At their lower end the beds wedge out below mean-annual stage, or grade into the sandy deposits of the Low Bench. Figure 7 shows the sectional geometry of the Middle and High Bench deposits and Figure 8 the internal organization within the Middle Bench.

In contrast to the Low Bench, cross-lamination is rare in the Middle and High Benches, however graded bedding is common. The grading may take any one of four configurations:

1) grading up from sand to mud (normal grading)   Simple

2) grading up from mud to sand (reverse grading)   Simple

Fig. 5.   Photograph of a Low Bench showing mud draped over low amplitude dunes (arrowed) accreting radially around the bench.

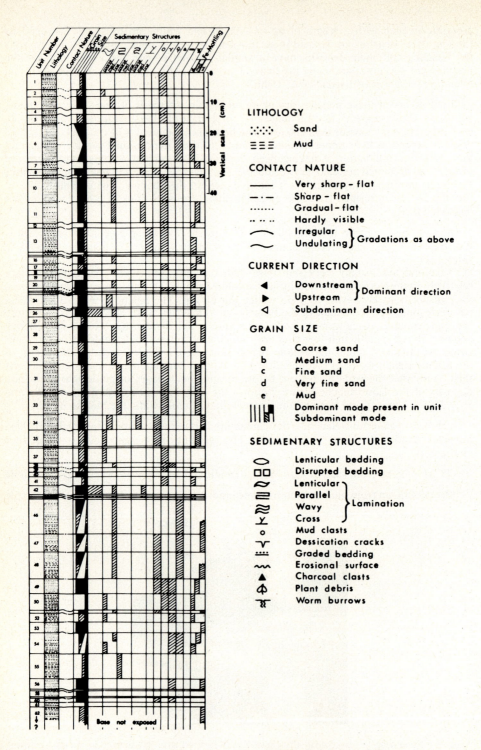

LITHOLOGY

:::::   Sand

≡≡≡   Mud

CONTACT NATURE

———   Very sharp – flat
—·—   Sharp – flat
·······   Gradual – flat
··· ···   Hardly visible
⌒ Irregular
∿ Undulating } Gradations as above

CURRENT DIRECTION

◀   Downstream } Dominant direction
▶   Upstream
◁   Subdominant direction

GRAIN SIZE

a   Coarse sand
b   Medium sand
c   Fine sand
d   Very fine sand
e   Mud
‖‖‖   Dominant mode present in unit
Subdominant mode

SEDIMENTARY STRUCTURES

◯   Lenticular bedding
◻◻   Disrupted bedding
∼   Lenticular
⩚   Parallel
∿   Wavy   } Lamination
Υ   Cross
o   Mud clasts
⊽   Dessication cracks
⸿   Graded bedding
∿∿   Erosional surface
▲   Charcoal clasts
Φ   Plant debris
⚏   Worm burrows

Fig. 6.   Graphic log of a portion of Middle Bench deposits.

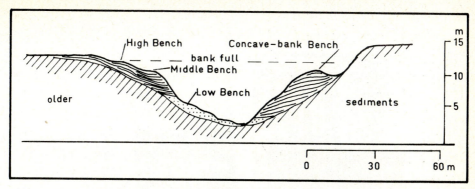

Fig. 7.   An idealised channel cross-section illustrating the positions of the Low, Middle, High and Concave-Bank Benches. Also see insets A and B Figure 1.

3) grading up from sand to mud to sand (waning-waxing grading)   Complex

4) grading up from mud to sand to mud (waxing-waning grading)   Complex

The most common configuration is (2), (4) is well represented in the sections examined but it is not common. (1) and (3) are rare. Within these four configurations the grading may be represented by either of the two following forms:

1) Regular transition from coarse to fine or fine to coarse up the bed. (continuous grading).

2) A sand (or mud) bed in which there is an increase in the number of fine (or coarse) laminae up through the beds, (laminated grading).

Laminated grading is the most common form in each of the configurations.

Overall the most common grading in these sediments is a laminated reverse graded bed (Fig. 9A, B) followed by laminated waxing-waning grading (Fig. 9C). The thickness of graded beds in the deposits of the Middle and High Benches is up to 12 cm.

The Middle and High Bench deposits contain many other primary structures. Parallel lamination within beds is common; it may be either flat or wavy and is sometimes disturbed. Flat parallel lamination is not extensively developed but when present usually passes laterally into wavy lamination. Both flat and wavy lamination may be disturbed although, on the whole, disturbed lamination is not very common. The flat lamination may be developed on flat depositional surfaces and in sand beds where the irregularities of the previous bed are smoothed. The wavy lamination forms in response to deposition over an already irregular surface. These irregularities are formed by several features including; vegetal growth, disturbance of the surface by animals, logs, human activity and by the swelling and contraction of the montmorillonite-rich sediments prior to any subsequent deposition. The waviness persists up-section, the amplitude of the waves gradually dying out until flat lamination again is the norm.

Disturbed bedding and lamination is not widespread or common in the unit as a whole but, on the steeply dipping portions of the unit nearest the river, disturbance, caused by slip of the beds, is common. Where the beds dip at greater than 30°, there is considerable disturbance of the bedding caused by slip, as a result of wetting of the sand beds during high stages. Disturbances on the lower parts of the unit caused by animals walking over the wet sediment are also quite common, (Fig. 10). There are also numerous minor irregularities in the bedding of unknown origin. Another form of bedding disruption is a

Fig. 8. Two photomosaics of sections through Middle Bench Deposits. A is from a hairpin bend and B from a 90° bend. Note the bedding, both wavy and flat, the wedgeouts and the conformity to bench surface with the beds dipping away towards the river.

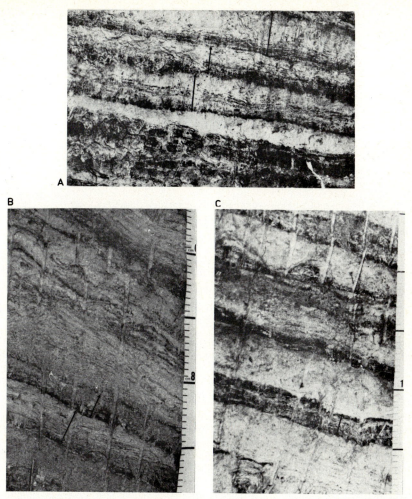

Fig. 9. Three photographs of graded bedding in Middle and High Bench deposits. A and B show (arrowed) reverse laminated grading and C waxing and waning laminated grading.

general 'mixing' or reduction in the definition of the bedding and lamination. This is caused by two factors; age and mineralogy. The deposits contain substantial volumes of montmorillonite which when wetted expands to about 1.25 - 1.5 times its original volume. The result of continual wetting and drying over a long period gradually disrupts the bedding. As well as the time factor in the wetting and drying process, time is also important with respect to pedogenesis. Although pedogenesis is not proceeding at a rate great enough to obliterate the bedding character of the sediments, it is none the less a continuous process which acts on the depositional surfaces whenever these surfaces are exposed. Hence during long periods of low flow (e.g. 1919-1920) some pedogenic modification may occur. Also as the benches accrete, they are exceeded less frequently and the time between depositional episodes increases allowing greater pedogenic modification.

Cross-lamination is not common in the Middle and High Benches. The most abundant form is small scale (less than 5 cm) ripple cross-lamination of fine sand separated by

Fig. 10. An example of strongly disturbed lamination in Middle Bench deposits. These features are interpreted as hoof prints with a raised rim (a) at the front and an undercut (b) at the front of the depression.

muddy laminae. They dip towards the river at 12° - 15°. No truncated sets have been found. The cross-bedding is most commonly developed on the less steeply dipping portions of the unit.

The cross laminated and many of the graded beds in the deposits exhibit elongation functions (Moss, 1972) typical of sediments deposited from bed-load.

In addition to these small scale cross-laminae there are larger sets about 8 cm thick which are developed at the slope break between the bench and the bank. The are tabular sets with well developed toe-set lamination of material finer than higher up the foresets, i.e. the cross-laminae are inversely graded. The laminae dip towards the stream at an angle of between 10° and 15° to the principal bedding.

Both types of cross-bedding indicate that the depositing current was flowing back into the channel, suggesting that the cross-bedding is perhaps formed during the falling stages of the flood. Although reverse flows over points along the Barwon have been recorded (Taylor *et al.,* 1972; Woodyer, 1975), there is not sufficient variability in the cross-beds of High and Middle Bench deposits to invoke reverse flow as the generator of the cross-lamination. Similarly it does not seem probable that they were generally formed by downstream bed-load motion, since there is little evidence of a downstream component in the cross-bedding.

Contacts between the beds in the Middle and High Bench deposits are for the most part depositional; erosional contacts are rare, localised and very minor. The major cause of erosion in the unit is scour around obstructions to flow, the most common of these being trees, logs and stumps. On the present depositional surface (the benches and banks) such scours are common phenomenena; however, they are not readily recognised in section since the material scoured out is sand and on the retreat of the flood the sand is draped with mud so that in section all that appears is a thinning of the sand, and to attribute this thinning to scour is difficult, particularly when they are not laminated. The presence of mud pellets is a further indication of erosion, the pellets having been formed by the erosion of thin slivers of mud from desiccated mud beds during the rising stages of the flood.

Concave-Bank Benches

These develop along the outer upstream bank in bends. Their surfaces are up to 10 m above the thalweg at the highest point and slope upstream. They have a shallow chute between the bench and bank and are crescentic in plan (Woodyer, 1975). In section the

Fig. 11. A recent bank collapse with a lobe of sediment protruding into the channel.

bench deposits are lensoid (Fig. 7) up to 8 m thick. The Concave-bank Benches are very mud rich (80%) with up to 60% less than 2 μm. They are not well bedded but some crude lamination is visible in the most recently deposited beds. Most of this lamination is produced by concentrations of organic debris which is abundant in these deposits. No other primary or secondary structures have been observed. If any were present, originally, it is highly likely that in such a clay rich deposit they would rapidly be destroyed by the continual wetting and drying process.

Bench Foundations

The deposits of the Middle and High Benches are dominatly suspended-load deposits and consequently only begin to deposit within the channel above the zone of bed-load activity. Hence they require a foundation on which to develop and this must be a minimum of 2.5 m above the thalweg. Along the study reach of the Barwon three types of footings are prominent.

At bends the Low Bench provides a pile of sandy sediment to the maximum height of bed-load movement. The deposits grade upward into the interbedded sands and clays of the Middle and High Point Benches.

On straight reaches the Middle and High Bench deposits begin to accumulate on bank slumps and behind ti-trees screens. Bank slumps occur after floods when steeply dipping, water laden, bank sediments slip into the channel (Fig. 11). These produce a small lobate sediment body in the channel. Because of the cohesive nature of the bank sediments and the extremely low energy conditions of the stream these lobes are not removed and hence provide a footing for suspended-load deposition.

Ti-trees (*Melaleuca linariifolia* Sm.) frequently inhabit the channel at times of low flow. Provided the river stays low enough for a year or two, these trees become established as low as 3 m above the thalweg. Once established they provide a screen behind which deposition can occur during high stages (Fig. 12). Once deposition behind the trees reaches a certain stage the trees are buried and die, leaving a bench devoid of ti-tree.

Spatial Variation in the Nature of the Middle and High Bench Deposits

There is systematic variation in the nature of deposits of the High and Middle Benches with the location of the deposit. The variations include the sand/mud ratio and the percentage of organic matter. The sand/mud ratio in these deposits is dependent on the turbulence in the channel and hence will be highest in the deposits on the inside of bends and least in deposits on straight reaches and on the outside of bends, Table 1. The ratio is highest for point benches on sharp bends and lowest for Concave-Bank Benches. The values for ti-trees and straight reach benches are intermediate between point and Concave-Bank Benches. The percentage of organic matter varies within the Middle and High Bench deposits with respect to the site of the deposit. The organic matter content is highest in the Concave-Bank Bench deposits (from field observation) where it imparts a spongy texture to the sediment, which persists at depth. The organic content is lowest at the point localities and intermediate at ti-tree and straight reach sites. This variation is again related to the energy conditions at the site at the time of deposition.

DISCUSSION

The Low Bench is deposited by bed-load activity at times of high flow. The majority of the material accumulates from the progression of low dune-like bedforms moving through the bend (Fig. 5) producing trough cross-beds and lenticular bedding. As the river falls the bed-load activity diminishes and the bedforms are eventually stabilised by a thin drape of mud deposited from suspension. The proportion of mud increases up-sequence due to the reduced bed-load activity higher in the channel.

Fig. 12. A ti-tree screen growing low within the channel and deposition of a bench beginning behind the screen.

At higher bench levels bed-load deposition gives way to suspended-load deposition typical of the Middle and High Benches. The internal geometry and the grading of beds in these deposits is evidence of their deposition from suspension. The mechanism of deposition of such fine muds in unresolved. Flocculation and physical contact of grains with the banks may play a role. The deposition of sand at heights of up to 12 m above the thalweg poses a further problem. No sand has been sampled in suspension in flows just below bankfull. Data from the spatial variations in sand/mud ratios at different angle bends shows that in general the tighter the bend the more sand there is in the deposits. This suggests that sand is temporarily suspended in the flow in localised regions of high turbulence and that it is this sand which is deposited on the Middle and High Benches.

Although the geometry and graded bedding evidence suggests that the deposits of the two higher benches are deposited from suspension the beds are frequently laminated. This suggests some post depositional remobilization of the sediments by the current to produce lamination. This is supported by elongation functions (Moss, 1972) which show typical lower-flow bed-load characteristics. The modification however, must be very slight as no ripples are formed.

The origin of the graded bed types within the Middle and High Bench deposits is the result of many factors. Some of the major factors include:

1) the nature of the hydrograph.

2) the energy gradients on the rising and receding limbs of the flood-wave.

3) the nature of the channel bed during the flood.

4) the suspended sediment concentration.

5) the temporal relationship between the sediment peak and the discharge peak.

6) the source of the water/sediment mix.

To the authors' knowledge there is no record of this style of grading in fluvial sediments, although it is expected that such grading would commonly occur in flood-plain

TABLE 1

Sand/Mud ratios of Middle and High Bench sediments from different sites along the Barwon River downstream of Walgett.

Bench Type	Bench Level	Sand/Mud	Bend Angle*
Concave+	—	0.2	20°
Concave+	—	0.25	20°
Ti-tree	—	0.9	straight reach
Ti-tree	—	0.6	straight reach
Point	Upper	0.3	150°
Point	Upper	3.7	20°
Point	Middle	0.5	120°
Point	Middle	1.5	120°
Point	Middle	1.3	80°
Point	Middle	1.5	20°

* Angle of the apex measured on the inside of the bend.
+ Concave = Concave-bank Bench.

or basin deposits. However, Kuenen & Migliorini (1950) suggest that in fluvial deposits the most common grading should be "inverted grading". Sanders (1965) and Costa (1974) record reverse grading in very coarse fluvial deposits. Fisk (1974) in a brief review of reverse grading comments that it is recorded from almost all types of sedimentary rocks, but that it is most obviously associated with conglomerates and breccias. In contrast we report grading at the opposite end of the size spectrum.

Due to the low gradients, velocities and turbulence, suspension and deposition of sand occurs mainly during the rising limb of the flood wave. Deposition of cohesive muds, below a critical shear force, increases with decreasing shear force and increasing sediment concentration (Partheniades, 1971). The shear force at the bench surface decreases with increasing depth. Thus, deposition of mud may occur near the flood peak or during high floods if the shear force is less than the critical shear force. The high percentage of montmorillonite in the suspended clay favours deposition because the critical shear force is relatively high. During high-flood levels short-circuiting of bends occurs reducing flow velocities and shear forces in bends. On this basis, also, high flood levels favour deposition of muds on point benches. However, deposition of mud may occur at about bankfull stage in the sharper bends. This deposition is associated with gentle reverse flows on the outside of bends, where channel widening occurs (Woodyer, 1978), or on the inside of bends due to the separation of flow from the bank. Near Walgett the sediment peak normally preceeds the flood peak by up to 14 days. If any of these conditions of low bed shear occur during the passage of the sediment peak then relatively rapid rates of mud deposition occur before the arrival of the flood peak and before the deposition of sands. Presumably this is the explanation for reverse grading. However conditions favouring the deposition of mud can occur at or after the flood peak, following the deposition of sand. This will also produce normal grading as is the case with the mud drapes on the Low Bench.

There are many possible origins for complex grading. The nature of the hydrograph from which the bed is deposited is the most obvious. A double peaked flood which continuously inundates the site could give rise to complex grading. The rising limb of the first peak could be associated with the deposition of sand followed by the deposition of mud, while the water levels are still high on the falling limb. This sequence could be repeated during the second rise and fall. This would explain the waxing-waning type of grading. The other type (i.e. waning-waxing) could result from a combination of the two types of simple grading. A mud may be deposited during the passage of the high sediment load ahead of the flood, a sand during the high discharge stage and another mud during the falling stages of the flood. Another possible cause of three-fold grading could be changes in the bed configuration during the passage of a flood-wave. Sand suspension may occur in eddies associated with dunes near the flood peak. This intermittent suspension and deposition of sand associated with continuous deposition of mud may produce grading. Another possible cause is the source of the waters in the flood-wave. The waters from different tributaries may contain different types and concentrations of sediment as well as temporally increasing energy gradients and since these waters may also arrive at the site of deposition at different times then a complexly graded bed could result.

Two levels of deposition, subject to 'constant' (Woodyer, 1968) frequency of inundation from stream to stream, suggests some fundamental control. However, Woodyer (1968) finds the case for the 'constant' inundation of the Middle Bench open to some doubt. On this basis the Middle Bench may be regarded simply as an intermediate stage of the construction of the High Bench. The Middle Bench and High Bench forming as the channel moves laterally.

Taylor (1976) contends that along the Barwon River there is little erosion and that lateral accretion features do not form solely in response to channel widening. The Middle and High Benches occur at a relatively constant height above the thalweg. This implies that they are forming in response to some relatively systematic parameter of the sediment/water flow.

Leopold *et al.* (1964) suggest that rivers develop their bank-full channel to cope with the normal range of flows and this gives rise to the constant bankfull frequency. Wolman and Miller (1959) suggested that the channel/flood plain junction was the result of a dynamic balance between the processes of deposition and transportation within the channel and on the flood plain and that this balance is relative to the most efficient means of sediment transport within the system.

Taylor (1976) suggests that the approximate coincidence of maximum sediment discharge (Fig. 2) with bank-full discharge on the Barwon River at Walgett supports the contention of Wolman and Miller (1959) and is a possible controlling factor in determining the level of the High Bench.

Taylor (1976) considers the Middle Bench level near Walgett is related to the maximum sediment concentration which proceeds the flood peak and occurs at about mean-annual stage which approximates the level of the Middle Bench. This is similar to a change in suspended sediment concentration reported by Wolman and Miller (1959), although they found no morphological evidence of this change.

Since the Middle Bench is flooded on the average two or three times as frequently as the High Bench and there is evidence that sedimentation rates on the Middle Bench are greater than on the High Bench it follows that the Middle Bench is accreting at a faster rate than the High Bench.

Eventually these two benches (Middle and High) merge at the bankfull or High Bench level, thus reducing the bankfull channel width. A reduction in channel width without any corresponding reduction in depth would cause an increase in flood-wave velocity along the channel (Rouse, 1950). In the case of overbank floods this would result in sediment concentration peaks occurring at higher stages up to bankfull (Woodyer *et al.,* in prep.). In the case of below bankfull floods the effect would be the same if the sediment peak leads the flood-wave peak. The available evidence for the study reach of the Barwon River indicates that this is the normal occurrence. If this relationship is preserved the only limit to the growth of the Middle Bench is the flood-plain (High Bench) level.

Continued accretion of benches in only one of many ways that the channel is obstructed along the Barwon River. Another possible obstruction is the growth of vegetation not only along the channel banks but also on small traction-clog islands during prolonged periods of low flow. These trap sediment, temporally divide the flow and with continued accretion coalesce with the bank severely reducing the channel width. The other major impedances to flow are local base level controls in the form of silcrete and calcrete reefs in the older sediments of the alluvial plains. These effectively dam the flow and cause deposition in the channel upstream of the reef.

Finally this continued accretion on the banks without significant compensatory erosion at all depositional sites, and the other forms of channel constriction, will lead to avulsion and the gradual development of a new channel. It is essentially this process of stepwise-migration of the channel as opposed to continuous lateral-migration which produced the anabranching river systems of central Southeastern Australia.

SUMMARY
Bank deposits within the Barwon River in general form a fining upwards sequence (Fig. 13). These deposits accrete both vertically and laterally. The low sandy-cross-bedded material grades vertically up into an interbedded sequence of sands and muds deposited from suspension. Channel restriction by deposition and other factors and the lack of erosion causes channel avulsion. The relict channel gradually fills with suspended material forming a clay plug (Fig. 13).

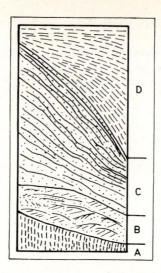

Fig. 13. An idealised vertical profile of channel sedimentation for suspended-load streams. A: older sediments, B: deposits of the Low Bench, cross-bedded sands with thin mud laminae, C: mud and sand interbeds deposited on the Middle and High Benches, D: dark clay plug of the abandoned channel fill.

ACKNOWLEDGEMENTS

The authors wish the thank Dr. K. A. W. Crook of the Australian National University and Dr. W. Mayer of the Canberra College of Advanced Education for critical review of this paper. We also thank the technical staff of the C.C.A.E. and C.S.I.R.O. for help in preparing some figures. One of us (G.T.) wish to acknowledge the financial support of the Department of Geology, Australian National University, Canberra, the research being carried out while he was on staff in the Department.

REFERENCES

Costa, J. E., 1974, Stratigraphic, morphologic and pedologic evidence of large floods in humid environments: Geology, v. 2, p. 301-303.

Einstein, J. A., 1950, The bedload function for sediment transporation in open channel flows: U.S. Dept. Agric. Tech. Bull., 1026, p. 70.

Fisk, L. H., 1974, Inverse grading as stratigraphic evidence of large floods — a comment: Geology, v. 2, p. 613-615.

Kuenen, Ph., H. and Migliorini, C. I., 1950, Turbidity currents as a cause of graded bedding: J. Geol., v. 58, p. 91-127.

Leopold, L. B., Wolman, M. G. and Miller, J. P., 1964, Fluvial Processes in Geomorphology: Freeman, 522p.

Lombard, A., 1963, Laminites: a structure of flysch sediments: J. Sediment. Petrol., v. 33, p. 14-22.

Moss, A. J., 1972, Bed-load sediments: Sedimentology, v. 18, p. 159-219.

Partheniades, E., 1971, Erosion and deposition of cohesive materials: *in* H. W. Shen, *ed.,* River Mechanics; Fort Collins, Colorado, p. 25-1 to 25-87.

Riley, S. J., 1973, Development of distributary channels with special reference to channel morphology: Unpub. Ph.D. thesis, Univ. Sydney.

———— and Taylor, G., 1978, The geomorphology of the Upper Darling River System with special reference to the present fluvial system: Proc. Roy. Soc. Vic., v.90, in press.

Rouse, H., 1950, Engineering hydraulics: John Wiley and Sons, New York p. 644-647.

Sanders, J. E., 1965, Primary sedimentary structures formed by turbidity currents and related resedimentation mechanisms: Soc. Econ. Paleont. Mineral. Spec. Pub. 12, p. 192-219.

Schumm, S. A., 1963, A tentative classification of alluvial channels: U.S. Geol. Survey, Circ. 477.

————, 1968, River adjustment to altered hydrologic regimen: Murrumbidgee River and paleochannels, Australia: U.S. Geol. Survey, Prof. Paper 598.

Taylor, G., 1976, The Barwon River, New South Wales — a study of a basin fill by a low gradient stream in a semi-arid climate: Unpub. Ph.D. thesis, Aust. Nat. Univ.

————, Crook, K. A. W., and Woodyer, K. D., 1972, Upstream dipping foreset cross-lamination: origin and implications for paleoslope analysis: J. Sediment. Petrol., v. 42, p. 178-181.

Wolman, M. G. and Miller, J. P., 1959, Magnitude and frequency of forces in Geomorphic processes: J. Geol., v. 68, p. 54-73.

Woodyer, K. D., 1968, Bankfull frequency in rivers: J. Hydrol., v. 6, p. 114-142.

————, 1975, Concave-bank benches on the Barwon River, N.S.W.: Aust. Geogr., v. 13, p. 36-40.

————, 1978, Sediment transport and deposition by the Darling River: Proc. Roy. Soc. Vic., v. 90, in press.

————, Taylor, G. and Crook, K. A. W., 1978, Sedimentation and benches in a very low gradient suspended-load stream: The Barwon River, New South Wales: Sediment. Geol., in press.

————, Taylor, G. and Watson, I., in prep: Wash load-flood peak relationships.

DISTINCTION OF AGGRADATIONAL AND DEGRADATIONAL FLUVIAL REGIMES IN VALLEY-FILL ALLUVIUM, TAPIA CANYON, NEW MEXICO

RUSSELL G. SHEPHERD[1]

ABSTRACT

An ephemeral tributary of the Rio Puerco northwest of Albuquerque, the present channel of Tapia Canyon is highly sinuous and deeply entrenched in the valley-fill alluvium. More than half of the drainage area is basalt. It produces gray or black gravel that constitutes almost all the bed-load of the stream. In contrast, the bedrock of the remaining area is almost all Cretaceous sandstone. It produces buff-colored sand that is transported principally as suspended load. The segregation of detritus in response to two source-rock types and modes of sediment transport enhances the distinction of different regimes of the past that are preserved in the entrenched channel walls. Furthermore, the geometry, distribution, and either lateral or vertical continuity of sedimentation units permit the identification of process-structure relations that distinguish episodes of aggradation and degradation.

Two basic types of deposits exist. Aggradational-regime deposits consist of sets of black or gray laminae of basalt gravel that are vertically persistent, comparatively uniform in grain size, and within which the individual laminae grade laterally but abruptly into sand laminae. Degradational-regime deposits are laterally persistent, they have sharply bounded, interfingering basalt and buff-colored sand units, and they overlap like cards thickening outward in a fanned-out deck. In comparison to degradational deposits, those resulting from aggradation suggest flow that was perennial and more uniformly distributed seasonally. Under these conditions, the deposition of basalt gravel effectively stablized the locus of thalweg deposition while the bed aggraded vertically. During degradational regimes, such as that at present, deposits reflect the seasonal, flashy nature of the flow. Lateral incision predominates. At one exposure of valley-fill alluvium in Tapia Canyon, radiocarbon and archeological dates temporally bracket episodes of vertical aggradation, lateral incision, and a neck cutoff.

INTRODUCTION

Episodes of fluvial scour and fill that occurred in channels millions of years ago are today commonly recognized in the stratigraphic record. Such processes of scour and fill only locally shape the beds of alluvial channels during individual runoff events. In contrast, the processes of aggradation and degradation operate over longer periods of time and involve the accumulation or removal of thicknesses of sediment that are frequently many times the maximum flood depth of a channel. Furthermore, aggradation or degradation may raise or lower an entire longitudinal profile while the overall gradient may either remain almost unchanged or change markedly.

In the stratigraphic record, ancient periods of aggradation are recorded by numerous, thick, stacked sequences of fluvial deposits of several different types, any of which may be well preserved. In contrast, degradational deposits would seem to have a lower preservation potential because a degrading stream removes sedimentary material and lowers the level of its bed.

In the past, sedimentologists have not had sufficient evidence to properly conclude whether a deposit is aggradational or degradational (Smith, 1973). However, distinctive aggradational and degradational deposits are preserved in the alluvial valley fill of Tapia Canyon, an ephemeral tributary of the Rio Puerco northwest of Albuquerque, New Mexico (Fig. 1).

---

[1]Willard Owens Associates, Inc., 7391 West 38th Avenue, Wheat Ridge, Colorado 80033

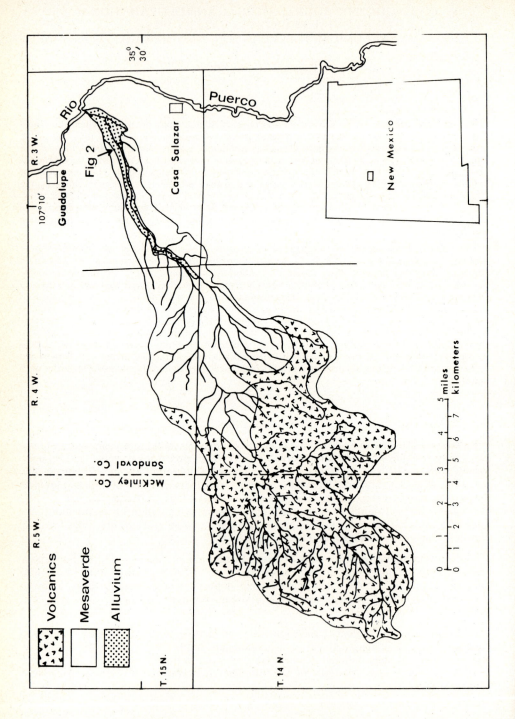

Fig. 1.   Map showing location of Tapia Canyon watershed, its bedrock lithology and the location of Figure 2.

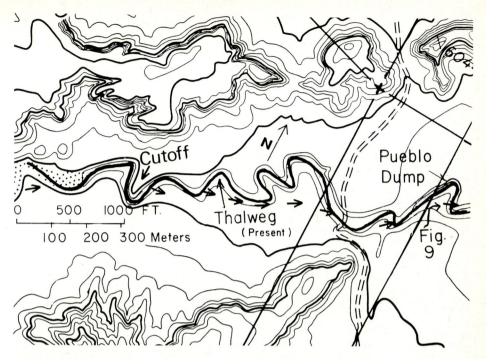

Fig. 2. Topographic map of Tapia Canyon a few kilometers upstream from its mouth, showing channel pattern, flow pattern during aggradation (arrows), slope (20 foot contour intervals) and locations of other sites discussed here.

The Rio Puerco is notorious for its high suspended load and ephemeral, flashy character (Nordin, 1963; Leopold, *et al.,* 1964). Its episodic arroyo cutting and filling are recorded in the walls of its presently-entrenched channel (Shepherd, 1976; Bryan, 1928). As a tributary to the Rio Puerco, the channel of Tapia Canyon is also currently entrenched, at places as much as ten meters into the valley-fill alluvium. However, the alluvial fill in Tapia Canyon provides a more comprehensible record of sedimentation because of the distinctive lithologic characteristics of the Tapia Canyon drainage.

TAPIA DRAINAGE CHARACTERISTICS

The Tapia Canyon watershed consists principally of two basic types of source rocks for fluvial sediment (Fig. 1). The upstream 90 km² of the 141 km² watershed consist of volcanic bedrock, mainly basalt and andesite (Bryan and Post, 1928; Hunt, 1938). The remaining downstream area of the watershed consists of sandstones (principally Gallup) and shales of the Mesaverde Group. A third type of bedrock, Mancos shale, crops out locally in the bed and banks of the entrenched channel in the valley-fill alluvium just upstream from the canyon mouth. This portion of the alluvial valley-fill (Fig. 2) provides the basis for most of the results presented here.

Although no records exist of flood events in Tapia Canyon, its channel morphology is characteristic of ephemeral streams in north-central New Mexico (Leopold and Miller, 1956; Nordin, 1963). Floods in arroyos in the region are usually "flashy" — stage or discharge hydrographs have steep slopes during both ascending and falling stages, and flow often ceases only a few hours after the channel was a torrent of water and mud. The climate is typically semi-arid.

PRESENT FLUVIAL SEDIMENTATION

To adequately understand the processes responsible for the sedimentary structures and sequences exposed in the entrenched-channel walls of Tapia Canyon, an examination was made of sedimentary processes currently operating in the existing channel.

The channel is, and has been for approximately the last ninety years, actively entrenching the sedimentary fill that was deposited during the last episode of alluviation (Bryan, 1928, p. 279). The existing channel is highly sinuous (sinuousity of 1.7; slope of 0.0076 in the 2 km reach shown in Fig. 2). During entrenchment, large alluvial blocks of the vertical channel walls slump into the channel between and during flood events (Fig. 3). They ultimately break up and partially account for the large amounts of load transported by the ephemeral flow. However, a significant volume of sediment is delivered to the main channel by tributaries draining the volcanic and sandstone source rock areas (Fig. 1).

The volcanic source rocks produce black and gray volcanic boulders and cobbles that reach the main channel as subrounded clasts, predominately of gravel size. In contrast, the sandstones of the Mesaverde Group produce sediment of sand and silt sizes that is buff-colored. Consequently, because the two distinct sizes and colors of clasts are available for transport, two distinct types of clasts occur in the main channel. The black and gray volcanic-derived gravel is coarser and is transported principally as rolling, sliding or saltating bed load, while the fine buff, sandstone-derived clasts are transported mainly by turbulent suspension near the bed.

The two types of clasts are segregated spatially (Fig. 4). The dark, volcanic gravel clasts are concentrated in the thalweg of the channel and in chute channels and flood paths that cross the upper parts of point bars at bends with large radii of curvature. The buff, fine-sand detritus is preferentially concentrated between gravel zones, in back water areas, and in other locations where tractive force is insufficient to transport gravel clasts but is high enough to transport fine sand.

Fig. 3. Photo showing entrenched, sinuous nature of present channel of Tapia Canyon. This point bar was cut off at the neck in the foreground the next year after this photo was taken.

Because existing point bars are incising laterally and simultaneously migrating downstream at rapid rates, virtually all aspects of point bar deposits are exposed. Numerous point bars, recently deposited, are now being cut at different angles (Fig. 5). Most conspicuous in these exposures are the lateral accretion surfaces and laterally persistent, interbedded sedimentation units of volcanic gravel and Mesaverde sand (Fig. 4). These are indicators of lateral channel migration during flashy, ephemeral-channel, degradational regimes similar to those of the present time.

The sedimentary structures of the two types of sedimentation units are quite different (Fig. 5). Ripple structures and flat lamination are the dominant sedimentary structures in the buff-sand units. The gravel units exhibit both curved and tabular, medium- and large-scale foreset cross-stratification, flat laminae, and prominent imbrication of the more platy clasts.

Valley Alluvium

Interpretation of the modes of deposition of the valley-fill alluvium are facilitated by the depositional segregation of bed load and suspended load. Two distinct basic types of deposits are well exposed in the entrenched-channel walls. One type is similar to the point bar deposits currently forming and being destroyed by meander migration in the present channel (Figs. 3-5). The other type is markedly different in geometry and continuity of sedimentation units. This type is the result of an aggradational fluvial regime that resulted in construction of much of the thick valley alluvium that is being removed today.

Aggradational Deposits

Aggradational deposits are distinctive because they have a vertical or near-vertical zone of dark, thalweg, basalt-gravel laminae that may be 7 to 10 m high (Figs. 6, 7). Some of these laminae may be truncated, the result of brief local scour after initial deposition,

Fig. 4. Photo showing machete (10 cm between lines) in suspended load deposit of buff sand, with basalt gravel on upper point bar in background. Finer, better-sorted gravel is in thalweg.

Fig. 5. Wall of scour-sliced point bar showing laterally persistent, vertically interbedded sedimentation units of basalt gravel and buff sand. Note lateral accretion surfaces, 10 cm between lines on machete.

but often sets of continuous, stacked laminae a meter thick exist (Fig. 6). Upwards the thalweg gravel laminae usually become slightly finer in mean grain size. For instance, a sandy, pebble gravel may exist at the base of a thalweg sequence, with a sandy, granule gravel near the top. The most impressive aspects of the thalweg gravel zones are their vertical persistence and continuity.

The thalweg-gravel laminae laterally grade abruptly into lamina sets of fine-sand, buff-colored, suspended load deposits (Figs. 6, 7). The border of the thalweg zone is characteristically a local, nearly-vertical zone of black, interfingering gravel and buff-colored sets of sand laminae (*see* loc. A in Fig. 7). Thalweg gravel zones range from less than a meter to 8 or 10 m in width, but average 4 to 5 m. Total bank-full channel widths of aggradational units, estimated from channel-form depositional surfaces including both thalweg and overbank laminae and sets, average approximately 7 m but may be twice or half that width in some exposures. The rather notable variation in these measurements is due to the varying angles, with regard to paleocurrent direction, at which the channel deposit has been cut and exposed in the valley walls.

Degradational Deposits

Degradational deposits contrast markedly with aggradational deposits, and are like point-bar deposits currently being built on the bottom of the entrenched channel. Buff-colored sand and dark gravel units are distinct, but in degradational deposits are laterally persistent and prominently intercalated vertically. Lateral accretion surfaces, the result of meander migration, are very prominent and spectacularly preserved (Fig. 8). Sedimentary structures in degradational deposits preserved in the valley fill are like those described previously from the modern point bars.

Degradational-regime deposits make up less of the valley-fill aluvium than do aggradational deposits. This is undoubtedly because they become eroded and partially

Fig. 6. Photo showing aggradational deposit with thalweg basalt gravel unit (behind the 4m rod) grading laterally and abruptly into the buff sand, overbank deposits.

Fig. 7. Photo showing another aggradational deposit with coarser gravel and a narrower thalweg zone. Note interfingering of gravel and sand to right of A.

Fig. 8. Photo of degradational deposit preserved in entrenched-channel walls. Note spectacular lateral accretion surfaces, basalt gravel at base, laterally-persistent units, and man at left for scale.

removed during flood events occurring after their deposition. Degradational-regime deposits are usually at the bottom of aggradational sequences and are usually scoured off at the top. However, an entire point bar sequence, probably the preserved remnant of the last stages of an ancient arroyo-cutting cycle, is preserved at one location in the valley-fill alluvium (Fig. 8).

SIGNIFICANCE OF RESULTS

Discovery of the aggradational or degradational nature of different portions of a valley-fill alluvial deposit is valuable in deciphering the climatic history of the area. In the middle Rio Puerco region including Tapia Canyon, approximately 1500 Pueblo archeological sites have been identified (Cynthia Irwin-Williams, 1975, oral communication). Numerous towns that were inhabited in the early part of this century are today abandoned along the valley of the Rio Puerco. The Puerco channel is deeply entrenced into the alluvium; the water table is as much as 12 to 15 m below the valley floor, and few livestock can be supported. Just as changes in climate and consequently stream regime have caused later settlers to leave the valley, so too were the Indians undoubtedly affected by entrenchment-alluviation cycles.

In Tapia Canyon numerous Pueblo archeological sites exist (Irwin-Williams, 1976) but one in particular is of especial interest. During the period between 1050 and 1130 A.D., immigrants from the Chaco Canyon area lived near the mouth of Tapia Canyon. Rubble from near one dwelling site fills a small gully on the surface of the alluvium (Fig. 2) and indicates that while the Indians lived there the stream was at the top of the valley fill. Possibly the Indians tried to prevent arroyo cutting after it began by filling the gully. Nevertheless, archeological dating results from the site show that the present valley floor was possibly being constructed as late as 1100 A.D. Also at this site, a radiocarbon date on a burned log buried in the alluvium approximately 6 m below the valley floor (Fig. 9) was obtained (dating done by Meyer Rubin of the U.S. Geological Survey). The age of the

Fig. 9. Photo showing aggradational deposit (below X), degradational deposit (below Y), and deposit possibly formed during the cutoff of a meander neck (z), in valley-fill alluvium.

wood was determined to be 2510 ± 250 years. The wood was just below the bottom of an aggradational-regime unit. Additional radiocarbon dates are necessary before any significant conclusions can be offered, but this one sample does temporally bracket a major, last period of alluviation in Tapia Canyon between 2500 and 1000 years before present.

The sedimentologic record of valley-fill sedimentary processes near the Pueblo dwelling site and dump indicates the types of fluvial regimes that were in effect during the last stages of final aggradation and the first stages of arroyo cutting. Figure 9 is a photograph of the entrenched-channel wall at location B, just downstream from the dwelling site (Fig. 2). The vertical, thalweg-gravel unit (X) is an aggradational deposit. The final aggradational channel was located there until arroyo cutting began, when lateral-accretion, laterally persistent units (Y) were deposited. These change laterally into the swale-fill, large-scale curved-laminae unit (Z) suggesting a large stream power such as might occur during an intense summer thunderstorm in the drainage area upstream in the present ephemeral channel. In addition, it is possible, but not proven, that deposit Z is the preserved channel deposit resulting from a neck cutoff during a flood. Such neck cutoffs are common in the present ephemeral channel. For example, the neck shown in Fig. 3 was cut through approximately one year after the photo was taken. The increased gradient through the cutoff (there are several feet of difference in thalweg levels across the neck in Fig. 3) could possibly cause deposition of such a massive unit as Z in Figure 9.

The vertical uniformity and regularity of grain size in the thalweg laminae of the aggradational deposits (Figs. 6, 7, 9) suggest that the stream flow responsible for the aggradational sequences was much less flashy than that of the present channel. Conditions of perennial flow, or at least more consistent, regular, and less flashy base flows, and flood hydrographs with rounded peaks are more likely for the aggradational deposits. It may be possible that a change to precipitation less seasonally distributed, without a significant change in the absolute quantity, could produce the difference in runoff characteristics. If so, then the increase in rainfall that Hall (1977, p. 1617) concluded as occurring beginning around 1100 A.D. at Chaco Canyon may not necessarily be in conflict with the conditions that caused the initiation of entrenchment in Tapia Canyon about that time.

Efforts to quantitatively compare the fluvial regimes of the aggradational deposits in the valley alluvium with the present, entrenched channel deposits have not yet provided results conclusive enough to report in final form. However, preliminary indications are that during the aggradational regime that resulted in the deposits shown in Figures 6 and 7, the channel had a lower width-depth ratio, more pronounced banks which were probably indicative of a regularly occurring bankfull stage, and the channel was much straighter than at present.

Deposition of basalt gravel during aggradation took place only in the thalweg of the aggrading channel, and provided a preferential locus of deposition for gravel that was subsequently carried into a reach along the bed. Only when the runoff character changed (to one more flashy) did the pattern change and then the continuity of vertical aggradation was disrupted.

In Tapia Canyon, the channel aggraded in response to a rise in base level as the main Puerco channel aggraded. However, the specific type of aggradational deposit was controlled by the runoff regime. Consequently, it is possible that point bar deposits could have been constructed during aggradation, if runoff conditions would have been different.

The results provided by Smith (1973, 1977) provide an interesting analogy to Tapia Canyon. Smith reported homologous aggradational deposits in glacial-fed delta-plain deposits in Banff Park, Canada. There, channel-bed gravel deposits are vertically persistent and bounded by mud. The processes operating are the same as those in Tapia Canyon, but the causes of aggradation are different-base level rise (Tapia Canyon) versus distributary-channel subsidence (Banff Park).

ACKNOWLEDGEMENTS

This work was begun with support from the U.S. Geological Survey's Sedimentary Processes program directed by Ed Clifton. Dave Macke and Carla Potter provided assistance during initial stages. Willard Owens Associates, Inc., provided field-work and manuscript-preparation assistance. Reviews were by Willard G. Owens and Richard Stirba.

REFERENCES

Bryan, K., 1928, Historical evidence on changes in the channel of the Rio Puerco, a tributary of the Rio Grande in New Mexico: J. Geol., v. 36, p. 265-282.

——— and Post, C. W., 1928, Erosion and control of silt on the Rio Puerco: Middle Rio Grande Conservancy District Report, 174p.

Hall, Stephen A., 1977, Late Quaternary sedimentation and paleoecologic history of Chaco Canyon, New Mexico: Geol. Soc. Am. Bull., v. 88, p. 1593-1618.

Hunt, C. B., 1938, Igneous geology and structure of the Mount Taylor volcanic field, New Mexico: U.S. Geol. Survey Prof. Paper 189-B, 29p.

Irwin-Williams, Cynthia, 1976, Archeological investigations in the area of the middle Puerco River, New Mexico, May 30 - December 31, 1975: Progress report to the Bureau of Land Management, 71p.

Leopold, L. B., and Miller, J. P., 1956, Ephemeral streams-hydraulic factors and their relation to the drainage net: U.S.Geol. Survey Prof. Paper 282-A, 37p.

———, Wolman, M. G., and Miller, J. P., 1964, Fluvial processes in geomorphology: W. H. Freeman and Company, San Francisco, 522p.

Nordin, C. F., 1963, A preliminary study of sediment transport parameters, Rio Puerco near Bernardo, New Mexico: U.S. Geol. Survey Prof. Paper 462-C, 21p.

Shepherd, R. G., 1976, Sedimentary processes and structures of ephemeral-stream point bars, Rio Puerco west of Albuquerque, New Mexico: Geol. Soc. Am. Abs. with Prog., v. 8, No. 6, p. 1103.

Smith, D. G., 1973, Aggradation of the Alexandra-North Saskatchewan River, Banff Park, Alberta: in Marie Morisawa, ed., Fluvial Geomorphology; Publications in Geomorphology, State University of New York, Binghampton, Chap. 9.

———, 1977, Fluvial sediments of Banff Park and adjacent areas, Field Trip Guide: First International Symposium on Fluvial Sedimentology, University of Calgary, 31p.

ANCIENT FLUVIAL SYSTEMS

ALLUVIAL AND DESTRUCTIVE BEACH FACIES FROM THE ARCHAEAN MOODIES GROUP, BARBERTON MOUNTAIN LAND, SOUTH AFRICA AND SWAZILAND

K. A. ERIKSSON[1]

ABSTRACT

The Moodies Group, approximately 3,300 m.y. in age, is the oldest relatively unmetamorphosed quartzitic assemblage of sediments presently known. This succession consists of a wide variety of sedimentary facies which accumulated *inter alia* in diverse alluvial and marginal marine depositional environments. A remarkable similarity is shown to exist between Holocene physical sedimentary processes and those operative during the Archaean.

A depositional model is proposed to relate the different alluvial sedimentary environments in space. Proximal alluvial plain deposits consist of matrix- and clast-supported conglomerates which were deposited, respectively, by mass flow and tractional processes. Mid-alluvial plain sandstones developed through vertical aggradation and mid-channel bar formation during falling stages of episodic floods. The most distal alluvial plain sediments comprise upward-fining channel-fill sequences enclosed within interlayered siltstones and shales of probable overbank origin. Vertical sequences of alluvial facies were determined by source area tectonics. During most active progradation alluvial plain sediments built directly onto shelf accumulations in the form of fan deltas. Marginal destructive swash bars developed on the fan deltas at the cessation of fluvial influx.

This investigation has shown that exposed granitic terrains existed greater than 3,000 m.y. ago, the erosion of which resulted in extensive subaerial alluvial sedimentation. The mineralogy of the fluvial sediments is indicative of an anoxygenic atmosphere at the time of their deposition. Braided fluvial processes predominated in the absence of levee stabilization by plant growth but a stable crust, on which prolonged reworking of sediments occurred, is indicated by the orthoquartzitic swash bar sandstones.

INTRODUCTION

Detailed palaeoenvironmental analyses on Archaean sediments are rare and have, to date, been confined almost exclusively to North America. Resedimented deep-water sequences predominate (*see* for example Donaldson and Jackson, 1965; Ojakangas, 1972; Turner and Walker, 1973; Hyde, 1975) but shallow-water alluvial deposits have also been identified (Turner and Walker, 1973; Hyde, 1975). In this account, an extensive assemblage of immature alluvial and intercalated mature beach sediments from the Moodies Group in the Barberton Mountain Land (Fig. 1) are discussed and related spatially in a single depositional model.

The recognition and palaeoenvironmental interpretation of Archaean alluvial and marginal marine sequences can provide important information pertaining to physical processes operative on the surface of the primitive earth. Extensive and diverse fluvial facies, furthermore, would indicate a moderately stable crust on which lateral channel migration could have occurred on broad alluvial plains. In addition, the presence or absence of iron oxide grain coatings in fluvial sediments holds one key to determining the composition of the Archaean atmosphere.

STRATIGRAPHIC FRAMEWORK AND AGE OF THE MOODIES GROUP

The Moodies Group is the upper of three subdivisions of the Swaziland Supergroup (Fig. 1). The basal Onverwacht Group consists predominantly of volcanics with only

---

[1]Geology Department, University of the Witwatersrand, Johannesburg, 2001, South Africa. Present address: Programs in Geosciences, University of Texas at Dallas, Richardson, Texas, 75080

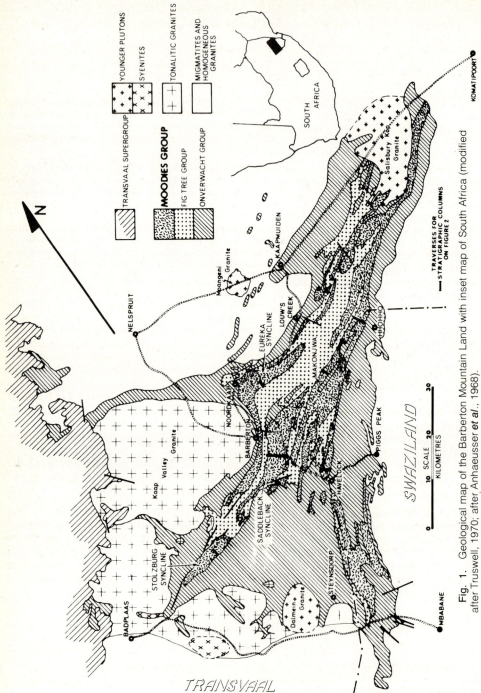

Fig. 1. Geological map of the Barberton Mountain Land with inset map of South Africa (modified after Truswell, 1970; after Anhaeusser *et al.*, 1968).

minor sediments (*see* Viljoen and Viljoen, 1969; Anhaeusser, 1973; amongst others) and is conformably overlain by immature clastic and chemical sediments of the Fig Tree Group (Reimer, 1975). In the southern parts of the Barberton Mountain Land, most notably in Swaziland, the Moodies Group rests directly on Onverwacht volcanics (Jones, 1969) and in the central parts unconformably on Fig Tree sediments. From the Makonjwa Range northwards (Fig. 1), as was established by Tomlinson (1967) and Anhaeusser (1969), a gradational contact exists between the Fig Tree and Moodies Groups.

No direct age determinations have been carried out on the Moodies Group, but these sediments can be dated relative to upper Onverwacht felsic volcanics, with an age of 3,360 ± 100 m.y. (Van Niekerk and Burger, 1969) at less than 3,400 m.y. A minimum age for the Moodies sediments is more difficult to determine. The 'intrusive' nature of the Kaap Valley Granite, which is dated at 3,360 ± 100 m.y. (Oosthuyzen, 1970), is now disputed and other more reliable evidence must be sought. As the Swaziland Supergroup is strongly deformed, the age of non-foliated or post-tectonic granites will provide a minimum age for the Moodies Group. Granites of this type include the intrusive Lochiel or Homogeneous Hood Granite, dated at 3,070 ± 60 m.y. (Allsopp *et al.*, 1962) and the intrusive Salisburykop Pluton which has an age of 3,060 ± 30 m.y. (Oosthuyzen, 1970). The oldest non-foliated granite is the Dalmein Pluton which is dated at 3,290 ± 80 m.y. (Oosthuyzen, 1970). This suggests that the Moodies Group is at least 3,300 m.y. in age which, in view of the conformable nature of the three groups within the Swaziland Supergroup, is not incompatible with the age of the upper Onverwacht. The Moodies Group is thus older than any of the sedimentary basins on the Kaap Vaal craton. The oldest of these is the Pongola Supergroup which rests on granites of the Lochiel-type dated at 3,000 ± 200 m.y. (H. L. Allsopp, pers. comm., 1975).

The Moodies Group attains its greatest thickness in the Eureka Syncline (Fig. 1) and is subdivisible into five stratigraphic units (MD1 - MD5; Fig. 2). A prominent amygdaloidal lava horizon at the base of unit MD4 allows for correlation between the Eureka, Saddleback and Stolzburg Synclines (Fig. 1). In the latter two synclines the Moodies Group is represented by stratigraphic units MD1 through MD4. South of the Saddleback Syncline no correlatable stratigraphy, apart from a basal conglomerate, can be recognized (Fig. 2).

Stratigraphic units discussed in this account are made up of an assemblage of sedimentary facies composed of conglomerates within thick impure sandstone sequences, subordinate shales and thin intercalated orthoquartzitic sandstones. In the southern parts of the Barberton Mountain Land, notably along the Swaziland border and northwards to the Makonjwa Range (Fig. 1), these are the only sedimentary facies present (*see* Fig. 2). In the Saddleback Syncline stratigraphic units MD1, the upper parts of MD3, and MD4 consist of this facies assemblage while further north representatives of this assemblage are less common and comprise only MD1, the upper part of MD4, and MD5 in the Eureka Syncline, and MD1 and the upper part of MD4 in the Stolzburg Syncline. Four sedimentary facies are distinguished in this assemblage and are characterized below in terms of their lithologies (Dott, 1964), sedimentary structures and vertical sequences which are then used to interpret specific shallow-water depositional environments with reference to Holocene depositional models from poorly-vegetated areas.

SEDIMENTARY FACIES

Conglomerate Facies

Occurrences of this facies are indicated on Figure 2, from which it is apparent that the thickest development of conglomerates occurs in the south, and specifically around Havelock. The conglomerates are very coarse, with an average mean pebble diameter of -5.10 ϕ (3.3 cm) (Table 1) but often contain boulders greater than 25 cm in diameter. Pebbles and boulders are generally spherical and rounded to well-rounded except in the

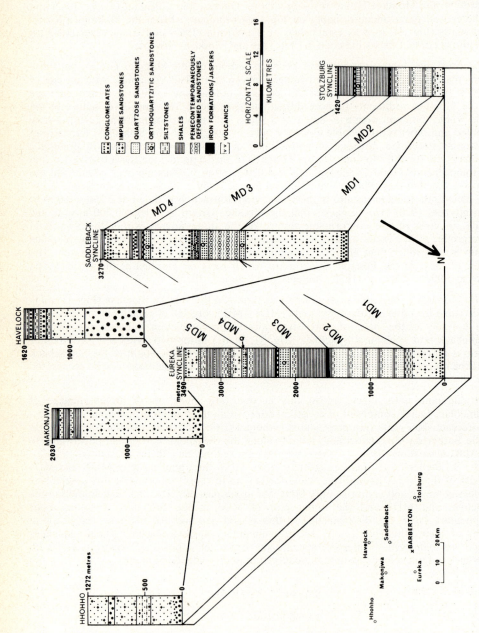

Fig. 2. Representative stratigraphic columns of the Moodies Group. Inset shows ground positions of column localities relative to Barberton. Orthoquartzitic sandstones contain greater than 95% quartz and quartzose sandstones between 90 and 95% quartz.

Table 1. STATISTICAL PARAMETERS FOR MOODIES CONGLOMERATES

Sample No.	E1	E2	E3	E4	S1	S2	S3	S4
Mean (ϕ)	−4.88	−5.25	−5.07	−4.56	−5.41	−5.62	−5.40	−4.65
Variance (ϕ)	1.03	1.41	1.82	1.17	1.37	2.40	1.58	.55
Std. Dev. (ϕ) (Dispersion)	1.01	1.19	1.35	1.08	1.17	1.55	1.26	.74
Max. Clast Size (cm)	12	30	50	24	33	45	38	16
Number of Clasts Measured	119	129	108	108	111	127	110	120

E1 - E4 : Basal Conglomerate, Havelock Area
S1 and S2 : Middle MD3 Conglomerate, Saddleback Syncline
S3 : Lower MD4 Conglomerate, Saddleback Syncline
S4 : Upper MD4 Conglomerate, Saddleback Syncline
(see Fig. 2 for stratigraphic positions).

Statistical parameters are calculated for grouped data with frequency count by class. Maximum diameters of all clasts >1 cm in diameter were measured at each locality.

Equations:

$$\mu \, (mean) \; = \; \frac{\sum\limits_{i=1}^{k} f_i \, x_i}{N}$$

$$\sigma^2 \; variance \; = \; \frac{\sum\limits_{i=1}^{k} f_i \, x_i^{\,2} \; - \; \dfrac{\left(\sum\limits_{i=1}^{k} f_i \, x_i\right)^{2}}{N}}{N - 1}$$

where :

x_i = Midpoint of Class Interval

f_i = Frequency in Each Class

k = Number of Classes

N = Number of Observations

(after Edwards, 1964, p. 41 and 59)

Hhohho area (Fig. 1) where a thin conglomerate with angular to subangular clasts occurs at the base of the succession. Sorting (dispersion) is commonly poor (>1.0 ϕ) and rarely moderate (.70-1.0 ϕ) (Folk, 1968, *see* Table 1). White, black and banded cherts, and jasper are the predominant clast types in the conglomerates (Table 2). Acid volcanic clasts, often silicified, are fairly abundant, but mafic varieties are rare. Granitic clasts, often greater than 25 cm in diameter, are confined to the basal conglomerate and occur most commonly in the Eureka Syncline.

Two conglomerate types occur in this facies; one is matrix-supported (Fig. 3) and the other clast-supported (Fig. 4). Matrix-supported types are most common lower down in the sedimentary succession, especially in the southern parts of the outcrop-belt. The basal conglomerate in the Havelock area, for instance, is predominantly matrix-supported with only occasional thin intercalated clast-supported units. Clast-supported conglomerates are

Table 2.   CLAST TYPES IN MOODIES CONGLOMERATES

Sample Number	E1	E4	S1	S3	S4
White Chert	10.5	1.9	25.2	45.5	8.1
Black Chert	26.8	48.2	30.0	1.8	58.5
Banded Chert	39.0	25.0	13.5	18.8	8.9
Vein Quartz	4.9	1.9	.9	1.8	4.1
Acid Volcanic	9.8	6.5	9.0	16.1	8.2
Mafic Volcanic	—	.9	3.6	8.1	1.6
Ironstone	3.2	2.8	17.1	6.5	—
Jasper	4.1	13.0	—	—	2.4
Brecciated Chert	1.6	—	.9	1.0	—

Figures in Percentages; Samples as in Table 1.

more abundant higher up in the stratigraphic succession, notably in unit MD4 in the Saddleback and Stolzburg synclines (Fig. 2). Contacts between sedimentary units, as defined by thin intercalated sandstones, are regular and non-erosional in the matrix-supported variety, while the clast-supported type frequently displays irregular basal contacts which may have a relief of up to 10 m. Imbrication is not apparent in the matrix-supported conglomerates but is sometimes seen in the clast-supported types (Fig. 4). The matrix-supported conglomerates display no size-grading of clasts, but an upward-fining is often noticeable in the clast-supported varieties. The relationship between bed thickness and maximum clast size for four different conglomerates is illustrated in Fig. 5. Correlation is best for conglomerates which are clearly matrix-supported (Havelock and MD2 Saddleback) and very poor for the lower MD4 clast-supported variety (Fig. 5). A fair correlation exists for the upper MD4 occurrence which contains both matrix- and clast-supported conglomerates.

The matrix of the conglomerates consists of medium- to coarse-grained, poorly- to very poorly-sorted lithic arenites and sublitharenites. The intercalated pebble-free sandstones are similar in composition to the matrix. These are most commonly plane-bedded and rarely greater than 1 m in thickness. They persist laterally for tens of metres where associated with the matrix-supported conglomerates, but tend to pinch out along strike where intercalated within the clast-supported variety (Fig. 6). Trough cross-bedded sandstones vertically separate some individual conglomerate units of the clast-supported type, and often grade laterally into these conglomerates.

The two end-member conglomerate types are best typified by the Havelock and lower MD4 conglomerates, respectively. Frequently, however, the matrix and clast-supported conglomerates display a rapid alternation in pairs varying between 4 and 10 m in thickness. This is particularly noticeable in the upper MD4 conglomerate of the Saddleback area (Fig. 2).

Trough Cross-Bedded Sandstone Facies

Sediments belonging to this facies are grouped together on Figure 2 as "impure sandstones". The sandstones vary in composition from sublitharenites and litharenites to lithic greywackes, are fine- to coarse-grained, poorly- to very-poorly sorted and are devoid of iron oxide pigmentation. Cumulative frequency curves (Fig. 7) and statistical parameters (Table 3) for a number of samples from this facies indicate their highly immature character. Framework — matrix relationships show that the clay minerals are primary and not diagenetic. Scattered pebbles, predominantly chert, are common but siltstone and shale layers constitute a very small proportion of this facies.

Fig. 3. Matrix-supported conglomerate: Havelock area (scale in 1 and 5 cm divisions).

Fig. 4. Imbricated, clast-supported conglomerate: Unit MD4; Saddleback Syncline.

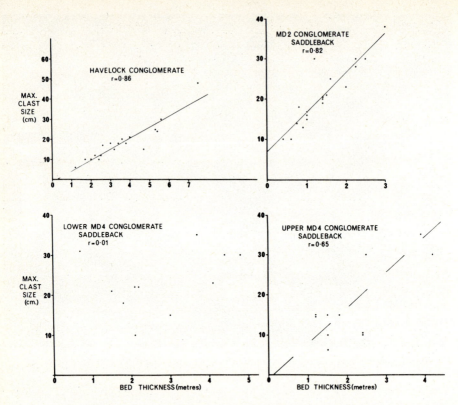

Fig. 5. Graphic plot of bed thickness against maximum clast-size for four different conglomerates (r = correlation coefficient). Upper two graphs are for matrix-supported, lower left graph for clast-supported, and lower right for interlayered clast and matrix-supported conglomerates.

Examples of vertical sequences of sedimentary structures developed in this facies are illustrated in Figure 8. Depositional cycles vary from 5 to 30 m in thickness, and often commence with a scattered pebble layer. Medium- and large-scale trough and occasional planar cross-beds (Figs. 9, 10) are the predominant sedimentary structures in overlying sandstones. Troughs are up to 6 m in width and vary from 20 cm to 2.5 m in thickness. Markedly erosional contacts, which may be occupied by shale-partings, separate the trough cross-bed sets. Shale-partings and flakes occur along some cross-bed foresets. Depositional cycles are capped by a variety of sedimentay structures including small-scale cross-bedding or ripple-cross-lamination, plane-bedding with or without primary current lineations, ripple-drift cross-lamination and desiccated shale-partings, shale drape-laminae and convolute lamination in siltstone. The thick sequences of 'impure sandstones' (Fig. 2) consist of stacked, overlapping and often scour-based partial or complete cycles of the types illustrated in Figure 8. Where not fully developed, the upper parts of cycles are missing, and thick uninterrupted intervals of trough cross-bedded sandstones are developed.

Palaeocurrent measurements on medium-scale trough cross-beds were made at a number of stations on different stratigraphic units. The cumulative results of these measurements for each stratigraphic unit are shown in Figure 11 (*see* Figs. 1, 2 for localities). A consistent northerly transport direction is indicated throughout the deposition of this facies.

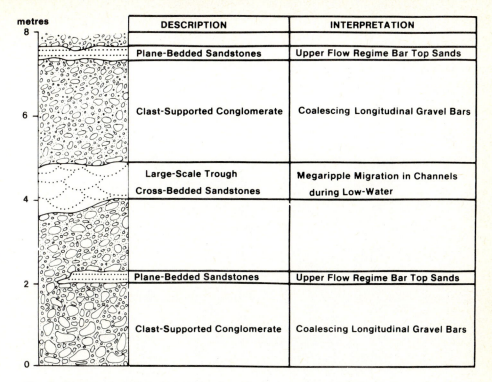

DESCRIPTION	INTERPRETATION
Plane-Bedded Sandstones	Upper Flow Regime Bar Top Sands
Clast-Supported Conglomerate	Coalescing Longitudinal Gravel Bars
Large-Scale Trough Cross-Bedded Sandstones	Megaripple Migration in Channels during Low-Water
Plane-Bedded Sandstones	Upper Flow Regime Bar Top Sands
Clast-Supported Conglomerate	Coalescing Longitudinal Gravel Bars

Fig. 6. Measured stratigraphic section through clast-supported conglomerates and associated sandstones: upper conglomerate; Unit MD4; Saddleback Syncline.

Impure Sandstone — Siltstone — Shale Facies

Sediments of this facies, although somewhat restricted in their vertical extent, occur as laterally persistent stratigraphic units under- and overlain by either of the two previously described facies. This facies is illustrated in Figure 2 as siltstones and shales in the Makonjwa and Havelock areas, part of unit MD1 and much of MD5 in the Eureka Syncline, limited portions of unit MD4 in the Saddleback Syncline and part of unit MD1 in the Stolzburg Syncline. Thin conglomerate-sandstone sequences, which always constitute less than 10% of measured stratigraphic intervals, occur as lenses, up to 100 m in strike length, within the predominant alternating siltstones and shales.

The coarse-grained sediments of this facies are arranged in 2 to 6 m thick upward-fining sequences (Fig. 12). Thin scattered pebble- and shale-flake conglomerates pass upwards into lithic greywackes structured by northerly-directed small- and medium-scale trough cross-beds. These sandstones decrease in grain-size upwards from coarse at the base to medium- and fine-grained at the top of the cycles. An upward decrease in cross-bed set thickness, from 30 to less than 10 cm, also occurs. The upper fine-grained sandstones occasionally display plane-bedding with primary current lineations, and are in turn overlain by siltstones with thin desiccated shale-partings. Single upward-fining cycles are enclosed within 20 to 50 m thick sequences of horizontally-laminated and less commonly ripple cross-laminated siltstones and horizontally-laminated shales (Fig. 13). The latter are commonly arranged in graded upward-fining units between 1 and 5 cm in thickness.

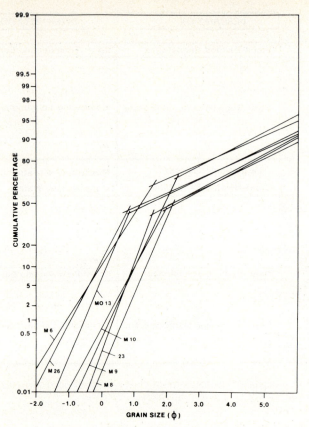

Fig. 7.   Grain-size cumulative frequency curves for impure sandstones. See Table 3 for sample localities. Curves drawn for converted sieve equivalent percentiles.

Table 3.   STATISTICAL PARAMETERS FOR MOODIES IMPURE SANDSTONES.

Sample	M8	M6	M013	23	M26	M9	M10
Mean (φ)	3.35	2.03	2.43	3.03	3.18	2.36	3.18
Variance (φ)	3.24	2.86	4.33	3.76	3.72	1.54	3.76
Std. Dev. (φ)	1.80	1.60	2.08	1.93	2.21	1.24	1.93
Number of Measurements	200	200	200	200	200	200	200

M8   : Unit MD1, Saddleback Syncline   M26 : Makonjwa Area
M6   : Unit MD3, Saddleback Syncline   M9   : Makonjwa Area
M013 : Unit MD4, Saddleback Syncline   M10 : Makonjwa Area
23   : Matrix to Lower Conglomerate in Unit MD4, Saddleback Syncline

(See Fig. 2 for stratigraphic positions)

Longest diameters of 200 grains were measured in thin section for each sample. Statistical parameters were calculated using the equations of Inman (1952) after converting cumulative frequency thin section percentiles to sieve equivalents (conversion equations after Harrell and Eriksson, in preparation).

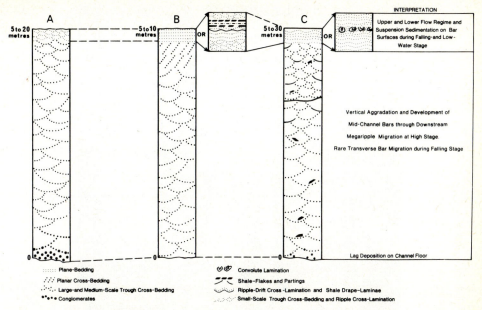

Fig. 8. Measured vertical sequences of lithologies and sedimentary structures in the trough cross-bedded sandstone facies.

PALAEOENVIRONMENTAL INTERPRETATION

General Alluvial Depositional Environment

A number of features exhibited by the three facies of this assemblage can be used in a general palaeoenvironmental interpretation. The most important of these are the immature character of the sediments, the strongly unimodal palaeocurrent patterns, and the widespread evidence of desiccation. These criteria indicate an environment of limited reworking which was influenced by unidirectional dispersal currents and subjected periodically to subaerial exposure, and are most compatible with an alluvial depositional setting. The shape of the cumulative frequency curves, specifically the high suspended population (Fig. 7; Visher, 1969) as well as the vertical sequences of lithologies and sedimentary structures developed in this facies assemblage (Figs. 6, 8 and 12) (see for example Klein, 1972; Boothroyd and Ashley, 1975), further support an alluvial depositional environment.

The three facies of this assemblage are now analyzed in the context of the broadly defined alluvial plain depositional environment.

Conglomerate Facies

The two readily distinguishable conglomerate types developed in this facies must have originated in different sedimentary subenvironments and as a result of contrasting depositional processes within the general alluvial setting.

The coarseness of the matrix-supported conglomerates, their poor sorting, lack of grading and stratification, and the general absence of associated cross-bedding as well as the consistent lateral thickness of individual beds, are all suggestive of a debris-flow mode of origin (Blissenbach, 1954; Bull, 1972; Walker, 1975). The arenaceous matrix in these conglomerates, however, argues against a true debris flow origin. Similar sand-supported

Fig. 9.   Large-scale trough cross-beds: Unit MD1: Saddleback Syncline.

conglomerates have been ascribed by Miall (1970) to 'debris flood' processes which are essentially sand debris flow phenomena. The less specific term 'mass flow' is preferred for these matrix-supported conglomerates which contain no direct evidence for a viscous, matrix-strength support mechanism during their transport (Middleton and Hampton, 1973). Individual matrix-supported conglomerate beds are thought to be related to separate mass flows, which were probably generated during successive floods. The bed thickness to maximum clast-size graphic plots (Fig. 5) provide further information as to the process involved in the deposition of the matrix-supported conglomerates. Bluck (1967) showed that conglomerates deposited by torrential floods exhibit a good correlation between those two variables. This is due to the fact that more intense floods can transport greater quantities of sediment and larger clasts which are deposited almost instantaneously as single bedding units. The excellent correlation between bed thickness and maximum clast-size in the matrix-supported conglomerates from this facies indicates that individual bedding units can be related to separate mass flows. The minor interbedded sandstones were probably formed during waning stages of individual floods.

Although coarse-grained and poorly-sorted, the clast-supported conglomerates of this facies are frequently well stratified and often display internal grading and weak

Fig. 10.   Planar cross-beds; Unit MD1; Saddleback Syncline.

imbrication. In addition, abundant channelling occurs, and the intercalated sandstones are frequently cross-bedded (Fig. 6). These features suggest that traction was the important process involved in the deposition of these alluvial conglomerates and associated sandstones. The poor correlation between clast-size and bed thickness in the lower MD4 conglomerate (Fig. 5) further supports this hypothesis.

The upper reaches of modern braided alluvial plains are frequently structured by longitudinal bars composed of pebbles, cobbles and boulders (*see*, for example, Doeglas, 1962; Boothroyd, 1972; Gustavson, 1974; Boothroyd and Ashley, 1975). These bars are believed by some workers to be initiated by deposition of the coarse bedload fraction of a stream as lags in the middle of the channel (Leopold and Wolman, 1957; Rust, 1972). The height of the bars may be determined by fluid and sediment discharge (Hein and Walker, 1977). If both remain high after deposition of the lag, the bar will grow downstream faster than it aggrades vertically. A rapid decrease in both fluid and sediment discharge conversely results in vertical aggradation of the bar. Exposed Holocene gravel bars are generally capped by sands which, depending on the flow regime, may be ripple-cross-laminated, small-scale cross-bedded or plane-bedded (Rust, 1972). Channels adjacent to the gravel bars frequently contain lower flow regime megaripples which continue to migrate at low flow stages after gravel movement has ceased (Williams and Rust, 1969; Boothroyd and Ashley, 1975). The clast-supported conglomerates and associated sandstones of this facies (Fig. 6) can be explained in terms of these Holocene processes. Open gravel frameworks developed during relatively high discharge with the matrix deposited during waning flow. The predominance of plane-bedding in the intercalated sandstones implies short-lived upper flow regime conditions across the surfaces of longitudinal gravel bars, and which was associated with rapid lowering of water level (Boothroyd and Ashley, 1975). Intercalated trough cross-bedded sandstones are thought to represent low-water channel deposits over which gravel bars migrated as they shifted laterally (Doeglas, 1962).

The alternating matrix- and clast-supported conglomerates can be interpreted as due to tractional, possible sheetflood (Bull, 1972) reworking of earlier mass flow deposits. The moderate correlation between bed thickness and maximum clast-size in the upper MD4 conglomerates of the Saddleback Syncline may be a reflection of these two depositional processes, and contrasts with the excellent and poor correlations of the end member Havelock and lower MD4 occurrences, respectively.

Trough Cross-Bedded Sandstone Facies

Upward-fining fluvial cycles form by channel-filling either through vertical aggradation (Moody-Stuart, 1966; Coleman, 1969; Cant and Walker, 1976; Cant, this volume) or lateral point bar accretion (Allen, 1970). The lack of evidence for point bar and overbank sedimentation in this facies favours vertical aggradation and consequent mid-channel bar formation as the dominant depositional process. Braided channels of the Brahmaputra River, for instance, are characterized by innumerable sandbars and mid-channel islands which are diamond-shaped in plan view, have their long axes parallel to the flow, and are covered on their longer downstream faces by ripples and larger bedforms. Vertical accumulations of cross-bedded sandstones, up to 18 m in thickness, are deposited in relatively short periods of time and are attributed to deposition by migrating sandbars during single flood cycles. As a result of rapid lateral migration of the thalweg, a large percentage of these sand units are preserved throughout the length of the river.

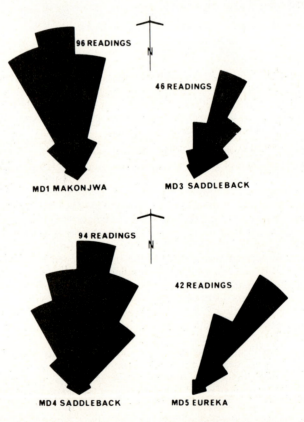

Fig. 11.   Cross-bed vector rose diagrams for the trough cross-bedded sandstone facies.

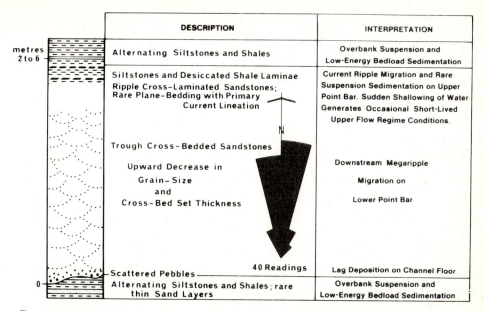

	DESCRIPTION	INTERPRETATION
metres 2 to 6	Alternating Siltstones and Shales	Overbank Suspension and Low-Energy Bedload Sedimentation
	Siltstones and Desiccated Shale Laminae Ripple Cross–Laminated Sandstones; Rare Plane–Bedding with Primary Current Lineation	Current Ripple Migration and Rare Suspension Sedimentation on Upper Point Bar. Sudden Shallowing of Water Generates Occasional Short-Lived Upper Flow Regime Conditions.
	Trough Cross–Bedded Sandstones Upward Decrease in Grain–Size and Cross–Bed Set Thickness	Downstream Megaripple Migration on Lower Point Bar
	40 Readings	
	Scattered Pebbles	Lag Deposition on Channel Floor
0	Alternating Siltstones and Shales; rare thin Sand Layers	Overbank Suspension and Low-Energy Bedload Sedimentation

Fig. 12. Measured stratigraphic section through impure sandstone-siltstone-shale facies: Makonjwa area.

Fig. 13. Finely-bedded alternating siltstones and shales: Unit MD4; Saddleback Syncline. Note the graded, upward-fining arrangement of siltstone-shale pairs especially above the hammer handle.

Similar depositional processes, including vertical aggradation and braid bar formation, can be invoked for the cross-bedded sandstone sequences in this facies (Fig. 8). The considerable vertical thicknesses of the cycles in this facies (Fig. 8) are, however, probably a result of stacking of partial depositional sequences. Migration of megaripples within channels, resulting in the formation of trough cross-beds, was the dominant depositional process. Downstream accretion of transverse bars during falling stage could account for the occasional planar cross-beds present in this facies (Fig. 8B). The abundance of trough cross-bedding suggests that the braided streams had a low-sinuosity (Moody-Stuart, 1966), a hypothesis which is supported by the strongly unimodal palaeocurrent patterns (Fig. 11). Low sinuosity streams also characteristically develop channel-fills by vertical aggradation, rather than lateral migration (Moody-Stuart, 1966). Basal conglomerates are interpreted as channel lags which formed during highest discharge and over which megarippples migrated as the stream velocity decreased.

Migration of large-scale bedforms occurs primarily during high stage. With decreasing discharge and resulting drop in waterlevel, a variety of hydraulic regimes are generated on bar surfaces. These are reflected in the diverse assemblage of sedimentary structures which cap depositional cycles. Plane-bedded sandstones (Fig. 8B) formed under upper flow regime conditions as a response to rapid lowering of the water-level (Boothroyd and Ashley, 1975). With gradually decreasing flow velocities, high-amplitude gave way to low-amplitude bedforms, leading to the development of small-scale cross-beds or ripple-cross-lamination at the top of depositional cycles (Fig. 8A). Capping ripple-drift cross-lamination (Fig. 8B, C) formed when abundant sand and silt were deposited rapidly from suspension. The gradation from type B to in-phase ripple-drift cross-lamination (Fig. 8C) reflects an increasing suspended load to bedload ratio (Jopling and Walker, 1968) under waning flow velocities (Gustavson, et al., 1975). Associated convolute lamination (Fig. 8C) is also indicative of rapid suspension sedimentation and developed during dewatering of the inherently saturated ripple-drift cross-laminated sands and silts. In-phase ripple-drift cross-lamination and associated shale drape-laminae are similar to overbank levee accumulations in modern braided alluvial plains (Boothroyd and Ashley, 1975) and represent the only such deposits in this facies. Overbank sedimentation was confined to abandoned reaches of braided alluvial plains.

Although vertical aggradation is considered to have been the most important process involved in the development of this facies, the persistence of the trough cross-bedded sandstones both along strike and down the palaeoslope indicates that downstream and lateral migration were likewise significant. Sandbars in the Brahmaputra River migrate downstream for up to 1700 m during a single flood, at rates of between 90 and 120 m per day. The same river has been found to migrate laterally for distances of over 700 m in short periods of time (Coleman, 1969). Even greater lateral migration rates occur for the Kosi River for which figures of up to 30 km per year have been measured (Fahnestock, 1963). Braided flood plains are thus highly active depositional regimes consisting of active and abandoned channels. Through constant shifting of stream courses, thick vertical accumulations of sediments are developed. Erosion during the lateral shifting of channels is responsible for the frequent removal of the bar surface deposits (Fig. 8) resulting in the formation of thick sequences of overlapping, scour-based trough cross-bedded sandstone lenses of the type present in this facies.

Impure Sandstone — Siltstone — Shale Facies

The character of the sediments in this facies indicates deposition in two contrasting fluvial subenvironments (see review by Allen, 1965). In terms of their textures, sedimentary structures and vertical sequences (Fig. 12), the conglomerates and sandstones are analogous to channel sediments of modern streams (Harms and Fahnestock, 1965; Sarkar and Basumallick, 1968). The enclosing finer-grained sediments closely resemble contemporary overbank floodplain deposits (McKee et al., 1967).

Their persistence along strike and limited thickness suggest that lateral rather than vertical accretion was the most important process responsible for the deposition of the conglomerates and sandstones of this facies. Meandering rivers most commonly undergo this type of lateral migration as a result of erosion of the outer and deposition on the inner bank. Vertical changes in depositional structures and grain-size within the coarse member can be attributed to decreasing bed shear stress (Allen, 1970). Basal scattered-pebble conglomerates represent channel lag deposits across which lower point bar coarse- and medium-grained trough cross-bedded sandstones accreted (*see* Fig. 12). A continuing decrease in flow velocity at shallower water depths resulted in the development of finer-grained sandstones, structured by small-scale trough cross-beds and ripple-cross-lamination, on the upper point bar. Occasional primary current lineations on plane-bedded fine-grained sandstones indicate short-lived upper flow regime conditions on the upper point bars as a result of a decrease in water depth.

The fine-grained member of this facies developed under waning bedload and suspension sedimentation processes on overbank floodplains and, as indicated by the absence of desiccation cracks, were probably maintained in a continually saturated state. The predominance of ripple-cross-lamination and plane-bedding in the siltstones and fine-grained sandstones, to the exclusion of large-scale sedimentary structures, is in accord with experimental findings. Willis *et al.* (1972) have shown that, for grain-sizes of less than 0.10 mm, increasing flow velocities result in ripples giving way directly to upper phase flat beds. Wolman and Leopold (1957) in turn found that high velocities are common during overbank flows.

In many respects the depositional model proposed for this facies coresponds to the high-sinuosity model of Moody-Stuart (1966). Although extensive large-scale planar cross-beds (epsilon cross-bedding of Allen, 1963) are not developed, lateral point bar migration was still an important process involved in the formation of the thin upward-fining cycles. Furthermore, the thickness of the coarse-grained units indicate that channel-depths varied between 2 and 6 m (Schumm, 1972; Leeder, 1973).

INTERRELATIONSHIPS OF THE DEPOSITIONAL ENVIRONMENTS

The Alluvial Facies

Having interpreted each alluvial facies in terms of local processes it is now necessary to develop a single depositional model for the assemblage of fluvial sediments. In general terms, braided alluvial plains are characterised by a downstream decrease in grain-size, especially for sediments which constitute bars (Smith, 1974; Boothroyd and Ashley, 1975), and a downstream variation in bar morphology. Channels in the upper reaches of braided alluvial plains generally contain longitudinal bars, while those in lower alluvial plains are more commonly structured by transverse or linguoid bars (Smith, 1970; Boothroyd and Ashley, 1975). Braided channels develop on steeper proximal slopes and meandering channels on low gradient, more distal parts of alluvial plains (Leopold and Wolman, 1957).

Following erosion of an emergent mixed granitic — volcanic source terrain to the south of the outcrop belt, sedimentation occurred under various hydrodynamic conditions on an extensive alluvial plain. Based on the preceding discussions, it can be inferred that the conglomeratic facies of this assemblage was deposited on the proximal, more southerly reaches, the thick sequences of trough cross-bedded sandstones on intermediate reaches, and the impure sandstone — siltstone — shale facies on the more northerly, distal reaches of the alluvial plain. The ubiquitously drab colour exhibited by the sediments of this fluvial assemblage indicates that the depositional and diagenetic environments were anoxygenic.

The exact nature of the upper alluvial plain, on which conglomerates were deposited by both mass flow and tractional processes, requires further clarification. Debris flows most characteristically occur on dry-region alluvial fans and result in the formation of muddy, matrix-supported conglomerates (Bull, 1972). Sandy, matrix-supported conglomerates of the type present in this facies are atypical of alluvial fans, and may have developed instead on the upper parts of humid-region alluvial plains during floods. Interrelationships between the clast-supported conglomerates and associated sandstones are similar to those for gravel bar and channel sediments on the upper reaches of the Scott and Yana fans in Alaska (Boothroyd and Ashley, 1975). Both conglomerate-types thus probably accumulated on upper alluvial plains but with the matrix-supported conglomerates deposited proximal to the clast-supported types (Hooke, 1967). The frequent interlayering of the two conglomerate types indicates that mass flow and tractional processes alternated through time on the upper alluvial plains.

Point bar and overbank deposits, of the type represented by the impure sandstone — siltstone — shale facies are rare in Precambrian sedimentary successions due to a lack of levee stabilization by vegetation at that time. (Schumm, 1968) Their occurrence in the Moodies Group may imply a broad, low-gradient distal alluvial plain on which high-sinuosity streams could have developed.

Vertical Stratigraphic Relationships

Vertical stacking of the depositional environments described here was determined by the intensity of terrigenous input (as controlled by source area tectonics) and the rate of basin subsidence which together regulated the landward or basinward migration of the facies belts.

Thus it can be suggested that stratigraphic unit MD1 (Fig. 2) throughout the Barberton Mountain Land developed as a result of the slow sourceward retreat of the alluvial plain

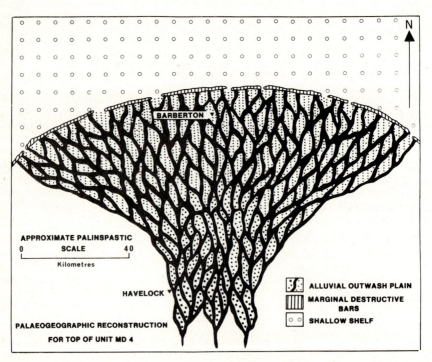

Fig. 14. Palaeogeographic fan delta model for actively prograding alluvial system: upper unit MD4, Moodies Group, Barberton Mountain Land.

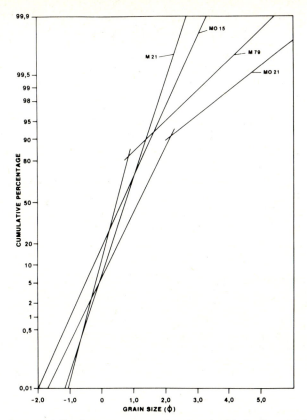

Fig. 15. Grain-size cumulative frequency curves for plane-bedded orthoquartzitic sandstones at the top of unit MD3, Saddleback Syncline. Curves drawn for converted sieve equivalent percentiles.

facies belts due to a diminution of the source area relief. This was in response to normal weathering processes and denudation in the source area, although pebble types in the basal conglomerate, such as the granitic clasts in the Eureka Syncline, often reflect derivation from local floor irregularities.

Active progradation, as controlled exclusively by source area tectonics, resulted in the development of the upper part of unit MD3 in the Saddleback Syncline (Fig. 2), while unit MD4 in the same syncline displays evidence of two periods of progradational sedimentation. The lower of the two upward-coarsening sequences consists of distal alluvial plain siltstones and shales, scoured into by proximal gravel bar deposits. Under conditions of maximum sediment input, mid-alluvial plain sandstones and conglomerates prograded directly onto interlayered sandstones, siltstones and shales, which are considered to be shallow shelf deposits (Fig. 14; Eriksson, 1977). This situation is analogous to the fan deltas along the coast of Alaska (Galloway, 1976) and is well illustrated at the top of unit MD4 in the Eureka and Stolzburg synclines (Fig. 2).

Marginal Destruction of the Alluvial Plain

Orthoquartzitic sandstone lenses up to 6 m in thickness and which may extend along strike for greater than 1 km occur at the top of unit MD4 in the Eureka and Stolzburg Synclines and cap units MD1 and MD4 in the Stolzburg Syncline (Fig. 2). Thin, well- to poorly-packed conglomerate lenses, composed of well-rounded resistant pebble types,

Table 4. STATISTICAL PARAMETERS FOR MOODIES ORTHOQUARTZITES INTERCALATED
WITHIN IMPURE SANDSTONES

Sample No.	M21	M015	M021	M79
Mean (ϕ)	0.77	0.67	1.18	0.54
Variance (ϕ)	0.28	0.52	0.56	0.21
Std. Dev. (ϕ)	0.53	0.72	0.75	0.46
Number of Measurements	200	200	200	200

All samples from top of Unit MD3, Saddleback Syncline. (See Fig. 2 for Stratigraphic Position).

Procedures for calculating the statistical parameters were the same as discussed in Table 3.

and separated by orthoquartzitic sandstones, are present within the upper MD3 and MD4 sandstones of the Saddleback and Stolzburg Synclines, respectively.

Cumulative frequency curves (Fig. 15) and statistical parameters (Table 4) for samples from one of these sandstones indicate them to be medium-grained and moderately- to well-sorted with suspension populations entirely absent from two of the samples. Plane-bedding is the dominant sedimentary structure in the sandstones while low-angle discordances and planar cross-bed intrasets, which have an easterly mode and vary from 15 to 80 cm in thickness, are also common.

The abundance of plane-bedding indicates a depositional environment that was subjected predominantly to upper flow regime conditions. Evidence of extensive reworking is provided by the well-sorted nature of the sandstones (Table 4). These criteria, coupled with the shape of the cumulative frequency curves (Fig. 15; Visher, 1969), are suggestive of a beach environment, and more specifically of a swash zone. The lateral impersistence of these sandstones favours their development on swash bars rather than along a foreshore. Associated planar cross-bedding is oriented at right angles to the fluvial palaeocurrent directions and probably reflects a longshore current influence. The intercalated conglomerates are similar to wave-worked deposits described from the west coast of the U.S.A. by Clifton (1973). Their lensoid nature and occurrence within clean sandstones serves to distinguish them from the fluvial gravels.

The plane-bedded orthoquartzitic sandstones are considered to have developed as a result of marginal reworking of alluvial plain sediments upon cessation of sediment supply (Fig. 14) and are indicative of a relatively stable crust. The poorly-sorted fluvial and mature beach (swash zone) sandstones are clearly separated on a plot of mean grain diameter against grain-size standard deviation for samples from the two inferred depositional environments (Fig. 16). The samples lie within or close to their respective fields as delineated by Moiola and Weiser (1968) using samples from known Holocene environments of deposition.

Similar destructive onlap relationships to those envisaged for this depositional system have been documented from the Mississippi Delta. Following abandonment of individual lobes and compaction of the sediment pile, clean sandstones are developed during transgressive wave reworking of the deltaic plain (Scruton, 1960; Kolb and Van Lopik, 1966; Morgan, 1970). The Gum Hollow fan delta (McGowen, 1970) and Santee River Delta (Stephens et al., 1976) display identical destructive relationships. Wave reworking of the Santee River Delta followed reduction and eventual termination of sediment supply through artificial damming-up of the river and its tributaries.

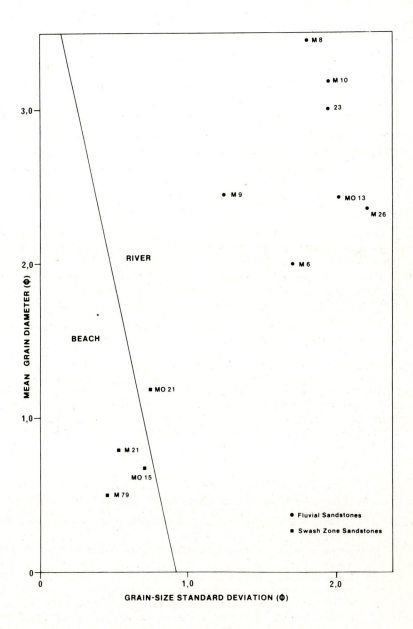

Fig. 16. Plot of mean grain diameter against grain-size standard deviation for immature fluvial and beach (swash zone) sandstones. Boundary between fluvial and beach fields after Moiola and Weiser (1968). Sample localities are described in Tables 3 and 4.

Conclusions

(a) The Moodies Group contains the oldest extensive and recognizable alluvial plain deposits.

(b) Braided fluvial sedimentary processes dominated and have analogues in a number of Holocene alluvial plains.

(c) The paucity of meander belt deposits is typical of the Precambrian and related to a lack of stabilization of levees by vegetation at that time. Poorly developed meander belt sequences in the Moodies Group probably formed on low-gradient distal alluvial plains.

(d) The quartzo-felspathic nature of arenaceous sediments in the Moodies Group indicates the widespread occurrence of granitic rocks at the time of deposition of these Archaean sediments. Unequivocal evidence in favour of emergent landmasses to the south of the outcrop belt exists, and contradicts the suggestion of Hargraves (1977) that a primordial sea covered the earth until 2,000 m.y. ago.

(e) The orthoquartzitic swash bar sandstones imply a relatively stable crust with slow subsidence at the time of deposition of the Moodies Group, on which prolonged reworking of sediments occurred.

(f) Vertical stratigraphic sequences illustrate the intimate relationship between source area tectonics and sedimentation.

(g) In contrast to the red colouration commonly exhibited by arenaceous fluvial sediments less than 2,000 m.y. in age (*see* for example Cloud, 1976), those in the Moodies Group are drab coloured, indicating an anoxygenic atmosphere at their time of deposition.

Acknowledgements

The Council of the University of the Witwatersrand, and the Council for Scientific and Industrial Research financially supported the field work. Helpful discussions were had with A. Button, D. K. Hobday and G. deV. Klein during preparation of the manuscript. N. D. Smith and B. R. Turner provided constructive comments leading to the final revision of the paper. Mrs. M. S. McCarthy is thanked for typing the manuscript and M. H. Hudson and M. Taylor provided photographic and cartographic assistance.

References

Allen, J. R. L., 1963, The classification of cross-stratified units with notes on their origin: Sedimentology, v. 2, p. 93-114.

———, 1965, A review of the origin and characteristics of recent alluvial sediments: Sedimentology, v. 5, p. 91-191.

———, 1970, Studies in fluviatile sedimentation: a comparison of fining-upward cyclothems with special reference to coarse-member composition and interpretation: J. Sediment. Petrol., v. 40, p. 298-323.

Allsopp, H. L., Roberts, H. R., Schreiner, G. D. L. and Hunter, D. R., 1962, Rb - Sr age measurements on various Swaziland granites: J. Geophys. Res., v. 67, p. 5307-5313.

Anhaeusser, C. R., 1969, The stratigraphy, structure, and gold mineralization of the Jamestown and Sheba Hills areas of the Barberton Mountain Land: Unpub. Ph.D. thesis, Univ. Witwatersrand, Johannesburg.

———, 1973, The evolution of the early Precambrian crust of South Africa. Phil. Trans. R. Soc. Lond., A. 273 - 359 - 388.

————, Roering, C., Viljoen, M. J. and Viljoen, R. P., 1968, The Barberton Mountain Land: a model of the elements and evolution of an Archaean fold belt: Trans. Geol. Soc. S. Afr., v. 71 (annex.), p. 225-253.

Blissenbach, E., 1954, Geology of alluvial fans in semi-arid regions. Geol. Soc. Am. Bull., v. 39, p. 465-484.

Bluck, B. J., 1967, Deposition of some Upper Old Red Sandstone conglomerates in the Clyde area: a study in the significance of bedding: Scott. J. Geol., v. 3, p. 139-167.

Boothroyd, J. C., 1972, Coarse-grained sedimentation on a braided outwash fan, northeast Gulf of Alaska: Coastal Research Division, Univ. South Carolina, Tech. Rept. No. 6 - CRD.

————, and Ashley, G. M., 1975, Processes, bar morphology and sedimentary structures on braided outwash fans, northeastern Gulf of Alaska, 193-222, *in* A. V. Jopling and B. C. McDonald, *eds.*, Glaciofluvial and glaciolacustrine sedimentation; Soc. Econ. Paleont. Mineral., Spec. Pub. 23, p. 193-222.

Bull, W. B., 1972, Recognition of alluvial fan deposits in the stratigraphic record 63-83, *in* J. K. Rigby and W. K. Hamblin, *eds.*, Recognition of ancient sedimentary environments; Soc. Econ. Paleont. Mineral., Spec. Pub. 16, p. 63-83.

Cant, D. J. and Walker, R. G., 1976, Development of a braided-fluvial facies model for the Devonian Battery Point Sandstone, Quebec: Can. J. Earth Sci., v. 13, p. 102-119.

Clifton, H. E., 1973, Pebble segretation and bed lenticularity in wave-worked versus alluvial gravels: Sedimentology, v. 20, p. 173-187.

Cloud, P. E., 1976, Major features in crustal evolution: Alex du Toit Mem. Lecture No. 14, Geol. Soc. S. Afr., 33p.

Coleman, J. M., 1969, Brahmaputra river: channel processes and sedimentation: Sediment. Geol., v. 3, p. 129-239.

Doeglas, D. J., 1962, The structure of sedimentary deposits of braided rivers: Sedimentology, v. 1, p. 167-190.

Donaldson, J. A. and Jackson, G. P., 1965, Archaean sedimentary rocks of North Spirit Lake area, N.W. Ontario: Can. J. Earth Sci., v. 2, p. 622-647.

Dott, R. H., 1964, Wacke, graywacke and matrix — what approach to immature sandstone classification? J. Sediment. Petrol., v. 34, p. 625-632.

Edwards, A. L., 1964, Statistical Analysis: Holt, Rinehart and Winston, New York, 234p.

Eriksson, K. A., 1977, A palaeoenvironmental analysis of the Archaean Moodies Group, Barberton Mountain Land, South Africa: Unpub. Ph.D. thesis, Univ. Witwatersrand, Johannesburg.

Fahnestock, R. K., 1963, Morphology and hydrology of a glacial stream: U.S. Geol. Survey, Prof. Paper 422-A.

Folk, R. L., 1968, Petrology of sedimentary rocks: Hemphills, Austin, Texas, 170p.

Galloway, W. E., 1976, Sediments and stratigraphic framework of the Copper River fan-delta, Alaska: J. Sediment. Petrol., v. 46, p. 726-737.

Gustavson, T. C., 1974, Sedimentation of gravel outwash fans, Malaspina Glacier Foreland, Alaska: J. Sediment. Petrol., v. 44, p. 378-389.

————, Boothroyd, J. C. and Ashley, G. M., 1975, Depositional sequences in glaciolacustrine deltas: *in* A. V. Jopling and B. C. McDonald, *eds.*, Glaciofluvial and glaciolacustrine sedimentation. Soc. Econ. Paleont. Mineral., Spec. Pub. 23, p. 264-280.

Hargraves, R. B., 1976, Precambrian geologic history: Nature, v. 193, p. 363-371.

Harms, J. C. and Fahnestock, R. K., 1965, Stratification, bedforms and flow phenomena (with an example from the Rio Grande): *in* G. V. Middleton, *ed.*, Primary sedimentary structures and their hydrodynamic interpretation; Soc. Econ. Paleont. Mineral., Spec. Pub. 12, p. 84-115.

Harrell, J. and Eriksson, K. A., in prep., Empirical conversion equations for thin-section and sieve derived size distribution statistics.

Hein, F. J. and Walker, R. G., 1977, Bar evolution and development of stratification in the gravelly, braided Kicking Horse River, British Columbia: Can. J. Earth Sci., v. 14, p. 562-570.

Hooke, R. L., 1967, Processes on arid-region alluvial fans: J. Geol., v. 75, p. 438-460.

Hyde, R. S., 1975, Depositional environment of Archaean exhalites, Kirkland Lake - Larder Lake area, Ontario: Geol. Soc. Am. North-Central Sectn., Abst. with Prog., p. 789.

Inman, D. L., 1952, Measures for describing the size distribution of sediments: J. Sediment. Petrol., v. 22, p. 125-145.

Jones, D. H., 1969, Geology and gold mineralization fo the Hhohho area, north-western Swaziland: Unpub. M.Sc. thesis, Univ. Witwatersrand, Johannesburg.

Jopling, A. V. and Walker, R. G., 1968, Morphology and origin of ripple-drift cross-lamination with examples from the Pleistocene of Massachusetts: J. Sediment. Petrol., v. 38, p. 971-984.

Klein, G. de V., 1972, Sedimentary model for determining paleotidal range: reply: Geol. Soc. Am. Bull., v. 83, p. 539-546.

Kolb, C. R. and Van Lopik, J. R., 1966, Depositional environments of the Mississippi River deltaic plain - southeastern Louisiana: in M. L. Shirley and J. A. Ragsdale, eds., Deltas in their geologic framework; Houston Geol. Soc., Houston, Tex., p. 17-62.

Leeder, M. R., 1973, Fluvial fining-upward cycles and the magnitude of palaeochannels: Geol. Mag., v. 110, p. 265-276.

Leopold, L. B. and Wolman, M. G., 1957, River channel patterns: braided, meandering and straight: U.S. Geol. Survey Prof. Paper 282-B.

McGowen, J. H., 1970, Gum Hollow fan delta: Nueces Bay, Texas: Rept. Invest. No. 69, Bur. Econ. Geol., Univ. Texas, Austin, Tex., 91p.

McKee, E. D., Crosby, E. J. and Berryhill, H. L., 1967, Flood deposits, Bijou Creek, Colorado: J. Sediment. Petrol., v. 37, p. 829-851.

Miall, A. D., 1970, Devonian alluvial fans, Prince of Wales Island, Arctic Canada: J. Sediment. Petrol., v. 40, p. 556-571.

Middleton, G. V. and Hampton, M. A., 1973, Sediment gravity flows: mechanics of flow and deposition: in G. V. Middleton and A. H. Bouma, eds., Turbidites and Deep-water Sedimentation; Pacific Section, Soc. Econ. Paleont. Mineral., Short Course, p. 1-38.

Moiola, R. D. and Weiser, D. D., 1968, Textural parameters and their evaluation: J. Sediment. Petrol., v. 38, p. 45-63.

Moody-Stuart, M., 1966, High and low sinuosity stream deposits with examples from the Devonian of Spitzbergen: J. Sediment. Petrol., v. 36, p. 1102-1117.

Morgan, J. P., 1970, Depositional processes and products in the deltaic environment: in J. P. Morgan, ed., Deltaic sedimentation, modern and ancient; Soc. Econ. Paleont. Mineral., Spec. Pub. 15, p. 31-47.

Ojakangas, R. W., 1972, Archaean volcanogenic greywackes of the Vermilion District, northeastern Minnesota: Geol. Soc. Am. Bull., v. 83, p. 429-442.

Oosthuyzen, E. J., 1970, The geochronology of a suite of rocks from the granitic terrain surrounding the Barberton Mountain Land: Unpub. Ph.D. thesis, Univ. Witwatersrand, Johannesburg.

Reimer, T. O., 1975, Untersuchungen über abtragung, sedimentation und diagenese im frühen Präkambrium am beispiel der Sheba Formation (Südafrika): Geologisches Jahrb., Reihe B., Heft, 17, 108p.

Rust, B. R., 1972, Structure and process in a braided river: Sedimentology, v. 18, p. 221-245.

Sarkar, S. K. and Basumallick, S., 1968, Morphology, structure and evolution of a channel island in the Barakar River, Barakar, West Bengal: J. Sediment. Petrol., v. 38, p. 746-754.

Schumm, S. A., 1968, Speculations concerning paleohydrologic controls of terrestrial sedimentation: Geol. Soc. Am. Bull., v. 79, p. 1573-1588.

————, 1972, Fluvial paleochannels: in J. K. Rigby and W. K. Hamblin, eds., Recognition of ancient sedimentary environments; Soc. Econ. Paleont. Mineral., Spec. Pub. 16, p. 98-107.

Scruton, P. C., 1960, Delta building and the deltaic sequence: in F. P. Shepard, F. B. Phleger and T. H. Van Andel, eds., Recent sediments, northwest Gulf of Mexico; Am. Assoc. Petrol. Geol., p. 82-102.

Smith, N. D., 1970, The braided stream depositional environment: a comparison of the Platte River with some Silurian clastic rocks, North-central Appalachians: Geol. Soc. Am. Bull., v. 81, p. 2993-3014.

————, 1974, Sedimentology and bar formation in the Upper Kicking Horse River, a braided outwash stream: J. Geol., v. 81, p. 205-223.

Stephens, D. G., van Niewenhuise, D. S., Mullin, P., Lee, C. and Kanes, W. H., 1976, Destructive phases of deltaic development: North Santee River Delta: J. Sediment. Petrol., v. 46, p. 132-144.

Tomlinson, R. S., 1967, The geology of the area between the Staircase Ridge and the Emlembe Range, Barberton Mountain Land: Unpub. M.Sc. thesis, Natal University, Durban.

Truswell, J. F., 1970, An Introduction to the Historical Geology of South Africa: Purnell, Cape Town, 167p.

Turner, C. C. and Walker, R. G., 1973, Sedimentology, stratigraphy and the crustal evolution of the Archaean greenstone belt near Sioux Lookout, Ontario: Can. J. Earth Sci., v. 10, p. 817-845.

Van Niekerk, C. B. and Burger, A. J., 1969, A note on the minimum age of the acid lava of the Onverwacht Series of the Swaziland System: Trans Geol. Soc. S. Afr., v. 72, p. 9-21.

Viljoen, M. J. and Viljoen, R. P., 1969, An introduction to the geology of the Barberton granite-greenstone terrain: Geol. Soc. S. Afr., Spec. Pub. 2.

Visher, G. S., 1969, Grain-size distribution and depositonal processes: J. Sediment. Petrol., v. 39, p. 1074-1106.

Walker, R. G., 1975, Conglomerate: Sedimentary structures and facies models: *in* J. C. Harms, J. B. Southard, D. R. Spearing and R. G. Walker, *eds.*, Depositional environments as interpreted from primary sedimentary structures and stratification sequences; Soc. Econ. Paleont. Mineral., Short Course, p. 133-161.

Williams, P. F. and Rust, B. R., 1969, The sedimentology of a braided river: J. Sediment. Petrol., v. 39, p. 649-679.

Willis, J. C., Coleman, N. L. and Ellis, W. M., 1972, Laboratory study of transport of fine sand: J. Hydraulics Div., Am. Soc. Civil Eng., v. 98, p. 489-501.

Wolman, M. G. and Leopold, L. B., 1957, River flood plains: some observations on their formation: U.S. Geol. Survey Prof. Paper 282-C.

PROTEROZOIC STREAM DEPOSITS:
SOME PROBLEMS OF RECOGNITION AND INTERPRETATION
OF ANCIENT SANDY FLUVIAL SYSTEMS

DARREL G. F. LONG[1]

ABSTRACT

Fluvial deposits of Proterozoic age (~2.5 - 0.6 Ga) have been reported from several continents. They include the products of deposition in alluvial fan, braided stream and possibly meandering stream environments. The braided stream deposits include both conglomeratic and non-conglomeratic sequences. Many of the non-conglomeratic, sandy, fluvial deposits, which may be several kilometres thick, are characterized by a paucity of fine-grained clastic material, whether as matrix or as associated mudrocks. In many cases, thick arenaceous sequences of Proterozoic age have been interpreted as fluvial by some authors and non-fluvial by others, with little conclusive evidence for either interpretation. Fossils are generally unavailable as evidence, except perhaps in the uppermost Proterozoic. Even stromatolites, where present, may have developed in nonmarine settings. Differentiation may in some cases be achieved by examination of grain size characteristics, in association with the type, scale and abundance of sedimentary structures, their directional attributes and vertical and lateral associations. Other criteria include sand-body geometry, lithofacies association, petrology and, in some examples, colour. A comparison can be made between Proterozoic and Phanerozoic fluvial environments, although the absence of vegetation must have had profound effects on the style of deposition in Proterozoic fluvial systems. The paucity of fines in many Proterozoic fluvial systems can be related to differences in hydraulic regime, dominance of bedload type streams, the extreme vulnerability of overbank flood deposits to later fluvial and aeolian erosion, and possibly to removal of fines from the fluvial systems as wash load.

INTRODUCTION

Thick arenaceous sequences are a conspicuous part of Proterozoic successions in many places in the world (*c.f.* Rankama, 1963, 1965, 1967, 1970). The great thickness of these sandstones (many exceeding 1 or 2 km) and the common lack of intimately associated mudrocks make their interpretation extremely difficult. This is especially true when attempting to match their depositional environments with models based on relatively thin Holocene sequences. Part of the problem stems from the general unavailability of fossils as direct evidence of depositional environment in Proterozoic successions. Even in sequences where stromatolites are present they cannot be considered as unequivocal evidence of deposition in a marine setting, as modern examples have been reported from both fresh and salt water lacustrine environments (Walter, 1976). In addition, the absence of prolific land vegetation must have had profound effects on the runoff characteristics in terrestrial environments (Schumm, 1968), possibly producing fluvial environments with no directly comparable modern analogues.

Similar sedimentary structures are known in deposits formed in radically different sedimentary environments; hence interpretations of the depositional milieu must be based on the relative abundance, type, scale, assemblage and sequence of lithologies and sedimentary structures within the rocks under investigation. Despite this seeming mass of difficulties, fluvial deposits have been recognized in Proterozoic sequences of several continents; they include the products of deposition in alluvial fan, braided stream and possibly meandering stream environments. Problems pertaining directly to the more sandy fluvial deposits will be considered first; followed by a review of conglomeratic and finer grained sequences, and possible depositional models which can be applied to the deposits of Proterozoic fluvial systems.

---

[1]Geological Survey of Canada, 3303 - 33rd St. N.W., Calgary, Alberta, Canada, T2L 2A7.

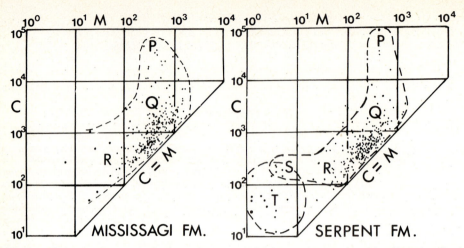

Fig. 1. CM plots (in microns) of two Proterozoic fluvial sandstones from the Huronian (lower Aphebian) of Ontario, Canada. Estimation of coarsest (C) and median (M) grain size based on field observations and estimates of least projection width (Folk, 1968, p. 11) of grains in thin section, P, Q, R, S, T, see text.

SANDY FLUVIAL DEPOSITS

Thick sandstone sequences are relatively common components of Proterozoic sequences in many parts of the world. The absence of abundant conglomeratic or muddy material in many places makes their interpretation extremely difficult. Despite this, several examples (some of which contain minor conglomeratic horizons) have been interpreted as the products of deposition in fluvial (?braided stream) environments. Examples include the Amagok (Cecile, 1976), Mississagi (Long, 1976, 1978), Serpent (Long, 1976), Lorrain (Wood, 1973), Kluziai (Hoffman, 1968), Athabasca (Fahrig, 1961; Ramaekers and Dunn, 1976), Hades Pass (Wallace, 1972) and Golneselv (Banks and Røe, 1974) Formations, the Bullo River Sandstone (Sweet et al., 1974) and parts of the Landersfjord Formation (Laird, 1972), Heavitree Quartzite (Clarke, 1976) and Katherine (Aitken et al., 1978), Wilson Island (Yeo, 1976) and Missoula (Winston, 1973, 1977) Groups. Of these, the Mississagi (Palonen, 1973), Serpent (Card et al., 1977, p. 9), Lorrain (Hadley, 1968) and Athabasca (Blake, 1956) Formations and the Heavitree Quartzite (Wells et al., 1970) have also been interpreted as marine. This dichotomy of interpretations is indicative of the basic problem of differentiating fluvial from marine sandstones in sequences where fossils are absent. Possible lines of evidence which might distinguish fluvial from marine facies in Proterozoic sandstones include grain size characteristics, petrography, sand-body geometry, lithofacies association and the type, scale, abundance and directional attributes of sedimentary structures.

Grain size characteristics

In most Proterozoic sequences, detailed observations of grain size distributions are impractical, due to the degree of lithification, and the vast amount of time required to undertake comprehensive size analysis using point counts on thin sections. Point counts require interpretation of grain to grain relationships in rocks which in many cases have undergone some recrystallization (at least of matrix components) and condensation of the framework grains (producing sutured contacts). The use of less specific techniques, such as plots of coarsest (C) and median (M) grain size (which can be approximated by using field observations in conjunction with thin section estimates) allows the construction of CM plots analogous to those of Passega (1957, 1964, 1977). CM plots of the Mississagi and Serpent Formations (Fig. 1; after Long, 1976) show several of the features Passega (1957) cited as typical of tractive current deposits such as rivers. In both examples, the bed load (PQ) and graded suspension (QR) segments of Passega's (1957) basic tractive current pattern are well defined. Poorly defined zones representing the zones of uniform (RS) and pelagic (T) suspension are also present in the CM plot of the Serpent Formation. The absence of well-defined zones of uniform and pelagic suspension in plots of the Mississagi Formation is similar to that observed by Bull (1972) in CM patterns of braided stream deposits in alluvial fan environments and can be explained by lack of favourable conditions (flood plains, protected channels) for deposition of finer grained sediments from suspension. Marine tractive current deposits show better developed uniform and pelagic suspension zones than fluvial deposits, with bed load and, to a lesser extent, graded suspension deposits being poorly developed (Passega, 1957, 1964).

The use of CM diagrams has its limitations (c.f. Passega, 1977) especially when the sequence under investigation includes deposits from more than one depositional environment, consequently this technique should only be used in conjunction with other evidence of depositional environments.

Sedimentary structures

Proterozoic sandy fluvial deposits are characterized by an abundance of planar and trough cross-stratification (Figs. 2-9), with lesser amounts of plane bedding and ripple cross-lamination. Unfortunately these stratification types are not restricted to any specific depositional environment. Solitary, straight-crested bedforms capable of producing planar cross-stratification are common in modern rivers (Sundborg, 1956; Allen, 1963; Coleman, 1969), especially braided rivers with sandy beds (Ore, 1964; Smith, 1970); they are also common in estuaries, beaches and shallow off-beach environments. Likewise, trains of ripples, megaripples and dunes with both straight and sinuous crests (producing planar and trough cross-stratification) are common in fluvial environments (Allen, 1963; Coleman, 1969), but are also present in the open sea (Allen, 1963; Imbrie and Buchanan, 1965).

The relative abundance of various sedimentary structures may be useful in defining depositional environment (Heckel, 1972; Picard and High, 1973). Heckel (1972, p. 243) noted that high-angle planar and trough crossbedding are features which occur only in marginal or nonmarine environments, hence the abundance of such forms might provide evidence against deposition in shallow-marine environments.

Scale of cross-stratification may be a useful indicator of depositional environment. Stratification types in marine settings are generally of smaller scale (i.e. dominated by ripple- and horizontal lamination) than most forms generated in fluvial environments. Exceptions do exist; for example large scale bedforms may be produced in open-marine settings (cf. Houbolt, 1968) and in aeolian environments. The internal geometry of large-scale bedforms generated in open-marine shelf settings may be more complex (cf. Hobday and Reading, 1972) than that of fluvial bedforms due to the influence of tidal reversals and storm events.

Sigmoidal and epsilon (Allen, 1963) cross-stratification, which might be considered indicative of lateral accretion, are very rare in Proterozoic sandy fluvial deposits. They have been recorded in sandstones of the Fugleberget (Hobday, 1974) and Golneselv (Banks and Røe, 1974) Formations, and from a single conglomeratic channel deposit in the Serpent Formation (Long, 1976). All three of these sequences have been interpreted as the products of deposition in braided stream environments. Large scale lateral accretion sets are present in the P3 sandstones of the Brock Inlier (Fig. 12), in the Katherine Group (Figs. 10, 11, 14) and in the Mississagi Formation (Fig. 13). Whereas those in the Brock Inlier can be attributed to deposition by lateral accretion in streams of moderate to high sinuosity (see later section), this may not be the case for the other examples which could have formed by lateral accretion on side- or inchannel-bars in a

Fig. 2. Trough cross-stratification in the Mississagi Formation.

Fig. 3. Trough cross-stratification superimposed on a solitary planar cross-stratified set, in the Mississagi Formation of Ontario. Note transitional boundary between bed types, indicating avalanching of smaller scale, sinuous crested, ripples over a larger scale transverse bar front.

Fig. 4. Small-scale trough cross-stratification in the Katherine Group, N.W.T. Scale in cm.

Fig. 5. Thinning-upward sequence of planar crossbeds in the Mississagi Formation of Ontario.

Fig. 6. Deformed planar crossbed, with scattered mud chips, in the Serpent Formation, Quirke Lake, Ontario.

Fig. 7. Thinning-upward sequence near the base of the Katherine Group, N.W.T.

Fig. 8. Recumbent-folded, deformed crossbedding in the upper part of the Katherine Group, N.W.T. Scale in cm.

Fig. 9. Thinning-upward sequence in P3 sandstones of the Brock Inlier, N.W.T. Planar sets are overlain by smaller scale ripple cross-lamination.

Fig. 10, 11. Large-scale cross-stratification in the Katherine Group (Mackenzie Mountains, N.W.T.). 10, producing a thick (50+ m) multistorey sand body exposed in Imperial Canyon. Note scoured base (arrow) of large composite set near base of cliff, and discontinuous character of mudstone beds (rare) in this section. 11, Detail of composite cross-stratified unit (7 m thick) with scoured base, exposed in canyon 10 km SE of Fig. 10.

Fig. 12. Large-scale composite cross-stratification in the P3 sandstone, Brock River Canyon, N.W.T. Note 5 - 7 m thick unit in middle of cliff (arrow) in which sandy lateral accretion sets are replaced laterally by red mudrock (? channel infill). Sinuosity calculation in text based on sandstone near base of this cliff.

Fig. 13. Compound, large-scale planar cross-stratified set in the Mississagi Formation, Louise Township, Ontario. The minor discordance (arrow) of foresets may indicate deposition by lateral accretion at a bar margin.

Fig. 14. Large-scale composite cross-stratified sets (2 - 5 m scale) exposed in a cliff face near Mount Eduni, N.W.T. (Katherine Gp.).

Fig. 15. Fining upward cycle in lower member of the Katherine Group, exposed in a section south of Keele River, N.W.T. Cycle grades from fine sand (1.18 m) at base, through red plane and ripple laminated very fine sandstone, to mudstone with minor ripple laminated sands (2.14 m).

braided stream environment. Williams (1966) used the strong divergence of paleocurrents indicated by planar and trough cross-stratification to infer that planar cross-stratified sets, in the Torridonian of part of northwest Scotland, were formed by lateral accretion on side- or inchannel-bars in a braided stream environment. Although evidence of lateral accretion may be used as supportive evidence for deposition in fluvial environments, it should be remembered that deposition by lateral accretion also occurs in tidal flat and tidal channel environments.

Evidence of penecontemporaneous deformation of cross-strata is common in Proterozoic fluvial sequences. It has been reported from the Athabasca (Fahrig, 1961), Dwaal Heuvel (Button, 1973), Golneselv (Banks and Røe, 1974), Fugleberget (Hobday, 1974), Glenelg (Miall, 1976a), Mississagi (Long, 1976, 1978) and Serpent (Long, 1976) Formations, from the Stonewall Sandstone (Plumb and Gemuts, 1976) and the Torridonian (Williams, 1966). The style of deformation of cross-strata in these sequences conforms with types a and b of Allen and Banks (1972, Fig. 1). Type a involves simple recumbent folds (Figs. 6, 8); while type b is slightly more complex with some disharmonic folds. These two styles of syndepositional deformation are considered by Allen and Banks (1972) to be especially common in rocks of fluvial origin and have been reported only rarely from rocks of shallow-marine origin.

Desiccation features, such as mudcracks (Fig. 16), provide evidence of at least temporary exposure in some Porterozoic sequences, as do rain print impressions which

Fig. 16. Subaerial desiccation cracks in thin mudstone unit, Serpent Formation, Ontario.

Fig. 17. Interbedded ripple-laminated sandstone and mudstone in a marine facies of the Katherine Group. Note presence of sand-filled syneresis cracks which project both upward and downward from the adjacent sandstone.

have been reported in the Jotnian Sandstone (Simonen and Kouvo, 1955) and the Athabasca (Fahrig, 1961, p. 3), Martin (Tremblay, 1972) and Fond du Lac (Morey, 1972) Formations. As these features require a muddy substrate to form, they are most common in intertidal environments. In many Proterozoic fluvial sequences, mudrocks are rarely preserved, with evidence of desiccation being limited to the presence of mud-chip conglomerates. Absence of such mud-chips does not preclude a fluvial origin as such clasts are readily broken down in fluvial environments within a few tens or hundreds of metres of their source (Smith, 1972a). Structures resembling mudcracks can be formed in subaqueous settings by syneresis (Donovan and Foster, 1972; Anderton, 1976) either at the sediment water interface, or during diagenesis. Syneresis cracks may be differentiated from subaerial desiccation cracks by their geometry. Syneresis cracks tend to have distinctly tapered terminations (Fig. 17) while, in plan view, mudcracks tend to be polygonal (Fig. 16).

Other (negative) evidence which might be used to support fluvial interpretations include the absence of sequences of alternating, bimodal-opposed, cross-stratification (herringbone cross-stratification), flaser bedding (Reineck and Wunderlich, 1968) and abundant symmetrical and asymmetrical ripple marks (Pettijohn et al., 1972, p. 494).

Structure sequence

A general upward tendency toward smaller bedforms and finer grain sizes is a characteristic feature of deposits formed in fluvial environments (Potter, 1967; Pettijohn et *al.*, 1972, p. 456). While fining-upward sequences are commonly used to support fluvial interpretations of many Phanerozoic sequences, they are comparatively rare in Proterozoic sequences (see earlier section), especially in those dominated by rocks of sand grade. Thinning-upward sequences are present locally in some Proterozoic sandstones such as the Amagok (Cecile, 1976) and Mississagi (Fig. 5; Long, 1976, 1978) Formations, the P3 sandstones (Fig. 9) and in the Katherine Group (Fig. 7) and can be used to support a fluvial origin for at least part of these successions. In many Proterozoic successions, distinct fining-upward and thinning-upward sequences are difficult to recognize, due to homogeneity of grain size and dominance of only one or two principal bedding types. One approach which may be used to circumvent this problem is to use a first order embedded Markov chain analysis (cf. Miall, 1973) in which different sedimentary structures or lithologies are used as facies states. Application of this method to rocks of the Mississagi Formation (Long, 1976, 1978) reveals a general tendency to an upward reduction in both grain size and scale of sedimentary structures. Examination of other Proterozoic sandstones in this fashion may provide a better basis for comparison with Phanerozoic successions.

Directional attributes of sedimentary structures

Perhaps the most useful approach which can be made to support fluvial interpretations of Proterozoic successions is the detailed examination of paleocurrent indicators. The shape of paleocurrent roses (after correction for tectonic dip) has been used for some time to support environmental interpretations of arenaceous sequences. Most fluvial systems are characterized by unimodal paleocurrent distributions (Potter and Siever, 1956; Selley, 1967, 1968; Wagner, 1977). Marine environments are characterized by diffuse (Wermund, 1965; Hofman, 1966; Spencer, 1971), unimodal (Klein, 1967; Pryor, 1971) or, in the majority of cases, bimodal-opposed paleocurrent distributions (Hülsemann, 1955; Hofman, 1966; Sedimentation Seminar, 1966; Hrabar *et al.*, 1971; Wagner, 1977). Bimodal-opposed paleocurrent distributions can also be expected in deposits of tidal flat environments (Klein, 1967; Selley, 1967, 1968), estuaries (Land and Hoyt, 1966) and lacustrine environments (Picard and High, 1972). Bimodal distributions with modes less than 180 degrees apart have been recorded from rocks formed in fluvial (Ore, 1964; High and Picard, 1974) as well as shallow marine (Picard and High, 1968) environments and hence cannot always be cited as evidence of marine influence.

The apparent shape of paleocurrent distributions may be strongly influenced by the sampling technique (Pettijohn, 1957), the scale, or hierachy (Miall, 1974) of the structures involved, and by the extent to which paleocurrent data are grouped both stratigraphically and areally. For example, grouping of data from paleocurrent observations in the Mississagi Formation led to the recognition of many apparently bimodal paleocurrent distributions (Long, 1976, 1978) which might be used (cf. Palonen, 1973) to support a marine origin for this formation. Examination of the same data at the outcrop level, for stations at which fifteen or more observations were recorded, showed that only 17% of the paleocurrent distributions had more than one principal mode (separated by more than 30°) which could be considered significant using the technique proposed by Tanner (1955). Only one of these was trimodal and one bimodal-bipolar. The remaining ten sets with bimodal distributions all had modes preferentially clustered at separations of 60° or 120° apart. These can be explained by local divergence of current systems within streams, perhaps augmented by changes in stream patterns during rising or falling flood stages and by shifting of stream systems leading to the production of multistorey sand bodies (cf. High and Picard, 1974). Alternatively these bimodal distributions may be related to the complex geometries of some fluvial bed forms, such as linguoid bars (Collinson, 1970), that might be produced by modification of transverse bars at high flow stage (Smith, 1972b, p. 625). The bimodal-opposed distribution pattern observed at one outcrop of the Mississagi Formation was due to opposed unimodal distributions of planar and trough cross-stratification (Long, 1976) and not to an alternating sequence of diametrically opposed cross-stratification as might be expected in a marine environment.

Detailed examination of paleocurrent distributions at the outcrop, or locality level, may provide a means to distinguish between marine and nonmarine components of sand-dominated sequences. For example changes from essentially unimodal to polymodal or diffuse paleocurrent roses have been used to distinguish fluvial from marine components of the Landersfjord Formation (Laird, 1972, p. 35), the Stirling Quartzite (Wertz, 1977) and parts of the Carr Boyd Group (Plumb and Gemuts, 1976, p. 46).

An alternative approach to the study of paleocurrent data is to use the degree of dispersion of cross-stratification as an indicator of depositional environment. This can be examined quantitatively using calculations of variance (square of standard deviation) about the vector mean in a circular distribution (Curray, 1956). Potter and Pettijohn (1963) indicated that, although some overlap is present, the most common variance for fluvial rocks is between 4000 and 8000, and for marine deposits is between 6000 and 8000. These values are based on calculations of variance where all the paleocurrent data for a formation have been grouped together; when the variance of paleocurrent data is examined at the outcrop level (as histograms of variance distributions) (Sedimentation Seminar, 1966; Button, 1973; Long and Young, in press) a slightly different picture emerges (Fig. 18). Marine deposits are characterized in more than 50% of the cases documented in Figure 18 by a modal variance class in the 6000 to 8000 range, while over 85% of the fluvial deposits have principal modes in the less than 4000 range. In all three of the marine deposits in Figure 18 with modal classes in the less than 4000 range there is a strong complementary peak in the 6000 to 8000 range. Unfortunately aeolian deposits, and some problematic marine or aeolian deposits such as the Navajo Sandstone (Fig. 18) cannot be clearly distinguished from fluvial deposits using this technique (Long and Young, in press.)

Other complementary evidence

Additional features which may be used to complement fluvial interpretations include petrography and sand body geometry (cf. Pettijohn *et al.*, 1972). As quartz arenites can be expected to form, even as first cycle sediments, in a marine, stable platform environment (Visser, 1975), the occurrence of quartz arenites might be considered as tentative evidence of deposition in a marine environment. As some Proterozoic fluvial sandsheets, such as the Athabasca Formation, are both remarkably thick and quartz rich (Fahrig, 1961), it is evident that this argument cannot be applied universally. The high quartz content of the Athabasca Formation may in part be related to reworking of older quartzite sequences (Ramaekers and Dunn, 1976, p. 306) and perhaps to intensive weathering conditions.

Local development of paleosoils provides evidence of emergent conditions, and hence can be used to support fluvial interpretations of adjacent rocks. Most examples cited in the literature appear to be developed on igneous or metamorphic basement rocks (Roscoe, 1969; Donaldson, 1967; Dodatko, 1975; Sochava *et al.*, 1975). Evidence of soil formation within sedimentary sequences is more limited. Thomson (1976) has suggested that colloidal rhythmic iron staining in the Pandurra Formation may indicate fossil weathering. Campbell and Cecile (1976, p. 372) have interpreted pisolitic quartzose dolomite and granular hematitic ironstone in the uppermost member of the Burnside River Formation as paleosoils developed during a period of quiescence prior to deposition of stromatolitic carbonates in the overlying Quadyuk Formation.

Red colouration of sedimentary rocks is often cited as supportive evidence of deposition in fluvial environments (cf. Simonen and Kouvo, 1955; Tremblay, 1972; Wallace, 1972, p. 365). One obvious problem is that the red colour of some sedimentary rocks may be due to later periods of diagenesis, rather than contemporary weathering.

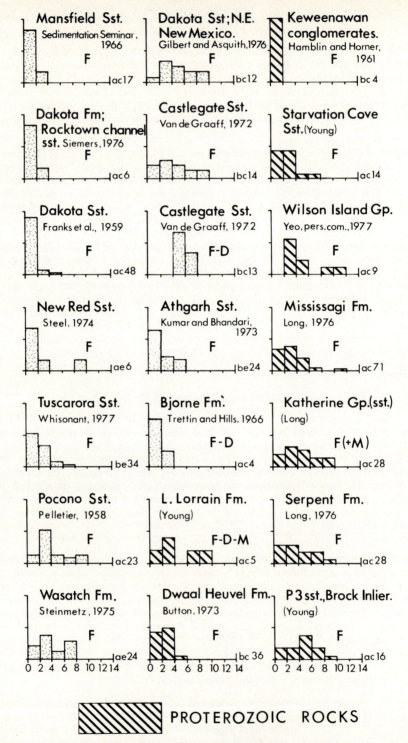

Fig. 18. Distribution of calculated values of variance in rocks deposited in marine (M), deltaic (D), fluvial (F), aeolian (A) and mixed environments. For explanation see Long and Young (in press).

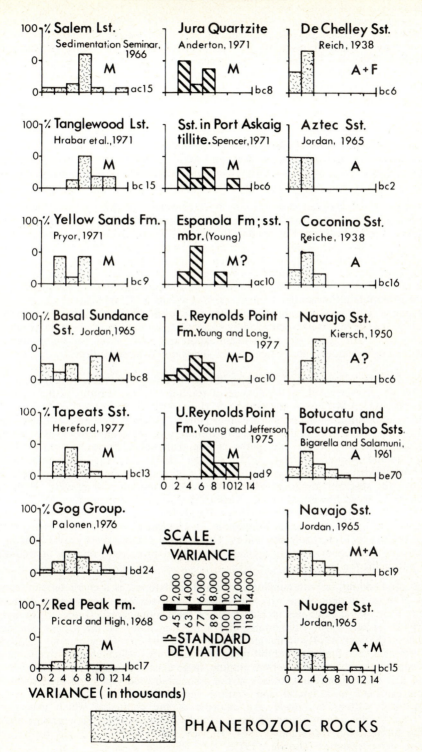

100 % Salem Lst.
Sedimentation Seminar, 1966
M
ac15

Jura Quartzite
Anderton, 1971
M
bc8

De Chelley Sst.
Reich, 1938
A+F
bc6

100 % Tanglewood Lst.
Hrabar et al.,1971
M
bc 15

Sst. in Port Askaig
tillite. Spencer,1971
M
bc6

Aztec Sst.
Jordan, 1965
A
bc2

100 % Yellow Sands Fm.
Pryor, 1971
M
bc9

Espanola Fm; sst.
mbr.(Young)
M?
ac10

Coconino Sst.
Reiche, 1938
A
bc16

100 % Basal Sundance
Sst. Jordan,1965
M
bc8

L. Reynolds Point
Fm.Young and Long, 1977
M–D
ac10

Navajo Sst.
Kiersch, 1950
A?
bc6

100 % Tapeats Sst.
Hereford,1977
M
bc13

U.Reynolds Point
Fm.Young and Jefferson, 1975
M
ad 9
0 2 4 6 8 10 12 14

Botucatu and
Tacuarembo Ssts.
Bigarella and Salamuni, 1961
A
be70

100 % Gog Group.
Palonen,1976
M
bd24

SCALE.
VARIANCE
2,000 4,000 6,000 8,000 10,000 12,000 14,000
0 45 63 77 89 100 110 118
STANDARD
DEVIATION

Navajo Sst.
Jordan, 1965
M+A
bc19

100 % Red Peak Fm.
Picard and High, 1968
M
bc17
0 2 4 6 8 10 12 14

Nugget Sst.
Jordan,1965
A+M
bc15
0 2 4 6 8 10 12 14

VARIANCE (in thousands)

PHANEROZOIC ROCKS

Likewise the absence of red pigmentation does not negate deposition in a fluvial environment as hematite pigments may be removed or altered during diagenesis under reducing conditions. The postulated absence of abundant free oxygen in the early Proterozoic atmosphere (Roscoe, 1969) could preclude the development of terrestrial red beds of earliest Aphebian age.

ALLUVIAL FAN DEPOSITS

Being typically coarse grained, alluvial fan deposits are the most easily recognized fluvial facies in Proterozoic rocks. They may be identified on the basis of grain size and geometry (Bull, 1972) even in intensely deformed sequences of high metamorphic grade, such as the Fleur de Lys metaconglomerates (DeWit, 1974) illustrated in Figure 19.

Proterozoic alluvial fan deposits are not restricted to any specific climatic zone; examples have been reported from inferred arid to semi-arid climates (Martin Formation: Tremblay, 1972; LaHood Formation: Boyce, 1975a, b), warm to moderately humid climates (Torridonian: Selley, 1965; Williams, 1968; Allen, 1972), wet or temperate climates (Van Horne Sandstone: McGowen and Groat, 1971; Witwatersrand: Vos, 1975) and in cold, glacial or paraglacial environments (Chibougamau Formation: Long, 1973, 1974; Mount Rogers Formation: Schwab, 1976; and possibly the Kaigas member of the Upper Stinkfontein Formation: Kröner, 1974). Two main assemblages can be recognized, representing end members of a complete spectrum of sedimentary structures and associations. The first is characterized by structures produced by deposition from traction currents (ie. stratified gravels, cross-stratification and imbrication) and the second by the absence of such features, and indications of deposition from mass-flows.

One of the best documented examples (though possibly not Precambrian) of an ancient fan deposit characterized by dominance of traction current features is the Van Horne Sandstone, which McGowen and Groat (1971) interpret as the product of deposition on a wet alluvial fan in a humid-temperate environment. Other Proterozoic examples include the Redstone River Formation (Fig. 20), uraniferous conglomerates in the Matinenda Formation (Roscoe, 1969) and parts of the Tarkwaian Series (Sestini, 1973) and Witwatersrand sequence (Vos, 1975). Clifton (1973) has suggested that, in some Phanerozoic sequences, high lateral continuity of conglomerate beds may be a useful criterion for discriminating between wave-worked and alluvial gravels. High lateral continuity is noted however in some parts of the Van Horne Sandstone (McGowen and Groat, 1971, p. 10), the Redstone River Formation (Fig. 20), and at least locally in conglomerates of the Matinenda Formation. This suggests problems in applying Clifton's (1973) criteria to Proterozoic sequences. High lateral continuity of gravel beds in some Proterozoic fan sequences may indicate sheet-flood processes, or lack of well-defined channels on the fan surfaces.

Massive bedding, in both conglomerates and sandstones, is a characteristic feature in many of those fan deposits in which mass-flow processes were important as depositional agents. Proterozoic examples include parts of the Chibougamau (Long, 1974) and Ring (Bjørlykke et al., 1976) Formations, the Kaigas member of the Upper Stinkfontein Formation (Kröner, 1974) and marginal facies of the Biskopås Conglomerate (Bjørlykke et al., 1976). As such fans may exhibit many of the features which characterize subaqueous fan deposits (ie. reverse and normal grading, large clasts "floating" in or near the top of beds), it is vitally important to look at the associated rock types and their sedimentary structures, both in a lateral and vertical sense, before any conclusions regarding depositional settings are made. An example of the confusion which may result is illustrated by the interpretation of the LaHood Formation (Belt Supergroup) both as a marine turbidite-fan (McMannis, 1963; Hawley, 1973) and as an arid region alluvial fan-fan delta complex (Boyce, 1975a, b). Possible evidence supporting Boyce's interpretation of this sequence as a subaerial fan complex comes from the presence of

Fig. 19. Fleur de Lys metaconglomerate, near Bear Cove, Newfoundland (Photo: E. W. R. Neale). Pebbles are elongated at a high angle to bedding (Parallel to hammer handle).

Fig. 20. Stratified conglomerates and conglomeratic sandstone in the Redstone River Formation, Mackenzie Mountains, N.W.T. (scale in cm; arrow indicates way up).

fining-upward sequences (involving upward transition from a pebbly base, through trough crossbeds, then plane beds, with a thin silt-shale layer rarely preserved at the top) which are repeated on a scale of 6-8 m in strata which directly overlie "Rubble beds", with clasts of up to 9 x 24 m, interpreted as debris flow deposits (Boyce, 1975b, p. 47).

Fan deposits with characteristics intermediate between these two end members are relatively common; examples include the South Channel (Donaldson, 1967), Hazel (Reid, 1974a, b) and Mount Rogers (Schwab, 1976) Formations, Keweenawan conglomerates (Hamblin and Horner, 1961; Babcock, 1974) and the Torridonian fan systems (Selley, 1965; Williams, 1968; Allen, 1972). In many of the above examples, fan deposits give way laterally to, and are intercalated with, conglomeratic sandstones interpreted as braided stream deposits.

CONGLOMERATIC (?BRAIDED) ALLUVIAL PLAIN DEPOSITS

Conglomerate-sandstone associations are relatively common in the Proterozoic, and are frequently interpreted in terms of deposition in braided stream environments. Evidence cited for such interpretations include marked lateral and vertical changes in grain size, pronounced channelling and/or scour surfaces, abundant cross-stratification with essentially unidirectional paleocurrents and, occasionally, the presence of red pigmentation. Examples include, in addition to those cited in the previous section, the basal parts of the Veidnesbotn (Banks *et al.*, 1974; Hobday, 1974), Thelon (Donaldson,

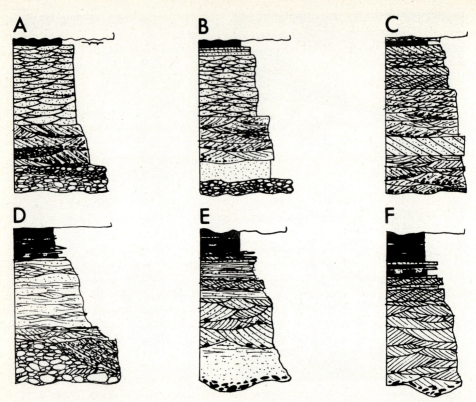

Fig. 21. Fining-upward cycles in Proterozoic rocks. A = Aphebian fining-upward cycle, Starvation Cove area of Victoria Island (proximal facies equivalent of B) based on Young (1974) and unpublished observations. Scale 3 - 5 m. B = Aphebian, "Ideal Complete Cycle" in the Burnside River Formation (after Campbell and Cecile, 1976). Scale 30 cm - 5 m (pers. com. Campbell and Cecile, 1977). C = Helikian, braided sandstone sequence in the Athabasca Formation at Manitou Falls (Ramaekers and Dunn, 1976). Scale 3.8 m. D = Hadrynian fining-upward cycle in the Mechum River Formation based on description by Schwab (1974). Scale 1 - 20 m. E = "Idealized minor cycle" in the Solor Church Formation (Upper Precambrian, Keweenawan) (Morey, 1974). Scale 0.9 - 15 m. F = Lower Aphebian fining-upward cycle in the upper sandstone member of the Espanola Formation, based on Young (1973). Scale 1.5 - 4.6 m.

1967), Athabasca (Ramaekers, 1975, 1976) and Guperas (Watters, 1974) Formations, parts of the Hornby Channel (Hoffman, 1968), Burnside River (Campbell and Cecile, 1976), Saunders Creek (Plumb and Gemuts, 1976), Eriksfjord (Poulsen, 1964) and Crystal Spring (Roberts, 1974) Formations, and proximal parts of the Bonner Quartzite (Winston, 1973, 1977).

One notable feature of some (but not all) of these conglomerate-sandstone sequences is the presence of fining-upward cycles (Fig. 21 A-D) which, in the Burnside River Formation (Fig. 21B; Campbell and Cecile, 1976) and its stratigraphic equivalents in the Starvation Cove area of Victoria Island (Fig. 21A; Young, 1974; Young and Jefferson, 1975) have been cited as evidence for deposition by lateral accretion. The "ideal complete cycle" in the Burnside River Formation (Fig. 21B) is essentially similar to the cycles observed in the more proximal facies developed on Victoria Island (Fig. 21A) except for its scale and the (very rare) presence of massive and plane-bedded sandstone facies. Cycles in the Athabasca Formation (Fig. 3C) have been compared (Ramaekers and Dunn, 1976; Ramaekers, pers. com., 1977) with cycles in the (Devonian) Battery Point Formation, which have been interpreted (Cant and Walker, 1976) as the product of deposition in relatively low sinuosity (braided) streams. Cycles in the Mechum River Formation (Fig. 21D) have been interpreted by Schwab (1974) as the product of deposition on point bars, presumably in a meandering stream environment. Distinctive lateral accretion surfaces (epsilon cross-stratification) which might support a point bar, or meandering stream, interpretation are not present in the Burnside River Formation (Campbell, pers. com., 1977) or in the Starvation Cove sequence, and have not been reported in the Mechum River and Athabasca Formations. These conglomeratic fining-upward cycles (Fig. 21 A-D) and possibly those reported by McGlynn (1971) in the Nonacho Group and by Sestini (1973) in the Tarkwaian Series may be examples of what Miall (1977) has termed "Donjek type" braided stream deposits.

Conglomerate sandstone sequences associated with glaciogene mixtite units are commonly interpreted as the products of deposition in fluvioglacial environments. Examples include the Skinner and Blackfellow Creek Sandstones (Sweet *et al*., 1974) and parts of the Ilfjord (Laird, 1972), Mount Rogers (Schwab, 1976), Mineral Fork (Varney, 1976) and Gowganda Formations. In the case of the Mount Rogers Formation, the conglomerates and associated matrix-poor sandstones are non-cyclical and contain evidence of deposition from traction currents in the form of plane bedding and trough cross-stratification (Schwab, 1976). In other glaciogene sequences, evidence of deposition from traction currents is rare; for example most of the Blackfellow Creek Sandstone (Sweet *et al*., 1974) is devoid of structures, and sedimentary structures are very rare in sandstones associated with the Mineral Fork (Varney, 1976) tillites. Sandstones in the Chibougamau (Long, 1973) and Ilfjord (Laird, 1972) Formations are distinctively massive and contain isolated boulders which appear to float in a matrix of medium or coarse sand grade. Lack of traction current features may indicate that deposition by mass flows (?debris flows) was a common feature of some Proterozoic fluvioglacial environments.

HIGH SINUOSITY (?MEANDERING) STREAM DEPOSITS

High sinuosity or meandering stream deposits have been recognized in very few Proterozoic sequences. Possible examples, characterized by the presence of distinct fining upward cycles, include the Fond du Lac (Morey, 1967, 1972), Red Castle (Wallace and Crittenden, 1969; Wallace, 1972) and Solor Church (Fig. 21E; Morey, 1974) Formations. Fining-upward cycles in the Espanola (Young, 1973) and Dwaal Heuvel (Button, 1973) Formations have been interpreted in terms of deposition in braided *or* meandering stream environments. Cycles within the above formations show, in addition to an upward reduction in grain size, a general upward trend toward reduction in scale of sedimentary structures (Fig. 21 E, F). This, together with the abundance of fines in the upper parts of the cycles, is consistent with deposition in a high sinuosity or meandering stream environment (cf. Moody-Stuart, 1966). One problem with this interpretation is that the lower (sandy) parts of the above cycles do not always contain evidence of deposition by lateral, rather than vertical, accretion. In none of the cases cited above has direct evidence been presented for the presence of lateral accretion surfaces analogous to those expected from migration of a point bar complex in a meandering stream environment. This may in part be due to the difficulty of recognizing lateral accretion sets where they have very low dips (Nijman and Puigdefabregas; this volume). Large-scale lateral accretion sets are rare in Proterozoic sequences; where present (Figs. 10-14) they do not always contain definite evidence of deposition in meandering streams, and may form as a component of more sandy braided fluvial systems (cf. Williams, 1966).

Fining-upward cycles in Proterozoic successions are not all of fluvial origin. Cycles superficially similar to fluvial examples may be generated in tidal channel (Johnson, 1975; Young and Long, 1977) and tidal flat (Klein, 1971, 1977) environments; consequently problems may arise in the interpretation of Proterozoic sandstone-mudstone cycles if close attention is not paid to paleocurrent indicators within each cycle. For example, a marine influence is indicted in at least some of the fining-upward cycles in the Espanola

Formation (Fig. 21F) by the presence of bimodal-opposed cross-bed distributions (Young *et al.*, 1977). The presence of what appears to be bimodal-opposed cross-stratification in the basal part of a fining-upward sequence (Fig. 15) in the Katherine Group (Aitken *et al.*, 1978) could be taken as evidence of a marine-tidal influence (cf. Klein, 1971, 1977). However this bimodality is actually related to opposed directions of cross-stratification on the flanks of trough cross-stratified sets which have essentially unidirectional trough-axis orientations, and not to a sequence of crossbeds with alternating, bimodal opposed, orientations (herringbone cross-stratification). Other features which may be useful in distinguishing tidal channels from fluvial cycles include the presence of abundant erosion surfaces (Johnson, 1975) or mud drapes (Young and Long, 1977) between crossbed sets, which may be related to tidal reversals, or tidally influenced fluctuations in current velocities. Reactivation surfaces, although present in some fluvial sequences (Banks, 1973), may be more abundant in tidally influenced environments.

Identification of high sinuosity stream deposits is exceptionally difficult in sequences with little or no mudrocks. Miller (1969) has suggested that a 90° divergence of paleocurrents, indicated by parting lineations from the directions indicated by associated crossbeds, may provide evidence of meandering stream conditions during deposition of the Auborus Formation. One potentially useful approach to evaluating stream sinuosity in Proterozoic deposits is the use of maximum angular range of mean channel azimuth, calculated by using a ten-point moving average of weighted crossbed azimuths, as measured in continuous vertical sequence (Miall, 1976b). This angle (Θ) is utilized to calculate sinuosity using the relation:

$$\text{Sinuosity} = \frac{1}{1 - (\Theta/252)^2} \text{ (Miall, 1976b, eq. 1)}$$

To date, attempts to apply this method to Proterozoic rocks have been limited. Values estimated for the Mississagi Formation range from 1.01 to 1.09 (Long, 1976, 1978) and for the Serpent Formation (Long, 1976) range from 1.01 to 1.48. Both of these units are interpreted as the products of deposition in (braided) stream systems with low to intermediate sinuosity. The sinuosity calculated for one set of observations (25 crossbeds) in the P3 sandstones of the Brock Inlier (Fig. 6) indicates a sinuosity of 1.47. This value is consistent with Young's (1977) interpretation of this section as part of a distal fluvial complex.

FLUVIAL MODELS

Attempts to match depositional environments of Proterozoic sequences with models based on Phanerozoic analogues is complicated by the apparent absence of abundant land plants. Schumm (1968) considered that, in the absence of vegetation, denudation and runoff rates would have been significantly higher, leading to larger floods and the dominance of what he terms bedload channels. The dramatic effect of vegetation on bank stabilization has been demonstrated by Smith (1976) who has shown that non-vegetated sediment can be 20,000 times less stable than comparable vegetated banks containing 16 to 18% by volume of plant roots. Experiments by Wolman and Brush (1961), backed by observations of natural stream channels, indicate that any increase in flow within a channel having non-cohesive banks is rapidly compensated by broadening of channels, with only a minor increase in flow depth above that required to initiate bed movement. If no vegetation is present to inhibit bank migration, increases in flow would have led to the production of beds with high width/depth ratios. Sheet erosion during rising flood stage may also explain the general deficiency of marked non-planar erosion surfaces in many sand-dominated fluvial sequences of Proterozoic age.

Attempts to subdivide Proterozoic fluvial sequences into those produced by braided and meandering stream systems are hindered by the absence of abundant mud-grade clastics (both as matrix and associated mudrocks) in many sequences of fluvial origin

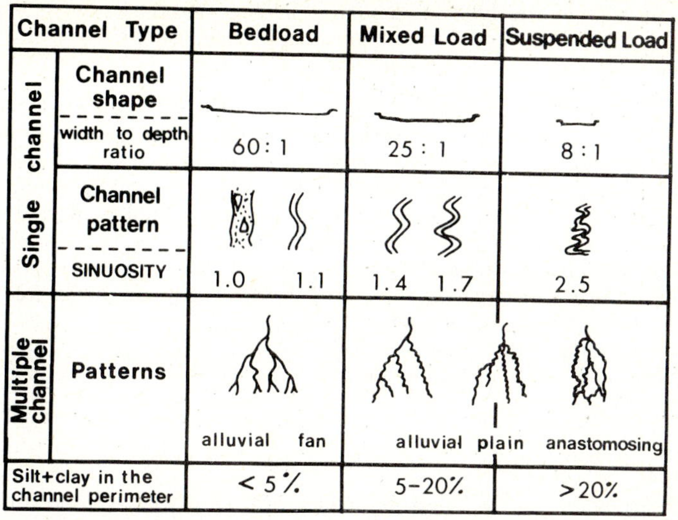

Channel Type		Bedload	Mixed Load	Suspended Load
Single channel	**Channel shape**			
	width to depth ratio	60 : 1	25 : 1	8 : 1
	Channel pattern			
	SINUOSITY	1.0 1.1	1.4 1.7	2.5
Multiple channel	**Patterns**			
		alluvial fan	alluvial plain	anastomosing
Silt+clay in the channel perimeter		< 5 %	5–20%	>20%

Fig. 22. Morphology of principal river types, as a function of their sediment load (modified from Schumm, 1968).

(Donaldson, 1967; Banks, 1973; Sweet *et al.*, 1974; Watters, 1974; Cecile, 1976; Ramaekers and Dunn, 1976). These deficiencies may be related to lower rates of production of fines (possibly related to more rapid rates of denudation; cf. Schumm, 1968) or to the removal of fine-grained material by contemporary fluvial and aeolian erosion. That fine-grained clastics were produced during the cycle of erosion in Proterozoic times is indicated by the abundance of fine-grained clastics in non-fluvial settings. For example, Cecile (1976, p. 3.35) has demonstrated that the paucity of mudstones in the fluvial facies of the Goulburn Group is compensated for by the abundance of mudrocks in the associated paralic-shallow marine facies. Fine-grained material may have effectively bypassed Proterozoic fluvial depositional settings if it was transported through the stream channels as wash load. The presence of mudflake conglomerates in some of the more sandy fluvial sequences attests to the extreme vulnerability of fine-grained sediments, deposited temporarily in overbank and protected channel environments, to later reworking.

Schumm (1968) proposed that rivers could be divided into three main types depending on the character of their sediment load. In this classification, the channels of bedload streams are characterized by less than 5% fines (less than 0.074 mm) in deposits forming the bank perimeter, mixed load stream deposits by 5% to 20% fines and suspension load stream deposits by greater than 20% fines. Other attributes of these stream types are summarized in Figure 22 (after Schumm, 1968). Using this classification, most documented examples of Proterozoic stream deposits can be considered as the products of deposition from bedload and, to a lesser extent, mixed-load streams. Schumm (1968) speculated that, in the absence of vegetation, the formation of meandering type streams, and the resulting generation of fining upward sequences, would be hindered by the lack of cohesive banks and increased runoff. The existence of large-scale accretion sets in a few Proterozoic fluvial deposits does not necessarily indicate a meandering stream pattern, as these structures may have been generated in braided stream environments by lateral migration of side- and inchannel bars. Fining-upward cycles do exist in Proterozoic rocks, although detailed studies of bedform geometry and sequence (cf. Fig. 21) must be made in many cases to ascertain whether these are the deposits of high sinuosity-meandering stream systems, or are the products of waning flood cycles. Detailed studies of vertical sequences of cross-stratification (Miall, 1976b) may lead to a better understanding of the relation of sinuosity to the type of deposit produced, although the technique requires rigorous testing on Holocene sequences produced by streams of known sinuosity before it can be applied with confidence to Proterozoic sequences.

While direct analogues for meandering streams may be difficult to find in the Proterozoic record, many of the thicker sandstone- (and conglomerate-sandstone-) dominated successions can be compared with models of braided stream deposits based on Phanerozoic examples. Using Miall's (1977; this volume) classification, many of the conglomerate bearing sequences can (at least in part) be related to deposition from Scott and Donjeck type streams, whereas the sandstone-dominated successions can be more closely related to Donjeck, South Saskatchewan and Platte type streams. Examples of Miall's (1977) Bijou Creek type deposits, consisting of superimposed flood cycles, are comparatively rare, although this mode of deposition has been suggested for some deposits in the Crystal Spring (Roberts, 1974, p. 52) and Ring (Bjørlykke *et al.*, 1976, p. 259) Formations and for a distal fluvial facies in the Belt Supergroup (Winston, this volume).

CONCLUSIONS

No single parameter can be cited as unequivocal evidence of deposition in a fluvial environment. Interpretation of Proterozoic clastic sequences must therefore rely on combined observations of sedimentary structures, paleocurrents, lithology and stratigraphy. In many cases, the basic concern is not to distinguish the type of fluvial environment, but to determine whether a formation is of fluvial, marine or mixed genesis. Although an attempt can be made to match Proterozoic fluvial deposits with models based on Recent examples, the behaviour of Proterozoic fluvial systems may have been drastically different from present-day fluvial systems (notably due to the lack of vegetation). More detailed regional studies of Proterozoic fluvial deposits are needed to see if, and how, these compare and contrast with more recent Phanerozoic deposits.

ACKNOWLEDGMENTS

I thank Andrew D. Miall, J. Ross McLean and Don Winston for their comments on an earlier version of this paper. Research on the P3 sandstones of the Brock Inlier, and sandstones of the Katherine Group in the North West Territories formed part of Geological Survey of Canada Project 730057. The author's understanding (?) of Proterozoic stream deposits has benefited greatly from discussions with G. M. Young, J. D. Aitken, F. H. A. Campbell, M. P. Cecile, P. P. Ramaekers and D. Winston.

REFERENCES

Aitken, J. D., Long, D. G. F., and Semikhatov, M. A., 1978, Progress in Helikian Stratigraphy, Mackenzie Mountains: *in* Current Research, Part A: Geol. Surv. Canada, Paper 78-1A, p. 481-484.

Allen, J. R. L., 1963, The classification of cross-stratified units with notes on their origin: Sedimentology, v. 2, p. 93-114.

———, and Banks, N. L., 1972, An interpretation and analysis of recumbent-folded deformed cross-bedding: Sedimentology, v. 19, p. 257-283.

Allen, P., 1972, Continental sandstones: the Torridonian of northwestern Scotland: *in* Arenaceous deposits: Sedimentation and diagenesis; 1972 National Conference on Earth Science, Dept. of Extension, Univ. of Alberta, and Alberta Soc. Petrol. Geol., p. 165-177 and p. 201.

Anderton, R., 1971, Dalradian paleocurrents from the Jura Quartzite: Scot. J. Geol., v. 7, p. 175-178.

———, 1976, Tidal-shelf sedimentation: an example from the Scottish Dalradian: Sedimentology, v. 23, p. 429-458.

Babcock, L. L., 1974, Jacobsville Sandstone: a lower-middle Keweenawan red bed sequence (Abstract): in P. E. Giblin, G. Bennett and E. J. Leahy, Program, Abstracts and Field Guides for the 20th Annual Institute on Lake Superior Geology, Sault Ste. Marie, Ontario, p. 1.

Banks, N. L., 1973, Falling stage features of a Precambrian braided stream: criteria for sub-aerial exposure: Sediment. Geol., v. 10, p. 147-154.

——, Hobday, D. K., Reading, H. G., and Taylor, P. N., 1974, Stratigraphy of the Late Precambrian 'Older Sandstone Series' of the Vrangerfjord area, Finnmark: Norges Geologiske Undersøkelse, Nr. 303 (Bull., 22), p. 1-15.

——, and Røe, S. L., 1974, Sedimentology of the Late Precambrian Golneselve Formation, Varangerfjord, Finnmark: Norges Geologiske Undersokelse, Nr. 303 (Bull., 22) p. 17-38.

Bigarella, J. J., and Salamuni, R., 1961, Early Mesozoic wind patterns as suggested by dune bedding in the Botucatú Sandstone of Brazil and Uraguay: Geol. Soc. Am. Bull., v. 72, p. 1089-1106.

Bjørlykke, K., Elvsborg, A., and Høy, T., 1976, Late Precambrian sedimentation in the central Sparagmite basin of south Norway: Norsk Geologisk Tidsskrift, v. 56, p. 233-290.

Blake, D. A. W., 1956, Geological Notes on the region south of Lake Athabasca and Black Lake, Saskatchewan and Alberta: Geol. Surv. Canada, Paper 55-33, 12p.

Boyce, R., 1975a, Depositional systems in the Ravalli Group: a conceptual model and possible modern analogue: Belt Symposium, 1, Univ. of Idaho, Moscow., ID, p. 139-158.

——, 1975b, Depositional systems in the LaHood Formation, Belt Supergroup, Precambrian, southwestern Montana: Unpub. Ph.D. thesis, Univ. of Texas at Austin, 248p.

Bull, W. B., 1972, Recognition of alluvial fan deposits in the stratigraphic record: in J. K. Rigby and W. K. Hamblin, eds., Recognition of Ancient sedimentary environments; Soc. Econ. Paleont. Mineral. Spec. Pub. 16, p. 63-83.

Button, A., 1973, A paleocurrent study of the Dwaal Heuvel Formation, Transvaal Supergroup: University of the Witwatersrand, Johannesburg, Economic Geology Research Unit, Information Circular No. 75, 17p.

Campbell, F. H. A., and Cecile, M. P., 1976, Geology of the Kilohigok Basin, Goulburn Group, Bathurst Inlet, District of Mackenzie: Geol. Surv. Canada, paper 76-1A, p. 369-377.

Cant, D. J., and Walker, R. G., 1976, Development of a braided-fluvial facies model for the Devonian Battery Point Sandstone, Quebec: Can. J. Earth Sci., v. 13, p. 102-119.

Card, K. D., Innes, D. G., and Debicki, R. L., 1977, Stratigraphy, sedimentology and petrology of the Huronian Supergroup in the Sudbury - Espanola area: Ontario Division of Mines, Geoscience Study 16, 99p.

Cecile, M. P., 1976, Stratigraphy and depositional history of the Upper Goulburn Group, Kilohigok Basin, Bathurst Inlet, N.W.T.: Unpub. Ph.D. thesis, Carleton Univ., Ottawa, Ontario, Canada, 204p.

Clarke, D., 1976, Heavitree Quartzite: in A. T. Wells, Geology of the Late Proterozoic - Paleozoic Amadeus Basin: 25th. Internat. Geol. Congr., Sydney, Australia, Excursion Guide No. 48A, p. 26-28.

Clifton, H. E., 1973, Pebble segregation and bed lenticularity in wave-worked versus alluvial gravel: Sedimentology, v. 20, p. 173-187.

Coleman, J. M., 1969, Brahmaputra River: channel process and sedimentation: Sediment. Geol., v. 3, p. 129-239.

Collinson, J. D., 1970, Bedforms of the Tana River, Norway: Geogr. Ann., Ser. A., v. 52, p. 31-56.

Curray, J. R., 1956, The analysis of two-dimensional orientation data: J. Geol., v. 64, p. 117-131.

DeWit, M. J., 1974, On the origin and deformation of the Fleur de Lys metaconglomerate, Appalachian fold belt, northwest Newfoundland: Can. J. Earth Sci., v. 11, p. 1168-1180.

Dodatko, A. D., 1975, Early Proterozoic continental sedimentation breaks and weathering epochs of the Russian Platform as shown by study of the Krivoy Rog district: Doklady Akad. Nauk. SSSR, v. 221, p. 216-218.

Donaldson, J. A., 1967, Two Proterozoic clastic sequences: a sedimentological comparison: Proc. Geol. Assoc. Canada, v. 18, p. 33-54.

Donovan, R. N., and Foster, R. J., 1972, Subaqueous shrinkage cracks from the Caithness Flagstone Series (Middle Devonian) of northeastern Scotland: J. Sediment. Petrol., v. 42, p. 309-317.

Fahrig, W. F., 1961, The geology of the Athabasca Formation: Geol. Surv. Canada, Bull. 68, 41p.

Folk, R. L., 1968, Petrology of sedimentary rocks: Hemphill's, Austin, Texas, 170p.

Franks, P. C., Coleman, G. L., Plummer, N., and Hamblin, W. K., 1959, Cross-stratification, Dakota sandstone (Cretaceous), Ottawa County, Kansas: Kansas Geol. Surv., Bull. 134, pt. 6, p. 223-238.

Gilbert, J. L., and Asquith, G. B., 1976, Sedimentology of braided alluvial interval of Dakota Sandstone Northeastern New Mexico: New Mexico Bureau of Mines Min. Res., Circular 150, 16p.

Hadley, D. G., 1968, Sedimentology of the Huronian Lorrain Formation, Ontario and Quebec, Canada: Unpub. Ph.D. thesis, Johns Hopkins Univ., Baltimore, 301p.

Hawley, D., 1973, Sedimentary environment of the LaHood Formation, a southeastern facies of the Belt Supergroup, Montana (Abstract): Belt Symposium, 1, Univ. of Idaho, Moscow, ID., p. 77.

Hamblin, W. K., and Horner, W., 1961, Sources of the Keweenawan conglomerates of northern Michigan: J. Geol., v. 69, p. 204-210.

Heckel, P. H., 1972, Recognition of ancient shallow marine environments: *in* J. K. Rigby and W. K. Hamblin, *eds.*, Recognition of ancient sedimentary environments: Soc. Econ. Paleont. Mineral. Spec. Pub. 16, p. 226-286.

Hereford, R., 1977, Deposition of the Tapeats Sandstone (Cambrian) in central Arizona: Geol. Soc. Am. Bull., v. 88, p. 199-211.

High, L. R., Jr., and Picard, M. D., 1974, Reliability of cross-stratification types as paleocurrent indicators in fluvial rocks: J. Sediment. Petrol., v. 44, p. 158-168.

Hobday, D. K., 1974, Interaction between fluvial and marine processes in the lower part of the Late Precambrian Vadsø Group, Finnmark: Norges Geologiske Undersøkelse, Nr. 303 (Bull. 22), p. 39-53.

———, and Reading, H. G., 1972, Fair weather versus storm processes in shallow marine sandbar sequences in the Late Precambrian of Finnmark, north Norway: J. Sediment. Petrol., v. 42, p. 318-324.

Hoffman, P. F., 1968, Stratigraphy of the lower Proterozoic (Aphebian), Great Slave Supergroup, East arm of Great Slave Lake, District of Mackenzie: Geol. Surv. Canada, Paper 68-42, 93p.

Hofman, H. J., 1966, Ordovician paleocurrents near Cincinatti, Ohio: J. Geol., v. 74, p. 868-890.

Houbolt, J. J. H. C., 1968, Recent sediments in the southern bight of the North Sea: Geol. Mijn., v. 47, p. 245-273.

Hrabar, S. V., Cressman, E. R., and Potter, P. E., 1971, Crossbedding of the Tanglewood Limestone member of the Lexington Limestone (Ordovician) of the Blue Grass region of Kentucky: Brigham Young University, Geology Studies, v. 18, p. 99-114.

Hülseman, J., 1955, Grosrippels und schrägschichtungs-gefüge im Nordsee-Watt und in der Molasse: Senckenbergiana Lethaea, v. 36, p. 359-388.

Imbrie, J., and Buchanan, H., 1965, Sedimentary structures in modern carbonate sands of the Bahamas: *in* G. V. Middleton, *ed.*, Primary sedimentary structures and their hydrodynamic interpretation: Soc. Econ. Paleont. Mineral. Spec. Pub. 12, p. 149-172.

Johnson, H. D., 1975, Tide- and wave-dominated inshore and shoreline sequences from the late Precambrian, Finnmark, north Norway: Sedimentology, v. 22, p. 45-73.

Jordan, W. M., 1965, Regional environmental study of the early Mesozoic Nugget and Navajo sandstones: Unpub. Ph.D. thesis, Univ. of Wisconsin, 206p.

Kiersch, G. A., 1950, Small scale structures and other features of the Navajo sandstone, northern part of San Rafael Swell, Utah: Am. Assoc. Petrol. Geol. Bull., v. 34, p. 923-942.

Klein, G. de V., 1967, Paleocurrent analysis in relation to modern marine sediment dispersal patterns: Am. Assoc. Petrol. Geol. Bull., v. 51, p. 366-382.

———, 1971, A sedimentary model for determining paleotidal range: Geol. Soc. Am. Bull., v. 82, p. 2585-2592.

———, 1977, Epilogue: Sediment. Geol., v. 18, p. 283-287.

Kröner, A., 1974, The Gariep Group, part 1, Late Precambrian Formations in the western Richersveld, northern Cape Province: University of Cape Town, Dept. of Geology, Chamber of Mines, Precambrian Research Unit, Bull. 13, 115p.

Kumar, S., and Bhandari, L. L., 1973, Paleocurrent analysis of the Athgarh Sandstone (Upper Gondwana), Cuttack District, Orissa (India): Sediment: Geol., v. 10, p. 61-75.

Laird, M. G., 1972, The stratigraphy and sedimentology of the Lakesfjord Group, Finnmark: Norges Geologiske Undersøkelse, Nr. 278, p. 13-40.

Land, L. S., and Hoyt, J. H., 1966, Sedimentation in a meandering estuary: Sedimentology, v. 6, p. 191-207.

Long, D. G. F., 1973, The stratigraphy and sedimentology of the Chibougamau Formation: Unpub. M.Sc. thesis, Univ. Western Ontario, London, Ontario, Canada, 305p.

———, 1974, Glacial and paraglacial genesis of conglomeratic rocks of the Chibougamau Formation (Aphebian), Chibougamau, Quebec: Can. J. Earth Sci., v. 11, p. 1236-1252.

———, 1976, The stratigraphy and sedimentology of the Huronian (lower Aphebian) Mississagi and Serpent Formations: Unpub. Ph. D. thesis, Univ. Western Ontario, London, Ontario, Canada, 291p.

———, 1978, Depositional environments of a thick Proterozoic sandstone: the (Huronian) Mississagi Formation of Ontario, Canada: Can. J. Earth Sci., v. 15, p. 190-206.

———, and Young, G. M., (in press), Dispersion of cross-stratification as a potential tool in the interpretation of Proterozoic arenites: J. Sediment. Petrol.

McGlynn, J. C., 1971, Stratigraphy, sedimentology and correlation of the Nonacho Group, District of Mackenzie: Geol. Surv. Canada, Paper 71-1A, p. 140-142.

McGowen, J. H., and Groat, C. G., 1971, Van Horn Sandstone, west Texas: an alluvial fan model for mineral exploration: Bur. Econ. Geol., University of Texas at Austin, Texas, Rept. Invest., No. 72, 57p.

McMannis, W. J., 1963, LaHood Formation, — a coarse facies of the Belt Series in southwestern Montana: Geol. Soc. Am. Bull., v. 74, p. 407-436.

Miall, A. D., 1973, Markov chain analysis applied to an ancient alluvial plain succession: Sedimentology, v. 20, p. 347-364.

———, 1974, Paleocurrent analysis of alluvial sediments: a discussion of directional variance and vector magnitude: J. Sediment. Petrol., v. 44, p. 1174-1185.

———, 1976a, Proterozoic and Paleozoic Geology of Banks Island, Arctic Canada: Geol. Surv. Canada, Bull. 258, 77p.

———, 1976b, Paleocurrent and paleohydrologic analysis of some vertical profiles through a Cretaceous braided stream deposit, Banks Island, Arctic Canada: Sedimentology, v. 23, p. 459-484.

———, 1977, A review of the braided-river depositional environment: Earth Sci. Rev., v. 13, p. 1-62.

Miller, R. McG., 1969, The Auborus Formation of the Bethanie district South West Africa: University of Cape Town, Dept. of Geology, Chamber of Mines, Precambrian Research Unit, Bull. 2, 32p.

Moody-Stuart, M., 1966, High and low sinuosity stream deposits, with examples from the Devonian of Spitzbergen: J. Sediment. Petrol., v. 36, p. 1102-1117.

Morey, G. B., 1967, Stratigraphy and petrology of the type Fond du Lac Formation, Duluth, Minnesota: Minnesota Geol. Survey, Rpt. Invest., 11, 191p.

———, 1972, Petrology of Keweenawan sandstones in the subsurface of southeastern Minnesota: *in* P. K. Sims and G. B. Morey, *eds.*, Geology of Minnesota: A centennial volume: Minnesota Geol. Survey, p. 436-449.

———, 1974, Cyclic sedimentation of the Solor Church Formation (Upper Precambrian, Keweenawan) southeastern Minnesota: J. Sediment. Petrol., v. 44, p. 872-884.

Ore, H. T., 1964, Some criteria for recognition of braided stream deposits: Univ. Wyoming, Dept. Geology, Contributions to Geology, v. 3, p. 1-14.

Palonen, P. A., 1973, Paleogeography of the Mississagi Formation and Lower Huronian cyclicity: *in* G. M. Young, *ed.*, Huronian Stratigraphy and Sedimentation: Geol. Assoc. Canada Spec. Paper 12, p. 157-167.

———, 1976, Sedimentology and stratigraphy of Gog Group sandstones in the southern Canadian Rockies: Unpub. Ph.D. thesis, Univ. Calgary, Alberta, Canada, 210p.

Passega, R., 1957, Texture as characteristic of clastic deposition: Am. Assoc. Petrol. Geol. Bull., v. 41, p. 1952-1984.

————, 1964, Grain size representation by CM patterns as a geological tool: J. Sediment. Petrol., v. 34, p. 830-847.

————, 1977, Significance of CM diagrams of sediments deposited by suspensions: Sedimentology, v. 24, p. 723-733.

Pelletier, B. R., 1958, Pocono paleocurrents in Pennsylvania and Maryland: Geol. Soc. Am. Bull., v. 69, p. 1033-1064.

Pettijohn, F. J., 1957, Paleocurrents of Lake Superior Precambrian Quartzites: Geol. Soc. Am. Bull., v. 68, p. 409-480.

————, Potter, P. E., and Siever, R., 1972, Sand and sandstone: Springer-Verlag, New York, N.Y., 618p.

Picard, M. D., and High, L. R., Jr., 1968, Shallow marine currents on the Early (?) Triassic Wyoming Shelf: J. Sediment. Petrol., v. 38, p. 411-423.

————, and ————, 1972, Criteria for recognising lacustrine rocks: in J. K. Rigby and W. K. Hamblin, eds., Recognition of ancient sedimentary environments: Soc. Econ. Paleont. Mineral, Spec. Pub. 16, p. 108-145.

————, and ————, 1973, Sedimentary structures of ephemeral streams: Developments in Sedimentology 17, Elsevier, New York, N.Y., 223p.

Plumb, K. A., and Gemuts, I., 1976, Precambrian geology of the Kimberley region, Western Australia: 25th. Intern. Geol. Congr., Sydney, Australia, Excursion Guide 44C, 69p.

Potter, P. E., 1967, Sand bodies and sedimentary environments: a review: Am. Assoc. Petrol. Geol. Bull., v. 51, p. 337-365.

————, and Pettijohn, F. J., 1963, Paleocurrents and basin analysis: Academic Press, New York, N.Y., 296p.

————, and Siever, R., 1956, Sources of basal Pennsylvanian sediments in the eastern Interior Basin, 1. Cross-bedding: J. Geol., v. 6, p. 225-244.

Poulsen, V., 1964, The sandstones of the Precambrian Eriksfjord Formation in south Greenland: Grønlands Geologiske Undersøgelse, Rapport Nr. 2, 16p.

Pryor, W. A., 1971, Petrology of the Permian Yellow Sands of northeastern England and their North Sea basin equivalents: Sediment. Geol., v. 6, p. 221-256.

Ramaekers, P. P., 1975, Athabasca Formation, southeast edge (74H): 1. Reconnaissance geological survey: in J. E. Christopher and R. Macdonald, eds., Summary of Investigations 1975: Saskatchewan Dept. Min. Res., p. 48-52.

————, 1976, Athabasca Formation, northeast edge (64L, 74I, 74P): Part 1 Reconnaissance geology: in J. E. Christopher and R. Macdonald, eds., Summary of Investigations 1976: Saskatchewan Dept. Min. Res., p. 73-77.

————, and Dunn, C. E., 1976, Geology and geochemistry of the eastern margin of the Athabasca Basin: in C. E. Dunn, ed., Uranium in Saskatchewan: Saskatchewan Geol. Soc., Spec. Pub. 3, p. 297-322.

Rankama, K., 1963, The Geologic Systems: The Precambrian, v. 1: Interscience Publishers — John Wiley and Sons Inc., N.Y., 279p.

————, 1965, The Geologic Systems: The Precambrian, v. 2: Interscience Publishers — John Wiley and Sons Inc., N.Y., 454p.

————, 1967, The Geologic Systems: The Precambrian, v. 3: Interscience Publishers — John Wiley and Sons Inc., N.Y., 325p.

————, 1970, The Geologic Systems: The Precambrian, v. 4: Interscience Publishers — John Wiley and Sons Inc., N.Y., 288p.

Reich, P., 1938, An analysis of cross-lamination in the Coconino Sandstone: J. Geol., v. 46, p. 905-932.

Reid, J. C., 1974a, Hazel Formation, west Texas; a Precambrian alluvial fan system (Abstract): Am. Assoc. Petrol. Geol., Soc. Econ. Paleont. Mineral., Ann. Mtg. Abstracts, v. 1, p. 74.

————, 1974b, Hazel Formation, Culderson and Hudspeth counties, Texas: Unpub. M.Sc. thesis, Univ. of Texas at Austin, Texas.

Reineck, H.-E., and Wunderlich, F., 1968, Classification and origin of flaser and lenticular bedding: Sedimentology, v. 11, p. 99-104.

Roberts, M. T., 1974, Stratigraphy and depositional environments of the Crystal Spring Formation, southern Death Valley region, California: Geol. Soc. Am., Cordilleran Sect., 70th Annual Meeting, March 29-31st, 1974, Las Vegas, Nevada: Field trip No. 1, Guidebook: Death Valley Region, California and Nevada, p. 49-57.

Roscoe, S. M., 1969, Huronian rocks and uraniferous conglomerates: Geol. Surv. Canada, Paper 68-40, 205p.

Schumm, S. A., 1968, Speculations concerning paleohydrologic controls of terrestrial sedimentation: Geol. Soc. Am. Bull., v. 79, p. 1573-1588.

Schwab, F. L., 1974, Mechum River Formation: Late Precambrian (?) alluvium in the Blue Ridge Province of Virginia: J. Sediment. Petrol., v. 44, p. 862-871.

———, 1976, Depositional environments, provenance, and tectonic framework: upper part of the Late Precambrian Mount Rogers Formation, Blue Ridge Province, southwestern Virginia: J. Sediment. Petrol., v. 46, p. 3-13.

Sedimentation Seminar, 1966, Cross-bedding in the Salem Limestone of central Indiana: Sedimentology, v. 6, p. 95-114.

Selley, R. C., 1965, Diagnostic characters of fluvial sediments of the Torridonian formation (Precambrian) of northwest Scotland: J. Sediment. Petrol., v. 35, p. 366-380.

———, 1967, Paleocurrents and sediment transport in nearshore sediments of the Sirte Basin, Libya: J. Geol., v. 75, p. 215-223.

———, 1968, A classification of paleocurrent models: J. Geol., v. 76, p. 99-110.

Sestini, G., 1973, Sedimentology of a paleoplacer: the gold-bearing Tarkwaian of Ghana: *in* G. C. Amstuts and A. J. Bernard, *eds.*, Ores in Sediments: Springer-Verlag, N.Y., p. 275-305.

Siemers, C. T., 1976, Sedimentology of the Rocktown Channel Sandstone, upper part of the Dakota Formation (Cretaceous), central Kansas: J. Sediment. Petrol., v. 46, p. 97-123.

Simonen, A., and Kouvo, O., 1955, Sandstones in Finland: Finland Commission Geologique, Geologien Toimikunta, Bull. 168, p. 57-87.

Smith, D. G., 1976, Effect of vegetation on lateral migration of anastomosed channels of a glacier meltwater river: Geol. Soc. Am. Bull., v. 87, p. 857-860.

Smith, N. D., 1970, The braided stream depositional environment: a comparison of the Platte River with some Silurian clastic rocks, north-central Appalachians: Geol. Soc. Am. Bull., v. 81, p. 2993-3014.

———, 1972a, Flume experiments on the durability of mud clasts: J. Sediment. Petrol., v. 42, p. 378-383.

———, 1972b, Some sedimentological aspects of planar cross-stratification in a sandy braided river: J. Sediment. Petrol., v. 42, p. 624-634.

Sochava, A. V., Savel'yev, A. A., and Shuleshko, I. K., 1975, Caliche in the middle Proterozoic sequence of central Karelia: Doklady Akad. Nauk. SSSR., v. 223, p. 173-176.

Spencer, A. M., 1971, Late Precambrian glaciation in Scotland: Geol. Soc. London, Mem. 6, 100p.

Steel, R. J., 1974, New Red Sandstone floodplain and piedmont sedimentation in the Hebridean Province, Scotland: J. Sediment. Petrol., v. 44, p. 336-357.

Steinmetz, R., 1975, Cross-bed variability in a single sand body: *in* E. H. T. Whitten, *ed.*, Quantitative studies in the geological sciences: Geol. Soc. Am., Mem. 142, p. 89-102.

Sweet, I. P., Pontifex, I. R., and Morgan, C. M., 1974, The geology of the 1:250,000 sheet area, Northern Territory (excluding Bonaparte Gulf Basin): Bur. Min. Res., Geology and Geophysics, Australia, Report 161, 98p.

Sundborg, A., 1956, The river Klaralven, a study of fluvial process: Geogr. Ann., Ser. A., v. 38, p. 125-316.

Tanner, W. F., 1955, Paleogeographic reconstruction from cross bedding studies: Am. Assoc. Petrol. Geol. Bull., v. 39, p. 2471-2483.

Thomson, B. P., 1976, Precambrian geology and tectonics of the Stuart Shelf and Torrens Hinge zone: *in* B. P. Thomson, B. Daily, R. P. Coats and B. G. Forbes, Late Precambrian and Cambrian geology of the Adelaide 'Geosyncline' and Stuart Shelf, South Australia: 25th Internat. Geol. Congr., Sydney, Australia, Excursion Guide No. 33A, p. 1-11.

Tremblay, L. P., 1972, Geology of the Beaverlodge mining area, Saskatchewan: Geol. Surv. Canada, Mem. 367, 265p.

Trettin, H. P., and Hills, L. V., 1966, Lower Triassic tar sands of northwestern Melville Island, Arctic Archipelago: Geol. Surv. Canada, Paper 66-34, 122p.

Van De Graaff, F. R., 1972, Fluvial-deltaic facies of the Castlegate Sandstone (Cretaceous), east-central Utah: J. Sediment. Petrol., v. 42, p. 558-571.

Varney, P. J., 1976, Depositional environments of the Mineral Fork Formation (Precambrian), Wasatch Mountains, Utah: in J. G. Hill, ed., Geology of the Cordilleran Hingeline: Rocky Mountain Assoc. Geol., 1976 symposium, p. 91-102.

Visser, J. N. J., 1975, The Table Mountain Group: a study in the depositon of quartz arenites on a stable shelf: Geol. Soc. South Africa, Trans., v. 77, p. 229-237.

Vos, R., 1975, An alluvial plain and lacustrine model for the Precambrian Witwatersrand deposits of South Africa: J. Sediment. Petrol., v. 45, p. 480-493.

Wagner, J. R., 1977, A comparison of environmental facies models for tidal and fluvial depositional systems (Abstract): Geol. Soc. Am., Abstracts with Programs, v. 9, p. 193.

Wallace, C. A., 1972, A basin analysis of the Upper Precambrian Uinta Mountain Group, Utah: Unpub. Ph.D. thesis, Univ. California, Santa Barbara, 412p.

————, and Crittenden, M. D., Jr., 1969, The stratigraphy, depositional environment and correlation of the Precambrian Uinta Mountain Group, Western Uinta Mountains, Utah: in J. B. Lindsay, ed., Geologic guidebook of the Uinta Mountains: Intermountain Association of Geologists, 16th. Ann. field conf., p. 127-142.

Walter, M. R., 1976, Stromatolites: Developments in Sedimentology 20, Elsevier, Amsterdam, 790p.

Watters, B. R., 1974, Stratigraphy, igneous petrology and evolution of the Sinclair Group in South West Africa: University of Cape Town, Dept. of Geology, Chamber of Mines, Precambrian Research Unit, Bull. 16, 235p.

Wells, A. T., Forman, D. J., Ranford, L. C., and Cook, P. J., 1970, Geology of the Amadeus Basin, Central Australia: Bureau Min. Res., Geology and Geophysics, Australia, Bull. 100, 222p.

Wermund, E. G., 1965, Cross-bedding in the Meridian sand: Sedimentology, v. 5, p. 69-79.

Wertz, W. E., 1977, Cross-bedding patterns and petrography of the Stirling Quartzite, Death Valley Region, California and Nevada (Abstract): Geol. Soc. Am., Abstracts with Programs, v. 9, No. 4, p. 526.

Whisonant, R. C., 1977, Lower Silurian Tuscarora (Clinch) dispersal patterns in western Virginia: Geol. Soc. Am. Bull., v. 88, p. 215-220.

Williams, G. E., 1966, Planar cross-stratification formed by the lateral migration of shallow streams: J. Sediment. Petrol., v. 36, p. 742-746.

————, 1968, Torridonian weathering, and its bearing on Torridonian paleoclimate and source: Scott. J. Geol., v. 4, p. 164-184.

Winston, D., 1973, The Precambrian Missoula Group of Montana as a braided stream and sea-margin sequence: Belt Symposium, 1, Univ. of Idaho, Moscow, ID., p. 208-220.

————, 1977, Alluvial fan, shallow water, and sub-wave base deposits of the Belt Supergroup near Missoula, Montana: in D. Winston and I. Lange, Field Guide No. 5, Rocky Mountain Section, 30th Ann. Meeting, p. 1-12.

Wolman, M. G., and Brush, L. M., Jr., 1961, Factors controling the size and shape of stream channels in coarse non-cohesive sands: in Physiographic and Hydraulic Studies on Rivers, 1956-1961: U.S. Geol. Survey, Prof. Paper 282-G, p. 183-210.

Wood, J., 1973, Stratigraphy and depositional environments of upper Huronian rocks of the Rawhide Lake - Flack Lake area, Ontario: in G. M. Young, ed., Huronian Stratigraphy and Sedimentation; Geol. Assoc. Canada, Spec. Paper 12, p. 73-95.

Yeo, G. M., 1976, Sedimentology and geochemistry of the Wilson Island Group, Northwest Territories (Abstract): Geol. Assoc. Canada, Program with Abstracts, v. 1 (Edmonton), p. 83.

Young, G. M., 1973, Origin of carbonate-rich Early Proterozoic Espanola Formation, Ontario, Canada: Geol. Soc. Am. Bull., v. 84, p. 136-160.

————, 1974, Stratigraphy, paleocurrents and stromatolites of Hadrynian (Upper Precambrian) rocks of Victoria Island, Arctic Archipelago, Canada: Precambrian Res., v. 1, p. 13-41.

———, 1977, Stratigraphic correlation of upper Proterozoic rocks of northwestern Canada: Can. J. Earth Sci., v. 14, p. 1771-1787.

———, and Jefferson, C. W., 1975, Late Precambrian shallow water deposits, Banks and Victoria Islands, Arctic Archipelago: Can. J. Earth Sci., v. 12, p. 1734-1748.

———, and Long, D. G. F., 1977, A tide-influenced delta complex in the upper Proterozoic Shaler Group, Victoria Island, Canada: Can. J. Earth Sci., v. 14, p. 2246-2261.

———, ———, and McLennan, S. M., 1977, Deltaic deposits in the upper Pecors, Espanola and Gowganda Formations (Huronian) (Abstract): 23rd Annual Institute on Lake Superior Geology, Thunder Bay, Abstracts with Proceedings, p. 46.

FLUVIAL SYSTEMS OF THE PRECAMBRIAN BELT SUPERGROUP, MONTANA AND IDAHO, U.S.A.

Don Winston[1]

Abstract

Fluvial quartzite and argillite wedges hundreds of meters thick extend from the fault-bounded south side of the Belt Basin out across the basin for hundreds of kilometers, forming parts of the late Precambrian Belt Supergroup of northwestern Montana, northern Idaho, northeastern Washington, and the Purcell of southeastern British Columbia. From investigations of the Bonner and Mount Shields Formations (upper Belt Missoula Group), five rock types can be identified that correspond to five major depositional tracts on alluvial fan, distal flat, and sea margin surfaces. The conglomeratic rock type contains coarse pebble lenses in matrix supported sand, and accumulated on the proximal parts of the fan surface close to the south bounding faults. Here braided streams deposited gravel in low, longitudinal channel bars and deposited poorly sorted muddy sand between the channels. The conglomeratic rock type passes down the fan to the coarse, crossbedded rock type, containing well sorted, coarse grained, feldspathic quartzose sand in 10 to 40 cm tabular unimodal crossbeds. This rock type represents mostly linguoid and transverse bars that migrated down broad, braided stream channels, working from side to side across the fan surface. Farther down fan, accumulated the fine, horizontally laminated rock type consisting of even beds 10 to 30 cm thick, within which are vertically graded sedimentation units of fine horizontally laminated sand that commonly passes up to rippled crossbedded sand capped by red argillite. This rock type accumulated from traction transported fine sand carried by shallow sheet wash floods across the distal parts of the fan. Suspended wash load continued out onto flats where flow slowed and ponded. Graded couplets of fine sand and silt to clay accumulated from the nearly standing water which repeatedly drained, forming the fine sand to clay couplets of the red argillite rock type. Seaward from the red argillite rock type, where the surfaces were not repeatedly drained, the sedimentary couplets are finer grained, thinner, and the clay was diagenetically altered to chlorite, forming the green argillite rock type. The model derived from the Mount Shields and Bonner appears to apply with minor modifications to other fluvial units of the Belt.

Introduction

The Belt Supergroup comprises an immense body of sedimentary rocks up to 16 km thick that extend from south-central Montana west into the states of Idaho and Washington, and north into British Columbia, where the Belt becomes the Purcell (Fig. 1). The Belt sediments accumulated in a basin that was filled for more than 600 million years (-1450 m.y. to -850 to -600 m.y.?) in response to downfaulting along its south side, where it is bounded by the east-trending Willow Creek fault system (McMannis, 1963). Rivers heading in the igneous and metamorphic terranes south of the Willow Creek fault, and possibly south of other complementary faults of uplifted terranes, flowed out across the Belt Basin forming alluvial fans that deposited sandstone wedges 100's to 1000's of meters thick and up to 300 kilometers across. At least five fluvial sandstone (quartzite) wedges are now recognized. They are: 1) the Revett Formation and associated facies; 2) the red sandstone member of the Mount Shields Formation (Mudge, 1972a, b); 3) the Bonner Formation; 4) a crossbedded quartzite unit high in the McNamara Formation (Bleiwas, 1977); and 5) the Pilcher and lower Flathead. Terrigenous mud settled in slow moving and standing waters distal to the alluvial fans.

The fluvial facies model described here, including both sand and mud components, was developed from stratigraphic and sedimentologic study of the Mount Shields and Bonner Formations (Fig. 2) of the middle part of the Missoula Group. Additional study of the upper McNamara (Bleiwas, 1977) and of the Revett (White *et al.*, 1977; White and Winston, 1977) help corroborate the model, with some modifications. Bowden (1977),

---

[1]University of Montana, Missoula, Montana U.S.A. 59812.

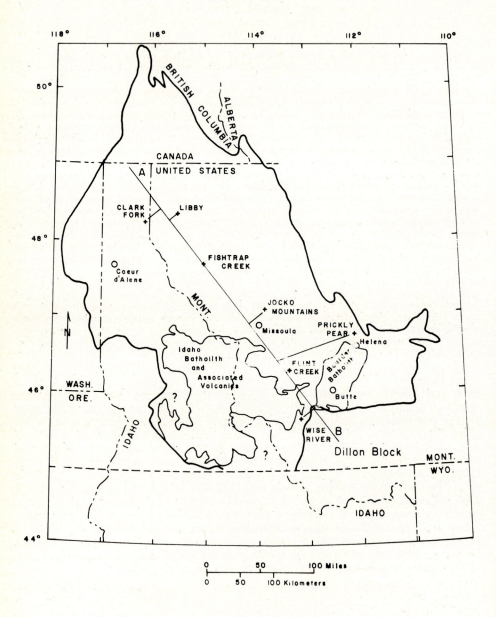

Fig. 1. Index map outlining the Belt Basin and locating measured sections and line of cross section. Modified from Harrison, 1972.

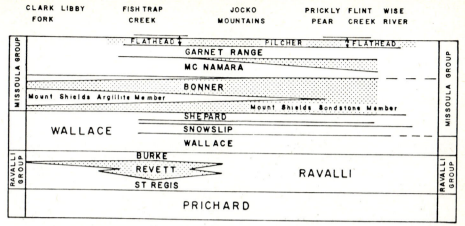

Fig. 2. Correlation diagram of Belt Supergroup formations along line A-B of Figure 1. Stippled pattern shows fluvial sandstone wedges. Thicknesses not to scale.

working completely independently, has developed a very similar interpretation of the Revett. Thus, the model developed for the middle Missoula Group appears to apply well to other parts of the Belt Supergroup.

Figure 3 schematically diagrams facies changes within the Mount Shields and Bonner Formations along line AB of Fig. 1. It shows that the fluvially related rocks can be subdivided into five rock types and that these rock types pass as facies one into another down the depositional slope. I interpret the conglomeratic rock type, the coarse, crossbedded rock type, and fine, horizontally laminated rock type, to represent facies of a huge alluvial fan complex that discharged suspended load onto distal flats represented by the red argillite and the green argillite rock types (Fig. 10).

Students at the University of Montana and I, (Winston, 1972, 1973a, 1973b, 1977; Winston and Jacob, 1977) have developed this model by: 1) identifying the basic rock types and describing sequences of sedimentary structures in them; 2) deducing the depositional processes based on the sedimentary structures of each rock type; and 3) integrating these depositional processes into an environmental model based on the regional stratigraphic framework (for an authoritative review of Belt stratigraphy, *see* Harrison, 1972). Notably absent from this method is direct comparison with recent sedimentary models. Analogous land forms on the scale of the Belt, if they exist in the recent, certainly have not been described. The advent of land vegetation, as pointed out by Schumm (1968) and Cotter (this volume) has drastically altered the earth's surface so that it is difficult to find words to express the geomorphology of the Precambrian earth.

Important to this analysis is the recognition of sequences of cycles of at least two scales, 1) the meter to millimeter scale representing single depositional events (sedimentation units), and 2) the 10's of meters scale, rarely identified, representing shifts of depositional environments. The thickness of the small scale cycles, or depositional events, decreases down the depositional slope from more than a meter in the conglomeratic proximal limit of the model, down to less than a centimeter far out in the silt and clay distal limit of the model (Fig. 9).

ROCK TYPES

Previous workers on the Belt have relied heavily on color, which is in part diagenetic, and on sedimentary texture and mineralogy in defining mappable rock units. By focusing

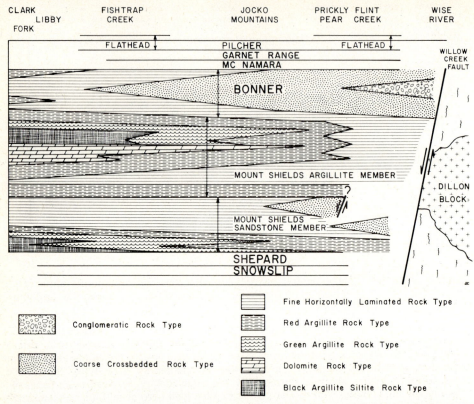

Fig. 3. Schematic cross sections of the Missoula Group along line A-B of Figure 1 showing fluvial facies changes down the depositional slope (right to left) within the Mount Shields and Bonner Formations. The generalized lateral sequence in a downstream direction is: conglomeratic rock type; coarse, crossbedded rock type; fine, horizontally laminated rock type; red argillite rock type; and green argillite rock type (dolomite and black argillite-siltite are not part of the fluvial sequence). Thicknesses are not to scale. Modified from Winston and Jacob, 1977.

on the types and associations of sedimentary structures along with the texture, mineralogy and color, we have defined rock types that are repeated through much of the Belt and form the major rock elements from which the model is constructed. Since the rock types are defined on the basis of texture, mineralogy, color, and sedimentary structures, they can be identified irrespective of their stratigraphic position. When analyzed within their stratigraphic framework, they commonly display lithofacies patterns which lead to the generalized facies tract interpretation given in Figure 10. Rock types commonly grade across the lithofacies boundaries, underscoring their genetic relationships. The principal rock types which belong to the Belt fluvial systems are: 1) the conglomeratic rock type; 2) the coarse, crossbedded rock type; 3) the fine, horizontally laminated rock type; 4) the red argillite rock type; and 5) the green argillite rock type.

Conglomeratic rock type

Description

The conglomeratic rock type consists of pebbles up to 10 cm in diameter, mostly of quartzite, quartz and microcline, supported in a matrix of coarse sand, in beds 0.5 to 1.5 m thick (Fig. 4). The long axes of the pebbles define the horizontal plane, but imbrication and grain support are not obvious. The quartzite pebbles are well rounded in contrast to microcline grains, which are characteristically angular. Some pebbles are quartz-feldspar mosaics.

Stratigraphic Setting

The coarsest and thickest units of the conglomeratic rock type crop out on thrust sheets of the Bonner Formation near the Big Hole River, close to the westward projection of the Willow Creek fault. At Wise River (Figs. 1, 3, 4), the quartzite pebbles reach 10 cm in diameter and form apparently lenticular intervals (Fraser and Waldrop, 1972) up to 15 m thick. Between the lenticular conglomeratic intervals are much thicker sequences of the coarse, crossbedded rock type up to 100 meters thick. Here the coarse crossbedded rock type is poorly sorted and crossbeds range from 30 cm thick down to large scale ripple crossbeds. Within the conglomerate and sand intervals are occasional, vaguely defined, fining upward sequences up to 5 m thick that grade upward from horizontally stratified, matrix supported pebbles to coarse crossbedded sand, to muddy sand with large ripple crossbeds.

To the north, at Flint Creek, the conglomeratic rock type of the Bonner forms an interval less than 10 m thick of well rounded quartzite pebbles and angular microcline grains up to one centimeter in diameter. The pebbles are supported in a coarse sand matrix and lie in horizontal concentrations and on foresets of low angle crossbeds. Above and below the conglomeratic interval are beds of the coarse, crossbedded rock type that are better sorted and more uniform than those at Wise River.

Fig. 4. Conglomeratic rock type from the Bonner Formation at Wise River. Quartzite pebbles 5 to 10 cm long are supported in a matrix of coarse, feldspathic sand. Jacobs staff scaled in feet.

Fine pebbles, less than one centimeter in diameter, occur in the upper part of the Mount Shields sandstone member high in the Prickly Pear section close to the southern, probably fault-bounded, limit of that member in the Helena area. Here also the pebbles are in coarse, crossbedded sands. However, they are too scattered to be considered a true conglomerate, and the rocks are placed in the coarse, crossbedded rock type.

Interpretation

The dominance of plane beds and low angle crossbeds, the matrix support, and the continuity of bedding across local outcrops near Wise River indicate that conglomeratic rock type near Wise River was probably deposited in the upper flow regime in the form of longitudinal bars on broad, probably braided, channel bottoms. The abrupt changes in grain size, the lenticular shape of the conglomeratic units, and the vaguely defined fining upward sequences suggests that both channels and overbank deposits shifted laterally and were locally preserved in the record. Rivers flowing from entrenched crystalline bedrock valleys may have formed poorly sorted immature channel and flanking overbank deposits at the heads of the alluvial fan sequences. However, farther down the fan, braided stream systems swept from side to side across the alluvial surfaces, reworking the sediments to leave a more mature, uniform record. This interpretation is consistent with the outcrops at Flint Creek, where the pebbles are smaller, and the sediments are more mature.

The textural inversion represented by the well rounded quartzite pebbles and the angular microcline grains of the same small size probably reflects a range of sources. Perhaps the rounded quartzite pebbles had longer transport histories before becoming incorporated into the Bonner.

Coarse, crossbedded rock type

Description

The coarse, crossbedded rock type is mostly well sorted, pink or reddish, quartzose, feldspathic (microcline and orthoclase) sandstone (now quartzite), crossbedded on a scale of 10 to 40 cm (Fig. 5). The characteristic pink or carmine color reflects minute hematite particles on the sand grains. Tabular, tangential crossbeds are at least four times as common as trough crossbeds. Inspection of single outcrops indicates that crossbedding directions are locally unimodal, so that the overall bedding appearance depends upon outcrop direction. Northwest trending outcrops that parallel flow direction have rather continuous, tabular, apparently unimodal cross strata that form continuous beds 20 to 100 cm thick that extend across outcrops 70 m wide. Outcrops transverse to flow direction display lenses 5 to 7 m long that are up to a meter thick, of apparent low angle crossbeds with both erosive and nonerosive lower contracts. Homogenous red argillite commonly forms angular mud clasts up to 5 cm in diameter on the toes and in foresets of crossbeds, and rarely forms lenses a meter or more thick. Fining upward sequences one to four meters thick, in which the scale of the sedimentary structures decreases along with the grain size, occur rarely in the coarse, crossbedded rock type. In the lower part of the Bonner in the Jocko Mountains and in the Mount Shields sandstone of the Prickly Pear section are cycles with scoured bases overlain by coarse crossbeds. These pass up to thinner tabular crossbeds, which in turn lead up to horizontally laminated muddy sand and climbing ripples, capped by red argillite couplets. These cycles form tabular sheets across outcrops, and lack apparent resedimentation surfaces and epsilon cross strata.

Stratigraphic Setting

The coarse, crossbedded rock type is most widespread in the Bonner Formation and forms about 200 m of the formation in the type area. To the south it includes beds of coarse, planar and low angle cross-stratified sandstone, that pass into lenses of the conglomeratic rock type. To the north, down the alluvial fan, it becomes interstratified with the fine grained, horizontally laminated rock type.

Fig. 5. Coarse, crossbedded rock type showing unimodal tabular and trough crossbeds in the Bonner at the type locality. Scale in centimeters and decimeters.

The rock type also occurs in the lower part of the Mount Shields sandstone in the Flint Creek section and in the uppermost part of the Mount Shields sandstone at the Prickly Pear section, where it is indistinguishable in hand specimen from the Bonner. At Prickly Pear, close to a source, it contains a few beds of quartz pebble sandstone, but no true conglomerates. The coarse crossbeds of the Mount Shields also pass downfan to the fine, horizontally laminated rock type that dominates the Mount Shields sandstone member.

Interpretation

The unimodal tabular crossbed upon crossbed-aspect of the coarse, crossbedded rock type most probably records accumulation from successive trains of linguoid or transverse bars and subordinate dunes (megaripples) on the beds of broad, flat, shallow braided streams (Platte River type of Miall, 1977, and this volume). The red argillite probably represents channel-fill mud that was only rarely preserved intact. Fining and thinning upward sequences or cycles a meter or more thick, probably record decreasing flow velocities of single floods, in which sediments accumulating in the broad channels, formed braided channel patterns.

Fine, horizontally laminated rock type

Description

The fine, horizontally laminated rock type consists of clean and well sorted, to poorly sorted and muddy, fine to medium grained, reddish or salmon colored, feldspathic sandstone (quartzite), that forms beds 10 to 30 cm thick with mostly fine horizontal internal laminations and common climbing ripple crossbeds in their upper parts (Fig. 6). The sand beds are commonly capped by a centimeter or two of red or purple mud. Magnetite and ilmenite grains mark many of the darker horizontal laminations. The rock type forms even beds that continue uniformly across outcrops up to 100 m long. The

Fig. 6. Fine, horizontally laminated rock type showing interbedded horizontal laminations and climbing ripple crossbeds in the Mount Shields sandstone member near Clearwater Junction, Montana. Pen for scale.

laminae resemble foreshore beach laminae, but nowhere do they lie at low angles to other bedding sets. By contrast, the beaches of the Belt formed beds and lenses less than a meter thick of coarse quartz and oolitic sandstone (quartzite) and lie in the red and green argillite rock types.

Stratigraphic Setting

The fine, horizontally laminated rock type comprises most of the 100 to 400 m thick Mount Shields sandstone member, that forms resistant outcrops of well sorted, evenly bedded, fine to medium grained clean quartzose feldspathic sandstone (quartzite). The rock type becomes interbedded with the coarse, crossbedded rock type locally in the southern, sourceward parts of the Mount Shields sandstone member. Elsewhere in the Mount Shields sandstone, calcareous nodules averaging more than 10 cm long and about one cm thick parallel the bedding and weather recessively. Calcite crystals in the nodules have both filled pore spaces and have replaced quartz grains. These nodules may represent incipient caliche.

The fine, horizontally laminated rock type in the Bonner contrasts with that in the Mount Shields member in being more poorly sorted and muddier. Where the coarse, crossbedded rock type of the Bonner passes northward to the fine, horizontally laminated rock type, the outcropping beds thin to 10 to 25 cm, and consist of graded beds of fine, horizontally laminated muddy sand capped by a few centimeters of sandy or silty red mud (argillite). Vertical sequences of horizontally laminated muddy sand that grade up into a single layer of ripples capped by horizontally laminated red mud are more characteristic of the Bonner, whereas clean, horizontally laminated sand passing up to 5 to 7 cm of climbing ripples is more characteristic of the Mount Shields. The fine, horizontally laminated rock type of the Bonner, like that of the Mount Shields, passes distally to the red argillite rock type.

Interpretation

The fine and medium sand of the horizontally laminated rock type ranges in grain size from 0.15 to 0.5 mm, in part too coarse to represent lower plane bed deposition. Instead, this rock type probably records deposition from currents that were flowing only fast enough to carry fine and medium sand, but nevertheless stayed in the upper flow regime, until velocity slowed enough to form ripples. Dunes and bars were probably not developed because the flow was too shallow to form them. The continuous, horizontally laminated beds probably represent broad, sheet flow, less than a meter deep, that deposited the fine to medium sand as traction load deposits. The wash load of very fine sand, silt and clay passed through and settled out on the distal flats from slowly moving or standing water masses.

The 20 to 30 cm vertical sequences characteristic of the Mount Shields comprise fine, horizontally laminated sand, passing up to climbing ripples, followed by thin, locally oscillation ripple-marked argillite; they represent decreasing flow velocities of single floods which left local ponds as a final event. Each flood deposited less traction load sand and more suspended silt and clay as it flowed down slope, so that distally each sequence passes into the red argillite rock type.

Thick sequences, more than 5 meters thick, of fine, horizontally laminated sand overlain by red argillite couplets probably represent shifts in the distributary lobes of the sandy sheetwash surfaces in the fan delta region.

The muddier, more poorly sorted, vertical sand to mud sequences more characteristic of the Bonner represent environments similar to those of the Mount Shields, in which the floods may have been more flashy and sudden.

Red argillite rock type

Description

The red argillite rock type (Fig. 7) consists of sedimentary couplets one to 5 cm thick of light colored fine sand and silt that grades up to red and purple clay (argillite). The lower fine sand and silt layers of the couplets are horizontally laminated or ripple cross laminated, and commonly contain lag concentrations of red mud chips. Ripples and plane beds of the lower parts of the couplets mostly grade up to the overlying mud and silt layers of the upper parts. Straight-crested oscillation ripples commonly cover the surfaces of the red mud layers and are, in turn, mostly cut by desiccation cracks. The thicker couplets are commonly dominated by sand, while most of the thinner couplets are dominated by mud. The red and purple pigment of argillite reflects minute hematite particles clinging to the terrigenous grains.

Stratigraphic Setting

The red argillite rock type becomes interbedded with the fine, horizontally laminated rock type in the distal, northwesternmost outcrops of the major sand wedges of the Mount Shields (Fishtrap Creek, Clark Fork) and Bonner (Clark Fork, Libby). Where the fine, horizontally laminated rock type grades laterally to the red argillite rock type, the red argillite rock type contains comparatively thick couplets up to 10 cm thick dominated by the lower sandy part. Where the red argillite rock type lies far from the fine, horizontally laminated rock type, the couplets are only one to 3 cm thick and are mostly silt and clay (argillite). Halite casts commonly dot the couplet surfaces in the Mount Shields argillite member. The red argillite rock type becomes interspersed with patches of green argillite that tend to coalesce in a "seaward" direction into continuous units of green argillite.

Fig. 7. Red argillite rock type showing repeated graded couplets of fine sand (white) up to red mudstone (dark). Pen for scale.

Interpretation

The red argillite rock type represents the wash load that was transported mostly in suspension, across the alluvial fan system out into standing or slow-moving water, and was deposited in small graded couplets. The oscillation ripple marks capping so many of the sedimentary couplets are the most convincing evidences of standing water. The abundant mud cracks, locally abundant salt casts, oxidized state of the iron, and dried mud chips indicate that the shallow water repeatedly drained from the flat mud surfaces, leaving them to desiccate. The evenness of the layers in individual outcrop and the extent of individual stratigraphic units indicates that the flat, mud surfaces must indeed have been extensive, stretching for hundreds of kilometers. The processes of flooding by turbid water, rippling by waves, and desiccation are common to both flood plains and to marginal marine mud flats. The absence in the Belt record of thick barrier beach sequences and the absence in the red argillite of epsilon crossbedding of migrating tidal channels indicate they are not mud flats generated by high marine tides. Perhaps the function of flood water storage, performed by flood plains flanking modern meandering rivers, was performed in the Belt by enormous mudflats distal to the alluvial sand fans. These distal flats, that may at times have stored turbid flood waters, graded imperceptably to the margin of the "Belt Sea." Winds probably alternately blew water from the "Belt Sea" across these same flats, leaving as a mark the salt casts. During times of fluvial progradation into the "Belt Sea", the demarcation between fluvial and sea margin environment was indeed vague. The "beach" may have been nothing more than discontinuous shoals, where breaking waves formed distinctive crescent-shaped assymetric ripples with their crescent points lined up the shoal in long rows. When the "Belt Sea" transgressed, destructive beaches formed, leaving sharply scoured lower boundaries and locally forming stromatolitic biostromes and coarse oolitic sand beds such as the high stand beach at the base of the Mount Shields argillite. The red hematite reflects a high original limonite content that was not reduced, and became hematite upon losing water.

Fig. 8. Green argillite rock type showing evenly laminated silt (dark) to mud (light) couplets cut by compacted syneresis cracks. Bonner-McNamara boundary, Blackfoot River. Scale marked in centimeters.

Green argillite rock type

Description

The green argillite rock type is the final fluvial-dominated rock type of the Belt facies tract, although much of it probably accumulated in "Belt Sea" margin environments. It is composed of pale green, fine sand and silt to clay couplets (Fig. 8).

Fine sand to clay couplets mostly range from 2 to 7 cm thick, while silt to clay couplets average a centimeter or less thick. Some sand and silt layers of the lower parts of the couplets tend to be sharply separated from the upper clay layers, and are not obviously graded. Oscillation ripples cover bedding planes and desiccation polygons, though common, are not so numerous as in the red argillite rock type. In addition, sinuous, plastically compacted syneresis cracks commonly cut the bedding. Transported flat green mud chips are locally incorporated in sands above scoured surfaces, but are not so abundant as in the red argillite rock type. Inverted cones and cracks up to 10 cm high and 2 or 3 cm across of dislodged argillite fragments probably represent water escape structures. The green color reflects chlorite flakes clinging to the larger terrigenous grains and filling the interstitial spaces in the absence of hematite. Small sand and argillite-filled channels up to 10 cm deep and 30 cm wide sporadically cut the green argillite beds. Within the green argillite sequences are occasional coarse, well rounded quartz sand beds less than 50 cm thick that are commonly cross stratified at high to low angles. Some beds are oolitic and some closely underlie stromatolite beds.

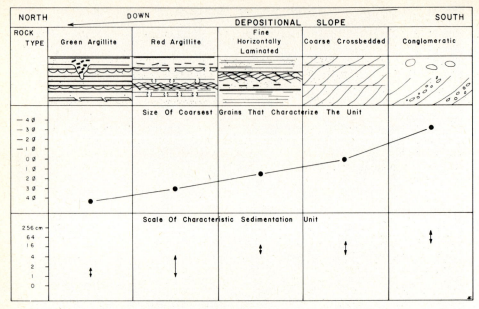

Fig. 9.   Diagram of lithic trends down the depositional slope.

Stratigraphic Setting

The green argillite occurs in patches within the dominantly red argillite sequences. Although the green to red boundaries commonly follow bedding planes, in places the color boundaries cut across bedding. Cross cutting relationships indicate that green beds have replaced potentially red beds. The green argillite intervals tend to become thicker at the expense of the red in a distal direction until some red sequences pass seaward to dominantly green sequences across distances tens of kilometers wide.

Interpretation

The patchy distribution of green argillite in red argillite sequences, and the mutual exclusion of hematite and chlorite indicates that the red to green transformation most probably records early diagenetic reduction of ferric iron, probably from limonite, to mobile ferrous iron, which then combined with smectitic clays to form chlorite minerals. McBride (1974) has recently proposed a similar interpretation of red and green colors in the Cretaceous Defunta Formation. Corroborating this interpretation is the evidence for more extended periods of water cover over the green argillite. Green argillite, in contrast to red argillite, contains fewer desiccation cracks, fewer dried mud chips, more subaqueous syneresis cracks and water escape structures. The sharper boundary between the lower silty parts of the sedimentary couplets and the upper clayey parts may reflect more thorough and repeated wave reworking of the lower parts of the couplets before suspended clay accumulated, forming the upper parts of the couplets. The seaward stratigraphic position of the green argillite compared to the red argillite is also consistent with the reduction in subaqueous sediments. Thus, fine grained, thinly laminated green argillite represents the most distal resting place of fluvially transported fine sediments. This environment probably represented the seaward continuation of the red argillite flats out into the "Belt Sea". Wind driven masses of water may have cut the small channels of the green argillite rock type, and the coarse quartz and oolitic sand beds may represent low stand beaches. The coarse quartz grains in these beach deposits migrated parallel to the low stand strand, not across the fine grained flats.

SEDIMENTOLOGIC SYNTHESIS

When the rock types of the Bonner and the Mount Shields are placed in their true stratigraphic framework (Fig. 3), consistent trends in both lithology and sedimentologic processes emerge from rock type to rock type (Fig. 9). Together they lead to an internally consistent regional interpretation (Fig. 10). This model was constructed from different levels of the cross section (Fig. 3) because not even the 330 kilometers from Wise River to Libby is great enough to illustrate the entire alluvial sequence at a single stratigraphic level. Lateral facies changes from the conglomeratic rock type to the coarse, cross bedded rock type, to the fine, horizontally laminated rock type, to the red argillite rock type are illustrated in the Bonner Formation and partly in the Mount Shields sandstone member. The fine, horizontally laminated rock type to red argillite to green argillite facies pattern is illustrated in the Mount Shields argillite member and in many other parts of the Missoula Group.

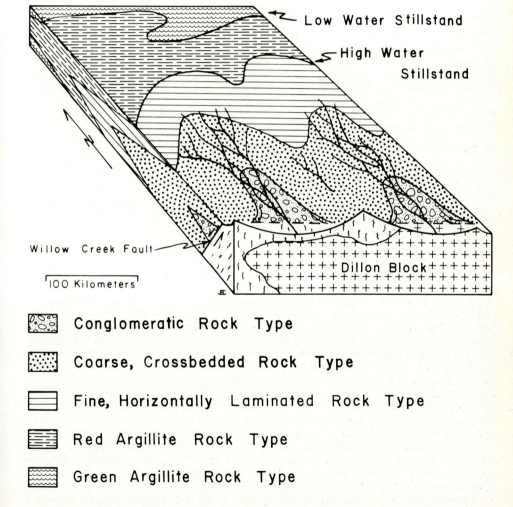

Fig. 10. Schematic block diagram of the fluvial facies model showing braided streams flowing from the Dillon Block northward to the "Belt Sea".

Thrusting

Before discussing the regional trends, it is necessary to point out that all of the measured stratigraphic sections lie in Laramide thrust plates that have translated the sections eastward or northeastward an unknown distance. Mudge (1972b) has estimated a minimum of 40 km of displacement for the thrust plates of the Sun River area. Wallace, *et al.*, (1976) have estimated up to 60 km for the sheets sliding off the Idaho batholith. The general figure of 40 km generally eastward, seems to be a reasonable approximation for most of the thrusting. The cross section in Figure 3 stretching 300 km from Wise River on the south to Libby on the north cuts perpendicularly to both depositional strike and to thrust direction. Therefore, the scale of the facies patterns compared to the scale of thrust offset parallel to depositional strike, appears great enough to discount the effect of thrusting at this broad, first order scale of interpretation. The stratigraphic framework now developed should allow us to estimate the offsets of the major thrusts, so that we can further refine our sedimentologic interpretations.

Lithic trends

The lithic trends are illustrated in Figure 9. Size of the largest grains in the Bonner, for instance, decreases consistently from 10 cm long pebbles in the southernmost section at Wise River to fine sand, silt and clay at Libby. Sedimentation units progressively thin from the conglomeratic rock type (1.0 to 1.5 m), to the coarse, crossbedded rock type (10 to 40 cm), to the fine, horizontally laminated rock type (10 to 30 cm), to the red argillite rock type (1.0 to 5 cm) to the green argillite rock type (1.0 to 2 cm). Correlative with this is the decrease in scale of the other sedimentary structures. Forms of the sedimentary structures also change progressively from south to north, from matrix supported conglomerate with plane beds and low angle crossbeds (conglomeratic rock type), to large scale crossbeds (coarse, crossbedded rock type), to plane beds and climbing ripples (fine, horizontally laminated rock type), to graded couplets (red and green argillite rock types). The continuity of this progression in these several characteristics indicates that the rock types form in response to a single depositional regime that transported sediments from south to north. This is in contrast to modern river-tidal flat regimes in which the tidal flats receive large volumes of sediment from the sea.

Sedimentologic trends

The interpretation of a great alluvial fan system that discharged flood water out onto distal flats (Fig. 10) likewise emerges by progressively integrating the inferred depositional processes of each rock type. Rivers emerging from fixed valleys in the crystalline terrane south of the Belt basin, flowed out onto the alluvial surface forming braided stream channels with long, longitudinal sand and gravel bars (conglomeratic rock type). Overbank deposits between the channel systems were mostly muddy, coarse sand. The gravel and plane bedded coarse sand gave way down fan to broad, shallow, coarse sandy, braided streams dominated by linguoid or transverse bars (coarse, crossbedded rock type). These broad channel systems worked back and forth across the alluvial fan surfaces reworking the sands and eroding most of the mud plugs deposited in the cut-off channels. Bed upon bed of unimodally crossbedded coarse sand reflects the persistence of the migrating lunguoid bars. The apparent unimodal flow indicates the channels were quite straight for long periods of time. As the channels dispersed across the fan surface, and as the surface flow infiltrated the sands, flow depth, velocity and competency decreased, forming the fine, horizontally laminated rock type. The continuity of the beds across outcrops indicates that the flow was nearly sheetlike, and too shallow to form transverse bars and dunes. Therefore, flow stayed in the upper regime until it decreased to the point where ripples formed in the lower flow regime. As flow of each flood event decreased, sand accumulated first in plane beds, then occasionally in climbing ripples. As the sheet wash passed beyond the distal limits of the fan, flow decreased still further until the water nearly ponded, depositing vertically graded sequences 3 to 5 cm thick of horizontally laminated sand overlain by ripple crossbedded sand that passes in turn up into horizontally laminated silt and clay (red argillite rock type). Waves rippled the tops of many of these mud surfaces. Water finally drained from the muddy surfaces leaving them to dry and crack. The dry mud polygons were commonly picked up and incorporated as mud chips in the sands of the succeeding flood. Farther out on the flats, the competency of the flood water diminished still further so that the graded couplets became silt to limonitic clay sequences one or two centimeters thick. Water from the "Belt Sea" was occasionally blown across these muddy surfaces, and upon drying contributed halite casts. Eventually these flood-dominated flats blended with the sea margin. Silt and clay were carried into the region in suspension and settled as silt to clay couplets mostly one centimeter or less thick (green argillite rock type). Windy periods alternating with calm periods divided the couplets into more sharply defined lower silty and upper clayey parts. Desiccation cracks gave way to syneresis cracks and dewatering structures. Ferric iron in the limonitic muds was reduced in the dominantly subaqueous sediments to ferrous iron, which became incorporated in the chlorite, imparting the green color to the argillite.

Paleohydrology

While the paleohydrology of the Belt fluvial systems is still well beyond our reach, a few constraints can be deduced. The graded couplets in the horizontally laminated rock type, the red argillite rock type, and the green argillite rock type indicate that the flow was sporadic, perhaps flashy or seasonal. Other lines of evidence also point to seasonality. On the one hand, the climate was too humid to preserve recognizable sand dunes on the alluvial sand fans or form bedded evaporites on the sea margin flats. On the other hand, ephemeral dry periods did form desiccation cracks and halite casts. Perhaps the climate had alternately wet and dry periods such as that in monsoon regions.

Conclusions

Although parts of this model are familiar, the model in its totality is foreign. There appear to be no comparable facies tracts in the recent, although individual rock types do have modern analogues. For instance, the conglomeratic rock type with its plane beds and low angle crossbeds compares closely to deposits of longitudinal gravel bars of many modern braided rivers. The downstream transition of the conglomeratic to coarse, crossbedded rock type has its modern counterpart in the Platte River (Smith, 1970) where longitudinal gravel bars pass down stream to transverse linguoid bars forming the Platte type of braided stream (Miall, 1977; this volume). The depositional processes of the fine, horizontally laminated rock type may be analogous to those of Bijou Creek (McKee *et al.*, 1967; Miall, 1977; this volume), but its down fan position relative to the crossbedded rock type in the Belt is not recorded in recent examples except possibly in the Gum Hollow fan delta (McGowen, 1971). Perhaps floods crossing the distributary lobes of the Missoula Group fan deltas behaved like modern floods in tributary creek channels such as the Bijou.

The environments of the red and green argillite rock types appear to have no modern analogue nor is their function fully understood. Their graded sequences recording decreasing flow velocity suggests they operated as distal flood plains (Bowden, 1977). However, their local high stand beaches at the fine, horizontally laminated sand - red argillite boundary and their low stand beaches close to the red to green argillite boundary suggest the red and green argillites may have functioned at times as sea margin flats. If the "Belt Sea" were an inland sea similar in size to the Caspian Sea with shallow margins like those of Lake Chad, then flood or flood seasons may have submerged miles of flats by raising the inland sea level only a few meters.

Belt sedimentary sequences in a vertical sense also differ from those of modern braided rivers and most post-Silurian meandering fluvial systems. Whereas most of the Panerozoic fluvial systems produce distinctive vertical sequences of lithologies and sedimentary structures at a scale of 10 to 20 m, the vertical sections of the Belt are remarkably uniform for hundreds of meters. Individual facies likewise carry laterally for hundreds of kilometers. Thus the scale of the Belt both in a vertical and lateral sense is an order of magnitude larger than most Phanerozoic fluvial deposits. Perhaps this is due in part to the tendency of facies changes in Phanerozoic rivers to occur laterally from the channels out into the flood plains, whereas in the Belt, facies are more widespread laterally and change distally down the fan. Possibly for some reason, the location of individual facies stayed in the same position for longer periods of time through repeated floods.

This is not to say that uniformitarianism does not apply to the Belt. The processes that produced the structures can easily be explained in a modern context. On the other hand, environments have not remained uniform since the Precambrian and, as pointed out earlier, probably the advent of land vegetation has greatly changed the earth. Thus, with proper mix of sedimentologic interpretation rigorously deduced from the sedimentary structures, of carefully described stratigraphy, and of controlled imagination, we can interpret Precambrian fluvial systems even in the absence of complete modern analogues. To do so, we must apply the principles of uniformitarianism, not the mirror of uniformity.

ACKNOWLEDGMENTS

I wish to thank the many students who, through the years, have helped open my eyes to the Belt rocks. They have made many critical observations and interpretations. These include Mike O'Connor, Frank Hall, Bob Brenner, and Harold Illich in the early 1960's, Jim Calbeck and Al Gensamer later on, and more recently Don Bleiwas, Dave Godlewski and Steve Lemoine. Phil Jacob helped measure most of the Missoula Group sections, and described large parts of them. His contribution is especially appreciated. The stratigraphic study would have been impossible without the regional mapping by geologists of the U.S. Geological Survey. Spirited discussions with Jack Harrison in particular have been stimulating, and I thank Jack for sharing so much of his data and for steering me in the right direction at times. Darrel Long and Jon Boothroyd kindly reviewed an early version of the manuscript, making helpful suggestions. Figures were drafted by John Cuplin; and Shirley Pettersen and Katie Marron typed and retyped the manuscript.

REFERENCES

Bleiwas, D. I., 1977, The McNamara-Garnet Range Transition (Missoula Group): Abs. Geol. Soc. Am. Regional Meeting, Missoula, Montana.

Bowden, T. D., 1977, Depositional processes and environments within the Revett Formation, Precambrian Belt Basin, northwestern Montana and northern Idaho: Abs. Geol. Soc. Am. Regional Meeting, Missoula, Montana.

Fraser, G. D., and Waldrop, H. A., 1972, Geologic map of the Wise River Quadrangle, Silver Bow and Beaverhead counties, Montana: U.S. Geol. Survey Map GQ-988.

Harrison, J. E., 1972, Precambrian Belt basin of northwestern United States — Its geometry, sedimentation, and copper occurrences: Geol. Soc. Am. Bull., v. 83, p. 1215-1240.

McBride, E. F., 1974, Significance of color in red, green, purple, olive, brown, and gray beds of Defunta Group, northeastern Mexico. J. Sediment. Petrol., v. 44, p. 760-773.

McGowen, J. H., 1971, Gum Hollow Fan Delta, Nueces Bay: Bureau Econ. Geol., Univ. of Texas, Rept. Invest. No. 69, 91p.

McKee, E. D., Crosby, E. J., and Berryhill, H. L., 1967, Flood deposits, Bijou Creek, Colorado: J. Sediment. Petrol., v. 37: p. 829-851.

McMannis, W. J., 1963, LaHood Formation — a coarse facies of the Belt Series in southwestern Montana: Geol. Soc. Am. Bull., v. 74, p. 407-436.

Miall, A. D., 1977, A review of the braided-river depositional environment: Earth Sci. Revs., v. 13, p. 1-62.

Mudge, M. R., 1972a, Pre-Quaternary rocks of the Sun River Canyon area, northwestern Montana: U.S. Geol. Survey Prof. Paper 663-A, 138p.

————, 1972b, Structural geology of the Sun River Canyon and adjacent area, northwestern Montana: U.S. Geol. Survey Prof. Paper 663B.

Schumm, S. A., 1968, Speculations concerning paleohydrologic controls of terrestrial sedimentation: Geol. Soc. Am. Bull., v. 79, p. 1573-1588.

Smith, N. D., 1970, The braided stream depositional environment: comparison of the Platte River with some Silurian clastic rocks, north-central Appalachians: Geol. Soc. Am. Bull., v. 81, p. 2993-3014.

Wallace, C. A., Harrison, J. E., Klepper, M. R., and Wells, J. E., 1976, Carbonate sedimentary breccias in the Wallace Formation (Belt Supergroup), Idaho and Montana, and their paleogeographic significance: Abs. Geol. Soc. Am. Program for Ann. Meeting, p. 1159.

Winston, Don, 1972, The Bonner Formation as a late Precambrian pediplain: Northwest Geology, v. 2, p. 53-58.

————, 1973a, The Bonner Formation (Precambrian Belt of Montana) as a braided stream sequence: Abs. Soc. Econ. Paleont. Mineral. Ann. Meeting, Anaheim, California.

————, 1973b, The Precambrian Missoula Group of Montana as a braided stream and sea-margin sequence: in P. T. Bishop and J. D. Powell, Co-chairmen, Belt Symposium; v. 1, Dept. of Geology, Univ. of Idaho, Moscow, Idaho.

————, 1977, Alluvial fan, shallow water, and sub-wave base deposits of the Belt Supergroup near Missoula, Montana: Rocky Mountain Section, Geol. Soc. Am. 30th Ann. Meeting, Field Guide No. 5, 41p.

————, and Jacob, J. P., 1977, Middle Missoula Group (Precambrian) units of Montana and Idaho: Abs. Geol. Soc. Am., Regional Meeting, Missoula, Montana.

White, B. G., and Winston, Don, 1977, The Revett/St. Regis "transition zone" near the Bunker Hill mine, Coeur d'Alene mining district, Idaho: Abs. Coeur d'Alene Field Conference, Wallace, Idaho, Nov. 3-5, 1977; Soc. Econ. Geol. and College of Mines, University of Idaho, Moscow, Idaho.

————, ————, and Jacob, P. J., 1977, The Revett Formation near Kellogg, Idaho: Abs. Geol. Soc. Am. Regional Meeting, Missoula, Montana.

THE EVOLUTION OF FLUVIAL STYLE, WITH SPECIAL REFERENCE TO THE CENTRAL APPALACHIAN PALEOZOIC

EDWARD COTTER[1]

ABSTRACT

Published interpretations of paleochannel patterns and my own interpretation of fluvial styles in the Paleozoic sequence of Pennsylvania both confirm an evolution of the style of rivers in mid-Paleozoic time. Of the more than one hundred published interpretations, virtually all pre-Silurian rivers were braided, whereas Silurian and younger rivers were either meandering or braided. Clouding this conclusion, however, are differences in tectonic, geomorphic, and climatic settings of the various depositional sites, the use of potentially different diagnostic criteria, and the paucity of interpretations of pre-Silurian rivers.

The Central Appalachian Paleozoic sequence exposed in central Pennsylvania contains eight formations of fluvial origin, ranging in age from Ordovician to Pennsylvanian. These accumulated on aggrading foreland plains in molasse phases of Appalachian clastic wedges. Consistent criteria were used to categorize the fluvial styles as fine-grained meanderbelt, coarse-grained meanderbelt, channeled-braided, or sheet--braided. The three older Paleozoic units (Bald Eagle, Juniata, Tuscarora) were deposited with a sheet-braided fluvial style; the five Devonian and Carboniferous units (Catskill, Pocono, Mauch Chunk Pottsville, Llewellyn) accumulated principally in fine- and coarse-grained meanderbelts, with downslope changes in fluvial style. In Early Paleozoic time only braided styles existed, under conditions that in the Later Paleozoic led to the development of meandering.

Likely cause of this middle Paleozoic fluvial style change was the advent of land vegetation, which altered flood peaks and sediment yields, enhanced production of finer sediment, and stabilized stream banks.

INTRODUCTION

About ten years ago, Schumm (1967, 1968) speculated that the evolution of land vegetation caused a change in the characteristic patterns of rivers. He thought that before land vegetation appeared in the middle of the Paleozoic Era, precipitation/sediment yield relations everywhere would have been much like those of arid regions today. Streams would transport great quantities of bed load, and river patterns characteristically would be braided.

The evolution of land vegetation would not only have encouraged the development of meandering river patterns, but it would also present a mechanism for producing cyclic fluvial sequences.

Schumm therefore thought that the geologic record ought to show a gradual transition in river patterns from essentially all braided to a more balanced mixture of braided and meandering, beginning in the middle of the Paleozoic Era.

I attempted to test this hypothesis with two different but related approaches. First, through compilation of published interpretations of river morphology for stratigraphic units organized according to age. And then by my own diagnosis of the style of Paleozoic rivers in the Pennsylvania part of the Appalachian Basin.

In referring to the morphology of ancient river systems I have elected to use the term "fluvial style", even though "channel pattern" is most commonly used for the trace in plan of modern rivers (Dury, 1969). Ancient rock sequences are related to channel pattern in plan only through abstract conceptual models. There is such a degree of uncertainty

---

[1]Department of Geology and Geography, Bucknell University, Lewisburg, Pennsylvania, U.S.A., 17837

about the validity of present conceptual models that I think it premature to presume that ancient channel patterns can be examined in the rock record. What can be examined is the internal organization of fluvial rock sequences, and this is what I call "fluvial style".

PUBLISHED INTERPRETATIONS OF FLUVIAL STYLE

I undertook a systematic survey of the literature of sedimentary geology for interpretations of fluvial style. Of particular help were the summaries of LeBlanc (1972), Allen (1970), and Miall (1977), as well as the series of Geo Abstracts, Section E — Sedimentology. Most of the interpretations are from English-language publications, and the rock sequences interpreted occur, in most cases, in North America and Europe. There are undoubtedly interpretations that were overlooked or missed in such sources as local and regional surveys. I did try more diligently to find Lower Paleozoic examples.

Interpretations were considered in the two basic categories of braided and meandering. The authors' interpretations had to be clearly stated in terms of these two categories; other varieties of interpretation, such as "high-velocity fluvial" could not be included. And I could not second-guess authors' interpretations or build my own interpretation from their descriptions. In short, an ancient fluvial system was braided or meandering if an author said it was.

There were a number of arbitrary decisions made about whether there was sufficient geographic and stratigraphic separation between interpreted units for them to be considered as separate entries in the tabulation or whether they should be lumped together as one entry. The nature of this problem can be appreciated if one considers the number of interpretations made of the Old Red Sandstone alone, both in the British Isles and in surrounding areas.

Before these arbitrary decisions reduced the number, there were a total of 136 useful published interpretations of fluvial style. Lumping together units that were too close geographically and stratigraphically reduced this number to 99,

    42 braided,
    34 meandering, and
    23 coexisting braided and meandering.

The coexisting braided and meandering interpretations were tabulated by assigning one to each of the separate braided and meandering categories for each of the 23 from the "both" category. This produced a total of 122 interpretations that were used in considering the evolution of fluvial style,

    65 braided, and
    57 meandering.

The 122 published interpretations are summarized in Tables 1 and 2, and the proportions of braided and meandering styles for each period are shown graphically in Figure 1.

There are a number of reasons why these data are difficult to interpret in the context of the effect of evolving vegetation on fluvial style. Precambrian and Early Paleozoic interpretations appear underrepresented. This could result from a combination of factors: Lower Paleozoic rocks are relatively less exposed for geological study (see Raup, 1976), and there was a relatively high proportion of nonterrestrial environments in Early Paleozoic time (see Sepkoski, 1976). It is also possible that some Lower Paleozoic fluvial sequences have been misinterpreted as shallow marine deposits (D. Winston, personal communication).

Another complication is that these interpretations were made by many different persons, using a variety of conceptual models, and probably using different diagnostic criteria.

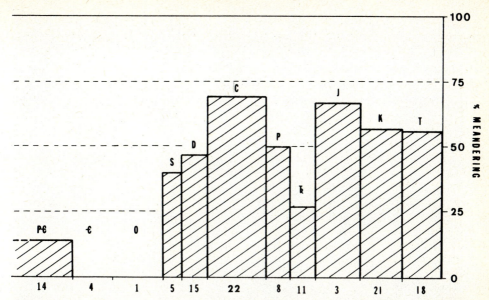

Fig. 1. Summary of published interpretations of fluvial style shown as percentage of streams of each period that were meandering. Total number of published interpretations for each period shown along base.

Perhaps the most important interpretive difficulty arises because these data represent an amalgamation of a great variety of sedimentary rock units that accumulated in a great variety of tectonic, geomorphic, and climatic settings.

Despite these complications, the published interpretations of fluvial style do show an increase in the proportion of meandering, beginning in the Silurian period, a time consistent with the advent of significant land vegetation (Gray and Boucot, 1977). However, it is troublesome to note that there is not a gradual increase in the proportion of meandering style resulting from the progressively more significant influence of more advanced forms of land vegetation, as suggested by Schumm (1967, 1968). The general trend ought to show significant effects of the evolution of flowering plants and grasses in the Cretaceous and Tertiary, respectively, and not such reversals of that general trend as the low proportion of meandering style for rivers of Triassic age.

The data presented here are consistent with some part of Schumm's hypothesis, yet it is difficult to isolate the single factor of evolving land vegetation as the cause of the tabulated change of fluvial style.

Certain of the complicating factors may be eliminated by examining the evolution of fluvial style in a single depositional basin. In this, one interpreter, using consistent diagnostic criteria, can judge the style of fluvial sequences that accumulated in similar tectonic, physiographic, and, insofar as possible, climatic settings. I undertook such a task in the interpretation of Paleozoic fluvial sequences in the Appalachian Basin strata exposed in central Pennsylvania.

PALEOZOIC SEQUENCE IN CENTRAL PENNSYLVANIA

Introduction

In the Folded Appalachians of central Pennsylvania the Paleozoic stratigraphic sequence is exposed in linear ridge and valley physiography, where essentially every

Fig. 2. Upper Ordovician Bald Eagle Sandstone adjacent to eastbound lanes of Interstate Highway 80 at Mile 181 between Loganton and Lock Haven Exits. Vertical drill marks spaced about 1 meter apart. Sheetlike genetic sandstone units with abandoned-channel mudstone in elongate lens in center.

major ridge is a fluvial sandstone. This sequence has traditionally been organized into cycles of flysch and molasse (Pettijohn, 1975). In the older molasse there are three formations of fluvial origin (Bald Eagle, Juniata, Tuscarora), and in the younger molasse there are five fluvial units (Catskill, Pocono, Mauch Chunk, Pottsville, Llewellyn).

Interpretations of the fluvial styles of these eight formations were based on extremely well-exposed sequences that were representative of the units. These exposures occur within an area bounded by a trapezoid defined by these Pennslyvania communities at its corners: Pottsville, Scranton, Port Matilda, and Newport. In general, outcrops of the three older Ordovician and Silurian units occur west of (farther from source) those of the five younger Devonian and Carboniferous units.

In order to be consistent with Schumm's hypothesis, the three older Paleozoic units, which accumulated before the evolution of significant land vegetation, ought to exhibit braided fluvial style. In contrast, the five younger Paleozoic units ought to have a greater proportion of meandering style.

The units and their fluvial style

Interpretation of ancient fluvial style has been made possible largely by the development of conceptual models that can serve as guides for ordering observations of vertical stratigraphic sequences. The basis of my judgments was the three-fold scheme of Brown (1973a) that divides fluvial depositional systems into categories of braided, coarse-grained meanderbelt, and fine-grained meanderbelt. As work progressed, I felt that it proved helpful to subdivide the braided category into "sheet-braided" and "channeled-braided". If individual genetic units were laterally very extensive, with high width/thickness ratios (greater than about 20/1) the braided style was termed sheet-braided. With individual genetic units more channel-shaped, making the formation complexly multichanneled in the manner of large-scale scour-and-fill, the style was termed channeled-braided. Thus, the Paleozoic fluvial units of central Pennsylvania were categorized as fine-grained meanderbelt, coarse-grained meanderbelt, channeled-braided, or sheet-braided.

Older Paleozoic units

Bald Eagle Formation. This unit is about 200 m thick in central Pennsylvania (Faill and Wells, 1974), although there is a great range of thickness because of basinward thinning, sourceward erosion, and the definition of the upper boundary in terms of a variably positioned color transition (A. M. Thompson, 1970). Its age is accepted as Late Ordovician (Cincinnatian) (Faill and Wells, 1974). The Bald Eagle is dominantly medium to coarse, subangular sandstone, with a medial conglomerate member; mudstone is rare, occurring in broad, isolated lenses (Fig. 2). Sedimentary structures comprise largely planar cross lamination, with subordinate trough cross lamination. Beds are medium to thick, and are grouped in laterally extensive, sheetlike genetic units (Fig. 2). Vertical sequences exhibit no definite pattern in grain size or sedimentary structures.

Guided by the conceptual framework of Brown (1973a) I consider these features to characterize a braided fluvial style, and the particular category to be sheet-braided.

Juniata Formation. Gradationally above the Bald Eagle, the Juniata is about 300 m thick (Faill and Wells, 1974, p. 18) and is traditionally assigned a Late Ordovician age (see Faill and Wells, 1974, p. 16). In comparison with the Bald Eagle, this unit has a somewhat finer grain size, slightly higher matrix content, and somewhat more common thin sheetlike to lenticular beds of mudstone. Mudstone is also characteristically present along the bases

Fig. 3. Upper Ordovician Juniata Formation adjacent to eastbound lanes of Interstate Highway 80 at Mile 182. Laterally extensive genetic units indicating sheet-braided fluvial style. Drill holes spaced about 1 meter apart.

Fig. 4. Lower Silurian Tuscarora Formation on east side of U. S. Route 22-322 one mile (1.6 km) north of Millerstown, Pa. Wooden staff on outcrop is 1. 5 m long. Extremely sheetlike genetic sandstone units separated by thin shale beds.

of beds and lining foreset laminae. Planar cross lamination is predominant in the sandstone, and even parallel lamination and trough cross lamination also occur. Broadly lenticular genetic units (Fig. 3) are composed of medium to thick beds of cross-laminated sandstone, and there are no discernible vertical patterns in lithology or structures.

The Juniata, also, has a sheet-braided fluvial style, although there is a slightly greater degree of channeling than in the Bald Eagle Formation.

Tuscarora Formation. Thickness of this unit in central Pennsylvania ranges from about 120 to 180 m (Faill and Wells, 1974, p. 27). Its age is reported by Berry and Boucot (1970) to be Early Silurian (Early Llandovery, Albion). Median grain size of the sandstone ranges from medium to fine, and the proportion of thin interbedded shales increases basinward (Yeakel, 1962; Smith, 1970). The thin to thick sandstone beds are characteristically planar and trough cross-laminated, and there are subordinate even, parallel laminated beds. Vertical sequences have no trends or patterns in lithology or structures. Genetic units are laterally very extensive sheets (Fig. 4); in places these units consist of a single cross-laminated sandstone bed between two very thin shale beds.

The Tuscarora, and its sourceward equivalent the Shawangunk Formation, have previously been interpreted to have formed from braided stream deposition (Smith, 1970; Epstein and Epstein, 1972; Epstein *et al.*, 1974; Faill and Wells, 1974). My observation of its characteristics and application of the criteria of Brown (1973a) lead to concurrence in that interpretation. The Tuscarora has a distinctly sheet-braided fluvial style.

Younger Paleozoic units

Catskill Formation. This Upper Devonian sequence is perhaps the most widely known of the units summarized in this paper. In central Pennsylvania the very thick (2000 to 2500

Fig. 5. Upper Devonian Catskill Formation, Duncannon Member, adjacent to southbound lanes of U. S. 22-322 at Buffalo Mountain, south of Millerstown, Pa. Fining-upward sequence shown is about 11 m thick and has small-displacement, low-angle reverse fault.

m. Faill and Wells, 1974, p. 120) nonmarine part is divided into two members. The lower Sherman Creek Member has symmetrical cycles consisting of red siltstone and shale alternating with thinner zones of very fine-grained, cross-laminated, red sandstone (Faill and Wells, 1974, p. 114). Above this, the Duncannon Member consists of asymmetric fining-upward cycles, grading upward from local basal conglomerates, through nonred, poorly sorted, planar and trough cross-laminated sandstone,to red silty mudstone and shale (Faill and Wells, 1974, p. 111) (Fig. 5).

The Catskill has been interpreted by a number of students to have accumulated in meandering stream systems (Allen and Friend, 1968; Allen, 1970; Woodrow *et al.,* 1973; Faill and Wells, 1974). In eastern Pennsylvania at more sourceward locations, the Catskill contains the deposits of braided streams interposed with those of meandering streams (Glaeser, 1974; Epstein *et al.*, 1974; Berg, 1975; Sevon, 1975a, b). However, in the area considered in this paper, the Catskill has features indicating that it formed in a fine-grained meanderbelt (Brown, 1973a).

Pocono Formation. Above a transitional Spechty Kopf Member, the Pocono is composed of medium-grained, poorly sorted sandstone and minor amounts of interbedded conglomerate and shale. It has medium to thick beds, principally containing planar cross lamination, arranged in complexly intercised channel-shaped genetic units (Fig. 6). There is no ordered pattern to vertical sequences. The age of the Pocono becomes Early Mississippian above the basal member (Read, 1955). The range of estimates of its thickness in central Pennsylvania is up to 500 m (Faill and Wells, 1974, p. 127).

In the conceptual framework of Brown (1973a) this unit formed in a braided fluvial depositional system. The shape of individual genetic units (Fig. 6) determines the category of its fluvial style as channeled braided.

Mauch Chunk Formation. This unit consists of thin to thick zones of gray to red, medium-grained sandstone alternating with zones of red siltstone and shale (Fig. 7). Mudstone zones commonly contain persistent horizons of caliche nodules. Conglomerates are present locally within the formation, and become laterally more persistent in the upper member that is transitional with the overlying Pottsville Formation. Sandstones are in thin to thick beds with common planar and trough cross lamination. In vertical sequences these zones of sandstone and mudstone are in cycles that are symmetric in the same sense as those of the lower member of the Catskill Formation. In some places, however, fining-upward sequences occur (Fig. 7; Meckel, 1970, Figs. 8 & 10).

The Mauch Chunk was interpreted to have had a fluvial origin by Barrell (1907), and more recent studies have supported this interpretation (Hoque, 1968; Meckel, 1970; Epstein *et al.,* 1974; Faill and Wells, 1974). In the four-fold scheme of this paper the Mauch Chunk has a fine-grained meanderbelt fluvial style. The conglomeratic upper member is considered to be part of the Mauch Chunk Formation because of the red color of the included shales. However, its fluvial style is that of a coarse-grained meanderbelt, closely related to that of the overlying Pottsville Formation.

Pottsville Formation. This unit consists largely of conglomerate and conglomeratic sandstone, with subordinate amounts of siltstone, shale, and coal. Beds are thin to very thick; certain bed sequences are laterally inclined away from depositional horizontal. The most common sedimentary structure in both conglomerates and sandstones is trough cross lamination. Fining-upward sequences occur at many localities (*see* Meckel, 1970, Figs. 9, 10); some of them are interrupted by multistoried superimposed cycles. The Pottsville Formation is approximately 400 m thick in the area studied (Wood *et al.,* 1969, Fig. 40). It is dated as Early and Middle Pennsylvanian (Wood *et al.,* 1969, p. 78).

Fig. 6.   Lower Mississippian Pocono Formation adjacent to north-bound lanes of U. S. Route 322 about 5 miles (8 km) north of village of Port Matilda, Pa. Lenticular genetic units of cross-bedded sandstone indicate channeled-braided fluvial style. Drill holes spaces about 1 m apart.

Fig. 7. Carboniferous Mauch Chunk Formation adjacent to west-bound lanes of Interstate Highway 80 near Nescopeck Creek at Mile 251. Alternating zones of drab sandstone and red shale accumulated in fine-grained meanderbelt fluvial style.

The features of the Pottsville fit the diagnostic criteria for deposition in a coarse-grained meanderbelt fluvial system, although at more easterly (sourceward) exposures the characteristics approach those of channeled-braided fluvial style.

Llewellyn Formation. The Llewellyn is characterized by fining-upward sequences that typically have fine conglomerate or coarse sandstone at the base and pass through trough cross-laminated sandstone into thinly interbedded very fine sandstone and siltstone. The uppermost parts of sequences are carbonaceous shales, underclays, and coals (Fig. 8). The unit is at least 800 m thick and is of Latest Middle Pennsylvanian and Late Pennsylvanian age (Wood *et al.,* 1969).

The Llewellyn Formation was deposited in a river system which had a fine-grained meanderbelt fluvial style (Brown, 1973a).

Summary. The fluvial styles of these eight Paleozoic units in central Pennsylvania are summarized in Figure 9. Note that the three pre-Devonian units (Bald Eagle, Juniata, and Tuscarora) are all assigned to the sheet-braided category, whereas none of the subsequent units are. Of the five Devonian and Carboniferous units, only the Pocono is braided, and the style is channeled-braided. The four other younger Paleozoic units were formed in fine-grained meanderbelts (Catskill, Mauch Chunk, and Llewellyn) and a coarse-grained meanderbelt (Pottsville).

Parallels in depositional settings

All eight fluvial formations in the central Pennsylvania Paleozoic stratigraphic sequence accumulated on aggrading foreland coastal plains at about the same distance from their source area. Rivers originated in orogenic highlands to the southeast and consistently

Fig. 8. Upper Carboniferous Llewellyn Formation adjacent to northbound lanes of Interstate Highway 81 at Mile 181 near Scranton, Pa. Multiple fining-upward sequences, typically capped by beds of coal, indicate fine-grained meanderbelt fluvial style.

flowed to the northwest (Meckel, 1970, Fig. 7; Pettijohn, 1975, Fig. 14-12). Detailed studies of the various dispersal systems have shown that the fall line and location of the source area did not change significantly through the period of accumulation of all eight units (summarized by Meckel, 1970, Fig. 6).

It might be possible to compare similar physiographic settings of deposition if one examines those parts of the formations that accumulated in closest proximity to the shoreline. In examination of basal continental deposits just above the transition with marine, and also the uppermost fluvial strata immediately below onlapping marine rocks, the older Paleozoic units (lowest Bald Eagle and upper-most Tuscarora) show that streams with sheet-braided fluvial style went right to the shoreline. In the same stratigraphic and physiographic setting the younger Paleozoic units (Catskill, Mauch Chunk, and Llewellyn) exhibit fine-grained meanderbelt fluvial style.

The relative proportion of silt and clay preserved in the Paleozoic stream deposits can serve as an index of the nature of the load transported by the streams and of the stream sinuosity (Cotter, 1971; Schumm, 1968, 1972). Much of the older Juniata and Tuscarora Formations has more silt and clay than the younger Pottsville and part of the Llewellyn Formations. Yet the Tuscarora and Juniata show sheet-braided fluvial style, and the Pottsville and Llewellyn meandered.

Modern meandering streams generally have lower gradients than braided streams (Leopold *et al.*, 1964, Fig. 7-39). Estimation of stream competence through observation of maximum and average grain size might serve as a measure of relative gradients of the Paleozoic streams. Only a very small proportion of the older units is conglomeratic, and the streams had sheet-braided fluvial styles. Coarsest sediment of all the units occurs in the uppermost Mauch Chunk and lower Pottsville Formations, yet these younger

Paleozoic units formed with coarse-grained meanderbelt fluvial style. These style differences are not merely a result of differences in distance from the source area; the meandering styles of the younger units existed much closer to the source than the braided styles of the older units. Younger Paleozoic units indicate downslope changes in fluvial style, with increased sinuosity downstream resulting in the contemporaneous existence of braided, coarse-grained meanderbelt, and fine-grained meanderbelt styles. Older Paleozoic units do not show such downslope style change.

Interpretations of paleoclimate are available only for a few of the younger Paleozoic units and for none of the older units. The Carboniferous Pottsville and Llewellyn Formations formed in relatively humid climates that enhanced the development of coastal plain swamps in which vegetation (coal) accumulated and was preserved. Both the Upper Devonian Catskill Formation and the Mississippian Mauch Chunk Formation have features indicating strong aridity, at least seasonally (Barrell, 1907; Allen and Friend,

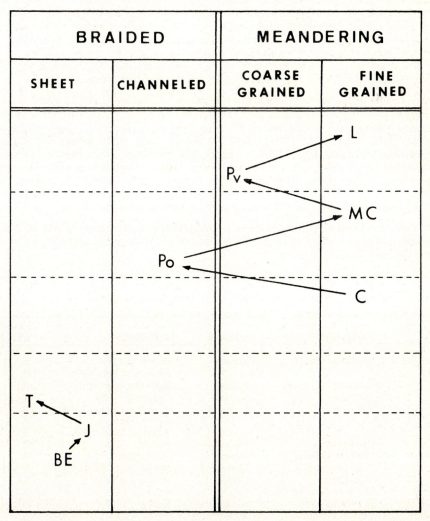

Fig. 9. Summary of fluvial styles of eight Paleozoic formations in the Appalachian Basin in central Pennsylvania. Formations symbolized by their initials.

1968; Woodrow *et al.*, 1973). It is likely that the coeval Pocono Formation also formed in at least a seasonally arid climate. Whether arid or humid these younger units, aside from the Pocono, had meandering fluvial styles. Comparison with the climatic settings of the older Paleozoic units is not possible; the very absence of land vegetation that helped to determine the braided fluvial styles adds to the inability to infer their paleoclimates.

SUMMARY AND CONCLUSIONS

Two approaches to examining secular change in fluvial style confirm that there was an evolution in the pattern of streams during the middle of the Paleozoic Era. Analysis of published interpretations of ancient fluvial style reveals a change from nearly all braided streams to a more balanced mixture of braided and meandering commencing in the Silurian Period. The same change is indicated by my own interpretation of the fluvial styles of eight Paleozoic units (Ordovician through Upper Carboniferous) in the central Pennsylvania part of the Appalachian Basin. Older Paleozoic units (Bald Eagle, Juniata, Tuscarora) all have sheet-braided fluvial styles; of the five younger Paleozoic units (Catskill, Pocono, Mauch Chunk, Pottsville, Llewellyn), only the Pocono is braided, and the others have varieties of meandering styles. There are so many parallels in the depositional settings of these Appalachian Paleozoic units that had some of the older braided units formed later in the Paleozoic they would have had meandering styles.

These results are consistent with the hypothesis of Schumm (1967, 1968) that the evolution of land vegetation should have resulted in changes in the characteristic patterns of streams. Vegetation has a significant effect on streams by retarding erosion, decreasing sediment yield, decreasing total runoff, discharge, and flood peaks for a given precipitation (Schumm, 1967, 1968; Gregory and Walling, 1973; Pearce, 1976); by decreasing bed load grain size and enhancing fine sediment production (Schumm, 1968; Ollier, 1969; Carroll, 1970; Birkeland, 1974; Douglas, 1976); and by increasing bank stability (Schumm, 1968; Smith, 1976). The direction of all these indicates that evolving land vegetation would enhance the tendency of streams to meander.

It is clear that there was a mid-Paleozoic change of fluvial style from nearly all braided to a mixture of both braided and meandering, and it is likely that the cause of this change was the advent of land vegetation.

Table 1.   SUMMARY — PUBLISHED INTERPRETATIONS OF FLUVIAL STYLE

AGE	ROCK UNITS	LOCALE	AUTHORS	FLUVIAL STYLE
Precambrian	Golneselv Fm.	Norway	Banks, 1973	B
	" "	"	Banks and Røe, 1974	B
	Sparagmite	Norway	Bjørlykke, *et al.*, 1976	B
	Huronian	Canada	Casshyap, 1968	M
	Torridonian	Scotland	Gracie and Stewart, 1967	B
	"	"	Selley, 1965	B
	"	"	Williams, 1966, 1969	B
	Vadso Group	Norway	Hobday, 1974	B
	Laksefjord Group	Norway	Laird, 1972	B
	Solor Church Fm.	Minnesota	Morey, 1972, 1974	B & M
	Missi Group	Ontario	Mukherjee, 1974	B
	Mount Rogers Fm.	Virginia	Schwab, 1976	B
	Grandfather Mtn.	N. Carolina	Schwab, 1977	B
	Dala & Jotnian	Scandinavia	Selley, 1970	B
	Tarkwaian	Ghana	Sestini, 1973	B
	Witwatersrand	S. Africa	Vos, 1975	B

Table 1 — Cont'd

AGE	ROCK UNITS	LOCALE	AUTHORS	FLUVIAL STYLE
Cambrian	Hardyston	New Jersey	Aaron, 1969	B
	Bowers Group	Antarctica	Andrews and Laird, 1976	B
	Tapeats Fm.	Arizona	Hereford, 1977	B
	Unicoi Fm.	Virginia	Schwab, 1972	B
Ordovician	Bald Eagle Fm.	Pennsylvania	Faill and Wells, 1974	B
Silurian	Bloomsburg Fm.	Pennsylvania	Epstein and Epstein, 1969	M
	Tuscarora Fm.	Pennsylvania	Smith, 1970	B
	Shawangunk Fm.	Pennsylvania	Epstein and Epstein, 1972	B
	Ringerike Group	Norway	Turner, 1974; Turner and Whitaker, 1976	B & M
	Clinch Fm.	Virginia	Whisonant, 1977	B
Devonian	Catskill	New York	Buttner, 1968	B
	"	Pennsylvania	Glaeser, 1974	B & M
	"	New York	Johnson and Friedman, 1969	M
	"	New York	McCave, 1968	M
	"	Pennsylvania	Ryan, 1965	M
	"	New York, Pa.	Allen and Friend, 1968	B & M
	Old Red Ss	Anglo-Welsh Basin	Allen, 1964	M
	" "	England, Wales	Allen, 1970	M
	" "	England	Allen, 1974	B & M
	" "	Scotland	Bluck, 1967	B
	" "	Eire	Graham, 1975	M
	" "	Scotland	Leeder, 1973	B & M
	" "	Shetland Is	Mykura, 1973	B
	" "	Norway	Nilsen, 1969	B
	" "	Scotland	Read and Johnson, 1967	M
	" "	Norway	Siedlecka, 1975	B
	Wood Bay Series	Spitsbergen	Friend, 1965	M
	" "	Spitsbergen	Friend & Moody-Stuart, 1972	B & M
	" "	Spitsbergen	Moody-Stuart, 1966	B & M
	Hervey Group	New So. Wales	Conolly, 1965	M
	Finke Group	Cen. Australia	Jones, 1973	B & M
	Mid.-Upp. Devonian	North Canada	Embry and Klovan, 1976	B & M
	Peel Sound Fm.	N. W. Terr.	Miall, 1970a, 1970b	B
	Hornelen Basin	Norway	Steel et al, 1977	B
	Grey Hoek Fm.	Spitsbergen	Worsley, 1972	M
	Battery Point Fm.	Quebec	Cant, 1973; Cant and Walker, 1976	B
Carboniferous	Kittaning Fm.	Pennsylvania	Beutner et al., 1967	M
	Cisco Group	Texas	Brown, 1973b	B & M
	Abbotsham/Northam	England	DeRaaf et al., 1965	M
	Conemaugh, Monongahela	W. Virginia	Donaldson, 1969	M
	Robinson Ss.	Illinois	Hewitt and Morgan, 1965	M
	Mauch Chunk Fm.	Pennsylvania	Hoskins, 1976	M
	Trenchard Group	England	Jones, 1972	B
	Rhondda Beds	England	Kelling, 1968	B & M
	Pottsville/Olean	New York, Pa.	Meckel, 1967	B
	Pottsville/Sharon	Ohio	Mrakovich & Coogan, 1974	B
	Carboniferous	Morocco	Padgett & Ehrlich, 1976	M
	Pocono Fm.	Pennsylvania	Sevon, 1975a, 1975b	B
	Kissinger Ss.	Texas	Shelton, 1973	B

Table 1. — Cont'd

AGE	ROCK UNITS	LOCALE	AUTHORS	FLUVIAL STYLE
	Pennsylvanian	Michigan Basin	Shideler, 1969	M
	Morrowan/Desmoinesian	Oklahoma	Visher et al., 1971	M
	Missourian	Oklahoma	Visher, 1965	M
	Cumberland Group	Nova Scotia	Way, 1968	M
	Llewellyn Fm.	Pennsylvania	Wood et al., 1969	M
	Carboniferous	Illinois Basin	Potter, 1963	B & M
	Yoredale	England	Elliott, 1976	M
Permian	Dunkard	West Virginia	Beerbower, 1964, 1969	M
	Cook	Texas	Bloomer, 1977	M
	New Red Ss.	Scotland	Bruck et al, 1967	B
	" "	England	Laming, 1966	B & M
	Prudhoe Bay Field	Alaska	Eckelmann et al., 1975	B
	Cedar Mesa	Utah	Mack, 1977	B
	Lower Gondwana	India	Casshyap, 1970, 1973	B & M
	" "	"	Casshyap & Qidwai, 1971	B & M
	" "	"	Niyogi, 1966	B
Triassic	Trujillo Ss.	Texas	Asquith & Cramer, 1975	B
	Hawkesbury Ss.	Australia	Conaghan & Jones, 1975	B
	Triassic	Spain	DeJong, 1971	B
	Chinle Fm.	Utah	Lupe, 1977	B
	"	Colorado Plat.	Stewart et al., 1972	M
	Triassic	Morocco	Mattis, 1977	B & M
	Stornoway Fm.	Great Britain	Steel & Wilson, 1975	B
	Bunter/L. Keuper	Great Britain	D. B. Thompson, 1970	B
	Verruca	Italy	Tangiorgi et al., 1977	B & M
Jurassic	Morrison Fm.	Colorado	Brady, 1969	M
	" "	New Mexico	Campbell, 1976	B
	" "	Utah	Derr, 1974	M
	" "	New Mexico	Flesch, 1974	B & M
	" "	Utah	Stokes, 1961	B & M
	Scalby Fm.	England	Nami, 1976	M
Cretaceous	Bida Ss.	Nigeria	Adeleye, 1974	B & M
	Fall River Ss.	Wyoming	Berg, 1968	M
	Tuscaloosa Fm.	Mississippi	Berg and Cook, 1968	B & M
	Mannville Group	Saskatchewan	Christopher, 1975	M
	Ferron Ss.	Utah	Cotter, 1971, 1975	M
	Dakota Fm.	Kansas	Franks, 1975	M
	"	New Mexico	Gilbert and Asquith, 1976	B
	"	Nebraska	Karl, 1976	B & M
	"	New Mexico	Owen, 1973	B & M
	Golden Spike Fm.	Montana	Gwinn and Mutch, 1965	B
	Hosston-Hensel	Texas	Hall and Turk, 1975	M
	J — interval	Nebraska	Harms, 1966	M
	Potomac Group	Maryland	Hansen, 1969	B & M
	Voznuevo Fm.	Spain	Jonker, 1972	B & M
	Isachsen Fm.	Arctic Canada	Miall, 1976	B
	Rocktown	Kansas	Siemers, 1976	M
	Cut Bank Ss.	Montana	Shelton, 1967	M
	Klikov Fm.	Czechoslovakia	Slanska, 1976	B
	Castlegate Ss.	Utah	Van De Graaff, 1972	B
	Arapahoe Fm.	Colorado	Weimer and Land, 1975	B

Table 1. — Cont'd

AGE	ROCK UNITS	LOCALE	AUTHORS	FLUVIAL STYLE
Tertiary	Aller	England	Edwards, 1973	B
	Tango Creek/Brother P.	Br. Columbia	Eisbacher, 1974	B & M
	Wilcox Group	Texas	Fisher & McGowen, 1969	M
	Kastri Fm.	Crete	Gradstein & Van Gelder, 1971	B
	Tongue River Fm.	North Dakota	Jacob, 1973	B & M
	,,	,,	Royse, 1970	M
	Oligocene	Spain	Nagtegaal, 1966	B
	Catahoula Fm.	Louisana	Paine & Meyerhoff, 1968	M
	Colton Fm.	Utah	Peterson, 1976	M
	Green River Fm.	Utah	Picard and High, 1972	M
	Miocene	Spain	Puigdefabregas, 1973	M
	Frio Group	Texas	Shelton, 1973	M
	Hoback Fm.	Wyoming	Spearing, 1969	B
	Pass Peak Fm.	Wyoming	Steidtmann, 1970	M
	Siwalik Series	India	Tandon, 1976	B
	Mirador-Misoa	Venezuela	Van Veen, 1971	M
	Vieja Group	Texas	Walton, 1977	B

B = braided, M = meandering

Table 2.   TABULATION OF PUBLISHED INTERPRETATIONS OF FLUVIAL STYLE

SYSTEM	STYLE	ALL REPORTS		SEPARATE UNITS		"BOTH" APPORTIONED EQUALLY	
		NO.	%	NO.	%	No.	%
PRECAMBRIAN	Braided	14	86	11	85	12	86
	Meandering	1	6	1	8	2	14
	Both	1	6	1	8	—	—
CAMBRIAN	Braided	4	100	4	100	4	100
	Meandering	0	0	0	0	0	0
	Both	0	0	0	0	—	—
ORDOVICIAN	Braided	1	100	1	100	1	100
	Meandering	0	0	0	0	0	0
	Both	0	0	0	0	—	—
SILURIAN	Braided	3	60	2	50	3	60
	Meandering	1	20	1	25	2	40
	Both	1	20	1	25	—	—
DEVONIAN	Braided	9	33	3	30	8	53
	Meandering	10	37	2	20	7	47
	Both	8	30	5	50	—	—
CARBONIFEROUS	Braided	5	25	4	21	7	31
	Meandering	12	60	12	63	15	69
	Both	3	15	3	16	—	—

Table 2. — Cont'd

SYSTEM	STYLE	ALL REPORTS		SEPARATE UNITS		"BOTH" APPORTIONED EQUALLY	
		NO.	%	NO.	%	NO.	%
PERMIAN	Braided	5	45	2	33	4	50
	Meandering	3	27	2	33	4	50
	Both	3	27	2	33	—	—
TRIASSIC	Braided	6	67	5	63	8	73
	Meandering	1	11	0	0	3	27
	Both	2	22	3	37	—	—
JURASSIC	Braided	1	17	0	0	1	33
	Meandering	3	50	1	50	2	67
	Both	2	34	1	50	—	—
CRETACEOUS	Braided	6	30	4	25	9	43
	Meandering	8	40	7	44	12	57
	Both	6	30	5	31	—	—
TERTIARY	Braided	6	35	6	38	8	44
	Meandering	9	53	8	50	10	56
	Both	2	12	2	12	—	—
TOTAL		136		99		122	

REFERENCES

Aaron, J. M., 1969, Petrology and origin of the Hardyston Quartzite (Lower Cambrian) in eastern Pennsylvania and western New Jersey: in Seymour Subitzky, ed., Geology of Selected Areas in New Jersey and Eastern Pennsylvania and Guidebook of Excursions; Rutgers Univ. Press, New Brunswick, N. J., p. 21-34.

Adeleye, D. R., 1974, Sedimentology of the fluvial Bida Sandstone (Cretaceous), Nigeria: Sediment. Geol., v. 12, p. 1-24.

Allen, J. R. L., 1964, Studies in fluviatile sedimentation: six cyclothems from the lower Old Red Sandstones, Anglo-Welsh basin: Sedimentology, v. 3, p. 163-198.

——, 1970, Studies in fluviatile sedimentation: a comparison of fining-upwards cyclothems, with special reference to coarse-member composition and interpretation: J. Sediment. Petrol., v. 40, p. 298-323.

——, 1974, Sedimentology of the Old Red Sandstone (Siluro-Devonian) in the Clee Hills area, Shropshire, England: Sediment. Geol., v. 12, p. 73-167.

——, and Friend, P. F., 1968, Deposition of the Catskill facies, Appalachian region: with notes on some other Old Red Sandstone basins: in G. deV. Klein, ed., Late Paleozoic and Mesozoic continental sedimentation, northeastern North America; Geol. Soc. Am. Spec. Paper 106, p. 21-74.

Andrews, P. B., and Laird, M. G., 1976, Sedimentology of a Late Cambrian regressive sequence (Bowers Group), northern Victoria Land, Antarctica: Sediment. Geology, v. 16, p. 21-44.

Asquith, G. B., and Cramer, S. L., 1975, Transverse braid bars in the Upper Triassic Trujillo Sandstone of the Texas Panhandle: J. Geol., v. 83, p. 657-661.

Banks, N. L., 1973, Falling-stage features of a Precambrian braided stream: criteria for sub-aerial exposure: Sediment. Geol., v. 10, p. 147-154.

Banks, N. L., and Røe, S.-L., 1974, Sedimentology of the Late Precambrian Golneselv Formation, Varangerfjorden, Finnmark: Norges Geol. Unders., v. 303, p. 17-38.

Barrell, Joseph, 1907, Origin and significance of the Mauch Chunk shale: Geol. Soc. Am. Bull., v. 18, p. 449-476.

Beerbower, J. R., 1964, Cyclothems and cyclic deposition mechanisms in alluvial plain sedimentation: Kansas Geol. Surv. Bull. 169, v. 1, p. 31-42.

———, 1969, Interpretation of cyclic Permo-Carboniferous deposition in alluvial plain sediments in West Virginia: Geol. Soc. Am. Bull., v. 80, p. 1843-1848.

Berg, R. R., 1968, Point bar origin of Fall River sandstone reservoirs, northeastern Wyoming: Am. Assoc. Petrol. Geol. Bull., v. 52, p. 2116-2122.

———, and Cook, B. C., 1968, Petrography and origin of lower Tuscaloosa sandstones, Mallalieu field, Lincoln County, Mississippi: Gulf Coast Assoc. Geol. Socs. Trans., v. 18, p. 242-255.

Berg, T. M., 1975, Geology and mineral resources of the Broadheadsville Quadrangle, Monroe and Carbon Counties, Pennsylvania: Pennsylvania Geol. Survey Atlas 205a, 60p.

Berry, W. B. N., and Boucot, A. J., 1970, Correlation of the North American Silurian rocks: Geol. Soc. Am. Spec. Paper 102, 289p.

Beutner, E. C., Flueckinger, L. A., and Gard, T. M., 1967, Bedding geometry in a Pennsylvanian channel sandstone: Geol. Soc. Am. Bull., v. 78, p. 911-916.

Birkeland, P. W., 1974, Pedology, weathering, and geomorphological research: Oxford Univ. Press, New York, 285p.

Bjørlykke, Knut, Elvsborg, A., and Hoy, T., 1976, Late Precambrian sedimentation in the central sparagmite basin of south Norway: Norsk Geologisk Tidsskrift, v. 56, p. 233-290.

Bloomer, R. R., 1977, Depositional environments of a reservoir sandstone in west-central Texas: Am. Assoc. Petrol. Geol. Bull., v. 61, p. 344-359.

Bluck, B. J., 1967, Deposition of some upper Old Red Sandstone conglomerates in the Clyde area: a study in the significance of bedding: Scott. Jour. Geol., v. 3, p. 139-167.

Brady, L. L., 1969, Stratigraphy and petrology of the Morrison Formation (Jurassic) of the Canon City, Colorado area: J. Sediment. Petrol., v. 39, p. 632-648.

Brown, L. F., Jr., 1973a, Cratonic basins: terrigenous clastic models: in L. F. Brown, Jr., et al., Pennsylvanian depositional systems in north-central Texas, a guide for interpreting terrigenous clastic facies in a cratonic basin: Bur. Econ. Geology, Univ. Texas at Austin, Guidebook No. 14, p. 10-30.

———, 1973b, Cisco depositional systems in north-central Texas: in L. F. Brown, Jr., et al., Pennsylvanian depostional systems in north-central Texas, a guide for interpreting terrigenous clastic facies in a cratonic basin; Bur. Econ. Geology, Univ. Texas at Austin, Guidebook No. 14, p. 57-122.

Bruck, P. M., Dedman, R. E., and Wilson, R. C. L., 1967, The New Red Sandstone of Raasay and Scalpay, Inner Hebrides: Scott. J. Geol., v. 3, p. 168-180.

Buttner, P. J. R., 1968, Proximal continental rhythmic sequences in the Genesee Group (lower Upper Devonian): in G. deV. Klein, ed., Late Paleozoic and Mesozoic continental sedimentation, northeastern North America; Geol. Soc. Am. Spec. Paper 106, p. 109-126.

Campbell, C. V., 1976, Reservoir geometry of a fluvial sheet sandstone: Am. Assoc. Petrol. Geol. Bull., v. 60, p. 1009-1020.

Cant, D. J., 1973, Devonian braided stream deposits in the Battery Point Formation, Gaspé, Est, Quebec: Maritime Sediments, v. 9, p. 13-20.

———, and Walker, R. G., 1976, Development of braided-fluvial facies model for the Devonian Battery Point Sandstone, Quebec: Can. J. Earth Sci., v. 13, p. 102-119.

Carroll, Dorothy, 1970, Rock weathering: Plenum Press, New York, 203p.

Casshyap, S. M., 1968, Huronian stratigraphy and paleocurrent analysis in the Espanola-Willisville area, Sudbury district, Ontario, Canada: J. Sediment. Petrol., v. 38, p. 920-942.

———, 1970, Sedimentary cycles and environment of deposition of the Barakar coal measures of Lower Gondwana, India: J. Sediment. Petrol., v. 40, p. 1302-1317.

———, 1973, Paleocurrents and paleogeographic reconstruction in the Barakar (Lower Gondwana) sandstones of peninsular India: Sediment. Geol., v. 9, p. 283-303.

——, and Qidwai, H. A., 1971, Paleocurrent analysis of Lower Gondwana sedimentary rocks, Pench Valley coalfield, Madhya Pradesh (India): Sediment. Geol., v. 5, p. 135-145.

Christopher, J. E., 1975, The depositional setting of the Mannville Group (Lower Cretaceous) in southwestern Saskatchewan: *in* W. G. E. Caldwell, *ed.,* The Cretaceous System in the Western Interior of North America: Geol. Assoc. Can. Spec. Paper 13, p. 523-552.

Conaghan, P. J., and Jones, J. G., 1975, The Hawkesbury Sandstone and the Brahmaputra: a depositional model for continental sheet sandstones: J. Geol. Soc. Australia, v. 22, p. 275-283.

Conolly, J. R., 1965, Petrology and origin of the Hervey Group, Upper Devonian, central New South Wales: J. Geol. Soc. Australia, v. 12, p. 123-166.

Cotter, Edward, 1971, Paleoflow characteristics of a Late Cretaceous river in Utah from analysis of sedimentary structures in the Ferron Sandstone: J. Sediment. Petrol., v. 41, p. 129-138.

——, 1975, Deltaic deposits in the Upper Cretaceous Ferron Sandstone of Utah: *in* M. L. S. Broussard, *ed.,* Deltas, Models for Exploration; Houston Geol. Soc., Houston, Texas, p. 471-484.

De Jong, J. D., 1971, Molasse and clastic-wedge sediments of the southern Cantabrian Mountains (NE Spain) as geomorphological and environmental indicators: Geol. Mijnbouw., v. 50, p. 399-416.

De Raaf, J. F. M., Reading, H. G., and Walker, R. G., 1965, Cyclic sedimentation in the Lower Westphalian of North Devon, England: Sedimentology, v. 4, p. 1-52.

Derr, M. E., 1974, Sedimentary structure and depositional environment of paleochannels in the Jurassic Morrison Formation near Green River, Utah: Brigham Young Univ. Geol. Stud., v. 21, p. 3-39.

Donaldson, A. C., 1969, Ancient deltaic sedimentation (Pennsylvanian) and its control on the distribution, thickness and quality of coals: *in* A. C. Donaldson, *ed.,* Some Appalachian coals and carbonates: models of ancient shallow-water deposition: West Virginia Geol. and Econ. Survey, p. 93-123.

Douglas, Ian, 1976, Erosion rates and climate: geomorphological implications: *in* E. Derbyshire, *ed.,* Geomorphology and Climate, Wiley, New York, p. 269-287.

Dury, G. H., 1969, Relation of morphometry to runoff frequency: *in* R. J. Chorley, *ed.,* Water, Earth, and Man; Methuen and Co., Ltd., London, p. 419-430.

Eckelmann, W. R., DeWitt, R. J., and Fisher, W. L., 1975, Prediction of fluvial-deltaic reservoir geometry, Prudhoe Bay field, Alaska: Ninth World Petrol. Congr. Proc., Tokyo, v. 2, Geology, p. 223-227.

Edwards, R. A., 1973, The Aller Gravels: Lower Tertiary braided river deposits in South Devon: Proc. Ussher Soc., v. 2, p. 608-616.

Eisbacher, G. H., 1974, Sedimentary history and tectonic evolution of the Sustut and Sifton Basins, north-central British Columbia: Geol. Surv. Can. Paper 73-31, 57p.

Elliott, T., 1976, The morphology, magnitude and regime of a Carboniferous fluvial-distributary channel: J. Sediment. Petrol., v. 46, p. 70-76.

Embry, Ashton, and Klovan, J. E., 1976, The Middle-Upper Devonian clastic wedge of the Franklinian Geosyncline: Bull. Can. Petrol. Geol., v. 24, p. 485-639.

Epstein, J. G., and Epstein, A. G., 1969, Geology of the Valley and Ridge province between Delaware Water Gap and Lehigh Gap, Pennsylvania: *in* S. Subitzky, *ed.,* Geology of selected areas in New Jersey and eastern Pennsylvania and Guidebook of Excursions: Rutgers Univ. Press, New Brunswick, N. J., p. 132-205.

——, 1972, Shawangunk Formation: U. S. Geol. Survey Prof. Paper 744, 45p.

Epstein, J. B., Sevon, W. D., and Glaeser, J. D., 1974, Geology and mineral resources of the Lehighton and Palmerton Quadrangles, Carbon and Northampton Counties, Pennsylvania: Pennsylvania Geol. Survey Atlas 195 cd, 460p.

Faill, R. T., and Wells, R. B., 1974, Geology and mineral resources of the Millerstown Quadrangle, Perry, Juniata, and Snyder Counties, Pennsylvania: Pennsylvania Geol. Survey Atlas 136, 276p.

Fisher, W. L., and J. H., McGowen, 1969, Depositional systems in Wilcox Group (Eocene) of Texas and their relationship to occurrence of oil and gas: Am. Assoc. Petrol. Geol. Bull., v. 53, p. 30-54.

Flesch, G. A., 1974, Stratigraphy and sedimentology of the Morrison Formation (Jurassic), Ojito Spring Quadrangle, Sandoval County, New Mexico: a preliminary report: New Mexico Geol. Soc. Guidebook, 25th Field Conf., p. 185-195.

Franks, P. C., 1975, The transgressive-regressive sequence of the Cretaceous Cheyenne, Kiowa, and Dakota Formations of Kansas: in W. G. E. Caldwell, ed., The Cretaceous System in the Western Interior of North America: Geol. Assoc. Can. Spec. Paper 13, p. 469-521.

Friend, P. F., 1965, Fluviatile sedimentary structures in the Wood Bay Series (Devonian) of Spitsbergen: Sedimentology, v. 5, p. 39-68.

————, and Moody-Stuart, M., 1972, Sedimentation of the Wood Bay Formation (Devonian) of Spitsbergen: regional analysis of a late orogenic basin: Oslo, Norsk. Polarinstitutt Skrifter 157, 77p.

Gray, Jane, and Boucot, A. J., 1977, Early vascular land plants: proof and conjecture: Lethaia, v. 10, p. 145-174.

Gilbert, J. L., and Asquith, G. B., 1976, Sedimentology of braided interval of Dakota Sandstone, northeastern New Mexico: New Mexico Bureau Mines Min. Res. Circ. 150.

Glaeser, J. D., 1974, Upper Devonian stratigraphy and sedimentary environments in northeastern Pennsylvania: Pennsylvania Geol. Surv. Bull. G 63.

Gracie, A. J., and Stewart, A. D., 1967, Torridonian sediments at Enard Bay, Ross-shire: Scott. J. Geol., v. 3, p. 181-194.

Gradstein, F. M., and Van Gelder, A., 1971, Prograding clastic fans and transition from a fluviatile to a marine environment in Neogene deposits of eastern Crete: Geol. Mijnbouw., v. 50, p. 383-392.

Graham, J. R., 1975, Deposits of a near-coastal fluvial plain — the Toe Head Formation (Upper Devonian) of southwest Cork, Eire: Sediment. Geol., v. 14, p. 45-61.

Gregory, K. J. and Walling, D. E., 1973, Drainage basin form and process: a geomorphological approach: Halsted Press, New York, 456p.

Gwinn, V. E., and Mutch, T. A., 1965, Intertongued Upper Cretaceous volcanic and nonvolcanic rocks, central-western Montana: Geol. Soc. Am. Bull., v. 76, p. 1125-1144.

Hall, W. D., and Turk, L. J., 1975, Aquifer evaluation using depositional systems: examples in north-central Texas: Ground Water, v. 13, p. 472-483.

Hansen, H. J., 1969, Depositional environments of subsurface Potomac Group in southern Maryland: Am. Assoc. Petrol. Geol. Bull., v. 53, p. 1923-1937.

Harms, J. C., 1966, Stratigraphic traps in a valley fill, western Nebraska: Am. Assoc. Petrol. Geol. Bull., v. 50, p. 2119-2149.

Hereford, Richard, 1977, Deposition of the Tapeats Sandstone (Cambrian) in central Arizona: Geol. Soc. Am. Bull., v. 88, p. 199-211.

Hewitt, C. H., and Morgan, J. T., 1965, The Frye in situ combustion test — reservoir characteristics: Jour. Petroleum Technology, v. 17, p. 337-342.

Hobday, D. K., 1974, Interaction between fluvial and marine processes in the lower part of the Late Precambrian Vadsø Group, Finnmark: Norges Geol. Unders., v. 303, p. 39-56.

Hoque, M. V., 1968, Sedimentologic and paleocurrent study of Mauch Chunk sandstones (Mississippian), south-central and western Pennsylvania: Am. Assoc. Petrol. Geol. Bull., v. 52, p. 246-263.

Hoskins, D. M., 1976, Geology and mineral resources of the Millersburg 15-minute Quadrangle, Dauphin, Juniata, Northumberland, Perry, and Snyder Counties, Pennsylvania: Pennsylvania Geol. Survey Atlas 146, 38p.

Jacob, A. F., 1973, Depositional environments of Paleocene Tongue River Formation, western North Dakota: Am. Assoc. Petrol. Geol. Bull., v. 57, p. 1038-1052.

Johnson, K. G., and Friedman, G. M., 1969, The Tully clastic correlatives (Upper Devonian) of New York State: a model for recognition of alluvial, dune (?), tidal, nearshore (bar and lagoon), and offshore sedimentary environments in a tectonic delta complex: J. Sediment. Petrol., v. 39, p. 451-485.

Jones, B. C., 1973, Sedimentology of the Upper Devonian to Lower Carboniferous Finke Group, Amadeus and Warburton Basins, central Australia: J. Geol. Soc. Australia, v. 20, p. 273-293.

Jones, P. C., 1972, Quartzarenite and litharenite facies in the fluvial foreland deposits of the Trenchard Group (Westphalian) Forest of Dean, England: Sediment. Geology, v. 8, p. 177-198.

Jonker, R. K., 1972, Fluvial sediments of Cretaceous age along the southern border of the Cantabrian Mountains, Spain: Leidse Geol. Meded., v. 48, p. 276-380.

Karl, H. A., 1976, Depositional history of Dakota Formation (Cretaceous) sandstones, southeastern Nebraska: J. Sediment. Petrol., v. 46, p. 124-131.

Kelling, G., 1968, Patterns of sedimentation in Rhondda Beds of South Wales: Am. Assoc. Petrol. Geol. Bull., v. 52, p. 2369-2386.

Laird, M. G., 1972, The stratigraphy and sedimentology of the Laksefjord Group, Finnmark: Norges Geol. Unders., v. 278, p. 13-40.

Laming, D. J. C., 1966, Imbrication, paleocurrents, and other sedimentary features in the lower New Red Sandstone, Devonshire, England: J. Sediment. Petrol., v. 36, p. 940-959.

LeBlanc, R. J., 1972, Geometry of sandstone reservoir bodies: *in* Underground Water Management and Environmental Implications: Am. Assoc. Petrol. Geol. Mem. 18, p. 133-190.

Leeder, M. R., 1973, Fluviatile fining-upwards cycles and the magnitude of palaeochannels: Geol. Mag., v. 110, p. 265-276.

Leopold, L. B., Wolman, M. G., and Miller, J. P., 1964, Fluvial processes in geomorphology: W. H. Freeman, San Francisco, 522p.

Lupe, Robert, 1977, Depositional environments as a guide to uranium mineralization in the Chinle Formation, San Rafael Swell, Utah: Jour. Res., U. S. Geol. Survey, v. 5, p. 365-372.

Mack, G. H., 1977, Depositional environments of the Cutler-Cedar Mesa facies transition (Permian) near Moab, Utah: Mountain Geologist, v. 14, p. 53-68.

Mattis, A. F., 1977, Nonmarine Triassic sedimentation, central High Atlas Mountains, Morocco: J. Sediment. Petrol., v. 47, p. 107-119.

McCave, I. N., 1968, Shallow and marginal marine sediments associated with the Catskill complex in the Middle Devonian of New York: *in* G. deV. Klein, *ed.,* Late Paleozoic and Mesozoic continental sedimentation, northeastern North America; Geol. Soc. Am. Spec. Paper 106, p. 75-107.

Meckel, L. D., 1967, Origin of Pottsville conglomerates (Pennsylvanian) in the central Appalachians: Geol. Soc. Am. Bull., v. 78, p. 223-258.

———, 1970, Paleozoic alluvial deposition in the central Appalachians: a summary: *in* G. W. Fisher, *et al., eds.,* Studies of Appalachian Geology: central and southern; Interscience Publ., New York, p. 49-81.

Miall, A. D., 1970a, Continental marine transition in the Devonian of Prince of Wales Island, Northwest Territories: Can. J. Earth Sci., v. 7, p. 125-144.

———, 1970b, Devonian alluvial fans, Prince of Wales Island, Arctic Canada: J. Sediment. Petrol., v. 40, p. 556-571.

———, 1976, Palaeocurrent and palaeohydrologic analysis of some vertical profiles through a Cretaceous braided stream deposit, Banks Island, Arctic Canada: Sedimentology, v. 23, p. 459-483.

———, 1977, A review of the braided-river depositional environment: Earth Sci. Revs., v. 13, p. 1-62.

Moody-Stuart, M., 1966, High and low sinuosity stream deposits, with examples from the Devonian of Spitsbergen: J. Sediment. Petrol., v. 36, p. 1102-1117.

Morey, G. B., 1972, Petrology of Keweenawan sandstones in the subsurface of southeastern Minnesota: *in* P. K. Sims, and G. E. Morey, *eds.,* Geology of Minnesota: a centennial volume: Minn. Geol. Survey, p. 436-449.

———, 1974, Cyclic sedimentation of the Solor Church Formation (Upper Precambrian, Keweenawan) southeastern Minnesota: J. Sediment. Petrol., v. 44, p. 872-884.

Mrakovich, J. V., and Coogan, A. H., 1974, Depositional environment of the Sharon Conglomerate Member of the Pottsville Formation in northeast Ohio: J. Sediment. Petrol., v. 44, p. 1186-1199.

Mukherjee, A. C., 1974, Some aspects of sedimentology of the Missi Group and its environment of deposition: Can. J. Earth Sci., v. 11, p. 1018-1019.

Mykura, W., 1973, The Old Red Sandstone sediments of Fair Isle, Shetland Islands: Geol. Survey Great Britain Bull., v. 41, p. 1-31.

Nami, M., 1976, An exhumed Jurassic meander belt from Yorkshire, England: Geol. Mag., v. 113, p. 47-52.

Nagtegaal, P. J. C., 1966, Scour-and-fill structures from a fluvial piedmont environment: Geol. Mijnbouw, v. 45, p. 342-354.

Niyogi, D., 1966, Lower Gondwana sedimentation in Saharjuri coalfield, Bihar, India: J. Sediment., v. 36, p. 960-972.

Nilsen, T. H., 1969, Old Red sedimentation in the Buelandet-Vaerlandet Devonian district, western Norway: Sediment. Geol., v. 3, p. 35-57.

Owen, D. E., 1973, Depositional history of the Dakota Sandstone, San Juan Basin area, New Mexico: *in* J. E. Fassett, *ed.*, Cretaceous and Tertiary rocks of the southern Colorado Plateau; Four Corners Geol. Soc. Memoir Book, p. 37-51.

Ollier, C. D., 1969, Weathering: American Elsevier, New York, 304p.

Padgett, G. V., and Ehrlich, R., 1976, Paleohydrologic analysis of a late Carboniferous fluvial system, southern Morocco: Geol. Soc. Am. Bull., v. 87, p. 1101-1104.

Paine, W. R., and Meyerhoff, A. A., 1968, Catahoula Formation of western Louisiana and thin-section criteria for fluviatile depositional environment: J. Sediment. Petrol., v. 38, p. 92-113.

Pearce, A. J., 1976, Geomorphic and hydrologic consequences of vegetation destruction, Sudbury, Ontario: Can. J. Earth Sci., v. 13, p. 1358-1373.

Peterson, A. R., 1976, Paleoenvironments of the Colton Formation, Colton, Utah: Brigham Young Univ. Geology Studies, v. 23, p. 3-35.

Pettijohn, F. J., 1975, Sedimentary rocks: Harper and Row, New York, Third Edition, 628p.

Picard, M. D., and High, L. R. Jr., 1972, Paleoenvironmental reconstructions in an area of rapid facies change, Parachute Creek Member of Green River Formation (Eocene), Uinta Basin, Utah: Geol. Soc. Am. Bull., v. 83, p. 2689-2708.

Potter, P. E., 1963, Late Paleozoic sandstones of the Illinois Basin: Illinois Geol. Survey Rept. Invest. 217, 92p.

Puigdefabregas, Cayo, 1973, Miocene point-bar in the Ebro basin, northern Spain: Sedimentology, v. 20, p. 133-144.

Raup, D. M., 1976, Species diversity in the Phanerozoic: An interpretation: Paleobiology, v. 2, p. 289-297.

Read, C. B., 1955, Floras of the Pocono Formation and Price Sandstone in parts of Pennsylvania, Maryland, West Virginia, and Virginia: U. S. Geol. Survey Prof. Paper 263, 32p.

Read, W. A., and Johnson, S. R. H., 1967, The sedimentology of sandstone formations within the Upper Old Red Sandstone and lowest Calciferous Sandstone Measures west of Stirling, Scotland: Scott. J. Geol., v. 3, p. 242-267.

Royse, C. F., Jr., 1970, A sedimentologic analysis of the Tongue River-Sentinel Butte interval (Paleocene) of the Williston basin, western North Dakota: Sediment. Geology, v. 4, p. 19-80.

Ryan, D. J., 1965, Cross-bedding formed by lateral accretion in the Catskill Formation near Jim Thorpe, Pennsylvania: Pennsylvania Acad. Sci. Proc., v. 38, p. 154-156.

Schumm, S. A., 1967, Paleohydrology: application of modern hydrologic data to problems of the ancient past: Internat. Hydrology Symp. Proc., Fort Collins, Colorado, v. 1, p. 185-193.

———, 1968, Speculations concerning paleohydrologic controls of terrestrial sedimentation: Geol. Soc. America Bull., v. 79, p. 1573-1588.

———, 1972, Fluvial paleochannels: *in* J. K. Rigby and W. K. Hamblin, *eds.*, Recognition of ancient sedimentary environments: Soc. Econ. Paleont. Mineral Spec. Pub. 16, p. 38-107.

Schwab, F. L., 1972, The Chilhowee Group and the Late Precambrian-Early Paleozoic framework in the central and southern Appalachians: *in* Peter Lessing *et al.*, *eds.*, Appalachian structures origin, evolution, and possible potential for new exploration frontiers; West Virginia Univ. and Geol. Survey, p. 59-101.

————, 1976, Depositional environments, provenance, and tectonic framework: upper part of the Late Precambrian Mount Rogers Formation, Blue Ridge Province, southwestern Virginia: J. Sediment. Petrol., v. 46, p. 3-13.

————, 1977, Grandfather Mountain Formation: depositional environment, provenance, and tectonic setting of Late Precambrian alluvium in the Blue Ridge of North Carolina: J. Sediment. Petrol., v. 47, p. 800-810.

Selley, R. C., 1965, Diagnostic characters of fluviatile sediments of the Torridonian Formation (Precambrian) of northwest Scotland: J. Sediment. Petrol., v. 35, p. 366-380.

————, 1970, Ancient sedimentary environments: Ithaca, New York, Cornell Univ. Press, 237p.

Sepkoski, J. J., Jr., 1976, Species diversity in the Phanerozoic: species-area effects: Paleobiology, v. 2, p. 298-303.

Sestini, G., 1973, Sedimentology of a paleoplacer: the gold-bearing Tarkwaian of Ghana: in G. C. Amstutz, and A. J. Bernard, eds., Ores in Sediments, Springer-Verlag, New York, p. 275-305.

Sevon, W. D., 1975a, Geology and mineral resources of the Hickory Run and Blakeslee Quadrangles, Carbon and Monroe Counties, Pennsylvania: Pennsylvania Geol. Survey Atlas 194 cd.

————, 1975b, Geology and mineral resources of the Christmans and Pohopoco Mountain Quadrangles, Carbon and Monroe Counties, Pennsylvannia: Pennsylvania Geol. Survey Atlas 195 ab.

Shelton, J. W., 1967, Stratigraphic models and general criteria for recognition of alluvial, barrier-bar, and turbidity current sand deposits: Am. Assoc. Petrol. Geol. Bull., v. 51, p. 2441-2461.

————, 1973, Models of sand and sandstone deposits: a methodology for determining sand genesis and trend: Oklahoma Geol. Survey Bull. 118, 122p.

Shideler, G. L., 1969, Dispersal patterns of Pennsylvanian sandstones in the Michigan basin: J. Sediment., v. 39, p. 1229-1237.

Siedlecka, Anna, 1975, Old Red Sandstone lithostratigraphy and sedimentation of the outer Fosen area, Trondheim region: Norges Geol. Unders., v. 321, p. 1-35.

Siemers, C. T., 1976, Sedimentology of the Rocktown channel sandstones, upper part of the Dakota Formation (Cretaceous), central Kansas: J. Sediment. Petrol., v. 46, p. 97-123.

Slanska, Jarmila, 1976, A red-bed formation in the South Bohemian Basins, Czechoslovakia: Sediment. Geol., v. 15, p. 135-164.

Smith, D. G., 1976, Effect of vegetation on lateral migration of anastomosed channels of a glacier meltwater river: Geol. Soc. Am. Bull., v. 87, p. 857-860.

Smith, N. D., 1970, The braided stream depositional environment, comparison of the Platte River with some Silurian clastics of the north-central Appalachians: Geol. Soc. Am. Bull., v. 81, p. 2993-3014.

Spearing, D. R., 1969, Stratigraphy, sedimentation and tectonic history of the Paleocene-Eocene Hoback Formation of Western Wyoming: Unpub. Ph. D. thesis, Univ. Michigan, Ann Arbor, Mich., 179p.

Steel, R. J., Maehle, S., Nilsen, H., Røe, S. L., and Spinnangr, Å., 1977, Coarsening-upward cycles in the alluvium of Hornelen Basin (Devonian) Norway: sedimentary response to tectonic events: Geol. Soc. Am. Bull., v. 88, p. 1124-1134.

Steel, R. J., and Wilson, A. C., 1975, Sedimentation and tectonism (? Permo-Triassic) on the margin of the North Minch Basin, Lewis: J. Geol. Soc. London, v. 131, p. 183-202.

Steidtmann, J. R., 1969, Environmental reconstruction from cross-stratification: an example from the Pass Peak Formation (Eocene), Western Wyoming: Contrib. Geology, v. 8, p. 168-170.

Stewart, J. H., Poole, F. G., and Wilson, R. F., 1972, Stratigraphy and origin of the Chinle Formation and related Upper Triassic strata in the Colorado Plateau region: U. S. Geol. Survey, Prof. Paper 690, 336p.

Stokes, W. L., 1961, Fluvial and eolian sandstone bodies in Colorado Plateau: in J. A. Peterson and J. C. Osmond, eds. Geometry of sandstone bodies; Am. Assoc. Petrol. Geol., p. 151-178.

Tandon, S. K., 1976, Siwalik sedimentation in a part of the Kumaun Himalaya, India: Sediment. Geology, v. 16, p. 131-154.

Tangiorgi, M., Rau, A., and Martini, I. P., 1977, Sedimentology of early-alpine, fluvio-marine, clastic deposits (Verrucano, Triassic) in the Monti Pisani (Italy): Sediment. Geology, v. 17, p. 311-332.

Thompson, A. M., 1970, Lithofacies and formation nomenclature in Upper Ordovician stratigraphy, central Appalachians: Geol. Soc. Am. Bull., v. 81, p. 1255-1260.

Thompson, D. B., 1970, Sedimentation of the Triassic (Scythian) red pebbly sandstones in the Cheshire Basin and its margins: Geol. J., v. 7, p. 183-261.

Turner, P., 1974, Lithostratigraphy and facies analysis of the Ringerike Group of the Oslo region: Norges Geol. Unders., v. 314, p. 101-131.

————, and Whitaker, J. H. M., 1976, Petrology and provenance of Late Silurian fluviatile sandstones from the Ringerike Group of Norway: Sediment. Geology, v. 16, p. 45-68.

Van De Graff, F. R., 1972, Fluvial-deltaic facies of the Castlegate Sandstone (Cretaceous) east-central Utah: J. Sediment. Petrol., v. 42, p. 558-571.

Van Veen, F. R., 1971, Depositional environments of the Eocene Mirador and Misoa Formations, Maracaibo Basin, Venezuela: Geol. Mijnbouw, v. 50, p. 527-546.

Visher, G. S., 1965, Fluvial processes as interpreted from ancient and recent fluvial deposits: *in* G. V. Middleton, *ed.*, Primary Sedimentary Structures and their Hydrodynamic Interpretation: Soc. Econ. Paleont. Mineral. Spec. Pub., p. 116-132.

————, Saitta, B. S., and Phares, R. S., 1971, Pennsylvanian delta patterns and petroleum occurrences in eastern Oklahoma: Am. Assoc. Petrol. Geol. Bull., v. 55, p. 1206-1230.

Vos, R. G., 1975, All alluvial plain and lacustrine model for the Precambrian Witwatersrand deposits of South Africa: J. Sediment. Petrol., v. 45, p. 480-493.

Walton, A. W., 1977, Petrology of volcanic sedimentary rocks, Vieja Group, southern Rim Rock Country, Trans-Pecos, Texas: J. Sediment. Petrol., v. 47, p. 137-157.

Way, J. H., Jr., 1968, Bed thickness analysis of some Carboniferous fluvial sedimentary rocks near Joggins, Nova Scotia: J. Sediment. Petrol., v. 38, p. 424-433.

Weimer, R. J., and Land, C. B., 1975, Maastrichtian deltaic and interdeltaic sedimentation in the Rocky Mountain region of the United States: *in* W. G. E. Caldwell, *ed.,* the Cretaceous System in the Western Interior of North America: Geol. Assoc. Can. Spec. Paper 13, p. 633-666.

Whisonant, R. C., 1977, Lower Silurian Tuscarora (Clinch) dispersal patterns in western Virginia: Geol. Soc. Am. Bull., v. 88, p. 215-220.

Williams, G. E., 1966, Paleogeography of the Torridonian Applecross Group: Nature, v. 209, p. 1303-1306.

————, 1969, Characteristics and origin of a Pre-Cambrian Pediment: J. Geology, v. 77, p. 183-207.

Wood, G. H., Jr., Trexler, J. P., and Kehn, T. M., 1969, Geology of the west-central part of the southern Anthracite field and adjoining areas: U. S. Geol. Survey Prof. Paper 602, 150p.

Woodrow, D. L., Fletcher, F. W., and Ahrnsbrak, W. F., 1973, Paleogeography and paleoclimate at the deposition sites of the Devonian Catskill and Old Red facies: Geol. Soc. Am. Bull., v. 84, p. 3051-3064.

Worsley, David, 1972, Sedimentological observations on the Grey Hoek Formation of northern Andrèe Land, Spitsbergen: Norsk Polarinst. Årbok for 1970, p. 102-111.

Yeakel, L. S., 1962, Tuscarora, Juniata and Bald Eagle paleocurrents and paleogeography in the central Appalachians: Geol. Soc. Am. Bull., v. 73, p. 1515-1540.

ALLUVIAL SAND DEPOSITION IN A RAPIDLY SUBSIDING BASIN (DEVONIAN, NORWAY)

RON STEEL[1] and SJUR M. AASHEIM[2]

ABSTRACT

A small (=< 2000 km^2), late-orogenic basin with a 25 km stratigraphic thickness of alluvium has been examined in an effort to define how the alluvial facies vary laterally between the basin walls and along its axis, and to determine the extent, if any, to which the rapid subsidence of the basin floor influenced the vertical organisation of the sedimentary pile.

The longitudinal infilling of the basin is the result of an alluvial plain or sandy fan delta, dominated by ephemeral, low sinuosity streams, prograding into floodbasin/lacustrine areas. There is a lateral facies change, either proximal/distal or axial/lateral involving coarse sandstones and conglomerates dominated by trough cross-strata, passing through finer sandstones characterised by planar cross-strata, into alternating fine sandstones, siltstones and mudstones dominated by ripple lamination.

The basin-fill is remarkably well organised into some 200 basinwide, upward coarsening cyclothems (100-200 m thick). These represent prograding sandstone bodies and are probably a response to major episodes of vertical (as a consequence of lateral) movement of the basin floor with respect to source areas. Each of these bodies is itself subdivided into laterally extensive upward coarsening sequences (10-20 m thick), dominant in the proximal reaches and more symmetrical sequences (2-10 m) in the distal reaches. Rapid subsidence and abundant sediment availability, leading to strongly prograding tendencies in the stream systems, is reflected even at this level of organisation in the alluvium.

THE BASIN

Hornelen basin is the largest of four, small, Caledonian, late-orogenic basins lying between Nordfjord and Sognefjord in western Norway (Kolderup, 1915; Bryhni, 1964, 1975; Steel, 1976; Steel *et al.*, 1977). It is bound by thrust (east), high-angle fault (north and south) and unconformable (west) margins (Fig. 1) and has an infill of probable middle Devonian age, as determined from plant fragments and fish remains found in the youngest sandstones. The basin succession is now gently folded and tilted to the east (Figs. 2 and 3), though with more severe, local folding against the margins, particularly in the north where the strata become vertical (Fig. 1). The surrounding basement rock, as also the basement horsts between each of the Devonian basins, has an east-west Caledonian grain and consists of Precambrian gneisses together with Cambro-Silurian metabasalts, schists, quartzites, metagreywackes, granodiorites and gabbros (Bryhni, 1964).

Hornelen Basin was filled largely with sand-sized alluvium, deposited on an alluvial plain or sandy fan delta growing to the west and north west (Figs. 1, 4). The margins of the basin are flanked by fanglomerates, which have been discussed elsewhere (Steel *et al.*, 1977; Larsen and Steel, 1978). Justification for suggesting the term 'fan delta' for the axial alluvium is found in the presence of a belt of fine sediment, largely of floodbasin but partly of lacustrine origin which envelopes successive prograding 'lobes' of sandy alluvium, both distally and laterally (Fig. 4). Along the northern margin particularly, there is evidence of periodic standing water, both from subaqueous debris flows derived from the marginal conglomeratic fans (Larsen and Steel, 1978) and from sandy density flows derived from the axial alluvial system, as discussed below.

Excellent exposure, clear signs that the present basin was largely the original basin (Fig. 2A), and the enormous stratigraphic thickness of the sandy fill ($\approx$25 km), provides

---

[1] Geological Institute, University of Bergen, Norway
[2] Statoil, Lagaardsvn. 78, Stavanger, Norway

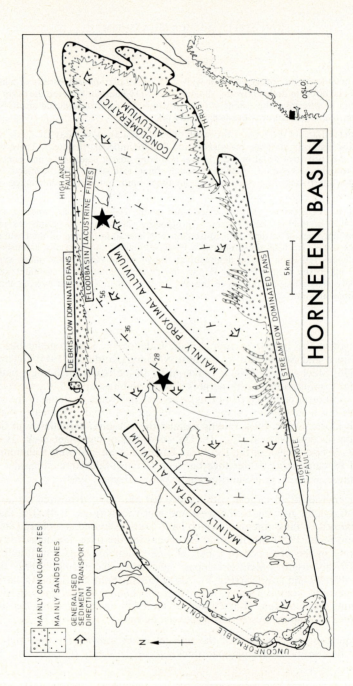

Fig. 1.   Simplified map of Hornelen Basin, showing distribution of main types of alluvial sediment.
Stars mark the Svelgen (axial) and °Alfotfjorden (marginal) areas where most of the detailed studies
were made.

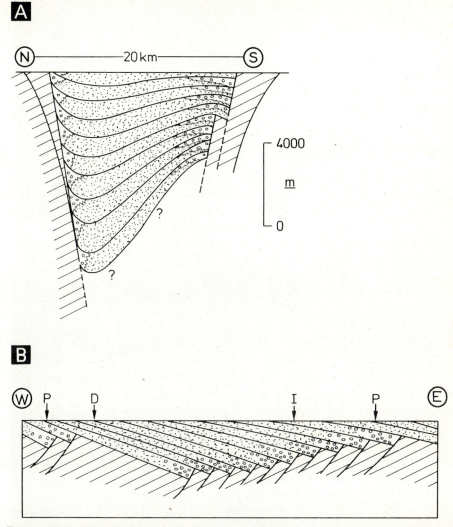

Fig. 2. (A) Simplified north-south section across Hornelen Basin, showing the basin asymmetry, the sheet-like nature of the upward coarsening sandstone bodies, and the cyclic organisation of the infill. (B) Speculative east-west section along part of Hornelen Basin showing repeated overlap of Devonian alluvium onto basement and a possible explanation of the present surface distribution of proximal (P), intermediate (I) and distal (D) alluvium across the basin (*see* Fig. 1). Not to scale.

an ideal opportunity to study the patterns of sedimentation generated within a small, rapidly subsiding alluvial basin. Of particular interest was the possibility of detecting any vertical organisation induced by rapid subsidence, over and above that normally inherent from processes within the alluvial system itself. The most obvious level at which tectonic influence can be detected, as argued by Steel *et al.* (1977), is that of the primary organisation of the succession into some 200, basinwide, upward coarsening cyclothems each of the order of 150 m thick (Figs. 2, 3). These represent an equivalent number of major progradational episodes of the alluvial system in response to periods of basin floor subsidence. Successive westward-prograding sandstone bodies, moreover, can be mapped as overlapping each other eastwards onto basement (Fig. 2B), suggesting that the

Fig. 3. View eastwards along the southern margin and axial region of Hornelen Basin. Strata dip eastwards and are folded along the margin. Note that the step-like topography (cyclicity on 100-200 m scale) is least well developed in the foreground (distal alluvium).

locus of the subsidence periodically shifted eastwards, probably as a result of right-lateral wrench fault movement along the northern edge of the subsiding area (*see* Steel, 1976, Fig. 9).

The discussion below centres on the organisation of sediment within the sandstone bodies. Lateral lithofacies variation in traverses parallel and at right angles to the basin axis, is used as a basis for the analysis, together with details of the vertical sequence at a number of localities along these traverses.

The Sandstone Bodies

Individual, basinwide sandstone bodies show facies variation from marginal fanglomerates through lacustrine/floodbasin fines to axial braided stream sandy alluvium (Fig. 4). This, together with the geometric relationship of successive bodies to each other (Fig. 2), makes it rather unlikely that climatic change was a primary control on the cyclicity. Internal drainage in a semi-arid climate setting with supply of sediment to a laterally migrating, subsiding area can most economically account for the organisation of the basin infill at this level.

The geometry, internal composition and interpretation of individual sandstone bodies is suggested in Figure 4. The north-south cross-section can be observed in the field by walking out individual cyclothems laterally. Within what appears to be simple, sheet-like sandstone bodies (Fig. 3), however, the following features are worth noting:

1. The northern and southern terminations of each 'sheet' are small fanglomerate wedges, showing that the present basin margins are more or less the original ones. Moreover, northern and southern margins contrast in fan geometry, in downfan grain-size gradient and in fan lithofacies (*see* Steel, 1976 for details).

2. The axial alluvial system was also somewhat asymmetric, in that fine-grained floodbasin and lacustrine sediments are best developed in a belt against the fans on the northern margin (Figs. 1, 4).

3. Individual bodies actually thicken considerably against the northern basin edge (Figs. 2A, 4A).

4. The bodies coarsen-upwards as a whole, both axially and in their conglomeratic extremities.

The east-west geometry and facies variation within individual sediment bodies is hypothetical. Figure 4B is a composite section, deduced from a study of many vertical sections from a variety of localities at various points along the east-west length of the bodies. In as much as the east-west variation in this composite section closely resembles the vertical trend within individual profiles, the former is probable a fairly reliable estimate of the actual east-west section through these prograding bodies. Because of the present eastwards tilt of strata, the western (now largely eroded) and eastern (deeply buried) extremities of the sheets are least certain. However, as shown in the model in Figure 2B, a probable eastwards rising of the basement causes proximal alluvium to dominate at the surface at the eastern end of the basin while distal alluvium is more common in the west (Fig. 1). In addition, the model predicts that any marked basement irregularities or major shifts in sediment dispersal pattern can cause unusually proximal or unusually distal sediments to appear on the surface. Details of the sedimentation sequence along this dimension are discussed below, but here it is pertinent to note the following gross east-west changes within sandstone bodies:

1. Although there is an upwards coarsening trend, particularly in the middle and proximal reaches of individual bodies, there is clearly a marked overall decrease in grain-size westwards.

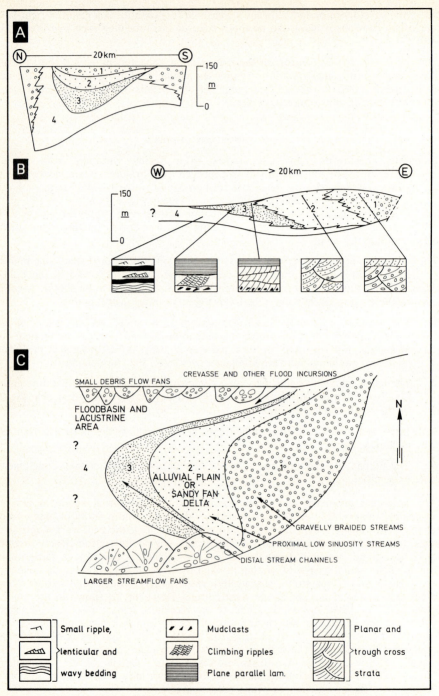

Fig. 4. Schematic north-south (A) and east-west (B) sections across a single sandstone body in Hornelen Basin, showing the dominant sedimentary structures in the 4 diachronous facies belts in (B). (C) Simplified palaeogeographic reconstruction for a time period early in the history of a single cycle.

2. Within each body there is a series of diachronous facies belts rising westwards, causing the proximal reaches (on the surface of the body) to be dominated by trough cross strata, the middle reaches by planar cross-strata and the distal reaches by ripple, wavy and plane-parallel lamination (Fig. 4B).

3. The east-west length of the bodies is difficult to determine, but most recent estimates suggest a minimum of some 20 km. The amount and rate of thinning of bodies westwards is also difficult to determine, because the cyclicity is least marked in the most distal facies (see Fig. 3).

The detailed studies, which are the basis of the sections in Figures 4A and 4B, supply evidence for the palaeogeographic scheme suggested in Figure 4C. In time, the dominant tendency within the system was for westwards and northwestwards progradation, with proximal alluvium coming to overlie distal alluvium, as shown so clearly by the upwards coarsening in individual sandstone bodies. The common thickening of the alluvium against the northern margin suggests a greater rate of subsidence here and probably an important northerly component to the palaeoslope. This component of tilt is consistent also with the tectonic model, in as much as the dominating element is the wrench fault bordering the basin's northern edge. Despite this it is important to note that the floodbasin/lacustrine belt along the northern margin appears to have been semi-permanent and only rarely the recipient of the much coarser axial alluvium.

Patterns Of Sedimentation Within Sandstone Bodies

Traverse parallel to direction of sediment transport

A series of 5 vertical profiles is presented here (Figs. 5, 6, 7, 9, 11). These are believed to show the internal details of any one of the Hornelen sandstones bodies along a proximal/distal traverse. The 3 intermediate profiles have been logged in the Svelgen area, while the most proximal and most distal have been taken from the outermost islands and from east of the Ålfotfjorden area respectively (Fig. 1). In general there is no difficulty in defining the base of a sandstone body in the field. The top is not always the coarsest portion of the sequence because in the northern half of the basin there is frequently a thin, relatively fine-grained capping portion, but the base of the overlying sequence is usually well marked by relatively thick units of siltstone.

It is emphasised that the various vertical sequences figured below actually contain a whole range of distal to proximal alluvium because of the progradational nature of the bodies. However, discussion centres on the upper portion of each sequence since this is representative of the alluvium present at that point when the progradation of the alluvial system was arrested.

Proximal conglomeratic alluvium

In their most proximal aspect the Hornelen sandstone bodies are conglomeratic. Unlike the basin margin fanglomerates, clasts here are generally relatively well sorted, well rounded (Fig. 5A) and of varied composition. The latter is particularly important in suggesting more distant and/or larger drainage areas for this axial system as compared with the basin margin fans. In addition, the conglomerates have channelled bases, and are most often apparently massive or show very low angle stratification, though both planar and trough cross-strata in sets up to 2 m are found. These features suggest channel scour and fill and longitudinal bar development in gravelly braided streams (Miall 1977). The conglomerates alternate with cross-stratified or plane parallel laminated sandstone (Fig. 5B) and, in some cases, these gradually overlie the conglomerates, suggesting conglomerate/sandstone couplets. Such couplets, together with alternations of coarse and fine grained conglomerates, often with filled and relatively open frameworks, may be a response to varying flow stage. In other cases, particularly where the conglomerates are

Fig. 5. (A) Conglomerates typical of the most proximal braided stream deposits. (B) Typical vertical sequence through these deposits. MPS = mean maximum particle size. See Figure 4 for legend.

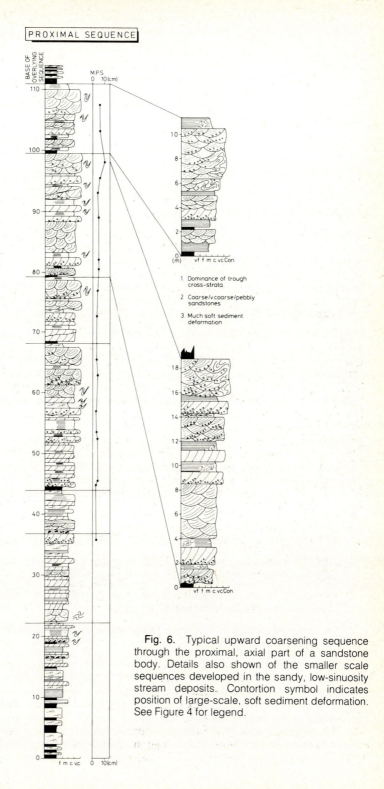

Fig. 6. Typical upward coarsening sequence through the proximal, axial part of a sandstone body. Details also shown of the smaller scale sequences developed in the sandy, low-sinuosity stream deposits. Contortion symbol indicates position of large-scale, soft sediment deformation. See Figure 4 for legend.

finer grained, they occupy thin sheets interstratified with sandstones, similar to the bar head bedding in the lateral bars of Bluck (1974, p. 534; 1976, p. 429, Figs. 9A, 14A.).

There are no obvious large-scale upward coarsening trends through this conglomeratic succession, and in some cases an upward increasing sandstone/conglomerate ratio indicates an upward fining sequence (Fig. 5B). In general, this most proximal alluvium has no characteristic cyclic organisation, although it is emphasised that very few of these sequences are exposed.

Proximal sandy alluvium

A complete section through a typical sandstone body in which the upper part is representative of a proximal sandy reach is shown in Figure 6. This alluvium, which is commonly very coarse grained or pebbly, is dominated by trough cross-strata (sets rarely thicker than 50 cm) and shows much soft sediment deformation, particularly asymmetric oversteepened or overturned cross-strata.

The facies in general probably represents channel scour and fill and dune/megaripple migration in straight or low sinuosity alluvial channels. As with the conglomeratic alluvium, the alternation of coarse cross-stratified sandstone with finer plane-parallel laminated sandstones could be interpreted in terms of waning flood conditions, producing fining-upward units with the plane-parallel laminae forming under very shallow water depths. Abundance of erosion surfaces, however, makes interpretation at this level difficult and many of the thicker units are likely to be multistorey.

One of the most notable features of this sequence is that it is clearly subdividable into upwards coarsening units of the order of 10-22 m in thickness, bounded by mudstone/siltstone horizons (Fig. 6). Sandy, planar or trough cross-strata generally give way upwards to pebbly sand trough cross-strata, while plane parallel lamination interbeds at all levels, as noted above. Because such units can be traced laterally for at least 1 km it is likely that they originate from progradation of a large segment of the alluvial plane.

Transitional sequence I

This type of alluvium is shown in the upper part of the succession in Figure 7A. It is dominated by planar cross-strata in fine/medium-grained sandstone and probably originated from linguoid or transverse bars in the middle and more distal reaches of Hornelen alluvial plain (e.g. Collinson 1970). In addition to the interbedded plane parallel laminated units there is also here the occasional presence of mudclast conglomerates.

Subcyclicity within the 100 m sequence is less obvious here, but there can be distinguished a series of laterally persistent (>0.75 km) units (5-10 m in thickness) bounded by mudstone/siltstone (Fig. 7B) which show the following internal features:

1. An initial upwards coarsening, followed by upwards fining, with coarsest levels in the middle or just below the middle of the unit (Fig. 7B, C).

2. An initial increasing then decreasing cross-stratal set thickness, with the thickest sets near the middle of the unit (Fig. 7B).

3. A common capping of low-angle trough cross-stratified sandstones comprising thin (<20 cm) sets (Fig. 7C).

This subcyclicity, as in the more proximal alluvium, probably originates from periodic progradation of substantial portions (? lobes) of the alluvial plain or fan delta. The near-symmetric nature of the motifs developed here, in contrast to those in Figure 6, suggests gradual waning from maximum flow strength and water depth, possibly due to less rapid abandonment of lobes in their distal reaches. As seen through six such lobes (Fig. 7B), palaeocurrents read from cross-strata show a consistent divergence from those derived from parting lineation (P.L.), suggesting systematic variation in current direction with bedform, probably as dependent on water depth and flow stage (*see* also Bluck 1974).

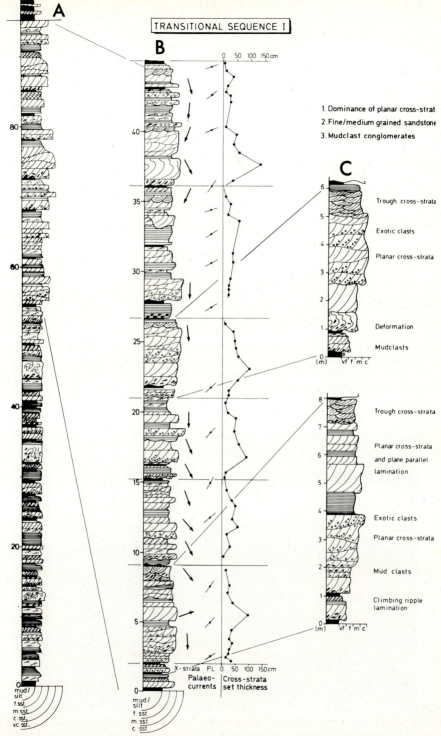

Fig. 7. (A) Typical upward coarsening sequence lateral (more distal) to that in Figure 6. (B) and (C) Details of the smaller scale, near symmetrical sequences within the upper part of the succession. Coarsest grained and thickest sets tend to occur in the middle of these sequences. See Figure 4 for legend.

Transitional sequence II

Figure 8 shows the upper half of a sequence which occupies a distal position in the sandstone body with respect to Transitional sequence I. This alluvium is organised into a series of units 3-7 m thick, each of which is bounded by mudstone, and one of which is shown in detail in Figure 9. In general the sediments are fine grained and there is an abundance of mudclast conglomerates and climbing ripple lamination. The 3-7 m units appear to be sheetlike and tend to be coarsest grained in the centre or just below centre, as was also the case for the units of this order of the size in Transitional sequence I. In contrast to the latter, however, there is here, in addition, a very clear small-scale organisation into upward fining units (~1 m thick) (Figs. 9, 10) which occupy very low channels and which have the following characteristics:

4. Mudstone cap

3. Ripple or climbing-ripple laminated, very fine sandstone

2. Plane-parallel laminated, very fine sandstone

1. Erosion surface overlain by cross-stratified (often deformed) fine sandstone with abundant mud flakes.

Large-scale soft sediment deformation, often as recumbent folds in the units up to 1 m thick, most commonly occurs in the coarsest, central portion of the 3-7 m units (Fig. 9). Lower in the sandstone body, associated with the folding there are often spectacular mudstone breccias in which clasts up to 1 m long occur.

Because of their geometry and internal structure the thin upward fining units do not resemble laterally accreted point bar sequences and it is suggested that they formed rather by vertical accretion in very broad shallow channels during the waning of flood periods. The larger scale, near symmetric package of units (e.g. Fig. 9) represents 10 or more such flood episodes which dispersed into lobes which themselves were prograding.

Fig. 8.   Upper 60 m of an upward coarsening sequence occuring distally (with reference to Fig. 7) in the sandstone body. Note the subdivision into some 10 smaller sequences. X marks the unit shown in detail in Figure 9. Svelgen.

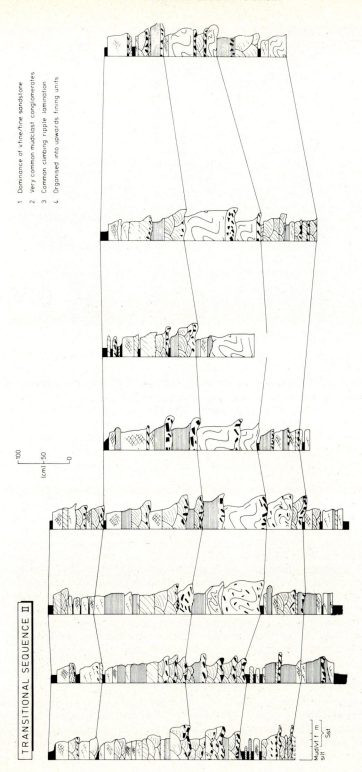

TRANSITIONAL SEQUENCE II

1 Dominance of v fine/fine sandstone
2 Very common mudclast conglomerates
3 Common climbing ripple lamination
4 Organised into upwards fining units

100
50
(cm) 0

Mud/vf f m
silt   Sst

Fig. 9. Details within the sequence marked X in Figure 8. Note that the sequence is near symmetrical, with the coarsest and thickest fining upward units occuring near the centre. Lateral distance shown is some 250 m. See Figure 4 for legend.

Fig. 10. Detail within one of the fining-upward units from Figure 9.

The abundance of mudclasts high in the body indicates ephemeral conditions there, while the mudstone breccias and folding may suggest significant slopes, possibly where the lobes advanced into areas of standing water. The general abundance of soft sediment deformation, and of recumbent folded cross-strata in particular, is remarkable in Hornelen Basin. Here, of all places, we can probably relate much of this deformation to the trigger action of earthquake shocks (*see* also Allen and Banks, 1972).

Floodbasin/lacustrine sequence

In their most distal segment the Hornelen sandstone bodies are dominantly siltstones and mudstones with very fine-grained sandstone interbeds showing rippled, wavy, and lenticular laminations (Fig. 11). As shown in Figure 4 these fine sediments appear to envelope the coarser alluvium laterally as well as distally. They are termed floodbasin deposits because of their fine grain-size and because they appear to have accumulated in basinal areas which remained relatively stable in position over time intervals represented by kilometers of sediment thickness. In addition, although mudcracks and mudclast conglomerates are common evidence of ephemeral flood conditions in some areas, there are many areas where such features are notably absent and where there is evidence of standing water into which subaqueous debris flows took place (*see* Larsen and Steel, 1978). For these reasons it is argued that some of the floodbasins were occupied by lakes, particularly those along the northern margin of the basin, and that the lobe-like axial alluvial system may periodically have been a sandy fan delta.

Traverse at right angles to the basin axis

Vertical sequences at three approximately laterally equivalent localities (Åskora, Sigdestad and Myklebustdalen in Fig. 12), along a traverse at right angles to the basin axis, are presented here. The sequence illustrated for these three localities is typical for the facies belts of proximal, transitional and floodbasin alluvium which, because of the lobate nature of the axial system, run subparallel to the basin margin here. Despite broad

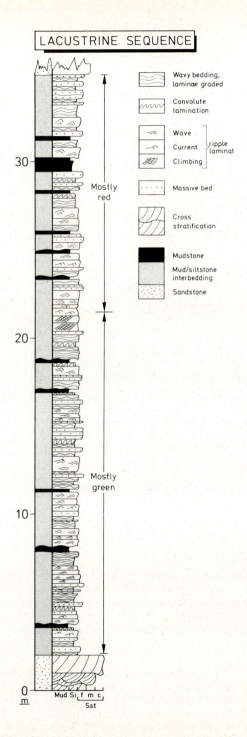

Fig. 11. Details through a fine grained lacustrine sequence, representative of the most distal reaches of the sandstone bodies. East of Ålfotfjorden.

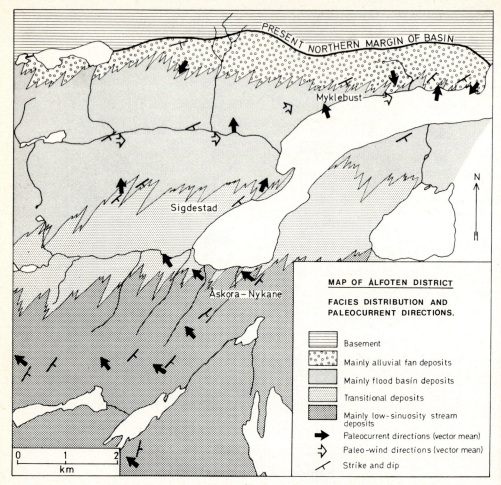

Fig. 12. Simplified facies map of the Ålfotfjorden area, showing the position of the Åskora-Sigdestad-Myklebust traverse, from proximal alluvium through transitional to floodbasin deposits.

lithological similarities and the similar outwards decrease of grain size (Fig. 12), the internal organisation of the sequence here differs in some important respects from that described for the sequences along the traverse parallel to the basin axis. In addition, a detailed study of the floodbasin facies has been made in this traverse, providing new information on a number of different types of flood-generated sedimentation unit.

Proximal sandy alluvium

This facies, as in the axial traverse, is very coarse-grained, often pebbly, and dominated by trough cross-stratification. It dominates the thick (? upwards coarsening) sequences illustrated from Åskora and Nykane (Fig. 13) and is divisible into units by the occurrence of mudstone/siltstone layers. A minority of these units show clear upwards coarsening or upwards fining trends of a type much more common in the transitional alluvium described below from Sigdestad. Most of the units, however, are some 8-12 m in thickness and are of more complex construction, showing no simple

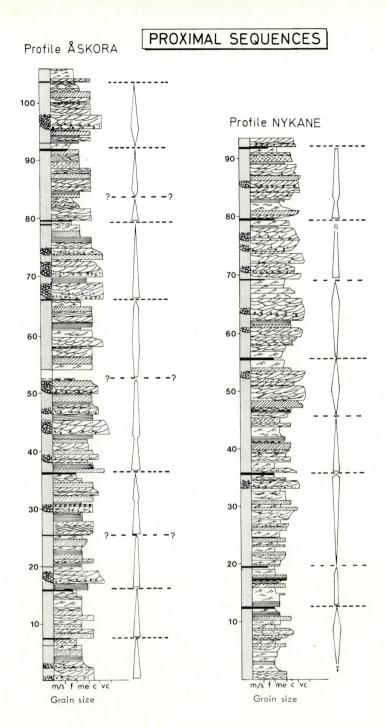

Fig. 13. Sequences through proximal alluvium at Askora and Nykane (see Fig. 12). Subdivision into smaller scale units is possible, but no systematic trends appear (compare with Fig. 6). See Figures 4 and 11 for legend.

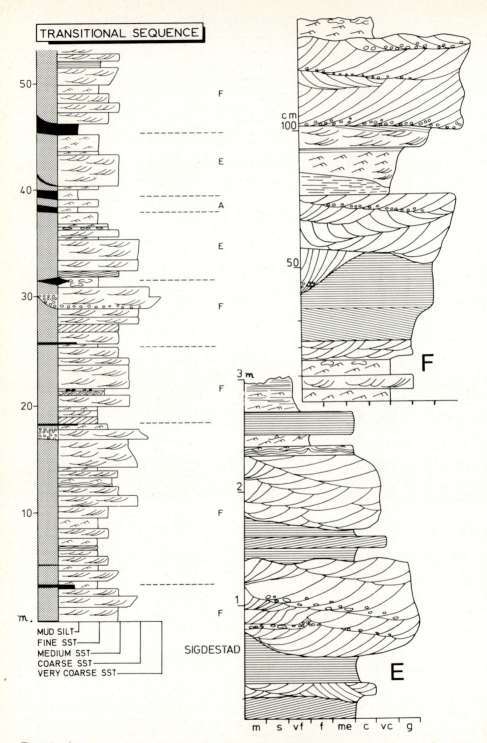

Fig. 14. Sequence through transitional alluvium from Sigdestad area. Subdivision into smaller scale sequences, which may be either upward coarsening (F) or upward fining (E), is possible.

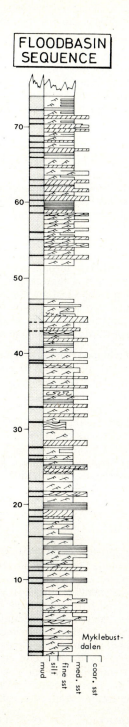

FLOODBASIN SEQUENCE

Myklebust-dalen

mud | silt | fine sst | med. sst | coar. sst

Fig. 15. Typical floodbasin sequence, measured in the Myklebust area.

upwards fining or coarsening but rather a vigorous repetition of sets or cosets, separated by local erosion surfaces and occasional thin draping mudstone layers.

The facies in general probably represents episodic growth and abandonment of braided stream bars and channels, with migrating dunes/megaripples as the dominating small-scale bedforms (eg. Bluck 1976). The somewhat random nature of the multistorey units here, in contrast to the well defined upwards coarsening units in the same facies along the traverse parallel to the basin axis (Fig. 6), may be explicable in terms of the marginal position of the braided alluvium here, in contrast to a more axial location (? site of optimum progradation) for the latter.

Transitional alluvium

Sequences through alluvium transitional between the coarse-grained braided stream sediments and the fine-grained floodbasin deposits are illustrated from Sigdestad (Fig. 14). Distinct upwards coarsening (usually with a fine capping) and less distinct upwards fining units, bounded by laterally persistent mudstone/siltstone, can be distinguished. The former units, usually only 2-5 m in thickness and occupying less than 30% of this alluvium, often show an upwards decrease in set thickness as well as grain size, and are never repeated vertically on each other (eg. Fig. 14, Type E unit). The latter units, generally 2-10 m thickness and occupying more than 70% of this type of alluvium, show upwards transition from ripple lamination through plane parallel or low angle cross-strata to trough cross-strata. There is usually a thin capping of wavy or ripple laminated fine sandstone/siltstone below the bounding siltstone/mudstone layer (Fig. 14, Type F unit) and sometimes there is upwards transition into a Type E unit. The thicker type F units are generally multistorey where individual sub-units thicken upwards, in addition to the overall upwards coarsening and upwards thickening of sandstone sets.

This alluvium probably also developed from low sinuosity stream systems, although the finer grain size and palaeogeographic position of this belt (Fig. 4C) indicate development in a distal position with respect to the coarser grained sequences. The upwards fining, Type E units may represent point-bar, bar tail/slough channel (eg. Bluck, 1976, Plate lb), or, more likely in this context, crevasse channel cut-off sequences. (*see* also Walker, 1976, Fig. 4) Individual upward coarsening units may have resulted from migration of bar heads over bar tails (eg. *see* medial bars of Bluck, 1976, Fig. 17). However, the general thickness and multistorey nature of these sequences suggests an origin from larger scale progradation of the alluvium, probably involving a prograding levee/crevasse channel complex (*see* also Elliott 1974) in this marginal region. It seems likely that in many cases the Type E unit is simply the upper portion of a near-symmetric Type F-Type E sequence which originated from progradation followed by gradual abandonment of a crevasse channel complex.

Floodbasin/lacustrine deposits.

This type of sequence is well illustrated along Myklebustdalen road section (Fig. 12), and Figure 15 shows a typical sample taken from the 1000 m succession studied here. In general the succession shows an alternation of thinly bedded fine sandstone, siltstone and mudstone. The occasional conglomeratic interbeds were derived, not from the axial alluvial system but from the alluvial fans which flank the nearby basin margin (Fig. 12). The good exposure and detailed examination of grain size and structure of some 4000 beds here allowed data treatment by Markov Chain analysis. The results from this, giving the vertical transition probabilities of various bed types, are summarised in Figure 16A. Full details of the analysis are given by Aasheim (1977). The analysis resulted in the erection of 4 models which summarise the most commonly occurring units in this succession (Fig. 16B). Some of the general features of these units (A-D), interpreted below as discrete flood-generated sedimentation units, are as follows:

1. Individual units are remarkably thin (4-25 cm), with the mean thickness of component beds varying from 3-63 mm for different bed types (Fig. 16B).

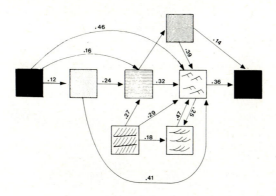

Fig. 16A

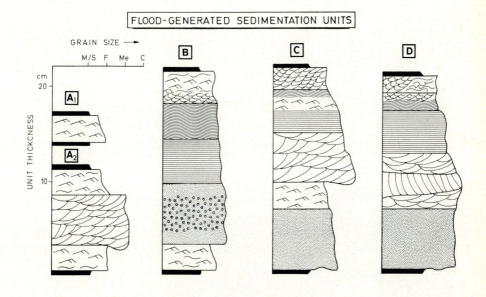

Fig. 16B

Fig. 16. (A) Flow diagram showing vertical transition porbabilities of various bed types within the Myklebust floodbasin succession (approximately 4000 transitions). (B) The 4 main models of flood-generated sedimentation units resultant from the analysis of the floodbasin succession. See Figures 4 and 11 for legend.

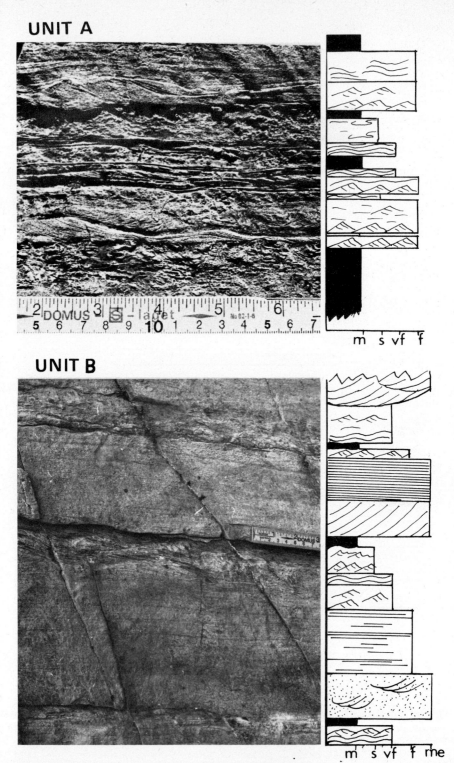

Fig. 17. Field examples of A and B type flood units from Myklebustdalen succession. Note the thickness of the unit.

UNIT C

UNIT D

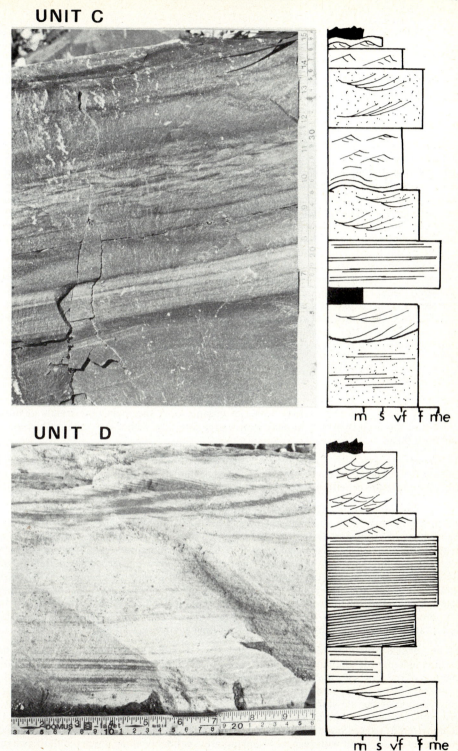

Fig. 18. Field examples of C and D type units from Myklebustdalen succession.

2. Although the flood units are dominated by an upwards fining trend, there is often a characteristic lowermost portion to the unit (2-6 cm) which shows an overall upwards coarsening (Fig. 16B, units A, B, C).

3. Individual bed-type boundaries are generally sharp, with the exceptional gradational boundaries where wavy overlies plane-parallel lamination and where small-ripple and climbing-ripple lamination alternate.

Units C and D are distinguished from A and B in being somewhat coarser grained (compare Figs. 18 and 17) and in occurring generally in coarser grained successions. In addition they have a more common development of medium-scale cross-strata (Fig. 18). They are interpreted as levee and small crevasse channel units respectively. The initial massive sandstone bed probably was deposited from flow with a high sediment concentration during flood stage, while the general upwards fining records waning flood conditions. The cross-stratified sets, particularly in Unit D which is notably less laterally persistent than Unit C, records dune migration in channellized flow. The occasional, initial upwards coarsening of units implies initial progradation of levee and small channel systems in the floodbasin. Unit B (Fig. 17) is more sheetlike than units C and D and is much less often repeated vertically on itself. It probably originated from subaqueous high concentration sheetflow as suggested by the initial massive or inverse/normal graded sandstone bed. The overlying plane parallel, wavy and ripple laminated portion may represent a later phase of lower sediment concentration turbidity flow. Unit A (two varieties, Fig. 16) is the most common, thinnest and finest grained (Fig. 17) of the 4 flood units recognised in the measured succession. It represents the most distal floodbasin deposits, and probably includes the distal ends of flood units B, C and D.

CONCLUSIONS AND DISCUSSION

Table 1 summarises the various scales of organisation, the main time trends at these scales and the way in which organisation varies laterally within the sandy alluvium of Hornelen Basin.

1. Large scale cycles. Analysis of the cyclicity on this scale involves (a) a geometrical/depositional model for a single cycle and (b) a basin model containing an explanation for the repetition of cycles.

In its three-dimensional form each cycle represents a sandstone body some 20 km wide (north-south) and more than 20 km long (east-west), within which there is lobe-like distribution of grain-size, fining both distally and laterally. In addition, within this lobe, belts of similar grain size rise diachronously westwards so that there is an overall upwards coarsening in the body. The sandstone represents alluvium which was deposited by fluvial systems dominated by a strong progradational tendency, probably a response to semi-arid climate, internal drainage and rapid subsidence of the basin floor. The progradation, together with the lobe-like distribution of the coarsest alluvium and the evidence for periodic standing water in the surrounding floodbasins may justify the term 'sandy fan delta' rather than 'alluvial plain' for the basin-filling system.

The vertical repetition of the sandstone bodies must be considered in relation to the small conglomeratic alluvial fan bodies, showing similar internal organisation, along both margins of the Hornelen Basin. As discussed more fully by Steel et al. (1977) the cyclicity is probably due to repeated lowering of the basin floor and the consequent repeated progradation of the alluvial system. In detail however, the tectonics were more complex than this, because a lateral eastwards overlap of successive sandstone bodies can be demonstrated, suggesting an important horizontal component of movement. This lateral movement of the basin with respect to upland area, probably along a wrench fault system, is one reason why the cyclicity is so prominent in the basin fill. Contrast between

Table 1. Summary of the scales and types of sequence present in the sandy alluvium of Hornelen Basin. C-U=coarsening upwards F-U=fining upwards.

	TRAVERSE PARALLEL TO BASIN AXIS					TRAVERSE ACROSS BASIN AXIS		
	Proximal				Distal	Proximal		Distal
	Congl. alluvium	Proximal	Trans.I	Trans.II	Flood/ lacustr.	Proximal alluvium	Trans.	Floodb/ lacustr.
CYCLICITY Large-scale 100-200 m	? F-U (few data)	Promi-nent, C-U	Promi-nent, C-U	Present C-U	Not obvious	Present, C-U	Present C-U	Not obvious
Smaller scale	Not obvious (few data)	10-22m, C-U (Fig. 6)	5-10m, Near symm. (Fig. 7)	3-7m, Near symm. (Fig. 7)	Few data	8-12m, No dominant trends (Fig. 13)	2-10m, C-U and Near symm. (Fig. 14)	Not present
FLOOD UNITS	? 1-6m, Congl/sst couplets (Fig. 5) or alternat-ing open and closed framework congl.	Difficult to define due to erosion surfaces		~0.2-1.5m F-U	?	Difficult to define, much erosion	1-4m, C-U or F-U	2-25cm F-U (Figs. 16-18)

proximal alluvium at the top of one body and the distal alluvium at the base of the succeeding one is exaggerated by the sourcewards overlap of bodies.

We know of no other alluvial basins showing this extraordinary repetition of upward coarsening cycles, although the number of basins documented as having developed under a strike-slip regime are relatively few as yet. Examples of individual sequences (eg. Leeder, 1973), or instances of a few repeated upward coarsening cycles on this scale or thicker (eg. Steel and Wilson, 1975; Heward in press and this volume) have been recorded, however, usually in the nature of basin-fill progradational sequences.

2. Smaller scale cycles. In Hornelen Basin there is also clear and abundant evidence of organisation of the alluvium into smaller scale upward coarsening sequences. These occur on a scale of 10-20 m in proximal reaches and probably become upward coarsening-fining sequences on a scale of 2-7 m in distal reaches. Although relatively few profiles have been illustrated here, these are typical of very many more which have been logged and examined.

Cycles of this type and of similar scale have not been commonly recorded either from modern or ancient braided alluvium. In modern braided alluvium limitation of exposure results in only small scale sequences, usually less than 2 m, being recorded (*see* Miall, 1977, Fig. 10). These generally show fining-upwards and result mostly from flood deposition or lateral accretion in channel bends. In ancient braided alluvium, sequences on a scale 2-20 m have been much more commonly recorded but upward fining trends (or more random patterns) are normal and are attributed to relatively local causes such as lateral accretion, channel aggradation, channel re-occupation or less locally, flooding (*see* Miall, 1977, Fig. 11). Some few instances of upward coarsening sequences have been documented, however. Examples have been given by Williams (1969) from Precambrian braided alluvium, by Bluck (1971, 1976) from both braided and meandering alluvium and from Costello and Walker (1972) from glacial outwash. In most of these cases the sequences are very local and originate from bar migration (coarse head over fine tail) or from re-occupation of abandoned channels. In the Precambrian case it is not clear how laterally extensive the 6-8 m upward coarsening sequences are, though it has been suggested that they originate from alternating sheetflood and braided bar deposition (Williams 1969, Table 1).

The Hornelen Basin sequences, despite their small size, are usually clearly multistorey, ie. not developed from single bars or channel fills, and are of great lateral extent compared to those described above. It is suggested that they originate simply from the strongly progradational tendencies of the alluvial system. They are likely to represent minor progradational episodes within the major (100-200 m thick) basin-fill response sequences, probably large depositional lobes, as suggested by the general proximal/distal decrease in the scale of the cyclicity (Table 1). In the general sense that these lobes were so subject to progradation, we tentatively suggest that their organisation, even on this small scale, reflects abundant sediment availability, together with the tectonic instability and rapid subsidence of the basin floor. It appears that the more predictable random or upward fining patterns of braided alluvium have been overprinted.

The abundance of upward coarsening sequences on a number of different scales here in braided stream deposits tends to support a suggestion of Costello and Walker (1972) that such trends could be important in braided alluvium. The important point, of course, is that these sequences are not necessarily forming as a direct response to the alluvial processes themselves (although the examples of Bluck 1971, 1976 and others seen by the present authors in Triassic Bunter Pebble Beds of England are convincing) but rather because braided streams complexes are commonly subject to progradation and form often on large alluvial lobes or sandy fans, whose existence may not be obvious in sequences of ancient alluvium. In the light of the above examples the suggestion of Miall (1977), that coarsening-upward sequences (eg. those of Costello and Walker 1972) may rather be a criterion of the glacial environment, is probably invalid.

The downslope change in organisation from upwards coarsening to more symmetric sequences is also of some interest. This many have been due to a tendency for the lobes to be more gradually abandoned in their distal as compared to their proximal reaches, or to a relatively slow rise and fall of discharge in distal areas, or it may be that the fine upper portion of sequences in the distal areas reflects waning sediment discharge from source, with this latest stage discharge of fines having bypassed the proximal reaches of the system. Table 1 and Figure 12 show that systematic progradation did not dominate in the more marginal reaches of the axial system. Rather more predictable random vertical patterns developed here. Farther towards the basin margin, however, upward coarsening sequences return, where coarser sediment was fed into the floodbasins. The sequences here, both in scale and in type, are not unlike those recorded from modern deltaic interdistributary bays, where flood-generated sediments are derived from the adjacent distributary channels (*see* Elliott, 1974).

3. Flood units. Definition of the smallest units of sedimentation (whether from discrete flood events or from varying flood stage within a river system) is generally difficult because of the abundance of erosion surfaces in braided stream alluvium. Table 1 shows that in Hornelen Basin these units can be clearly picked out only in the more distal alluvium, where the mudstone capping to units has been most easily preserved. Only in the floodbasin sequence has there been any rigorous analysis of these units, showing that rarely more than 20 cm of sediment were deposited by single flood episodes. Similar types of thin sequence have been recorded as ephemeral flood cycles (eg. McKee *et al.,* 1967; Williams, 1971), from the distal parts of outwash fans (eg. Boothroyd and Ashley, 1975) and from levee/crevasse complexes marginal to braided rivers (eg. Coleman, 1969; Singh 1972), but, as far as we are aware, the initial, fine-grained portion to these units has not been described previously. Thicker (~1 m), more distinct fining-upwards flood units occur in lobes from the Transitional II sequence profile, in the distal reaches of the axial traverse. The large amounts of climbing ripple lamination in the upper portion of these flood units is worth noting; it suggests a widening of the flow or dispersal of the sediment in sheets in front of the channels in these distal reaches of the alluvial system. In the more proximal belts of alluvium small-scale units showing either upwards fining or upwards coarsening can be distinguished, but the significance of these is not clear.

Acknowledgements

We thank our colleagues in Bergen and Stavanger for enthusiastic discussions during the course of the West Norway Devonian Project, and are particularly grateful to Ånon Spinnangr and E. Undersrud for unpublished data used in Figures 5 and 10. We thank Norges Tekniske Naturvitenskaplig Forskningsråd and the University of Bergen for supporting the work financially, Brian Bluck and David Thompson for reading and commenting on the manuscript, and Ellen Irgens, Masaoki Adachi and Jan Lien for the drafting and photograhic work.

References

Aasheim, S., 1977, Devonian sediments along a segment of the northern margin of Hornelen Basin, with emphasis on flood-generated cyclic sedimentation in the Ålfoten area, western Norway: Cand. Real. Thesis, Univ. of Bergen, 542 p.

Allen, J.R.L., and Banks, N.L., 1972, An interpretation and analysis of recumbent-folded, deformed cross-bedding: Sedimentology, v. 19, p. 257-83.

Bluck, B.J. , 1971, Sedimentation in the meandering River Endrick: Scott. J. Geol., v. 7, p. 93-138.

——, 1974, Structure and directional properties of some valley sandur deposits in southern Iceland: Sedimentology, v. 21, p. 533-554.

——, 1976, Sedimentation in some Scottish rivers of low sinuosity: Trans. Roy. Soc. Edin., v. 69, p. 425-456.

Boothroyd, J.C., and Ashley, G.M., 1975, Process, bar morphology and sedimentary structures on braided outwash fans, northeastern Gulf of Alaska: in A.V. Jopling and B.C. McDonald, *eds*. Glaciofluvial and glacio-lacustrine sedimentation; Soc. Econ. Paleontol. Mineral. Spec. Pub. 23, p. 193-222.

Bryhni I., 1964, Relasjonen mellom senkaledonsk tektonikk og sedimentasjon ved Hornelens og H°asteinens devon: Norges Geologiske Undersokelse, v. 223, p. 10-25.

————, 1975, The West Norwegian Basins of Old Red Sandstone: Proc. IXieme Congress Int. Sed., Nice 1975.

Coleman, J.M., 1969, Brahmaputra River: Channel Processes and sedimentation: Sediment. Geol., v. 3, p. 129-239.

Collinson, J.D., 1970, Bedforms of the Tana River, Norway: Geogr. Ann., v. 52A, p. 51-36.

Costello, W.R., and Walker, R.G., 1972, Pleistocene sedimentology, Credit River, southern Ontario. A new component of the braided river model: J. Sediment. Petrol., v. 42, p. 389-400.

Elliott, T., 1974, Interdistributary bay sequences and their genesis: Sedimentology, v. 21, p. 61-62.

Heward, A.P., in press, Alluvial fan and lacustrine sediments from the Stephanian A and B (La Magdalena, Cinera-Matallana and Sabero) coalfields, northern Spain: Sedimentology.

Larsen, V. and Steel, R.J., 1978, The sedimentary history of a debris flow-dominated, Devonian alluvial fan - a study of textural inversion: Sedimentology, v. 25, p. 37-59.

Leeder, M.R., 1973, Sedimentology and palaeogeography of the Upper Old Red Sandstone in the Scottich Border Basin: Scott. J. Geol., v. 9, p. 102-119.

Kolderup, C.F., 1915, Vestlandets devonfelter og deres plantefossiler: Naturen, p. 217-232.

McKee, E.D., Crosby, E.J., and Berryhill, H.L., 1967, Flood deposits, Bijou Creek, Colorado: J. Sediment. Petrol., v. 37, p. 829-851.

Miall, A.D., 1977, A review of the braided river depositional environment: Earth Sci. Rev., v. 13, p. 1-62.

Singh, I.B., 1972, On the bedding in the natural levee and point bar deposits of the Gomti River, Uttar Pradesh, India: Sediment. Geol., v. 7, p. 309-317.

Steel, R.J., 1976, Devonian basins of western Norway - sedimentary response to tectonism and varying tectonic context: Tectonophysics, v. 36, p. 207-224.

Steel, R.J. and Wilson, A.C., 1975, Sedimentation and tectonism (?Permo-Triassic) on the margin of the North Minch Basin: Geol. Soc. Lond. J., v. 131, p. 183-202.

Steel, R. J., Maehle, S., Nilsen, H., Roe, S. L., and Spinnangr, Å., 1977, Coarsening-upward cycles in the alluvium of Hornelen Basin (Devonian) Norway. Sedimentary response to tectonic events: Geol. Soc. Am. Bull., v. 88, p. 1124-1134.

Walker, R.G., 1976, Facies models 3. Sandy fluvial systems: Geoscience Canada, v. 3, p. 101-109.

Williams, G.E., 1969, Characteristics and origin of a Precambrian pediment: J. Geol., v. 77, p. 183-207.

————, 1971, Flood deposits of the sand-bed ephemeral streams of central Australia: Sedimentology, v. 17, p. 1-40.

FLUVIAL DEPOSITS OF THE ECCA AND BEAUFORT GROUPS IN THE EASTERN KAROO BASIN, SOUTHERN AFRICA

DAVID K. HOBDAY[1]

ABSTRACT

The Permian-Triassic Ecca and Beaufort Groups of the eastern Karoo Basin display several distinct fluvial facies, including channels of the delta plain, mixed-load high-sinuosity channel units and braided bedload stream deposits. Ecca sediment influx was mainly from the north and east, and shoal water deltas prograded across a broad, shallow platform which deepened southward into a flysch basin. The lower delta plain was traversed by multiple branching channels up to 4 m deep. These channels filled both by the downstream migration of solitary, large-scale bedforms, and by deposition from progressively smaller bedforms representing waning flow. Channels of the upper delta plain are up to 15 m deep and display point bar lateral accretion structures, with intrasets decreasing in scale upward. Associated topstratum and interfluvial deposits are silty and carbonaceous with irregular crevasse-splay sandstones, and are commonly overlain by coal of probable allochthonous origin. Along the updip basin margin similar coal-bearing cycles of the alluvial plain pinch out laterally against basement ridges.

The succeeding Beaufort basin was characterized by the development of a subsiding foreland related to the tectonically-active Cape Fold Belt to the south. A molasse-type wedge attained a maximum thickness of 1,000 m along the southern depo-axis, thinning northward where it prograded across a lacustrine shelf. The more distal portions of this clastic wedge resembled smaller lobes which developed along the northeastern basin margins, and comprise mixed-load, high sinuosity channel deposits. The coarser grained proximal channel facies consist of (1) braided stream deposits which are dominated by solitary sets of planar cross-beds, with associated plane-beds and trough cross-bed cosets, and (2) deposits made up entirely of upper flow regime plane-beds, probably recording flash floods in ephemeral streams.

INTRODUCTION

The Ecca and Beaufort Groups (Permian-Triassic) of the eastern Karoo Basin (Fig. 1) comprise shallow-water platform, deltaic, fluvial, and lacustrine deposits. These accumulated during the episode of climatic amelioration which followed Dwyka continental glaciation. Ice melting was accompanied by a brief marine connection in the west (McLachlan and Anderson, 1973). Although there is local mineralogical and micropalaeontological evidence of marine conditions in the postglacial Ecca basin which covered much of the subcontinent (Hart, 1964), brackish conditions are generally thought to have prevailed (McLachlan, 1973). The Beaufort Group contains fish, amphibians, reptiles, invertebrates and plants, all of which indicate a freshwater environment (Van Dijk *et al.*, in press).

ECCA GROUP

The Ecca basin was deep in the south, where flysch-type deposition occurred (Truswell and Ryan, 1969), with a broad, gently-subsiding platform to the north (Fig. 2). Arkosic sediments, shed by isostatically-rejuvenated, deglaciated highlands to the north and east, prograded intermittently across the platform. These regressive fluvio-deltaic deposits (Fig. 3) intertongue offshore with argillaceous sediments.

The deltas were of the shoal-water variety as described by Fisk (1955) and Gould (1970). Progradation generated 30 - 150 m thick vertical sequences which grade from sandy siltstone of the prodelta, through alternating siltstone and sandstone (distal mouth bar), into ripple cross-laminated sandstones overlain by coarser, trough cross-bedded

---

[1]Department of Geology, University of Natal, Pietermaritzburg, South Africa.
Present address: Bureau of Economic Geology, University of Texas at Austin, Austin, Texas 78712.

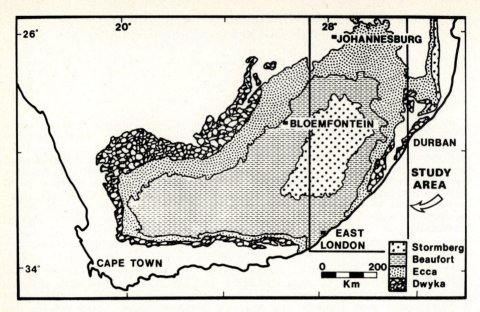

Fig. 1.   The Karoo Basin, with study area indicated.

sandstones (proximal mouth bar). The delta plain is dominated by 2 - 25 m thick "bay-fill" deposits of carbonaceous mudstones and siltstones with lenticular crevasse-splay sandstones and small upward-coarsening subdelta lobes (Hobday, 1973). Transgressive deposits, which accompanied delta abandonment and decay, consist generally of 1 - 2 m thick organic-rich, sideritic mudstones.

Three fluvial subenvironments are recognized in the southward-building Ecca clastic wedge: lower delta plain, upper delta plain, and alluvial plain (Fig. 4). A distinctive channel variety is associated with each of these.

Channels of the lower delta plain

Cutting into the upper parts of the deltaic sequences are small, closely-spaced, bifurcating and mutually-truncating channels (Fig. 4). Maximum depths are 4 m, with widths generally between 8 and 60 m. Most channelling is concentrated in the upper surface of the proximal mouth bar sandstones, but some steep-sided, narrow channels are erosive into finer-grained bay-fill deposits.

The channels are filled by either a solitary, large-scale trough cross-bed set or by cosets of trough cross-beds. In some cases the sets show an irregular upward decrease in thickness from 40 - 75 cm near the base to 10 cm or less at the top. Others show no vertical changes in texture or internal geometry, with scour surfaces and lenses of small-pebble intraformational conglomerate at all levels.

The three-dimensional geometry of the channels and their relationship to the underlying deltaic sequences suggest a system of multiple, branching, shallow distributaries of the lower delta plain. An analogous but larger-scale pattern existed in the Lafourche Delta (Gould, 1970) and is seen in subdeltas of the modern Mississippi (Coleman et al., 1969). According to Coleman and Wright (1975) factors which favour this geometry include a high subsidence rate, low wave action, a gentle offshore profile, and a small tidal range, conditions which are thought to have prevailed over much of the northeastern Karoo Basin (Hobday, 1973).

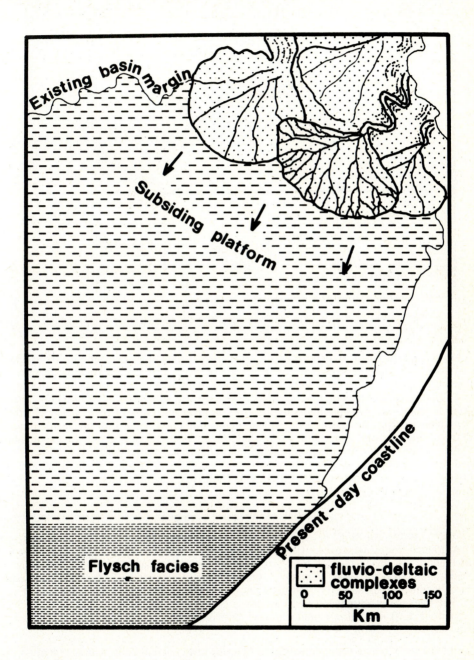

Fig. 2. Schematic reconstruction of Ecca palaeogeography showing fluvio-deltaic progradation across a subsiding platform.

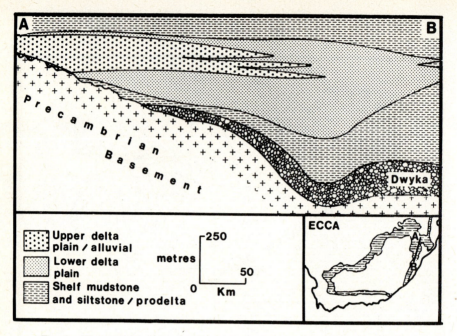

Fig. 3.   Ecca clastic wedge interfingering offshore with siltstones and mudstones.

Most channels filled by deposition on downcurrent-migrating, discontinuously-crested bedforms. The largest sets indicate bedforms of over 4 m in height; these probably resembled the lunate bars or sandwaves recorded in modern channels by Coleman (1969) and Collinson (1970) among others. Some of the vertical sequences of channel-fill cosets record waning energy, possibly occasioned by gradual diversion of flow. Other channel units indicate sporadic high energy bedload deposition intervening with the accumulation of fine sediment from suspension. Levee deposits are not preserved above the channel-fill units, and overbank deposition was largely restricted to ponds and embayments. Avulsion and the establishment of new channel courses was favoured by the gradient advantage offered by flow into inter-channel depressions (Ferm and Cavaroc, 1968).

Channels of the upper delta plain

Traced northward (Fig. 3) and vertically (Fig. 5) the proportion of sandstone to finer-grained rocks increases. The transition into a coal-bearing, largely arenaceous interval in the upper part of Fig. 5 is taken to mark a change to upper delta plain conditions (Hobday and Mathew, 1975). Lateral accretion structures (epsilon cross-bedding of Allen, 1963) are well developed in the upper delta plain sandstones (Fig. 6). The thicker sandstones (up to 15 m) erode deeply through the distributary mouth bar deposits (Fig. 7) in a pattern similar to that described in the modern Guadalupe Delta by Donaldson *et al.*, (1970).

The erosively-based vertical sequences fine upward concomitantly with a reduction in scale of sedimentary structures. A sporadic basal conglomerate, made up of well-rounded extrabasinal pebbles and boulders and crudely-imbricated intraformational mudclasts, ranges between 5 cm and 120 cm in thickness. Horizontal, coalified tree trunks up to 30 m long (Mathew, 1974, p. 40) occur at the base of some channels, and are orientated parallel to the general southerly paleoslope.

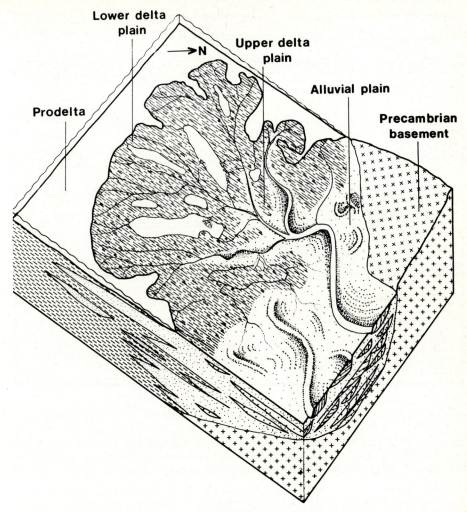

Fig. 4. Idealized subdivision of Ecca fluvio-deltaic system into lower delta plain, upper delta plain and alluvial plain.

Massive sandstone with scattered pebbles, corresponding to the rapidly-dumped "axial wedge" of Kelling (1968), commonly intervenes between the basal conglomerate and the overlying trough cross-bedded sandstones. The latter are generally medium- to coarse-grained and arkosic, and grade upward into finer-grained, microtrough cross-laminated sandstones. Foreset azimuths are strongly unimodal within individual channel sequences, and deviate from the inclination direction of the large-scale side-fill structures by angles of up to 40 degrees. Plane-beds are rare, but occur as thin intercalations at all levels.

The uppermost part of the average sequence displays ripple-drift cross-lamination in fine-grained silty sandstone, and alternating subhorizontal laminae of carbonaceous siltstone and sandstone. The proportion of sandstone decreases upward, with an increasing amount of finely-macerated plant debris. Sideritic concretions are locally abundant. Sparse rootlet beds and invertebrate burrows are present. Where coal is

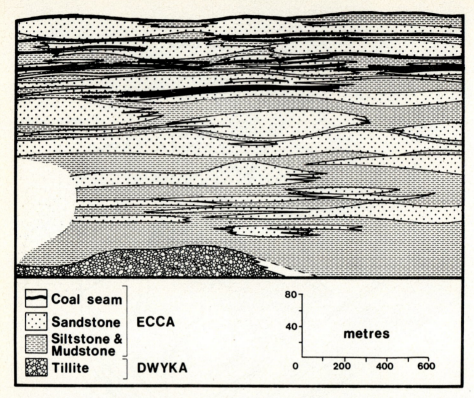

Fig. 5. Detailed stratigraphic section from northern Natal showing vertical gradation from facies of the lower delta plain (below coal-bearing interval) to upper delta plain (coal-bearing interval).

present it terminates the cycle. Most coal seams are well-banded and have a high inorganic content. There are no seatearths, and the coals frequently lie directly upon carbonaceous, clayey siltstone with some lamination preserved.

The full sequence from conglomeratic base to coal is commonly not developed, either because of non-deposition, or erosion of the upper parts by subsequent channels. Nevertheless, Markov analysis confirms the preferred vertical arrangement of conglomeratic and massive sandstone⟶ cross-bedded sandstone⟶ ripple cross-laminated sandstone⟶ alternating sandstone and siltstone (Hobday et al., 1975). This cyclic pattern corresponds closely with the meandering floodplain models of several workers (e.g. Allen, 1965, 1970; Visher, 1972), and sequences of this type have been described from other Gondwana continents (Casshyap, 1970).

A secondary Markov chain of siltstone/mudstone⟶ massive sandstone⟶ siltstone/mudstone represents another commonly observed pattern of the upper delta plain. These predominantly fine-grained units are up to 10 m thick and are lateral facies equivalents of major channel sandstones. The argillaceous component is rich in plant detritus, with *Glossopteris* leaves sometimes visible on bedding surfaces. A general upward-coarsening pattern results from the intercalation of sandstone units in the upper part. These range from persistent sandy laminae to erosively-based, internally structureless sandstone lenses up to a metre thick. Coal seams continue across the top of the siltstones from the adjacent channel cycles. Deposition in an interfluvial floodbasin is indicated, with the introduction of suspended clastics and rafted organic material by overbank flooding, probably accompanied by *in situ* contribution of vegetal matter.

Fig. 6. Ecca channel-fill unit of the upper delta plain, 8 m thick, showing side-fill structures. There is a gradation from erosively-based conglomerate through sandstone into rooted siltstone at the top.

Fig. 7. Channel sandstones up to 15 m thick incised into deltaic deposits.

Crevasse splays occurred during the latter stages of infilling (Hobday and Mathew, 1975). Most crevasses were ephemeral, but occasionally they led to total diversion of the discharge and the initiation of a new channel course.

The coals contrast with those of the Northern Hemisphere Carboniferous in the absence of seatearths and the rarity of rootlet beds. An abundance of clay minerals, the predominance of macerals of the inertinite group, mainly semifusinite, and the absence of cleat led Hoffman and Hoehne (1960) to propose an allochthonous origin. Despite support for an autochthonous origin (Wyberg, 1925; Plumstead, 1966), the laminated substrate and the well-developed layering in the coal, often accentuated by detrital partings, provides additional evidence of a rafted origin (cf. Cohen, 1970).

Channels of the alluvial plain

Basement valleys around the northern margin of the Karoo Basin were only partly filled by Dwyka glaciogene deposits, and thus continued to exercise control over Ecca drainage patterns. The valley-fill deposits are highly varied. Some of the lowermost coals overlie varved shales and Gilbert-type deltaics (K. A. Eriksson, pers. comm., 1977). Similar basal deposits are observed by B. C. Cairncross (pers. comm., 1977) in an adjacent coalfield, and these are overlain by elongate channel sandstones up to 16 m thick. In addition, Eriksson has recognized braided stream deposits in this area. Up to five superimposed, coal-capped, upward-fining alluvial cycles are developed locally, pinching out against basement highs. In the Witbank coalfield, for example, the floodplain was bounded on one side by a felsite ridge, which was gradually buried by aggradation so that the uppermost coals override the crest (Haughton, 1969, p. 405).

Steyn (1977) distinguishes between active channel-fill sequences, in which the coal overlies upward-fining arkosic sandstones, and abandoned channel-fill coals, which are associated with thick carbonaceous siltstones.

Beaufort Group

The Beaufort basin fluctuated markedly in outline, with a progressive reduction in size. Shrinkage of the basin was accompanied by segmentation into a series of shallow lakes (Van Dijk *et al.*, in press). Initial influx was mainly along the southwestern margin, where a prograding alluvial flood-plain developed (Kübler, 1977; B. R. Turner, this volume). Quiescent lacustrine conditions prevailed in the Lower Beaufort of the eastern Karoo Basin, with local deposition of arkosic sediments derived from erosion of uplifted granitic areas in the northeast (Theron, 1975). J. R. Turner (1977) recognizes westward-building fluvio-deltaic complexes which resemble patterns in the Ecca. One difference is the development in the Lower Beaufort upper delta plain of low-energy point bar sequences in which the accretionary surfaces are siltstone-draped, and enclose cosets of microtrough cross-lamination.

Middle Beaufort influx into the southeastern segment of the basin was probably generated by erosion of a tectonically-active source beyond the present-day coastline. A sandstone-dominated wedge thins northwards from 1,000 m in the foreland basin along the southern coast to less than 100 m in northern Natal. Two distinct fluvial end members are recognized (Fig. 8), a distal alluvial floodplain facies, which makes up the bulk of the lower half of the succession, and a coarser, more proximal component, which occurs towards the top.

Distal mixed-load channel facies

Deposits of the lower alluvial floodplain consist of overlapping lenticular units characterized by repetitive vertical sequences of: erosive base ⟶ conglomerate or

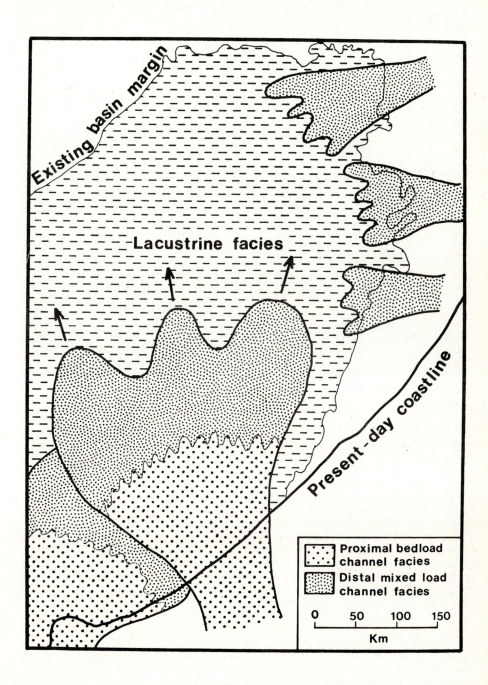

Fig. 8. Schematic reconstruction of Beaufort palaeogeography showing major fluvial influx from a tectonically-active southerly source, with minor contributions from the east and northeast.

Fig. 9.   Bedding plane surface in upper fine-grained part of mixed-load channel facies, showing current-oriented trace fossils.

massive sandstone➝ trough cross-bedded sandstone➝plane-bedded and/or microtrough and ripple-drift cross-laminated sandstone➝ maroon or green siltstone or mudstone. This arrangement is confirmed by Markov analysis (H. D. Greenshields, pers. comm., 1977).

The basal erosion surfaces are highly irregular, with scours, flutes, grooves and load structures, but are broadly concave-upward. Approximately 90% of the clasts are locally derived, and include plastically-deformed mudstone slabs up to 40 cm long, reworked calcareous and siliceous concretions, and rolled root- or burrow-casts. Water-worn limb-bones, tusks and skulls of synapsids (mammal-like reptiles) are common. Vein quartz, granite, metasediment and sandstone pebbles are well-rounded and up to 8 cm in diameter. The sandstone pebbles are of characteristic Cape Supergroup (Early Palaeozoic) lithology, thus suggesting derivation from an eastward extension of the Cape Fold Belt, which prior to disruption of Gondwanaland would have been situated to the south.

The massive sandstone unit is poorly sorted and contains scattered pebbles. Complex overturned foresets and water escape structures occur in the cross-bedded sandstones. The capping siltstones and mudstones vary from well-laminated, grey and sandy to structureless and green-grey or maroon with colour mottling. Rootlet beds, branching subvertical invertebrate burrows, desiccation cracks and calcareous and siliceous concretions are present. The concretions are probably a product of soil-forming processes.

One burrow type, which is useful as palaeocurrent indicator, is preserved as elongate to elliptical traces (Fig. 9) aligned parallel to the current direction as indicated by the underlying cross-beds. Spreiten record upcurrent mining by the organism. The burrows are wall-like in vertical section, extending to a maximum depth of 25 cm, with a width of up to 6 cm.

The vertical sequence of textures and structures resembles the deposits of high-sinuosity, mixed-load, low-gradient streams of aggrading alluvial plains (Allen, 1965; Schumm, 1972; Galloway, 1977). The absence of discernible point-bar accretion surfaces may be due to deposition on very low-angle point-bar profiles. Channel lag deposits were probably introduced largely by cutbank slumping during falling river stage. The overlying sequence records the decrease of megaripple amplitude up the point-bar surface, and the development of plane-beds resulting from increased Froude number due to shallowing (Visher, 1965). Upper point-bar and levee deposits accumulated mainly from suspension, and were subject to subaerial exposure, with the development of pedogenic concretions indicating prolonged stability (Leeder, 1975). The deposits were generally oxidised and red, but locally reducing conditions were brought about by the incorporation of abundant organic material. Progressive Triassic aridification (Haughton, 1924) may explain the absence of coal at the top of the southern Beaufort fluvial sequences, which in many other respects resemble the Ecca.

Crevasse sandstones occur above thick (2 - 12 m), upward-coarsening siltstone units (Fig. 10) which probably accumulated in lakes and bays at the basinward edge of the prograding fluvial complex. The basal siltstones are wave rippled and contain fossil plants, insects, freshwater fish, and almost complete skeletons of the therapsid *Lystrosaurus*, indicating *in situ* preservation (Van Dijk *et al.*, in press). Nearest to the trunk channels the crevasse sandstones are massive to poorly-bedded, markedly erosive, and

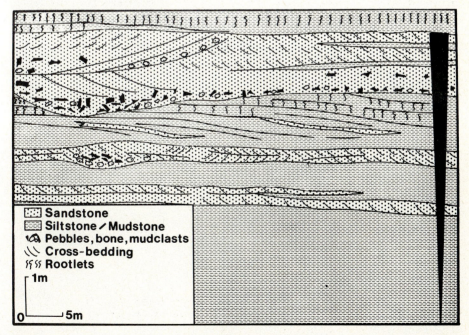

Fig. 10. Upward coarsening deposits of lower alluvial floodbasin origin, contributed by overbank flooding and crevassing. The two lower sandstones are distal crevasse splays; the upper thick sandstone cuts through levee deposits and is interpreted as a proximal crevasse channel.

contain pebbles and water-worn bone fragments. Distally the crevasse sandstones become progressively more conformable with bay siltstones, and show internal structures indicative of wave and current reworking.

Proximal bedload channel facies

The coarser-grained upper part of the southern Middle Beaufort succession comprises the following lithofacies: (1) conglomerate, (2) planar cross-bedded sandstone, (3) trough cross-bedded sandstone, (4) plane-bedded sandstone, and (5) rare siltstone. The average fully-developed vertical sequence is: erosive base overlain by conglomerate $\longrightarrow$ cross-bedding$\longrightarrow$ plane-bedding$\longrightarrow$ siltstone. However, the upper lithological units are invariably truncated to some degree, so that many sequences preserve only conglomerate and cross-bedded sandstone.

The channels overlap one another laterally and are superimposed to thicknesses of over 100 m (Fig. 11). At one exposure the composite stratigraphic thickness of 74 stacked channels is 81 m, giving an average preserved thickness of 1.1 m per channel. Width to depth ratios of between 45 and 60 were determined for those channels whose widths it was possible to ascertain; but these are minimum values, with most channels being more sheet-like.

Channel floors are uneven, with shallow scour pits and troughs. The conglomerate comprises extrabasinal pebbles up to 12 cm in diameter, locally-eroded argillaceous clasts ranging from small pebbles to a slab 52 cm long and 10 cm thick, reworked concretions, and fragmentary bone. The proportion of extrabasinal to intrabasinal clasts varies markedly from one locality to another. The conglomerate is commonly only a single layer

Fig. 11. Fine-grained, mixed-load channel deposits (lower right) overlain by a thick succession of superimposed coarser-grained proximal bedload channel facies.

of pebbles, the tabular varieties showing good imbrication in accord with a northerly palaeoslope. Thicker conglomerate units (20 - 50 cm) show normal grading, inverse grading, or inverse - to - normal grading (Walker, 1975). Most clasts are matrix-supported, and in this variety there is no stratification or imbrication.

Gradationally overlying the conglomerate is medium- to coarse-grained impure arkose with scattered mudclasts. These deposits commonly show a solitary planar cross-bed set between 25 and 80 cm thick. The foresets are normally sigmoidal, becoming progressively less steeply inclined towards the top of the set, where they merge into plane-beds with current lineation. Downstream gradation from planar cross-bedding into plane-beds is also observed. Cosets of trough and, less commonly, planar cross-beds occur locally.

Penecontemporaneous deformation structures, including complexly overturned foresets, often with imbricate microthrusts (Fig. 12), convolute lamination, water-escape pipes and sand volcanoes, are ubiquitous in the sandstones. In places all vestiges of bedding are obliterated. In cosets, frequently only one set is involved in the deformation, the overlying foresets smoothing out irregularities in the surface of the deformed set. Lateral displacement is always directed down the palaeoslope. This suggests that the deforming force was current friction (or gravity) acting on a fluidized or liquidized sand (Coleman, 1969; Allen and Banks, 1972). Excess pore water pressures may have been a consequence of rapidly falling water level (J. H. Barwis, pers. comm., 1977).

A layer of green-grey siltstone up to 10 cm thick is infrequently preserved at the top of the channel-fill. Erosion of this unit provided the intraclasts of the basal conglomerate of succeeding channels.

Fig. 12. Penecontemporaneous deformation in proximal channel facies. Note imbricate microthrusts in lower part, convolute lamination above, and upward decrease in intensity of deformation.

The high channel width to depth ratios, the dominance of traction current sedimentation, and the high proportion of planar cross-bedding together suggest that deposition occurred in a bedload fluvial environment (Schumm, 1972; Galloway, 1977). The channel-fill characteristics conform to some extent with the "Platte-type" braided stream of Miall (1977), documented in detail by Smith (1970, 1971) except that cyclic upward-fining character and the basal gravel unit are more similar to the "Donjek type" (Williams and Rust, 1969; Miall, 1977, this volume) or "South Saskatchewan type" (Cant, this volume; Miall, this volume).

Thin basal pebble layers probably represent a winnowed lag, whereas the thicker, matrix-supported conglomerates resemble the debris flood deposits of Miall (1977). Solitary sets of planar cross-beds possibly developed by downstream accretion on the lee faces of transverse bars (Smith, 1970), but the Beaufort deposits differ from this Platte River analogue in having sigmoidal foresets, in contrast to the angular basal contacts developed in the Platte River. These curving, asymptotic foresets probably indicate a well-developed separation eddy (Collinson, 1970). Smith (1970) has shown that in the Platte River, planar cross-beds usually grade upstream into trough cross-beds. Furthermore, in very shallow water, sediment is transported over the top of the bar as plane-beds. These observations may account for the arrangement of sedimentary structures in the Beaufort sandstones. Siltstones may represent suspension deposition in abandoned channels. The low preservation potential of these beds (Cant, 1976) is a consequence of constantly shifting channel patterns.

Erosively-based units comprising only plane-bedding (Fig. 13) occur interspersed with the more common cross-bedded channel-fill sandstone sequences. These upper flow regime deposits may have been a response to flashy discharge in ephemeral channels within the braided floodplain system. These flood-dominated channel sediments conform with the "Bijou Creek type" of Miall (1977, this volume), based on the studies by McKee *et al.*, (1967).

Fig. 13.   Stacked, erosively-based plane-bedded units.

Although showing a general upward-coarsening tendency, fluvial facies in the southern foreland basin are complexly interlayered, reflecting minor shifts in the balance between sediment influx and basinal subsidence. Farther basinward only the more distal, mixed-load channel facies are represented.

Conclusions

A comparison of the Ecca and Beaufort successions reveals contrasted tectono-sedimentary terrains, with marked differences in the style of fluvial sedimentation.

The postglacial Ecca intracratonic basin received sediment eroded from isostatically-uplifted deglaciated highlands to the north and east. Shoal water deltas prograded across a broad, gently subsiding platform which was bounded in the south by a deep, narrow flysch basin. The lower delta plain of the southward-building complex was traversed by numerous small, bifurcating distributaries, whereas the upper delta and alluvial plains had larger meandering channels which generated upward-fining, coal-bearing sequences.

The southern flysch trough was a shortlived feature, and after infilling was transformed into a syntectonic foredeep related to the Cape Fold Belt. The resultant northward-building clastic wedge dominated Beaufort sedimentation. More proximal parts were characterized by bedload deposition in a braided stream environment. Grain size decreased downstream and was accompanied by a change to a meandering condition.

The molasse complex graded northward into a lacustrine environment, and together with smaller clastic lobes in the northeast, led to progressive segmentation and filling of the basin.

Acknowledgments

I thank Ken Eriksson and Victor von Brunn for refereeing this paper and for suggesting improvements. I gratefully acknowledge a travel subsidy from the organizers of the symposium.

References

Allen, J. R. L., 1963, The classification of cross-stratified units, with notes on their origin: Sedimentology, v. 2, p. 93-114.

———, 1965, A review of the origin and characteristics of recent alluvial sediments: Sedimentology, v. 5, p. 89-191.

———, 1970, Studies in fluviatile sedimentation: a comparison of fining-upward cyclothems, with special reference to coarse-member composition and interpretation: J. Sediment. Petrol., v. 40, p. 298-323.

———, and Banks, N. L., 1972, An interpretation and analysis of recument-folded deformed cross-bedding: Sedimentology, v. 19, p. 257-283.

Cant, D. J., 1976, Selective preservation of flood stage deposits in a braided fluvial environment: Geol. Assoc. Canada, Program Abstracts, v. 1, p. 77.

Casshyap, S. M., 1970, Sedimentary cycles and environment of deposition of the Barakar coal measures of Lower Gondwana, India: J. Sediment. Petrol., v. 40, p. 1302-1317.

Cohen, A. D., 1970, An allochthonous peat deposit from southern Florida: Geol. Soc. Am. Bull., v. 81, p. 2477-2482.

Coleman, J. M., 1969, Brahmaputra River: channel processes and sedimentation: Sediment. Geol., v. 3, p. 129-239.

———, and Wright, L.D., 1975, Modern river deltas: variability of processes and sand bodies: *in* M. L. Broussard, *ed.*, Deltas, models for exploration; Houston Geol. Soc., Houston, Texas, p. 99-150.

————, Gagliano, S. M., and Morgan, J. P., 1969, Mississippi River subdeltas: natural models of deltaic sedimentation: Coastal Studies Bulletin Number 3, Louisiana State Univ. Press, p. 23-28.

Collinson, J. D., 1970, Bed forms of the Tana River, Norway: Geogr. Ann., v. 52A, p. 31-55.

Donaldson, A. C., Martin, R. H., and Kanes, W. H., 1970, Holocene Guadalupe Delta of Texas Gulf Coast: *in* J. P. Morgan, *ed.*, Deltaic sedimentation, modern and ancient; Soc. Econ. Paleont. Mineral. Spec. Pub. 15, p. 107-137.

Ferm, J. C., and Cavaroc, V. V., Jr., 1968, A nonmarine model for the Allegheny of West Virginia: *in* G. deV. Klein, *ed.*, Late Paleozoic and Mesozoic continental sedimentation, northeastern North America; Geol. Soc. Am. Spec. Paper 106, p. 1-19.

Fisk, H. N., 1955, Sand facies of the Recent Mississippi Delta deposits, Proc. Fourth World Petroleum Congr., Rome: Section 1C, p. 377-398.

Galloway, W. E., 1977, Catahoula Formation of the Texas Coastal Plain: depositional systems, composition, structural development, ground-water flow history, and uranium distribution: Bur. Economic Geology, Univ. Texas at Austin, Rept. Invest. No. 87, 59p.

Gould, H. R., 1970, The Mississippi Delta complex: *in* J. P. Morgan, *ed.*, Deltaic sedimentation, modern and ancient; Soc. Econ. Paleont. Mineral. Spec. Pub. 15, p. 3-30.

Hart, G. F., 1964, Where was the Lower Karroo sea? Scientific South Africa, v. 1, p. 289-290.

Haughton, S. H., 1924, Fauna and stratigraphy of the Stormberg Series: Annals of the South African Museum, v. 12, p. 128-142.

————, 1969, Geological history of southern Africa: Geol. Soc. South Africa, Johannesburg, 535p.

Hobday, D. K., 1973, Middle Ecca deltaic deposits in the Muden-Tugela Ferry area of Natal: Trans. Geol. Soc. South Africa, v. 76, p. 309-318.

————, and Mathew, D., 1975, Late Paleozoic fluviatile and deltaic deposits in the northeast Karroo Basin, South Africa: *in* M. L. Broussard, *ed.*, Deltas, models for exploration; Houston Geol. Soc., Houston, Texas, p. 457-470.

————, Tavener-Smith, R., and Mathew, D., 1975, Markov analysis and the recognition of palaeoenvironments in the Ecca Group: Trans. Geol. Soc. South Africa, v. 78, p. 75-82.

Hoffman, H., and Hoehne, K., 1960, Petrographische Eigenschaften und rheologisches Verhalten von permischen Steinkohlen, insbesondere der Gondwanaformation; III die Gondwanakohlen von Südafrika: Brennst. Chem., v. 41, p. 142-162.

Kelling, G., 1968, Patterns of sedimentation in the Rhondda Beds of South Wales: Am. Assoc. Petrol. Geol. Bull., v. 52, p. 2369-2386.

Kübler, M., 1977, The sedimentology and uranium mineralization of the Beaufort Group, Cape Province: Unpub. M.Sc. thesis, Witwatersrand Univ., Johannesburg, 106p.

Leeder, M. R., 1975, Pedogenic carbonates and flood sediment accretion rates: a quantitative model for alluvial arid-zone lithofacies: Geol. Mag., v. 112, p. 257-270.

Mathew, D., 1974, A statistical and palaeoenvironmental analysis of the Ecca Group in northern Natal: Unpub. M.Sc. thesis, Natal Univ., Pietermaritzburg, 158p.

McKee, E. D., Crosby, E. J., and Berryhill, H. L., 1967, Flood deposits, Bijou Creek, Colorado, June, 1965: J. Sediment. Petrol., v. 37, p. 829-851.

McLachlan, I. R., 1973, Problematic microfossils from the Lower Karroo Beds in South Africa: Palaeontologia Africana, v. 15, p. 1-12.

————, and Anderson, A., 1973, A review of the evidence for marine conditions in Southern Africa during Dwyka times: Palaeontologia Africana, v. 15, p. 37-64.

Miall, A. D., 1977, A review of the braided river depositional environment: Earth Sci. Revs., v. 13, p. 1-62.

Plumstead, E. P., 1966, The story of South Africa's coal: Optima, v. 16, p. 186-213.

Schumm, S. A., 1972, Fluvial paleochannels: *in* J. K. Rigby and W. K. Hamblin, *eds.*, Recognition of ancient sedimentary environments; Soc. Econ. Paleont. Mineral. Spec. Pub. 19, p. 98-107.

Smith, N. D., 1970, The braided stream depositional environment: comparison of the Platte River with some Silurian clastic rocks, north-central Appalachians: Geol. Soc. Am. Bull., v. 81, p. 2993-3014.

————, 1971, Transverse bars and braiding in the lower Platte River, Nebraska: Geol. Soc. Am. Bull., v. 82, p. 3407-3420.

Steyn, P. P. A., 1977, Die sedimentologie van die Davelsteenk oolveld in die Oos-Transvaal met spesiale verwysing na die Vryheid Formasie van die Karoo Supergroep: Geokongres 77 Abstracts, Geol. Soc. South Africa, Johannesburg, p. 105-108.

Theron, J. C., 1975, Sedimentological evidence for the extension of the African continent southwards during the late Permian - early Triassic times: *in* K. S. W. Campbell, *ed.*, Gondwana Geology; Australian National Univ. Press, p. 61-71.

Turner, J. R., 1977, Palaeoenvironmental study of the Lower Beaufort in the northeastern Karoo Basin: Unpub. M.Sc. thesis, Natal Univ., Pietermaritzburg, 138p.

Truswell, J. F., and Ryan, P. J., 1969, A flysch facies in the Lower Ecca Group of the southern Karroo and a portion of the Transkei: Trans. Geol. Soc. South Africa, v. 72, p. 151-158.

Van Dijk, D. E., Hobday, D. K., and Tankard, A. J., in press, Permo-Triassic lacustrine deposits in the eastern Karoo Basin, South Africa: *in* M. E. Tucker and A. Matter, *eds.*, Modern and ancient lake sediments, Int. Assoc. Sedimentologists, Spec. Pub.

Visher, G. S., 1965, Fluvial processes as interpreted from ancient and recent fluvial deposits: *in* G. V. Middleton, *ed.*, Primary sedimentary structures and their hydrodynamic interpretation; Soc. Econ. Paleont. Mineral. Spec. Pub. 12, p. 116-132.

————, 1972, Physical characteristics of fluvial deposits: *in* J. K. Rigby and W. K. Hamblin, *eds.*, Recognition of ancient sedimentary environments; Soc. Econ. Paleont. Mineral. Spec. Pub. 16, p. 84-97.

Walker, R. G., 1975, Generalized facies models for resedimented conglomerates of turbidite association: Geol. Soc. Am. Bull., v. 86, p. 737-748.

Williams, P. F. and Rust, B. R., 1969, The sedimentology of a braided river: J. Sediment. Petrol., v. 39, p. 649-679.

Wyberg, W. J., 1925, Coal Resources of the Union of South Africa: South African Geological Survey Mem. 19, 157p.

CHANGING CHANNEL MORPHOLOGY AND MAGNITUDE IN THE SCALBY FORMATION (M. JURASSIC) OF YORKSHIRE, ENGLAND

M. Nami[1] and M. R. Leeder[1]

Abstract

An upward change in palaeochannel morphology from low sinuosity, possibly braided, to high sinuosity channels set in a matrix of overbank fines is deduced for the Scalby Formation. The resulting sandbodies are of sheet and ribbon morphologies, respectively. The early low sinuosity channels are thought to have carried much higher discharges than the later high sinuosity channels. The deduced upward changes in channel morphology and magnitude are thought to be due to initial hinterland upwarp causing high valley gradients and initial fluvial advance into the marine Yorkshire basin followed by rising sea level causing extensive alluviation, lowered gradients and final marine transgression of the Cornbrash sea over the defunct alluvial plains.

Introduction And Stratigraphy

During Middle Jurassic times in the North Sea Basin periods of upwarp associated with the formation of graben/horst complexes (Fig. 1) gave rise to several widespread sedimentary regressions due to the growth of terrestrial drainage systems (Kent, 1975). The deposits left by these regressions now form important hydrocarbon reservoirs. Middle Jurassic outcrops in fluvial facies occur most notably in Yorkshire where they are magnificently exposed in structurally simple coastal outcrops.

The Scalby Formation (some 60 m thick) is the youngest of three regressional episodes in the otherwise wholly marine Jurassic succession of Yorkshire. The regressive sequences form part of the Ravenscar Group (Fig. 1) as defined by Hemingway and Knox (1975). Black (1928, 1929, 1934) divided the Formation, then known as the Upper Estuarine Series, into two major groups, the Current-Bedded and Level-Bedded series. The former consisted of the basal Moor Grit and the Current Bedded Sandstone. Black interpreted the Moor Grit, which comprises coarse, cross-stratified, quartz-arenites, as delta foreset beds. The Current-Bedded Sandstone, consisting of impure silty sandstone, was interpreted as forming in a shallow water environment with variable current action behind the delta front. The Level-Bedded series, consisting of laminated siltstones and mudstones cut by channel deposits, or 'washouts' as Black termed them, were ascribed to deposition in freshwater lagoons which received fine sediment from a delta distributary system.

Black's environmental interpretation of the Scalby Formation was not developed further until one of us (MN) discovered that Black's Current Bedded Sandstone was in fact a complex stream meander belt faithfully preserving point bar lateral accretion deposits in a large number of truncated meander loops (Nami, 1976a, 1976b).

Sandbodies And Palaeochannel Morphology

Within the Scalby Formation two major lithofacies may be distinguished: 1) coarse, ribbon and sheet sandstone bodies set in 2) fine siltstones and mudstones. The ribbon or sheet sandstone members may be further divided into those that show lenticular internal bedding (vertically accreted) or sigmoidal internal bedding (laterally accreted).

Lenticular bedded sheet sandstone body
(braided, low-sinuosity channels)

This is the Moor Grit Member, at the base of the Formation, which has a sharp and erosive contact with the underlying nearshore marine Scarborough Formation (Fig. 2).

---

[1]Department of Earth Sciences, University of Leeds, Leeds LS2 9JT, W. Yorkshire, U.K.

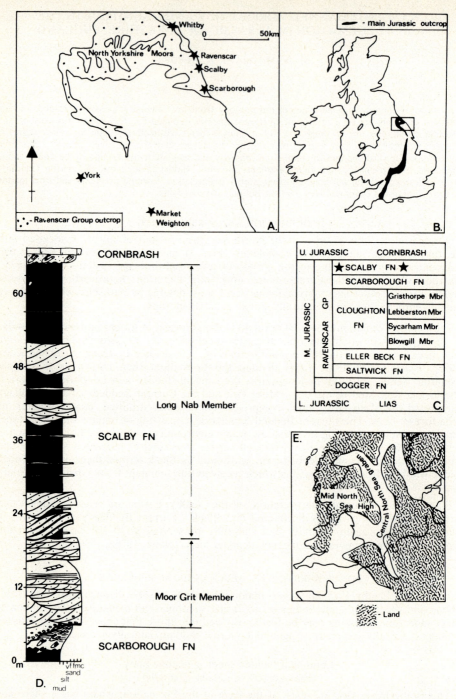

Fig. 1. 1A, B — location map. Cromer Point, Long Nab and Hundale Point lie respectively, 1.5 km, 3.25 km and 4 km north of Scalby. 1C — Stratigraphic divisions of the Middle Jurassic in Yorkshire (after Hemingway and Knox, 1973). 1D — Generalised stratigraphic log of Scalby Formation. 1E — Generalised distribution of bedrock land areas during the Middle Jurassic of northwest Europe (after Sellwood and Hallam, 1974).

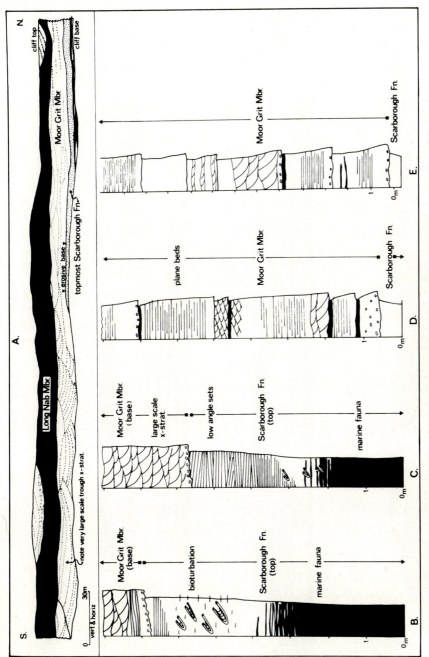

Fig. 2. 2A — Section to show very large scale trough cross stratification in the Moor Grit member as developed in the cliffs south of Hundale Point. 2B-E — Logs to show sections through the topmost nearshore marine facies of the Scarborough Formation and the Moor Grit Member. 2B — 300 m south of Hundale Point, 2C — north of Beast Cliffs near Ravenscar, 2D — 300 m south of Hundale Point, 2E — White Nab, 3 km south of Scarborough.

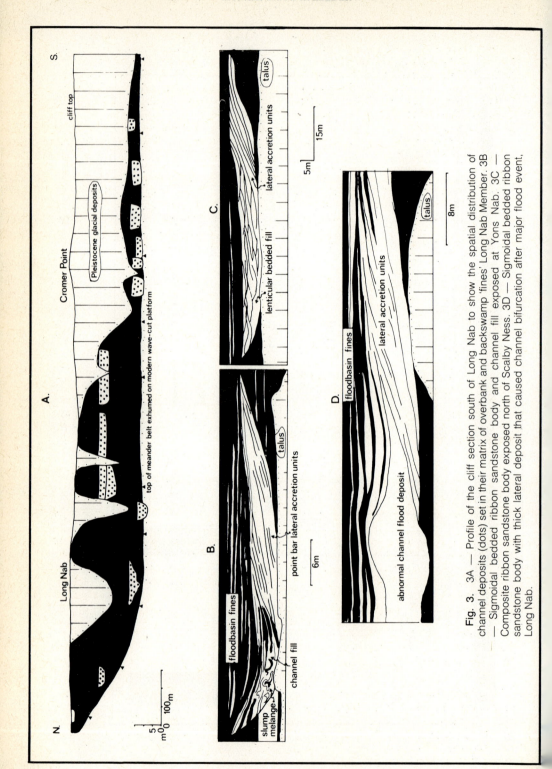

Fig. 3. 3A — Profile of the cliff section south of Long Nab to show the spatial distribution of channel deposits (dots) set in their matrix of overbank and backswamp 'fines' Long Nab Member. 3B — Sigmoidal bedded ribbon sandstone body and channel fill exposed at Yons Nab. 3C — Composite ribbon sandstone body exposed north of Scalby Ness. 3D — Sigmoidal bedded ribbon sandstone body with thick lateral deposit that caused channel bifurcation after major flood event, Long Nab.

The sheet deposit, with unimodal palaeocurrent trends, may be up to 12 m thick and is traceable over an outcrop width normal to palaeocurrent trend of about 70 km. The fine to coarse-grained sometimes pebbly sandstone may be multistorey in character with 0.5 - 2.5 m sandstone units separated by thin siltstones and mudstones that occasionally bear small dinosaur footprints (Fig. 2). Each unit shows a basal scoured surface with quartz pebbles, mudstone and ironstone intraclasts and carbonised plant debris. Internally the units may show upper phase plane beds, large scale trough and tabular cross stratification and climbing ripple small scale cross stratification. South of Hundale Point the whole member comprises spectacular giant sets of large scale trough cross stratification up to 10 m thick (Fig. 2).

This frequently multistorey sheet sandstone is attributed to dominantly bedload deposition in a lateral succession of broad and sometimes deep channel reaches. Dune bedforms up to 10 m high are indicated by the giant trough sets and are considered analagous to the channel sand waves recorded in the braided Brahmaputra River (Coleman, 1969). There is no evidence of deposition by lateral accretion and it is inferred that the channels were of low sinuosity and possibly braided.

Lenticular bedded ribbon sandstone bodies
(low-sinuosity, ?non-braided channel or reach)

These occur in the Long Nab member as isolate sandstone ribbons surrounded by fine-grained floodbasin/backswamp deposits. The ribbons range from 20 - 140 m wide and from 3 - 8 m thick. The fine to medium grained sandstones with unimodal currents show internal lenticular bedding with 0.1 - 0.6 m thick sets of large scale trough stratification, small scale cross lamination and parallel lamination. Fining upwards sometimes occurs. Lateral accretion bedding is absent.

The deposits are thought to be due to infilling of symmetrical, mud-bank channels containing dune fields. Minimal channel migration, low channel sinuosity and channel fill by vertical accretion are indicated.

Simple sigmoidal bedded ribbon sandstone bodies
(high sinuosity meandering reaches)

These also occur in the Long Nab member as isolate deposits surrounded by fine-grained floodbasin deposits (Fig. 3). Several examples in the coast section south of Cromer Point show basal erosive contacts with the complex sigmoidal bedded deposit of the exhumed meander belt (q.v.). Lateral interconnectedness between the sandstone bodies is not common. The sandbodies range from 3 - 9 m thick and from 30 - 500 m wide. Internally they are dominated by sigmoidal dipping sets of lateral accretion bedding (Fig. 3) which show diagnostic palaeocurrent flow normal to the dip of the sets. Individual lateral accretion units have sharp bases, often with abundant intraclasts, and invariably fine upwards from fine/medium sandstone to coarse siltstones and mudstones. Small scale cross lamination is the commonest internal structure but some large scale sets also occur. Possible sets of scroll bar origin show migration directions up the dip direction of the epsilon sets. Rarely, whole channel sections are seen together with cut banks and channel fill (Fig. 3).

These sandstone bodies are confidently ascribed to point bar deposits of meandering reaches cut into cohesive, mud-grade banks. The cluster of such deposits just above the exhumed meander belt deposit south of the cliff section at Cromer Point (Fig. 3) may possibly represent a vertical section through a meander belt normal to general palaeoflow. At other localities only isolate ribbon sandbodies occur, indicating the absence of extensive meander belts.

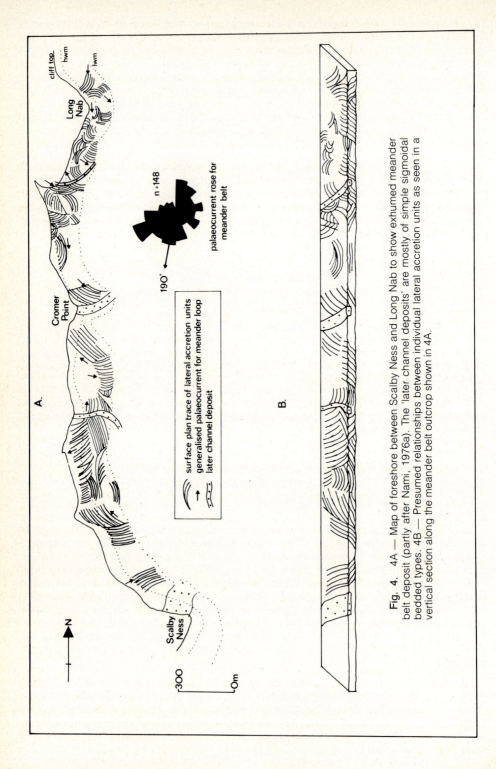

Fig. 4. 4A — Map of foreshore between Scalby Ness and Long Nab to show exhumed meander belt deposit (partly after Nami, 1976a). The 'later channel deposits' are mostly of simple sigmoidal bedded types. 4B — Presumed relationships between individual lateral accretion units as seen in a vertical section along the meander belt outcrop shown in 4A.

Complex sigmoidal bedded sheet sandstone body
(high sinuosity stream meander belt)

This unique deposit erosively overlies the Moor Grit Member on the broad intertidal platform between Scalby Ness and Long Nab (Nami, 1976b; Fig. 4). Perfect exposures up to 300 m wide, display the curvilinear lateral accretion sets of an exhumed meander belt in plan view. Vertical sections through the deposit are complex (Fig. 4). The individual lateral accretion sets comprise fine sandstones grading upwards into mudstones. Small scale cross laminations are the dominant internal structure. A variety of large scale point bar slump deposits and small scale dewatering structures occur.

Composite sigmoidal/lenticular bedded ribbon
sandstone bodies

These show that certain reaches started life as meandering reaches producing lateral accretion deposits of point bar origin. The final stages of deposition were of vertical accretion type so that lenticular bedded sandstone units were deposited. The reason for such changes is not clear to us. In one example a period of deep scouring followed by an abnormally thick deposit of lateral accretion origin seems to have effectively split the former channel into separate anabranches where a lenticular bedded fill subsequently developed.

CHANGING PALAEOCHANNEL MORPHOLOGY AND MAGNITUDE

The coarse grained sheet sandstone of low sinuosity ? braided origin at the base of the Scalby Formation passes upwards into finer grained, generally isolate ribbon sand bodies, usually of high sinuosity origin, set in a matrix of overbank fines. These gross upward changes may be best explained by postulating an initial period of high slope and river entrenchment followed by general basin alluviation and river aggradation due to rising sea level (Fig. 5). A well-known effect of the early Holocene sea level rise was massive alluviation giving rise to successions such as those of the Mississippi Valley and many other alluvial plains in which coarse-grained, braided stream, valley-fill sequences are succeeded by finer grained alluvium comprising high sinuosity meander belt sands set in a matrix of overbank fines. If such a model is applicable to the Scalby Formation then the final marine transgression over the Jurassic alluvial plains is represented by the deposits of the Upper Cornbrash (the nearshore marine facies at the top of Fig. 5) of Callovian age, which time was a period of world wide transgression. Fairly rapid deposition of the floodbasin fines which are volumetrically dominant in the Long Nab Member is supported by the absence of coal and soil horizons in a climate that was obviously conducive to peat formation since numerous coal horizons occur in equivalent successions in the North Sea hydrocarbon province. Humid subtropical soil formation is further indicated by the very abundant kaolinitic clay mineral fractions we have analysed from the Long Nab Member mudstones.

The cause of the increase in valley slope which eventually led to deposition of the Moor Grit sheet sand over a prominent erosion surface is unknown; sea level fall or hinterland uplift seem obvious alternatives. Since it is known that the middle Jurassic in the North Sea was a time of marked vertical crustal movements the latter hypothesis may apply.

The frequent occurrence of lateral accretion surfaces in the ribbon sand bodies described above enables estimates to be made of the magnitude of the channels responsible. Using Method A of Leeder (1973), with suitable units, the magnitude of the 'mean' high sinuosity channel from the Long Nab Member may be calculated (Fig. 6). It is evident that the channels were generally of fairly small size and discharge magnitude, a point confirmed by a general lack of large scale stratification. Whether or not the channels were distributive is unknown. To judge by the abundance of mudstone/siltstone laminae, many of the channels must have carried very substantial amounts of suspended material.

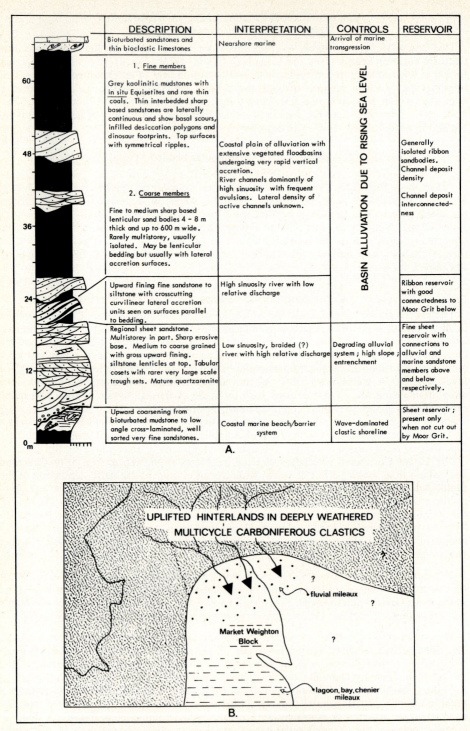

	DESCRIPTION	INTERPRETATION	CONTROLS	RESERVOIR
	Bioturbated sandstones and thin bioclastic limestones	Nearshore marine	Arrival of marine transgression	
	1. Fine members\n\nGrey kaolinitic mudstones with in situ Equisetites and rare thin coals. Thin interbedded sharp based sandstones are laterally continuous and show basal scours, infilled desiccation polygons and dinosaur footprints. Top surfaces with symmetrical ripples.\n\n2. Coarse members\n\nFine to medium sharp based lenticular sand bodies 4 - 8 m thick and up to 600 m wide. Rarely multistorey, usually isolated. May be lenticular bedding but usually with lateral accretion surfaces.	Coastal plain of alluviation with extensive vegetated floodbasins undergoing very rapid vertical accretion.\nRiver channels dominantly of high sinuosity with frequent avulsions. Lateral density of active channels unknown.	BASIN ALLUVIATION DUE TO RISING SEA LEVEL	Generally isolated ribbon sandbodies.\nChannel deposit density\n\nChannel deposit interconnected-ness
	Upward fining fine sandstone to siltstone with crosscutting curvilinear lateral accretion units seen on surfaces parallel to bedding.	High sinuosity river with low relative discharge		Ribbon reservoir with good connectedness to Moor Grit below
	Regional sheet sandstone. Multistorey in part. Sharp erosive base. Medium to coarse grained with gross upward fining. siltstone lenticles at top. Tabular cosets with rarer very large scale trough sets. Mature quartzarenite	Low sinuosity, braided (?) river with high relative discharge	Degrading alluvial system ; high slope ; entrenchment	Fine sheet reservoir with connections to alluvial and marine sandstone members above and below respectively.
	Upward coarsening from bioturbated mudstone to low angle cross-laminated, well sorted very fine sandstones.	Coastal marine beach/barrier system	Wave-dominated clastic shoreline	Sheet reservoir ; present only when not cut out by Moor Grit.

A.

UPLIFTED HINTERLANDS IN DEEPLY WEATHERED MULTICYCLE CARBONIFEROUS CLASTICS

fluvial mileaux

Market Weighton Block

lagoon, bay, chenier mileaux

B.

Fig. 5. 5A — Summary log to show interpretation of Scalby Formation alluvium. 5B — Generalised palaeogeography of N. England during Scalby Formation times.

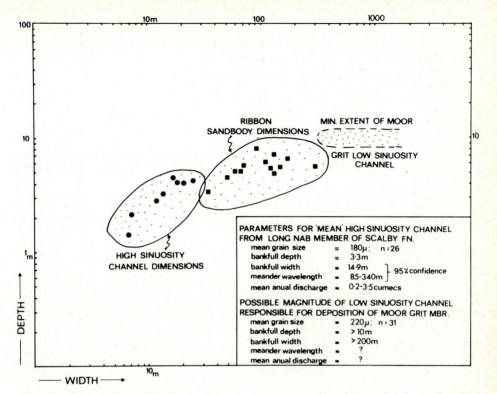

Fig. 6. Graph to show width/depth relations and palaeohydrology deduced for Scalby Formation Channels (method A of Leeder 1973).

It is evident from general consideration of the maximum grain sizes and of the giant scale of some of the cross stratification present in the Moor Grit sheet deposit that the low sinuosity channels responsible were considerably greater in their discharge magnitude and flow power than the later high sinuosity channels. The lack of preserved channel cross sections, due probably to frequent channel wanderings, prevents any quantitative estimate of the magnitudes involved.

Our final model for the changing morphology of the Jurassic alluvial plains of Yorkshire bears great resemblance to that put forward by Allen (1976) in his reinterpretation of the Cretaceous Wealden alluvium of southeast England. In both areas it may become vital to disentangle the possible influences of tectonism, climate and eustasy in causing such vertical changes in alluvial successions. At the present time global tectonics encourage our tectonic bias. Perhaps Schumm's recent text (1977) will go some way towards correcting the balance towards climatic and intrinsic basin interpretations.

ACKNOWLEDGEMENTS

MN gratefully acknowledges financial support from the Ministry of Science and Higher Education in Iran. The paper was presented at the Calgary Fluvial Symposium with the aid of grants from the Symposium sponsors, the Royal Society, and the University of Leeds, all of whom we gratefully thank. We thank Peter Friend and John Collinson for comments on an earlier version of the manuscript.

REFERENCES

Allen, P., 1976, Wealden of the Weald: a new model: Geol. Assoc. London Proc., v. 86, p. 389-437.

Black, M., 1928, 'Washouts' in the Estuarine Series of Yorkshire: Geol. Mag., v. 65, p. 301-325.

——, 1929, Drifted plant-beds of the Upper Estuarine Series of Yorkshire: Quart. J. Geol. Soc. London, v. 85, p. 389-439.

——, 1934, Sedimentation of the Aalenian rocks of Yorkshire: Yorks. Geol. Soc. Proc., v. 22, p. 265-279.

Coleman, J. M., 1969, Brahmaputra River: channel processes and sedimentation: Sediment. Geol., v. 3, p. 129-239.

Hemingway, J. E., and Knox, R. W. O'B., 1973, Lithostratigraphical nomenclature of the Middle Jurassic strata of the Yorkshire basin of N.E. England: Yorks. Geol. Soc. Proc., v. 39, p. 527-535.

Kent, P. E., 1975, Review of North Sea Basin developments: J. Geol. Soc. London, v. 131, p. 435-468.

Leeder, M. R., 1973, Fluviatile fining-upwards cycles and the magnitude of palaeochannels: Geol. Mag., v. 110, p. 265-276.

Nami, M., 1976a, Sedimentology of the Scalby Formation (Upper Deltaic Series) in Yorkshire: Unpub. Ph.D. thesis, University of Leeds, 195 p.

——, 1976b, An exhumed Jurassic meander belt from Yorkshire: Geol. Mag., v. 113, p. 47-52.

Schumm, S. A., 1977, The fluvial system: Wiley, 338 p.

Sellwood, B. W., and Hallam, A., 1974, Bathonian volcanicity and North Sea rifting: Nature, v. 25, p. 27

CYCLICITY, TECTONICS AND COAL:
SOME ASPECTS OF FLUVIAL SEDIMENTOLOGY
IN THE BRAZEAU-PASKAPOO FORMATIONS, COAL VALLEY AREA,
ALBERTA, CANADA

J. Ross McLean[1] and Tomasz Jerzykiewicz[2]

ABSTRACT

The Upper Cretaceous - lower Tertiary Brazeau-Paskapoo Formations in the central Foothills of Alberta encompass an estimated thickness of 3600 m of nonmarine sediments of alluvial plain origin. They were deposited in a foreland, or molasse, basin accompanying orogeny in the Cordillera to the southwest.

Three orders of cyclicity are observed: (1) first order cycles representing mappable lithostratigraphic units; (2) second order cycles representing successive recurrences of fluvial channels at a particular location; and (3) third order cycles representing high- and low-energy overbank deposits on a flood plain.

The main allocyclic control on sedimentation was tectonics. Progressive, but sporadic, encroachment of the thrust belt of the Rocky Mountains on the foreland basin produced sporadic loading and subsidence by isostatic compensation. Coarser members of first order cycles reflect major thrust loading events that initially produced maximum aggradation on the southwestern flank of the foreland basin and then was transmitted progressively eastward. Continued subsidence due to sediment loading produced further aggradation but at a reduced rate, resulting in the finer members of first order cycles which are characterized by a greater proportion of overbank to channel deposits. Second order cycles are predominantly of autocyclic origin but with an allocyclic override. Third order cycles are entirely autocyclic.

Rapid marine transgressions are related to rapid (10^4 - 10^5 years) thrust loading and widespread subsidence of a flexurally rigid lithosphere. Slower transgressions are a function of sediment starving at the coastal plain by preferential sedimentation at the proximal edge of the foreland basin, or by creation of local base levels.

Calculated rates of deposition suggest that long periods of nondeposition must have occurred.

Modern river classifications appear to be inadequate for application to ancient fluvial systems because they lack the perspective of time. Present models for ancient fluvial sequence interpretation are premature and often misleading.

Channel sands, interspersed with various thicknesses of overbank deposits, have characteristics of both braided and meandering rivers, suggesting that both of these morphological types were present. Many channel deposits, composed of horizontally stratified, coarse-grained sandstones, are interpreted to have a flood origin.

The inferred tectonic - sedimentation framework for the Brazeau-Paskapoo sequence is analogous in many respects to the modern Indogangetic Plain of northern India and Bangladesh.

Thick coal deposits are of alluvial plain origin and are associated with a climatic change and widespread floral extinction at the Cretaceous-Tertiary transition. They also coincide with a major orogenic event producing widespread subsidence and marine transgression.

INTRODUCTION

The foreland basin (or foredeep, exogeocline, retroarc basin, or molasse basin), along the eastern edge of the Canadian Rocky Mountains, was the site of active subsidence

[1]Institute of Sedimentary and Petroleum Geology, 3303 - 33rd St. N.W., Calgary, Alberta, Canada T2L 2A7.
[2]Institute of Geological Sciences, University of Wroclaw, ul Cybulskiego 30, 50 - 205 Wroclaw, Poland.

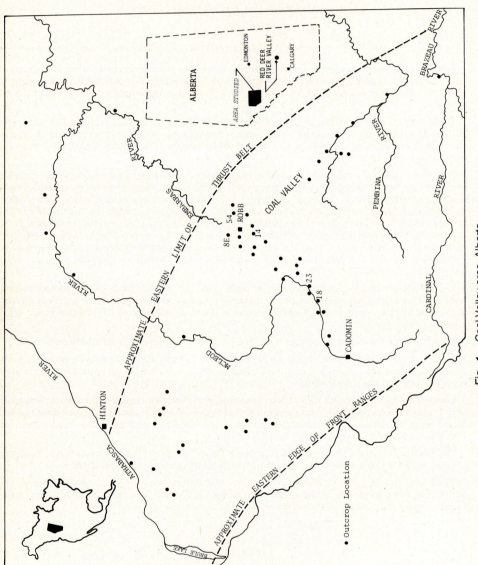

Fig. 1.   Coal Valley area, Alberta.

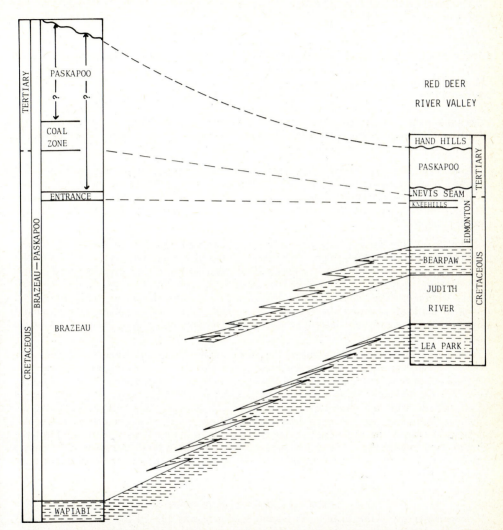

Fig. 2. Lithostratigraphic nomenclature and schematic cross section, Coal Valley to Red Deer River Valley. Dashed pattern indicates marine shales. Tentative correlation of the Entrance Conglomerate with the Kneehills Tuff (Eliuk, 1969), and of the coal zone at Coal Valley with the upper Edmonton coal group (Srivastava, 1970; Sweet, personal comm.) is based on palynological studies.

adjacent to an area of active orogenic uplift, in which were deposited synorogenic clastic sediments ranging in age from Late Jurassic to mid-Tertiary (Bally *et al.*, 1966; Price and Mountjoy, 1970; Eisbacher *et al.*, 1974). The entire clastic wedge thickens markedly towards the mountains.

The Upper Cretaceous - lower Tertiary segment in the Coal Valley area (Fig. 1), with which this paper is concerned, comprises an estimated thickness of 3600 m (12,000 ft) of predominantly nonmarine, alluvial sediments. Complete outcrop sections are unknown, and the succession as a whole usually is not well exposed. Several lithostratigraphic nomenclatural schemes have been proposed (Malloch, 1911; Allan and Rutherford, 1923,

1934; Mackay, 1930; Lang, 1947) and there has been a determined effort to match lithostratigraphic units with the time-stratigraphic Cretaceous-Tertiary boundary, which the facts have steadfastly resisted, but which has resulted in much nomenclatural inexactitude and confusion (cf. Eliuk, 1969). The lithostratigraphic nomenclature is under review and recommendations will be forthcoming in Jerzykiewicz and McLean (in preparation). Brazeau and Paskapoo each have formational status (Fig. 2) but the lower boundary of the Paskapoo Formation in the Foothills is being disputed at present. A group name is required for the entire undivided, nonmarine sequence from the top of the Wapiabi marine shales to the present erosional surface. However, we feel that this is not the place to introduce revised lithostratigraphic nomenclature and we will, therefore, use the dual name Brazeau-Paskapoo when referring to this undivided sequence. The Brazeau-Paskapoo spans the same interval referred to as "Belly River-Paskapoo Assemblage" by Eisbacher *et al*. (1974).

Coal seams were observed during early reconnaissance geology at the end of the last century and mining was active in a few areas until about 1950, particularly in the Coal Valley area. Renewed interest in coal has led to recent exploration in this area. The coal is generally high volatile "C" bituminous, with 10 to 25 percent ash, and less than 0.5 percent sulphur (dry basis).

The Coal Valley area is situated near the northeastern edge of the Foothills thrust belt. Sediments of comparable age are over-ridden by the McConnell Thrust at the front of the Front Ranges (Fig. 1) so that it seems unlikely that any thrusting was taking place any closer than that during the Late Cretaceous. Using the palinspastically restored cross-section of Price and Mountjoy (1970, Fig. 2-3), the closest potential source area (the front of the leading thrust sheet) would be 40 km distant. This is probably a minimum distance but certainly the leading thrust sheet was no farther than twice this distance away.

The paleogeographic setting during deposition of the Brazeau-Paskapoo indicates that the closest marine waters were probably no closer than 80 km from the area of maximum subsidence of the foredeep in the Coal Valley area (Williams and Burk, 1964). As a result, much of the sediment being supplied from the rising Cordillera never reached a coastal environment, but was deposited well inland in an alluvial plain setting. This is quite distinct from most major modern depositional settings which have been studied in detail, where the bulk of the sediment is being transported to the coast to be deposited in large deltas or submarine fans. The Indogangetic Plain of northern India and the upper Amazon Basin are probably modern analogues of the foreland basin, but they have not been extensively studied (*see* later section).

SEDIMENTOLOGICAL OBSERVATIONS

Detailed study of the Brazeau-Paskapoo sequence was carried out in the Coal Valley area (Fig. 1) where exposures are good, but discontinuous, and less than 20 percent of the entire section is exposed. Details of sections and facies analysis are given in Jerzykiewicz and McLean (in preparation). Representative segments of columnar sections are shown in Figure 3. Lithofacies sequence diagrams for coal-bearing and non-coal-bearing segments of the section are shown in Figure 4, together with composite vertical sequences. Results of paleocurrent studies are shown in Figure 5.

Several observations are of particular importance:

(1) There is well-developed cyclicity at two levels observable in single outcrops and a grosser cyclicity observed in the total Brazeau-Paskapoo succession.

(2) Proportion of channel deposits to flood basin deposits varies (Fig. 3) but the latter are usually of greater cumulative thickness than the former. However, in some sections (8E of Fig. 3) vertically stacked channel sands were observed. Vertically stacked flood basin deposits are common.

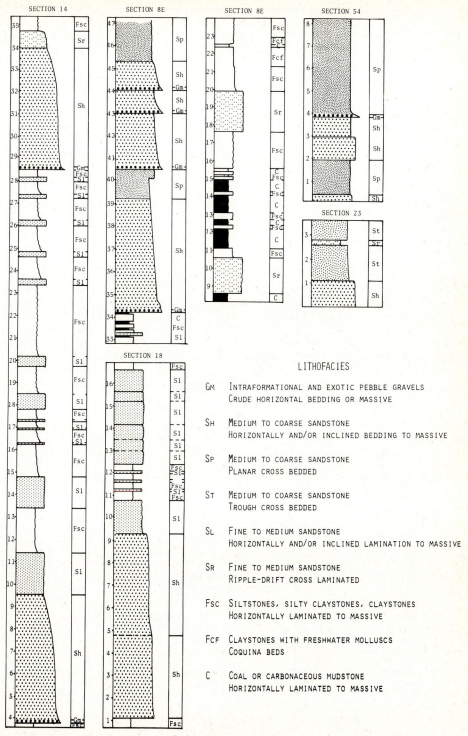

Fig. 3. Representative segments of columnar sections from the Brazeau-Paskapoo. Facies symbols modified after Miall (1977a, and this volume). Thickness in metres. Sections are segments of longer sections given in Jerzykiewicz and McLean (in preparation).

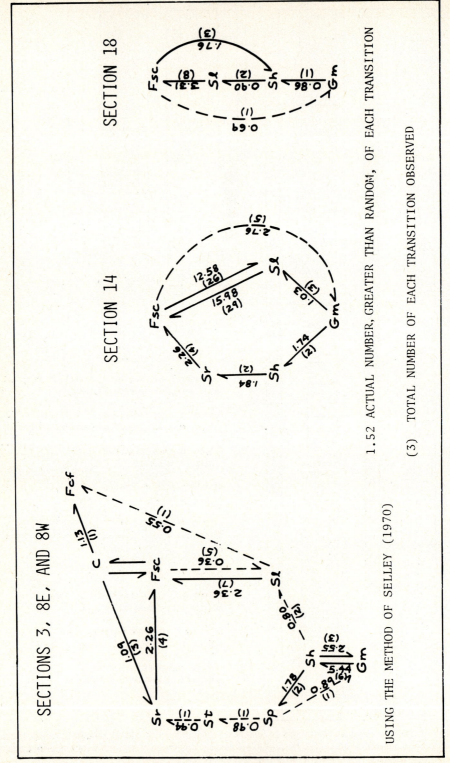

1.52 ACTUAL NUMBER, GREATER THAN RANDOM, OF EACH TRANSITION

(3)   TOTAL NUMBER OF EACH TRANSITION OBSERVED

USING THE METHOD OF SELLEY (1970)

Fig. 4.   Lithofacies sequence diagrams.

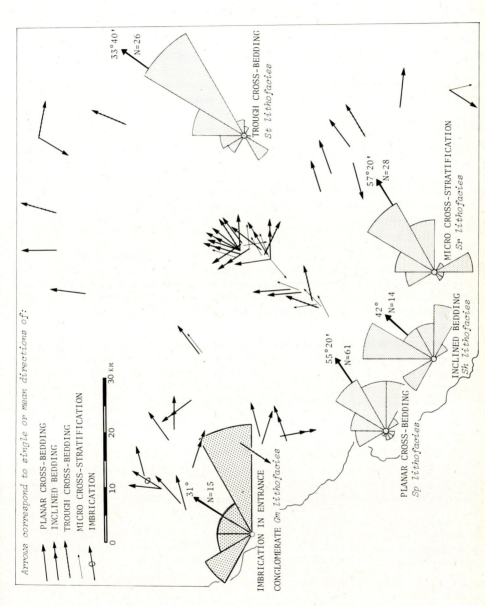

Fig. 5. Paleocurrent data for the Coal Valley area.

(3) Medium- to coarse-grained sandstones with horizontal bedding to low-angle crossbedding (facies *Sh*) are very common, and often comprise almost all of the coarse-grained members of fining upward sequences.

(4) Planar crossbedding is important in some fining-upward sequences; trough crossbedding is less common.

(5) Paleocurrents are strongly unimodal.

(6) Thick coal seams occur in only one part of the section (Fig. 2).

All evidence indicates that the observed sequences are of alluvial plain origin, with a diverse and variable flood basin character including, at times, lakes and extensive swamps.

CYCLICITY AND CONTROLS ON PRESERVED ALLUVIAL VERTICAL SEQUENCES

Cycle Types and Controls

Cyclic sedimentation is natural to any alluvial plain (Beerbower, 1964, p. 41) and would be conspicuous by its absence. Cyclicity is of two basic types: autocyclic and allocyclic. The former are mechanisms which require no change in the total energy and material input into a sedimentary system but involve simply the redistribution of these elements within the system (Beerbower, 1964, p. 32). Normal fluctuations in discharge are encompassed by this category. Mechanisms include lateral migration of channels, avulsion, crevassing, and subsidence due to compaction or due to isostatic adjustment to loading (the latter is not stated by Beerbower, but is implicit). Allocyclic mechanisms are those which result from changes in supply of energy or material and include such factors as eustatic variations in sea level, climatic changes, irregular elevation of the source area, and spasmodic depression of the basin (subsidence other than that due to sediment loading).

A well developed cyclicity is present in the Brazeau-Paskapoo succession as shown by vertical sequence analysis (Fig. 4) and by observations of columnar sections (Jerzykiewicz and McLean, in preparation). The two levels of cyclicity observable in these sections are: (1) larger scale (metres to tens of metres thick), with alternations of coarser channel facies (*Gm, Sh, St, Sp*) (symbols as in Fig. 3) with the finer overbank facies (*Fsc, Sl, Fcf, C*); (2) smaller scale (centimetres to metres thick); within the overbank facies, there are usually well developed couplets composed of: (a) facies *Fsc* and *Sl* in non-coal-bearing sections (sections 14 and 18, Fig. 3), and (b) facies *Fsc* and *C*, and *Sl* and *C* in coal-bearing sections (section 8E, Fig. 3).

A third cyclicity can be seen on the much larger scale (hundreds of metres thick) of the whole Brazeau-Paskapoo sequence. Figure 6 shows the subdivisions of the succession observed by Douglas (1958), Lang (1947) and Ollerenshaw (1968).

A sequence from the Siwaliks of the Himalayan Foredeep (Johnson and Vondra, 1972) is shown for comparison and will be discussed later. The cycles represent variations in proportions of sand and gravel to finer clastics, which have geomorphic expression as prominent ridges and valleys, and are used as mappable divisions.

For the purposes of discussion, the largest cycles are called first order cycles; the smallest, third order cycles.

First Order Cycles and Tectonics

Allocyclic origin

The first order cycles can only be of allocyclic origin as their scale is not reconcilable with energy distributions on an alluvial plain. Beerbower (1964, p. 38) suggested that such

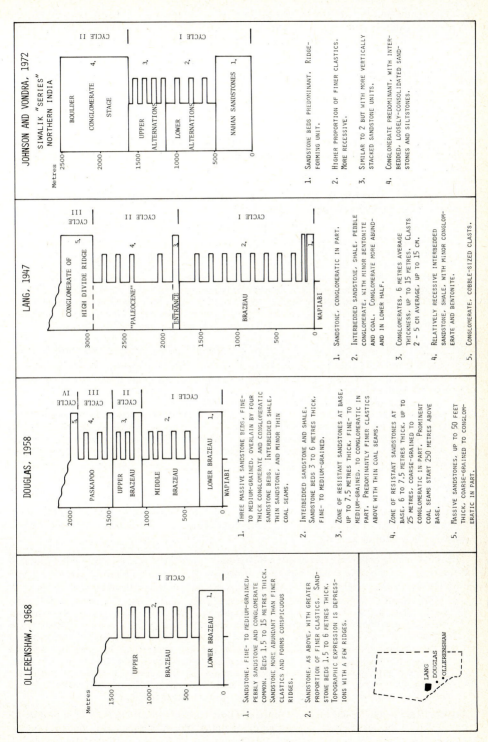

Fig. 6. Schematic representation of first order cycles in the Brazeau-Paskapoo sequence and Siwalik "Series" of northern India.

cycles are generated by periods of disequilibrium, when the alluvial plain morphology and the energy system acting on that plain are out of balance, interspersed with periods of equilibrium. The fluvial system out of equilibrium will always tend towards the equilibrium state (Mackin, 1948, p. 38). Beerbower (1964, p. 38) suggested that:

"Since sediments deposited during a disequilibrium period must differ from those laid down during the subsequent as well as prior steady state periods, an alternation of sedimentary types is inevitable. To the extent that sediments formed during the two steady states may be more similar than the intervening ones deposited during transition, the sequence will display at least a crude cyclicity."

This type of mechanism is suggested for the large cycles of the Brazeau-Paskapoo sequence. The controlling mechanism was tectonic.

Tectonic mechanism

The orogenic process in Western Canada, as visualized by Price and Mountjoy (1970, p. 7), is stated most succinctly in their own words:

"The distinctly layered and strongly anisotropic mass of rock deposited within the Cordilleran miogeosyncline and on the shelf along the margin of the craton has been stripped along the layering from its crystalline basement. The mass has moved upward and up to at least 125 miles northeastward along an array of discrete, discontinuous, interleaved, and overlapping thrust faults, out of the region now occupied by plutonic rocks of the Eastern Crystalline Belt of the Cordillera, to where it is now stacked up on the flank of the craton in the thrust sheets typical of the Rocky Mountains. This process of northeasterly foreshortening involved more than 8 km of crustal thickening that is restricted to the suprastructure of bedded sedimentary rocks above a passive infrastructure of crystalline basement rocks."

The motive force was envisioned as follows (Price and Mountjoy, 1970, p. 18):

". . . it is our thesis that the progressive northeasterly development of a succession of discrete, subparallel, overlapping and interleaved thrust faults in the Rocky Mountains should be correlated step by step, with upwelling and lateral spreading of tongues of hot gneissic rocks in the infrastructure of the Western Cordillera. We view the emplacement of these tongues of hot mobile rocks as a continuing process, in which earlier tongues are eventually uplifted and domed above later tongues that move in and spread out beneath them. Each local pulse in this continuing process is expressed in the Rocky Mountains as a corresponding local surge in thrusting. The thrusts accordingly encroach on the eastern edge of the geosyncline and adjacent parts of the cratonic cover."

Thrusting and Subsidence

Thrust faulting took place spasmodically over a period of about 140 million years and proceeded from southwest to northeast. Thus there was progressive, but not continuous, loading of the crust to the northeast, resulting in discontinuous episodes of subsidence by isostatic adjustment to loading.

The inextricable relationship between thrust fault generation and isostatic adjustment in the foredeep, or molasse basin, has been noted by Bally *et al.* (1966), Price (1973) and Elliott (1977). Price's (1973, p. 500) observations are particularly pertinent to our study:

"The record of fluctuating rates of sedimentation and of intermittent erosion in the clastic wedge deposits of the foredeep is a sensitive gauge of variations in the intensity of deformation within the whole of the orogenic belt. The time delay in isostatic adjustment to supracrustal loads is geologically very short. The lateral gravitational spreading that occurred in conjunction with each pulse of upwelling in the core zone of the orogenic belt led to a commensurate and essentially contemporaneous subsidence of the foredeep . . . Thus intervals of rapid concurrent intense deformation in the foredeep are evidence of concurrent intense deformation in the orogenic belt, and, conversely, lulls in sedimentation in the foredeep are evidence of relative tectonic quiescence in the orogenic belt."

Price's deductions, based on this theory of large-scale gravitational flow of supracrustal rocks, refer to the large-scale intertonguing of major clastic wedges and marine tongues but, we believe, are applicable, as well, to subdivisions within the major clastic wedges.

Walcott (1970, Fig. 5) showed that isostatic response of the crust to unloading due to de-glaciation takes place exponentially, with uplift most rapid initially, and slowing with time. Reversing this, the initial rate of subsidence due to simple loading would be relatively rapid and would slow with time. Some 200 m of post-glacial isostatic uplift has occurred during the past 7000 years in central Canada and is still proceeding. Walcott (1970, p. 719) calculates a total uplift of about 550 m with a "relaxation time" of 1 to 2 x 10^4 years. Certainly, this is very rapid in geological terms.

Walcott (1970, 1973) and Murrell (1976) showed that the continental lithosphere has a high flexural rigidity and reacts in an elastic manner to loading so that a load applied on the southwestern edge of the craton will produce subsidence extending several hundred kilometres laterally from the load. The effects of long term loading (10^6 or more years) are complex and characterized by creep deformation (Murrell, 1976, p. 10). The effects of this are beyond the scope of the present paper.

Examination of structure contour maps of the Precambrian basement (Burwash *et al.*, 1964, Fig. 2-3) and of a prominent Cretaceous marker bed, the First White Specks (Williams and Burk, 1964, Fig. 12-15) indicate a more or less continuous increase in the amount of depression of these surfaces from northeast to southwest, perpendicular to the mountain front, as would be expected with a load applied to the southwest edge of a crustal layer with high flexural rigidity.

Lateral continuity of many major clastic wedges among the synorogenic deposits suggests that many major pulses of orogenic activity occurred more or less synchronously over much of the Rocky Mountain belt, but the occurrence of clastic wedges confined to only part of the foredeep (cf. Stelck, 1975, Figs. 3, 8) suggests that, at times, pulses were perhaps more localized. It seems probable that, masked within the larger clastic wedges, many sub-wedges exist which mark the effect of discrete local thrust movements in particular parts of the thrust sheet entity. Applied stress, from the southwest, would be relieved preferentially along certain thrusts which then would remain inactive while others, in their turn, would be activated, before further movement occurred on the same fault. Noncontinuous, nonuniform thrust movement would result in noncontinuous, nonuniform loading and, thus, noncontinuous, nonuniform subsidence of the foredeep. Overall, most of these small inequities have been averaged out within the framework of the larger clastic wedge and are not seen in the gross geometry of the synorogenic succession.

It is our contention that the nonuniform thrusting, loading and subsidence are manifested in the first and second order cyclicity observed in the Brazeau-Paskapoo sequence and, by extrapolation, in other entirely nonmarine synorogenic successions in the western part of the foredeep, although many of the latter are not preserved due to their inclusion in later thrusting as the foredeep migrated, and their subsequent erosion.

Orogeny and movement on thrust faults is sporadic rather than continuous, as indicated in previous quotations. However, the actual rate of movement of thrust sheets is not known. Crude time limits can be placed on some sheets by the age of sediments which the fault cuts (Royce *et al.*, 1975), but these do not indicate rate of movement. Elliott (1976, p. 293) estimated an average rate of movement on the McConnell Thrust of the Canadian Rocky Mountains at 5 mm/year over 8 million years, but this does not represent the more realistic, sporadic propagation rate, which may have been much greater over short periods (10^3 - 10^4 yrs) of time. It is of interest that most modern rivers which drain the eastern slopes of the Rocky Mountains cut directly across thrust sheets rather than being diverted by them. They were thus able to maintain their positions (antecedent streams) by erosion as the sheets were uplifted and moved northeastward. This has also been observed in the southern slopes of the Himalayas (Holmes, 1965, p. 595) where many rivers have maintained their pre-uplift courses, heading in the Tibetan Plateau, north of the highest peaks. The Indus River has downcut 5150 m through the rising mountain chain during the past 5 to 10 million years. This represents a very slow average downcutting rate of 1 to 0.5 mm/year but, as in the Rocky Mountains, orogenic movements were probably sporadic rather than continuous so that downcutting was probably greater at times. Uplift may have exceeded downcutting ability at times and damming of the rivers may have occurred, such as with the Arun River of Nepal (Holmes, 1965, p. 598) or the Colorado River of Nevada (Holmes, 1965, p. 592). However, infilling from upstream and headward erosion from downstream would eventually eliminate the lake and the river would resume its old course.

Subsidence and Cyclicity

Tectonic control: The interaction between an orogenic belt and its foreland basin is undoubtedly very complex in detail. However, unnecessary detail will only tend to obscure basic relationships and, thus, we have used an oversimplified approach assuming that the general mechanism is correct, and that deductions drawn from the general case will shed light on the relationship between the complex system and its sedimentary record.

If we take a hypothetical situation of a thrust faulting event producing 300 m (1000 ft) of additional load on the crust, and use a density of 2.7 for the material of the thrust plate and 3.3 for the mantle material which must be displaced, then the 300 m load would produce 2.7/3.3 x 300 = 245 m of subsidence directly beneath it and directly in front of it at the western edge of the foreland basin. If the continental crust is considered homogenous, the elastic deformation will produce a progressively smaller downward warping of the crust laterally outwards from the thrust belt, possibly for hundreds of kilometres.

If, as mentioned above, we assume an equilibrium condition prior to thrust movement, then upon thrusting the situation changes from equilibrium with no net degradation or aggradation to disequilibrium with net aggradation. It seems probable in an active tectonic area, that equilibrium in a stream profile might never be achieved, but that there were only variations in the degree of departure from the ideal equilibrium state.

Sediment infill continues as subsidence is taking place and three general states combining these two factors can be attained: (1) subsidence rate greater than infilling rate; (2) subsidence equal to infilling; and (3) subsidence less than infilling. All cases involve aggradation but rates are variable. In the first state, departures from the equilibrium gradient profile would increase, with rivers continually trying to build up the profile but continually falling farther behind. Such a situation would produce maximum alluviation in the proximal locality, but minimum alluviation distally. The second state, which is least likely because it represents a single combination of factors rather than a range, is similar to the first case in that it would produce maximum aggradation, but with uniform disequilibrium. The third state would produce aggradation until the river had established an equilibrium profile, adding only enough new detritus to maintain this profile. The remainder would bypass the proximal area, to be deposited more distally. Such a situation would produce maximum alluviation distally, and possibly internal erosion and reworking proximally.

Where loading is relatively rapid or essentially instantaneous in geological terms, a situation is possible wherein sedimentation proceeds from state 1 to state 3 with time. Initial rapid subsidence due to isostatic compensation produces a state 1 situation. Without renewed loading, sedimentation overtakes subsidence and a state 2 or 3 situation is achieved. However, subsidence continues due to the sediment loading, so that aggradation continues although at a slower rate, and a state 3 situation could be maintained for a long period of time.

The initial 245 m downwarp, as it infills with clastic sediments of density about 2.3, would undergo further isostatic compensation of 170 m. A chain reaction of subsidence and loading is established, which taken to the twelfth event would produce a total subsidence of about 800 m. Effects of compaction have been ignored but would increase this figure by a substantial amount. Only four such thrust loading events could account for the entire 3600 m thickness estimated for the Brazeau-Paskapoo sequence. It must be emphasized, however, that the 300 m taken as the thickness of the thrust sheet was only an arbitrary figure and incremental loading was probably of less magnitude.

Bloom (1967) indicated that even relatively small loads will be compensated isostatically, so that subsidence due to sediment loading would be expected to be relatively continual, although not uniformly so, variations being due to changes in rate of loading.

The sand-rich portions of the first order cycles (Fig. 6) are interpreted as deposits of state 1, resulting from uplift of the source combined with subsidence of the depositional area. The coarsest clasts are found in these units, reflecting periods of highest gradients. Lateral migration was sufficiently rapid that only a subordinate amount of overbank sediments was preserved. The present Kosi River alluvial fan (Gole and Chitale, 1966; *also see* Holmes, 1965, p. 542) may be analogous in part. This much cited, but little studied, river is known to have migrated over 100 km during the past 240 years. The present Ganges and Brahmaputra rivers (Coleman, 1969) might also be analogous.

Infilling of a depression and construction of a graded profile (Mackin, 1948) begins at the proximal edge and moves progressively distally. If the energy and sediment input to the system does not diminish, coarse clastics will be introduced progressively farther out into the foredeep, and a coarsening-upward sequence will be developed. Such coarsening upward cycles (10 to 100 m thick) were observed by Steel *et al.* (1977) in a Norwegian structural basin, and by Fuchtbauer (1967) in the Alpine Molasse. Distinct coarsening-upward cycles have not been recognized in the Brazeau-Paskapoo sequence as yet, but may be present in the sand-rich portions of the first order cycles. Alternatively, if energy input to the system does diminish with time, a fining upward sequence would result. Heward (this volume) documents examples of fining-upward, coarsening-upward, and combined coarsening-, then fining-upward sequences in proximal alluvial fan settings, produced by variations in controlling mechanisms.

The portions of the first order cycles containing a higher proportion of overbank sediments were deposited during periods of greater quiescence, between major thrust events, and are the product of states 2 or 3. Lateral energy gradients were insufficient to rework and homogenize overbank sediments before subsidence could remove them below the erosive zone. Aggradation was active and subsidence was probably of the sediment loading type.

Thus, an allocyclic mechanism — the initial loading produced by thrust sheet progradation — has initiated a process whereby an autocyclic mechanism — subsidence due to sediment loading — combined with other autocyclic mechanisms of energy distribution across an alluvial plain, can produce all of the features observed in the Brazeau-Paskapoo sequence. However, this explanation cannot be considered conclusive and further field work is planned by one of us (JRM) to further test this hypothesis. It is, however, a possibility which is consistent with the tectonic setting and which has not been adequately considered in this and other areas of the world.

Other controls: It is highly unlikely that, over a period of several million years, other allocyclic mechanisms — eustacy and climatic change — would be inoperative. They would act to modify the pattern basically controlled by tectonically induced subsidence.

Eustatic sea level changes cannot be justified as the primary cause of cyclicity. Jeletzky (1971, p. 75) concluded that the Late Cretaceous inundations and emergence of the Canadian Western Interior basins were chiefly caused by oscillatory movements of the tectonically active Cordilleran region, because they are not coincident from one region to another and are too localized and variable in their location in time and space to be caused by worldwide or continental epeirogenic movements or by eustatic oscillations of sea level.

There is no evidence in the paleontological record (Russell, 1977) for any major fluctuations or oscillatory changes in climate which might account for the major cyclicity observed in the Brazeau-Paskapoo sequence. A catastrophic change is postulated for the worldwide change in fauna and flora at the Cretaceous-Tertiary boundary (Russell and Tucker, 1971; Kent, 197) but other authors (Srivastava, 1970; Tschudy, 1970; Sweet, personal comm.) state that it was a gradual, albeit geologically rapid, change.

It is suggested that climate and eustacy may have produced modifications in channel morphology and variations in the character and thickness of the second order cycles in the Brazeau-Paskapoo sequence, but were not of sufficient magnitude to have produced the larger first order cycles.

Distal Loading Effects

Marine Transgressions: An additional feature which might accompany the thrust loading-subsidence model is of interest. If, as is believed, the crust behaves with flexural rigidity and the zone of subsidence acts several hundred kilometres northeastward from the area of loading then, if the sea is within this radius of subsidence, it will transgress across the subsiding area to sea level elevation on the modified alluvial plain surface. Such a transgression could be rapid by geological standards, 10^3 - 10^5 years, and would be essentially isochronous in terms of biological evolution of even rapidly evolving organisms such as ammonites (Gill and Cobban, 1966, Table 2).

If state 1 exists, sedimentation of coarser clastics will be concentrated in the proximal foredeep area and the coastal areas will receive only fine clastics and may be essentially sediment starved. This might allow further marine transgression by continuing subsidence and compaction of the coastal areas and by marine reworking of previously deposited nonmarine sediments.

The nature of the Bearpaw sea transgression (Fig. 2) is in harmony with these conditions. Over much of the southern Alberta Plains, the base of the Bearpaw Formation is essentially isochronous (Russell, 1939) and is interpreted to be the result of thrust loading and relatively rapid subsidence [less than 5×10^5 years, which is the level of resolution using ammonites (Gill and Cobban, 1966)] leading to relatively rapid marine transgression. In the Foothills of southwestern Alberta, Wall and Rosene (1977, p. 852) have shown that the base of the Bearpaw Formation is several ammonite zones younger than the Bearpaw Formation farther east, indicating that the boundary is distinctly diachronous. This is interpreted as the product of slower transgression accompanying sediment starving and slower subsidence due to sediment loading. Once state 3 was achieved in the foredeep, more and coarser sediment was transported to the coast and slow infilling of the sea took place producing the observed prominent diachronous upper contact of the marine formation.

There is no evidence of the Bearpaw Sea in the Coal Valley area. This area contains the maximum preserved thickness of Upper Cretaceous-lower Tertiary sediments in what has been called the Central Alberta Basin (Williams and Burk, 1964, Fig. 12-15; Taylor et al., 1964, Fig. 13-1). The basin itself is an interesting feature suggesting some segmentation of the clastic wedge. In view of the previous discussion, it may be a manifestation of a zone of more intense thrust sheet loading and consequent greater overall subsidence, perhaps associated with a major paleofluvial system, (possibly the ancestral Athabasca River), which probably drained a large area west of the present continental divide in Late Cretaceous-early Tertiary time.

Local Base Levels: Although a homogeneous lithosphere is assumed for modelling purposes (Walcott, 1970, 1973), the continental crust is obviously very heterogeneous, so that different segments will react to loading in different manners, producing a nonuniform subsidence profile which will contribute locally to the overall disequilibrium.

A situation can be visualized wherein the sea might not be within the area of influence of a subsidence event, but where a portion of the original river profile to sea level might be depressed so that a lake is formed and a local base level is created. Aggradation and infilling of the lake from the west with headward erosion by streams draining the lake would eventually eliminate this temporary base level, after which erosion would cause the river to erode back through the lake deposits to an equilibrium profile position. While infilling of the lake is taking place, all but the finest sediment would be cut off from the sea. If an appreciable length of time is involved, transgression by reworking of coastal sediments might occur.

Some probable lacustrine deposits have been identified in the Coal Valley area (Jerzykiewicz and McLean, in preparation) but these were relatively small lakes on the flood plain. However, the tectono-depositional setting, with the largest quantities of detritus being deposited well inland of the sea where maximum subsidence was occurring, is conducive to the establishment of local base levels, and more lake deposits may be identified as study of this sequence progresses.

Second Order Cycles

Introduction

The intermediate scale cycles, the largest observable in individual outcrops, represent the record of the arrival of a river at that particular site, its erosion into older flood plain sediments, and its departure from that locality by lateral migration or abandonment, followed by a period during which only overbank sediments were deposited before a river once again occupied this site. A great deal of variability is obviously possible in the thickness and content of these cyclothems. Some of the variations in the overbank sediments will be discussed below. Channel deposits can vary from entirely facies *Sh*, to entirely facies *Sp*, to mixtures of facies *Sh, Sp* and *St* (Fig. 3; and Jerzykiewicz and McLean, in preparation).

Second order cycles may be of (1) purely autocyclic origin, (2) a combination of autocyclic and allocyclic mechanisms or (3) purely allocyclic origin. Allen (1974) investigated possible models combining various types of cyclicity to explain the pedogenic carbonates which he observed associated with fining-upward cycles in alluvial plain deposits. The allocyclic mechanisms considered were climatic fluctuations and base level changes. It would appear that he assumed a subsidence rate equal to depositional rate. Certainly an exclusively tectonic control was not investigated. However, as discussed above, we believe that tectonics played a key role in influencing deposition in the Brazeau-Paskapoo succession. Rate of subsidence and rate of sedimentation, strongly influenced by thrust faulting and subsequent loading of the foredeep, modified the character of the channel deposits and of the ratio of channel to overbank sediments. Autocyclic mechanisms produced the second order cycles but were, at least in part, ultimately controlled by tectonics, an allocyclic mechanism.

The second order cycles are the fundamental building blocks of fluviatile sequences and the classification of ancient fluvial deposits rightfully has been concentrated here. However, attempts to apply current classifications and models of fluviatile sedimentation to ancient fluvial sequences, such as the Brazeau-Paskapoo sequence, are often unsuccessful. The reasons for this are investigated in the following section.

Classification of Ancient Fluvial Deposits

Factors affecting vertical profiles: Systematic study of ancient fluvial systems has lagged behind that of modern rivers, but has been the subject of an ever increasing amount of investigation over the past 10 to 15 years. Naturally, classification of ancient fluvial deposits has attempted to follow that of modern rivers. Attempts have been concentrated on the channel facies of fluvial fining-upward sequences, and particularly on the interpreted morphology of the channel (Allen, 1965; Walker, 1975; Cant and Walker, 1976; Miall, 1977a).

However, the preserved vertical succession of alluvial plain sedimentation is the sum of all of the numerous factors which influence that sedimentation, both channel and overbank. Such deposits are the record of alluvial aggradation, but rates and character of aggradation can vary greatly. Mackin (1948, p. 503) observed:

> "Contrasts between aggradational fills of different types depend largely on contrasts in the rate of lateral shifting of the depositing stream relative to the rate of upbuilding of the deposit and, especially, on the mechanism of lateral shifting, whether by meander swing and sweep processes . . . or by avulsion as in the braided stream. These habits depend in turn on slope, discharge, load, channel characteristics, and the hydrographic regime of the stream, resistance of the valley floor materials to lateral cutting, vegetative cover on the valley floor, distance from the source of the detritus and from the mouth of the stream, relief and erosional processes on adjacent uplands, and other factors which cannot be evaluated here."

Channel morphology is just one of many factors affecting the nature of the aggradation fill and attempts to relate vertical sequences to this one factor may be too simplistic to be of general utility. The whole vertical profile, showing both channel and overbank deposits and reflecting the sum of all controlling factors, should be considered in any classification of ancient fluvial systems.

An aggrading stream is a stream out of equilibrium, but one which is always attempting to achieve the ideal graded profile (Mackin, 1948). However, in a tectonically active area, its efforts may be thwarted by changes in source area elevation, irregular subsidence of the basin, changes in sea level, climatic changes, and all the interconnected effects of these on stream discharge, load, and channel characteristics.

Observations of modern alluvial plain sedimentation, such as those by Leopold and co-workers, who preferentially chose rivers which appeared to be in equilibrium, with no net aggradation or degradation, may not be applicable to the aggradation setting. For example, Wolman and Leopold (1957, p. 106) stated that, although the relative proportions of channel and overbank deposits varies, on the average the amount of overbank material appears to be small. Examination of the geological record, in particular the deposits of the Brazeau-Paskapoo sequence, suggests that this observation cannot be taken as a valid generalization and that the overbank deposits may be much thicker than the channel deposits. The relative proportions of the two depends fundamentally on the balance between lateral energy gradient efficiency in homogenizing a flood plain and the combined effects of subsidence and compaction which act to remove the heterogenous section below the level of erosion and thus preserve it. Other factors certainly contribute but these are the two principal contesting factors.

Such major external factors as sporadic subsidence associated with tectonism, rapid eustatic change of sea level, or rapid fluctuations of climate may profoundly affect the characteristics of the vertical sequence. Allen (1974b) has led the way here, as has happened so often in the past decade and a half, with some very interesting observations on the factors responsible for fluvial cyclothems containing pedogenic carbonates. However, we feel that he has not given sufficient attention to the role of tectonics, as discussed above.

Model Concept: Recent attempts to establish models, which can be used as 'norms' for comparison of fluvial vertical sequences (Walker, 1975; Miall, 1977a, b, and this volume), are premature in our opinion. Too much emphasis has been placed on channel facies, and very little on overbank facies, in particular their thickness relative to associated channel facies. Thick overbank sequences have been observed in the Brazeau-Paskapoo, associated with thinner channel facies which have characteristics suggestive of both meandering and braided origin. Although current models suggest that only thin overbank sequences are associated with braided stream deposits, we suggest that the absolute and relative thicknesses of facies have no intrinsic meaning but depend on other factors as well as channel type. The preservability of vertical sequences under different tectonic and climatic conditions has not been sufficiently investigated, nor has the effect of unusually energy-intensive conditions such as major floods. Channel facies of probably flood origin were observed in the Brazeau-Paskapoo sequence and will be discussed in a later section.

Other authors in this volume have dealt with the model concept in detail and we will not repeat their arguments. Suffice it to say that our feeling is that at the present, relatively primitive stage in study of ancient fluvial systems, general models are often more misleading than helpful in that they give the impression of a uniformity which is not observed in practice. Objective description of a great many fluvial sequences is needed, from as many different major settings as exist, so that realistic models can be developed which will show the variations possible with variations in each controlling factor. In the meantime, we should be content to compare and contrast specific vertical profiles without trying to produce a universal vertical profile or model. Concentration should be on the differences, rather than glossing them over in favour of apparent similarities.

Classification of Brazeau-Paskapoo deposits

The nature of the fluviatile succession in the Brazeau-Paskapoo sequence is shown in Figures 3 and 4. The common, thick sequences of finer grained sediments, interpreted as overbank deposits, have been considered indicative of a meandering, or high sinuosity, river (cf. Moody-Stuart, 1966; Cant and Walker, 1976), but this now appears untenable as a universal rule. The fact that most Brazeau-Paskapoo channel sands are seated in mudstones, claystones and siltstones, and were observed locally to be scoured into these sediments, suggests that banks were relatively cohesive. The common occurrence of rootlets, which tend to stabilize banks (Schumm, 1968) also favours this interpretation. Such a situation favours high sinuosity, meandering rivers (Fisk, 1952; Schumm, 1968). The occurrence of stacked channel sands, as in section 8E, indicates that channels were sometimes floored in less cohesive sediment, favouring a low sinuosity (braided) channel.

Other channel sand bodies in the sections studied contain a high proportion of planar crossbeds (section 54, Fig. 3). Miall (1977b, p. 10) suggested that this facies is abundant in many braided rivers but scarce to absent in most meandering rivers. If this is true, then the deposits of braided rivers may be common in the Brazeau-Paskapoo sequence. However, Jackson (this volume) disputes this suggestion by Miall.

Epsilon crossbedding and fine-grained channel fills, suggested as diagnostic of meandering river channel deposition by Allen and Friend (1968, p. 54), have not been positively identified in the outcrop of the Coal Valley area where large exposures are uncommon.

Flood origin of facies Sh

There is a growing feeling among geologists that, what is observed in the sediment surface of an alluvial plain at a particular instant in the present may not be the key to the past. Uniformitarism of product depends very much on the quantum of time considered. Most studies of modern rivers deal with only the two horizontal dimensions and a small segment of the vertical dimension in a four-dimensional problem. The results are often not translatable into the two dimensions most frequently encountered by geologists — vertical succession and time. Geological time, the fourth dimension, often involves thousands to millions of years and the products of these spans are not observable in the fleeting moment of a single lifetime. The slow, progressive accumulation of 1000 years may be entirely removed and replaced by the products of a single, exceptionally aggressive flood, the rare event in geology (Gretener, 1967).

The common occurrence of facies *Sh* as the predominant facies in the coarse member of fining upward cyclothems is unusual. This facies, in the Brazeau-Paskapoo sequence, is composed of medium- to coarse-grained, massive, horizontally bedded, or low-angle crossbedded sandstone. Massive beds may actually be bedded or crossbedded but lack the particles of different size, colour, or composition necessary to accentuate the bedding. Also, horizontal bedding may, in some cases, contain low angle crossbeds not visible in outcrop, as described by Coleman (1969, p. 215) from the Brahmaputra River.

The occurrence of primary current lineation, the scoured base, and the occasional upward transition to dune crossbedding indicate formation under upper flow regime conditions (Simons et al., 1965). Such sequences have not been commonly reported, but are not unknown.

Allen (1970, p. 318) suggested that predominantly horizontal bedding was typical of one end-member in a series of coarse-grained members of fining upward point bar sequences. This end member represented higher energy deposits — higher slope and lower sinuosity — than those deposits composed of crossbedded sands formed under lower flow regime conditions. Allen (1970, p. 319) suggested that such deposits were deposited closer to the stream source than those characterized by other bedforms.

McKee et al. (1967), in a study of the results of an unusually severe flood in Bijou Creek, Colorado, observed a preponderance (90 - 95%) of horizontally bedded sandstones, both in the channel and as overbank accumulations. Grain size is fine- to coarse-grained but the study also showed that objects such as steel beams and concrete slabs from bridges were also transported and thus the competency of flow was much greater than necessary to transport coarse-grained sand. Much larger clasts would have been transported had they been available.

Miall (1977a) proposed the Bijou Creek deposits as one of four generalized vertical profile models for braided stream deposition. He (1977b) visualizes this model as the product of ephemeral streams with deposition occurring during flash floods, with upper flow regime conditions attained. Models based on modern stream environments, such as these, must be viewed very critically as they attempt to extrapolate the present into the past with short term uniformitarian thinking. Miall's models are build-ups of hypothetical repetitive events similar to those forming the modern braided stream succession. They may be valid and, indeed, they can be applied to certain ancient fluvial sequences as shown by Miall, but they may prove to be applicable to only a small proportion of the range of braided stream profiles preservable in nature.

Smith (1970, p. 2999) observed that longitudinal bars, characterized in part by horizontal stratification, are predominant in the upper reaches (about 125 km) of the braided South Platte River, Colorado. His Figure 11 suggests that not more than 50 percent of any bar was composed of horizontal stratification, which differs from our observations for the sequence at Coal Valley.

The segment of section 18 shown in Figure 4 shows predominantly horizontally bedded sandstone both in the interpreted channel deposit (facies Sh) and in the overbank deposits (facies Sl). This is very similar to the situation observed by McKee et al. (1967). The segment of section 14 (Fig. 4) shows prominent horizontal bedding in the channel but much less in the overbank sediments. In both cases, channels are scoured into fine-grained overbank sediments. The segment of section 8E (Fig. 4) shows stacked channel sands with facies Gm, Sh, and Sp but no preserved overbank sediments.

It is known that, as the water depth in a channel increases, the velocity required to produce upper flow conditions on the floor increases (Southard, 1975, Figs. 2-4, 2-6; Reineck and Singh, 1975, p. 13). Many channel sands of facies Sh show no discontinuity in sedimentation, suggesting that they may be the product of a single depositional event. The thickness of the unit should thus be an approximate indication of channel depth during flood. Thicknesses of up to 7 m (23 ft) were observed. Comparing these values to data given by Reineck and Singh (1975, Table 3), maximum velocities of 8 m/s (26.5 f/s) would be required to produce the upper flow regime.

These values are high in comparison to values for rivers in the United States (Leopold *et al.*, 1964, p. 167) where mean velocities of rivers in flood vary from 2 to 3 m/s (6 - 10 f/s), few exceed 4 m/s (13 f/s), and values above 9.1 m/s (30 f/s) are unknown. Values calculated for the Foothills Group may be too high, the method of calculation being open to question as the tabulated values of Reineck and Singh are unsubstantiated. However, values of this order of magnitude do suggest flood conditions and, perhaps, unusually strong floods.

The sand bodies may represent periods of intense channel scouring followed by rapid deposition during falling stages of the flood. Depths of scour of up to 38 m are reported by Leopold *et al.* (1964, p. 229) but there is some doubt (Lane and Borland, 1954) as to whether or not significant scouring occurs in the channel other than in constricted areas. Certainly, greater depth of scour can occur in these areas, but Leopold *et al.* (1964, p. 234) suggest that scour can occur simultaneously through relatively long reaches, both in pools and over bars.

It is not known whether these flood deposits are of braided or meandering stream origin. Floods are common to most rivers and are primarily a function of seasonal variations in precipitation and, consequently, runoff. The frequency of flooding is nearly the same in regions of very diverse runoff, from tropics to semi-arid regions (Leopold *et al.*, 1964, p. 326). Scouring is observed in both river types (Wolman and Leopold, 1957, p. 105; Coleman, 1969, p. 173). Unfortunately, there is no documentation of the nature of the bedforms preserved below the zone of active alluviation, but only on the nature of surficial bedforms. Coleman (1969, p. 214) observed some horizontally bedded deposits but also many small to very large bedforms, the relative preservability of which are unknown.

Cant (1976, p. 77) noted that, during flood stage in the braided South Saskatchewan River, ripples, sand waves, and small dunes and transverse bars were replaced by large (up to 1.5 m high) transverse bars and, locally, a flat bed, and that trenching showed large crossbed sets and upper flow regime parallel laminations. He concluded, "An infrequent very large flood event therefore creates sedimentary structures which are selectively preserved compared to those created during periods of normal sedimentation." A similar probability of selective preservation of high discharge deposits has been investigated by Blodgett and Stanley (this volume).

Third Order Cycles

The smallest cycles, as illustrated in columnar sections 14 and 8E (Fig. 3), represent stacking of vertical accretion deposits and are undoubtedly of autocyclic origin. The $Sl \rightarrow Fsc/Sl \rightarrow Fsc$ sequence of section 14 consists of abrupt based, fining upward couplets which result from higher energy crevasse splay or levee sands being deposited during the active phase of flooding, and low energy siltstone, mudstone, and claystone deposited from suspension during and after the declining phase of the flood.

Facies *C* deposits consist of: (a) coal, which was deposited in floodplain swamps, and (b) carbonaceous mudstone to sandstone, which could be deposited in swamps subject to frequent influx of clastic detritus, or in a higher energy environment where thin beds and laminae of transported carbonaceous debris were interbedded with sand, silt or mud.

Chronology of Cycles

Estimates of rate and duration of deposition can be made which shed some light on the nature of sedimentation and tectonic events.

Flood plain depositional rates obviously can vary greatly from one river to another, and from one part of the flood plain to another. Thus, the higher energy channel and associated levee facies will usually aggrade more rapidly than low energy regions far from an active channel, although this is extremely variable. Leopold *et al*. (1964, p. 433-34) tabulated average flood basin deposition rates ranging from 9 to 46 mm per year for both small and very large rivers. Coleman (1969, p. 232) measured a maximum thickness of 25.4 mm (1 inch) deposited by a single flood on the flood plain of the Brahmaputra River. Leeder (1975) summarized observations by many researchers on accretion rates for flood plain deposition and found a range in values from 1 to 100 mm. A 3 m thick unit of facies *Fsc* would, if we use a figure of 25.4 mm/flood, take 118 flood events to accumulate. Using an average flood recurrence interval of 2.33 years (Leopold *et al*., 1964, p. 64), it would take 275 years for this unit to accumulate. This accumulation rate is rapid compared to the results of Ritter *et al*. (1973, p. 374), who calculated an accretion rate of 0.14 to 1.15 mm/year for deposits of the Delaware River. Alexander and Prior (1971, p. 361) reported as much as 460 mm of sedimentation on a natural levee during a single flood event, but showed that average rates were closer to 10 to 16 mm/year due to alternating erosion and deposition on the levees. It seems probable that multiple depositional events were responsible for deposition of each *Fsc* facies unit. This is reflected in lamination due to compositional and colour variations in this facies. Diastems are abundant but usually are not distinguished individually as they are obscured by pedogenic and organic mechanisms operating on a sediment thickness greater than the average small increment per flood event.

Facies *Sl*, interpreted as crevasse splay or levee deposits, varies greatly in thickness and some are composed of several recognizable depositional events (Sect. 18, Fig. 3). Coleman (1969, p. 232) recorded crevasse splays on levees from 1 to 3 m (few feet to over 10 ft) thick. As mentioned above, nearly 0.5 m was deposited by a single flood of the Ohio River (Alexander and Prior, 1971), McKee *et al*. (1967, p. 829) recorded deposits 0.6 to 3.5 m thick on the flood plain of Bijou Creek, as the result of the major 1965 flood. Certainly, individual *Sl* facies units may be the result of single depositional events. Stacked units of facies *Sl* (as in Sect. 18, Fig. 3) probably represent proximal levee deposits with the deposits of several successive flood events represented.

Facies *C*, coal and carbonaceous-rich beds, are present in different parts of the Brazeau-Paskapoo sequence in different localities laterally along the foredeep. The major beds occur in a zone about 185 m (600 ft) thick as shown in Figure 2. Seams up to 10 m thick are known. Ten metres of bituminous coal represent about 50 m of original peat (Stach *et al*., 1975, p. 18). Average accumulation rates for peat of 1 mm/year for the Klang delta of Malaysia (Coleman *et al*., 1970), 1 mm/year for the Mississippi River (Fisk, 1960), and 3 - 4 mm/year in northwest Borneo (Anderson, 1964) have been reported. At 1 mm/year, 50,000 years would be required to accumulate this thickness, or 12,500 years at 4 mm/year. The implications of this will be discussed in a later section.

In-channel deposition rates are difficult to ascertain as a river may occupy the channel for a much longer time than is required to deposit the preserved sequence. Allen (1974, Table 2) estimated a maximum rate of deposition of 0.4 m/year within cyclothems with recurrence frequencies of 1 to 2 x 10^4 years. Flood deposits, such as our facies *Sh* deposits, are believed to have accumulated several metres of sediment during a single event. Coleman (1969, Fig. 3) shows 8 abandoned channel positions for the Brahmaputra River, some braided, other meandering, which have been formed during the Holocene (about 10,000 years). This is an average of one avulsion event in 1250 years. As discussed below, the Himalayan foredeep is analogous in many ways to the Rocky Mountain foredeep of Late Cretaceous-early Tertiary time, so that this rate may have some significance for our interpretation.

The Brazeau Formation, represents deposition from about 82.5 m.y. to 66 m.y. (Stott, 1963, p. 112; Obradovich and Cobban, 1975, Table 1; Shafiqullah *et al*., 1964). The total thickness of the formation is about 1650 m (Lang, 1947), so that the average depositional rate is 0.1 mm/year. This is markedly less than the expected rates calculated above. This might be due to the expected rates being grossly in error, which seems unlikely since the

Rocky Mountain foredeep was an area of active aggradation, and cited rates of accumulation even in relatively stable areas are never as low as 0.1 mm/year. More likely, it represents long periods of erosion or stability interspersed with periods of active aggradation as suggested above in the discussion of the relative rates of subsidence and sedimentation. These figures suggest that periods of nondeposition may have exceeded, in total, periods of deposition. The nondepositional periods would favour peat accumulation if other factors were favourable.

THE HIMALAYAN MOLASSE BASIN — AN ANALOGUE

The tectono-depositional setting visualized for western Alberta in Late Cretaceous-early Tertiary time is analogous in part to that of the Himalayan molasse basin of Northern India at the present time. Certainly, the plate tectonic model for the Himalayas differs from that of western North America (cf. Dewey and Bird, 1970) and the Rocky Mountains may never have achieved the lofty elevations of the main Himalayan ranges, but there are useful analogies in the tectonic and depositional settings. "The Siwaliks represent a thick, comformable sequence of Upper Cenozoic, predominantly fluviatile, molasse-like deposits derived from the erosion of the rising Himalaya Mountains" (Vondra and Johnson, 1968, p. 308). A schematic representation of their section is shown in Figure 6.

The lower part of the Himalayan succession, the Nahan sandstones, consists of a high proportion of channel sands, often vertically stacked with subordinate vertical accretion deposits. They were interpreted (Johnson and Vondra, 1972, p. 130-133) as the product of rapid aggradation of "loosely" sinuous streams in a rapidly subsiding basin, with rapid lateral shifting and little preservation of overbank sediments. This is analogous to the sand-rich segments of the Brazeau-Paskapoo succession.

The Lower Alternations is characterized by single-story channel sands with multi-storied vertical accretion deposits interpreted as the product of more sinuous channels with prominent development of thick overbank sediments. This unit is similar to the middle to upper Brazeau exemplified by sections 14 and 18 (Fig. 3).

The Upper Alternations is sedimentologically between the Nahan sandstones and the Lower Alternations with some multistoried channel sands, as well as single storied channel sands and multistoried overbank sediments. An analogy may be the section shown in section 8E (Fig. 3).

The Boulder Conglomerate Stage is interpreted as an alluvial fan deposit resulting from uplift along the Main Boundary Fault of the Himalayan front (Johnson and Vondra, 1972). The Entrance Conglomerate and other conglomeratic beds observed in the Brazeau-Paskapoo sequence (Allan and Rutherford, 1934; Lang, 1947; Douglas, 1958; Ollerenshaw, 1968) may represent distal facies of alluvial fans, the proximal portions of which are not present in the Coal Valley area.

Johnson and Vondra (1972, p. 133) compare the Siwalik sedimentation to the present Bhabar (the modern coalesced alluvial apron at the foot of the Siwalik Hills) and the Indogangetic Plain, where rapid subsidence of the Plain is occurring simultaneously with uplift of the outermost Siwalik Hills along an inferred thrust fault. In the area of subsidence, the entire alluvial sequence is preserved, with individual sand bodies reflecting rapid aggradation. "Continual subsidence and thrusting results in a subsequent movement outward, away from the uplift, of the zone of maximum sediment accumulation in the Indogangetic Plain — a process not unlike the development of the foredeep itself." This is exactly analogous to the tectonic and sedimentary development of the Rocky Mountain foredeep.

Coleman (1969, p. 159) observed of the Ganges-Brahmaputra river system: "A delicate balance exists between the rapid subsidence of the land and the up-building of the land by flood-lain deposits." An interesting observation is that, despite its proximity to the

Himalayas, the Ganges River is essentially a large meandering river (Coleman, 1969, p. 135). The Brahmaputra River was, until recently, also a meandering river but, after capture of a smaller river with a heavier silt load, changed to its present braided pattern (Coleman, 1969, p. 143). Both rivers are characterized by well-developed levees and they have, in part, very extensive overlapping floodplains. For example, the flood of 1955 inundated about 1/3 of Bangladesh covering an estimated 46,000 km² (Coleman, 1969, Fig. 10). Because of the numerous major and minor channels which are actively migrating, flood plain deposits along the Brahmaputra are presently forming only a thin veneer capping channel sands and silts. The current alluvial plain sequence may be similar to that of the Nahan Sandstone, although this is only speculation. One can conceive of a situation in which subsidence was more rapid than deposition of the coarse channel facies, where meandering and braided rivers might coexist side by side, and thick flood basin deposits could form and be preserved. That is, there is the potential for thick flood basin deposits to be associated with braided stream deposits.

The morphological metamorphosis of the Brahmaputra is of further interest. The present braided river is flowing across a flood plain dominated by older meander belt deposits, so that it will be subject to the same lateral constraints, provided by clay-plugged, abandoned meander loops, as would a meandering river. A braided belt is established analogous to a meander belt. If this situation can be maintained for a relatively long period of time, significant relief on the alluvial plain could be developed by channel aggradation, and thick overbank deposits could form and be preserved before avulsion relocated the channel. Constrained braided channels may be a significant factor in fluvial deposition in certain settings, such as proximal molasse basins, and could produce a very different vertical profile from that of an unconstrained braided channel as observed on alluvial fans (Boothroyd and Ashley, 1975; Gole and Chitale, 1966).

Evidence for both braided and meandering rivers in the Brazeau-Paskapoo sequence suggests that conditions for coexistence of the two major river types may have existed in the Coal Valley area during Late Cretaceous-early Tertiary time.

COAL IN THE ALLUVIAL ENVIRONMENT

Most commercial coal seams are associated with protected coastal environments, lower and upper delta plains, and lower alluvial plains (Weimer, 1976). However, there is no evidence of marine proximity in the Brazeau-Paskapoo sequence of the Coal Valley area. Facies associations and environmental interpretation of sediments associated with the coal beds of the Brazeau-Paskapoo succession strongly indicate that the peat accumulated on an alluvial plain.

Alluvial plain peats accumulate as (1) areally restricted back-levee swamps that parallel channels, (2) more extensive flood basin swamps which are commonly associated with lacustrine deposits (Weimer, 1976, p. 9) or (3) abandoned channel fills up to 1000's of metres wide and many kilometres long. Fluvial channel sands occur below and above, as well as within, the 180 m (600 ft) thick coal-bearing section of the Brazeau-Paskapoo sequence. Some deposits believed to be of lacustrine origin have been differentiated in two outcrops (Jerzykiewicz and McLean, in preparation) but may be small lakes on the flood plain, rather than a large lake forming a temporary base level. There is no evidence of large scale lacustrine delta building or lacustrine shoreline deposits as yet but, as suggested earlier, this possibility needs further investigation. Outcrops are too sparse and company borehole information was not available to determine the detailed geometry of the seams and associated sediments. The common interbeds of facies *Fsc, Sl* and, occasionally, *Sr* suggest that peat-accumulating areas were usually not very far distant from active channels.

The more proximal alluvial flood plain setting (not lower alluvial plain) for peat accumulation is perhaps less common than others mentioned above, but is certainly not unique. Duff *et al.* (1967) discuss numerous coal-bearing sequences in the southern hemisphere associated with fluvio-lacustrine deposits, Stach *et al.* (1975, p. 16) noted that peat formation ". . . is connected with the later phase of foredeep formation, i.e. with a time when a continental milieu already predominates," and Weimer (1976, p. 9) noted that Upper Cretaceous to lower Tertiary coal deposits of the Western Interior of the United States are related to fluvial systems draining across fresh water-dominated intermontane basins.

Certain physical conditions are dictated by the presence of thick coal seams:

(1) The area must have been free of eroding currents for periods of about 50,000 years, as calculated above, and possibly much longer. This is a relatively long period relative to the estimated rate of accumulation of flood derived overbank sediments. However, the previously discussed rates of sedimentation relative to the time available suggested that long periods of nondeposition of coarse clastics probably existed during deposition of the Brazeau-Paskapoo succession.

(2) Since preservation of peat requires a water table high enough to cover the decomposing vegetation but sufficiently low that it does not drown the vegetation, either subsidence was taking place at a rate equal to vegetative accumulation rate, or the water table was slowly rising. Probably a combination of factors was operative, varying from one area to another, and one time to another. The ground water table would aggrade as the flood plain aggraded and subsidence due to loading and compaction took place. The 50 m of peat required to form 10 m of bituminous coal (Stach *et al.*, 1975, p. 18) would require 19 m of subsidence if the water table level remains constant and if there is a continuous reduction in the volume of the peat during deposition from 1:1 at the surface to 3:1 at the lowest layer (Stach *et al.*, 1975, p. 18). The 19 m of subsidence over 50,000 years is an average of 0.4 mm/year, a very moderate rate.

The fact that thick coal seams are not more common in the Brazeau-Paskapoo sequence but, at Coal Valley, are grouped into a single 180 m interval, indicates that a combination of suitable conditions was not often attained. The causes of this are uncertain. Energy gradients may have been such that any peat formed was later destroyed, but the presence of thick clastic overbank sequences dictates against this. Possibly the rate of subsidence was usually too great and plant growth could not be maintained for the necessary periods of time. This may be valid in that coal seams are common in much of the laterally equivalent Judith River Formation and Edmonton Group in the Plains area (Fig. 2) where subsidence would be less active and alluviation less intense. Thin coal seams and abundant woody and carbonaceous debris throughout the Brazeau-Paskapoo sequence suggest that this may have been a factor.

A tectonic influence, or control, of peat accumulation is possible. There is a growing body of evidence that a very widespread coal-bearing interval occurs just above the Cretaceous-Tertiary boundary, having been documented by Brown (1962, p. 9) in the northern Great Plains of the United States, by Sweet (personal comm.) in southern Saskatchewan, and by Srivastava (1970, p. 272) in the Plains of central Alberta. The 180 m thick coal interval at Coal Valley occurs in a 900 m interval known to contain the Cretaceous-Tertiary boundary (Eliuk, 1969) and, by comparison with the aforementioned reports, we believe that the boundary most likely lies just at the base of the coal-bearing interval. A drastic change in flora and fauna, accompanying a climatic change, occurred across the Cretaceous-Tertiary boundary although the cause is a moot point (Russell and Tucker, 1971; Srivastava, 1970; Tschudy, 1970).

Some evidence suggests that there may have been a change to a higher energy environment in the early Tertiary, although not coincident with the Cretaceous-Tertiary boundary as suggested by Russell (1977). The fluvial sandstones of the Paskapoo Formation of the central Alberta Plains (Fig. 2) lie approximately 35 m above the Nevis coal seam which is just above the Cretaceous-Tertiary boundary (Srivastava, 1970, p. 272) and are distinctly coarser grained than underlying sediments (Gibson, 1977, p. 7). At Coal Valley, the top of the coal seam group is overlain abruptly by stacked sandstone channels (Section 8E, Fig. 3) which lie some 180 m above the proposed location of the Cretaceous-Tertiary boundary. A possible explanation of this, in light of previous discussion, is that the peat was being deposited during an early phase of our state 1, following a period of thrust faulting and pronounced subsidence of the foreland area. Active alluviation would take place on the western edge of the basin but, to the east, the depressed profile with reduced gradients and less coarse clastic influx would be more suitable for development of extensive swamps. The eastward migration of coarse clastic sedimentation would eventually result in destruction of the peat-forming environments, depositing the fluvial channel sands at Coal Valley and, later, in the central Plains area.

An additional bit of evidence which is in harmony with this hypothesis is the approximate synchroneity of the time of deposition of the widespread coal interval with the transgression of the Cannonball Sea of North Dakota which Leffingwell (1970, Fig. 5) shows just above the Cretaceous-Tertiary boundary. There is no evidence that this marine transgression was ever closer than 1000 km southeast of the Coal Valley area, suggesting that nearly synchronous tectonic events may have occurred along the Cordilleran far to the south at this time, perhaps as part of a global event which also produced the worldwide faunal and floral changes observed at the Cretaceous-Tertiary boundary.

ACKNOWLEDGEMENTS

The authors are indebted to A. D. Miall and D. G. F. Long for suggested improvements to an earlier draft of this paper. We are grateful to the National Research Council of Canada for providing a postdoctoral fellowship which allowed T. Jerzykiewicz to work at the Institute of Sedimentary and Petroleum Geology for 14 months, and to the University of Wrocław for granting him a leave of absence. Mr. P. Ward assisted ably during field investigations.

REFERENCES

Alexander, C. S., and Prior, J. C., 1971, Holocene sedimentation rates in overbank deposits in the Black Bottom of the lower Ohio River, southern Illinois: Am. J. Sci., v. 270, p. 361-372.

Allan, J. A. and Rutherford, R. L., 1923, Saunders Creek and Nordegg coal basins, Alberta, Canada: Scientific and Indust. Research Council of Alberta, Rept. No. 6, 66p.

———, 1934, Geology of Central Alberta: Geol. Div., Research Council of Alberta, Rept. No. 30, 41p.

Allen, J. R. L., 1965, Studies in fluviatile sedimentation: six cyclothems from the Lower Old Red Sandstone, Anglo-Welsh Basin: Sedimentology, v. 3, p. 163-198.

———, 1970, Studies in fluviatile sedimentation: a comparison of fining-upwards cyclothems, with special reference to a coarse-member composition and interpretation: J. Sediment. Petrol., v. 40, p. 298-323.

———, 1974, Studies in fluviatile sedimentation: implications of pedogenic carbonate units, Lower Old Red Sandstone, Anglo-Welsh outcrop: Geol. J., v. 9, p. 181-208.

——— and Friend, P. F., 1968, Deposition of the Catskill facies, Appalachian region: with notes on some other Old Red Sandstone Basins: in G. de V. Klein, ed., Late Paleozoic and Mesozoic Continental Sedimentation, Northeastern North America; Geol. Soc. Am. Spec. Paper 106, p. 21-74.

Anderson, J. A. R., 1964, The structure and development of the peat swamps of Sarawak and Brunei: J. Trop. Geogr., v. 18, p. 7-16.

Bally, A. W., Gordy, P. L. and Stewart, G. A., 1966, Structure, seismic data, and orogenic evolution of southern Canadian Rocky Mountains: Bull. Can. Petrol. Geol., v. 14, p. 337-381.

Beerbower, J. R., 1964, Cyclothems and cyclic depositional mechanisms in alluvial plain sedimentation: in D. F. Merriam, ed., Symposium on Cyclic Sedimentation; State Geol. Surv. Kansas, Bull. 169, p. 31-42.

Bloom, A. L., 1967, Pleistocene shoreline: a new test of isostacy: Geol. Soc. Am. Bull., v. 78, p. 1477-1494.

Boothroyd, J. C. and Ashley, G. M., 1975, Processes, bar morphology, and sedimentary structures in braided outwash fans, northeastern Gulf of Alaska: in A. V. Jopling and B. C. McDonald, eds., Glaciofluvial and Glaciolacustrine sedimentation; Soc. Econ. Paleont. Mineral. Spec. Pub. 23, p. 264-280.

Brown, R. W., 1962, Paleocene flora of the Rocky Mountains and Great Plains: U.S. Geol. Survey Prof. Paper 375, 106p.

Burwash, R. A., Baadsgaard, H., Peterman, Z. E., and Hunt, G. H., 1964, Precambrian: in R. G. McCrossan and R. P. Glaister, eds., Geological History of Western Canada; Alta. Soc. Petrol. Geol., p. 14-19.

Cant, D. J., 1976, Selective preservation of flood stage deposits in a braided fluvial environment: Geol. Assoc. Canada, 29th Ann. Mtg., Edmonton, Prog. with absts., p. 77.

—— and Walker, R. G., 1976, Development of a braided-fluvial facies model for the Devonian Battery Point Sandstone, Quebec: Can. J. Earth Sci., v. 13, p. 102-119.

Coleman, J. M., 1969, Brahmaputra River: channel processes and sedimentation: Sediment. Geol., v. 3, p. 129-239.

——, Gagliano, S. M. and Smith, W. G., 1970, Sedimentation in a Malaysian high tide tropical delta: in J. P. Morgan, ed., Deltaic sedimentation: modern and ancient; Soc. Econ. Paleont. Mineral. Spec. Pub. 15, p. 185-197.

Dewey, J. F., and Bird, J. M., 1970, Mountain belts and the New Global Tectonics: J. Geophys. Res., v. 75, p. 2625-2647.

Douglas, R. J. W., 1958, Chungo Creek map-area, Alberta (83C/9): Geol. Surv. Canada, Paper 58-3, 45p.

Duff, P. McL. D., Hallam, A. and Walton, E. K., 1967, Cyclic sedimentation: Elsevier, New York, 280p.

Eisbacher, G. H., Carrigy, M. A. and Campbell, R. B., 1974, Paleodrainage pattern and late orogenic basins of the Canadian Cordillera: in W. R. Dickinson, ed., Tectonics and Sedimentation; Soc. Econ. Paleont. Mineral. Spec. Pub. 22, p. 143-166.

Eliuk, L. S., 1969, Correlation of the Entrance Conglomerate, Alberta, by Palynology: Unpub. M.Sc. thesis, Univ. Alberta, Edmonton, Canada, 128p.

Elliott, D., 1976, The energy balance and deformation mechanisms of thrust sheets: Phil. Trans. Roy. Soc. London, v. 283, p. 289-312.

——, 1977, Some aspects of the geometry and mechanics of thrust belts: Calgary, Alberta; Can. Soc. Petrol. Geol., 8th Ann. Seminar Notes.

Fisk, N. H., 1952, Mississippi River Valley geology relation to river regime: Am. Soc. Civil Eng., Trans., v. 117, p. 667-682. (Reprinted in S. A. Schumm, ed., River Morphology; Benchmark papers in Geology, Stroudsburg, Penn., Dowden, Hutchinson and Ross).

——, 1960, Recent Mississippi River sedimentation and peat accumulation: Compte Rendu, 4th Carboniferous Cong., Heerlin, 1958, v. 1, p. 187-199.

Fuchtbauer, H., 1967, Die Sandsteine in der Molasse nordlich der Alpen: Geol. Runds., v. 56, p. 266-300.

Gibson, D. W., 1977, Upper Cretaceous and Tertiary coal-bearing strata in the Drumheller-Ardley region, Red Deer River Valley, Alberta: Geol. Surv. Canada, Paper 76-35, 41p.

Gill, J. R. and Cobban, W. A., 1966, The Red Bird section of the Upper Cretaceous Pierre Shale in Wyoming: U.S. Geol. Survey, Prof. Paper 393-A, 73p.

Gole, C. V. and Chitale, S. V., 1966, Inland delta building activity of Kosi River: J. Hydraulics Div., Am. Soc. Civil Eng., v. 92, HY2, p. 111-126.

Gretener, P. E., 1967, Significance of the rare event in geology: Am. Assoc. Petrol. Geol. Bull., v. 51, p. 2197-2206.

Holmes, A., 1965, Principles of Physical Geology: Nelson, London, Gt. Britain, 1288p.

Jeletzky, J. A., 1971, Marine Cretaceous biotic provinces and paleogeography of Western and Arctic Canada: illustrated by a detailed study of ammonites: Geol. Surv. Canada, Paper 70-22, 92p.

Jerzykiewicz, T. and McLean, J. R., in preparation, Sedimentology of coal-bearing Cretaceous-Tertiary sequence, Coal Valley area, Alberta: Geol. Surv. Canada Paper.

Johnson, G. D. and Vondra, C. F., 1972, Siwalik sediments in a portion of the Punjab re-entrant: the sequence at Haritalyangar, District Bilaspur, H. P.: Himalayan Geol., v. 2, Wadia Inst. of Himalayan Geology, Delhi, p. 120-144.

Kent, D. V., 1977, An estimate of the duration of the faunal change at the Cretaceous-Tertiary boundary: Geology, v. 5, p. 769-771.

Lane, E. W. and Borland, W. M., 1954, River-bed scour during floods: Am. Soc. Civil Eng. Trans., v. 119, p. 1069-1079. (Reprinted in S. A. Schumm, ed., River Morphology; Benchmark Papers in Geology; Stroudesurg, Penn., Dowden, Hutchinson, and Ross).

Lang, A. H., 1947, Brûlé and Entrance map-areas, Alberta: Geol. Surv. Canada Mem. 244, 65p.

Leeder, M. R., 1975, Pedogenic carbonates and flood sediment accretion rates: a quantitative model for alluvial arid-zone lithofacies: Geol. Mag., v. 112, p. 257-270.

Leffingwell, H. A., 1970, Palynology of the Lance (Late Cretaceous) and Fort Union (Paleocene) Formations of the type Lance area, Wyoming: in R. M. Kosanke and A. T. Cross, eds., Symposium on Palynology of the Late Cretaceous and Early Tertiary; Geol. Soc. Am. Spec. Paper 127, p. 1-64.

Leopold, L. B., Wolman, M. G. and Miller, J. P., 1964, Fluvial processes in geomorphology: W. H. Freeman and Co., San Francisco, 522p.

Mackay, B. R., 1930, Stratigraphy and structure of bituminous coalfields in the vicinity of Jasper Park, Alberta: Trans. Canadian Inst. Mining Metall., v. 33, p. 473-509.

Mackin, J. H., 1948, Concept of the graded river: Geol. Soc. Am. Bull., v. 59, p. 463-511.

Malloch, G. S., 1911, Bighorn Coal Basin, Alberta: Geol. Surv. Canada, Memoir 9-E, 66p.

McKee, E. D., Crosby, E. J. and Berryhill, H. L., Jr., 1967, Flood deposits, Bijou Creek, Colorado, June 1965: J. Sediment. Petrol., v. 37, p. 829-851.

Miall, A. D., 1977a, A review of the braided river depositional environment: Earth Sci. Revs., v. 13, p. 1-62.

———, 1977b, Fluvial sedimentology: Can. Soc. Petrol. Geol., Notes to accompany lecture series on fluvial sedimentology.

Moody-Stuart, M., 1966, High and low sinuosity stream deposits, with examples from the Devonian of Spitsbergen: J. Sediment. Petrol., v. 36, p. 1102-1117.

Murrell, S. A. F., 1976, Rheology of the lithosphere — experimental indications: Tectonophysics, v. 36, p. 5-20.

Obradovich, J. D. and Cobban, W. A., 1975, A time-scale for the late Cretaceous of the Western Interior of North America: in W. G. E. Caldwell, ed., The Cretaceous System in the Western Interior of North America; Geol. Assoc. Canada, Spec. Paper 13, p. 31-54.

Ollerenshaw, N. C., 1968, Preliminary account of the geology of Limestone Mountain map-area, southern Foothills, Alberta: Geol. Surv. Canada, Paper 68-24, 37p.

Price, R. A., 1973, Large-scale gravitational flow of supracrustal rocks, southern Canadian Rockies: in K. A. DeJong and R. Scholten, eds., Gravity and Tectonics; J. Wiley and Sons, New York, p. 491-502.

———, and Mountjoy, E. W., 1970, Geologic structure of the Canadian Rocky Mountains between Bow and Athabasca Rivers — a progress report: in J. O. Wheeler, ed., Structure of the South Canadian Cordillera, Geol. Assoc. Canada Spec. Pub. 6, p. 7-25.

Reineck, H. E. and Singh, I. B., 1975, Depositional sedimentary environments with reference to terrigenous clastics: Springer-Verlag, New York, 439p.

Ritter, D. F., Kinsey, W. F., III, and Kauffman, M. E., 1973, Overbank sedimentation in the Delaware River Valley during the last 6000 years: Science, v. 179, p. 374-375.

Royce, F., Warner, M. A. and Reese, R. L., 1975, Thrust belt structural geometry and related stratigraphic problems, Wyoming-Idaho-Northern Utah: in D. W. Bolyard, ed., Deep Drilling Frontiers of the Central Rocky Mountains; Rocky Mountain Assoc. Geol., Denver, Colorado, p. 41-54.

Russell, D. A., 1977, A vanished world: the dinosaurs of Western Canada: National Museums of Canada, Natural History Series No. 4, 142p.

——, and Tucker, W., 1971, Supernovae and the extinction of the dinosaurs: Nature, v. 229, p. 553-554.

Russell, L. S., 1939, Land and sea movements in the Late Cretaceous of Western Canada: Trans. Roy. Soc. Canada, Ser. 3, v. 33, sec. 4, p. 81-99.

Schumm, S. A., 1968, Speculations concerning paleohydrologic controls of terrestrial sedimentation: Geol. Soc. Am. Bull., v. 79, p. 1573-1588.

Shafiqullah, M., Folinsbee, R. E., Baadsgaard, H., Cumming, G. L. and Lerbekmo, J. F., 1964, Geochronology of Cretaceous-Tertiary boundary, Alberta, Canada: Proc. 22nd Internat. Geol. Congr., Pt. III, sec. 3, p. 1-20.

Simons, D. B., Richardson, E. V. and Nordin, C. F. Jr., 1965, Sedimentary structures generated by flow in alluvial channels: *in* G. V. Middleton, *ed.*, Primary sedimentary structures and their hydrodynamic interpretation; Soc. Econ. Paleont. Mineral. Spec. Pub. 12, p. 34-52.

Smith, N. D., 1970, The braided stream depositional environment: comparison of the Platte River with some Silurian clastic rocks, north-central Appalachians: Geol. Soc. Am. Bull., v. 81, p. 2993-3014.

Southard, J. B., 1975, Bed configurations: *in* Depositional environments as interpreted from primary sedimentary structures and stratification sequences; Soc. Econ. Paleont. Mineral., Short Course No. 2, p. 5-43.

Srivastava, S. K., 1970, Pollen biostratigraphy and paleoecology of the Edmonton Formation (Maestrichtian), Alberta, Canada: Palaeog., Palaeocl., Palaeoec., v. 7, p. 221-276.

Stach, E., Markowsky, M.-Th., Teichmuller, M., Taylor, G. H., Chandra, D. and Teichmuller, R., 1975, Stach's Textbook of Coal Petrology; 2nd ed., Gebruder Borntraeger, Berlin, 428p.

Steel, R. J., Maihle, S., Nilsen, H., Roe, S. L. and Spinnangr, A., 1977, Coarsening-upward cycles in the alluvium of Hornelin Basin (Devonian) Norway: sedimentary response to tectonic events: Geol. Soc. Am. Bull., v. 88, p. 1124-1134.

Stelck, C.R., 1975, Basement control of Cretaceous sand sequences in western Canada: *in* W. G. E. Caldwell, *ed.*, The Cretaceous System in the Western Interior of North America; Geol. Assoc. Canada Spec. Paper 13, p. 427-440.

Stott, D. F., 1963, The Cretaceous Alberta Group and equivalent rocks, Rocky Mountain Foothills, Alberta: Geol. Surv. Canada, Memoir 317, 306p.

Taylor, R. S., Mathews, W. H. and Kupsch, W. O., 1964, Tertiary: *in* R. G. McCrossan and R. P. Glaister, *eds.*, Geological History of Western Canada; Alta. Soc. Petrol. Geol., Chapt. 13, p. 190-194.

Tschudy, R. H., 1970, Palynology of the Cretaceous-Tertiary boundary in the northern Rocky Mountain and Mississippi Embayment regions: *in* R. M. Kosanke and A. T. Cross, *eds.*, Symposium on palynology of the Late Cretaceous and Early Tertiary; Geol. Soc. Am. Spec. Paper 127, p. 65-112.

Vondra, C. F., and Johnson, G. D., 1968, Stratigraphy of the Siwalik deposits of the sub-Himalayan Gumber-Sakarghat fault block in Himachal Pradesh India (Abs): Geol. Soc. Am., Ann. Mtg. (1968, Mexico City), Prog. with abst., p. 308.

Walcott, R. I., 1970, Isostatic response to loading of the crust in Canada: Can. J. Earth Sci., v. 7, p. 716-726.

——, 1973, Structure of the Earth from glacio-isostatic rebound: Ann. Rev. Earth Planet. Sci., v. 1, p. 15-37.

Walker, R. G., 1975, From sedimentary structures to facies models: example from fluvial environments: *in* Depositional environments as interpreted from primary sedimentary structures and stratification sequences; Soc. Econ. Paleont. Mineral., Short Course No. 2, p. 63-79.

Wall, J. H. and Rosene, R. K., 1977, Upper Cretaceous stratigraphy and micropaleontology of the Crowsnest Pass - Waterton area, southern Alberta Foothills: *in* M. S. Shawa, *ed.*, Cordilleran geology of southern Alberta and adjacent areas; Can. Soc. Petrol. Geol. Bull., v. 25, p. 842-867.

Weimer, R. J., 1976, Stratigraphy and tectonics of western coals: *in* D. K. Murray, *ed.*, Geology of Rocky Mountain Coal; Colo. Geol. Surv. Resource Ser. 1, p. 9-27.

Williams, G. D., and Burk, C. F., Jr., 1964, Upper Cretaceous: *in* R. G. McCrossan and R. P. Glaister, *eds.*, Geological History of Western Canada: Alta. Soc. Petrol. Geol., p. 169-189.

Wolman, M. G. and Leopold, L. B., 1957, River flood plains: some observations on their formation: U.S. Geol. Survey Prof. Paper 282-C, p. 86-109.

MEANDERING STREAM DEPOSITS FROM THE TERTIARY OF THE SOUTHERN PYRENEES

Cayo Puigdefabregas[1] and Arthur Van Vliet[2]

Abstract

Three examples of meandering stream deposits from the Tertiary of the Southern Pyrenees show different types of sandbody. Individual active channel sequences in the three examples fit reasonably well the "classical" point-bar model. Lateral accretion bedding is clearly developed and indicates substantial discontinuity in point-bar growth. Accretion surfaces can be traced either throughout entire channel sequences or only through their upper portions. In two of the examples the point-bars are preserved as isolated bodies in a mudstone matrix. In the third example, stacked amalgamated point-bars form more extensive sandbodies which are interpreted as meanderbelts. These different modes of preservation of meandering stream deposits are thought to be dependent on: aggradation rate of the floodbasin, life-time of the channels, stability of position of meanderbelts in the floodbasin and presence of additional floodplain-mud suppliers other than the channels themselves.

Introduction

Aggradation of meandering stream systems after a certain period of time result in a volume of sediment consisting of two basic types of deposits: a) deposits of active and abandoned channels and b) floodplain deposits. The spatial arrangement and relative quantities of these basic types of deposits in a fluvial sediment body as a whole depends on the interaction of a complex set of variables that all together can be summarized as "the longer term behaviour of channels in a floodbasin". Important variables in this respect are, for example, the aggradation rate of the system, the frequency of avulsive events and hence the stability of position of meanderbelts in a floodbasin, the life-time of channels and the rate and mode of meander migration in the belts.

The large scale properties of modern and ancient fluvial systems received much less attention in literature than the smaller (channel) scale processes and deposits in fluvial environments (see e.g. Miall, 1977, for a review).

Active channel sequences of meandering streams in ancient rocks have been described by, among others, Allen (1964, 1965, 1970) from the Old Red Sandstone of England and Wales. Such active channel or point-bar sequences are characterised by the combination of the following features:

a) A scoured basal contact.

b) An upwards change from higher energy to lower energy stratification types.

c) An upwards fining of the dominant grainsize.

d) A large scale inclined bedding (epsilon-cross-stratification described by Allen, 1965) passing through the sequence and reflecting the lateral accretion of the point-bar.

---

[1]Geologisch en Mineralogisch Instituut der Rijksuniversiteit, Garenmarkt 1b, Leiden, The Netherlands.
Present address: Departamento de Estratigrafia, Universidad Autónoma de Barcelona, Bellaterra (Barcelona), Spain.
[2]Koninklijke/Shell Exploratie en Produktie Laboratorium, Volmerlaan 6, Rijswijk (Z-H), The Netherlands.

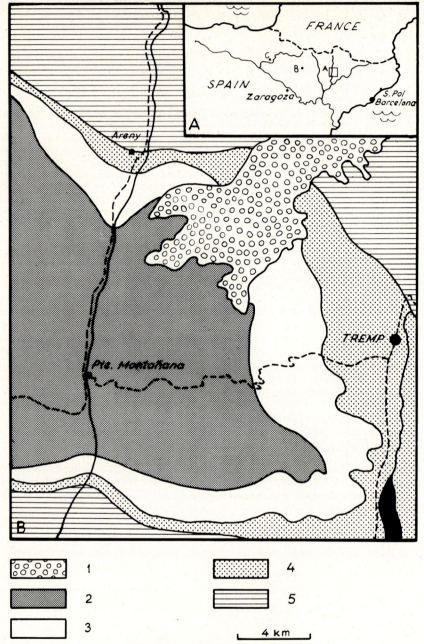

Fig. 1. A: Location map. a — Puente Montañana (L. Eocene), b — Puerto de Monrepos (Oligocene), c — Murillo and Santacara (Miocene).

B: Geological sketchmap of the area around Puente Montañana. 1) Collegats Formation, postorogenic fanglomerates (U. Eocene - Oligocene). 2) Montañana Group, fluvio-deltaic deposits (L. Eocene). 3) Ager Group, shallow marine deposits (U. Paleocene - L. Eocene). 4) Tremp Formation, fluvial deposits (U. Maastrichtian Paleocene). 5) Undifferentiated U. Cretaceous, marine deposits.

Fig. 2. General view of exposure north of Puente Montañana (Barranco de Montañana). Note the two active channel sequences (arrows mark their bases) and floodplain fines in between.

The value of these characteristics for the identification of ancient meandering stream deposits was confirmed by observations in some exposures of exceptional quality, where similar vertical sequences could directly be related to horizontal point-bar geometries (Puigdefabregas, 1973; Nami, 1976).

In this paper three examples of meandering stream deposits from Tertiary formations of the Southern Pyrenees are discussed (Fig. 1A). The first example comes from the Lower Eocene Montañana Group. The other two, already published examples (Puigdefabregas 1973, 1975), from Miocene and Oligocene formations in the same area are briefly reconsidered for the sake of a comparison.

THE MONTAÑANA GROUP

Setting and previous work

The Lower Eocene Montañana Group of the Southern Central Pyrenees crops out in a roughly E-W trending zone approximately 60 km long and 15-25 km wide. Tectonic deformation within the predominantly subhorizontal sequence is limited, thus providing an excellent opportunity for regional sedimentological analysis. Maximum preserved thickness is in the order of 1500 m. The eastern extremity of the Montañana Group outcrop is shown in Fig. 1B. The fluvio-deltaic origin of the group was first recognized by van Eden (1970). Nijman and Nio (1975) presented a more elaborate regional sedimentological study.

Marine influence in the Montañana Group gradually increases from east to west. In the eastern part of the outcrop, in the area of Puente de Montañana (Fig. 1B), the sequence consists almost exclusively of alluvial plain deposits, except for its basal part.

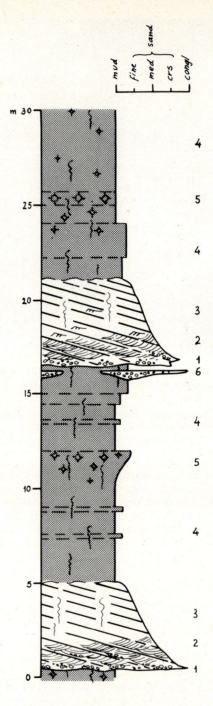

Fig. 3. Generalised vertical section of exposure of Fig. 2. Legend: 1) scoured surface (locally) overlain by lag deposits. 2) trough crossbedded (dm - scale) sandstone. 3) inclined bedded homogenised muddy sandstone. 4) mottled mudstone. 5) calcrete horizon in mottled mudstone. 6) small channelised conglomerate body. 7) dark, non-mottled mudstone and fine sandstone.

Approximately 50 km to the west, the group is almost entirely made up of lower deltaic plain and delta front platform deposits. The overall fluvial transport direction was roughly westward with locally important supply via alluvial fans bordering the deltaic plain in the north.

The Montañana Group was deposited directly on top of the Central South Pyrenean Unit (see Seguret, 1970) that was tectonically emplaced in the earliest Eocene. The group overlies the shallow marine Ager Group with a rather abrupt facies change and locally subtle angular unconformities (see Nijman and Nio, 1975). In the eastern part of the outcrop the Montañana Group is unconformably overlain by the post-orogenic fanglomerates of Late Eocene-Oligocene age.

Meandering stream deposits

The facies of the Montañana Group as described below is characteristic of the lower part of the group (Montllobat Fm. of Nijman and Nio, 1975) in the vicinity of the village of Puente de Montañana. Many of the features described and illustrations shown come from excellent exposures ca. 1 km north of the village along the main road Lérida-Viella.

Description

In the exposure two sandstone bodies are found intercalated in a muddy succession (Figs. 2, 3). Both sandstone bodies are almost 5 metres thick and show a well developed fining upward trend of the dominant grainsize together with an increase in mud content. Bases of the bodies are scoured surfaces locally overlain by patches of gravel. A distinct inclined (15-20°) accretionary bedding of decimetre scale units is observed within the bodies (Fig. 5). Accretion surfaces are marked by mud or silt drapes and can be traced down almost to the bases of the bodies. The only depositional structure normally observed within the accretionary units is decimetre scale trough-crossbedding in their lower parts (Fig. 7A), the upper parts have been homogenized by rootlets. Very rarely ripple cross-lamination is preserved in the higher levels of the bodies. Similar but thinner bodies in nearby exposures are homogenized throughout. Sometimes the accretionary units are grouped in alternating muddier and sandier bundles (Fig. 5B). Minor internal discordances may be visible between accretionary units. A series of strike measurements of accretion surfaces along a line of ca. 250 m in the upper body of the exposure shows a gradual change in dip direction of ca. 30 degrees. In other exposures the curvature of the accretionary units can actually be observed in plan view (Fig. 9). Both sandstone bodies of Fig. 2 can be traced parallel to the dip direction of the accretion surfaces. At the termination of the bodies the last depositional slope is overlain unconformably by a mudstone with thin siltstone and sandstone beds representing the post-abandonment channel-fill or clay plug (Figs. 6A, B). As with the upper sequence of Fig. 2 the cut bank is also exposed (Fig. 8) the bankfull width of the channel could be determined at approximately 30 m.

Sandstone bodies as described above are scattered in a mudstone matrix and form mostly less than 25% (but up to 40%) of the total lithology of the formation. Occasionally mudstone sequences up to ca. 50 m without sandstones interbedded were observed. The mudstones are mottled by rootlets and sometimes show a minor development of calcrete nodules (Fig. 7B) which are either dispersed or organized in discrete horizons (Figs. 3, 4). The nodules are incipient calcite concretions showing a concentric growth. Fine concentrically arranged (?shrinkage) cracks filled with sparry calcite are found in a "matrix" with a "turbate" microtexture. Quartz grains are extensively replaced by calcite. The nodules are found frequently reworked as small pebbles in the basal portions of the sandstone bodies thereby indicating their early pedogenetic origin.

In the exposure described small channelized conglomeratic bodies also occur (Fig. 5A). In a northward direction in the outcrop area of the Montañana Group (Fig. 1) conglomerate channels and sheets become progressively more important.

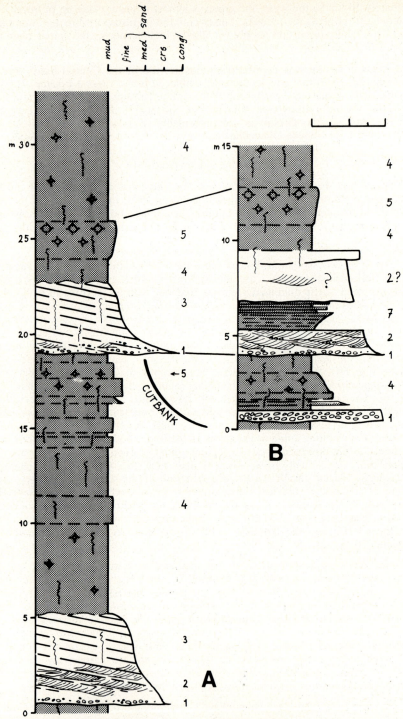

Fig. 4. Generalised vertical sections of exposures in the Barranco de Romeria (ca. 250 m south of exposure of Fig. 2). For legend see caption of Fig. 3.

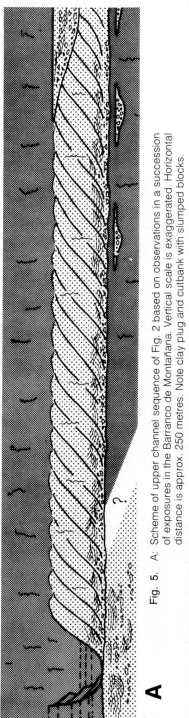

Fig. 5. A: Scheme of upper channel sequence of Fig. 2 based on observations in a succession of exposures in the Barranco de Montañana. Vertical scale is exaggerated. Horizontal distance is approx. 250 metres. Note clay plug and cutbank with slumped blocks.

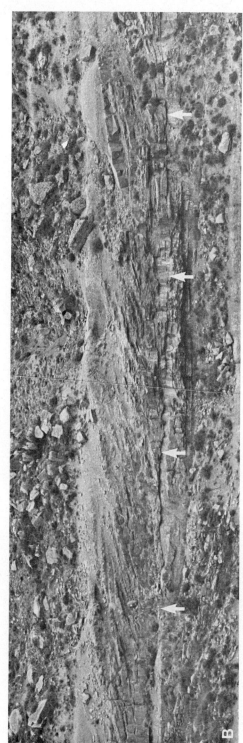

B: Detail view of upper channel sequence of Fig. 2. Note the well developed accretion bedding and the presence of muddier and sandier bundles of accretionary units. Base of the channel sequence is marked by arrows.

Fig. 6.   Clay plug of lower channel sequence of Fig. 2. Arrows mark the point-bar-slope at the moment of channel abandonment. A distinct paleosol (S) is present in the top of the clay plug.

Fig. 7. A: Trough crossbedding in basal part of lower channel sequence of Fig. 2.
B: Calcrete nodules.

Fig. 8. A: Cutbank associated with upper channel sequence of Fig. 2. Abandoned channel-fill is at the left and parallel-bedded floodplain mudstone with paleosols is at the right.
B: Detail view of cutbank. Note erosional surface and slumped blocks marked by arrows.

Interpretation

The observations in the exposures near Puente de Montañana strongly suggest that the described sandstone bodies result from point-bar deposition in small meandering streams from a mud-rich source. Channel depths were approximately 5 m and bankfull widths were in the order of a few tens of metres. The well developed accretionary bedding that can be traced throughout the point-bar sequence and the mud-draped accretion surfaces indicate that point-bar growth was highly discontinuous and that the channels were probably alternating wet and dry. Temporary subaerial exposure of abandoned channels is also suggested by the rootlets in the clay plugs. The interpreted abandoned channel fill in Fig. 4B (5.20-6.80 m) consists of dark laminated mudstone rich in macerated plant material. Mottling did not affect these deposits as they remained submerged and became rapidly buried under a thick sandstone bed deposited during an attempt at channel reactivation.

The alternation of muddier and sandier bundles of accretionary units could point to larger scale fluctuations in discharge. The intensive rootlet mottling and minor calcrete development in the mudstones indicate a subhumid climate which is compatible with the discontinuous discharge interpreted from the accretionary structure of the point-bars. The active channel sequences near Puente de Montañana closely resemble those described by Bluck (1971, Fig. 22) from the lower reach of the River Endrick.

A slightly different type of point-bar sequence from the Lower Montañana Group is shown in Fig. 10. This type of sequence is characteristic of the thicker (5-10 m) sandstone bodies in the area of Puente de Montañana. The basic difference with the earlier described point-bars is that lateral accretion bedding is only observed in the upper part of the active channel sequence. The "massive" lower part is a few metres thick, comprising a coset of trough-crossbedded sets (dm-m scale). Apparently the accretion was only discontinuous in the upper part of the point-bar, probably in the zone between bankfull and (?seasonal) low water level. Most likely the larger channels in the Lower Montañana Group were not alternating wet and dry but continued to carry water during the periods of minimum discharge in the system.

Considering the larger scale arrangement of sandstone bodies and mudstones in the succession, two important features can be observed:

1) Sandstones only form a minor part of the sequence.

2) Mostly isolated single active channel sequences occur, with only exceptional more complex sandbody build-ups.

It is hardly conceivable that the mudstone as a whole results exclusively from overbank processes related to the small channels. Though intensive mottling obliterates the observation of depositional features the mudstone could have been deposited partly from muddy sheetfloods in a distal fan environment. Evidence for the presence of an alluvial

Fig. 9. Plan view of an upper point-bar surface showing scroll-bar relief. The exposure is located a few kilometers east of Puente Montañana and is related to a similar channel sequence as the ones observed in the exposure of Fig. 2.

fan or apron bordering the Montañana floodbasin in the north is provided by the conglomerate channels and sheets appearing in the northern part of the outcrop. Another possibility is that the small channels were associated with a main channel as dominating supplier of overbank sediment. The existence of such a main channel, however, cannot be demonstrated in the available exposures.

The isolated, single point-bar sequences indicate that the channels were rather short-lived features created by avulsive events. A longer term stable position of meanderbelts in the floodbasin was not reached. Rapid aggradation of the floodplain caused early burial of the pointbars preventing them from being (partially) eroded by younger channels.

COMPARISON WITH OTHER EXAMPLES FROM THE SOUTH PYRENEAN AREA

Miocene of the Ebro Basin

Active channel sequences produced by meandering streams in the Miocene of the Ebro Basin were recognized and described by Puigdefabregas (1973). Excellent exposure conditions near the villages of Murillo el Fruto and Santacara (Navarra province) allow the relationship between vertical sections through the subhorizontally lying point-bars and their lateral geometry to be established. The sandstone bodies are composed of a series of curved accretionary units that can be distinguished on aerial photographs. A plan view of a point-bar complex drawn from aerial photographs is presented in Fig. 11A. The thickness of the point-bar sequences ranges from one to two metres. Mud-draped accretion surfaces are distinct and can be traced down to the bases of the point-bars (Fig. 11B). Depositional structures within the accretionary units, large (dm) and small (cm) scale crossbedding, are better preserved than in the examples from the Montañana Group due to the effective absence of rootlets and pedogenesis. The point-bars are again found

Fig. 10. Channel sequence few kilometers north of Puente Montañana. Note that accretion bedding only occurs in the upper half of the channel sequence. The top of the sequence shows scroll-bar relief (arrows).

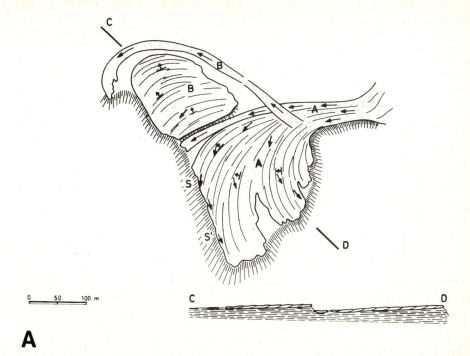

Fig. 11.　A: Plan view (drawn after aerial photographs) and cross-section of a point-bar complex with its accretionary structure. Miocene of the Ebro Basin. (modified after Puigdefabregas, 1973) "S" is topographic slope.

B: Lower part (circa one metre) of similar channel sequence as shown in 11A. Total thickness of such sequences ranges from 1 to 2 m. Note that the accretion planes continue downward to the channel-floor (arrow).

as isolated single channel sequences in a mudstone matrix. The sandstones generally make up less than 10% of the succession. The facies association is interpreted as a distal alluvial fan deposit in which the mudstones largely originated from muddy sheetfloods. Occasionally small meandering channels, with a highly discontinuous discharge, were cut in the distal fan mud-plain. Apart from their smaller dimensions, those channels were basically similar to the ones interpreted from the Montañana Group. Evidence for more proximal fan facies in the Miocene nearby is provided by Puigdefabregas (1973).

Oligocene of the Jaca Basin

Along the road from Sabiñanigo to Puerto de Monrepos a 2000 m thick Oligocene succession of fluvial sediments is exposed. Like the other examples described the sequence consists of sandstone bodies alternating with mudstones. Sandstone forms about 35 percent of the bulk lithology. The exposure discussed here (Fig. 12) is located near the Pardina de Escusaguás. Total thickness of the sandstone body shown in Fig. 12 is ca. 8 m. Two internal scoured surfaces overlain by gravel patches divide the body into three units that largely consist of cosets of trough-crossbedded sets (cf. Fig. 13C), with an average set height of 20 cm. Paleocurrent directions in these units appear to be different from each other: N330E for the lower, N250E for the middle and N20E for the upper unit. The uppermost unit shows in its upper part a fining upward trend and a distinct development of lateral accretion bedding. The top surface of the sandstone body displays a beautifully preserved scroll-bar topography.

With its obvious multistorey character the sandstone body is totally different from the single isolated point-bars discussed so far. We suggest that such sandstone bodies represent entire fossil meanderbelts and that the units within them are individual active channel sequences. Only the uppermost channel sequence was completely preserved, the ones below are incomplete (top absent) as a result of erosion. The substantial difference

Fig. 12.   Multistorey channel sequence near Puerto de Monrepos (Oligocene). The sequence is 8 m thick and consists of three amalgamated units of which only the uppermost was entirely preserved. Note the scroll-bar topography on the top surface of the sandstone body.

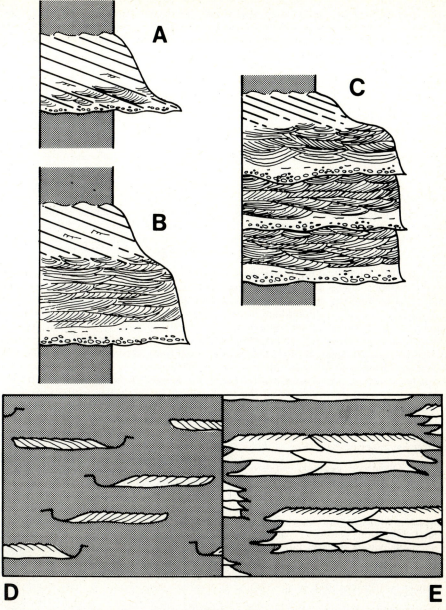

Fig. 13. Scheme showing the different types of channel sequences and their different ways of preservation as discussed in this paper.

A: Single active channel sequence as observed in the Lower Montañana Group (smaller channels) and the Miocene of the Ebro Basin.

B: Single active channel sequence as observed in the Lower Montañana Group (larger channels).

C: Amalgamated active channel sequences as observed in the Oligocene near Puerto de Monrepos.

D: Isolated point-bars in a mudstone matrix (Montañana Group and Miocene).

E: Multistorey sandbodies (meanderbelts) in a mudstone matrix (Oligocene).

in paleocurrents between the units might indicate that the multistorey aspect of the sandstone body largely results from downstream meander migration in the belt rather than by cut-offs followed by re-establishment of the sinuosity of the channels (see Bluck, 1971, p. 133-134).

In contrast to the other examples, sandstone bodies of this type are likely to be formed by meandering channels that are more persistent and develop a stable position of meanderbelts in a floodbasin. A relatively slow rate of floodplain aggradation will prevent the complete preservation of individual point-bars except for the youngest ones that become inactive with abandonment of the whole meanderbelt.

CONCLUSIONS

Three examples of meandering stream deposits have been described and compared. In terms of individual active channel sequences the deposits fit reasonably well the "classical" point-bar model, characterized by a distinct sequential arrangement of dominant grainsize and sedimentary structures. Exposure conditions, though generally very good, do not permit the study of variability between vertical sections of different parts of a single point-bar (cf. Bluck, 1971; Jackson, 1976).

The smaller channels in the Lower Montañana Group and the Miocene examples show a well developed lateral accretion bedding throughout the channel sequence (Fig. 13A). Point-bar accretion was highly discontinuous and the channels were probably alternating wet and dry. The thicker point-bars in the Lower Montañana Group and the Oligocene example show lateral accretion bedding only in their upper parts (Fig. 13B) most likely in the point-bar zone between bankfull level and (?seasonal) low water level. With respect to the larger scale arrangement of channel deposits and fines, the examples from the Lower Montañana Group and Miocene are basically similar; individual point-bars are preserved as isolated bodies in a mudstone matrix (Fig. 13D). Keeping in mind the small size of the channels, it is hard to visualize that the thick mudstone sequences entirely result from overbank deposition from those channels. To preserve meandering stream deposits in a way as illustrated by these two examples the following conditions are required: short life of the channels, frequent avulsive events, high rate of floodplain aggradation and an additional supplier of the fines.

In the Oligocene example sandstone/mudstone ratio is slightly higher. Here the sandbodies are more continuous (km scale), composed of stacked amalgamated point-bars (Fig. 13C, E) and are interpreted as whole meanderbelts. Apparently the channels here were more persistent features and the meanderbelts had a stable position in the floodbasin for longer periods of time. Aggradation of the floodplain was, with respect to the rate of channel migration, probably slower than in the other examples. It is, however, questionable whether aggradation of a "simple" depositional system, a meandering channel flanked by areas of its own overbank deposition, can ever be so fast that meander migration in the belt is "slow" enough to preserve complete individual point-bar sequences (see Bluck, 1971, p. 133-134).

The examples discussed in this paper refer to small "fine grained" point-bars as opposed to the coarse grained point-bars with large bar upbuilding on a lower point-bar platform. A model of coarse grained point-bar deposits from the Montañana Group is described separately in this volume by Nijman and Puigdefabregas.

ACKNOWLEDGEMENTS

We thank Drs. B. J. Bluck, P. L. de Boer, M. C. Budding, P. F. Friend, P. J. C. Nagtegaal and W. Nijman for review of the manuscript and Mr. S. van der Meulen for showing us the exposure of figure 9.

References

Allen, J. R. L., 1964, Studies in fluviatile sedimentation: Six cyclothems from the Lower Old Red Sandstone, Anglo Welsh Basin: Sedimentology, v. 3, p. 163-198.

————, 1965, The sedimentation and paleogeography of the Old Red Sandstone of Anglesey, North Wales: Yorks. Geol. Soc. Proc., v. 35, p. 139-185.

————, 1970, Studies in fluviatile sedimentation: A comparison of fining-upward cyclothems, with special reference to coarse member composition and interpretation: J. Sediment. Petrol., v. 40, p. 298-323.

Bluck, B. J., 1971, Sedimentation in the meandering River Endrick: Scott. J. Geol., v. 7, p. 93-138.

Eden, J. G. van, 1970, A reconnaissance of deltaic environment in the Middle Eocene of the South Central Pyrenees, Spain: Geol. Mijnbouw, v. 49, p. 145-157.

Jackson, R. G., 1976, Depositional model of point-bars in the lower Wabash River: J. Sediment. Petrol., v. 46, p. 579-594.

Miall, A. D., 1977, Fluvial Sedimentology: Notes to accompany a lecture series on fluvial sedimentology held in Calgary, 19th October 1977, Can. Soc. Petrol. Geol., 111p.

Nami, M., 1976, An exhumed Jurassic meander belt from Yorkshire, England: Geol. Mag., v. 113, p. 47-52.

Nijman, W. and Nio, S. D., 1975, The Eocene Montañana delta: *in* The sedimentary evolution of the Paleogene South Pyrenean Basin, Guide book of excursion 19, IXth Internat. Congr. Sedimentology, Nice, July 1975.

Puigdefabregas, C., 1973, Miocene point bar deposits in the Ebro Basin, Northern Spain: Sedimentology, v. 20, p. 133-144.

————, 1975, La sedimentacion molásica en la cuenca de Jaca: Monografias del Instituto de Estudios Pyrenaicos, jaca, v. 104.

Seguret, M., 1970, Etude tectonique des nappes et séries décollées de la partie centrale du versant sud des Pyrénées: Thesis, Montpellier.

COARSE-GRAINED POINT BAR STRUCTURE IN A MOLASSE-TYPE FLUVIAL SYSTEM, EOCENE CASTISENT SANDSTONE FORMATION, SOUTH PYRENEAN BASIN.

W. Nijman[1] and C. Puigdefabregas[2]

ABSTRACT

A large coarse-grained point bar lithosome in the upper part of a multi-storey fluvial sheet sandstone, the Castisent Formation, is described. This sandstone forms a major fluvial system within the Eocene Montañana Deltaic Complex in the Tremp-Graus section of the South Pyrenean Basin. The point bar unit has a crescent-shaped wedge geometry, shows two forms of lateral accretion units, and is characterised by a vertical sequence of scour pool conglomerate, lower point bar trough-fill cross-strata, a well-defined transverse bar facies restricted to the downstream half of the point bar body, upper point bar festoon-beds and an accretionary bank with meander scroll structure, interrupted by a flat-based chute. Perfect outcrop conditions allow the documentation of both the vertical sequence and the point bar relief forms, the cut bank and abandonment fill included.

A comparison is made with some modern examples of coarse-grained point bars.

INTRODUCTION

At present only a few detailed descriptions of coarse-grained point bars are available. McGowen and Garner (1970) were the first to describe this type of point bar from the Recent Amite River in Louisiana and the Colorado River in Texas. They compared these examples with sequences preserved in the Pleistocene Colorado River terraces and in the Eocene Simsboro Sandstone, Texas. In 1971 Bluck described similar bars in the River Endrick in Scotland, paying special attention to their preservation under conditions of downstream meander migration. Jackson (1975, 1976a, 1976b) gives detailed information on structure, sequential order and hydrodynamic regime of coarse-grained point bars of the Lower Wabash River, Illinois.

Data from these researches have in common that this type of point bar does not show an obvious fining-up trend in the vertical sequence except in the uppermost part, transitional to the topstratum and, secondly, that the vertical profile is characterized by an interval of bar foresets. The examples from the Recent relate to rivers with a sinuosity of 1.4 - 2.3 and with either a fluctuating water-level or a strong velocity variation.

In September 1977 we examined a fluvial sandstone lithosome of Eocene age which, because of the preservation of the cut bank, its crescent-shaped wedge geometry and the overall presence of lateral accretion surfaces is interpreted as a huge point bar, although with respect to the grain size, paleocurrent variability and the structural properties of most of its vertical profile it would easily tally with an interpretation as a braided, low-sinuosity, fluvial body.

The point bar lithosome forms part of the Castisent Formation in the Eocene Montañana Deltaic System in the Southern Pyrenees.

---

[1]Dept. of Sedimentology, Leiden and Utrecht State Universities, Garenmarkt 1B, Leiden, Netherlands.
[2]formerly same address; at present: Departamento de Estratigrafia, Universidad Autónoma de Barcelona, Bellaterra, Barcelona, Spain.

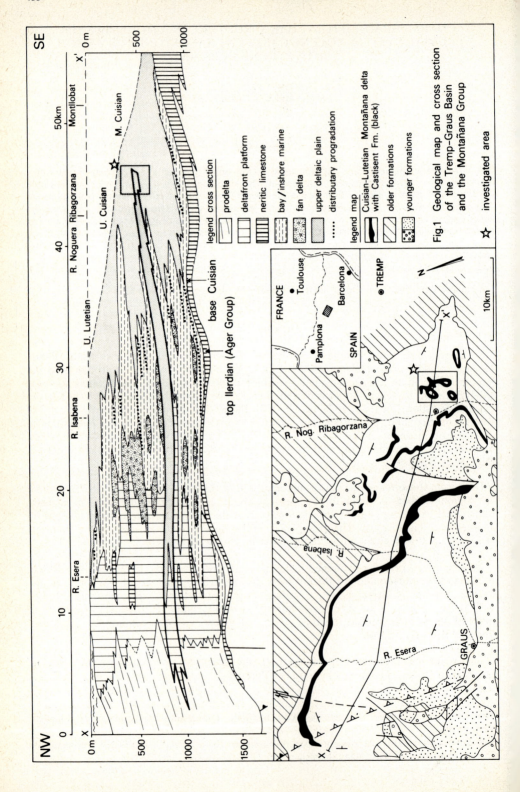

Fig.1   Geological map and cross section of the Tremp–Graus Basin and the Montañana Group

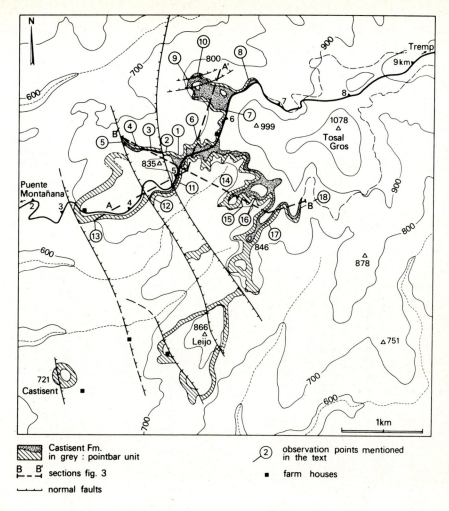

Fig. 2: Outcrop map of the Castisent Sandstone in its type area.

THE CASTISENT SANDSTONE

Geologic setting and gross sedimentologic features

During the Lower Tertiary the South Pyrenean Basin (Mutti *et al.*, 1972; Puigdefabregas *et al.*, 1975, Fig. 1) was an elongated embayment along the southern flank of the rising Pyrenean Axial Zone. The basin opened towards the northwest into a former eastern extension of the Gulf of Biscay. In that direction the basin was filled by prograding fluvial and deltaic mega-sequences. The general sedimentation pattern therefore consists of a northwesterly transition of molasse-type (Van Houten, 1974) fluvial and deltaic deposits into deeper water sediments. Sedimentation was strongly controlled by intense tectonic activity of the Pyrenean orogene (Séguret, 1970; Garrido, 1973; Arthaud *et al.*, 1977, p. 157). The Tremp-Graus Basin forms the central sector of the South Pyrenean Basin and is the depocentre of the Montañana Deltaic System (van Eden, 1970; Nijman and Nio, 1975) during the Lower and Middle Eocene, Cuisian and Lutetian stages (Fig. 1).

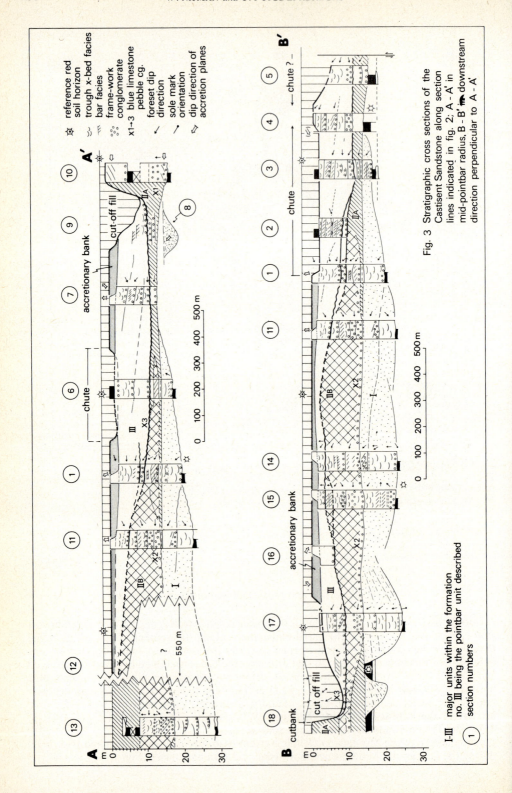

Fig. 3 Stratigraphic cross sections of the
Castisent Sandstone along section
lines indicated in fig. 2; A - A' in
mid-pointbar radius, B - B' ➤ downstream
direction perpendicular to A - A'.

I-III   major units within the formation
        no. III being the pointbar unit described
①   section numbers

Infilling of this basin was accomplished by alluvial systems, draining hinterlands to the north, east and south. The coarse grain size of the bed load material transported into the area and its high content of limestone clasts point to nearby sources and a relatively high gradient. The alluvium from the northern Axial Zone displays the more proximal facies in the form of voluminous fan systems. These coalesce with alluvial and deltaic plain deposits of rivers which followed the basin axis, i.e. in a direction parallel to the orogenic strike. Interference of fan and river deposits contributes considerably to the complexity of the depositional systems and to the immaturity of the material. According to palynological investigations of Haseldonckx (1973) climatic conditions were humid-tropical for the lower part of the Montañana Group while upwards in the sequence a more pronounced seasonal aridity can be inferred.

The Castisent Sandstone Formation represents a major fluvial feeder system of the Montañana delta. The area of investigation is situated in its easternmost outcrop (outline in Fig. 1), which, on the basis of paleocurrent measurements, is considered to be also the point of entry of the system into the Tremp-Graus Basin. A detailed outcrop map is shown in Fig. 2, together with indications of the relevant observation points numbered (1) to (18) throughout this paper.

In its type area (the village of Castisent is in the southwest corner of Fig. 2) the Castisent Sandstone is a multi-storey fluvial channel system, up to 40 m thick, with a width-thickness ratio of about 100 (Fig. 3). Beautifully exposed in a ragged erosion cliff (Fig. 4) it appears as a greyish white sandstone sheet between the darker, yellow-grey coloured flood plain deposits of the Montañana upper deltaic plain. Its dominant lithotype is a calcareous lithic arenite with a high content of quartz and mostly calcareous rock fragments, up to 15% feldspar and a calcite cement. Conglomerates are polymict with variable amounts of limestone, sandstone, quartz, black chert, mud pebbles, calcrete nodules and some intrusive rock components. Limonitized wood fragments are frequently observed and sometimes vertebrate ossicles.

Immediately westward of the survey area the Castisent Sandstone splits into several distributary units, at the same time admixing with alluvial fan material from the north (Vlugt, unpublished M.Sc. thesis, 1975). Within a distance of 40 km it attains a littoral facies at only a few km from its corresponding prodelta slope (Fig. 1, cross-section). Another aspect in this western sector is that the Castisent Formation, in which for reasons of mappability the lateral equivalent transitional and delta front deposits are included, represents a major progradational phase between two, most probably tectonically induced, marine transgressions.

Throughout the fluvial part of the formation a red soil marker bed is used as a reference for correlation.

Subdivision of the Castisent Sandstone in its type area.

In cross-section (Fig. 3) the Castisent Sandstone has a three-storey composition. This paper mainly deals with the upper unit. Although investigation of the complex is still in progress, we will briefly characterize the lower two intervals:

Unit 1: an up to 8 m thick sheet, consisting of imbricated channel fills of coarse- to granule-sized, white sandstones. It contains gravel lags often concentrated at the base of huge, 5 to 10 m wide trough-fill sets, which are the dominant primary structures. Large bar sets are a subordinate feature and because of the dimensions sometimes difficult to distinguish from the trough-fill structures. Paleocurrent bearings around 340° to 360° indicate that the river system entered the basin in this area from the south-southeast. The channel system is confined to a 4 km wide area between the village of Castisent and the hill of Tosal Gros (Fig. 2) its axis coinciding with the figured graben structure. There the channel fills coalesce to form a continuous sheet, much the same as in the example described by Campbell (1976). However, along the northeastern side of the unit separate bodies of very coarse sandstone fill up, to over brim full, channel forms which are deeply (7 m) erosive into the substratum (locs. 8, 18, Fig. 3). These fills have a pronounced lateral accretion, dipping away from the main body, and in cross-section show much resemblance with examples described from Spitsbergen by Moody-Stuart (1966, Fig. 5), although in our case a high sinuosity of these side channels is to be excluded.

Geometry, structure and paleocurrent variability within and between the channels probably indicate a braided, fluvial system which, because of the contrast with the substratum, rather abruptly entered the basin over a newly established paleoslope.

Unit II: In all the section points of Fig. 3, at 15 to 18 m below the top of the formation, the Castisent Sandstone markedly changes both in composition and structure. This second interval is made up of two facies groups.

One — IIA — is a renewed extension of the type of floodplain deposit that underlies the formation, with intercalated channel fills some of which attain the epsilon-stratified sheet form of high-sinuosity streams, as in the examples described by Puigdefabregas and Van Vliet (this volume). Lag conglomerates, however, are relatively thick, reach cobble size and are dominantly composed of bluish grey Mesozoic limestone and reworked calcrete nodules which also form the main constituents of many sandstone (X-1 cg. in Fig. 3, loc. 10).

The second facies type — IIB — is the normal Castisent-type sandstone, somewhat admixed with limestone material producing a more greyish colour. The sandstone unit is underlain by a lag conglomerate with abundant bluish limestone pebbles wherever in contact or near-contact with unit IIA, suggesting a reworking (X-1 into X-2, Fig. 3). With respect to unit I, conglomerates are more abundant (locs. 11, 13), bar foreset structures, alternating with trough-fill cross-stratified sets, are more conspicuous, and cosets are frequently separated by clay layers (Fig. 4) rich in lignitized wood fragments with limonite crusts. Logs and mud pebbles also occur in the sandstones. Paleocurrent bearings vary around a mean of 275°, a deviation of some 70° with respect to unit I. The relations between the facies groups within interval II are not yet clear. Different structural style and reworking of the limestone pebbles, for which a southern provenance was observed in IIA, suggest two systems successive in time, the second eroding the first completely in many places. However, in locs. (13) and (18) IIA-type sediments also overly IIB sandstone bodies.

A tentative interpretation may be that over unit I a vegetated flood plain was established through which the Castisent river renewed its course.

The third unit is the Castisent point bar wedge to be described in the next section. Its areal extent is depicted in Fig. 2. In the southern part of the map area an analogous unit was found immediately overlying unit IIB and is itself overlain by a fourth sandstone sheet that gains in importance to the west of the map area.

THE CASTISENT POINT BAR

External geometry.

Outline of the point bar body.

The two cross-sections of Fig. 3 illustrate the wedge shape of the third Castisent unit. In plan view — see also Fig. 19A — it has a crescent form with a radius of about 1.5 km. A maximum thickness of 11 m is measured in section (7). The basal scoured contact with lag conglomerate deeply incises unit II and can be traced almost continuously in the field. The inner boundary of the wedge is not easy to detect. In some places, sections (4) and (12), it merely consists of a mottled topstratum above unit IIB-sandstones; identification,

Fig. 4A: Exposed SE-NW cross-section of the Castisent Sandstone at sites (1), centre, and (2), utmost right. Compare Figs. 3A and 18.

Fig. 4B: Enlargement of Fig. 4A, showing point bar section at site (1).

Fig. 4 (combined): I, II and III, bases of respective Castisent units; s , preserved scroll bars with high-angle accretion planes, ➜ low-angle accretion surface; a, b, d and e, correspond to facies within point bar unit III, mentioned in text. Paleocurrent generally to the right.

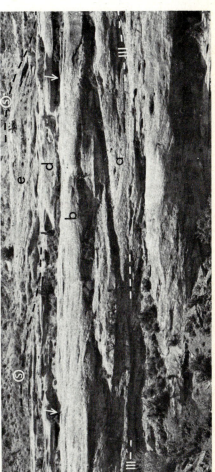

(Photographs by A. van Vliet)

then, completely depends upon the possibility of tracing the beds laterally toward the wedge depo-centre. Most of the wedge surface is level with respect to the red soil marker bed. The outer boundary could be localized at two places, (10) and (18), where the wedge tapers off against the corresponding cut bank.

Cut bank and abandonment fill (Fig. 5).

In the two last-mentioned exposures the channel lag conglomerate is abruptly overlain by a dark grey mud with intercalated sandstone layers (Fig. 5A). Thickness of this deposit amounts 8 to 9 m. The sandstones are medium to very fine-grained, have an irregular base and sharp to gradual contact with the overlying muds. Primary structures are small-scale current ripples (Fig. 5B). Some wave ripples were found and often the primary structure is destroyed by bioturbation. Sometimes mud-crack patterns were observed.

This sequence abuts against unit II sandstones and flood plain muds. At site (18) the contact strikes to N 340° E (Fig. 5A) and in loc. (10) the contact could be traced as a juxtaposition of the dark-grey mud with the more vividly coloured, pedogenetically altered, flood plain marls over a distance of more than 100 m. In both cases the steeply dipping contact is parallel to the local paleocurrent directions in the wedge.

The structure is therefore interpreted as the preserved cut bank and abandonment fill of the wedge and considered to be a strong argument in favour of the interpretation of the latter as a huge point bar. Both cut bank sections are immediately covered by the red soil marker bed, hence are in the same stratigraphic position, and with the use of the general trend in paleocurrent measurements can be easily connected underneath the cover of younger deposits in Tosal Gros (Figs. 2, 19A).

Internal structure and surface morphology.

Facies within the point bar wedge and their morphologic expression in the bar surface.

The morphologic features of the point bar surface are exceptionally well preserved, apparently because of rapid burial by abandonment fill sediment and muds of the accreting flood plain. Where erosion removed this fine-grained cover, large segments of the exhumed bar surface and slope are now open to observation (Figs. 7, 8, 12).

The bar surface displays a step-wise topography, the steps being closely related to distinct facies within the bar body. These can be correlated with sub-environments as the scour pool, lower and upper point bar, accretionary bank and chute.

a. Conglomeratic lag. Pebbly to cobbly framework conglomerates, 0.5 - 1.5 m thick, cover the basal scoured surface and, as such, are typical of the scour pool environment. Bluish limestone pebbles form a dominant constituent of the lag (X-3 in Fig. 3) only in places where either unit IIA or IIB conglomerates of that type have been reworked, for instance contact X3-X2 at site (18), X3-X1 in section (9). Repetitions of lag conglomerates are frequently observed at higher levels of the vertical profile with gradually decreasing maximum pebble size (Fig. 18). Pebbles are well rounded, often rather flat and imbricated. At only one site, (2) in Fig. 18, imbrication was in the upstream direction. Primary structures include trough-fill cross-bedding, bar foreset stratification and low-angle planar bedding (Fig. 4B).

Alternating framework gravels and cross-stratified sands are reported by Bluck (1971, p. 108-109, Fig. 8) to be characteristic of riffles. Being deposited just upstream of or into the pool, they are closely associated with the scour pool sediments. Part of the conglomerate facies here described may indeed represent the riffle environment. The fact, however, that the conglomerates are mostly linked to scoured surfaces which pass into lateral accretion planes (see section on vertical facies distribution), renders an interpretation of the repetitive lag deposits as laterally offlapping scour pool gravel sediments plausible.

Fig. 5A: Cut bank and cut-off fill at site (18). 1, cut bank contact transitional into basal scoured surface at lower right; 2, unit II-sandstones in the cut bank; 3, cut-off fill; 4, position of overlying red paleosol marker bed.

Fig. 5B: Detail of cut-off fill of Fig. 5A, showing alternating fine-grained current-rippled sandstones and dark grey claystone. Ruler 15 cm.

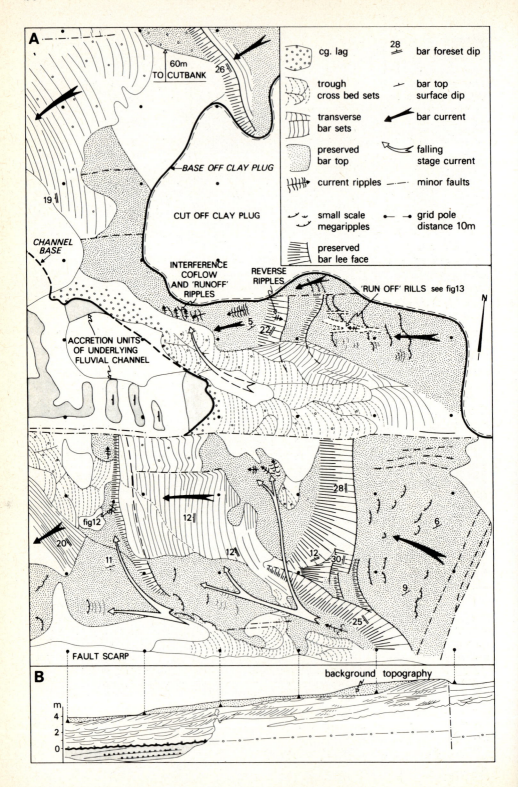

b. Trough-fill cross-bedded (conglomeratic) sandstone. In gradational contact with the underlying lag deposit, this facies is mainly exposed at deep levels of the point bar profiles (Fig. 6). Grain size is very coarse and the common structure is trough-fill cross-stratification of large dimension. Set width is in the order of 3 to 8 m attaining heights of about 50 to 75 cm and a length/width ratio of about 3. Within the sets lamination is often at very low angles near the set base and upward fining to medium sand is a common feature. Pebble strings occur at or near the set base.

The same type of structure, which is also characteristic for the underlying Castisent units, is reported by several authors from high- or low-sinuosity streams in comparable sequential position (Frazier and Osanik, 1961, Fig. 4; Harms *et al.*, 1963, p. 570-575; McGowen and Garner, 1970, p. 84, 98 and many others) and from braided streams (see Miall, 1977, for a review). For our example the genetic interpretation suggested by Harms *et al.* (1963) seems most appropriate: the trough-fill stratification is the result of current erosion producing the troughs, which are subsequently infilled by avalanching. Whether or not this process is related to migration of large bedforms, like the dune facies of Jackson (1975, 1976b), is not clear (see also next section). There is no evidence for subaerial exposure of the interval such as for the next facies to be described. For that reason the facies is related to the lower point bar interval.

c. Transverse bar facies. This facies is very distinctive for the Castisent point bar. Due to the exceptional quality of the exposure both the vertical profile and the corresponding original topography could be observed over a surface of 5000 m² at site (9), and related to the cut bank in the nearby exposure at site (10). Fig. 6A gives a detailed plan view, reconstructed with the aim of a 10 m-grid over the outcrop, and Fig. 6B a cross-section along the paleocurrent direction. Photographs further illustrate the bar morphology (Figs. 7-9) and minor bed forms on the bar crest (Fig. 13) and in inter-crest areas (Fig. 12).

Both Figures 6 and 7 show a large, step-wise surface form, corresponding to bars advancing over each others' stoss-sides. Where the original surface is preserved it is coloured brown by iron oxides. Crests are sinuous and spaced at about 30 m. Crestal height is 60 to 150 cm, composed of one or two superimposed lee faces. Correction for a slight northwestern tilt brings the bar top surfaces in Fig. 6 in an almost horizontal position (compare also the parallel position of bar top and channel base in Fig. 6B) and reduces the maximum indicated foreset dip to about 25°. The foresets are tangential at the base. By contrast a sharp contact exists between the bar foreset and a 10 to 20 cm thick top-coset of small-scale sinuous to lunate mega-ripples found all over the exposed bar surface as preserved bed forms (Figs. 6A, 9).

Immediately in front of the bar lee faces small-scale current ripple trains occur with two distinct foreset orientations. One is in the bar foreset dip direction, which here as in all the other locations does not deviate significantly from the channel orientation measured at cut banks, pebble imbrications and lower point bar trough-beds. It consists of 5 cm-spaced straight-crested ripples, passing into 10 cm-spaced linguoid ripples and finally into 1 m-spaced small mega-ripples away from the bar front, thus marking a gradual increase in dimension towards the top-set mega-ripples near the crest of the next bar.

The foreset dip of the other current ripple system is oriented at right angles to the bar fronts with foreset direction towards the channel axis. Sometimes these ripples deflect against the bar toe (Fig. 12), elsewhere they interfere with the ripples of the bar top-set as closely spaced current ripples within wider mega-ripple troughs (Fig. 6A, centre). The channel-wards oriented ripples can, because of the same unique fine grain size, be related to sediment veneering the floor of re-entrants in the bar crests, which, at least in our observation site, are in the same position in successive bar fronts. All the other structures belonging to the bars are medium to coarse-grained.

Fig. **6A**:   Plan view of the preserved point bar surface with transverse bars at site (9); explanation in text.

Fig. **6B**:   Corresponding cross-section in paleocurrent direction, exposed at southern fault scarp of Fig. 6A.

Fig. 7-9: Transverse bar facies at site (9). Compare Fig. 6A. Fig. 7: View from north on two successive bars, a and b, forming down-current step-wise point bar surface; Fig. 8: opposite view of front of bar (a); at (x) cut-off fill, covering transverse bar exposed in cross-section behind minor fault indicated by arrows in Figs. 7 and 8; Fig. 9: 1.5 m high lee face of bar (a) with sharp upper contact to mega-rippled top set.

Fig. 10: Downstream view of transverse bar at site (3). Arrow points to flat chute base contact.

Fig. 11: Detail of bar facies in section (3); compare Fig. 18. Note tangential bar toe lamination and, at (x), intercalated coset of low-angle laminated toe sets transitional into small mega-ripple top sets.

A further feature are rills disrupting the top-set mega-ripple crests (Fig. 13, located in Fig. 6A), merging into micro-deltas at the ripple slip face. These structures, although of limited dimension, are identical to those recorded by Collinson (1970, p. 43), Rust (1972, Fig. 14b) and Singh and Kumar (1974, Fig. 11) in Recent fluvial bars where they characterize falling stage conditions.

In cross-section (Fig. 10) individual bars can be traced over some tens of meters. At the down-current end the bar sets flatten at the same time as small-scale mega-ripple sets intercalate (Fig. 11). Clay drapes over the bars may occur, but mainly in connection to low-angle accretion planes (Figs. 6B, 14).

With respect to the structure and the orientation at right angles to the thalweg, the bars are very similar to the transverse bars described by Jackson (1976b) from the Lower Wabash River and the undulatory mega-ripples recorded from the Benares point bar in the Ganga River by Singh and Kumar (1974). For a review on transverse bars in braided rivers, see Miall (1977, p. 14). In Fig. 6A two current systems are inferred to have been operative in the bar formation. One, the bar current, responsible for the generation of the bars, oriented at right angles to the migrating top-set mega-ripples and to the bar lee faces (solid arrows). The other, confined to bar crest re-entrants and deflected towards the channel axis representing the falling stage current (open arrows). The existence of the latter implies an original dip of the inter-crest areas towards the thalweg. Minor structures related to the last phases of waning current are called 'run-off'' ripples and rills on the analogy of those on emerging tidal sand bars (Klein, 1970). No inter-bar delta lobes facing the channel, as recorded by Singh and Kumar (1974), were found.

It is furthermore important to note that no deep scours immediately in front of the bar face were observed, so that a direct relation between the bars and the underlying trough-fill facies could not be established. The latter facies, moreover, also occurs outside areas of bar preservation.

d. Medium- to small-scale festoon-beds. At the next higher structural level trough cross-stratification of a type comparable to facies (b) but on a smaller scale is the dominant structure of the well sorted medium-grained sandstones of this interval. Set width generally does not surpass 2 m and diminishes everywhere to about 50 cm upwards in the profile (Fig. 18). Length/width ratios sometimes surpass 10.

Although scour-and-fill could be the general mechanism here also, the distinctive properties with respect to the lower interval are more suggestive for an origin by dune migration. The more general term festoon is here used merely as an easy distinction with regard to facies (b). From its position in the sequence it is referred to as the upper point bar interval.

e. Accretionary bank facies. The medium to fine-grained sandstones of this facies comprise the uppermost two meters of most of the sections of the point bar wedge (Figs. 3, 18) and have a gradational contact to both the underlying facies and the overlying flood plain fines. They show a well defined fining-up trend and are generally colour-mottled by pedogenesis.

Fig. 12: 'Run-off' current ripples at right angles to and in front of a transverse bar lee face. Location indicated in Fig. 6A.

Fig. 13: 'Run-off' rills incising small-scale sinuous mega-ripples of a transverse bar top set and developing micro-deltas (b) in front of the mega-ripple slip face. Ruler: 15 cm. Location indicated in Fig. 6A.

Fig. 14: Upstream view at section site (7). Cut bank position 400 m to the left. Preserved scroll topography between high-angle accretion surfaces (arrows); at (x) single transverse bar set, facing the viewer, at right angles to the dip direction of the accretion; low-angle accretion surface immediately over the transverse bar. Near left arrow, geologist in accreted position for scale.

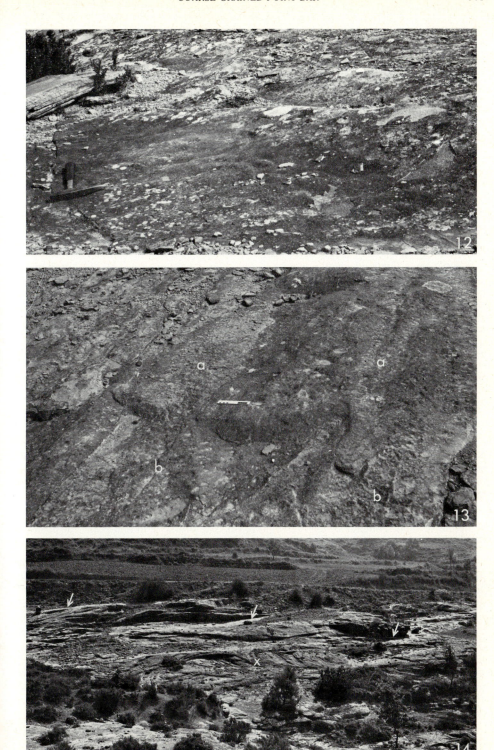

The main structural feature is its arrangement in definite accretionary units, separated by sharply expressed accretion planes dipping everywhere at right angles to the trough axes of the underlying facies (Figs. 4B, 14). Internal bedding of the accretionary units parallels the major bounding accretion planes. The units form crescentic ribbons, even visible on aerial photographs, with a maximum height of 1.5 m and a spacing in the order of 15 m (Fig. 19A). The dip angle of the accretion varies around 12° with a maximum observed value of 18° at site (15). Minor structures are seldom preserved e.g. at site (1) where they are composed of small-scale ripple sets in an interrupted wavy lamination, building perpendicularly to the accretionary dip. The so-defined facies is very similar to that of the accretionary bank of the River Endrick (Bluck, 1971, p. 107) and it is beyond doubt that the ribbons are typical meander scrolls.

f. Chute facies. In some areas, for instance sites (2) to (4), the accretionary bank facies is absent and facies (d) ends with a sharp contact and abrupt grain size change into clays, rich in plant material, with intercalated, burrowed, fine sandstones similar to those of the channel cut-off fill. The area around location (4) (Fig. 18) is especially interesting in showing that the contact represents a flat, channel-like depression, floored with straight to sinuous mega-ripples bearing linguoid current ripples on the stoss side (Fig. 16), and with a concave sideward profile against a ridge-like elevation (Fig. 15), the latter being characterized by a sequence of evenly laminated sandstones with parting lineation and dispersed pebbles passing into current ripple sets. The sets dip slightly towards the main channel (Fig. 17).

The entire complex is interpreted as a chute, down to the base level of the accretionary bank, with a chute-fill being probably formed at the time of channel abandonment. The elevation in location (4) is best explained as a ridge between two chute areas or between chute and thalweg.

The point bar lithosome is finally covered by flood plain deposits with several paleosol horizons, one of which is the red soil marker bed.

Spatial relations of the point bar facies.

Within the point bar wedge the spatial distribution of the facies follows a systematic pattern. Figure 18 illustrates a typical example of the vertical distribution in a correlation scheme of a series of profiles; Figure 19A compiles the data of the horizontal distribution of the more prominent facies types: i.e. the trough-fill cross-bedded and transverse bar facies, together with that of the point bar surface features.

Vertical facies distribution. Profile (7), corresponding to the location of Fig. 14, reveals a complete vertical sequence of the facies (a) to (e). The profile is in mid-point bar position (Fig. 19A), as is profile (1). In the latter, however, no intermediate bar facies is found.

The series of profiles (1) to (5) constitutes a detailed cross-section from a mid-point bar position in the down-current direction. Excellent outcrop conditions (part of the outcrop is shown in Fig. 4) allow tracing in the field of the facies changes. For the lower and intermediate point bar level the facies change between sections (1) and (5) is the same as

---

Fig. 15-17: Chute facies at sites (2) to (4). Compare Figs. 18 and 19A. Fig. 15: Upstream, i.e. eastward, view of chute cross-section. Arrows indicate erosive chute base, with concave sideward profile against chute-ridge with levee-type facies of Fig. 17 at (x). The chute-fill consists of dark grey clay- and siltstone with macerated plant material.

Fig. 16: Straight-crested and sinuous, westward facing megaripples, capped by linguoid current ripples, on the flat chut base, some distance upstream of Fig. 15.

Fig. 17: Interchute-ridge facies of even-laminated medium-grained sandstone with dispersed pebbles and parting lineations, gradational upward into fine-grained small-scale ripple sets. Ruler 1 m.

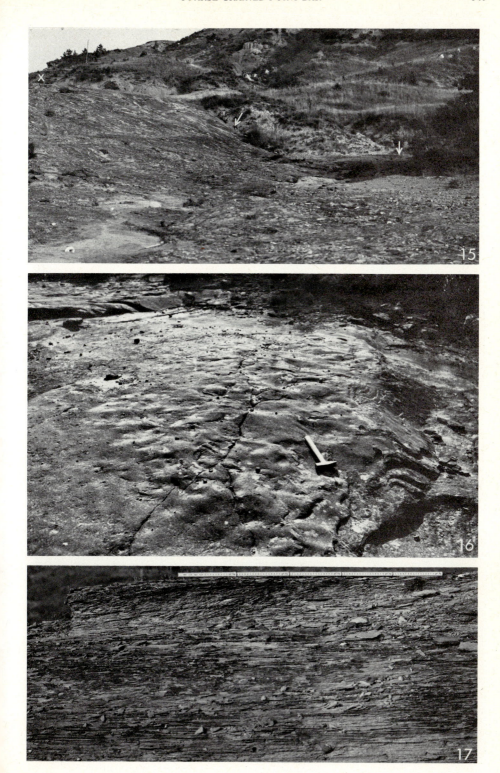

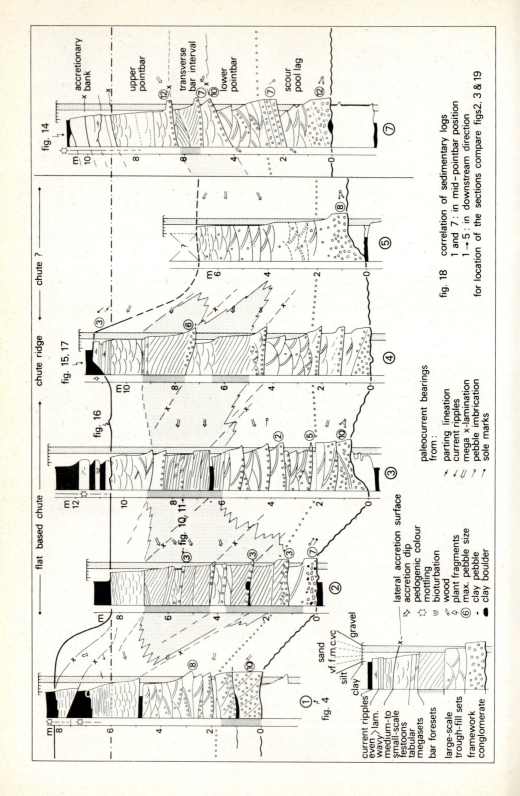

fig. 18   correlation of sedimentary logs
          1 and 7: in mid-pointbar position
          1→5 : in downstream direction
          for location of the sections compare figs 2, 3 & 19

observed between sections (7) and (9); compare Fig. 6B. From Figure 18 it is apparent than the transverse bar facies is limited in extent to the downstream reaches of the point bar, where it even may comprise the major part of the vertical profile from the top of the scour pool facies up to 1.5 m below the base of the accretionary bank (2). Maximum bar set height is attained in the downstream profile (4). A second feature is that the general fining-up trend in the profiles is often composed of repeated upward fining intervals with more or less complete successions from lag to upper point bar facies. Compare for instance the 4.4 - 7.6 m interval of log (4) and the repetition of lag/trough-fill facies in log (7). Repetitions of this kind are invariably linked to the presence of major low-angle accretion planes (Fig. 4B), often covered by a thin clay layer, which are to be distinguished from the high-angle accretion planes of the accretionary bank, although both of them are related to growth stages of the point bar (see discussion). This is an important fact to realize in the interpretation of incompletely preserved sequences in aggradational fluvial sandstone bodies of this type. From photographs the dip of these larger accretion surfaces can be estimated to amount about 5°.

Facies changes from (a) to (d) are observed within individual accretion units. The occurrence of large-scale trough-fill sets at 9 m in section (3) probably indicates an early phase of development and filling of the chute. Profile (5) shows an incomplete sequence (a)-(b)-(d), the trough-fill facies being laterally equivalent to the transverse bar interval of profile (4) and the festoon-bedded upper interval identical to the upper point bar interval. The section probably represents a deeper chute incision in the downstream end of the point bar near its exit to the main channel (compare Fig. 19A).

Horizontal facies distribution. The horizontal distribution of the transverse bar facies (Fig. 19A) corroborates the conclusion from the vertical profiles of Fig. 18. It is restricted to the downstream half of the point bar body except for the occurrence at sites (17) and (18) in the southeast of the area. The areal extent of the chute base features in the point bar surface, however, leads to the inference in Fig. 19A of the existence of a major flat-bottomed chute, dissecting the point bar over its entire length, eroding the older scrolls between sites (1) and (14) to (16). The bars between (17) and (18) then occupy a position where, during flood stage, river flow was split into a pool current and a chute current, and in that case are analogous to the bar head sand deposits of the River Endrick (Bluck, 1971, p. 108 and his fig. 24).

Paleocurrent variability. The reconstruction of Fig. 19A is largely based on the regular swing of 90° in paleocurrent bearings. Systematic changes in paleocurrent measurements also occur in the vertical sense (Fig. 3 and 18). Profiles (1), (2) and (7) show a deflection towards the west in the upward direction; (3) to (5) an additional eastward rotation in the upper part of the profile while in profile (17) the trend is slightly to the east, being somewhat more pronounced in combination with profiles (14) and (15). These rotations are to be related with the migrational behaviour of the Castisent point bar (see next section).

INTERPRETATION AND DISCUSSION

In Table 1 the observed and inferred parameters of the Castisent point bar are tabulated together with those of four examples of Recent coarse-grained point bars from the literature. The Castisent bar is considered a typical fossil representative of this class of fluvial deposits. As compared to the criteria for distinction between meandering and braided fluvial systems, recently summarized by Miall (1977, p. 52), the intermediate character of the Castisent point bar is evident from:

— a well defined, though repeated, fining-up trend;

— concurrence of well developed lateral accretion with vertical accretion forms like the large-scale transverse bars;

— abundance of large-scale trough-bedded conglomerates and very coarse sandstones in scour pool and lower point bar intervals;

— the abandonment by abrupt cut-off or avulsion and the subsequent development of fine-grained fill deposits in sharp contact with the point bar surface;

— development of a large chute, but without the formation of corresponding chute bars on the higher bar platform.

We will further elaborate these conclusions in the following comparative analysis. The presence of a general fining-up trend implies that the major bar foreset interval in the vertical sequence is not coarser than the underlying trough-bedded sandstones, as is the case in the modern examples cited in the table. Closely related to this difference, the structural sequence in the Castisent point bar shows a reversal with respect to that reported from the Amite River and the Eocene Simsboro Sandstone by McGowen and Garner (1970, p. 98, 106), i.e. in the latter two examples there is a sequence of (1) large-scale trough-fill cross-beds, (2) small to medium-scale foreset and trough cross-beds, (3) large-scale bar foresets overlain by (4) accretionary bank deposits, while in our example the sequence is 1-3-2-4, indicative of vertical accretion in the transitional zone between lower and upper point bar, i.e. on the point bar slope.

As to the two types of lateral accretion units, mentioned before and visualized in the cross-sectional model of Fig. 19B, there appears to be no correspondance between spacing and number of low-angle and high-angle accretionary units, the latter being more closely spaced and higher in number. The meander scrolls of the Castisent point bar nowhere exhibit foresetting towards the convex bank as reported by Jackson (1976b, p. 594) from the Wabash River, or as in the embryonic point bars of the River Klarälven (Sundborg, 1956, p. 274-276). Both examples are in medium-grained sand while the Castisent scrolls are fine-grained, which may account for this difference. According to Hickin and Nanson (1975) scroll spacing (which in their examples from the Beatton River averages 11.6 m in a somewhat smaller river dimension than in the Castisent case) is related to the rate of lateral migration of the bend, itself being a function of several parameters among which are the relative curvature and the discharge, for instance bankfull to near-bankfull discharge fluctuations. On the other hand, the observed low-angle accretion planes are probably related to longer periods of interruption of the accretionary process. The clay drapes observed on these surfaces indicate conditions of prolonged low velocities or flow stagnation, possibly combined with low water stands. As a consequence the probability of formation is smaller for low-angle accretion units, reflecting exceptional conditions, than for scroll units which belong to the normal process of meander migration.

Morphologically, the Castisent point bar fits best with the Columbus point bar of the Colorado River (McGowen and Garner, 1971, p. 91). Sinuosity, relative curvature and general dimensions are of the same order. Both bars correspond in having flat-based chutes without chute bars, a feature which, according to these authors, is conterminous with a dominance of transverse bar foresets and trough-fill cross-strata in this type of coarse-grained point bar. The Columbus bar, moreover, has a levee on the scour pool side of the chute with foresets deflected towards the pool, analogous to the chute ridge at site (4) (Fig. 19A).

The Castisent transverse bars are developed over a much wider range than their equivalents in the Wabash point bars, where they occur at mean low-water level (Jackson, 1976a, Figs. 16-18). This high range is related to a higher ratio of water level range to bankfull depth in our example (Table 1). The occurrence of the transverse bars in the Castisent model corresponds to the reach downchannel from the site where the streamlines separate from the convex bank to head for the concave bank (Leopold and Wolman, 1960, Fig. 5). Compare Jackson's (1976a, p. 589) zone of fully developed depositional facies. The transverse bars in the Wabash River do not show much influence

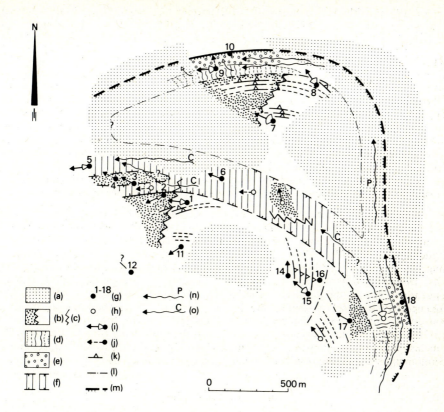

Fig. 19A: Reconstructed plan of the Castisent point bar and river bend. Legend: (a) not exposed, (b) transverse bar facies below point bar surface, (c) trough cross-bedded facies only, below the surface, (d) transverse bar facies exposed in point bar slope surface or below cut-off fill, (e) scour pool conglomerate below cut-off fill, (f) exposed chute base and/or chute-fill deposits, (g) locations of sections, (h) additional observation points, (i) paleocurrent bearings from large-scale structures, (j) the same, from small-scale structures, (k) scroll accretions with dip direction, (l) outward extension of accretionary bank deposits, (m) cut bank, observed and interpreted, (n) inferred pool current and (o) inferred chute current at bankfull stage.

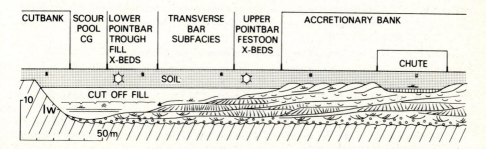

Fig. 19B: Cross-sectional model of Castisent point bar along radius in downstream point bar segment. Transverse bar foresets dip towards viewer.

of spiral motion within the water body. Curvature of the Castisent bend tallies with that of the Bozeman bend where, according to Jackson (1975, p. 1515), spiral motion is quite low at all stages and is even directed towards the thalweg during flood periods (his fig. 5).

This brings us to the point of the observed gradual change in paleocurrent directions in the vertical succession. If crestlines of primary structures are not deflected towards the convex bank due to helicoidal flow, as is often the case, the following distinct pattern of shift of paleocurrent bearings can be expected, moving upward in the same section: in the case of downstream meander migration, most of the sections in an anticlockwise bend would be expected to show an eastward rotation; however, in a situation of lateral extension with increasing curvature, such as indicated in the scroll pattern of the Castisent meander, such a bend will display an eastward rotation in the upstream sector of the point bar and a westward rotation in the downstream sector. This seems to be the case in our example. The aberrant eastward rotation in the top of sections (3) to (5) probably reflects the influence of the chute topography in that particular area.

Our final remark concerns the vertical facies succession to be expected in an aggrading system of the Castisent type. If aggradation leads to a multi-storey sandstone sheet, the accretionary banks and upper point bar intervals have little chance to survive (see Puigdefabregas and Van Vliet, this volume). This results in a vertical repetition of trough-fill and transverse bar facies or, in the case of upward stacking of point bar apex units, trough-fill cosets upon each other, separated by gravel lags. Both situations frequently occur in the Castisent unit II (Fig. 3) and await further analysis. Such a sequence would closely resemble the Donjek or South Saskatchewan type of braided river depositional profile, distinguished by Miall (1977, p. 46; this volume). Distinction between such a braided fluvial system and one of the type of the sinuous Castisent River in its last phase of development, would depend mainly on the ability to distinguish lateral accretion units and on the results of detailed paleocurrent analysis.

ACKNOWLEDGEMENTS

This study forms part of a research programme in the South Pyrenean Basin of our department. Some information was also provided by Mrs. W. P. Vlugt, F. de Reuver and H. Welsink from their M.Sc. studies. We gratefully acknowledge Ds. B. J. Bluck and A. D. Miall for critically reviewing the manuscript. Thanks are due to Mr. W. Bosman for field assistance and to Mrs. M. L. Brittijn, J. Bult and J. A. Schiet for preparation of the figures.

REFERENCES

Arthaud, F., Mégard, F., and Séguret, M., 1977, Cadre tectonique de quelques bassins sédimentaires: Bull. Centres Rech. Explor.-Prod. Elf-Aquitaine, v. 1, p. 147-188.

Bluck, B. J., 1971, Sedimentation in the meandering River Endrick: Scott. J. Geol., v. 7, p. 93-138.

Campbell, C. V., 1976, Reservoir geometry of a fluvial sheet sandstone: Am. Assoc. Petrol. Geol. Bull., v. 60, p. 1009-1020.

Collinson, J. D., 1970, Bed forms of the Tana River, Norway: Geogr. Ann., v. 52A, p. 31-55.

Eden, J. G. van, 1970, A reconnaissance of deltaic environment in the Middle Eocene of the South-Central Pyrenees, Spain: Geol. Mijnbouw, v. 49, p. 145-157.

Frazier, D. E. and Osanik, A., 1961, Point bar deposits, Old River Locksite, Louisiana: Gulf Coast Assoc. Geol. Soc. Trans., v. 11, p. 121-137.

Garrido, A., 1973, Estudio geológico y relacion entre tectónica y sedimentacion del Secundario y Terciario de la vertiente meridional Pirenaica en su zona central: Tesis de Doctorado, Universidad de Granada, 395p.

Harms, J. C., McKenzie, D. B. and McCubbin, D. G., 1963, Stratification in modern sands of the Red River, Louisiana: J. Geol., v. 71, p. 566-580.

Table 1. Comparison of the Castisent model with some modern coarse-grained point bars

	McGowen and Garner (1970)		Bluck (1971)	Jackson (1975, 1976a, b)	this paper
	Amite River Magnolia point b.	Colorado River Columbus point b.	River Endrick (reach B)	lower Wabash River Helm bend	Castisent point bar
River parameters					
river sinuosity	1.4 - 1.7	1.5 - 1.75	[±2]	2.3	estim. 1.5 - 2
arc of curvature	[90°]	[90°]	[max. 250°]	[180°]	min. 110°
radius of curvature, Rc	[115 m]	[675 m]	[60 - 135 m]	[500 m]	850 m
rel. curvature Rc/width	[±5]	[±7.5]	[2.5 - 5.5]	1.8	6
water level range/ bankfull depth	[0.7]	no data	[max. 6.5]	[0.4 - 0.5]	0.75
Point bar structures					
lateral accretion	no data	no data	high-angle channel-ward bedding in accretionary bank; low-angle in lower point b. (reach C)	scroll migration by high-angle foresetting towards convex bank	high-angle channel-ward bedding in accretionary bank; low-angle in lower to upper point b
scroll spacing	—	—	[5 - 10 m]	10 - 40 m	±15 m
chutes	present	present	no indication	no indication	present
bar foreset forms in upper point bar	linguoid chute bars	transverse bars	linguoid bars	longitudinal (scroll) bars	—
in intermediate level	—	—	—	transverse bars	transverse bars
bar foreset forms in lower point bar	small transverse bars	—	—	sandwaves	—
grain size trend	no FU trend; grain size maxima in scour pool and upper point bar		FU and CU, maxima in riffle and upper point bar head; FU accretionary bank.	indifferent or CU in upper point bar; FU accretionary bank	overall FU, composed of repeated FU units

N.B. Data between [ ] are calculated from figures in the quoted literature and therefore sometimes are not precise.

Haseldonckx, P., 1973, The palynology of some Palaeogene deposits between the Río Esera and the Rio Segre, Southern Pyrenees, Spain: Leidsche Geol. Med., v. 49, p. 145-165.

Hickin, E. J. and Nanson, G. C., 1975, The character of channel migration on the Beatton River, Northeast British Columbia, Canada: Geol. Soc. Am. Bull., v. 86, p. 487-494.

Jackson, R. G. II, 1975, Velocity — bedform — texture patterns of meander bends in the lower Wabash River of Illinois and Indiana: Geol. Soc. Am. Bull., v. 86, p. 1511-1522.

——, 1976a, Depositional model of point bars in the lower Wabash River: J. Sediment. Petrol., v. 46, p. 579-594.

——, 1976b, Large-scale ripples of the lower Wabash River: Sedimentology, v. 23, p. 593-624.

Klein, G. deV., 1970, Depositional and dispersal dynamics of intertidal sand bars: J. Sediment. Petrol., v. 40, p. 1095-1127.

Leopold, L. B. and Wolman, M. G., 1960, River meanders: Geol. Soc. Am. Bull., v. 71, p. 769-794.

McGowen, J. H. and Garner, L. E., 1970, Physiographic features and stratification types of coarse-grained point bars: modern and ancient examples: Sedimentology, v. 14, p. 77-111.

Miall, A. D., 1977, A review of the braided-river depositional environment: Earth Sci. Revs., v. 13, p. 1-62.

Moody-Stuart, M., 1966, High- and low-sinuosity stream deposits, with examples from the Devonian of Spitsbergen: J. Sediment. Petrol., v. 36, p. 1102-1117.

Mutti, E., Luterbacher, H., Ferrer, J. and Rosell, J., 1972, Schema stratigrafico e lineamenti di facies del Paleogeno marino della zona centrale sudpirenaica tra Tremp (Catalogna) e Pamplona (Navarro): Mem. Soc. Geol. It., v. 11, p. 391-416.

Nijman, W. and Nio, S. D., 1975, The Eocene Montañana Delta, Tremp - Graus Basin, Southern Pyrenees: IXth Internat. Sed. Congr., Nice, France, Guide book Exc. 19, part B, 56 pp.

Puigdefabregas, C., Rupke, N. A. and Sole Sedo, J., 1975, The sedimentary evolution of the Jaca Basin: IXth Internat. Sed. Congr., Nice, France, Guide book Exc. 19, part C, 33 pp.

Rust, B. R., 1972, Structure and process in a braided river: Sedimentology, v. 18, p. 221-245.

Séguret, M., 1970, Etude tectonique des nappes et séries décollées de la partie centrale du versant sud des Pyrénées: Thèse Fac. Sci. Université de Montpellier, Publications USTECA 1972.

Singh, I. B. and Kumar, S., 1974, Mega- and giant ripples in the Ganga, Yamuna and Son Rivers, Uttar Pradesh, India: Sediment. Geol., v. 12, p. 53-66.

Sundborg, Å., 1956, The River Klarälven, a study of fluvial processes: Geogr. Ann., v. 38, p. 125-316.

Van Houten, F. B., 1974, Northern Alpine Molasse and similar Cenozoic sequences of Southern Europe: in R. H. Dott Jr., and R. H. Shaver, eds., Modern and ancient geosynclinal sedimentation, Soc. Econ. Paleont. Mineral., Spec. Pub. 19, p. 260-273.

FLUVIAL FACIES OF THE PLIO-PLEISTOCENE KOOBI FORA FORMATION, KARARI RIDGE, EAST LAKE TURKANA, KENYA

CARL F. VONDRA[1] AND DANIEL R. BURGGRAF, JR.[1]

ABSTRACT

Four major lithofacies have been described and interpreted (Vondra and Bowen, 1976) as a complex of lacustrine, transitional lacustrine, deltaic and fluvial environments of an embayment of paleo Lake Turkana. Fluvial environments are recorded by the lenticular conglomerate, sandstone and mudstone facies of the Upper Member of the Koobi Fora Formation which consists of four interfingering subfacies: 1) the interbedded basalt conglomerate and pebbly mudstone, 2) the lenticular basalt clast conglomerate, 3) the polymictic conglomerate and sandstone, and 4) the interbedded sandstone and tuffaceous siltstone subfacies. These are well exposed along the Karari Escarpment and represent 1) alluvial fan channel and debris flow; 2) high-energy channel — bar core, gravel sheet; 3) lower energy channel — bar-side, transverse bar, and point bar; and 4) floodplain depositional environments, respectively.

Body geometries, sequences of primary structures and textural and compositional variations by which the subfacies are interpreted are well exemplified by the Upper Member. Lateral upstream-downstream (longitudinal) and channel-floodplain (transverse) as well as vertical subfacies relationships indicate that a series of ephemeral streams entered the basin from the volcanic highlands to the northeast and east, forming a belt of alluvial fan deposits along the basin margin. A high-energy, low sinuosity braided river system flowed southwestward across a broad alluvial plain toward paleo Lake Turkana. This system gave way vertically to a low energy, high sinuosity meandering river system as the embayment was filled with sediment by a prograding delta complex.

Episodic volcanism surrounding the Lake Turkana basin is manifest throughout the sedimentary sequence by interbedded volcaniclastic units, and reflects the tectonically active Plio-Pleistocene setting of the region.

INTRODUCTION

During the past eight years anthropological, archaeological, and geological studies in the northwestern quarter of Kenya have provided evidence regarding the occurrence of early man in East Africa and the environments in which he lived (Coppens et al., 1976). The area lies along the northeastern margin of Lake Turkana[2] (Fig. 1) and consists of a series of intertongued lacustrine, transitional-lacustrine, deltaic, and fluvial sediments of Pliocene to Recent age. They include abundant volcaniclastic debris, recording a period of geologic activity in which the lake level rose and fell as the land surface responded to tectonic pressures associated with crustal fragmentation which produced the East African rift system.

The strata of the Plio-Pleistocene Koobi Fora Formation record a general westward regression of the paleo-Lake Turkana shoreline. They have been divided into a Lower Member, consisting primarily of lacustrine, transitional-lacustrine, and deltaic sediments, and an Upper Member which includes about 45 m of conglomerate, sandstone, mudrock, and tuff of fluvial origin.

Attention has been focused on the fluvial sediments because they contain abundant evidence of early man, including cranial and post cranial fossil remains (Leakey, 1976; Walker, 1976; Wood, 1976) and at least two distinct stone artifact assemblages (Isaac et al., 1976; Isaac, 1976). Detailed geologic studies consisting of bed-by-bed correlation and mapping at a scale of 1:6,000, measurement and description of over 200 critical exposures

---

[1]Dept. of Earth Sciences, Iowa State University, Ames, Iowa, 50011
[2]Formerly Lake Rudolf, Lake Turkana was renamed by the Kenyan Government early in 1975. The extreme northern portion of the lake lying in Ethiopia is still referred to as Lake Rudolf.

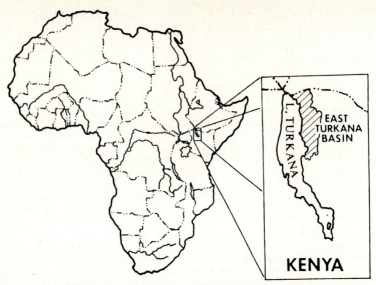

Fig. 1.   Location map of the East Turkana Basin.

integrated with sample collection, and laboratory analyses of grain-size distribution, mineralogy, and geochemistry, have been completed for an area of approximately thirty-six square kilometers.

GEOLOGIC SETTING

Lake Turkana in northwestern Kenya, lies in a depression bounded on the west by a normal fault and on the east by a monoclinal flexure in a structurally complex area between the Kenyan and Ethiopian domes (Baker *et al.*, 1972). The southern tip of the lake occupies a depression to the northeast of, and in trend with, the Suguta Valley, which is considered to be the main trend of the Gregory rift in this region. The remainder of the lake occurs in an associated fault-bounded block which may represent an inactive rift, abandoned as tectonic stresses migrated to the east (Cerling and Powers, 1977).

A succession of Pliocene through Recent sediments (up to 325 m thick) unconformably overlies Miocene and Pliocene volcanics which, farther east, uphold the gently westward dipping backslope of a cuesta — the Suregei — and form the eastern and southern boundaries of the East Turkana Basin. Regional study of the sediments has led to the establishment of four lithostratigraphic units (Fig. 2) termed the Kubi Algi Formation (Pliocene), the Koobi Fora Formation (Plio-Pleistocene), the Guomde Formation (late Pleistocene), and the Galana Boi Beds (Holocene), (Bowen and Vondra, 1973; Vondra and Bowen, 1976, 1977). Each of these units consists of portions of up to four major facies (Fig. 3) which are the: 1) laminated siltstone facies; 2) arenaceous bioclastic carbonate facies; 3) lenticular fine-grained sandstone and lenticular bedded siltstone facies; and 4) lenticular conglomerate, sandstone and mudstone facies. These represent, respectively; 1) prodelta and shallow-shelf lacustrine; 2) littoral-lacustrine beach, barrier beach and lagoon; 3) delta plain-distributary channel and interdistributary flood basin, and and 4) fluvial channel and flood basin, environments of deposition.

SUBFACIES OF THE LENTICULAR CONGLOMERATE, SANDSTONE
AND MUDSTONE FACIES

Within the lenticular conglomerate, sandstone and mudstone facies occur the fluvial sediments of the Upper Member, Koobi Fora Formation. Through detailed sedimentologic

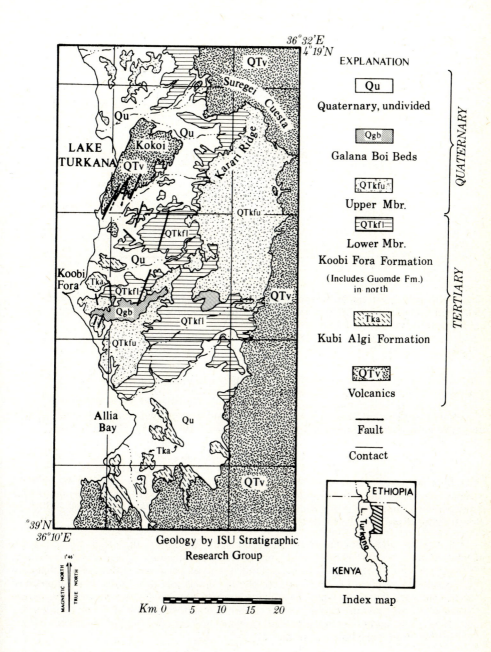

Fig. 2. Geologic map of the East Turkana area.

Major Lithofacies of the East Turkana sediments:

1) Laminated Siltstone   *(Prodelta; shallow-shelf lacustrine)*

2) Arenaceous Bioclastic Carbonate   *(Littoral-lacustrine beach, barrier beach,*
                                                                    lagoon

3) Lenticular Fine-grained Sandstone and   *(Delta plain-distributary channel*
    Lenticular-bedded Siltstone   *interdistributary flood basin*

4) Lenticular Conglomerate, Sandstone and *(Fluvial channel and flood basin)*
    Mudstone

Fig. 3.

and stratigraphic investigations, four fluvial subfacies (Fig. 4) have been distinguished and described. These are the: 1) interbedded basalt cobble conglomerate and pebbly mudstone subfacies; 2) lenticular basalt clast conglomerate subfacies; 3) polymictic conglomerate and sandstone subfacies; and 4) interbedded sandstone and tuffaceous siltstone subfacies.

Interbedded basalt cobble conglomerate and pebbly mudstone subfacies

The interbedded basalt cobble conglomerate and pebbly mudstone subfacies occurs along the basin margin in the northern portion of area 129[1] (Fig. 5). This subfacies (Fig. 6) attains a thickness of as much as 15 m and is characterized by subrounded to rounded pebbles and cobbles of light olive gray (5Y5/2) to medium gray (N5) basalt and ignimbrite in a matrix of pale yellowish brown (10YR6/2) to moderate yellowish brown (10YR5/4) litharenite or grayish brown (5YR3/2) arenaceous mudstone (Fig. 6). Individual beds range from 0.4 m thick clast and/or matrix-supported units with indistinct upper and lower bedding planes, to 2.0 m thick clast-supported, lenticular bodies exhibiting distinct downcutting of the basal surfaces (Frank, 1976).

The pebbles and cobbles show no preferred orientation in either the clast or matrix-supported conglomerates, and while primary sedimentary structures are generally lacking, crude horizontal stratification and indistinct large-scale planar cross-stratification are sometimes visible in the clast-supported conglomerates (Fig. 6). Mudstones interbedded with them are typically arenaceous with abundant volcanic rock fragments and isolated pebbles of basalt and ignimbrite composition. Additionally, limonitic concretions, and carbonate nodules and stringers are common in the mudstones while clay balls occur in some horizons of the conglomerates.

---

[1] A portion of the region that contained abundant and well-preserved fossils was divided into subequal areas (based on physiographic boundaries such as streams and drainage divides) each of which was assigned a number to facilitate the documentation of fossil discoveries.

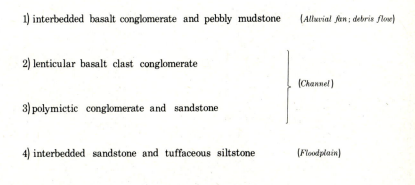

Fig. 4.

The sequence of conglomerates and pebbly mudstones comprises a complex, wedge-shaped body of detritus thinning to the west from the present exposure of the Miocene and Pliocene volcanics of the Suregei Cuesta. It grades into conglomeratic sandstones and narrow ribbons of basalt clast conglomerates. The interbedded basalt cobble conglomerate and pebbly mudstone subfacies of the Koobi Fora Formation has characteristics of the general fluvial facies *"Gm"* and *"Gp"* of Miall (1977, this volume), which include massive or crudely bedded gravel and planar cross-bedded gravel, respectively.

Depositional environment

The geographic positions of the strata of this facies with respect to the volcanic source area directly to the east, coarseness of the sediments, occurrence of matrix-supported conglomerates interbedded with the clast-supported units, and westward intertonguing with conglomerates and sandstones toward the basin center suggest that the sediments of the interbedded basalt cobble conglomerate and pebbly mudstone subfacies represent an ancient alluvial fan system (Fig. 6), the characteristics of which are similar to those described by Blissenbach (1954), Hooke (1967), Bull (1964a, b; 1968; 1972) and Walker (1975).

Bull (1972) describes two fundamental types of deposits which occur in coarse-grained alluvial fans. These include sheets and lenses of water-laid gravel, sand, and silt and interbedded lobes of matrix-supported conglomerates that represent debris-flow deposits. Both are well represented in the strata of the interbedded basalt cobble conglomerate and pebbly mudstone subfacies. Matrix-supported basalt clast conglomerates and arenaceous mudstones with occasional granules and/or pebbles of basalt are interpreted to be debris-flow deposits which indicate environmental factors of low annual precipitation with irregular and brief intervals of heavy rainfall and probably, steep slopes with little vegetative cover (Bull, 1972, p. 69). The coarse clastic debris and the abundant fine-grained material which constitutes the matrix of the interbedded arenaceous mudstones and pebbly mudstones were derived from older Pliocene basalts, ignimbrites and paleosols exposed in the adjoining volcanic highlands.

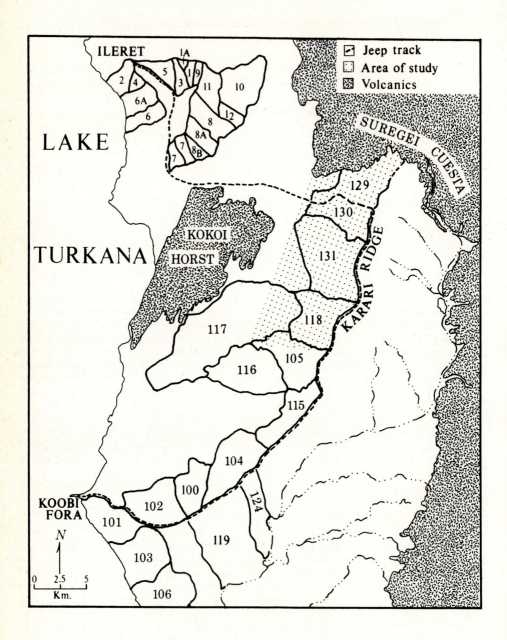

Fig. 5. Map showing fossil collecting localities.

Interbedded basalt conglomerate and pebbly mudstone subfacies

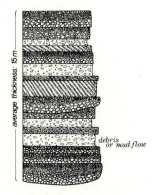

Primary Structures: horizontal stratification to indistinct large scale planar x-stratification

Bedding: indistinct to thick-bedded

Basal Contacts: sharp, erosional; rarely transitional

Fig. 6.

Clast-supported conglomerates are interpreted to have been deposited by ephemeral streams which migrated across the surface of the alluvial fan during periods of high rainfall. The massive to thick-bedded, tabular deposits represent deposition as sheetfloods across the fan, while the lenticular, more poorly sorted, and crossbedded conglomerates represent stream-channel deposits similar to those described by Bull (1964a, 1968, 1972).

Lenticular basalt clast conglomerate subfacies

The lenticular basalt clast conglomerate subfacies is characterized by dominantly structureless, loose to poorly indurated, laterally discontinuous, lenticular beds up to 4.3 m thick. Individual beds consist primarily of well-rounded and subspherical pebbles of light olive gray (5Y5/2) basalt in a matrix of pale yellowish brown (10YR6/2) to moderate yellowish brown (10YR5/4) conglomeratic sand consisting of subangular clasts of quartz, feldspar, gneiss, granite, and ignimbrite (Fig. 7). These sediments are typically poorly to very poorly sorted and strongly fine-skewed with leptokurtic grain-size distribution, reflecting the presence of a relatively better sorted coarse-grained mode, and a poorly sorted fine-grained tail. Graphically, (grain size versus fraction weight) this relationship is seen as a bimodal distribution with the coarser mode including a much greater percentage of the sample weight than the finer one. More will be said of this relationship under environmental interpretations.

Commonly, portions of the gravel beds are cemented by calcite with the degree of induration varying from very dense, highly cemented units usually less than 20 cm thick, to porous, partially cemented horizons less than 50 cm thick. Occasionally, the gravel sized material will include well-rounded clasts of calcite-cemented sand or basalt pebbles which indicate reworking of a previously cemented (caliche) deposit.

Primary sedimentary structures are commonly limited to indistinct horizontal stratification and fining-upward graded bedding, but cross-stratified conglomerates occur as well. Horizontally stratified beds averaging about 20 cm in thickness constitute the bulk of the facies. These are occasionally capped by large-scale planar cross-stratification composed of two individual sets. The lower set attains a thickness of about 50 cm and is characterized by high angle foresets while the upper set is thinner and consists of low angle foresets. The cross-stratification occurs in cobble conglomerates with a matrix of very coarse-grained conglomeratic sandstone which almost invariably grades laterally and vertically into sandstones and conglomeratic sandstones.

The lenticular basalt clast conglomerate subfacies commonly includes interbedded wedges and/or lenses of small-scale trough cross-stratified, coarse-grained sand. Individual outcrops normally are composed of two or more sequences consisting of basalt cobble conglomerate, overlain by coarse-grained sand and conglomeratic sandstone. The sequences range from 40 cm to 4 m in thickness. The sandstones commonly contain carbonate-cemented root casts and crotovina and in some cases may be entirely cemented by a lace-like calcium carbonate framework (caliche). In many instances, fragmented and abraded mammalian skeletal remains are found associated with the clastic debris, and occasionally, stone artifact assemblages occur on the upper surface of individual gravel units.

In map view, lenticular basalt cobble conglomerate outcrops can sometimes be traced as narrow, isolated, low-sinuosity ribbons for distances of nearly a kilometer. They commonly attain widths no greater than 50 m. Laterally they grade into the conglomeratic sandstone of the polymictic conglomerate and sandstone subfacies.

Lenticular basalt clast conglomerate subfacies

avg. thick.
4 m

Primary Structures: *horizontal stratification; large scale planar and trough x-stratification*

Bedding: *thin- to thick-bedded; often graded*

Basal Contacts: *erosional; sharp to transitional*

Grain size frequency statistics:

	Average	Range
Graphic Mean (M_z)*	-2.70 φ	-4.01 - -1.81 φ
Inclusive Graphic Standard Deviation (σ_i)*	2.03 φ	1.45 - 2.76 φ
Inclusive Graphic Skewness (Sk_i)*	0.47	-0.03 - 0.76
Graphic Kurtosis (K_G)*	1.52	0.64 - 2.47

*after Folk and Ward (1957)

Fig. 7.

Depositional environments

Deposits of the lenticular basalt clast conglomerate subfacies are interpreted to represent major channels draining the alluvial fans and adjacent volcanic highlands. Many features of the strata of this subfacies are similar to those of braided outwash deposits described by numerous workers, including Eynon and Walker (1974), Rust (1975), and Boothroyd and Ashley (1975). The structureless to crudely horizontally stratified gravels are similar to the gravel braid bars of Rust (1975, p. 245) and to the upper fan, coarse-gravel, longitudinal bars of Boothroyd and Ashley (1975, p. 202). The lack of internal cross-stratification is due, apparently, to relatively shallow water depth and the failure of the migrating bar form to develop recognizable slip-faces. On the other hand, large-scale planar cross-stratified gravels indicate the development of bars with higher relief in a system having energy great enough to transport the pebble-sized material as bed load with the development of gravel bars as stable primary bedforms (Rust, 1975, p. 246). Eynon and Walker (1974) describe tabular and trough cross-stratified gravels in sets up to 35 cm thick occurring in the "bar front and bar stoss-side facies," and also in the "shallow-braided stream facies" which overlies either the "sandy side channel facies" or the "bar top facies."

While it is unlikely that discharge from the channels in which the sediments of the lenticular basalt clast conglomerate subfacies were deposited was large enough to develop the very large scale bar forms described by Eynon and Walker (1974), it is probable that similar facies represent genetically similar bar positions. Thus, in the Upper Member of the Koobi Fora Formation, the structureless gravels overlain by large scale planar cross-stratified sandy gravel, and interfingering laterally with large-scale trough cross-stratified and horizontally stratified conglomeratic sandstone represent the "bar core — bar top — side channel facies" association of Eynon and Walker (1974).

One point of deviation between the sediments comprising the lenticular basalt clast conglomerate subfacies and the braided gravel outwash deposits of other workers is the apparent lack of imbrication of the gravel clasts in the former. This is due to the fact that the majority of the clasts in this subfacies are at least moderately spherical, making the recognition of any imbrication difficult. Also, because the deposits are, for the most part, loose to poorly indurated and possess little cohesive strength between adjacent clasts, in outcrop, they readily cave or slump, destroying any preferred orientation that may have existed. However, even in those outcrops in which cross-stratification could be observed, only indistinct imbrication was noted.

The bimodal distribution of grain sizes alluded to above suggests that during conditions of waning flow, sand-sized material was deposited and sifted into the underlying gravels. The occurrence of root casts and caliche horizons in the finer bar top sediments suggests that the streams were ephemeral in nature. It is also possible, then, that wind-blown fine-grained material may have been deposited upon the gravels during dry periods. The poor sorting of the fine-grained component of the lenticular basalt clast conglomerate subfacies is probably due to the addition of finer grained sieve deposits from waning-flow sand and wind-blown silt and fine sand.

Polymictic conglomerate and sandstone subfacies

Roughly the lower half of the Upper Member of the Koobi Fora Formation consists of moderately sorted to very poorly sorted, grayish orange (10YR7/4) to moderate yellowish brown (10YR5/4) feldspathic sandstone, arkose, and conglomeratic litharenite which comprise the polymictic conglomerate and sandstone subfacies (Fig. 8). This subfacies occurs throughout the Karari Escarpment and intertongues with each of the three other subfacies. Although it does pinch out along the northern extremity of the Karari Escarpment it attains a thickness of over 20 m to the south in the central portion of area 131 (Fig. 5); its average thickness over much of the study area is 15 m. The boundary between the Lower and Upper Members of the Koobi Fora Formation (Vondra and Bowen, 1977) is marked by the erosional basal contact of this subfacies with the underlying transitional-lacustrine sediments.

Individual beds of this subfacies are variable in thickness because of their lenticular nature. Their average thickness is about 1.5 m. They generally possess an erosional base and consist of a fining-upward sequence of massive pebble gravel and coarse-grained sand grading vertically into cross-stratified coarse- to fine-grained sand, occasionally capped with a siltstone or claystone.

Because of the poor induration of these sediments, good exposures are confined to areas of bank-caving along present day ephemeral streams and to the actively eroding badlands area adjacent to the Karari Ridge. In these outcrops primary structures consist of large- and small-scale planar and trough cross-stratified, and horizontally stratified medium- to coarse-grained sandstone, with lesser amounts of festoon cross-stratified and ripple-laminated silty sandstone. Three groups, or sequences, of structures are evident.

The first group consists of solitary sets or cosets of large-scale trough cross-stratified sands. Individual sets reach up to 1 m in thickness, 2 m in apparent width and 3 to 4 m in length. They possess erosional basal contacts which are often marked by a thin layer of granules. This sequence ranges in thickness from that of an individual set up to cosets as much as 3 m thick.

The second group is composed of large-scale planar cross-stratified sands in sets ranging in thickness from 5 to 30 cm. Sets are tabular to wedge-shaped in cross-section and commonly include reactivation surfaces. Cosets of planar cross-stratified sands with individual sets being separated by very thin horizontally laminated sand commonly reach thicknesses of greater than 1 m. Each successive set often is thinner than the preceding one. Frequently the uppermost set is overlain by up to 50 cm of horizontally laminated sands which in turn grade upward into 50 cm of either small-scale festoon cross-stratified or ripple-laminated fine-grained sand.

Finally, the third primary structure sequence consists of a lower unit of poorly sorted structureless, sandy pebble conglomerate or a conglomeratic sand grading upward into cosets of large-scale trough cross-stratified sand. Individual sets range up to 70 cm in thickness and possess erosional basal surfaces. The uppermost set is truncated and overlain by horizontally stratified fine- to medium-grained sand which, in turn, is overlain by one or more sets of large-scale trough cross-strata also composed of fine- to medium-grained sands. The sequence fines upward into small-scale ripple cross-stratified, fine-grained sand and silt as much as 1 m thick. This, in turn, grades upward into a veneer of mudrock. The entire sequence may reach a thickness of about 5 m, but quite commonly, erosion prior to the deposition of a succeeding sequence, has removed the upper portions, thus the complete sequence is often not preserved.

The strata of the polymictic conglomerate and sandstone subfacies contain a variety of other structures as well. Root casts and crotovina cemented by calcite are abundant in the finer grained rocks of the subfacies and clay balls and rip-up clasts commonly are present in the sandstones. In addition, vertebrate remains occur within some of the medium-grained deposits and usually are preserved as bone splinters or badly abraded limb bones. On no occasion was an articulated specimen found in the deposits of this subfacies.

Depositional environment

The deposits of the polymictic conglomerate and sandstone subfacies are interpreted as representing large-scale dune and transverse, longitudinal, and point bar environments of deposition. Ore (1963); Allen (1965b, 1970); Williams and Rust (1969); Williams (1971); Kessler (1971); Rust (1972); Smith (1970, 1971, 1974) and many others have reported on braided river systems and have discussed bar forms and corresponding stratification

Polymictic conglomerate and sandstone subfacies

average thickness: 20 m

Primary structures: *small and large scale planar and trough x-stratification, horizontal stratification*

Bedding: *thick-bedded, indistinct; graded*

Basal Contacts: *sharp to transitional; erosional; lower 0.2 m often cemented with carbonate*

Grain size frequency statistics:

	Average	Range
Graphic Mean $(M_z)^*$	1.32 ϕ	-0.91-3.87ϕ
Inclusive Graphic Standard Deviation $(\sigma_1)^*$	1.19ϕ	0.59-2.67ϕ
Inclusive Graphic Skewness $(Sk_1)^*$	0.14	-0.55-0.61
Graphic Kurtosis $(k_G)^*$	1.27	0.57-2.22

*after Folk and Ward (1957)

Fig. 8.

variations. Miall (1977, this volume) has summarized deposits of sand-sized material in braided-rivers as consisting of various combinations of fine "sandy facies" based on grain size and the dominant primary sedimentary structures present. These are trough crossbedded sand (*St*), planar crossbedded sand (*Sp*), ripple cross-laminated sand (*Sr*), horizontally-bedded sand (*Sh*), and scour-fill sand (*Ss*).

The first sequence of the polymictic conglomerate and sandstone subfacies, consisting of singular sets of large-scale trough cross-stratified sand, corresponds to Miall's (1977) "Facies *St*" and represents a scour and fill feature. This cross-stratification type (theta cross-stratification of Allen (1963) includes large-scale arcuate forsets which are discordant with the lower bounding surface of the set. Williams (1971) and Williams and Rust (1969) report the occurrence of scour features in modern braided rivers where they form at the intersection of minor channels, at re-entrants along the leading edge of a large-scale bed form, or downstream from a large object (e.g. a log) in a stream bed where turbulent flow conditions occur.

Sequences of large-scale planar cross-stratified sands (alpha type of Allen, 1963) in tabular to wedge-shaped sets as occur in the second sequence described are formed by downstream migration of sand waves with avalanching of individual grains down the slip-face. Such deposits (Facies *Sp* of Miall, 1977) represent transverse bars formed as stable primary bed forms in the upper lower flow regime (Harms and Fahnestock, 1965) of a sandy braided river (Smith, 1974). An individual bar form includes the large-scale planar cross-stratification and often, a capping veneer of horizontally stratified sands representing stoss-side deposition. Thus, grouped sets of planar and horizontally stratified sands in this subfacies document aggradation in a braided river with stacking of consecutive transverse bars as downstream transport of a sand-laden bedload proceeds.

The third stratification sequence observed in the polymictic conglomerate and sandstone subfacies is most common in the stratigraphically, higher portions of the subfacies. The combination of a fining-upward sequence with a thin basal unit of massive pebble conglomerates or conglomeratic sands overlain by large-scale trough cross-stratified sands, horizontally stratified sand, and an upper unit of small-scale trough cross-stratified fine sand grading upward into silt and argillaceous siltstone ("Facies *Gm, St, Sh, Sr* and *Fm*" respectively of Miall, 1977), and thick, laterally equivalent fine-grained fine-grained overbank sediments, suggests a point bar environment of deposition. These same characteristics were described by Bernard and Major (1963) for point bar deposits of the Brazos River near Richmond, Texas. Also, Visher (1965) proposed a general fluvial model based on a similar combination of a fining-upward cycle and the sequence of large-scale trough crossbedding, current lamination, and ripple lamination. Later workers have continued to study and formalize models for stratification in fluvial deposits (Allen, 1970; Visher, 1972; Steel, 1974) and have applied these models to the identification of ancient meandering river systems preserved in the geologic record.

It should be pointed out, however, that similar fining-upwards sequences can also be found in deposits of braided rivers, notably of the Donjek and South Saskatchewan types (Miall, 1977, p. 46-48; Miall, this volume; Williams and Rust, 1969, p. 669-670; Cant, this volume). The authors believe that it is unlikely though, that a fine-grained overbank sequence (the interbedded sandstone and tuffaceous siltstone subfacies) as thick as that found in the Upper Member of the Koobi Fora Formation would be preserved in a braided-river system.

In summary, the deposits of the polymictic conglomerate and sandstone subfacies record a gradual change in the dominant type of fluvial system during the early Pleistocene in the area of the present Karari Ridge. The lower strata indicate that deposition occurred in a low-sinuosity braided river system with the development of longitudinal and transverse bar forms. Scour-fills are interpreted to have formed in re-entrants along with the leading edge of a migrating bed form similar to those described by Williams (1971, p. 137) for large-scale lunate ripples of an ephemeral stream in South Australia.

Strata of the upper portion of the facies suggest that the dominant fluvial system was a high-sinuosity meandering river with associated point bar development and overbank flooding (see next subfacies) on a broad low relief floodplain. Thicker cyclic stratification sequences suggest a smaller width to depth ratio for the channel dimensions. The relative lack of root casts, crotovina, and caliche horizons in the sand channel-fills of the upper portion of the polymictic conglomerate and sandstone subfacies as compared to the lower portion suggests that the flow was more persistent during the later stages of deposition of the subfacies. Thus, perennial flow, or at least a less transient channel system would have prevented the development of vegetation and caliche horizons in the channel-fill deposits. The fine-grained overbank sediments by comparison, are replete with structures indicative of subaerial exposure with root casts being abundant.

Interbedded sandstone and tuffaceous siltstone subfacies

Overlying and intertonguing with the thick accumulation of sands and conglomerates of the polymictic conglomerate and sandstone subfacies is a series of medium- to fine-grained sandstones, siltstones, tuffs, tuffaceous siltstones and mudstones (Fig. 9). Along the Karari Ridge, this series of strata averages about 18 m in total thickness and comprises the interbedded sandstone and tuffaceous siltstone subfacies; the base of which is often marked by a light gray (N7) tuff bed ranging up to 1.8 m thick.

While typified by uniform bed thickness and relatively gradational contacts, this subfacies exhibits the greatest lithologic variability of the fluvial subfacies. It includes moderately to poorly sorted grayish orange (10YR7/5) to dark grayish orange (10YR6/4) fine- to medium-grained sandstones and tuffaceous siltstones interbedded with tabular to lenticular pale brown (5YR5/2) to pale yellowish brown (10YR5/2) mudstones and light gray tuffs. Individual beds are usually less than 2 m thick and occur as tabular bodies which are often laterally continuous for as much as a kilometer. Along the Karari Ridge, the interbedded sandstone and tuffaceous siltstone subfacies consists of a series of fining-upward sequences with occasional interbedded tuffs. However, in many cases deviations from a general fining-upward cycle occur such as coarsening-upward sandstone beds up to 70 cm thick and very poorly sorted, and complexly interbedded sandstones, siltstones, and pumice clast conglomerates.

In the general case, the lower portion of each sequence consists of fine- to medium-grained sandstone up to 2 m thick with occasional lenses and partings of arenaceous to argillaceous siltstone less than 20 cm thick. The sandstones are often horizontally stratified or ripple laminated but vary from massive, structureless units to indistinctly large-scale cross-stratified deposits with individual sets less than 15 cm thick. The sands consist of moderately to poorly sorted subangular clasts of quartz and feldspars with fragments of granite, quartzite, gneiss, and ignimbrite. They include a fine-grained matrix of micas, quartz, and feldspar with abundant montmorillonite and mixed-layer montmorillonite-illite clays. Also, in many instances, lenses up to several meters in length and less than 50 cm thick occur, consisting of granules and pebbles of pumice in a sand matrix.

The sandstones grade upward into massive, parallel laminated and ripple laminated arenaceous and tuffaceous siltstones usually several meters thick. They invariably contain abundant small, calcareous root casts (less than 1 cm thick and several centimeters long) accompanied by nodular carbonate concretions up to several centimeters in diameter. In most cases the upper 10 to 20 cm of the siltstone are very argillaceous and contain more root casts and nodular carbonate concretions than the underlying portions. In other localities, the siltstone grades upward into a silty mudstone as much as 2 to 3 m in thickness, and typified by abundant white (N9) carbonate concretions, occasional green (pale olive, 10Y6/2) mottling, and blocky fracture. The mudstones are usually massive, only rarely displaying an indistinct horizontal stratification or ripple lamination.

Tuff units and tuffaceous sediments can occur at practically any stratigraphic level within the interbedded sandstone tuffaceous siltstone subfacies but are most likely to occur near the lower and upper boundaries. They consist primarily of light gray (N7) angular glass shards with varying amounts of included clastic debris. Although, in many cases they occur as uniformly thick, laterally continuous sheets they may also occur as lentils less than 10 m in width and 50 cm thick, or as irregular fingers which superficially appear to truncate the bedding of lateral siltstones.

The tuffs very often exhibit internal sedimentary structures and may consist of as many as three stratification zones forming a sequence in the thicker tuffs. This sequence includes a lower, massive zone up to 50 cm thick overlain by an intermediate zone of parallel laminated to small-scale planar or trough cross-stratified tuff up to 1.0 meter thick; and an uppermost, thin (less than 20 cm) zone of ripple laminated to massive tuff. A corresponding fining-upward of shard size is usually evident in units displaying this stratification sequence.

In several instances, tuff deposits occur as concave-up lenses about 75 cm thick and up to 25 m wide. They rest on and are overlain by horizontal, parallel-stratified tuffaceous siltstone or sandstone indicating that structural deformation does not account for the concave-up nature of the tuff. These concave-up tuff lenses are interpreted to represent a viscous ash-water slurry which surged through a swale and flowed upward and outward over the banks. The water was quickly lost through filtration to the underlying loose sediment, and collapse of the fluidized sediment flow left the ash as a mantle conforming to the concavity of the small depression.

Depositional environment

The fine-grained and tabular nature of the strata of the interbedded sandstone and tuffaceous siltstone subfacies and their interfingering relationship with coarse-grained lenticular deposits of the polymictic conglomerate and sandstone subfacies suggests that deposition took place on a broad, low relief floodplain. Many authors, including Wolman and Leopold (1957), Allen (1964, 1965a, 1965b, 1970), Leopold *et al*. (1964), Andersen and Picard (1974) and Steel (1974) have reported on ancient as well as modern floodplain deposits, pointing out the abundance of massive to horizontally- or ripple-laminated siltstone, trace fossils, mottling, and overall fining-upward sequences that typify this type of environment. In particular, Allen (1970, p. 302-303) notes the characteristic blocky fracture, bioturbation, scattered trace fossil content, and mottling of argillaceous and arenaceous siltstones of the fining-upward cyclothems comprising the Catskill Formation in the Appalachian Mountains of North America. Andersen and Picard (1974, p. 176) report that fine-grained floodplain sediments of the early Tertiary Uinta and Duchesne River Formations in the Uinta Basin of northeastern Utah, consist of abundant, indistinctly-bedded, fine-grained deposits typically showing evidence of bioturbation, mottling, and interstratified "thin, but laterally persistent sandstone beds." In his study of clastic sediments of New Red Sandstone age in western Scotland, Steel (1974, p. 351) identifies a fine member in the D2 cyclothem which consists of "siltstone with ripple-drift lamination and bioturbation," and also reports the occurrence of poorly developed caliche horizons in the D2 cyclothem. These deposits are attributed to overbank flooding of a high-sinuosity (meandering) river (Steel, 1974, p. 352).

The strata of the interbedded sandstone and tuffaceous siltstone subfacies almost certainly include some eolian deposits as well. Thin, very fine grained sandstone and siltstone beds with a relatively high degree of sorting occur sporadically within the sequence and may represent deposition by wind.

Vertical accretion by overbank flooding is suggested as the origin of the bulk of the strata of this facies. It is likely that vegetation covering the floodplain surface aided in trapping and holding either wind-blow or water-born sediment as reported by Schumm and Lichty (1963), due to the abundance of root casts and apparent bioturbation structures throughout the sequence of fine-grained sediments. Such vertical accretion has been advocated for other ancient floodplain deposits as well (Andersen and Picard, 1974, p. 176).

PALEOGEOGRAPHY

The recognition of particular types of fluvial depositional environments represented by the strata of the Upper Member of the Koobi Fora Formation, coupled with the geographic distribution and intertonguing relationships between each subfacies makes it possible to reconstruct the paleogeography during the time the sediments were deposited.

Widespread fluvial deposition in the eastern portion of the East Turkana Basin began shortly after deposition of the KBS Tuff[1] (Frank, 1976; Burggraf, 1976), a prominent marker bed in the uppermost few meters of the deltaic sequence of the Lower Member (Bowen and Vondra, 1973; Vondra and Bowen, 1976). Subsidence in the west led to

---

[1]The KBS Tuff has been dated by several different laboratories and techniques. Radiometric step-heating methods by Fitch *et al*. (1976) have yielded a date of 2.44 m.y. B.P.; fission-track dating by Hurford *et al*. (1976) supports an age of about 2.4 m.y. B.P. Conventional potassium-argon techniques by Curtis *et al*. (1975), however, point to an age of 1.6 to 1.8 m.y. B.P. for the KBS Tuff in different parts of the basin.

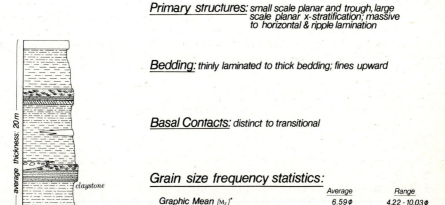

Interbedded sandstone, tuffaceous siltstone subfacies

Primary structures: small scale planar and trough, large scale planar x-stratification; massive to horizontal & ripple lamination

Bedding: thinly laminated to thick bedding; fines upward

Basal Contacts: distinct to transitional

Grain size frequency statistics:

	Average	Range
Graphic Mean $(M_z)^*$	6.59ø	4.22 - 10.03ø
Inclusive Graphic Standard Deviation $(\sigma_1)^*$	2.79ø	0.76 - 4.77ø
Inclusive Graphic Skewness $(Sk_1)^*$	0.10	-0.45 - 0.59
Graphic Kurtosis $(K_G)^*$	1.13	0.08 - 1.75

*after Folk and Ward (1957)

average thickness: 20 m

claystone

siltstone

sand

tuff

Fig. 9.

relative elevation of the basin margins to the north and east with subsequent initiation of erosion of these areas. Concurrently, a climatic change toward less humid conditions was occurring as suggested by oxygen-isotope studies of carbonates, and by the apparent increase in alkalinity of paleo Lake Turkana; recorded by the occurrence of zeolites in the sediments, molluscan extinction, and change in authigenic mineral content of the KBS Tuff and later sediments (Cerling *et al.*, 1977).

Adjacent to the volcanic highlands bordering the basin along the east and southeast, broad alluvial fans accumulated as wedge-shaped gravels of the interbedded basalt cobble conglomerate and pebbly mudstone subfacies. A major southwestwardly flowing stream brought abundant clastic debris from plutonic igneous and metamorphic terranes north and east of the volcanic highlands as well (Mathiesen, 1977), and entered the basin along its northeast margin. The shift to a high energy fluvial system resulted in initial erosion of the uppermost portion of the Lower Member of the Koobi Fora Formation. In the northern portion of area 105 (Fig. 5) for example, a maximum downcutting of 11 meters has been recorded (Burggraf, 1976).

As rapid erosion of the surrounding highlands took place a massive influx of gravel through fine sand and silt occurred and resulted in the development of a shallow, braided alluvial plain throughout the area presently occupied by the Karari Ridge. Low sinuosity channels migrated across the plain depositing coarse sands and gravels as longitudinal and transverse bars and gravel sheets. Aggradation in the basin continued with the development of a thick succession of sands and gravels with minor amounts of fine-grained strata. Depositon appears to have been largely by sand bed ephemeral streams. Numerous caliche horizons and abundant root casts throughout the sand channel-fills indicate that flow in the channels was intermittent and transient so that vegetation and mature caliche profiles could develop. However, in the southernmost portion of the study area there is evidence that a perennial braided stream system existed (Greenwood, personal comm. regarding fish fauna, in Mathisen, 1977). Clay balls and rip-up clasts are commonly present in the lower portions of subsequent sandstone deposits, indicating that thin veneers of silt and clay-sized material which blanketed the coarser grained sediments during waning flood conditions were torn up and incorporated into the sediments of a succeeding flood cycle. Erosion during lateral channel migration with dissection of longitudinal and transverse bars during waning flood conditions resulted in the removal of most of the fine-grained sediment. Such braided river deposits are recorded in the lenticular basalt clast conglomerate subfacies and in the lower portion of the polymictic conglomerate and sandstone subfacies.

The beginning of widespread fine-grained sediment deposition in the form of the interbedded sandstone, tuffaceous siltstone subfacies is marked in the Karari Ridge area, by a thick (to 1.8 m) tuff known as the Okote Tuff. While the transition from the predominantly coarse-grained deposits of the lower half of the Upper Member to the predominantly fine-grained sediments of the upper half appears to be an interfingering, conformable relationship, the coincident occurrence of this change with the volcaniclastic Okote Tuff suggests that basin-margin volcanism and concomitant subsidence may be responsible, in part, for the changing sedimentation pattern. In any case, a decrease in the relative relief difference between the basin and surrounding source terranes led to the development of a meandering (high-sinuosity) drainage system. This is suggested by the occurrence of thick point-bar sand sequences and laterally equivalent massive to laminated arenaceous siltstones, siltstones and argillaceous siltstones. In this sequence, caliche horizons are less well developed than in the underlying subfacies but calcareous concretions and root casts abound. Virtually the entire upper half of the Upper Member contains abundant tuffaceous debris and thin, lenticular tuff units occur in swales of the former low relief floodplain. Associated argillaceous siltstone horizons and silty mudstones which commonly occur capping thick siltstone sequences may represent ancient soil horizons. Their blocky fracture and otherwise massive nature suggests extensive bioturbation as does the occurrence of nodular carbonate concretions and root casts (Allen, 1970 and White, 1976).

The uppermost unit of the interbedded sandstone and tuffaceous siltstone subfacies is the Karari Tuff (up to 2 m thick) which has been radiometrically dated at 1.32 m.y. B.P. (Fitch and Miller, 1976). It represents the end of widespread basin aggradation and fluvial deposition in the Karari Ridge area, and was followed by a period of erosion and intermittent sedimentation. Toward the basin center, at least two subsequent episodes of lacustrine transgression and deposition occurred and are represented by the Guomde Formation and the Galana Boi beds (Bowen and Vondra, 1973).

A short episode of local fluvial sedimentation in the Karari Ridge is evidenced by a thin sequence of mid to late Pleistocene sandstones and conglomerates which truncated the Karari Tuff and underlying strata in many places; and which, in turn have undergone active erosion since the last phase of regional uplift during the latest Pleistocene.

Today the area is drained by a series of shallow, low-sinuosity ephemeral streams with bed material which consists almost wholly of fine sand and coarser clastic debris.

SUMMARY

Fluvial sedimentation during the early Pleistocene is represented in the vicinity of the Karari Ridge by four subfacies belonging to the lenticular conglomerate, sandstone, and mudstone facies of Vondra and Bowen (1976). They comprise the Upper Member of the

Koobi Fora Formation and indicate changing depositional conditions through the middle Pleistocene. The lower half of the sequence indicates that a braided river dominated the eastern portion of the East Turkana Basin and was replaced by a high-sinuosity meandering river system. Each system has been interpreted from an integrated study of bed geometries, grain-size distributions, internal primary structures and facies relationships. This work is a detailed study of a small portion of the East Turkana Basin. As such it represents the first step toward developing a comprehensive understanding of the environmental conditions of the entire basin during the Pliocene and Pleistocene epochs. Such information is critical to the accurate documentation and interpretation of the rich Koobi Fora fauna — an assemblage which includes some of the earliest known and most prolific and significant occurrences of early man.

ACKNOWLEDGMENTS

We are deeply indebted to R. E. F. Leakey and the National Museums of Kenya without whose cooperation this work would not have been possible. Many members of the Iowa State University Cenozoic Stratigraphic Research Group have assisted in this study in various ways, and acknowledgment for their contribution is extended to Bruce Bowen, Hal Frank, Howard White, Russ Bainbridge, Mark Mathisen and Hoyt Acuff. We are grateful to the various members of the East Turkana Research Group for invaluable discussions during all phases of the study. Finally, we wish to thank Robert Palmquist and Howard White of the Iowa State University Department of Earth Sciences for critically reviewing the contents of this manuscript. However, we accept full responsibility for any errors which may appear in either the text or figures. This research was funded by National Science Foundation Grants GS-37813 and GA-25684 to Carl F. Vondra.

REFERENCES

Allen, J. R. L., 1963, The classification of cross-stratified units with notes on the origin: Sedimentology, v. 2, p. 93-114.

———, 1964, Studies in fluviatile sedimentation: six cyclothems from the Lower Old Red Sandstone, Anglo-Welsh Basin: Sedimentology, v. 3, p. 163-198.

———, 1965a, Fining-upwards cycles in alluvial successions: Geol. J., v. 4, p. 229-246.

———, 1965b, A review of the origin and characteristics of recent alluvial sediments: Sedimentology, v. 5, p. 89-191.

———, 1970, Studies in fluviatile sedimentation: a comparison of fining upwards cyclothems, with special reference to coarse-member composition and interpretation: J. Sediment. Petrol., v. 40, p. 298-323.

Andersen, D. W., and Picard, M. D., 1974, Evolution of synorogenic clastic deposits in the intermontane Uinta Basin of Utah: in W. R. Dickinson, ed., Tectonics and Sedimentation; Soc. Econ. paleont. Mineral. Spec. Pub. 22, p. 167-189.

Baker, B. H., Mohr, P. A., and Williams, L. A. J., 1972, Geology of the Eastern Rift System of Africa: Geol. Soc. Am. Spec. Paper 136, 67p.

Bernard, H. A., and Major, C. F. Jr., 1963, Recent meander belt deposits of the Brazos River: An alluvial "sand" model (abs.): Am. Assoc. Petrol. Geol. Bull., v. 47, p. 350-351.

Blissenbach, E., 1954, Geology of alluvial fans in semiarid regions: Geol. Soc. Am. Bull., v. 65, p. 175-190.

Boothroyd, J. C., and Ashley, G. M., 1975, Processes, bar morphology, and sedimentary structures on braided outwash fans, northeastern Gulf of Alaska: in A. V. Jopling and B. C. McDonald, eds., Glaciofluvial and glaciolacustrine sedimentation: Soc. Econ. Paleont. Mineral. Spec. Pub. 23, p. 193-222.

Bowen, B. E., and Vondra, C. F., 1973, Stratigraphical relationships of Plio-Pleistocene deposits, East Rudolf, Kenya: Nature, v. 242, p. 391-393.

Bull, W. B., 1964a, Alluvial fans and near-surface subsidence in western Fresno County, California: U.S. Geol. Survey Prof. Paper 437-A, 70p.

———, 1964b, Geomorphology of segmented alluvial fans in western Fresno County, California: U.S. Geol. Survey Prof. Paper 352-E, p. 89-129.

———, 1968, Alluvial fans: J. Geol. Ed., v. 16, p. 101-106.

———, 1972, Recognition of alluvial fan deposits in the stratigraphic record: *in* J. K. Rigby and W. K. Hamblin, *eds.*, Recognition of ancient sedimentary environments; Soc. Econ. Paleont. Mineral. Spec. Pub. 16, p. 63-83.

Burggraf, D. R., 1976, Stratigraphy of the Upper Member, Koobi Fora Formation, southern Karari Escarpment, East Turkana Basin, Kenya: Unpub M.S. thesis, Iowa State University, Ames, Iowa, U.S.A., 116p.

Cerling, T. E., Hay, R. L., and O'Neil, J. R., 1977, Isotopic evidence for dramatic climatic changes in East Africa during the Pleistocene: Nature, v. 267, p. 137-138.

———, and Powers, D. W., 1977, Paleorifting between the Gregory and Ethiopian Rifts: Geology, v. 5, p. 441-444.

Coppens, Y., Howell, F. C., Isaac, G. L1, and Leakey, R. E. F., 1976, Earliest Man and environments in the Lake Rudolf Basin: Univ. Chicago Press, 615p.

Curtis, G. H., Drake, R. E., Cerling, T. E., Cerling, B. W., and Hampel, 1975, Age of KBS Tuff in Koobi Fora Formation, northern Kenya: Nature, v. 258, p. 395-398.

Eynon, G., and Walker, R. G., 1974, Facies relationships in Pleistocene outwash gravels, southern Ontario: A model for bar growth in braided rivers: Sedimentology, v. 21, p. 43-70.

Fitch, F. J., Hooker, P. J., and Miller, J. A., 1976, $^{40}Ar/^{39}Ar$ dating of the KBS Tuff in Koobi Fora Formation, East Rudolf, Kenya: Nature, v. 263, p. 740-744.

———, and Miller, J. A., 1976, Conventional Potassium - Argon and Argon - 40/Argon-39 dating of volcanic rocks from East Rudolf, *in* Y. Coppens, F. C. Howell, G. L. Isaac, and R. E. F. Leakey, *eds.*, Earliest Man and Environments in the Lake Rudolf Basin: University of Chicago Press, p. 123-147.

Frank, H. J., 1976, Stratigraphy of the Upper Member, Koobi Fora Formation, northern Karari Escarpment, East Turkana Basin, Kenya: Unpub. M.S. thesis, Iowa State University, Ames, Iowa, U.S.A., 118p.

Harms, J. C., and Fahnestock, R. K., 1965, Stratification, bed forms, and flow phenomena (with an example from the Rio Grande); *in* G. V. Middleton, *ed.*, Primary sedimentary structures and their hydrodynamic interpretation: Soc. Econ. Paleont. Mineral. Spec. Pub. 12, p. 84-115.

Hooke, R. L. B., 1967, Processes on arid-region alluvial fans: J. Geol., v. 75, p. 438-460.

Hurford, A. J., Gleadow, A. J. W., and Naeser, C. W., 1976, Fission-track dating of pumice from the KBS Tuff, East Rudolf, Kenya: Nature, v. 263, p. 738-740.

Isaac, G. Ll., 1976, Plio-Pleistocene artifact assemblages from East Rudolf, Kenya: *in* Y. Coppens, F. C. Howell, G. Ll. Isaac, and R. E. F. Leakey, *eds.*, Earliest Man and environments in the Lake Rudolf Basin; Univ. Chicago Press, p. 552-564.

———, Harris, J. W. K., and Crader, D., 1976, Archeological evidence from the Koobi Fora Formation: *in* Y. Coppens, F. C. Howell, G. Ll. Isaac, and R. E. F. Leakey, *eds.*, Earliest Man and environments in the Lake Rudolf Basin; Univ. Chicago Press, p. 533-551.

Kessler, L. G., 1971, Characteristics of the braided stream depositional environment with examples from the south Canadian River, Texas: Wyoming Geol. Assoc. Earth Science Bull., March, p. 25-35.

Leakey, R. E. F., 1976, An overview of the Hominidae from East Rudolf, Kenya: *in* Y. Coppens, F. C. Howell, G. Ll. Isaac, and R. E. F. Leakey, *eds.*, Earliest Man and environments in the Lake Rudolf Basin: Univ. Chicago Press, p. 476-483.

Leopold, L. B., Wolman, M. G., and Miller, J. P., 1964, Fluvial processes in geomorphology: Freeman, 522p.

Mathisen, M. E., 1977, A provenance and environmental analysis of the Plio-Pleistocene sediments in the East Turkana Basin, Lake Turkana, Kenya: Unpub. M.S. thesis, Iowa State University, Ames, Iowa, U.S.A., 233p.

Miall, A. D., 1977, A review of the braided river depositional environment: Earth Sci. Revs., v. 13, p. 1-62.

Ore, H. T., 1963, Some criteria for recognition of braided stream deposits: Contr. to Geol., v. 3, p. 1-14.

Rust, B. R., 1972, Structure and process in a braided river: Sedimentology, v. 18, p. 221-245.

———, 1975, Fabric and structure in glaciofluvial gravels: in A. V. Jopling, and B. C. McDonald, eds., Glaciofluvial and glaciolacustrine sedimentation; Soc. Econ. Paleont. Mineral. Spec. Pub. 23, p. 238-248.

Schumm, S. A., and Lichty, W., 1963, Channel widening and floodplain construction along Cimarron River in S. W. Kansas: U. S. Geol. Survey Prof. Paper 352-D, p. 71-88.

Smith, N. D., 1970, The braided stream depositional environment: Comparison of the Platte River with some Silurian clastic rocks, northcentral Appalachians: Geol. Soc. Am. Bull., v. 81, p. 2993-3014.

———, 1971, Transverse bars and braiding in the lower Platte River, Nebraska: Geol. Soc. Am. Bull., v. 82, p. 3407-3420.

———, 1974, Sedimentology and bar formation in the upper Kicking Horse River, a braided outwash stream: J. Geol., v. 81, p. 205-223.

Steel, R. J., 1974, New Red Sandstone floodplain and piedmont sedimentation in the Hebridean province, Scotland: J. Sediment. Petrol., v. 44, p. 336-357.

Visher, G. S., 1965, Fluvial processes as interpreted from ancient and recent fluvial deposits: in G. V. Middleton, ed., Primary sedimentary structures and their hydrodynamic interpretation; Soc. Econ. Paleont. Mineral. Spec. Pub. 12, p. 116-132.

———, 1972, Physical characteristics of fluvial deposits, in J. K. Rigby and Wm. K. Hamblin, eds., Recognition of ancient sedimentary environments; Soc. Econ. Paleont. Mineral. Spec. Pub. 16, p. 84-97.

Vondra, C. F., and Bowen, B. E., 1976, Plio-Pleistocene deposits and environments, East Rudolf, Kenya: in Y. Coppens, F. C. Howell, G. Ll. Isaac, and R. E. F. Leakey, eds., Earliest Man and environments in the Lake Rudolf Basin; Univ. Chicago Press, p. 79-93.

———, 1977, Stratigraphy, sedimentary facies, and paleoenvironments, East Rudolf, Kenya: in Geol. Soc. London, Spec. Pub. 6, in press.

Walker, R. G., 1975, Conglomerate: Sedimentary structures and facies models: in Depositional environments as interpreted from primary sedimentary structues and stratification sequences; Soc. Econ. Paleont. Mineral. Short Course 2, p. 133-161.

Walker, A., 1976, Remains attributable to Australopithecus in the East Rudolf succession, in Y. Coppens, F. C. Howell, G. Ll. Isaac, and R. E. F. Leakey, eds., Earliest Man and environments in the Lake Rudolf Basin; Univ. Chicago Press, p. 484-489.

White, H. J., 1976, Stratigraphy of the Lower Member, Koobi Fora Formation, southern Karari Escarpment, East Turkana Basin, Kenya: Unpub. M. S. thesis, Iowa State University, Ames, Iowa, U.S.A., 134p.

Williams, G. E., 1968, Formation of large-scale trough cross-stratification in a fluvial environment: J. Sediment, Petrol., v. 38, p. 136-140.

———, 1971, Flood deposits of the sand-bed ephemeral streams of central Australia: Sedimentology, v. 17, p. 1-40.

Williams, P. F., and Rust, B. R., 1969, The sedimentology of a braided river: J. Sediment. Petrol., v. 39, p. 649-679.

Wolman, M. G. and Leopold, L. B., 1957, River flood plains: some observations on their formation: U.S. Geol. Survey Prof. Paper 282-C, p. 87-107.

Wood, B. A., 1976, Remains attributable to Homo in the East Rudolf succession: in Y. Coppens, F. C. Howell, G. Ll. Isaac, and R. E. F. Leakey, eds., Earliest Man and environments in the Lake Rudolf Basin; Univ. Chicago Press, p. 490-506.

FLUVIAL FACIES MODELS

DISTINCTIVE FEATURES OF SOME ANCIENT RIVER SYSTEMS

P. F. Friend[1]

Abstract

Features of some Old Red Sandstone (Devonian) and Tertiary river systems are described that distinguish them from many of the present-day river systems that have been studied. Their sequences and structures suggest that the sediment accumulated in areas of down-stream decrease of river size. Channel incision was not widespread, and the alluvium was built up into convex-upward lobes in some areas. All these features suggest the presence of terminal fans, similar to those described from the Indo-Gangetic plain. Incision is presumed to be more important in present-day studies because of Quaternary events. Lack of vegetation cover and proximity of orogenic activity were important controls of the ancient fluvial systems.

Introduction

This Conference has helped to demonstrate the considerable progress that has been made in relating the sedimentary structures of ancient fluvial sequences to those being formed by present-day rivers. And it is perhaps inevitable that much less progress has been made in relating the larger features of fluvial systems. This is partly due to their size. It is difficult to investigate ancient features bigger than those contained within one exposure, not just because they are incompletely exposed, but also because it is not usually possible to correlate exposure with exposure with any precision. But the problem is also one of sampling. I wish to suggest here that, in many instances, present-day river systems have been compared with ancient systems that were even more different than is usually realised.

My examples will mainly be taken from four of the ancient fluvial stratigraphic units (or sequences of units) best known to me (Fig. 1):

1) Wood Bay Formation (Devonian, Old Red Sandstone) of Spitsbergen (Friend and Moody-Stuart, 1972).

2) Vilddal Supergroup (Devonian, Old Red Sandstone) of East Greenland (Friend *et al.*, 1976a, b. in prep.; Alexander-Marrack and Friend, 1976).

3) Franz Joseph Complex, composed of the Kap Kolthoff Supergroup, Kap Graah Group and Mount Celsius Supergroup (Devonian, Old Red Sandstone) of East Greenland (Friend *et al.*, 1976a, b, in prep.; Yeats and Friend, 1977; Nicholson and Friend, 1976).

4) Molasse-like sediments (Oligocene and Miocene) Ebro basin, Spain (Van Houten, 1974; Riba *et al.*, 1967; Friend *et al.*, in press).

All these units are thick (thousands of metres, Table 1) and accumulated in relatively small basins (tens of km across), near uplifting source mountains. In each case they accumulated at the end of periods of complex major crustal deformation, and they differ in this respect, from other ancient fluvial sequences that form parts of relatively stable, conformable, successions.

Downstream Decrease Of River Depth

Patterns of palaeocurrents and major facies trends within these four fluvial units show general constancy during their accumulation. It is therefore possible, in spite of the lack

---

[1]Department of Geology, University of Cambridge, Sedgwick Museum, Downing Street, Cambridge CB2 3EQ, England.

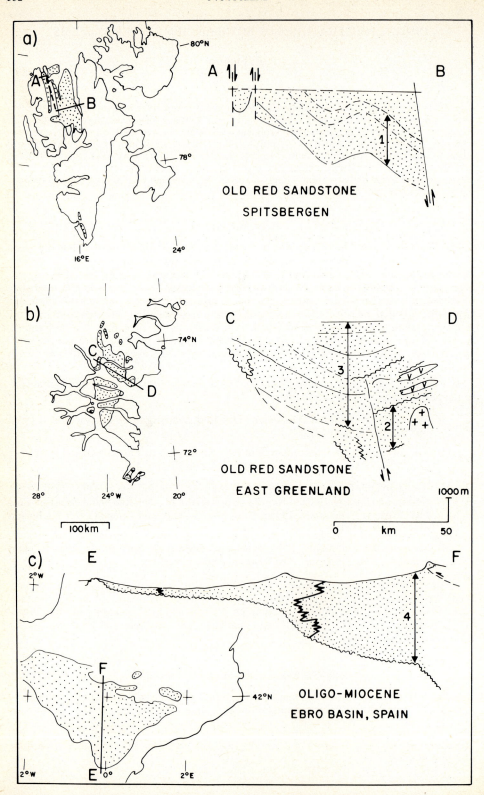

a)

A

B

A   B

1

OLD RED SANDSTONE
SPITSBERGEN

80°N

78°

24°

16°E

b)

C

D

C   D

3

2

OLD RED SANDSTONE
EAST GREENLAND

74°N

72°

28°   24°W   20°

100km

1000m

0   km   50

c)   E   F

4

OLIGO-MIOCENE
EBRO BASIN, SPAIN

2°W

F

E

2°W   0°   2°E

42°N

of precise internal correlation, to arrange detailed sediment logs to show generalised downstream trends of variation of sedimentary features.

The generalised downstream trend for the Wood Bay Formation in Spitsbergen is shown in Fig. 2. The proximal sediment is composed largely of sandstone, with very minor amounts of siltstone. The sandstone is generally cross-stratified, although some flat-bedding is usually present. These sediments were deposited by a bed-load type of river (Schumm, 1968), probably of the braided "Platte" type (Miall, 1977, this volume). The distal sediment in Spitsbergen is composed almost entirely of siltstone, with minor and thin sandstones and calcareous marlstones. I interpret this distal type as the result of long-term accumulation in a clay-playa (Friend and Moody-Stuart, 1972). The intermediate types show downstream trends of decreasing 1) grain-size of sandstones, 2) thickness of sandstone bodies, 3) thickness of cross-stratification sets, and increasing proportion of 4) siltstone, 5) small-scale cross-stratification and 6) flat-bedding. These features together are best explained as the result of sedimentation by a river system with decreasing river depth and flow strength that extended into, and ended in, an area of clay flats.

In Greenland, a similar downstream trend exists in the Vilddal Supergroup (Fig. 2), and also higher in the succession (Kap Kolthoff Supergroup). The proximal sandstones are very similar to those of Spitsbergen, but the distal type is somewhat coarser and more varied than the Spitsbergen clay-playa deposits. However the same trends of overall decrease in grain-size, and in cross-stratification set thickness are present, coupled with a tendency for the amount of flat bedding and small-scale cross-stratification to increase.

The thick succession of the north side of the Ebro Basin, in Spain, also shows a similar downstream trend. Massive coarse conglomerates occur, forming distinct bodies along the northern (Pyreneean) edge of the alluvium. The centre of the basin is occupied by a thick development of limestone and gypsum essentially lacking in siliciclastic material. Between these two extremes, sandstone bodies are present, but become thinner, and generally less coarse, distally.

The idea that the channels of these particular fluvial units decreased in depth downstream, is, at first sight, a surprising one. In Miall's (1977, p. 49) review of some aspects of fluvial sedimentation, the observation that planar cross-stratification units decrease in thickness downstream was reported from three publications. Miall suggested that this decrease might reflect loss of river competency downstream, with a corresponding decrease in the scale of bed-forms. However he pointed out that "depth generally increases downstream and would be expected to be reflected in larger bed-forms." My suggestion is that, in the particular cases being considered here, depth decreased downstream, resulting in smaller bed-forms, and thinner cross-stratification units. It is certainly not possible with present knowledge, to use the thickness of sets of cross-stratification to specify numerically the depth of water flow, but the general suggestion of shallowing based on this feature, is also supported by the tendency for sandstone bodies to thin and fine downstream, and, in some formations, to die out altogether.

Fig. 1. Location and cross-sectional geometry of four stratigraphic units. Stipple indicates the outcrop area and cross-sectional extent of a) Old Red Sandstone of Spitsbergen, b) Old Red Sandstone of East Greenland, and c) Oligo-Miocene molasse-like sediments of the Ebro basin, Spain. Internal subdivisions are indicated in the cross-sections. Arrows in each cross-section indicate 1) Wood Bay Formation, 2) Vilddal Supergroup, 3) Kap Kolthoff Supergroup to Mount Celsius Supergroup, 4) Oligo-Miocene Molasse-like sequence.

Table 1. THICKNESS AND RATES OF ACCUMULATION OF FLUVIAL STRATIGRAPHIC
SEQUENCES

1)   Spitsbergen, **Wood Bay Formation,** 2000 m, Seigenian to early Eifelian, 20 ma, 0.1
     m/1000 years.

2)   Greenland, **Vilddal Supergroup,** 2000 m, Givetian, 4 ma, 0.5 m/1000 years.

3)   Greenland, **Kap Kolthoff Supergroup to Mount Celsius Supergroup,** 5000 m,
     Givetian to Famennian, 16 ma, 0.3 m/1000 years.

4)   Spain, Ebro basin, **Molasse-like sediments,** northern succession, 4000 m, late
     Oligocene to early Miocene, 10 ma, 0.3 m/1000 years.

5)   Texas, **Kisinger Sandstone,** 50 m, Pennsylvanian, basal Virgilian, say 1 ma, 0.05
     m/1000 years.

6)   Pakistan, **Upper Siwalik Group,** 1000 m, latest Miocene to present, 2.6 ma, 0.4
     m/×1000 years (Keller *et al.*, 1977).

   Time intervals estimated from London Geological Society Phanerozoic Time Scale (Harland *et al.*,
1964).

ABSENCE OF ALLUVIAL INCISION

   Two scales of alluvial incision phenomena will now be discussed. The first involves the
extent and depth of development of channels. The second involves the larger-scale, but
related, erosion of "valleys" within the alluvium.

Incision of channels

   Scour structures large enough to indicate episodes of channel downcutting (incision)
have often been reported from ancient fluvial units. But the occurrence of these features
varies considerably from one fluvial succession to another.

   The southern flank of the Ebro Basin is remarkable for the presence of a number of
areas of exposures of narrow, elongate ("ribbon") sandstone bodies (Riba *et al.*, 1967;
Friend *et al.*, in prep.). These bodies are a few tens of metres in width, and a few metres
in height (Fig. 3 a-c). Each was formed by an episode of channel incision followed by
aggradation with sediment. In some cases, this erosion-deposition alternation was
repeated several times at the same location, giving rise to the multi-storey sandstone
bodies seen today. These narrow bodies provide evidence of deep incision by channels,
and of the restriction of lateral channel movement within the incision.

   The northern flank of the Ebro basin does not display the same evidence of deep
incision and restriction. Although some of the narrow elongate sandstone bodies are
present, the most abundant sandstone bodies have sheet form, and result from lateral
channel movement as can be seen from their large cross-stratification and the local
presence of channel-scale scours on their bases. More distally, sheet sandstones are still
the commonest, although they are usually thinner and have rather planar bases, lacking
local downcutting on a channel-scale.

   Fig. 2.   Logs illustrating generalised downstream trends of sedimentation, a) from Wood Bay
Formation, Spitsbergen (letter-plus-number label identifies locality), b) from Vilddal Supergroup,
Greenland (number-plus-letter label identifies sequence type in the classification scheme used in
Greenland, Friend *et al.*, 1976a).

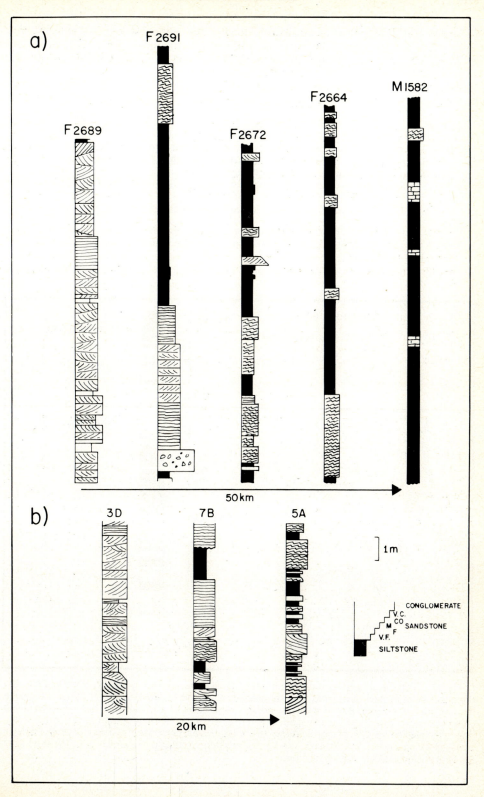

a)

F 2691

F 2689

F 2672

F 2664

M 1582

50 km

b)

3D

7B

5A

1 m

CONGLOMERATE

V.C.
CO SANDSTONE
M
F
V.F.

SILTSTONE

20 km

Evidence of downcutting by channels is notably rare in the Old Red Sandstone of both Spitsbergen and Greenland. In Spitsbergen, the sandstone bodies, where separated from each other by distinct siltstone intervals, generally lack basal irregularities of greater relief than a few tens of centimetres. In Greenland, the lack of sandstone-siltstone alternations makes it difficult to distinguish major from minor scour-surfaces, but channel-scale scour features are certainly not common, and there is no evidence of deep incision episodes.

The lack of evidence for incision in the ancient fluvial sequences of Spitsbergen and Greenland raises the further possibility that the sediment depositing flows may not have been as clearly channelised as most present-day flows that have been studied. Much of the deposition may have been from sheet floods.

Absence of valley incision and flood-plain restriction

The Kisinger Sandstone of Pennsylvanian age, outcropping in north-central Texas, U.S.A., has been interpreted (Shelton, 1973, p. 18) as the result of river aggradation in an alluvial valley that had been eroded in slightly older Pennsylvanian limestones and shales. The valley is typically about 3 km wide and 50 m deep (Fig. 4a), and must have formed at

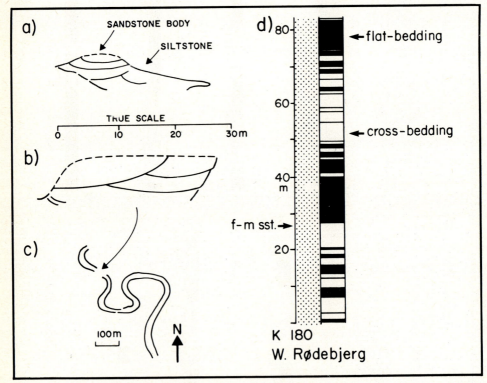

Fig. 3. Diagram illustrating strongly incised channel deposits, and cyclic sequence that resulted from lack of restriction of braided-river sedimentation. a) and b) are sketches of ribbon sandstone bodies from the Ebro Basin showing incision and "multi-storey" internal structure indicating alternations of erosion and sedimenttion. c) is a map of the highly sinuous form of ribbon b). d) is a log of a section in Greenland, showing two megacycles composed of varying proportions of cross-stratification and flat-bedding. The cycles are interpreted as the result of lateral movements of an unrestricted braided channel complex.

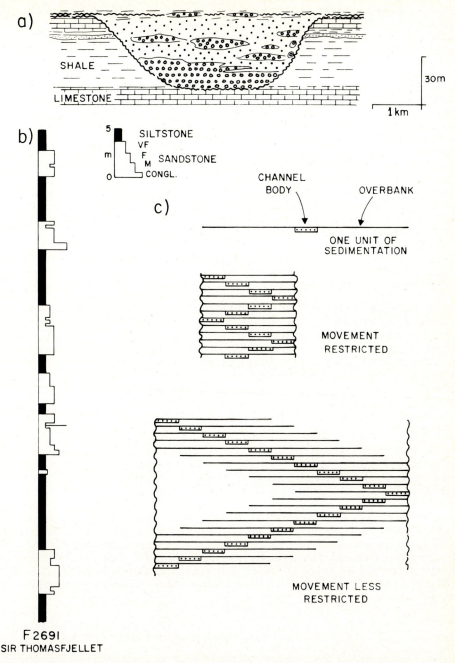

Fig. 4. Diagram illustrating incision of floodplain sediments, and effect of lack of lateral restriction on cyclic sequence. a) Cross-section (Shelton, 1973) through Kisinger Sandstone, north central Texas; b) typical sandstone-siltstone cyclic sequence, Wood Bay Formation, Spitsbergen (F 2691, Sir Thomasfjellet); c) highly simplified model illustrating effect of restriction of lateral movement on floodplain sedimentation: upper sketch shows unrestricted overbank and channel sediment unit: middle sketch shows in vertical section effect of simple, regular lateral movement of channel complex, to produce cycles in vertical profile. This assumes that marginal overbank material is flushed out of the system due to lateral restriction: lower sketch shows how less restriction allows more lateral movement and the construction of cycles with more overbank silt.

a time of pronounced relative lowering of sea-level. The Sandstone itself consists of conglomerates and sandstones, often in fining-upwards sequences. Estimates of the width of the river, presumably braided, suggest that it varied from about 70 m up to perhaps 300 m in width (Shelton, 1973, p. 24).

This example has been summarised, because it illustrates clearly the erosion of a valley, followed by aggradation of a river floodplain sequence. Some of Allen's (1974), models of alluvial 'architecture' include aggraded alluvial valleys, and the restriction of a floodplain by local valley walls is common in many of the present-day river systems that have been studied. Yet no examples of this have been found in any of the ancient river formations being discussed in this paper, except in the most marginal deposits, where, in Greenland and Spain, conglomerates do occupy valley forms eroded in pre-alluvial bed-rock. It may be argued that ancient aggraded valleys can only be detected where the valley walls are lithologically quite distinct from the alluvial fill, as is the case in the examples just quoted. Yet careful observations of these ancient deposits, both walking along individual strata, and examining from a distance well exposed, near-vertical, Arctic mountain faces, have failed to reveal "internal" unconformities. I suggest that this does reflect a genuine difference in erosion-deposition patterns.

Sandstone-siltstone alternations, the typical fining-upwards cycles of some Old Red Sandstone areas, including Spitsbergen, provide further evidence of the freedom of floodplain sedimentation to move extensively.

Fig. 4b shows a typical section of sandstone-siltstone alternations from the Wood Bay Formation in Spitsbergen. The proportion of suspended-load siltstone to dominantly bed-load sandstone is high, and puzzling in terms of the thicknesses of suspended-load silt usually reported from present-day floodplains (Allen, 1965). Large accumulations of overbank material are favoured by large lateral movements of the channel area, as shown in Fig. 4c, provided that sufficient lateral spreading of the suspended fine-grain sediment occurs at times of flood. The presence of sandstones and thick siltstones may be explained in a number of ways, eg. high input of fine sediment, periodic restriction of channel movement, but it is also favoured by lack of close restriction of the floodplain by 'valley' walls.

Another form of cyclicity seems also to be favoured by lack of restricting valley walls. In Greenland, some of the sandstone successions show a large-scale cyclicity (megacyclicity of Yeats and Friend, 1978) due to variations in the proportions of cross-stratification and flat-bedding (Fig. 3d). This cyclicity is a feature of Greenland sediments that are broadly of Miall's (1977, this volume) Platte-type of braided river sediment sequence. The cross-stratification appears to have formed by the movement of megaripples or bar faces within the braided channels, whereas the flat-bedding seems to have formed in shallower water (Smith, 1971). The large-scale cycles may have been built by a long-period change of environment from channel complex centre to channel complex margin. Regular alternations of these environments must reflect the freedom of the complex to move, and thus the lack of incision. Most present-day braided-river are incised to some extent.

Convex-Upwards, Lobate Topography Of River Systems

The western part of the Kap Kolthoff Supergroup in Greenland was deposited by a number of river systems, each one distinguished by the grain-size, colour and palaeocurrent pattern of the major sediment bodies that resulted (Yeats and Friend, 1978). Conglomerates and sandstones are the predominant types of sediment. The only exceptions to this are the siltstones that occur in local units, metres to tens of metres thick, close to the basal unconformity, and at the boundaries between conglomerate and major sandstone bodies, and between major sandstone bodies (Fig. 5a).

An explanation of this situation has been found (Yeats and Friend, 1978) in terms of the contemporaneous topography (Fig. 5b), and is based on observations of the Icelandic sandur, particularly Skeiderarsandur. The different river systems probably built convex-upwards lobes of sediment, each somewhat higher in the centre than at the margins, due to the loss of flow strength marginally as water percolated into the sediment. Local hollows between these sediment lobes, and also close to the outcropping bed-rock, accumulated sediments of silt-grade, because only rather weak streams emerged from the main river systems into the hollows.

This explanation of the presence of fine-grained marginal facies requires that each main river system had a convex-upwards lobate topographic form.

TERMINAL FAN MODEL

I have shown that many of the ancient fluvial formations discussed in this paper provide evidence for 1) downstream decrease of river depth, 2) absence of river incision, and 3) lobate topographic form of river systems.

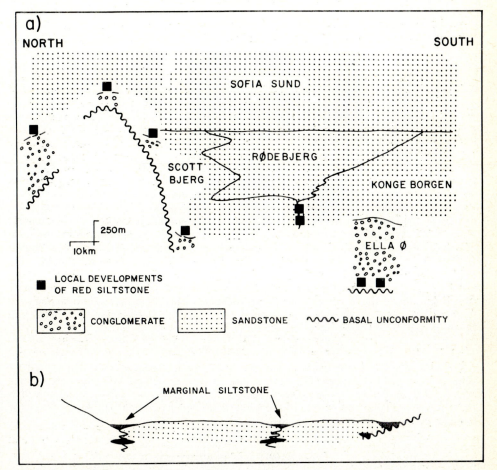

Fig. 5. Occurrence of siltstone marginal to conglomerate and sandstone fluvial systems. a) diagrammatic section through the major sediment bodies of the western edge of the Kap Kolthoff Supergroup, East Greenland, b) hypothetical sketch illustrating probable mode of accumulation of silts in hollows between morphologically higher sand and gravel depositing systems.

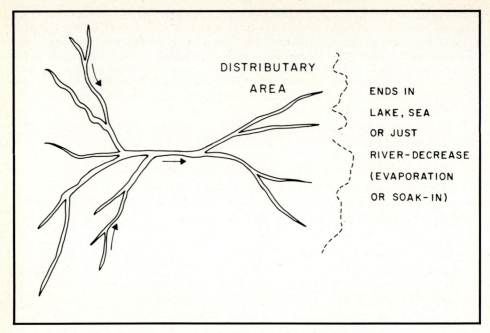

Fig. 6.   Sketch diagram illustrating a distributary area.

I suggest that all these features are characteristic of the "terminal fans" recently described by Mukerji (1976) from the Indo-Gangetic Plain. Rivers emerging from the foothills of the Himalayas suggest a decrease in discharge due to a number of factors including 1) the loss of water into permeable alluvium, and 2) evapo-transpiration coupled with a lower precipitation on the plains than in the mountains. This results in a channel pattern of distributive (Allen, 1965) type, and the ultimate disappearance of all channels (Fig. 6). The fans have a distinctly lobate, convex-upward, topographic form, and the examples described by Mukerji are a few kilometres in width and in downstream length. Geddes (1960), however, describes much larger "cones" as the main morphological units of the entire Indo-Gangetic Plain, each cone corresponding to a major river system. Williams (1970) describes the sediments of a Saharan alluvial piedmont that has many similar features to the terminal fans. These sediments vary downstream in a similar way to those described above from the ancient sequences in Spitsbergen, Greenland and Spain.

The incision question

The lack of evidence of incision in many of the ancient sequences is not shared by some of these present-day terminal fans. It may be that, compared with Quaternary times (West, 1968), earlier times saw steadier regimes of river grade and climate, coupled with a steadier and/or higher supply rate of sediment, that may have minimised the tendency for periodic (Schumm, 1975) channel incision.

COMPARING PRESENT WITH PAST

Vegetation factors

Schumm (1968) has pointed out the major influence of the evolution of vegetation cover on erosion and fluvial processes. The lack of sediment-binding vegetation during the accumulation of the Old Red Sandstone may have decreased the cohesion of channel banks and discouraged incision of the channels, and it must also have increased the rate of run-off and instantaneous ability to transport and deposit sediment. Both these effects are likely to have favoured the terminal fan situation described above.

The Ebro basin appears to have been at least seasonally arid at the time of deposition of the Oligo-Miocene molasse-like sediment. Vegetation was probably not a significant factor in controlling the sediment features in this case either.

Orogenic factors

The high rates of sediment supply necessary to feed the terminal fans ultimately imply high relief in the river source areas.

The long life of most of the fans also implies continuing uplift to maintain this relief.

Differential tectonic movement, particularly tectonic basin formation, can often be seen to be generating alluvial fans, by changing the gradients of the fluvial systems, i.e. forming sediment traps.

The orogenic nature of the four fluvial units discussed in this paper is clearly indicated by their regional tectonic histories. It is also implied by their great thicknesses, and high average accumulation rates (Table 1). In contrast, the strongly incised Kisinger Sandstone (see above) has a more stable tectonic setting.

It is my suggestion that the four stratigraphic units of this paper show important differences from many Quaternary deposits at least partly because they formed under orogenic conditions. Studies of Quaternary sediments of orogenic environment are not common. A promising possibility comes again from the Indo-Gangetic plain. The Upper Siwalik Group (Pliocene and Pleistocene) river sediment has accumulated there from the rising Himilayas, and provides a comparable thickness and average accumulation rate to those of the older orogenic units (Table 1, Keller *et al.*, 1977).

ACKNOWLEDGEMENTS

I thank Professor J. R. L. Allen (Reading Univesity) and Dr. M. R. Leeder (University of Leeds) for reading a draft of this manuscript, Dr. J. D. Marshall (Shell) for helpful advice, and the School of Environmental Sciences, University of East Anglia, for its hospitality during a sabbatical term.

REFERENCES

Alexander-Marrack, P. D. and Friend, P. F. 1976, Devonian sediments of East Greenland III, The eastern sequence, Vilddal Supergroup and part of the Kap Kolthoff Supergroup: Medd. om Grønland, Bd. 206, 3, 121p.

Allen, J. R. L., 1965, A review of the origin and characteristics of Recent alluvial sediments: Sedimentology, v. 5, p. 89-191.

——, 1974, Studies in Fluviatile sedimentation: implications of pedogenic carbonate units, Lower Old Red Sandstone, Anglo-Welsh outcrop: Geol. J., v. 9, p. 181-208.

Friend, P. F., Alexander-Marrack, P. D., Nicholson, J., and Yeats, A. K., 1976a, Devonian sediments of East Greenland I, Introduction, classification of sequences, petrographic notes: Medd. om Grønland, Bed. 206, 1, 56p.

——, ——, —— and ——, 1976b, Devonian sediments of East Greenland II, Sedimentary structures and fossils: Medd. om Grønland, Bd. 206, 2, 91p.

——, ——, —— and ——, in prep., Devonian sediments of east Greenland VI, Summary and significance: Medd. om Grønland, Bd. 206, 6.

—— and Moody-Stuart, M., 1972, Sedimentation of the Wood Bay Formation (Devonian) of Spitsbergen: regional analysis of a late orogenic basin: Norsk Polarinstitutt Skrifter, Nr. 157, 77p.

——, Slater, M. J. and Williams, R. C. in press., Vertical and lateral building of river sandstone bodies, Ebro basin, Spain: Geol. Soc. London, J.

Geddes, A., 1960, The alluvial morphology of the Indo-Gangetic plain: its mapping and geographical significance: Inst. British Geogr., Trans., Pub. 28, p. 253-276.

Harland, W. B., Smith, A. Gilbert, Wilcock, B., 1964, The Phanerozoic Time-scale: Geol. Soc. London, Quart. J., v. 120s, 458p.

Keller, H. M., Tahirkheli, R. A. K., Mirza, M. A., Johnson, G. D., Johnson, N. M., Opdyke, N. D., 1977, Magnetic polarity stratigraphy of the Upper Siwalik Deposits, Pabbi Hills, Pakistan: Earth Planet. Sci. Letters, v. 36, p. 187-201.

Miall, A. D., 1977, A review of the Braided-River depositional environment: Earth Sci. Revs., v. 13, p. 1-62.

Mukerji, A. B., 1976, Terminal fans of inland streams in Sutlej-Yamuna Plain, India: Zeits. Geomorphologie, Neue Folge., Bd. 20, p. 190-204.

Nicholson, J. and Friend, P. F., 1976, Devonian sediments of East Greenland V. The central sequence, Kap Graah Group and Mount Celsius Supergroup: Medd. om Grønland, Bd. 206, 5, 117p.

Riba, O., Villena, J. and Quirantes, 1967, Nota preliminar sobre la sedimentación en paleocanales terciarios de la zona de Caspe-Chiprana (Provincia de Zaragoza): Ann. Edafol. Agrobiol., T. 26, p. 617-634.

Schumm, S. A., 1968, Speculations concerning palaeohydrologic controls of terrestrial sedimentation: Geol. Soc. Am. Bull., v. 79, p. 1573-1588.

——, 1975, Episodic erosion: a modification of the geomorphic cycle: in W. N. Melhorn and R. C. Flemal, eds., Theories of Land Form Development; Proc. 6th ann. Geomorph. Symp., Pub. in Geomorph., State University of New York, Binghampton, p. 69-85.

Shelton, J. W., 1973, Models of sand and sandstone deposits: a methodology for determining sand genesis and trend: Oklahoma Geol. Surv. Bull. v. 118, 122p.

Smith, N. D., 1971, Pseudo-planar stratification produced by very low amplitude sand waves: J. Sediment. Petrol., v. 41, p. 69-73.

Van Houten, F. B., 1974, Northern Alpine molasse and similar Cenozoic sequences of Southern Europe: Soc. Econ. Palaeont. Mineral., Spec. Pub. 19, p. 260-273.

West, R. G., 1968, Pleistocene geology and biology: Longman Group Limited, 379p.

Williams, G. E., 1970, Piedmont sedimentation and late Quaternary chronology in the Biskra region of northern Sahara: Zeits. Geomorphologie, Supplement band 10, p. 40-63.

Yeats, A. K. and Friend, P. F., 1978, Devonian sediments of East Greenland IV, The western sequence, Kap Kolthoff Supergroup of the western areas: Medd. om Grønland, Bd. 206, 4, 112p.

PRELIMINARY EVALUATION OF LITHOFACIES MODELS FOR MEANDERING ALLUVIAL STREAMS

ROSCOE G. JACKSON II[1]

ABSTRACT

Recently completed and ongoing studies of meandering streams disclose a complexity of sedimentary facies not found in standard fining-upward models and later models from the Endrick and Wabash Rivers. The following features are usually deemed diagnostic of low-sinuosity undivided or braided streams but typify many meandering streams: negligible mud, large facies changes over short distances laterally, vertical sequences that do not denote an upward decrease in flow regime, predominance of gravel (sometimes quite coarse), scarcity of cross-stratification, absence of natural levees, and sheetlike geometry of channel facies. Many other widely used criteria for meandering streams do not occur in all meandering streams or else can exist in non-meandering streams.

The ubiquitous assumption of a close correspondence between channel pattern and lithofacies in rivers presents additional complications to fluvial facies models. The morphology of many streams varies greatly from one short reach to another, can be drastically modified by great floods, or can be transitional between recognized types of channel pattern.

Five lithofacies classes of meandering streams are identified tentatively: muddy fine-grained streams, sand-bed streams with modest fine member, sand-bed streams without mud, graveliferous sand-bed streams, and streams with coarse gravel and little sand. The last three classes include vital features ordinarily deemed typical of non-meandering streams.

These considerations imply that the reconstruction of depositional environment of ancient fluvial sediments requires detailed knowledge of the three-dimensional lithofacies. Failure to appreciate this fact may be one reason why environmental interpretations of ancient fluvial sediments commonly are equivocal. A second important factor may be the present ignorance of the manner and potential of preservation of individual fluvial units when stacked upon one another.

INTRODUCTION

The past several years have witnessed a renewed interest in actualistic facies models of alluvial streams. Substantial attention has been devoted during this time to "braided" streams (e.g., Collinson, 1970; Smith, 1971, 1974a; Bluck, 1974, 1976; Jopling and McDonald, 1975). Miall (1977, this volume) has summarized much of this recently acquired knowledge about lithofacies of these non-meandering streams. During this period the investigation of meandering streams has continued in a general effort to refine some of the basic facies models — in particular, the fining-upward cycle — and to come to a quantitative understanding of the depositional processes responsible for recognized facies styles (e.g., Allen, 1970; Bridge, 1975). Concomitantly, the hydromorphological relations of Leopold and Wolman and their colleagues (e.g., Leopold and Maddock, 1953; Leopold and Wolman, 1957; Leopold et al., 1964) and of Schumm (1960, 1963, 1968, 1972) have begun to be employed for the paleohydraulic interpretation of ancient fluvial deposits, particularly those of inferred meandering-stream origin (e.g., Cotter, 1971; Leeder, 1973a).

Despite this welcome reinvigoration and diversification of geological research on meandering streams, there is yet to appear a thorough appraisal of the significant recent developments. The latest such effort was Allen (1965a), but his review appraised facies models only very briefly. Since 1965 a number of Holocene meandering streams have been examined in varying degrees of detail from hydraulic, geomorphological, and sedimentological points of view. These studies offer profound implications to actualistic fluvial facies models, and a review of their salient findings is one rationale for this paper.

[1]Department of Geological Sciences, Northwestern University, Evanston, Illinois 60201 U.S.A.

The present paper attempts a *preliminary* evaluation of lithofacies models of meandering alluvial streams. Several explicit aims are entertained. Firstly, I shall demonstrate that standard concepts of meandering facies contain many faulty generalizations based upon what now can be documented to be unreliable, superficial observations. In the course of this critique I shall summarize and compare the main features of facies models of meandering streams, with special attention paid to results from some quite recent studies. Thirdly, I shall summarize briefly the main geomorphic and sedimentary features of Holocene meandering streams which I surveyed during 1976 and 1977. The bewildering variety of features in these streams run contrary to nearly all the widely accepted criteria for meandering streams. Some implications of this perversity of the Holocene for facies models will be identified tentatively, and a consensus of reliable and unreliable facies attributes will be offered in a similar tentative vein. The final topic to be approached is manner and potential of preservation of fluvial sediments. I hope to demonstrate that this subject is a key uncertainty of actualistic facies models of meandering streams.

GEOLOGICAL PERSPECTIVE OF MEANDERING-STREAM FACIES MODELS

At the outset of an appraisal of lithofacies models, it seems essential to place them into proper spatial and temporal framework, for the simple reason that the process of preservation may selectively remove or alter salient sedimentary characteristics of Holocene environments. In this writer's opinion a major deficiency of the body of studies of fluvial facies is a prevalent neglect of the subject of preservation. Allen (1967) voiced a like complaint.

Figure 1 illustrates the geological setting of the valley fill of the modern lower Mississippi River. This and many other environmental reconstructions in Fisk (1944, 1947) constitute today the most thorough three-dimensional documentation of a Holocene alluvial sequence. The only other comparable achievement comes from the Brazos River (Bernard *et al.*, 1970). Three important, but widely neglected, features of Figure 1 bear profound implications to the rational use of actualistic facies models for the objective interpretation of depositional environments of ancient fluvial sediments. Firstly, sediments of the modern meandering river (natural levee, backswamp, clay plug, and point bar) merely comprise a veneer over the main valley fill of sands and gravels, laid down by the braided early Holocene river of much greater stream discharge and competence. From the other panels of Plate 5 of Fisk (1947) and the pertinent text discussions, it can be deduced that a return to glacial conditions of increased discharge and competence would cause substantial alteration, and possibly complete removal, of this sedimentary veneer of the modern meandering phase.

Another noteworthy aspect of Figure 1 is the comparatively small volume of sediment in the point-bar deposits of the modern meander belt, when viewed in the context of the total valley fill. Point bars are the customary targets of investigations of Holocene streams (e.g., Sundborg, 1956; Bernard *et al.*, 1970; Jackson, 1976a), yet in the lower Mississippi River they constitute only a modest fraction of the present meander belt. This active meander belt is but a small portion of the sedimentary package of the Holocene meandering phase, which in turn includes only about one-half of the volume of the total valley fill. The history and mechanics of deposition of the largest portion of the valley fill is not considered in virtually all published investigations of Holocene meandering streams. Besides Fisk (1944, 1947) and Bernard *et al.* (1970) the sole exceptions to this sweeping indictment are Bluck (1971) and Jackson (1976a), both of whom considered only the influence of rates and directions of bend migration upon lithofacies preservation of channel deposits. Bridge (1975) has considered theoretically the effect of bend migration and aggradation rates upon the lithofacies preservation of point-bar deposits of fining-upward cycles.

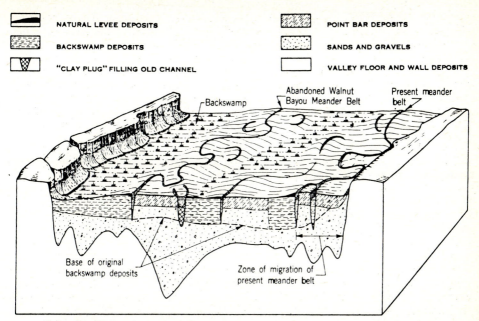

	NATURAL LEVEE DEPOSITS		POINT BAR DEPOSITS
	BACKSWAMP DEPOSITS		SANDS AND GRAVELS
	"CLAY PLUG" FILLING OLD CHANNEL		VALLEY FLOOR AND WALL DEPOSITS

Present Stage-Mississippi, diverted from Walnut Bayou Meander Belt west of Vicksburg, established present course in lowland east of the Cocodrie Meander Belt Ridge.

Fig. 1. Geological setting of Holocene alluvium of lower Mississippi River, near Vicksburg, Mississippi. Shows main features of floodplain and valley fill. In this figure "point-bar deposits" are the sediment that ". . . consists of an alternation of sand bar ridges, capped with thin topstratum, and swales underlain by clay plugs . . .". This definition of point bar (Fisk, 1944, p. 20) is effectively the same as the "upper point bar" of Sundborg (1956; p. 273-277). "Point bar" in the present paper is defined in the more conventional manner as "The total body of sediment, along the inner bank of a bend, from the thalweg to the lowest limit of the fine member." (The fine member consists of the natural levee deposits, backswamp deposits, and "clay plug" in this illustration.) Therefore, the sands and gravels that lie above the base of original backswamp deposits in this figure are included in the point-bar deposits discussed in the test. From panel 6 of Plate 5 of Fisk (1947).

The proportion of fine member (largely deposits of natural levees, backswamp, and clay plugs) to coarse member (point-bar deposits) in the meandering-phase sediments of Figure 1 varies greatly across the valley. Other cross sections in Fisk (1947) show commensurate variability in the valley fill. The total thickness of meandering-phase sediments likewise displays a large, erratic variation across the valley, most strikingly in cross sections containing local areas yet to be reached by the meandering phase of the river (Fisk, 1944, pls. 1, 7A, and 16).

These three cursory observations make evident the highly simplified approximations intrinsic to a lithofacies model like the fining-upward cycle. Equally apparent are the major variations in space and time of the preservation potential of the valley fill and the unpredictability of this variation. Consider now in this light the numerous thick deposits of ancient fluvial sediments described so frequently in the literature (e.g., Allen, 1964, 1965c, 1970; Friend and Moody-Stuart, 1972; Steel *et al.*, 1977). These deposits can consist of many alluvial sequences, or cycles, stacked one upon the other. To anyone interpreting ancient sediments, the issue of manner and potential of preservation of a Holocene subenvironment, be it point bar or backswamp or whatever, is crucial and complex. Yet the topic of preservation continues to be virtually ignored in facies studies of Holocene streams and rarely considered in adequate detail in examinations of ancient sediments of presumed meandering-stream origin.

The valley fill in Figure 1 consists of a couplet of a basal coarse unit of sand and gravel deposited by the early Holocene braided river and an overlying finer-grained unit of sand and mud from the subsequent meandering stream. Nami and Leeder (this volume) and Pryor (this volume) each has identified a similar couplet in an ancient deposit. Their interpretations constitute the sole documentation in the rock record of this style of preservation, which one would expect to find often in fluvial sediments deposited during periods of extensive continental glaciation. Leeder's (this volume) stochastic model of alluviation in an alluvial setting like that of the modern lower Mississippi River is a welcome first attempt to simulate numerically the genesis of a thick sequence of Mississippi-type fluvial cycles.

The vital geological perspective derived from even these superficial considerations of preservation points emphatically to the abyss between the lithofacies models from modern meandering streams and the ancient fluvial sediments to which these models are applied. Besides the obvious question of how representative modern fluvial environments are of the past, one critical factor in this methodological gap is the ultimate preservation of any Holocene fluvial environment; that is, how much (vertically and areally) of the modern sediment escapes erosion by succeeding sediments stacked upon it. The precise modes of lithofacies preservation in streams remain essentially unknown, mainly because of the difficulty of determining channel behaviour through long periods of time, say thousands of years, and of estimating long-term preservation potential of subenvironments within modern stream valleys.

Owing to this regrettable lack of understanding of preservation, the remainder of this paper will concentrate on the better-known "veneer" of alluvium — the channel and immediately proximal sediments of the modern stream — which form the heart of all fluvial facies models. The complex, largely undeciphered fates that this depositional veneer can suffer during its incorporation into the rock record should be kept in mind throughout the following discussions.

Status of Facies Models For Meandering Alluvial Streams

The traditional depositional model for meandering streams is the familiar fining-upward cycle. It arose from the work of Nanz (1954) and Bersier (1958) in ancient sediments and from observations in modern stream valleys by Fisk (1944, 1947), Sundborg (1956, p. 273-276), Bernard and Major (1963), and Bernard *et al*. (1970). Allen (1963a) presented the first complete statement of the fining-upward cycle as a definitive vertical sequence of grain size and primary sedimentary structures. He provided an explanation of the sequence in terms of the flow-regime classification of bedforms and the assumed properties of bend flow. His model has been refined successively over the years (Allen, 1970; Bridge, 1975) and remains today a facies standard for ancient sediments.

Starting with McGowen and Garner (1970) and Bluck (1971), later investigators have begun to realize that some Holocene meandering streams conform poorly to the fining-upward cycle. These recent findings will be discussed later, but they have exerted minimal influence to date on the interpretation of depositional environment.

Assumption of the facies models

The implicit assumption common to all fluvial facies models proposed to date is an intuitive belief that each channel pattern must yield a distinctive lithofacies produced by distinctive depositional processes. This assumption lies behind the usual categorization of fluvial facies models into models for meandering streams, braided streams, and (occasionally) low-sinuosity streams (cf. Moody-Stuart, 1966; Miall, 1977).

The descriptive classification of rivers that has been most widely adopted by fluvial scientists was proposed by Leopold and Wolman (1957). These authors defined rivers on the basis of channel pattern, namely the number of channels and the sinuosity of each channel. Strict adherence to their classification scheme leads to the recognition of four river types (Miall, 1977, Fig. 1; Rust, 1978 and this volume): low-sinuosity single-channel (straight), high-sinuosity single-channel (meandering), low-sinuosity multiple-channel (called braided by Miall, 1977), and high-sinuosity multiple-channel (anastomosing in Miall's usage). Leopold and Wolman (1957) considered both versions of divided stream to be braided. However, Schumm (1968) and Smith (1973, 1976) have described fundamental differences in channel behavior and stability between some anastomosing rivers and the more intensively examined low-sinuosity divided streams. These differences suggest that the full fourfold discrimination should be followed, and this paper will do so.

This rational classification of stream morphology commonly is abused by sedimentologists. There is a pervasive tendency to consider streams as either meandering or "braided" (e.g., Allen, 1965a, 1970; Miall, 1977). Such "braided" streams include all three types of non-meandering streams defined above. This confusion in morphological terminology creates ambiguity in the definitions of facies models, as Miall (1977, p. 1-5) has discussed. The terminological difficulty may arise in part from several factors that conspire to complicate enormously the simple scheme of Leopold and Wolman (1957). Although Kellerhals *et al.* (1976) mention many of these obfuscating aspects of natural streams, these complications are pertinent enough to lithofacies models based upon channel pattern that an analysis of them is given below.

Complexities of channel pattern

Many streams at low flow are divided, whereas high-flow configurations are undivided. Sand-bed plains rivers commonly display a low-sinuosity highly braided pattern during low flow, when sediment-transport rates are minimal, but show a low-sinuosity undivided channel at high, channel-forming discharges. Four examples of this morphological variation with stream discharge are the Brahmaputra (Coleman, 1969, p. 145-151), Platte (Smith, 1970, 1971; Blodgett, 1974; Blodgett and Stanley, this volume), South Canadian (Kessler, 1971), and Tana Rivers (Collinson, 1970). Indisputably meandering streams such as the lower Mississippi (Fisk, 1947, pls. 2, 3) and lower Wabash Rivers (Jackson, 1976a, Fig. 2) often are locally braided in crossings during low flows. Smaller meandering streams of less uniform discharge often appear thoroughly braided at low flow. A low-sinuosity single channel at formative stream discharges seems to be the case for the Brahmaputra, Platte, South Canadian, Tana, and numerous other "braided" streams. Such discharge-related variability of channel pattern presents many implications to preservation potential of lithofacies described at low flow and to paleocurrent patterns (e.g., Bluck, 1974; Rust, 1975; Banks and Collinson, 1974; Smith, 1974b; Schwartz, this volume).

Morphological classification becomes an acute issue in glacial outwash (sandur) streams. These streams tend to show a single channel of extraordinarily large width-depth ratio at peak flood (when sediment discharges are greatest) but become braided at only slightly lower flows, due to the emergence of low-relief bars. Examples include streams on the Baffin Island sandurs (Church, 1972, p. 97-101) and on Iceland sandurs (Nummedal *et al.*, 1974; Rust, 1975). Other sandur stream, in contrast, show divided channels even during peak flood (e.g., Boothroyd and Ashley, 1975, Figs. 3A, 26). Consequently, a complete continuum in channel pattern (and thus presumably lithofacies) likely exists between divided low-sinuosity streams and true undivided low-sinuosity streams. End members of this continuum may well differ appreciably in lithofacies styles. Similar morphological continua between other pairs of stream type exist as well. Examples of morphological transitions from meandering to low-sinuosity undivided streams include the two in Figure 2, several described by Leopold and Wolman (1957), the Cimarron river (Shelton and Noble, 1974), and the Red River of Texas and Oklahoma (Schwartz, 1977). The latter two rivers contain a mixture of sedimentary features of meandering streams

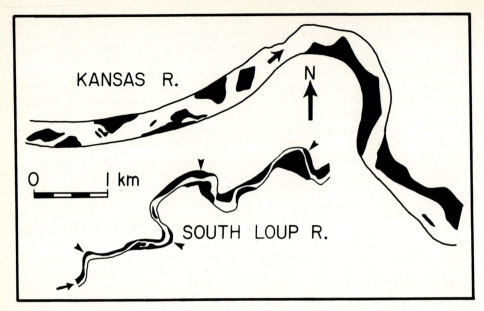

Fig. 2.   Channel pattern at low flow of Kansas River near Williamstown, Kansas (39°02'N 95°18'W), and of South Loup River near Boelus, Nebraska (41°02½'N 98°43'W). Sand bars exposed during low flow are shown in solid black. Each arrowhead denotes a bar on the outside of a bend. Maps traced from Williamstown and Boelus 1:24,000 topographic maps published by the U.S. Geological Survey.

(point bars and accretion ridges) with features most usually associated with "braided" streams (longitudinal and transverse bars and a lack of mud units). Transitions from ideal meandering to ideal anastomosing appear in the Barwon (Taylor, 1976) and in the Little Sugar and Little Wind Rivers.

Even when the channel pattern of a stream during dominant, channel-forming discharges is known, one can find many Holocene streams that do not fit within this fourfold classification (Leopold and Wolman, 1957, p. 40, 59-63). Figure 2 illustrates two means by which streams can depart from ideal form. The Kansas River near Williamstown contains short undivided reaches (at bankfull flow) of alternately high and low sinuosity; one such pair of reaches appears in Figure 2. Other rivers of this ilk include the Calamus (Brice, 1964, Fig. 24), middle Mississippi, Missouri, Brazos, and upper Wabash Rivers. The straight reaches are not due to recent bend cutoffs, and some of these streams suffer little or no channel modification by man. Leopold and Wolman (1957, p. 59-63) described other types of abrupt transition of channel form. Many transitions apparently arose from subtle changes in gradient or sediment size along the valley. Schumm (1960, p. 25-28) emphasized the role of changes in sediment load as a cause of downstream variations in channel morphology.

The South Loup River in Figure 2 presents an undivided sand-bed channel of high sinuosity, but the distribution of alternate (side) bars around bends is similar to that in many low-sinuosity streams. The location of alternate bars on the outsides of bends is most atypical of the normal bed topography of meandering streams (e.g., Sundborg, 1956, Fig. 41; Jackson, 1975a, Figs. 8-9). Sharply curved bends in the Little Wind River (Tables 5-6) show coarse-gravel bars attached to the outer banks.

A third deviation from ideal channel form is the existence of compound, or multiple, meanders (Brice, 1974), in which thalweg wavelength can be a small fraction of meander wavelength. Distributions of sediment size and bed topography around compound meander bends is probably more complex than distributions around simple bends. The Calamus River at 25-35 km NW of Burwell, Nebraska, displays a combination of all three departures from ideal form and, as might be expected, scarcely resembles any of the four basic classes of channel pattern.

If lithofacies style reflects channel pattern, each of these three examples of morphologically ''hybrid'' streams must contain a facies type that is a mixture of two or more form-defined facies. This confusing amalgamation throws substantial doubt upon the assumption that channel morphology *must* reflect distinctive lithofacies and formative physical processes, inasmuch as there is every reason to suppose a continuum of channel pattern between recognized types.

Another caveat of channel pattern concerns the occurrence in some Holocene streams of drastic changes in morphology that are produced by infrequent, great floods. Pre-existing channel pattern and valley features can be obliterated by such catastrophic events. Sand-bed rivers in arid or semiarid climates are very susceptible to this phenomenon (Schumm and Lichty, 1963; Kessler, 1971; Schumm, 1971). Some graveliferous streams suffer channel modification of like magnitude (e.g., Baker, 1977; Patton and Baker, 1977). In general, though, gravelly streams in humid regions and mud-bed streams appear to possess considerable channel stability through major floods (see Gardner, 1977, and references therein). The incomplete documentation of channel changes by infrequent events, whether floods or prolonged droughts, prevents any precise estimation of the stability of a given stream (Wolman and Miller, 1960, p. 70-72). The huge volumes of sediment transported in unstable rivers during rare floods must influence greatly the lithofacies that is ultimately preserved (e.g., McKee *et al.*, 1967).

A final complication of channel pattern in some streams involves substantial changes in channel processes with normal variations in stream discharge. In the lower Wabash River, for instance, gravel on the upper point bars in transitional flow zones is moved only during bankfull and flood discharges, during which times channel activity in fully developed flow zones can be much less striking (Jackson, 1975a, p. 1514-1517). During times of much lower stream discharge the situation reverses; then many of the largest bedforms of significant preservation potential (especially scroll bars) develop in the fully developed flow zones. This temporal variability in bed activity along a single bend is duplicated across the channel. The resulting complex spatial variation of channel processes with time in a short reach appears also in the meandering River Endrick (Bluck, 1971, p. 98-100, Fig. 12) and in many non-meandering streams (e.g., Rust, 1972; Bluck, 1974, 1976; Lewin, 1976).

This assessment of channel morphology indicates that one can expect a diversity of lithofacies types preserved in many fluvial environments. The variation in facies styles must consist of a continuum including those diagnostic of each of the four recognized channel patterns. Recent reconstruction of the Holocene depositional history of the Colorado River of Texas (Baker and Penteado-Orellana, 1977; Looney and Baker, 1977) and Fisk's (1944, 1947) work support this contention. The studies of Jahns (1947), McKee *et al.* (1967), Williams (1971), and Baker (1977) illustrate the wide range in depositional processes and facies expectable from great floods. The modification of facies and channel processes from prolonged drought remains virtually unknown, although there exist unpublished observations of a major reduction of channel capacity by vegetation.

Genetic classification of rivers

Schumm's (1963, 1968, 1972) classification considers sediment load and mud content of channel periphery as mutually dependent genetic factors of channel pattern. He derived empirical equations to relate these factors to such important morphological parameters as width-depth ratio of channels and sinuosity. His classification and attendant empirical equations have found enthusiastic acceptance from sedimentologists who have strived to reconstruct paleohydraulic regime (e.g., Cotter, 1971; Elliott, 1976). Inherent errors in and limitations of Schumm's approach have been voiced by Schumm himself and others (Schumm, 1960, p. 18-19; Cotter, 1971, p. 132-134; Leeder, 1973a, p. 271-272; Nummedal, 1973; Riley, 1975; Ethridge and Schumm, this volume; Baker, this volume).

In a separate paper I shall explore in depth the limitations of Schumm's classification and empirical equations as well as other hydromorphological relations. Suffice it to say here that many Holocene meandering streams, including several in Tables 3-6 subsequently, show width-depth ratios, mean velocities, gradients, and mud contents that violate severely many commonly used hydromorphological relations. Major causes of deviation include (1) a high variation in mud content from cross section to cross section along the channel, (2) stabilization of stream banks by vegetation (particularly effective in small streams), (3) discharge regime, (4) large variation in width-depth ratio around a single bend and between bends and crossings, and (5) size of sediment available to the stream. These discrepancies are of sufficient magnitude and commonness to place strong reservation on the broad reliability of hydromorphological relations in paleohydraulic analyses.

Criteria of channel pattern

Criteria to distinguish meandering streams from non-meandering streams can be categorized as geomorphological (Table 1) or as sedimentological (Table 2). The sedimentological criteria for meandering streams appear, often implicitly, in meandering-stream facies models proposed prior to 1971. These facies models even today form the methodological basis for inferring the meandering-stream origin of ancient sediments. Inferences of paleohydraulic regime ordinarily incorporate one or more geomorpholocial criteria or hydromorphological relations derived therefrom.

The criteria in Tables 1 and 2 come from a diversity of geomorphological, hydraulic, and sedimentological observations, many of which are demonstrably incomplete or erroneous. This section will examine each criterion briefly and mention Holocene streams which provide counterexamples to each.

Geomorphological criteria

Gradient, discharge rate, and bank stability as criteria in Table 1 come largely from the work of Leopold and Wolman (e.g., Leopold and Wolman, 1957, 1960; Leopold *et al.*, 1964). The load criterion is due to Schumm (1968, Table 5; 1972, Table 1) and is derived directly from the sand-mud ratio of the deposit (Schumm, 1968, Fig. 27). The sand-mud criterion and the two immediately below it in Table 1 are in essence sedimentological but originate from geomorphological considerations. The criterion on deposit width arises from the universal observation of a greater width-depth ratio of channels in modern divided streams than in the single channels of undivided streams of comparable mean annual discharge. The assessment of natural levees in Table 1 likewise is difficult to attribute to a single source.

Perhaps the most objectionable criterion in Table 1 is the one on channel flow. It is virtually without meaning from two standpoints. Not only do all unentrenched rivers show "unconfined" flow (in not being restricted to well-defined channels) during flood, but also many divided streams display remarkably explicit channels. Examples of the latter aspect include distributary networks on many deltaic plains (e.g., Allen, 1965b, p. 569-572; Axelsson, 1967, p. 99-106; Smith, 1973, 1976) and such diverse non-meandering streams as the Kosi (Gole and Chitale, 1966), Knik and Matanuska (Fahnestock and Bradley, 1973), Yellow (Chien, 1961), Brahmaputra, Donjek, Platte, and Tana Rivers. Streams which are weakly confined to channel systems appear to be restricted to Arctic sandurs (e.g., Church, 1972; Nummedal *et al.*, 1974; Boothroyd and Ashley, 1975) that are utterly free of bedrock bounds.

TABLE 1. COMMONLY CITED GEOMORPHOLOGICAL CRITERIA FOR ALLUVIAL STREAMS

	Braided	Coarse-grained meander belt	Fine-grained meander belt	Straight distributary
Gradient	High------------------------------------Low			
Channel flow	Unconfined------------------------------Confined			
*Discharge rate	Flashy----------------------------------Continuous			
*Bedload/suspended load	High------------------------------------Low			
*Sand/mud (deposit)	High------------------------------------Low			
*Sand body (deposit)	Wide-------Multilateral----Multistoried---Narrow			
*Natural levees	Slight----------------------------------Prominent			
Bank stability	Slight----------------------------------Great			

Note: Adapted from Brown (1973, his Fig. 8B).

* Criterion is not generally valid.

Of the other criteria in Table 1, most are not generally valid, in the sense that one can find many streams in which one or more can be disobeyed. To start with, consider discharge rate. Some meandering streams, especially ephemeral ones, possess undeniably "flashy" discharges: Little Dry Creek (Tables 5-6), Bagwell Creek (38°47'N 88°05½'W), Ninnescah River (37°25'N 97°20'W), and Red River (Harms *et al.*, 1963). Alternatively, many small non-meandering streams near or within the Sand Hills of Nebraska show very uniform discharges; excellent examples include the Platte (Smith, 1974b), Loup Rivers (Brice, 1964, p. D1), and Calamus River (Fig. 2). The non-meandering Brahmaputra and Ganges Rivers (Coleman, 1969) rise steeply to peak flood but rather uniform discharges prevail during the summer floods. Although Smith (1974b) has demonstrated that "braided" streams need not show highly irregular discharges, the discharge-rate criterion in Table 1 remains too inadequately examined to allow objective conclusions on its reliability.

The load criterion comes solely from five measurements reported by Schumm (1968, his Fig. 27), who emphasized their obvious limitations. One counterexample to this criterion comes again from the Sand Hills. Cedar Creek (Table 5) contains no suspended load during the prevalent low flows, nor do the deposits of the modern stream show any appreciable deposition from suspension. Yet this stream meanders freely and possesses a higher sinuosity than many streams of substantial mud content and suspended load (Table 5).

The sand-mud criterion is contradicted by the several meandering streams in Tables 3-5 which contain minimal mud in their vertical sequences. The banks of all these small streams are stabilized solely by vegetation. While it thus is certain that many meandering streams need not conform to this criterion, the suggestion in Table 1 of high sand-mud ratios in braided-stream deposits may be better founded. Few examples of Holocene braided-stream deposits with appreciable thicknesses of mud units are known, excluding anastomosing streams (Schumm, 1968; Smith, 1976), which are rarely considered by fluvial scientists. Nonetheless, this criterion is frequently abused by the call upon it to infer a braided-stream origin for ancient fluvial sequences which contain little mud. This practice is doubly regrettable because, in Holocene streams, mud units usually occur in the higher parts of the vertical section, in precisely the position most vulnerable to erosion by later events of deposition of overlying units.

The inference of sand-body geometry in Table 1 is due to implicit extrapolation of two generalizations: (1) most modern braided streams are wider than meandering streams of commensurate mean discharge and (2) many braided streams migrate laterally over great distances (e.g., Kosi and Brahmaputra Rivers), whereas many (by no means all) meandering streams are constrained to narrow meander belts, outside of which muds accumulate in floodbasins (Fig. 1).

On the other hand, meandering streams with vertical sequences of much lower sand-mud ratios than the lower Mississippi River (cf. Tables 3-5) present a different picture. Some of these streams show much lower densities of meander scars on their floodplains than does the lower Mississippi. For example, the lower Wabash River flows in a single meander belt that contains surprisingly few oxbow lakes and meander scars (Fidlar, 1948, p. 85-88, pl. 1). The mud-free meandering streams in Tables 3-5 can show sand-gravel deposits whose width equals the total width of the modern valley. Furthermore, anastomosing streams can possess floodplain features and a channel behaviour which indicate relatively narrow sand bodies. Many "braided" streams are confined within narrow permanent valleys, and their deposits are correspondingly narrow: Bijou Creek (McKee *et al.*, 1967), Platte River (Smith, 1971), South Canadian River (Kessler, 1971), most of the larger streams of the Nebraska Sand Hills, and graveliferous valley-sandur streams such as the Kerlingordalsá (Krigström, 1962, Fig. 1), White of Washington state (Fahnestock, 1963), Donjek, Knik and Matanuska Rivers.

In sum, the criterion of sand-body geometry may be reasonable for streams in exceedingly wide valleys but becomes suspect for "low-mud" meandering streams and for all streams in narrow valleys.

Many meandering streams display negligible natural levees: the Endrick, lower Wabash, and all streams in Tables 3-5 of thickness ratio less than about 0.3. The non-meandering Brahmaputra shows modest natural levees (Coleman, 1969, p. 230-232), and the low-sinuosity undivided Yellow shows prominent natural levees (Chien, 1961). Church (1972, p. 110-112) has described gravelly "sandur levees" in the non-meandering South River. These and other reports suggest that natural levees of a great range in physical scale and sediment size can exist in all streams whose floodplains suffer periodic inundation by sediment-laden water.

From this analysis it appears that the only geomorphological criteria of possible broad validity involve gradient and bank stability. The former originated from the classic Figure 46 of Leopold and Wolman (1957) and consequently is valid only for (1) comparisons of streams of equivalent bankfull discharge and (2) the distinction between divided streams and meandering streams. Low-sinuosity streams in their Fig. 46 fall haphazardly into the fields of both of the preceding channel patterns and thus do not possess a gradient sensitivity in the manner implied for straight distributaries in Table 1. Although the experimental work of Schumm and Khan (1972) and Parker's (1976, Fig. 4) theoretical analysis are compatible with the gradient relation in Table 1, both studies suggest that additional factors besides gradient determine channel pattern.

The many limitations of geomorphological criteria make appropriate the mention of the following fact. Published inferences of depositional environment or of paleohydraulic regime of ancient fluvial sediments invariably rely upon one or more of the dubious criteria of Table 1.

Sedimentological Criteria

Many criteria for distinguishing sediments of meandering streams from deposits of non-meandering streams are employed widely in the examination of ancient sediments. These criteria (Table 2) come from a variety of sources, which, for the sake of brevity, will not be listed here. Each criterion has been employed to varying degrees of frequency, for the inference of a meandering-stream origin. Most published analyses use several of

TABLE 2.   COMMONLY CITED SEDIMENTOLOGICAL CRITERIA FOR FLUVIAL DEPOSITS

	Meandering	Non-meandering
Vertical sequence of lithofacies	Fining-upward cycles (of grain size and sed. structures)	No consistent sequence
Fine member	Normally common and appreciably thick	Uncommon and thin
Rock gravel in coarse member	Small amounts; few large clasts	Can be abundant, with large clasts
Scroll bars	Common	Absent
Epsilon-cross stratification	Common	Absent
Scouring surfaces in coarse member	Uncommon	Abundant
Channel-fill mud deposits	Common, esp. in muddy streams; long and arcuate	Minor; short
Chute-fill and chute bars	Expected in "coarse-grained" streams	Uncommon
Natural levees	Often prominent	Minor
Dispersion of current indicators	Large, often >180°	Small, often <90°
Exhumed meander belt	Can be expected in proper sections	Absent
Continuity of sand and gravel beds (in coarse member)	Often great, with little lateral change in texture	Beds often lenticular and discontinuous

these criteria and not just one. The following paragraphs will indicate the extreme unreliability of the first six criteria and the last one in Table 2.

Channel-fill deposits of mud abound on the floodplains of many meandering streams but are much less frequent (in terms of area of meander scars per unit area of the floodplain) in meandering streams with little mud. Floodplains of some braided streams contain short, abandoned anabranches (e.g., Williams and Rust, 1969; Boothroyd and Ashley, 1975, Fig. 3; Church and Gilbert, 1975, p. 76, Fig. 44c). Such channel fills commonly contain much sand and gravel, in contrast to the typical mud fills of muddy meandering streams (LeBlanc, 1972, Fig. 15), but are not so dissimilar to channel fills in mud-free meandering streams (mainly sand with a thin topstratum of decayed plant debris — author's unpublished observations). These facts argue against the indiscriminate use of this criterion without consideration of the mud content of the vertical sequence.

One instance in which channel fill indicates a sinuous channel, if not a meandering stream *per se*, is a complete exposure normal to the local channel axis of a bend. This sort of section reveals the typical asymmetrical triangular cross profile of the channel at the time of abandonment. Glowacz and Horne (1971, their Fig. 7) have provided two examples of such ideal sections. Each example was doubly persuasive in showing the immediately adjacent fill of the channel, which contained epsilon-cross-stratification (Table 8).

As discussed earlier, many meandering streams contain insignificant natural levees, whereas some non-meandering streams can display large levees. This conclusion extends naturally to levee deposits.

Paleocurrent dispersion is now appreciated to be a more complicated phenomenon than once thought. One complication is the general increase in directional variance with decrease in physical scale of the current indicator (Allen, 1966, 1967; Miall, 1974). Recent work in modern and ancient fluvial sediments has revealed a consistently lower dispersion in certain current indicators, most notably (1) imbrication of discoidal gravel clasts (Bluck, 1974; Rust, 1975) and (2) orientation of trough axes of largescale cross-strata (Harms *et al.*, 1963, p. 572-573; Dott, 1973; Michelson and Dott, 1973). Fair correspondence of mean channel orientation with the orientation of either indicator requires many directional readings at representative sites across the channel. Cross-stratification contains an inherently high dispersion in many non-meandering streams (Smith, 1972; Bluck, 1974; Rust, 1975; Cant, this volume) of a magnitude commensurate with that measured in meandering streams (e.g., Harms *et al.*, 1963, Fig. 7; Bluck, 1971, Figs. 12, 14c, 25; Bluck, 1976, p. 454). Caution is therefore warranted in the inference of channel pattern from comparative differences in paleocurrent direction.

One indisputably dependable criterion for highly sinuous channels is the presence of exhumed meander belts, consisting of the accretionary ridge-and-swale topography so common on floodplains of Holocene meandering streams (e.g., Fisk, 1947, Fig. 1, pls. 2, 4; Bluck, 1971, p. 125-126; Hickin and Nanson, 1975). The excellent exposures, parallel to overall bedding planes, required to reveal this feature are not often met with in the rock record; but Puigdefábregas (1973), Nami (1976), Nami and Leeder (this volume), and Padgett and Ehrlich (1976) describe such fortunate cases. The first two authors also found epsilon-cross-stratification in vertical sections through the resepctive deposits (Table 8).

One cautionary note on exhumed meander belts is the necessity of an appreciable planwise curvature in the accretionary topography in order to assume safely a highly sinuous channel. Some low-sinuosity streams form a broadly arcuate accretionary topography by the lateral migration of bends of low curvature; examples include the Cimarron River (Shelton and Noble, 1974), the upper Wabash River near Newport, Indiana (author's unpublished observations), and the Volga (Shantzer, 1951, as cited by Allen, 1970, p. 313). In like manner low-sinuosity streams can show occasional bends of appreciable curvature (e.g., upper Wabash R.; Gustavson, 1977, Figs. 1, 22), so numerous, intersecting sets of accretionary topography are needed in order to infer unequivocally a meandering-stream origin. Finally, anastomosing streams can yield an accretionary topography (Axelsson, 1967; Allen, 1965b, p. 569). This fact emphasizes once again the point that many criteria of channel form indicate channel sinuosity and not channel pattern *per se*.

CRITICISMS OF SPECIFIC LITHOFACIES MODELS

Lithofacies studies of Holocene meandering streams have revealed facies relationships that either contradict or do not appear in the fining-upward cycle (Tables 3 and 4). Some

TABLE 3. PUBLISHED REPORTS OF MEANDERING STREAMS WHOSE VERTICAL SEQUENCES OF LITHOFACIES DEPART FROM THE FINING-UPWARD MODEL

Reference	Stream	$\overline{Q}$ (m³/s)	$\overline{D}/D_{99}$ (mm)	Thick-ness (m)	Fine/coarse member
Fisk (1944)	Mississippi	16,100	0.4/ 40	20	~0.5
*Bluck (1971)	Endrick	7	~1 />10	4	0.3-2.0
*Jackson (1975a, 1976a)	Wabash	760	0.8/ 12	10	0.1-0.5
Shepherd (1976)	Rio Puerco	~1	0.05/ ?1	4	<0.1?

* Report includes measurements of spatial trends in mean grain size of surficial sediments on point bars.

TABLE 4. PUBLISHED REPORTS OF MEANDERING STREAMS WHOSE SURFICIAL (POINT-BAR) SEDIMENTS SHOW VARIATIONS IN MEAN SIZE CONTRARY TO THOSE IMPLIED BY THE FINING-UPWARD MODEL

Reference	Stream	$\overline{Q}$ (m³/s)	$\overline{D}/D_{99}$ (mm)	Thick-ness (m)	Fine/coarse member
McGowen and Garner (1970)	Amite	52	~0.8/~10	8	<0.1
	Colorado	94	~0.8/~10	8?	<0.1
*Wiethe (1970)	Whitewater	45	1 / 50	5	<0.5
Nilsson and Martvall (1972)	Öre	32	~0.8/>10	5	?
Erkek (1973)	Oker	11	0.8/>5	2	?
Bridge and Jarvis (1976)	South Esk	5	1.1/ 10	3	<0.1
*Levey (1976)	Congaree	260	0.6/ 35	~6	0.3-1?
Nanson (1977)	Beatton	30	1 / 10	7	~1

* Only the subaerial portion of a point bar on one bend was examined.

Note: In Tables 3-5, $\overline{D}$ and D_{99} are the mean size and the coarsest one percentile of the coarse member.

of the reports in Table 4 are compromised in part by their restriction to surficial point-bar sediments, which need not represent long-term depositional trends in the stream. For example, the reach of the Colorado River examined by McGowen and Garner (1970) contains areal trends of mean grain size on point bars which resemble closely the textural patterns from subaerial portions of point bars in the lower Wabash River (Jackson, 1975a, 1976a), once the veneers of mud and sand deposited in both rivers during low flow (and commonly removed by succeeding high flows) are eliminated from consideration. Point bars in the Brazos River near the Richmond site of Bernard *et al.* (1970), the one Holocene type example for the fining-upward model, show the same facies characteristic. These results for the Brazos and Colorado come from my reconnaissance of these streams in June 1973.

Indeed, in the formal literature on meandering streams there is no explicit documentation of point-bar lithofacies which conforms unequivocally to the fining-upward cycle. Three reasons may explain this troubling state of affairs.

First of all, sediments of most of the streams in Tables 3 and 4 do not resemble well the numerous ancient fluvial deposits in which the fining-upward cycle attributed to meandering rivers is reported so frequently. Part of this poor correspondence between the modern and the (inferred) ancient consists of a much smaller ratio of thickness of fine member to coarse-member thickness in Holocene streams (less than 1, commonly near zero, versus ratios from about 1 to in excess of 10 in ancient sediments) and of a substantially coarser sediment in the coarse member. That is, most of the streams in Tables 3 and 4 contain appreciable amounts of gravel, whereas most of the ancient examples contain coarse members of medium-sand size or finer (e.g., Allen, 1970, p. 300-303). Indeed, many ancient deposits with significant gravel and little mud are routinely termed "braided-stream" deposits simply on the basis of their coarse grain size and lack of mud (e.g., Allen and Friend, 1968, p. 57; Shelton, 1973, p. 18-24; Campbell, 1976, p. 1017).

The facies studies of Table 3 include two recently proposed facies models, each of which refines substantially the fining-upward cycle. The Wabash model of Jackson (1976a) appears compatible with the lithofacies of point bars in the Beatton, Brazos, Colorado, Congaree, Öre, S. Esk, and Whitewater in Tables 3 and 4. All these rivers present remarkably similar ratios of thickness of fine and coarse members, mean grain size of coarse member, and spatial trends in mean size of point-bar sediment. The River Endrick (Bluck, 1971) differs from this depositional style primarily in the possession of a substantial gravel-riffle facies and a lesser proportion of cross-stratified sand.

The third point of departure of streams in Tables 3-4 from the fining-upward cycle lies in the occurrence of specific sedimentary features which do not appear in the ideal model. For instance, the Wabash River displays prominent transverse bars and scroll bars (Jackson, 1976b) and inner-bank accretionary deposits of leveelike appearance (Jackson, 1975b, p. 237-239; facies 4 of Fig. 14 of Jackson, 1976a); the Endrick contains prominent gravel riffles in crossings (Bluck, 1971). Paleocurrent directions and bed topography of bends are much more variable in both streams than the fining-upward cycle would imply.

Regrettably for the sake of simplicity, two streams in Table 3 differ greatly from both the fining-upward cycle and the Wabash and Endrick variants. The Rio Puerco contains a predominance of upper-regime sedimentary structures, chute cutoffs, and very fine grain size of the coarse member. It appears to be a distinct facies type in itself. The lower Mississippi River shows some facies similarities to the lower Wabash (especially in the possible existence of transitional and fully developed textural zones — see, e.g., Fisk (1947, p. 65, sects. 182, 183) — but differs from the Wabash in possessing prominent chute channels, frequent chute cutoffs, and thick natural levees (on both banks of each bend).

TABLE 5. MEANDERING STREAMS SURVEYED IN 1976 AND 1977

Stream	$\overline{Q}$ (m³/s)	$\overline{h}$ (m)	$\overline{D}/D_{99}$ (mm)	Fine/ coarse member	Valley slope (m/km)	Sin.
Seeley Creek	0.1	1.2	<0.05/0.2	>10	0.85	2.4
Little Sugar R.	2.	2.0	0.1/1	1-2	0.60	1.8
Baraboo River	3.5	3.5	0.2/2	1.	0.46	1.7
Little Dry Cr.	<0.1	1.0	0.3/5	0.2	5.0	1.4
Cedar River	1.	1.3	0.2/1.0	0.0	1.7	2.1
Elkhorn River	4.5	2.	0.3/4	<0.05	0.91	1.6
Vermilion R.	28.	5.	0.8/20	<0.3	0.47	1.7
Little Wind R.	4.	2.0	10/200	0.6	4.6	2.0

Note: In Tables 3-5, $\overline{D}$ and D_{99} are the mean size and the coarsest one percentile of the coarse member; $\overline{h}$ is the average thickness of the vertical sequence, from thalweg to floodplain. Sin. is the average channel sinuosity.

The preceding superficial examination suggests the existence of at least four basic facies styles in meandering streams as evidenced from the literature: the traditional fining-upward model (not documented in the Holocene), the Endrick and Wabash facies models, and the bizarre Rio Puerco.

The streams in Tables 3-4 violate nearly all the common criteria, reviewed in the preceding section, for meandering streams. Some contain virtually no fine member, many contain much gravel, some lack natural levees and chutes entirely, some occupy wide Mississippi-type valleys with many abandoned meander belts while others occupy narrow valleys, and so on down the list of Tables 1 and 2.

LITHOFACIES OF MEANDERING STREAMS SURVEYED IN 1976 AND 1977

In an effort to assess lithofacies variabiliy of Holocene meandering streams, the writer has investigated some additional streams. These streams plus those of Tables 3-4 span a broad spectrum of discharge regime (from ephemeral to virtually constant discharge), grain size of coarse member, and mud content (specifically, the thickness ratio of fine member to coarse member). The newly examined streams meander freely, lack obvious man-made modification, are either at temporary grade or else are aggrading, and are sufficiently small to allow direct examination of the entire vertical sequence in most cases.

Table 5 lists the main features of several streams which have been examined in some detail. Some 30 other freely meandering streams were examined cursorily; they will be cited only to exemplify particular facies types. One-to-three bends in each of five streams

TABLE 6.   GEOMORPHIC AND SEDIMENTARY FEATURES OF 5 MEANDERING STREAMS

Little Sugar River (near Monticello, Wisconsin; 42°44'N 89°31'W)

1. Scroll bars and rock gravel are absent.

2. Vertical sequence consists mainly of levee-like cm- and dm-thick interbeds of sandy silt and fine-medium quartz sand. Basal 0.5-1.0 m of the 2-m thick sequence consists mainly of clean fine sand, likely rippled bedded.

3. Present channel at low flow locally contains abundant mud granules, rotting leaves, and muddy sand. The intraformational granules likely are preserved in part through succeeding high flows.

4. Inner banks of both bends, which are tightly (but typically) curved, are dominantly erosional, in a very confusing fashion; little possibility of epsilon-cross-stratification (ECS). Broader bends contain inner banks with gentler transverse slopes and sediment deposition.

5. Rapid lateral migration of channel despite seemingly resistant banks.

6. Most of the mud in the vertical sequence is deposited within channel and is not true "overbank" mud.

Cedar Creek (15 km NW of Ericson, Nebraska; 41°54'N 98°49½'W)

1. Very low suspended load at low, prevailing flow. Active small-scale ripples predominate at low flow. One point dune (Hickin, 1969) per bend. No evidence of other flow-regime bedforms.

2. Bed material almost exclusively fine-medium quartz sand. Vertical sequence consists of 90% beds of clean sand and 10% beds of sand-silt, often with much plant debris. No consistent vertical trends in grain size and primary sedimentary structures, but surficial bed material displays transitional and fully developed textural zones.in subtle fashion.

3. Good development of transitional and fully developed flow zones at low flow.

4. Possible ECS on upper point-bar surface is defined by thin layer of sandy plant debris.

5. Channel-fill deposits consist largely of fine sand, with thin topstratum of decayed plant material; very little mud in channel fill.

Little Dry Creek (7 km SE of Mountain View, Wyoming; 41°15'N 110°15½'W)

1. Ephemeral: summer floods show flashy discharges, high suspended loads with much mud, and much scour-and-fill of bed. Spring flood due to snow melt.

2. Many chute cutoffs; each bend typically shows multiple chutes in varying stages of development. No chute bars; rare neck cutoffs.

3. Abundant presence and preservation of lower-regime bedforms in dominantly sand bed; upper-regime bedforms absent. Occasional, subdued scroll bars. No evidence of ECS.

4. General, erratic upward fining of vertical sequences. Dune cross-strata common near base and small-scale cross-strata prevail near top. Surficial sediment usually not indicative of preserved lithofacies.

5. Fine member ranges from sandy silt to fine sand with scattered silt beds and does not show cross-stratification; thickness ranges from <0.1 m to >0.8 m--extremely variable along a single bend.

TABLE 6 (CONT.). GEOMORPHIC AND SEDIMENTARY FEATURES OF 5 MEANDERING STREAMS

Vermilion River (3 km WNW of Cayuga, Indiana; 39°58'N 87°29'W)

1. Prominent gravel riffles in crossings are armored with rock clasts 2-5 cm in diameter. Bends show same patterns of bed topography and bed-material size observed in lower Wabash River. Excluding the riffles, lithofacies of these two streams is quite similar. Modest inner accretionary banks.

2. The downstream one-third of the middle bend (of the three mapped) has assumed a braided aspect, at low flow, since 1973.

3. Prominent scroll bars on two bends; several small ones on third bend.

4. In many portions of each bend the bedforms and sediments on point-bar surfaces are, as in the lower Wabash River, thoroughly unrepresentative of depositional conditions during the high, channel-forming discharges.

Little Wind River (3 km E of Ethete, Wyoming; 43°02'N 108°44'W)

1. Prominent gravel riffles in crossings and at irregular intervals in bends produce an erratic cross-sectional channel profile. Riffles armored with equant rock clasts 5-20 cm in diameter.

2. Sudden facies transition from coarse gravel on lower point bar to levee-like silt-sand on upper point bar; latter corresponds to inner accretionary bank of Bluck (1971). Pools contain only slightly finer bed material than do riffles. Sand common only near downstream end of each "point bar", which resembles in morphology and texture the lateral bars of gravel outwash plains (e.g., Bluck, 1976) more closely than the point bars of finer-grained meandering streams.

3. Fair, but erratic, development of transitional and fully developed textural zones in surficial bed material. No bed-material movement at typical low flows.

4. Negligible occurrence of flow-regime bedforms, owing to general low content of sand in channel bed.

5. Fine member comprises 0% to 73% of local thickness of vertical sequence and consists of levee-like alternations of fine sand, silty sand, and silt. Bed material of coarse member is mainly an unstratified pebble-cobble gravel, of sand content 0-50%.

in Table 5 were surveyed thoroughly for bed topography, spatial distributions of surficial sediments and bedforms, and vertical sequence (revealed in cores and trenches). Final results from this study, for which the analysis is not yet completed, will be presented elsewhere.

Main features of five streams

Table 6 summarizes the salient features of these streams. Ancillary information for each stream follows below.

Little Sugar River

Two bends were mapped by plane table on 24-28 August 1976, and cores were procured on 31 August. The modern stream reworks thin lacustrine silt that was deposited in a Wisconsinan proglacial lake. Several abandoned meander belts exist in the modern valley. The vertical sequence of this perennial stream is incompletely known, because cores could not be taken nor trenches excavated below the mean low-water level, roughly at the boundary between the thick fine member and the thin coarse member. However, observations of grain size of surficial channel sediments indicate that the vertical sequence probably resembles the fining-upward cycle much more closely than do the four streams below and those in Tables 3-4.

Cedar Creek

One bend was mapped on 11-15 August 1977. Six other bends and the floodplain were examined briefly during this time. This stream flows in a valley of about ½-km width in the Nebraska Sand Hills, which consist of eolian sand dunes produced during a Pleistocene glacial episode (Ahlbrandt and Fryberger, 1976; Warren, 1976, and references therein). The entire drainage basin resides within the Sand Hills; consequently sediment in the modern stream is almost entirely fine and medium quartz sand. Ground water provides a uniform discharge the year around, and the highly permeable sand of the Sand Hills yields an appreciable surface runoff only during unusually intense rainstorms or rapid melting of a thick snow cover.

Little Dry Creek

One meander bend in this ephemeral, basin-and-range stream was mapped during three days, starting 31 August 1977, when the stream channel was dry. Little Dry Creek occupies a very narrow meander belt on a broad lacustrine (?) plain upon which exist several abandoned meander belts.

Only in the predominance of chute cutoffs over neck cutoffs does this stream resemble Shepherd's (1976) Rio Puerco.

Vermilion River

Three successive bends were mapped by plane table in August 1976. The reach lies within the Pleistocene sluiceway valley of the upper Wabash River. Its climatic regime and range in sediment size are very much like those of the lower Wabash. The main lithofacies difference between the two streams is the presence in the Vermilion of gravel riffles, which is likely due to its slightly coarser grade of gravel. Point-bar lithofacies of the Vermilion also resembles rather closely the channel sediments of the Endrick (Bluck, 1971). The main sedimentary dissimilarities between the Endrick and the Vermilion are the relatively minor development of inner accretionary banks and the possibly greater abundance of flow-regime bedforms in the Vermilion. The latter disparity may be due to a slightly lesser proportion of sand in the Endrick.

In terms of sediment size and bedforms the Vermilion bridges the small "lithofacies gap" between the lower Wabash and the Endrick.

Little Wind River

This stream is one of many graveliferous streams of varying channel pattern that drain the Wind River Mtns. near Lander, Wyoming. Two consecutive regular bends of rather low curvature were mapped on 19-28 August 1977. Several other bends in nearby reaches downstream to the confluence with the Popo Agie River were visited briefly during this period. The climate in the study area is strongly continental and dry (average annual precipitation being on the order of 25 cm), but there is a significant spring flood from snow melt in the mountainous head waters.

The supreme sedimentological attraction of the Little Wind is the very coarse gravel of its coarse member. Few coarser-grained streams of any channel pattern have been described in the literature, and none meander. [Several such gravelly streams are the most proximal reaches of the Donjek (Rust, 1972, Fig. 13B; Rust, 1975, p. 240), Knik (Bradley *et al.,* 1972, p. 1270-1271), and White (Fahnestock, 1963, p. A23-A26) Rivers, and of the Baffin Island sandurs (Church, 1972, Figs. 20, 69, 80)].

In spite of abundant oxbow lakes and meander scars in the single meander belt, migration rates of bends are low (on the order of 1 m/yr) for the size of stream. Several bends further downstream show a puzzling wavy bed topography arising from closely spaced coarse-gravel riffles separated by pools of sandy gravel to sand.

LITHOFACIES CLASSES OF MEANDERING STREAMS

The analysis to this point suggests a need for revision of the standard lithofacies approach to meandering streams. That discussion has revealed the existence of great variability in lithofacies of modern meandering streams. This section first will define, in quite tentative fashion, five classes of lithofacies of meandering streams. Then some limitations and implications of this initial classification will be given. Table 7 lists the diagnostic features of each facies class.

Many of the muddy fine-grained streams of Table 7 occupy broad, flat valleys which were lake bottoms during Pleistocene glacial episodes. Their valley fills are composed mainly of lacustrine silts and clays, with minimal sand and gravel. The modern stream merely reworks this fine material. Examples of muddy fine-grained meandering streams include Little Sugar River (Tables 5-6), Baraboo River (near La Valle and Reedsburg, Wisconsin; approx. 43°35'N 90°05'W), and Seeley Creek (43°26'N 89°55'W) in Wisconsin, and Fox and Little Wabash Rivers (near Mount Erie, Illinois; approx. 38°33'N 88°10'W) and Patoka River (Fidlar, 1948, Fig. 3) of the Wabash River basin. Two examples from Australia are the Barwon (Taylor, 1976; Taylor and Woodyer, this volume) and the Murrumbidgee (Schumm, 1968). The latter stream contains a somewhat coarser channel material (median size 0.40-1.30mm) than do the other examples but otherwise typifies this class of meandering stream. The modern channel of each of the first three streams shows a surprisingly rapid rate of lateral migration, in spite of seemingly resistant bank materials, and many avulsions. The Little Wabash, Patoka, and Murrumbidgee migrate much more slowly. The Barwon shows little or no Holocene lateral migration, despite its high sinuosity.

Streams of class #2 in Table 7 include Little Dry Creek (Tables 5-6), lower Mississippi River between Memphis and Vicksburg (Fisk, 1944), and a local reach of Honey Creek (43°16½'N 89°50'W) immediately downstream from a 10-km-long meandering reach of facies type #1. Streams of class #3 display some of the widest departures from conventional facies models for meandering streams. Holocene examples include Cedar Creek (Tables 5-6) and Elkhorn River near O'Neill, Nebraska (Table 5). All known streams of class #3 display vertical-sequence thicknesses of less than 3 m, banks stabilized by vegetation, and rather uniform discharges.

Facies class #4 has received the most intensive study by sedimentologists. These graveliferous sand-bed streams are common on coastal plains of the Atlantic and Gulf Coasts of the U.S. (Amite, Brazos, Colorado, and U. Congaree Rivers of Tables 3-4; Watts Branch: Wolman and Leopold, 1957, p. 92-96; Leopold, 1973; Appalachicola and Pascagoula Rivers), in the U.S. mid-Continent (lower Wabash and many tributaries; Iowa, Chippewa, and Minnesota Rivers), and elsewhere (Beatton, Oker, and Öre Rivers in Table 4). All the larger and most of the smaller streams in Tables 3-4 are perennial and occur in humid-to-subhumid climates. Some are virtually free of mud (Amite, Colorado, and S. Esk).

The final facies class contains the most startling departures from the standard facies models for meandering streams. The streams of this class exist near or within mountain ranges and flow on comparatively steep slopes. Examples include Little Wind River (Tables 5-6), Popo Agie River near Hudson, Wyoming (Leopold and Wolman, 1957, Figs. 43-44, Apps. E, F, H), and other graveliferous Wyoming streams mentioned in Leopold and Wolman (1957) (e.g., Hams Fork, Little Popo Agie River, and Wind River). Although all Holocene examples come from severely continental climates, there seems to be no genetic influence of climate upon the lithofacies style, and class #5 streams likely occur in other climates.

TABLE 7. 5 PRELIMINARY LITHOFACIES CLASSES OF HOLOCENE MEANDERING STREAMS

#1: Muddy fine-grained streams (lack rock gravel)

1. Channel shows small width-depth ratio, commonly a very high sinuosity, and an erratic cross-sectional profile. Steep point-bar slopes often exceed 20°.

2. Transitional and fully developed textural zones in coarse member usually not recognizable. Upward fining of coarse member common but not ubiquitous.

3. Fine member equals or exceeds coarse member in thickness, but its mud need not be "overbank", even in part. Locally abundant intraformational mud chips in coarse member. Bed material usually is of medium-sand size or finer.

4. ECS possible in point-bar deposits of gently curved bends and likely consists of thin laminae of fine sand and silt. Prominent natural levees and channel-fill mud deposits expected. Scroll bars and chutes absent.

#2: Sand-bed streams with modest thickness of fine member

1. Channel shows variable width-depth ratio, prominent point bars of modest transverse slope, the usual asymmetrical triangular cross section in bends, and substantial scour-and-fill during major floods.

2. Upward fining of coarse member common but not ubiquitous. Generally poor development of transitional and fully developed textural zones.

3. Fine member is thinner than, but comparably thick to, coarse member; its mud is largely "overbank".

4. ECS likely in coarse member and can dip steeply in small streams. Prominent natural levees and channel-fill mud deposits common. Scroll bars and chutes can be both common and prominent.

5. Rates of channel migration uniformly large; both chute and neck cutoffs espected.

6. Chutes, chute cutoffs, and major scouring surfaces (which truncate any ECS) predominate in ephemeral streams.

#3: Sand-bed streams lacking mud and rock gravel

1. Channel displays larger width-depth ratio than muddier meandering streams of comparable channel size and the normal asymmetrical triangular cross section.

2. Good development of transitional and fully developed flow and textural zones in each bend. Textural zonation becomes subtle in streams with a narrow size range of sand. Coarse member can contain all sand sizes or only a narrow size range, depending upon provenance.

3. Natural levees, channel-fill mud deposits, and inner accretionary banks are minor to absent entirely. No riffles (in the textural sense). Scroll bars can be common.

TABLE 7 (CONTINUED): 5 PRELIMINARY CLASSES
OF LITHOFACIES OF HOLOCENE MEANDERING STREAMS

#4: Graveliferous sand-bed streams

1. Deposits usually contain comparatively little mud, especially "overbank"
   mud. Some streams (e.g., S. Esk) are virtually mud free. Predominance
   of coarse sands and fine gravel in coarse member.

2. Prominent development and preservation of transitional and fully
   developed textural and flow zones for each bend, in much the same manner
   of the lower Wabash River (Jackson, 1975a, 1976a). Textural zonation
   enhanced by the large size range in coarse member and the regular bed
   topography around bends. Prominent point bars.

3. Riffles (in the textural sense), ECS in coarse member, natural levees,
   and major scouring surfaces are rare to absent entirely. Scroll bars and
   channel-fill mud deposits are common, but latter comprise a comparatively
   minor volume of alluvial fill.

4. Levee-like, accretionary deposits of inner bank can be prominent
   (local thickness as much as 1/3 of total vertical sequence) in upstream
   portion of each bend and can show ECS.

#5: Streams with coarse gravel and little sand

1. Substantial channel sinuosity, commonly about 1.8. Bends tend not to be
   tightly curved but can be very long. Highly erratic bed topography arises
   from presence of prominent gravel riffles (with armoring) in crossings
   and in bends.

2. Bed-material transport limited to high flows, in contrast to finer-grained
   streams. Suspended loads can be very high in muddy streams susceptible
   to flash floods.

3. Fair development of transitional and fully developed textural zones often
   is obscured by spectacular pool-and-riffle sequences.

4. Fine member is highly variable in thickness (along a bend and from river
   to river in a small region) and consists largely of levee-like
   alternations of sand and silt. This levee-like deposit resembles well
   in morphology and structure the inner accretionary bank of Bluck (1971).
   Little true "overbank" mud.

5. Flow-regime bedforms, ECS in coarse member, major scouring surfaces, and
   scroll bars are rare to absent entirely, largely owing to paucity of sand.

On the basis of the results developed thus far in this paper, I see no merit to the
suggestion (McGowen and Garner, 1970, Table 1; Brown, 1973, p. 13-15; Fig. 1 herein)
that ''coarse-grained'' meandering streams are transitional in any way between ''braided''
streams and ''fine-grained'' meandering streams.

Limitations of the facies classification

The reader must appreciate the incomplete status of the preceding analysis of
lithofacies. With the possible exception of graveliferous sand-bed streams the facies
classes of Table 7 are poorly documented. The first two classes received only cursory
attention in this study, which seems to be the first published survey of classes 1, 3, and 5.
A thorough description of channel and overbank sediments, bend migration, flow
conditions, and lithofacies preservation in meandering streams of each class must be
obtained in order to appraise authoritatively the suitability of this initial categorization.

Another reservation concerns the fact that several published studies have revealed meandering streams whose lithofacies conform poorly to any of the five types in Table 7. For instance, the Vermilion River (Tables 5-6) and the River Endrick (Bluck, 1971) fit awkwardly between classes 4 and 5. Each stream contains all vital aspects of class #4 except for the lack of gravel riffles and differs from class #5 mainly in the presence of flow-regime bedforms and substantial sand. The Rio Puerco (Shepherd, 1976) is distinctly unlike any of the five facies classes. Class #2 is largely an artifact to include the lower Mississippi River and similar streams whose facies seem intermediate between classes 1 and 3. It appears both likely and reasonable to suppose a complete continuum exists between each adjacent pair of the five classes. The five facies types are specific points in this continuum of hydrological regime, sediment size, and mud content, but by no means cover all the possible main styles of lithofacies. To illustrate this latter point, ephemeral streams in the total body of investigation of meandering-stream facies are represented by the Rio Puerco and Little Dry Creek, each of which offers spectacularly distinctive sedimentary features.

A meandering stream can show a downstream gradation among these facies types just as some streams display a downstream transition in channel pattern. Two prime examples of downstream gradation in facies type come from the Colorado and Brazos Rivers of Texas. The Colorado is a graveliferous meandering stream near Austin (Baker and Penteado-Orellana, 1977), becomes a Wabash-like graveliferous sand-bed stream near Columbus and Garwood (McGowen and Garner, 1970; author's unpublished observations), and finally develops into a muddy sand-bed stream on the plain of its delta (Kanes, 1970, p. 81-84, Fig. 24). The Brazos varies in like manner from Calvert (Folk and Ward, 1957) to San Felipe and Richmond (Bernard et al., 1970; author's unpublished observations) to its deltaic plain (A. W. Walton, written comm.). This concept of gradation is compatible with standard notions of the downstream evolution of major streams (e.g., Brown, 1973, Figs. 8-9). However, in these two cases (and in many others among the streams cited in this paper) the stream remained freely meandering throughout the transitions in sediment size and valley slope along its course.

Implications of the facies classification

Facies classes 3-5, and to a large extent #2, incorporate sedimentary features usually deemed typical of "braided" streams (cf. Tables 1-2): thin-to-absent fine member, abundance of gravel, presence of coarse gravel, vertical sequences which do not always fine upwards and which show great spatial variability, absence of epsilon-cross-stratification and natural levees, large width-depth ratio of channel, and lack of flow-regime bedforms and sedimentary structures. The great majority of ancient fluvial sediments of inferred meandering-stream origin resemble most closely facies type #1, whereas ancient deposits resembling types 3-5 ordinarily are interpreted routinely as "braided-stream" deposits. The marked similarity of facies classes 4-5 to alleged sedimentary features of "braided" streams suggests that many ancient graveliferous deposits have been interpreted mistakenly to be of non-meandering origin.

The astonishingly large variability in lithofacies of Holocene meandering stream points to the desirability of a different approach to environmental analyses of ancient fluvial sediments. Instead of reconstruction of paleochannel pattern, the emphasis would seem to lie more profitably in estimates of paleodepositional conditions based upon grain sizes, sedimentary structures, and geometry of the deposit. This recommendation is strengthened by the earlier conclusions on the unreliability of standard hydro-morphological relations and on the complexity of channel pattern in natural streams. streams.

The final implication is perhaps self evident by now. If the relatively simple physical framework of the meandering alluvial stream (unidirectional flow, often quasi-steady, in a single explicit channel) displays such diversity of lithofacies, then one can expect a commensurate, if not greater, lithofacies variability in the geometrically more complex divided streams. The sad state of present affairs, well illustrated in Miall (1977), is that vertical sequences of deposits of Holocene non-meandering streams remain very poorly known. Until such studies are completed, considerable skepticism on the use of broad criteria to distinguish meandering streams from non-meandering streams would seem to be in order.

ROLE OF EPSILON-CROSS-STRATIFICATION (ECS)
AS A FACIES CRITERION

The original definition of ECS came from Allen (1963b), who later (1965c, p. 155) suggested that it was produced by lateral accretion of a point bar. Numerous authors (Table 8) subsequently have reported ECS in ancient fluvial sediments. There has been an increasing tendency to assume that the presence of ECS indicates deposition in a meandering, point-bar stream (e.g., Cotter, 1971, p. 131; Leeder, 1974, p. 138; Elliott, 1976, p. 70), following Allen's suggestion. The unanimity of depositional environment in Table 8 arises largely from the authors' assumptions that the presence of ECS and fining-upward cycles denote a meandering-stream deposit. Even the absence of ECS from a fluvial sequence has been proposed on occasion to indicate deposition in a non-meandering stream. Several observations from modern and ancient fluvial environments cast doubt on the general validity of these two inferences concerning ECS, particularly the latter.

All 18 published documentations of ECS in ancient fluvial deposits (Table 8) revealed ECS in beds of maximum mean grain size less than 1 mm, excluding local conglomerates of intraformational mud chips. Thickness ratios of mudstone to ECS-sandstone were, where mentioned, commonly not less than 1. The few deposits with much lower ratios exhibited extensive multistorying. Most of these ECS-bearing deposits thus seem very similar to facies class #1 in Table 7. Many examples of ECS consisted of steeply dipping interbeds of fine sandstone and mudstone, reminiscent of the vertical sequence observed in Little Sugar River (Table 6) and expected in facies class #1. The ECS in beds of maximum grain size less than coarse sand invariably was steeply dipping, typically 10-20°. Similar transverse slopes of inner banks of bends in muddy fine-grained streams are common (e.g., Baraboo and Little Sugar Rivers), whereas streams of lower mud content — and higher width-depth ratios (Schumm, 1963) — ordinarily present lower point-bar slopes (see also Allen, 1970, p. 312). Allen's point that very low point-bar slopes in large streams will render difficult the recognition of ECS in deposits from such streams is likely valid for facies classes 2-5 in Table 7, but less appropriate for the muddy fine-grained streams.

This paper has indicated the probable absence of ECS in many modern meandering streams. Unpublished reports of EXS exist from the Rio Puerco (Shepherd, 1976) and from the Brazos River (A. W. Walton, written comm.). In the ECS-containing reaches of both streams, the bed material is finer than coarse sand, and the bed and banks contain substantial amounts of mud. The numerous reports of ECS from meandering channels in intertidal mudflats (e.g., Bridges and Leeder, 1976, p. 533, 548-550) pertain to a muddy environment once again of considerable lithological similarity to facies class #1. They do not answer satisfactorily the question of occurrence of ECS in other meandering streams.

The author's examination of Holocene meandering streams in the field and in the literature has failed to turn up a single genuine example of ECS beyond the two unpublished reports cited above. Admittedly one can find in modern streams an occasional fine-grained bed or local bedding surface parallel to the point-bar surface (e.g., Cedar Creek; Leopold and Wolman, 1960, their Fig. 6), but a description of true ECS appears yet to be achieved. The earlier remarks on facies classes make clear the low likelihood of encountering ECS in graveliferous streams (facies classes 4-5). Not only do their stepped cross-channel profiles from floodplain to thalweg (e.g., Harms *et al.*, 1963,

TABLE 8.   PUBLISHED REPORTS OF EPSILON-CROSS-STRATIFICATION (ECS) IN ANCIENT FLUVIAL DEPOSITS

Reference	Deposit	Bed-material size of ECS unit	Number of ECS units	Thickness of ECS (m)	Fine member/coarse member	Depositional environment of ECS unit (proposed by Ref.)
Allen (1965c, p. 151-154)	Old Red Sandstone of Anglesey	basal intraform. gravel to silt and sand at top	10	1.1-2.7	about 1	point bar in meandering stream
Moody-Stuart (1966)	Wood Bay Series (Devonian)	ditto above; mean sizes <0.5 mm	51	0.5-6.3	<0.2 *	high-sinuosity stream
Beutner et al. (1967)	Kittaning Fm. (Pennsylvanian)	medium sand with 20% matrix	1	≤12	>1	inside of meander bend
Allen and Friend (1968, p. 44-47)	Catskill Formation (Devonian)	basal coarse-medium sand to silt and fine sand in upper 2/3 †	1	7	about 1	point bar in meandering stream
Read (1969, p. 339-340)	Namurian of central Scotland	very fine to medium sand	2 [5]	2-3	0.5-1	meandering stream
Cotter (1971)	Ferron Sandstone (U. Cretaceous)	fine sand	"large number"	about 8	not given (likely ~1)	point bar in highly sinuous stream
Glowacz and Horne (1971)	Catahoula and Fleming Fms.) (Early Miocene)	silty sand	2	about 5	indeterminate	sinuous channel on upper deltaic plain
Leeder (1973b, p. 123-124)	Upper Old Red Sandstone	fine and medium sand	3	0.7-1.1	<0.5	point bar in meandering stream
Molenaar (1973, p. 90)	Torrivio Ss. Mr. (U. Cretaceous)	fine and medium sand	1	about 5	indeterminate	point bar in distributary channel
Puigdefàbregas (1973)	Miocene fluvial point-bar deposits	sand, grains <1 mm	>1	1-2	about 10	point bar in meandering stream #
Donaldson (1974, p. 59-62)	Lower Mahoning Ss. (Pennsylvanian)	sand	1	about 10	not given	point bar in meandering stream
Horne et al. (1974, p. 106-107) (Carboniferous)	Breathitt Fm.	basal coarse sand to fine sand at top	?	<15	likely >1	point bar in meandering stream on deltaic plain
Leeder (1974, p. 138)	Brewcastle Fm. (Tournaisian)	basal fine sand to silt at top	2	about 1	near 0.0	point bar of meandering stream

TABLE 8 (CONTINUED).   PUBLISHED REPORTS OF EPSILON-CROSS-STRATIFICATION (ECS) IN ANCIENT FLUVIAL DEPOSITS

Elliott (1976)	Yoredale cyclothem (Carboniferous)	sand (probably fine or medium)	1	7.5	indeterminate	point bar in meandering distributary channel
Nami (1976); Nami and Leeder (this vol.)	Scalby Formation (Middle Jurassic)	basal coarse-medium sand to silty clay at top	"common"	3-4	likely about 1	point bar in meandering alluvial stream #
Padgett and Ehrlich (1976, their Fig. 3)	U. Carboniferous of southern Morocco	basal pebbly sand to silty sand at top; 5% mud content	? **	5 **	likely <1	point bar in meandering channel #
Peterson (1976, p. 8 and pl. 4)	Colton Formation (Lower Eocene)	basal fine sand to silt at top	2	2	likely about 1	point bar in meandering channel
Mossop (this vol.)	McMurray Fm. (L. Cretaceous)	thick beds of fine sand with thin partings of silt	>1	up to 25	not given	inner bank or point bar of deep sinuous channel

The following papers describe cross-stratification interpreted to represent lateral-accretion surfaces but which is more likely attributable to the cross-channel migration of small bars.

Williams (1966)	Torridonian Ss. (Precambrian)	coarse sand to pebble gravel	18	0.20-2.50	likely near 0.0	likely, point bar in braided stream
Picard and High (1973, Fig. 133)	Duchesne River Fm. (Eocene-Oligo.?)	fine and medium sand, likely	1	0.2	not given	point bar in meandering stream

* Friend and Moody-Stuart (1972, their Fig. 14)

† ECS prominent only in upper 2/3 of deposit.

§ Not true ECS, according to Read, but rather the beta-cross-stratification of Allen (1963b).

Exhumed meander belt, consisting of accretionary ridge-and-swale topography with high planwise curvature of ridges, is present on ground surface.

** Despite sketch of prominent ECS in their Fig. 3B, there is no mention of ECS elsewhere in their paper.

p. 568; McGowen and Garner, 1970, Figs. 3, 4, 13; Bluck, 1971, p. 98-99, Figs. 21-24; Jackson, 1975b, Fig. 4.2, in conjunction with Jackson, 1976c, Figs. 7-8) make a smooth sigmoidal ECS improbable, but the (previously discussed) differential activity of the channel bed with changes in stream discharge and position along the bend also is incompatible with the formation of ECS. Furthermore, these streams invariably display stable bed topographies through major floods, which means that any major scour-and-fill of the bed (one apparent means to generate ECS) is associated only with the erratic migration of large flow-regime bedforms (e.g., Jackson, 1976c, Figs. 7-8).

The final reservation about ECS concerns non-uniqueness. Granted the reasonable concept of ECS as a product of lateral accretion, one would expect ECS in *any* stream showing surfaces of lateral accretion. The large lateral shifts in various sorts of "braid" bars observed in many decidedly non-meandering streams (e.g., Coleman, 1969, p. 177-183; Allen, 1970, p. 313; Bluck, 1976, p. 449-451) suggest the possibility of ECS in non-meandering streams. Because some such bars can rise nearly to the elevation of the floodplain, ECS generated this way could be as thick as the coarse member — as ordinarily assumed in meandering streams.

Some Holocene meandering streams present a lateral-accretion analogous to ECS. The inner accretionary banks of the River Endrick contain a steeply dipping stratification defined by alternations of thin mud and sand beds (Bluck, 1971, p. 107, 128-132). This cross-stratification appears in the lower portion of the fine member, and its thickness often is measurably different from the thickness of the underlying coarse member and markedly less than the thickness of the point-bar complex. Sundborg (1956, Fig. 47, p. 276) described the same bedding feature from a terrace deposit of the meandering River Klarälven. He concluded that it defined former surfaces of the upper point bar, which is equivalent to the inner accretionary bank of Bluck. Inner accretionary banks, where well developed, in the Vermilion and Little Wind Rivers (Tables 5-6) display an accretionary stratification which at least superficially is identical to that in the Endrick. In October 1977 I observed an homologous inclined bedding in deposits of the inner accretionary bank of the lower Wabash River (Jackson, 1975b, p. 237-239, described this subenvironment).

This sort of accretionary bedding is not present in all meander bends. Some mud-free meandering streams (e.g., Cedar Creek) lack inner accretionary banks entirely, although others contain modest ones (e.g., S. Esk: Bridge and Jarvis, 1976, Figs. 3, 26). Typical, sharply curved bends in many muddy fine-grained streams (e.g., Baraboo and Little Sugar Rivers) lack an inner accretionary bank. Thus this lateral-accretion feature may be widespread in meandering streams but is not ubiquitous. Puigdefábregas and van Vliet (this volume) describe a possible ancient homolog from Eocone point-bar deposits of the Southern Pyrenees.

These remarks suggest considerable caution in the use of either the presence or the absence of ECS in fluvial deposits as an indicator of paleochannel pattern.

SUMMARY OF LITHOFACIES CRITERIA

The preceding sections demonstrated that virtually all the criteria of Tables 1-2 suffer major shortcomings which make them undependable indicators of channel pattern. Only the first and last criteria in Table 1 and the next-to-last criterion in Table 2 appear to be both meaningful and broadly valid. The criteria in Table VI of Miall (1977) also seem to be quite unreliable in general.

The few facies criteria which *may* be broadly reliable are the following. Low paleocurrent variance of imbrication of large pebbles and cobbles is expected in low-sinuosity streams (Bluck, 1974; Rust, 1975), which may be divided or undivided. Exceedingly coarse gravel — say, of mean length of intermediate axes greater than 0.5 m

— likely is diagnostic of a very proximal non-meandering stream. Substantial mud content in coarse member, thick fine member (cf. facies class #1), and asymmetric channel-fills with much mud (one of the oldest criteria) are strongly indicative of a meandering stream. Exhumed meander belts with many intersecting sets of highly curved accretionary topography point to a meandering-stream origin. In the strictest sense, of course, the latter four criteria imply only a highly sinuous channel with lateral migration. The channel pattern could be either meandering or anastomosing; however, the former seems to be much the more common in nature today.

PRESERVATION OF LITHOFACIES

The section on geological perspective introduced preservation potential as the key methodological gap between actualistic facies models and environmental interpretation of fluvial rock units. The largely unexplored realm of the mode and potential of preservation of the sedimentary features described from Holocene streams limits the analysis of this topic to broad generalizations.

Several factors appear at first glance to be pertinent to the incorporation of a fluvial lithofacies into a stacked sequence of individual deposits: (1) rate and duration of basin subsidence, (2) rates and magnitudes of sea-level changes irrespective of subsidence, (3) proximity to non-fluvial environments, (4) short-term channel behavior (e.g., stability of channel pattern and avulsion), (5) number and width of meander belts (for meandering streams), and (6) tectonic activity during sediment accumulation. These factors, many of which are mutually interdependent, remain to be quantified or even defined rigorously, although Leeder (this volume) presents an initial attempt at modeling stochastically factors 1 and 4.

The Wabash depositional model (Jackson, 1976a) can illustrate the conceivable variety of stacking scenarios. The artificial sequence at the left in Figure 3 would be expected during moderate subsidence with the river valley remaining well above sea level. This sequence exemplifies the type of mud-free coarse-grained deposit that, in the rock record, routinely is interpreted to be of ''braided-stream'' origin. The thin fine member of each individual deposit could well be removed entirely by erosion during the deposition of overlying members. The scene at the right in Figure 3 postulates close proximity to sea level, much as in the modern lower Mississippi River. This scenario incorporates the behavior of the river during a Pleistocene-type glaciation. The valley fill of the lower Wabash River near Grayville resembles the valley fill of the lower Mississippi (Fisk, 1944, 1947) in consisting of a basal graveliferous sand underlying the finer-grained muddier deposits of the Holocene meandering phase (Fidlar, 1948, p. 20-23, 96-100). This correspondence supports the main aspects of the scene envisioned in Figure 3.

In view of the discussions in preceding sections the drastic oversimplification of Figure 3 scarcely bears mention. In effect, only factor 3 above was varied. The large lateral variability in lithofacies of the point-bar deposits (Jackson, 1976a, Figs. 13-16) and of the entire valley fill (cf. Fig. 1) was ignored.

Perhaps the major unknown element in preservation of Holocene meandering-stream deposits is a comprehension of the possible occurrence of particular facies styles in settings which are unusually conducive to ultimate preservation. This issue involves the preservation factors 1-3 and 6 above. My analysis to date suggests that the coastal plains of the U.S. Gulf Coast, a province that likely enjoys a high long-term preservation potential for fluvial sediments, contains modern meandering streams of all facies types in Table 7. Relative proportions of each facies class are not known, but most large meandering streams (mean annual discharge greater than $50 m^3/s$) appear to be of class #2 or #4. Other sites of significant preservation potential remain to be considered.

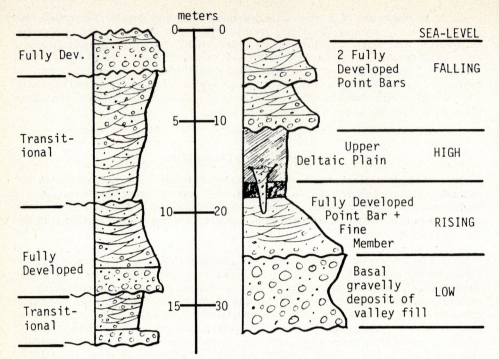

Fig. 3. Two hypothetical examples of stacking sequence of deposits of the modern lower Wabash River. See text for discussion.

MAJOR RESEARCH NEEDS

This report has described the many gaps in our understanding of fluvial lithofacies. There is critical need for additional documentation of the three-dimensional lithofacies of meandering streams, and especially non-meandering streams. This assessment contains an implicit pessimistic note of the extreme variability in lithofacies already encountered in meandering streams. It appears certain, therefore, that there is little hope of a single diagnostic facies model, or even a small suite of reliable general models, for meandering streams. The substantial overlap of facies of graveliferous and mud-free meandering streams with facies of non-meandering streams argues further against the hope of universally reliable facies criteria for meandering streams.

There is an urgent need for methods of estimating paleohydraulic and paleogeomorphic conditions in ancient fluvial systems. Approaches are needed which do not rely upon the hydromorphological relations employed widely today. The possible usefulness of grain sizes, sedimentary structures, and channel geometry in paleohydraulic analysis has been demonstrated tentatively (Nummedal, 1973; Bridge, 1976, 1977, this volume); but the complexity of process and response in modern meandering streams does not encourage prospects for precise methods.

Future studies of fluvial facies should perhaps strive less towards the erection of facies models but should consider more thoroughly than has been the wont so far the style and potential of preservation of characteristic sedimentary features. The pioneering works of Fisk, LeBlanc, Bernard, and their colleagues indicate the fruitful results that may be forthcoming from a renewed analysis of preservation.

ACKNOWLEDGMENTS

Research Grant EAR 76-10748 from the National Science Foundation helped support research during 1976 and 1977. Field research during summer 1976 was funded largely by a faculty research grant from Northwestern University. J. L. Carr assisted in the field during summer 1976. D. P. Eberhart and M. A. Lynch helped reduce field data into draftable form. The manuscript benefitted from critical reviews by J. S. Bridge, G. E. Gerner-Adams, R. T. Getzen, R. A. Levey, and A. D. Miall.

REFERENCES

Ahlbrandt, T. S., and Fryberger, S. G., 1976, Structures and textures of eolian deposits in the Nebraska Sand Hills, U.S.A.: 25th Internat. Geol. Congr., Abstracts, v. 3, p. 829.

Allen, J. R. L., 1963a, Henry Clifton Sorby and the sedimentary structures of sands and sandstones in relation to flow conditions: Geol. Mijnbouw, v. 42, p. 223-228.

——, 1963b, The classification of cross-stratified units, with notes on their origin: Sedimentology, v. 2, p. 93-114.

——, 1964, Studies in fluviatile sedimentation: Six cyclothems from the Lower Old Red Sandstone, Anglo-Welsh Basin: Sedimentology, v. 3, p. 163-198.

——, 1965a, A review of the origin and characteristics of Recent alluvial sediments: Sedimentology, v. 5, p. 89-191.

——, 1965b, Late Quaternary Niger delta and adjacent areas: Sedimentary environments and lithofacies: Am. Assoc. Petrol. Geol. Bull., v. 49, p. 547-600.

——, 1965c, The sedimentation and palaeogeography of the Old Red Sandstone of Anglesey, North Wales: Yorks. Geol. Soc. Proc., v. 35, p. 139-185.

——, 1966, On bedforms and paleocurrents: Sedimentology, v. 6, p. 153-190.

——, 1967, Notes on some fundamentals of palaeocurrent analysis, with reference to preservation potential and sources of variance: Sedimentology, v. 9, p. 75-88.

——, 1970, Studies in fluviatile sedimentation: A comparison of fining-upwards cyclothems, with special reference to coarse-member composition and interpretation: J. Sediment. Petrol., v. 40, p. 298-323.

——, and Friend, P. F., 1968, Deposition of the Catskill facies, Appalachian region: With notes on some other Old Red Sandstone Basins: *in* G. deV. Klein, *ed.*, Late Paleozoic and Mesozoic continental sedimentation, northeastern North America; Geol. Soc. Am. Spec. Paper 106, p. 21-74.

Axelsson, V., 1967, The Laitaure delta: A study of deltaic morphology and processes: Geogr.Ann., v. 49A, p. 1-127.

Baker, V. R., 1977, Stream-channel response to floods, with examples from central Texas: Geol. Soc. Am. Bull., v. 88, p. 1057-1071.

——, and Penteado-Orellana, M. M., 1977, Adjustment to Quaternary climatic change by the Colorado River in central Texas: J. Geol., v. 85, p. 395-422.

Banks, N. L., and Collinson, J. D., 1974, Discussion of "Some sedimentological aspects of planar cross-stratification in a sandy braided river": J. Sediment. Petrol., v. 44, p. 265-267.

Bernard, H. A., and Major, C. F., 1963, Recent meander belt of the Brazos River: An alluvial "sand" model (abs.): Am. Assoc. Petrol. Geol. Bull., v. 47, p. 350.

——, Major, C. F. Jr., Parrot, B. S., and LeBlanc, R. J. Sr., 1970, Recent sediments of southeast Texas: Bur. Econ. Geology, Univ. Texas, Guidebook 11.

Bersier, A., 1958, Séquences détritiques et divagations fluviales: Eclog. Geol. Helvetiae, v. 51, p. 854-893.

Beutner, E. C., Flueckinger, L. A., and Gard, T. M., 1967, Bedding geometry in a Pennsylvanian channel sandstone: Geol. Soc. Am. Bull., v. 78, p. 911-916.

Blodgett, R. H., 1974, Comparison of Oligocene and modern braided stream sedimentation on the High Plains: Unpub. M.S. thesis, Lincoln, Nebraska Univ., 58p.

Bluck, B. J., 1971, Sedimentation in the meandering River Endrick: Scott. J. Geol., v. 7, p. 93-138.

————, 1974, Structure and directional properties of some valley sandur deposits in southern Iceland: Sedimentology, v. 21, p. 533-554.

————, 1976, Sedimentation in some Scottish rivers of low sinuosity: Royal Soc. Edinburgh Trans., v. 69, n. 18, p. 425-456.

Boothroyd, J. C., and Ashley, G. M., 1975, Processes, bar morphology, and sedimentary structures on braided outwash fans, northeastern Gulf of Alaska: in A. V. Jopling and B. C. McDonald, eds., Glaciofluvial and glaciolacustrine sedimentation; Soc. Econ. Paleont. Mineral. Spec. Pub. 23, p. 193-222.

Bradley, W. C., Fahnestock, R. K., and Rowehamp, E. T., 1972, Coarse sediment transport by flood flows on Knik River, Alaska: Geol. Soc. Am. Bull., v. 83, p. 1261-1284.

Brice, J. C., 1964, Channel patterns and terraces of the Loup Rivers in Nebraska: U.S. Geol. Survey Prof. Paper 422-D, 41p.

————, 1974, Evolution of meander loops: Geol. Soc. Am. Bull., v. 85, p. 581-586.

Bridge, J. S., 1975, Computer simulation of sedimentation in meandering streams: Sedimentology, v. 22, p. 3-43.

————, 1976, Mathematical model and FORTRAN IV program to predict flow, bed topography and grain size in open-channel bends: Computers and Geosciences, v. 2, p. 407-416.

————, 1977, Flow, bed topography, grain size and sedimentary structure in open-channel bends: A three-dimensional model: Earth Surf. Proc., v. 2, p. 401-416.

————, and Jarvis, J., 1976, Flow and sedimentary processes in the meandering River South Esk, Glen Cova, Scotland: Earth Surf. Proc., v. 1, p. 303-336.

Bridges, P. H., and Leeder, M. R., 1976, Sedimentary model for intertidal mudflat channels with examples from the Solway Firth, Scotland: Sedimentology, v. 23, p. 533-552.

Brown, L. F. Jr., 1973, Cratonic basins: Terrigenous clastic models: in L. F. Brown Jr., A. W. Cleaves III, and A. W. Erxleben, Pennsylvanian depositional systems in north-central Texas: A guide for interpreting terrigenous clastic facies in a cratonic basin: Bur. Econ. Geology, Univ. Texas, Guidebook n. 14, p. 10-30.

Campbell, C. V., 1976, Reservoir geometry of a fluvial sheet sandstone: Am. Assoc. Petroleum Geologists Bull., v. 60, p. 1009-1020.

Chien, N., 1961, The braided stream of the Lower Yellow River: Scientia Sinica, v. 10, p. 734-754.

Church, M., 1972, Baffin Island sandurs: A study of arctic fluvial processes: Geol. Surv. Can. Bull. 216, 208p.

————, and Gilbert, R., 1975, Proglacial fluvial and lacustrine environments: in A. V. Jopling and B. C. McDonald, eds., Glaciofluvial and glaciolacustrine sedimentation; Soc. Econ. Paleont. Mineral. Spec. Pub. 23, p. 22-100.

Coleman, J. M., 1969, Brahmaputra River: Channel processes and sedimentation: Sediment. Geol., v. 3, p. 129-239.

Collinson, J. D., 1970, Bedforms of the Tana River, Norway: Geogr. Ann., v. 52A, p. 31-56.

Cotter, E., 1971, Sedimentary structures and the interpretation of paleoflow characteristics of the Ferron Sandstone (Upper Cretaceous), Utah: J. Sediment. Petrol., v. 41, p. 129-138.

Donaldson, A. C., 1974, Pennsylvanian sedimentation of central Appalachians: Geol. Soc. Am. Spec. Paper 148, p. 47-78.

Dott, R. H. Jr., 1973, Paleocurrent analysis of trough cross stratification: J. Sediment. Petrol., v. 43, p. 779-783.

Elliott, T., 1976, The morphology, magnitude and regime of a Carboniferous fluvial-distributary channel: J. Sediment. Petrol., v. 46, p. 70-76.

Erkek, C., 1973, Investigation on the transport of sediment in plain rivers: Proc. 15th Congress Int. Assoc. Hydraulic Research, v. 1, p. 411-417.

Fahnestock, R. K., 1963, Morphology and hydrology of a glacial stream — White River, Mount Rainier, Washington: U.S. Geol. Survey Prof. Paper 422A, 70p.

————, and Bradley, W. C., 1973, Knik and Matanuska Rivers, Alaska: A contrast in braiding: in M. Morisawa, ed., Fluvial geomorphology; SUNY-Binghamton, Publs. in Geomorphology, p. 221-250.

Fidlar, M. M., 1948, Physiography of the Lower Wabash Valley: Indiana Dept. Conservation, Div. Geology, Bull. 2, 112p.

Fisk, H. N., 1944, Geological investigation of the alluvial valley of the lower Mississippi River: Mississippi River Comm., Vicksburg, Miss., 78p.

——, 1947, Fine-grained alluvial deposits and their effects on Mississippi River activity: Mississippi River Comm., Vicksburg, Miss., 82p.

Folk, R. L., and Ward, W. C., 1957, Brazos River bar: A study in the significance of grain size parameters: J. Sediment. Petrol., v. 27, p. 3-26.

Friend, P. F., and Moody-Stuart, M., 1972, Sedimentation of the Wood Bay Formation (Devonian) of Spitsbergen: Regional analysis of a late orogenic basin: Norsk Polarinstitutt Skrifter n. 157, 77p.

Gardner, J. S., 1977, Some geomorphic effects of a catastrophic flood on the Grant River, Ontario: Can. J. Earth Sci., v. 14, p. 2294-2300.

Glowacz, M. E., and Horne, J. C., 1971, Depositional environments of the Early Miocene as exposed in the Cane River Diversion Canal, Rapides Parish, Louisiana: Gulf Coast Assoc. Geol. Soc. Trans., v. 21, p. 379-386.

Gole, C. V., and Chitale, S. V., 1966, Inland delta building activity of Kosi River: J. Hydraulics Div., Am. Soc. Civil Eng., v. 92, n. HY2, p. 111-126.

Gustavson, T. C., 1978, Bed forms and stratification types of modern gravel meander lobes, Nueces River, Texas: Sedimentology, v. 25 (in Press).

Harms, J. C., MacKenzie, D.B., and McCubbin, D. G., 1963, Stratification in modern sands of the Red River, Louisiana: J. Geol., v. 71, p. 566-580.

Hickin, E. J., 1969, A newly-identified process of point bar formation in natural streams: Am. J. Sci., v. 267, p. 999-1010.

——, and Nanson, G. C., 1975, The character of channel migration on the Beatton River, northeast British Columbia, Canada: Geol. Soc. Am. Bull., v. 86, p. 487-494.

Horne, J. C., Ferm, J. C., and Swinchatt, J. P., 1974, Depositional model for the Mississippian-Pennsylvanian boundary in northeastern Kentucky: Geol. Soc. Am. Spec. Paper 148, p. 97-114.

Jackson, R. G. II, 1975a, Velocity — bed-form — texture patterns of meander bends in the lower Wabash River of Illinois and Indiana: Geol. Soc. Am. Bull., v. 86, p. 1511-1522.

——, 1975b, A depositional model of point bars in the lower Wabash River: Unpub. Ph.D. thesis, Univ. Illinois, Urbana, 269p.

——, 1976a, Depositional model of point bars in the lower Wabash River: J. Sediment. Petrol., v. 46, p. 579-594.

——, 1976b, Largescale ripples of the lower Wabash River: Sedimentology, v. 23, p. 593-623.

——, 1976c, Unsteady-flow distributions of hydraulic and sedimentologic parameters across meander bends of the lower Wabash River, Illinois-Indiana USA: Internat. Symp. Unsteady Flow Open Channels Proc., Univ. Newcastle-upon-Tyne, Brit. Hydromechanics Res. Assoc., Paper G4, 14p.

Jahns, R. H., 1947, Geologic features of the Connecticut valley, Massachusetts as related to recent floods: U.S. Geol. Survey Water-Supply Paper 996, 158p.

Jopling, A. V., and McDonald, B. C., eds., 1975, Glaciofluvial and glaciolacustrine sedimentation: Soc. Econ. Paleont. Mineral. Spec. Pub. 23, 320p.

Kanes, W. H., 1970, Facies and development of the Colorado River delta: in J. P. Morgan, ed., Deltaic sedimenation, modern and ancient; Soc. Econ. Paleont. Mineral. Spec. Pub. 15, p. 78-106.

Kellerhals, R., Church, M., and Bray, D. I., 1976, Classification and analysis of river processes: J. Hydraulics Div., Am. Soc. Civil Eng., v. 102, p. 813-829.

Kessler, L. G. II, 1971, Characteristics of the braided stream depositional environment with examples from the South Canadian River, Texas: Wyoming Geol. Assoc. Earth Sci. Bull., v. 4, p. 25-35.

Krigström, A., 1962, Geomorphological studies of sandur plains and their braided rivers in Iceland: Geogr. Ann., v. 44A, p. 328-346.

LeBlanc, R. J. Sr., 1972, Geometry of sandstone reservoir bodies: Am. Assoc. Petrol. Geol. Memoir 18, p. 133-189.

Leeder, M. R., 1973a, Fluviatile fining-upwards cycles and the magnitude of palaeochannels: Geol. Mag., v. 110, p. 265-276.

————, 1973b, Sedimentology and palaeogeography of the Upper Old Red Sandstone in the Scottish Border Basin: Scott. J. Geol., v. 9, p. 117-144.

————, 1974, Lower Border Group (Tournaisian) fluvio-deltaic sedimentation and palaeogeography of the Northumberland Basin: Yorks. Geol. Soc. Proc., v. 40, pt. 2, p. 129-180.

Leopold, L. B., 1973, River channel change with time: An example: Geol. Soc. Am. Bull., v. 84, p. 1845-1860.

————, and Maddock, T., 1953, The hydraulic geometry of stream channels and some physiographic implications: U.S. Geol. Survey Prof. Paper 252, 57p.

————, and Wolman, M. G., 1957, River channel patterns: Braided, meandering and straight: U.S. Geol. Survey Prof. Paper 282-B, p. 39-85.

————, and ————, 1960, River meanders: Geol. Soc. Am. Bull., v. 71, p. 769-794.

————, ————, and Miller, J. P., 1964, Fluvial processes in geomorphology: W. H. Freeman and Co., San Francisco, 522p.

Levey, R. A., 1976, Characteristics of coarse-grained point bars, Upper Congaree River, South Carolina: in M. O. Hayes and T. W. Kana, eds., Terrigenous clastic depositional environments; Coastal Research Div., Geol. Dept., Univ. South Carolina, Tech. Rept. n. 11-CRD, p. II-38 to II-51.

Lewin, J., 1976, Initiation of bed forms and meanders in coarse-grained sediments: Geol. Soc. Am. Bull., v. 87, p. 281-285.

Looney, R. M., and Baker, V. R., 1977, Late Quaternary geomorphic evolution of the Colorado River, Inner Texas Coastal Plain: Gulf Coast Assoc. Geol. Soc. Trans., v. 27, p. 323-333.

McGowen, J. H., and Garner, L. E., 1970, Physiographic features and stratification types of coarse-grained point bars: Modern and ancient examples: Sedimentology, v. 14, p. 77-111.

McKee, E. D., Crosby, E. J., and Berryhill, H. L., 1967, Flood deposits, Bijou Creek, Colorado: J. Sediment. Petrol., v. 37, p. 829-851.

Miall, A. D., 1974, Paleocurrent analysis of alluvial sediments: A discussion of directional variance and vector magnitude: J. Sediment. Petrol., v. 44, p. 1174-1185.

————, 1977, A review of the braided-river depositional environment: Earth Sci. Revs., v. 13, p. 1-62.

Michelson, P. C., and Dott, R. H. Jr., 1973, Orientation analysis of trough cross stratification in Upper Cambrian sandstones of western Wisconsin: J. Sediment. Petrol., v. 43, p. 784-794.

Molenaar, C. M., 1973, Sedimenary facies and correlations of the Gallup Sandstone and associated formations, northeastern New Mexico: in J. E. Fassett, ed., Cretaceous and Tertiary rocks of the southern Colorado Plateau; Durango, Colo., Memoir, Four Corners Geol. Soc., p. 85-110.

Moody-Stuart, M., 1966, High- and low-sinuosity stream deposits, with examples from the Devonian of Spitzbergen: J. Sediment. Petrol., v. 36, p. 1102-1117.

Nami, M., 1976, An exhumed Jurassic meander belt from Yorkshire, England: Geol. Mag., v. 113, p. 47-52.

Nanson, G. C., 1977, Channel migration, floodplain formation, and vegetation succession on a meandering-river floodplain in N.E. British Columbia, Canada: Unpub. Ph.D. thesis, Simon Fraser Univ., Vancouver, B.C., Canada, 371p.

Nanz, R. H. Jr., 1954, Genesis of Oligocene sandstone reservoir, Seeligson Field, Jim Wells and Kleberg Counties, Texas: Am. Assoc. Petrol. Geol. Bull., v. 38, p. 96-117.

Nilsson, G., and Martvall, S., 1972, The River Öre and its meanders: Uppsala Univ., UNGI Rept. 19, 154p. (in Swedish, with English summary and figure subcaptions).

Nummedal, D., 1973, Paleoflow characteristics of a Late Cretaceous river in Utah from analysis of sedimentary structures in the Ferron Sandstone: A discussion: J. Sediment. Petrology, v. 43, p. 1176-1179.

————, Hine, A. C., Ward, L. G., Hayes, M. O., Boothroyd, J. C., Stephen, M. F., and Hubbard, D. K., 1974, Recent migrations of the Skeidarásandur shoreline, southeast Iceland: Coastal Research Div., Geol. Dept., Univ. South Carolina, Final Report for Contract N60921-73-C-0258, 183p.

Padgett, G. V., and Ehrlich, R., 1976, Paleohydraulic analysis of a late Carboniferous fluvial system, southern Morocco: Geol. Soc. Am. Bull., v. 87, p. 1101-1104.

Parker, G., 1976, On the cause and characteristic scales of meandering and braiding in rivers: J. Fluid Mechanics, v. 76, p. 457-480.

Patton, P. C., and Baker, V. R., 1977, Geomorphic response of central Texas stream channels to catastrophic rainfall and runoff: *in* D. Doehring, *ed.*, Geomorphology of arid and semi-arid regions: SUNY-Binghamton, Publs. in Geomorphology, p. 189-217.

Peterson, A. R., 1976, Paleoenvironments of the Colton Formation, Colton, Utah: B.Y.U. Geology Studies, v. 23, p. 3-35.

Picard, M. D., and High, L. R., 1973, Sedimentary structures of ephemeral streams: Developments in Sedimentology 17, Elsevier, Amsterdam, 223p.

Puigdefábregas, C., 1973, Miocene point-bar deposits in the Ebro Basin, northern Spain: Sedimentology, v. 20, p. 133-144.

Read, W. A., 1969, Fluviatile deposits in Namurian rocks of central Scotland: Geol. Mag., v. 106, p. 331-347.

Riley, S. J., 1975, The channel shape-grain size relation in eastern Australia and some paleohydraulic implications: Sediment. Geol., v. 14, p. 253-258.

Rust, B. R., 1972, Structure and process in a braided river: Sedimentology, v. 18, p. 221-246.

————, 1975, Fabric and structure in glaciofluvial gravels: *in* A. V. Jopling and B. C. McDonald, *eds.*, Glaciofluvial and glaciolacustrine sedimentation: Soc. Econ. Paleont. Mineral. Spec. Pub. 23, p. 238-248.

————, 1978, Application of modern fluvial processes to the interpretation of ancient alluvial successions: 5th Guelph Symp. Geomorphology Proc., Norwich, England, Geo Abstracts Ltd. (in press).

Schumm, S. A., 1960, The shape of alluvial channels in relation to sediment type: U.S. Geol. Survey Prof. Paper 352-B, p. 17-30.

————, 1963, Sinuosity of alluvial channels on the Great Plains: Geol. Soc. Am. Bull., v. 74, p. 1089-1100.

————, 1968, River adjustment to altered hydrologic regimen — Murrumbidgee River and paleochannels, Australia: U.S. Geol. Survey Prof. Paper 598, 65p.

————, 1971, Fluvial geomorphology: Channel adjustment and river metamorphosis: *in* H. W. Shen, *ed.*, River mechanics, Ft. Collins, Colo., v. I, ch. 5, 22p.

————, 1972, Fluvial paleochannels: *in* J. K. Rigby and W. K. Hamblin, *eds.*, Recognition of ancient sedimentary environments, Soc. Econ. Paleont. Mineral. Spec. Pub. 16, p. 98-107.

————, and Lichty, R. W., 1963, Channel widening and flood-plain construction along Cimarron River in southwestern Kansas: U.S. Geol. Survey Prof. Paper 352-D, p. 71-88.

————, and Khan, H. R., 1972, Experimental study of channel patterns: Geol. Soc. Am. Bull., v. 83, p. 1755-1770.

Schwartz, D. E., 1977, Flow patterns and bar morphology in braided-to-meandering transition zone — Red River, Texas and Oklahoma: Am. Assoc. Petrol. Geol. and Soc. Econ. Paleont. Mineral., Abstracts, June 12-16, 1977 meeting, p. 107.

Shantzer, E. V., 1951, Alluvium of plains rivers in a temperate zone and its significance for understanding the laws governing the structure and formation of alluvial suites: Trav. Inst. Sci. Akad. Nauk, v. 135, p. 1-271 (in Russian).

Shelton, J. W., 1973, Models of sand and sandstone deposits: A methodology for determining sand genesis and trend: Oklahoma Geol. Survey Bull. 118, 122p.

————, and Noble, R. L., 1974, Depositional features of braided-meandering stream: Am. Assoc. Petrol. Geol. Bull., v. 58, p. 742-749.

Shepherd, R. G., 1976, Sedimentary processes and structures of ephemeral-stream point bars, Rio
    Puerco near Albuquerque, New Mexico: Geol. Soc. Am. Abstracts Prog. (Ann. Mtg.), v. 8, p.
    1103.

Smith, D. G., 1973, Aggradation of the Alexandra-North Saskatchewan River, Banff Park, Alberta:
    in M. Morisawa, *ed.,* Fluvial geomorphology: SUNY-Binghamton, Publs. in Geomorphology,
    p. 201-219.

——, 1976, Effect of vegetation on lateral migration of anastomosed channels of a glacier
    meltwater river: Geol. Soc. Am. Bull., v. 87, p. 857-860.

Smith, N. D., 1970, The braided stream depositional environment: Comparison of the Platte River
    with some Silurian clastic rocks, north-central Appalachians: Geol. Soc. Am. Bull., v. 81, p.
    2993-3014.

——, 1971, Transverse bars and braiding in the lower Platte River, Nebraska: Geol. Soc. Am.
    Bull., v. 82, p. 3407-3420.

——, 1972, Some sedimentological aspects of planar cross-stratification in a sandy braided river: J.
    Sediment. Petrol., v. 42, p. 624-634.

——, 1974a, Sedimentology and bar formation in the upper Kicking Horse River, a braided
    outwash stream: J. Geol., v. 82, p. 205-223.

——, 1974b, Some sedimentological aspects of planar cross-stratification in a sandy braided river:
    A reply to N. L. Banks and J. D. Collinson: J. Sediment. Petrol., v. 44, p. 267-269.

Steel, R. J., Maehle, S., Nilsen, H., Røe, S. L., and Spinnangr, A., 1977, Coarsening-upward cycles
    in the alluvium of Hornelen Basin (Devonian) Norway: Sedimentary response to tectonic
    events: Geol. Soc. Am. Bull., v. 88, p. 1124-1134.

Sundborg, A., 1956, The River Klarälven, a study in fluvial processes: Geogr. Ann., v. 38, p.
125-316.

Taylor, G., 1976, A depositional model for suspended-load, low gradient streams: 25th Internat.
    Geol. Congr., Abstracts, v. 3, p. 836.

Warren, A., 1976, Morphology and sediments of the Nebraska Sand Hills in relation to Pleistocene
    winds and the development of aeolian bedforms: J. Geol., v. 84, p. 685-700.

Wiethe, J. D., 1970, Textural parameters of a modern point bar: Compass Sigma Gamma Epsilon, v.
    47, p. 110-118.

Williams, G. E., 1966, Planar cross-stratification formed by the lateral migration of shallow streams:
    J. Sediment. Petrol., v. 36, p. 742-746.

——, 1971, Flood deposits of the sand-bed ephemeral streams of central Australia: Sedimentology,
    v. 17, p. 1-40.

Williams, P. F., and Rust, B. R., 1969, The sedimentology of a braided river: J. Sediment. Petrol., v.
    39, p. 649-679.

Wolman, M. G., and Leopold, L. B., 1957, River flood plains: Some observations on their
    formation: U.S. Geol. Survey Prof. Paper 282-C.

——, and Miller, J. P., 1960, Magnitude and frequency of forces in geomorphic processes: J.
    Geol., v. 68, p. 54-74.

VERTICAL SEQUENCE AND SAND BODY SHAPE IN ALLUVIAL SEQUENCES

J. D. COLLINSON[1]

ABSTRACT

The type of channel which deposited coarse member sand bodies in alluvial sequences has largely been deduced from coarse member - fine member ratios, palaeocurrent patterns and the internal organisation of grain size and sedimentary structures. The shape of the sand bodies has seldom been considered an important piece of evidence and has more often been a predictive target. Work on modern rivers shows that no one vertical sequence typifies a particular type of channel but that different channel types generate diverse suites of sequences which give scope for ambiguity in interpretations. For highly sinuous streams, the maximum lateral extent of the sand bodies generated by point bar migration can be predicted by empirical equations. Comparisons of predicted maximum lateral extents with the dimensions of ancient sandstone bodies shows that some sandstones, previously interpreted as of point bar origin, are more likely to be the products of low sinuosity streams or of sheet floods.

INTRODUCTION

Prediction of the extent and shape of sandstone bodies in an alluvial sequence depends largely on the ability to identify the type of alluvial system responsible for it. This approach forms the basis of much current thinking and much work has recently gone into establishing facies models of different fluvial systems with this aim. Such work, however, seems to ignore the possibility that sand body shape itself may often be powerful evidence for arriving at such a reconstruction in spite of the pioneering work in this direction by Moody-Stuart (1966). Further, it may be that without this information, interpretations based on, say, the vertical sequence of internal structures may be ambiguous or even misleading.

This paper reviews some of these ambiguities, and examines the extent to which the range of possible interpretations may be reduced by using sand body shape in a more active way to establish environmental models.

APPROACHES TO THE PROBLEM:

Continental or paralic sequences of inter-bedded sandstone and siltstone/-mudstone-dominated units in which the sandstones are sharp-based and often upwards-fining are well described in terms of vertical sequence of sedimentary structures and grain size. They are commonly divided into 'coarse members' and 'fine members' and these subdivisions are then equated with 'channel' and 'interchannel' (overbank) deposition respectively (e.g. Allen, 1965a; Allen & Friend, 1968; Leeder, 1973). Detailed descriptions of both coarse and fine members have then been used to characterise the environments more fully. Most attempts to characterise the river type have, naturally, concentrated largely on the coarse member, whilst fine members which may contain pedogenic units have been used to characterise the prevailing climate and to estimate frequencies of channel migration and shifting (Allen, 1974a; Leeder, 1975).

Essentially four main properties of a sequence have been considered as reflecting channel type and helping to place it within or even outside the spectrum of channel types known at the present-day: the four properties are:

1) Ratio of thickness of coarse member to fine member in the overall sequence.

2) Palaeocurrent patterns.

[1]Department of Geology. The University, Keele, Staffs. ST5 5BG, England.

3)   Internal organisation of the coarse member sand body.

4)   Shape of the coarse member sand body.

It is intended here to concentrate on the last two aspects, though brief statements and comments on the first two seem appropriate, particularly as a combination of all four types of evidence should ideally be used in any interpretation.

COARSE MEMBER/FINE MEMBER THICKNESS RATIO

This gross property of the sequence has been taken to reflect the nature of the channel systems active during deposition (e.g. Thompson, 1970). The argument stems from observations that the depth-width ratio of channels increases with the proportion of cohesive material in their banks (Schumm, 1960) and that sinuosity increases with higher suspended loads, other parameters being equal (Schumm & Kahn, 1972). It has thence been argued that sequences with a large proportion of fine material are more likely to have involved sinuous streams while those with little preserved fine material were produced by streams of low sinuosity. This approach seems rather naive as it largely ignores the question of how material is preserved in the deposition of the overall sequence. The critical factor in preserving overbank sediment in a vertical section is the nature and frequency of channel migration over a particular point on the alluvial plain in relation to overall subsidence rates (see Leeder, this volume). In so far as meandering streams tend to move their position by a combination of short migrations and avulsion rather than by steady migration over long distances, there may be some truth in the approach. However, the frequency of this process at a point on an alluvial plain is likely to have more to do with the width of the plain and the prevailing rate of subsidence than with the nature of the bank material and the suspended load.

PALAEOCURRENT PATTERNS:

It was once considered that the dispersion of palaeocurrents measured in alluvial sand bodies gave a good indication of channel sinuosity, with high dispersion relating to high sinuosity (e.g. Pelletier, 1958; Selley, 1965, 1968). Detailed work both on modern rivers and ancient alluvial sequences has shown that the picture is extremely complex (Miall, 1974). Low sinuosity sandy braided streams are able to produce highly dispersed cross-bedding within a restricted reach due to the tendency of transverse bars to migrate at high angles to the downcurrent trend under conditions of reduced flow (Smith, 1972). The nature of the river hydrograph probably plays an important part in determining the precise pattern of dispersion in any one case (Jones, 1977). While gross patterns of palaeocurrent distribution probably have little value, it is possible that a detailed integration of palaeocurrent measurements and particular measured sections will yield real insight (e.g. Cant & Walker, 1976). Also a combination of low within-section variance and high between-section variance for cross-bedding directions from a particular sequence could be an argument for high sinuosity.

INTERNAL ORGANISATION OF THE COARSE MEMBER:

The organisation of grain sizes and bedding within the coarse members has probably been the main criterion for reconstructing channel type. The approach has centred around the equation of a fining upwards unit with lateral accretion on a point bar in a meandering river system (Bernard & Major, 1963; Allen, 1964, 1965a). This classical model, largely based on single vertical sequences, has been developed to a high degree of sophistication (Allen 1970a, 1970b). It is, however, based on very little data from present-day meanders, and recently described point bars show much greater complexity than that envisaged by the model (e.g. Jackson, 1976). The classical point bar model was of great interpretative power in that it explained differences in vertical sequence in terms of differing channel parameters such as slope and sinuosity. It is now clear that different sequences can be generated by the same point bar with channel parameters held constant, the differences being due to a downstream segregation of both flow pattern and bed response through the operation of a spatial lag effect. That being the case, some of the interpretative power of the classical model is lost. Where more laterally extensive exposure is available these problems are greatly reduced as it may then become possible to see not only a suite of vertical sequences but also their spatial relationships. It is clearly more desirable to develop two- to three-dimensional models but these are of little direct help to those who have only one-dimensional data (e.g. boreholes or stream sections).

These rather pessimistic reservations apply at least equally to low sinuosity sandy streams and their alleged ancient analogues (cf. Miall, 1977). The recent description of the Battery Point Sandstone by Cant and Walker (1976) is one of the few attempts at recognizing ancient sandy low sinuosity stream deposits. It suffers, however, from the conceptual problems outlined above for point bars and is of highly dubious statistical validity. A sandy braided river has even greater scope than a meandering river for lateral variability in the vertical sequences which it can generate, as has been so clearly shown for the S. Saskatchewan River (Cant, this volume). To attempt, therefore, to categorise the deposits in terms of one vertical sequence seems unrealistic and again a three-dimensional model is required. For the deposits of a sandy braided stream the spatial distribution of the different types of sequence is less predictable than for the deposits of a meandering stream, adding yet another dimension to the scope for ambiguity in interpretation. It would have been more helpful, in my view, to have been told something of the observed lateral variability within the Battery Point Sandstone rather than to have this information evaporated in the 'distillation' process. For the distillation by Markov analysis to have statistical validity the number of observed transitions must increase on a steepening curve as the number of 'states' increases. To reduce the number of states is usually to throw away valuable hydrodynamic information. It is better to try to characterise a sequence not by a single distilled sequence but by some description of the variability of *observed* sequences, ideally with known spatial relationships.

No single vertical sequence categorises a particular channel type. Different channel types each generate suites of vertical sequences with greater or lesser degrees of spatial organisation and with varying relative abundances and preservation potential. With a continuum of channel types, there will be a continuum of subtly changing suites and discrimination of channel type based on one-dimensional exposure must ultimately depend on sophisticated statistical and optimisation techniques which match the variabilities of sequence and define levels of sampling appropriate to confident interpretation.

Epsilon Cross-Bedding

The foregoing pessimism is somewhat ameliorated by the existence of epsilon cross-bedding or lateral accretion bedding. This is now widely recognised in alluvial coarse members and is interpreted as reflecting the lateral shift of a channel depositional bank, usually taken to be a point bar (Allen, 1965b). Its occurrence therefore leads to the reconstruction of a channel of high sinuosity and this is exceptionally confirmed when scroll-bar topography on bedding planes coexists with the lateral accretion bedding (e.g. Puigdefabregas, 1973; Nami, 1976; Nami and Leeder, this volume). The converse does not apply and the absence of epsilon cross-bedding cannot be taken to indicate that point bars were not involved. There are many well rehearsed reasons why a point bar sand body should not always display epsilon cross-bedding (Allen, 1970a).

SAND BODY SHAPE

The shape and extent of sand bodies is commonly a prediction target rather than a factor which bears heavily on the interpretation of channel type. This is inevitable in the context of hydrocarbon or coal exploration, and one aspect of the problem has been examined by Leeder (this volume). However, in order to apply a model such as Leeder proposes, it is first necessary to establish that a meandering river system was involved and that it behaved avulsively. In the light of the foregoing discussion of vertical sequences, it is clear that the recognition of channel type may be far from unambiguous.

Where exposure of the sand body is extensive or where there are closely spaced boreholes, the recognition of sand body shape is possible, and this can give further evidence of the channel type if internal structures are unhelpful or ambiguous.

There have been several attempts at using the cross-section shape of the sand body to determine channel type. Moody-Stuart (1966) recognised two main sand body shapes in the Old Red Sandstone of Spitzbergen. Rather steep sided flat bottomed sand bodies were associated with epsilon cross-bedding and therefore attributed to meandering streams whilst more lenticular concave upwards based units without epsilon cross-bedding were attributed to low sinuosity streams which generated the sand body largely by a process of vertical accretion. A similar approach has been followed, at least as far as low sinuosity streams are concerned, in several other cases (e.g. Sykes, 1974; Campbell, 1976). In addition to these two laterally restricted types of sand body there also exist rather extensive flat-based sheet sandstones which lack epsilon cross-bedding, which have been variously interpreted as the products of both high and low sinuosity streams. It seems appropriate to ask whether there is any basis in our knowledge of present day streams to substantiate the use of sand body shape as a guide to channel type.

CHANNEL BEHAVIOUR

To use shape in this way we must understand something of channel behaviour. For meandering streams this is quite well understood. Free meanders in a wide floodplain develop in discrete meanderbelts which are stabilised by clay plugs generated by cut-off events. Channel migration and point bar accretion is confined within this belt and larger movements tend to take place by avulsion (see Leeder, this volume). Abandoned meander belts therefore comprise a complex of laterally restricted sand bodies and intervening clay plugs (Fig. 1B). The maximum lateral extent of the sand bodies is determined by the wavelength and amplitude of the meanders, which in turn are functions of channel size. There are a number of empirical equations of varying exactness which relate channel width, channel depth, meander belt width and discharge: —

$$Wm = 65.8 \, \overline{Q}^{0.47} \quad \text{(Carlston 1965)} \tag{1}$$

$$W = 7 \, \overline{Q}^{0.46} \quad \text{(approx.) (Carlston 1965)} \tag{2}$$

$$\text{For sinuosity} > 1.7 \quad W = 6.8 \, h^{1.54} \quad \text{(Leeder, 1973)} \tag{3}$$

$$\lambda = 10.9 \, W^{1.01} \quad \text{(Leopold \& Wolman, 1960)} \tag{4}$$

Where Wm is width of meander belt, W is bankfull channel width, h is bankfull channel depth, λ is meander wavelength and $\overline{Q}$ is mean annual discharge.

Combining these and ignoring compound errors gives the following relationships: —

$$Wm = 64.6 \, h^{1.54} \tag{5}$$

$$\lambda = 74.1 \, h^{1.54} \tag{6}$$

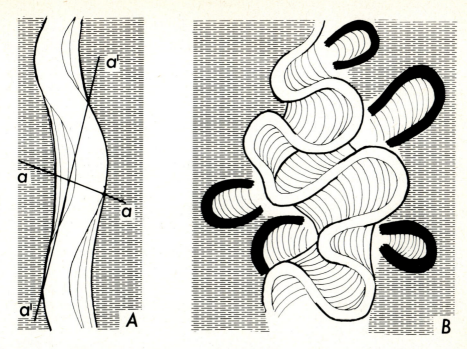

Fig. 1. Schematic plan views of the sand bodies generated by A) low sinuosity and B) high sinuosity streams. In A, if the channel is abandoned before sinuosity develops and the abandonment is gradual, an elongate sand body with no internal erosion surfaces results. The lateral extent of such a sand body, seen in vertical section is clearly dependant upon the orientation of the section in relation to the current (cf. a-a and a'-a'). In B, the frequent cut-off of meander loops and the development of clay plugs (black) in abandoned channels leads to a series of restricted sand bodies separated either by erosion surfaces (heavy lines) or clay plugs. The final abandoned channel may be sand or clay filled. The orientation of a vertical section through such sand bodies has little influence on their observed lateral extent.

These equations can only be considered approximate, though they are probably of the right order of magnitude. Their errors cannot be calculated but are probably rather high. It would clearly be desirable to establish similar relationships directly from the raw data rather than by combination of other empirical equations. Meanderbelt width (Wm) and roughly half the meander wavelength (λ) should control the maximum lateral extent of any point bar sand body. In reality, a randomly positioned and oriented vertical section should show a sand body somewhat less extensive than these limits (Fig. 1B). There may, in some cases, be lateral coalescence of several point bar sands to give more extensive sand bodies but these are not likely to be very common and it should be possible to recognize erosion surfaces between the various point bar units.

Channels which were potentially meandering with a wavelength predicted by equation (6) may sometimes have been abandoned at an early stage in their development and while still of low sinuosity. In such cases, mechanisms of meander cut-off and the development of clay plugs will not have occurred and if abandonment was gradual an elongate sand body may have resulted (Fig. 1A). There is then clearly a wide discrepancy between the apparent lateral extent of the sand body in sections oriented parallel (line a' - a') and perpendicular to the current direction (line a - a). This is not the case with high sinuosity stream deposits where apparent extent is largely independent of orientation of the section due to the cross-cutting of the meander belts by erosion surfaces and clay plugs (Fig. 1B).

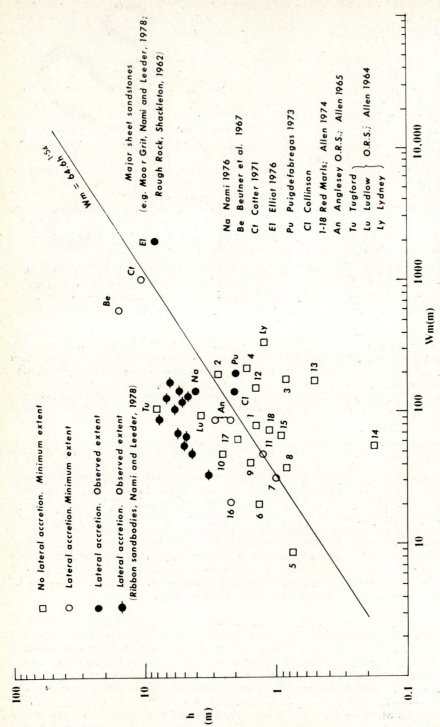

Fig. 2. Lateral extents and thicknesses of alluvial sand bodies, showing the predicted line of meanderbelt width (Wm) in relation to depth (h). Sources of data are indicated. 'Collinson' is a generalised value for channel sandbodies in the Upper Carboniferous of the Spanish Pyrenees. 'Nami and Leeder, 1978' refers to their paper in this volume.

Using this rationale, it should be possible to predict the likely maximum lateral extent of a sand body from a knowledge of its thickness if it is the product of a high sinuosity channel, sand body thickness being taken as equivalent to channel depth.

SAND BODY EXTENT IN RELATION TO PREDICTED VALUES

In order to test this procedure, the dimensions of several flat-based sand bodies have been plotted in relation to the line generated by equation (5) (Fig. 2). The line predicting Wm is plotted as it gives the largest value of lateral extent and we are interested in maximum extents. $\lambda/2$ and Wm/2 are both a more realistic estimate of lateral extent but using Wm accounts for the unlikely possibility of a meander loop sometimes migrating over the full width of the meander belt. Many of the sand bodies plotted have been interpreted, either explicitly or by implication, as lateral accretion deposits, some on the basis of epsilon cross-bedding and others on the basis that they were fining upwards units. For most of the examples, neither margin or only one margin was observed and therefore only minimum values of lateral extent are available. Except for those examples which show lateral accretion bedding and whose exposures are therefore, presumably, oriented at a high angle to the palaeocurrent, the orientation of the plane of observed lateral extent in relation to current direction is not recorded.

From the data it is clear that all the sand bodies for which there is independent evidence of point bar sedimentation (i.e. lateral accretion bedding) plot fairly close to the expected line. Those with lateral accretion bedding but for which only a minimum lateral extent is available, plot on or to the left of the line. Their true extents would clearly shift their positions to the right by some unknown distance.

The extent values plotted for the ribbon sand bodies of the Yorkshire Jurassic (Nami & Leeder, this volume) are the widths of the ribbons. The values of lateral extent recorded (i.e. the widths) are much less than what could have been developed had a high sinuosity course had chance to develop. As discussed above, the observed extents of such sand bodies in two-dimensional vertical exposures would depend very much on the orientation of the exposure with respect to palaeocurrent.

The sand bodies which lack lateral accretion bedding show a rather more widely scattered distribution, on either side of the predicted line. As all these have minimum values of lateral extent, their true positions should also lie further to the right. Some of the sand bodies lie too far to the right of their expected position to be due to the result of meandering stream deposition. Units 3, 13, 14 and *Ly* are nearly an order of magnitude wider than predicted. Some of the others could also move into this area, but it is not possible to rule out a meandering stream origin for them. It seems unlikely, however, that units 15, 18, 12, 4 at least, should be regarded as of high sinuosity origin on the basis of sand body shape.

While this approach does not enable us to identify the sand bodies of high sinuosity streams, it does give us a method of eliminating certain units which on the evidence of their internal organisation alone could be compatible with point bar sedimentation.

THE PROBLEM OF SHEET SANDSTONES

Elimination of these units from the realm of point bar sedimentation immediately raises the question of what they actually are. For some sand sheets which are several metres thick, (e.g. Moor Grit: Nami & Leeder, this volume; Rough Rock: Shackleton, 1962) the tendency seems to be to interpret them as the product of low sinuosity channels, mainly on the grounds that they are not of high sinuosity origin. The implication of this is that low sinuosity streams are able to migrate laterally and continuously over long distances to develop sheet sandstones. While such behaviour has been hinted at in publications (e.g.

Allen, 1965c, Fig. 35c) there is no present-day river which has been described as behaving in this way. Such an interpretation is also at odds with the examples quoted earlier where concave upwards lenticular sand bodies are interpreted in terms of low sinuosity streams which are cut and filled by vertical accretion. With such behaviour, a sheet sandstone could be generated by lateral coalescence of lenticular bodies (e.g. Campbell, 1976), but the erosion surfaces between the individual channels should be recognisable. This model has a modern analogue in the behaviour of the Kosi River which appears to migrate in discrete steps (Gole and Chitale, 1966). The variation in shape and lateral extent within alleged low sinuosity stream channel deposits therefore suggests two contrasting modes of large-scale channel behaviour. If the reasons for these differences in behaviour were understood they could offer considerably more insight into environmental reconstruction and possibly to predictive ability.

Thinner sand bodies which are removed from the ambit of point bar deposition by their lateral extent present a different problem. Their flat basal scour surfaces and their internal organisation which commonly shows upwards fining of grain size and a change in sedimentary structures from cross-bedding to parallel or cross-lamination record a generally waning flow. The point bar model would interpret this waning as the result of spatial segregation of the various flow fields combined with lateral migration. As the point bar model is ruled out by the shape there seems no reason why the waning should not have resulted from temporal segregation, for instance in a waning sheet flood (e.g. McKee *et al.*, 1967, Bull. 1972). This alternative interpretation gathers some support when some of the actual sequences are examined in more detail. The sand body which has been described as a channel deposit in the Old Red Sandstone at Lydney, for example, is the thickest of several similar ones in the same sequence. Its selection as a 'channel' unit seems somewhat arbitrary when a sheetflood or crevasse splay interpretation would probably be considered appropriate for its thinner neighbours. One should not, however, loose sight of the possibility of the exposure giving a longitudinal section through a ribbon channel sandstone. On the basis of shape, if it is a channel deposit, it cannot be due to a high sinuosity system, whatever the orientation of the exposure. Similar arguments can be applied to the examples from the Red Marls in Pembrokeshire where individual sand bodies show quite marked lateral changes both of thickness and of internal organisation (Allen, 1974b). When the thicknesses of these sandstone units fall as low as 10 cm (unit 14) a channel interpretation is difficult to sustain.

The question which arises is 'When is a sand body of channel origin and when should it be regarded as a sheetflood deposit on the floodplain or on the distal part of an alluvial fan?' When channel margins and lateral accretion bedding are absent, the safest policy at present is to question whether so called 'coarse members' thinner than 1-2 m need be equated with channel deposition at all.

Acknowledgments

I would like to thank Bernard Besly, Trevor Elliott and Mike Leeder for their helpful comments on the manuscript.

References

Allen, J. R. L., 1964, Studies in fluviatile sedimentation: Six cyclothems from the Lower Old Red Sandstone, Anglo-Welsh Basin: Sedimentology, v. 3, p. 163-198.

———, 1965a, Fining upwards cycles in alluvial successions: Geol. J., v. 4, p. 229-246.

———, 1965b, The sedimentation and palaeogeography of the Old Red Sandstone of Anglesey, North Wales: Yorks. Geol. Soc. Proc., v. 35, p. 139-185.

———, 1965c, A review of the origin and characteristics of Recent alluvial sediments: Sedimentology, v. 5, p. 89-191.

————, 1970a, Studies in fluviatile sedimentation: A comparison of fining-upwards cyclothems with special reference to coarse-member composition and interpretation: J. Sediment. Petrol., v. 40, p. 298-323.

————, 1970b, A quantitative model of grain size and sedimentary structures in lateral deposits: Geol. J., v. 7, p. 129-146.

————, 1974a, Studies in fluviatile sedimentation: implications of pedogenic carbonate units, Lower Old Red Sandstone. Anglo-Welsh outcrop: Geol. J., v. 9, p. 181-208.

————, 1974b, Studies in fluviatile sedimentation: lateral variation in some fining upwards cyclothems from the Red Marls, Pembrokeshire: Geol. J., v. 9, p. 1-16.

———— and Friend, P. F., 1968, Deposition of the Catskill Facies; Appalachian region: with notes on some other Old Red Sandstone Basins: *in* G. de V. Klein, *ed.*, Late Paleozoic and Mesozoic Continental Sedimentation, northeastern North America, Geol. Soc. Am. spec. paper 106, p. 21-74.

Bernard, H. A. and Major, C. F., 1963, Recent meander belt deposits of the Brazos River: an alluvial "sand" model: Am. Assoc. Petrol. Geol. Bull., v. 47, p. 350.

Beutner, E. C., Flueckinger, L. A. and Gard, T. M., 1967, Bedding geometry in a Pennsylvanian channel sandstone: Geol. Soc. Am. Bull., v. 78, p. 911-916.

Bull, W. B., 1972, Recognition of alluvial-fan deposits in the stratigraphical record, *in* K. J. Rigby and W. K. Hamblin, *eds.*, Recognition of Ancient Sedimentary Environments, Soc. Econ. Paleont. Mineral. Spec. Pub. 16, p. 63-83.

Campbell, C. V., 1976, Reservoir geometry of a fluvial sheet sandstone: Am. Assoc. Petrol. Geol. Bull., v. 60, p. 1009-1020.

Cant, D. J. and Walker, R. G., 1976, Development of a braided-fluvial facies model for the Devonian Battery Point Sandstone, Quebec: Can. J. Earth Sci. v. 13, p. 102-119.

Carlston, C. W., 1965, The relation of free meander geometry to stream discharge and its geomorphic implications: Am. J. Sci., v. 263, p. 864-885.

Cotter, E., 1971, Paleoflow characteristics of a Late Cretaceous river in Utah from analysis of sedimentary structures in the Ferron Sandstone: J. Sediment. Petrol., v. 41, p. 129-138.

Elliott, T., 1976, The morphology, magnitude and regime of a Carboniferous fluvial-distributary channel: J. Sediment. Petrol., v. 46, p. 70-76.

Gole, C. V. and Chitale, S. V., 1966, Inland delta building activity of the Kosi River: J. Hydraulics Div., Am. Soc. Civ. Engrs., v. 92, p. 111-126.

Jackson, R. G., II. 1976, Depositional model of point bars in the Lower Wabash River: J. Sediment. Petrol., v. 46, p. 579-594.

Jones, C. M., 1977, The effects of varying discharge regimes on bed form sedimentary structures in modern rivers: Geology, v. 5, p. 567-570.

Leeder, M. R., 1973, Sedimentology and palaeogeography of the Upper Old Red Sandstone in the Scottish Border Basin: Scott. J. Geol., v. 9, p. 117-144.

————, 1975, Pedogenic carbonate and flood sediment accretion rates: a quantitative model for alluvial arid-zone lithofacies: Geol. Mag., v. 112, p. 257-270.

Leopold, L. B. and Wolman, M. G., 1960, River meanders; Geol. Soc. Am. Bull., v. 71, p. 769-794.

McKee, E. D., Crosby, E. J. and Berryhill, H. L. Jr., 1967, Flood deposits, Bijou Creek, Colorado, June, 1965: J. Sediment. Petrol., v. 37, p. 829-851.

Miall, A. D., 1974, Paleocurrent analysis of alluvial sediments: A discussion of directional variance and vector magnitude: J. Sediment. Petrol., v. 44, p. 1174-1185.

————, 1977, A review of the braided-river depositional environment: Earth Sci. Revs., v. 13, p. 1-62.

Moody-Stuart, M., 1966, High- and Low-sinuosity stream deposits, with examples from the Devonian of Spitsbergen: J. Sediment. Petrol., v. 36, p. 1102-1117.

Nami, M., 1976, An exhumed Jurassic meander belt from Yorkshire, England: Geol. Mag., v. 113, p. 47-52.

Pelletier, B. R., 1958, Pocono paleocurrents in Pennsylvania and Maryland: Geol. Soc. Am. Bull., v. 69, p. 1033-1064.

Puigdefabregas, C., 1973, Miocene point bar deposits in the Ebro Basin, Northern Spain: Sedimentology, v. 20, p. 133-144.

Schumm, S. A., 1960, The shape of alluvial channels in relation to sediment types: U.S. Geol. Survey Prof. Paper 352-B, p. 17-30.

———— and Kahn, H. R., 1972, Experimental study of channel patterns: Geol. Soc. Am. Bull., v. 83, p. 1755-1770.

Shackleton, J. S., 1962, Cross-strata of the Rough Rock (Millstone Grit Series) in the Pennines: Geol. J., v. 3, p. 109-118.

Selley, R. C., 1965, Diagnostic characters of fluviatile sediments of the Torridonian Formations (Precambrian) of northwest Scotland: J. Sediment. Petrol., v. 35, p. 366-380.

————, 1968, A classification of paleocurrent models: J. Geol., v. 76, p. 99-110.

Smith, N. D., 1972, Some sedimentological aspects of planar cross-stratification in a sandy braided river: J. Sediment. Petrol., v. 42, p. 624-634.

Sykes, R. M., 1974, Sedimentological studies in Southern Jameson Land, East Greenland. 1. Fluviatile sequences in the Kap Stewart Formation (Rhaetic-Hettangian): Geol. Soc. Denmark Bull., v. 23, p. 203-212.

Thompson, D. B., 1970, Sedimentation of the Triassic (Scythian) Red Pebbly Sandstones in the Cheshire Basin and its margins: Geol. J., v. 7, p. 183-216.

A QUANTITATIVE STRATIGRAPHIC MODEL FOR ALLUVIUM, WITH SPECIAL REFERENCE TO CHANNEL DEPOSIT DENSITY AND INTERCONNECTEDNESS

M. R. Leeder[1]

Abstract

Channel deposit density and interconnectedness within a 'matrix' of overbank fines in stratigraphic sections through alluvium are inversely related to the rates of contemporary channel avulsion and floodplain accretion and to the width of the alluvial plain. Simulated sections through alluvium generated by random channel avulsions and residence intervals reveal spurious local concentrations of channel deposits. Such concentrations may be given misleading non-random interpretations in stratigraphic or exploration schemes based upon incomplete exposures or sparse borehole data.

Introduction

To the student of present day sedimentary processes and landforms the geologically rapid oscillations of sea level during the Pleistocene have given both advantages and disadvantages. Advantages include a vivid appreciation of the effects of eustatic changes upon sedimentary and geomorphic systems; disadvantages include the departure of accessible Pleistocene and Holocene systems from a 'steady state'.

Nowhere are such disadvantages better illustrated than in the alluvial environment. Periods of deep incision during sea-level 'lows' or during arid periods alternated with depositional cycles caused by subsequent aggradation during sea-level rises and during humid periods. Pleistocene to Holocene alluvium is thus noteworthy for its rapid vertical facies changes and changing channel regimes. Although such changes must have occurred periodically in the geological past, it is also apparent that many thick successions of ancient alluvium probably accumulated during periods of relative 'steady state' or at least when the intervals of sea-level or hinterland gradient changes were comparable with the relaxation times of the alluvial system.

Apart from the purely scientific interest concerning the dynamics of alluvial stratigraphy there are pressing economic problems such as the prediction of reservoir positions, reservoir extent and interconnectedness and the prediction of channel courses that contain mineral resources (uranium, copper, placers) or which interfere with the economic extraction of resources (channel bodies 'cutting-out' coals). With these problems in mind, this paper outlines a simple quantitative approach to the simulation of stratigraphic sections through alluvial deposits. The first fruits of this approach link process studies of modern alluvial systems with the generation of regional stratigraphic sequences that may be compared with sections measured through ancient alluvium.

Discussion of Variables

The proportional divisions of alluvium between coarse channel deposits and overbank fines is controlled by channel type and magnitude, rates of lateral channel migration, avulsion and vertical floodplain accretion. These variables depend upon those climatic and geological factors that determine discharge magnitude and distribution, valley slope, type of sediment load and sediment yield.

---

[1]Department of Earth Sciences, University of Leeds, Leeds LS2 9JT, W. Yorkshire, U.K.

Channel Deposit Size In Static Channel Systems

The total area of a channel deposit in a vertical section normal to valley slope will depend upon channel sinuosity. In a high sinuosity system the maximum area of deposits around an active channel will be given by the product of the local mean meander belt width (Mw) and the channel bankfull depth (h). These meander belt deposits will comprise a number of interconnecting individual point bar deposits interbedded with arcuate screens of clay/silt grade 'clay plugs' (Fig. 1). In a static low sinuosity system the area of channel deposits will be given by the product of bankfull channel width (w) and some fraction of the bankfull channel depth. Channel widths range over 10-1000 h, with no relationship between the two parameters. Low sinuosity channel deposits comprise mesoform bedform deposits in sand or gravel with occasional lensoid screens of fines deposited in anabranches of the braided channel system during falling and low stage.

Channel Migration And Avulsion

It is known that high-sinuosity streams may be confined to their meander belt for long periods, but that a sudden flood-induced shift, known as an avulsion, will transfer the whole channel system to another part of the floodplain. Such a process leaves the abandoned meander belt in a state of perfect preservation and it is gradually buried by subsequent flood deposits. Such a model is one of the chief ways of producing distinct alluvial fining upwards cycles. Many rivers including the Meander (Russell, 1954), Rufiji (Anderson, 1961), Brazos (Bernard et al., 1970) and Mississippi (Fisk, 1944) show abandoned meander belts on their present floodplains and it is evident, particularly from Mississippi evidence, that realistic mean avulsion periodicity may be $1 - 2 \times 10^3$ years.

Two possible modes of avulsion are possible. Consider a funnel-shaped alluvial plain marked by one major channel system issuing from an upland drainage basin through the neck of the alluvial funnel (Fig. 1). In the first mode, the channel may always avulse at or close to the node of the alluvial plain, where a gradient change may exist between the confined valley upstream of this point and the free meanders downstream. Such a node is present in the Mississippi alluvial basin at Thebes Gap where the Mississippi River enters, and also near Cairo, Illinois, where the Ohio River joins the Mississippi. Several avulsions of both the Mississippi and Ohio occurred from these nodes between 5500 and 2500 BP (Fisk, 1944). Nodal avulsion will cause the number of channel bodies to be constant with respect to distance downvalley from the node. In the second mode of avulsion the channel can divert randomly along the length of its alluvial plain. This has occurred 4 times in the Mississippi basin since the Thebes Gap diversion and the post-glacial sea level still stand (Fisk, 1944). Random downvalley avulsion may cause the number of channel bodies to increase regularly downstream from the node, but may not if the channels rejoin the existing channel course at some point downstream. In the following analysis we shall assume for simplicity that in this second mode of avulsion the new channel position is not influenced by previous abandoned meander belt alluvial ridges on the river floodplain. It seems probable, however, that some degree of control is likely since an upstanding meander belt ridge may constitute a physical barrier until it subsides by compaction and is gradually buried by flood sedimentation. The effectiveness of alluvial ridges in influencing new channel positions will depend upon original relief, rate of subsidence and the avulsion periodicity. It is also not impossible that abandoned segments could become re-occupied by a later channel.

By way of contrast to high sinuosity channels, low sinuosity, braided channels are marked by greater instability which causes both short term channel migrations of a quasi-continuous nature and avulsions with a much shorter recurrence interval. There is thus a greater tendency for channel deposits to overlap both laterally and vertically.

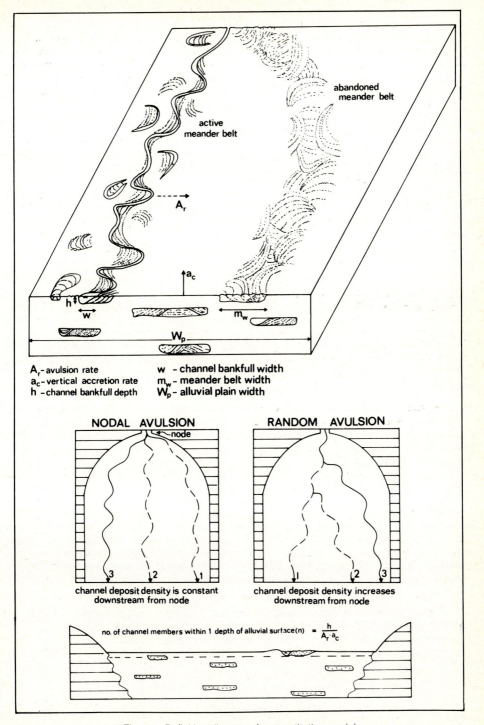

A$_r$ – avulsion rate
a$_c$ – vertical accretion rate
h – channel bankfull depth

w – channel bankfull width
m$_w$ – meander belt width
W$_p$ – alluvial plain width

NODAL AVULSION

channel deposit density is constant
downstream from node

RANDOM AVULSION

channel deposit density increases
downstream from node

no. of channel members within 1 depth of alluvial surface(n) $= \dfrac{h}{A_r \cdot a_c}$

Fig. 1. Definition diagrams for quantitative model.

Vertical Floodplain Accretion

The lateral movements of channels to and fro across an alluvial plain will be accompanied by a net upward accretion of the whole depositional surface due to flood sediment accretion in backswamps. If the accretion rate balances the tectonic subsidence rate and the valley slope remains constant then it is clear that the balance between the horizontal movement and vertical accretion will determine the resultant distribution of channel deposits within any given stratigraphic interval of alluvium.

As expected, values for the vertical accretion rate (a_c) vary widely as a consequence of channel size, floodplain width, flood sediment concentrations, overbank flood periodicity etc. Long term mean values of between 0.5 - 1.5 mm yr^{-1} are recorded from C_{14} and archaeological studies (Summary in Leeder, 1975). Working backwards from the known thickness of top stratum fines in the Mississippi basin that have accumulated over about 5000 years yields values of between 3 - 6 mm yr^{-1}. Such calculations ignore compaction (q.v.). Recent data from the 1973 4-year flood in the Mississippi basin points to an uncompacted minimum accretion rate of about 2.75 mm yr^{-1} for the alluvial backswamps (Kesel et al., 1974). The range of values quoted above are for backswamp accretion, it is well known that higher rates pertain in areas close to the channel such as levees and point bar tops.

The thickness of overbank deposits produced by a particular accretion rate will decrease with depth due to compaction as accretion proceeds. Thus even with a constant surface accretion rate balanced by tectonic subsidence there will exist, after a lengthy time interval, a downward trend towards increasing channel deposit density. Channel deposit interconnectedness on the other hand will not be affected by compaction. We should thus distinguish a surface accretion rate from a fully compacted, nominal accretion rate (a'_c). In the subsequent discussion we will make no attempt to integrate a_c with respect to depth and will rely on a'_c to determine the deep stratigraphy of alluvial sediments.

The problem of estimating compaction rates in alluvial overbank fines is likely to be a tricky one since they are likely to be greatly influenced by rates of surface dewatering and pedogenesis on the exposed alluvial surface. Compaction is likely to greatly affect accretion rates derived from C_{14} dating of alluvial deposits, a point largely ignored by workers in this field.

Width Of Alluvial Plain

All of the effects so far discussed take place on the surface of an alluvial plain of definite width. The side-limits to avulsion and the effects of vertical accretion will thus be drawn at the marginal interfluves. In all modern alluvial plains, interfluves are usually prominent bluff-lines resulting from incision and part-burial due to Quaternary sea-level changes. We can obviously have little idea of the nature of bounding areas between ancient alluvial systems on wide coastal plains of alluviation that remained in a 'steady state'.

All other variables being constant, we can say that the narrower the alluvial plain the greater tendency there will exist for channel interconnectedness, culminating in the production of a multistorey, valley-fill deposit when the width of the alluvial plain approximates to the width of the channel system or meander belt.

DERIVATION OF THE QUANTITATIVE MODEL

Fractional area of channel deposits

Consider an alluvial plain containing a single major channel system that periodically undergoes avulsion of nodal type. Assuming steady subsidence, over a period of time, t, the number of individual channel deposits will equal t/A_r where A_r is the mean channel residence interval in years.

These channel deposits may be set in a 'matrix' of fine-grained backswamp or topstratum sediments. The total area of sediment in a section across the alluvial plain normal to the valley slope will be $a'_c \cdot t \cdot W_p$, where a'_c = 'compacted' vertical accretion rate and W_p = width of alluvial plain.

The total area of channel deposits in the section will be $C_a t/A_r$ where C_a is the area of a single channel deposit. The mean fractional area of channel deposits in a section will thus be

$$F_a = \frac{C_a}{A_r \cdot a'_c \cdot W_p} \tag{1}$$

For high sinuosity channels, $C_a = M_w \cdot h$, where M_w = meander belt width and h = channel bankfull depth. Thus

$$F_a = \frac{M_w \cdot h}{A_r \cdot a'_c \cdot W_p} \tag{2}$$

For low sinuosity channels, $C_a = W \cdot xh$ where W = channel bankfull width and $x = <1$. Thus

$$F_a = \frac{W \cdot xh}{A_r \cdot a'_c \cdot W_p} \tag{3}$$

However, the above derivations do not take into account the point that during its residence interval at a particular site on the alluvial plain the channel will undergo some degree of aggradation in response to the gradual buildup of the alluvial surface. This is the channel's response to floodplain accretion, enabling it to maintain its equilibrium channel cross section. The degree of channel aggradation will be determined by the product of the uncompacted accretion rate and the residence interval of the channel at a particular site. Thus the thickness of an aggraded channel deposit after a mean residence interval will be $h + (A_r \cdot a_c)$. Substituting into Eq. 2 and 3 gives the final expressions.

$$Fa = \frac{M_w \cdot (h + A_r \cdot a_c)}{A_r \cdot a'_c \cdot W_p} \quad \text{for high sinuosity channels} \tag{4}$$

$$Fa = \frac{W \cdot (xh + A_r \cdot a_c)}{A_r \cdot a'_c \cdot W_p} \quad \text{for low sinuosity channels} \tag{5}$$

Mean interconnectedness ratio

Consider a channel system on an alluvial plain which periodically avulses in nodal fashion by shift distances which are whole multiples of its deposit width. The probability of the channel bracketing on either side or directly coinciding with previous abandoned channel deposits on the alluvial surface (number = n) is, ignoring possible multiple contacts,

$$P = \frac{3n}{W'_p} \tag{6}$$

where W'_p is the total floodplain width expressed as a dimensionless multiple of whole channel deposit widths. Now n, the number of abandoned channel deposit targets on the alluvial surface (Fig. 1) is equal to the number of channels produced by avulsion in the time taken for the alluvial surface to accrete one whole channel deposit thickness (C_h). Thus ignoring compaction effects within one channel deposit thickness of the alluvial surface,

$$n = \frac{C_h}{A_r \cdot a_c} \qquad (7)$$

with n expressed to the nearest whole number value.

Substituting for n in Equation 6 yields the expression for the interconnectedness ratio, that is, the ratio of touching channel deposits to total channel deposits in an alluvial section. This approximate solution ignores the problems of complex probability theory arising due to multiple channel contacts. The approach is strictly valid only when Fa <0.5.

$$I_r = \frac{3C_h}{A_r \cdot a_c \cdot W'_p} \qquad (8)$$

Where for high sinuosity channels, $C_h = h$ and for low sinuosity channels $C_h = xh$ (x <1).

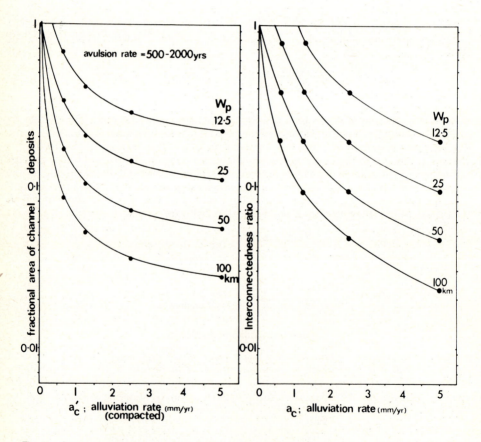

Fig. 2.   Graphs to show fractional area of high sinuosity channel deposits and interconnectedness ratio of channel deposits as a function of alluviation rate and width of alluvial plain. Nodal avulsion rate considered random between 500-2000 years with a mean periodicity of 1250 years. Discussion in text. Meander belt width = 1 km; channel depth = 5 m. Alluviation rate for F_a considered fully compacted ($a'_c = 0.5a_c$). Curves calculated from equations 4 and 8.

Discussion

Derivation of Equations 1 to 8 above both assume nodal avulsion in which F_a and I_r will both be constant along the length of the alluvial plain. In random downvalley avulsion, however, both parameters will decrease in value upstream. The decrease in values upstream will be controlled by the ratio L_s/L where L_s is the length of the alluvial plain down to the particular plane of section and L is the total length of the system.

In Figure 2, Equations 4 and 8 have been solved for high sinuosity channel deposits with various possible values of a'_c; a_c and W_p, and a mean avulsion periodicity of 1250 years (range 500-2000 yrs).

SIMULATION OF ALLUVIAL STRATIGRAPHY

It is hoped that in the future, Equations 1-8 may be used as the basis for computer simulation of alluvial stratigraphy. For the moment an example of manual simulation is presented in Figure 3, using random number tables to determine the channel position and the residence interval of a channel at a particular site.

A particular consequence of the simulation models is the apparent concentration of channel deposits in certain areas of the hypothetical sections (Figure 3). The effect of such apparent concentrations is best illustrated by examining randomly chosen 2 km wide sequences across an alluvial plain section (Figure 4). Sequences A-F show how the nature of the complete alluvial section might be deduced from available outcrop evidence. Sections 4 and 5 both reveal no channel deposits and such a result might lead the stratigrapher or drilling crew to concentrate on the ground between sections 1 and 2. A similar decision might follow the results achieved after sequence E where the observer might infer the existence of 3 distinct fluvial systems separated by areas of ground containing no channel deposits. The true position if continuous exposure was available is shown in sequence F. The correct procedure for prediction in such circumstances is to assume a random distribution of channels and to base a stratigraphic classification or exploration strategy upon this assumption. If there was some reason to doubt a random model, then a *test* for spatial randomness of the data should be applied. Such conclusions might apply to the subsurface evaluation of the two distinct fluvial systems deduced for the Mt. Pleasant Series of Texas by Fisher and McGowen (1969). Sections such as those illustrated in Figure 3 clearly have some use in subsurface reservoir modelling.

CONCLUSIONS

The fractional concentration of channel deposits and the interconnectedness ratio of channel deposits in alluvial successions are given by two simple equations which relate these properties to parameters such as the avulsion rate, vertical accretion rate, channel geometry and the width of the alluvial plain. Simulation of stratigraphic sections through alluvium is demonstrated and reveals that random models of channel avulsion give rise to certain spurious concentrations of channel deposits that might be misleading to stratigraphers or economic geologists. Future work should set out to investigate the effect of compaction in alluvial successions and to use the computer to simulate alluvial stratigraphy employing more complex models in which periodic tectonic subsidence, compaction, lateral variability in accretion rates and alluvial ridge influences upon channel avulsion paths might be taken more fully into account.

594 M. R. LEEDER

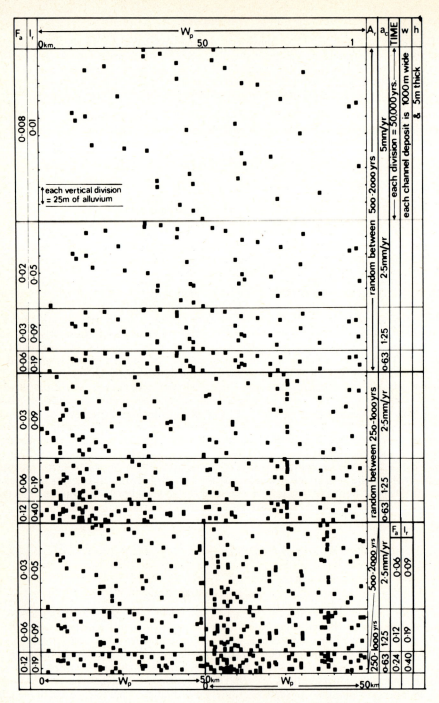

Fig. 3. Simulations of subsurface alluvial stratigraphy using random numbers for channel residence interval and flood plain position illustrating the effects of changing the controlling variables avulsion rate, alluviation rate and width of alluvial plain. All sections normal to palaeoslope of alluvial plain. Note $a_c = 2a'_c$. Solid squares are the channel deposits (to scale), unshaded is fine-grained alluvium.

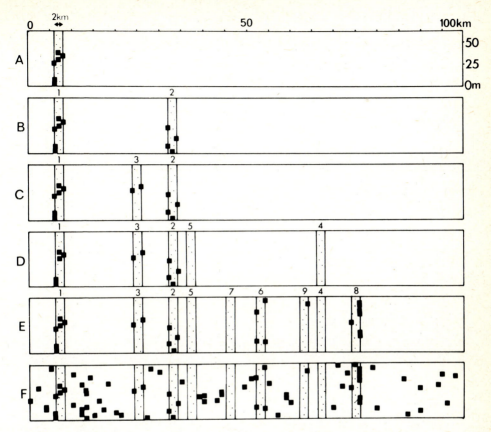

Fig. 4.   Sections to illustrate the effects of sampling random 2 km outcrops of a 100 km wide, 60 m thick stratigraphic interval through alluvium with channel deposits. Section F is the actual whole section. Alternatively the vertical lines bounding the 2 km sections may be taken to represent borehole locations.

ACKNOWLEDGEMENTS

I thank John Bridge, John Collinson, Alan Fowler, Peter Friend, Roscoe Jackson II, Tessa de Mowbray and Andy Gardiner for help and useful comments at various stages of this work. The paper was presented at the Calgary Fluvial Symposium with the aid of grants from the Symposium sponsors, the Royal Society and the University of Leeds, all of whom I gratefully thank.

REFERENCES

Anderson, B., 1961, The Rufiji Basin, Tanganyika: Vol. 7. Soils of the main irrigable areas: FAO rept. to Government of Tanganyika.

Bernard, H. A., Major, C. F. Jr., Parrott, B. S. and LeBlanc, J. R. Sr., 1970, Recent sediments of southeast Texas: Univ. of Texas at Austin, Bureau of Economic Geology, Guidebook 11, 16p.

Fisher, W. L. and McGowen, J. H., 1969, Depositional systems in the Wilcox Group (Eocene) and their relation to the occurrence of oil and gas: Am. Assoc. Petrol. Geol. Bull., v. 53, p. 30-54.

Fisk, H. N., 1944, Geological investigations of the alluvial valley of the Lower Mississippi River: Mississippi River Comm., Vicksburg, 78p.

Kesel, R. H., Dunne, K. C., McDonald, R. C., Allison, K. R. and Spicer, B. E., 1974, Lateral erosion and overbank deposition on the Mississippi River in Louisiana caused by 1973 flooding: Geology v. 2, p. 461-464.

Leeder, M. R., 1975, Pedogenic carbonates and flood sediment accretion rates: a quantitative model for alluvial arid-zone lithofacies: Geol. Mag., v. 112, p. 257-270.

Russell, R. J., 1954, Alluvial morphology of Anatolian rivers: Ann. Assoc. Am. Geogr. v. 44, p. 363-387.

LITHOFACIES TYPES AND VERTICAL PROFILE MODELS IN BRAIDED RIVER DEPOSITS: A SUMMARY

ANDREW D. MIALL[1]

ABSTRACT

This article serves as an introduction to the papers dealing with braided river deposits in this volume.

A lithofacies code erected earlier by the writer is expanded to include matrix-supported gravel, low-angle cross stratified sand, erosion surfaces with intraclast conglomerates, and massive mud deposits.

The four vertical profile models erected by the writer are expanded to six. A new model, the "Trollheim type" is proposed, to include gravelly deposits characterized by abundant debris flows. The Donjek sequence type is restricted to gravel-dominated cyclic deposits and a new model, the "South Saskatchewan type", is erected for sand dominated cyclic deposits. The Scott, Platte and Bijou Creek models remain essentially unchanged.

INTRODUCTION

In a recent review of braided river depositional environments I attempted to summarize published data on lithofacies types and facies associations by erecting a code system for lithofacies descriptions and four vertical profile models which, it was suggested, represent the most commonly occurring facies associations (Miall, 1977). Subsequently, work by other writers, particularly that reported in this volume, has shown that the code system and the vertical profile models can both be usefully expanded to include a wider range of depositional variability. The following brief discussion is offered as an attempt to integrate these new data into the review published earlier, and as an introduction to the more detailed papers on braided depositional models which follow.

LITHOFACIES CODES

A revised listing of the lithofacies types identified in braided river deposits is given in Table 1. The following are added to those originally proposed by Miall (1977):

Gms: proposed by Rust (this volume) for massive, matrix supported gravel, both clasts and matrix characterized by very poor sorting; interpreted to be of debris flow origin.

Se: erosional scours with a lag deposit of silt or mud intraclasts, proposed by Rust (this volume) and identical to facies SS of Cant and Walker (1976).

Sl: low angle (< 10°) cross-stratified sand, described by Cant and Walker (1976; their facies G), Rust (this volume) and McLean and Jerzykiewicz (this volume).

Facies *Sse*, *She* and *Spe* are proposed by Boothroyd and Nummedal (this volume) as eolian equivalents of facies *Ss*, *Sh* and *Sp*. However, detailed criteria for distinguishing these facies from those of water-laid origin are not provided.

Fm: originally suggested for mud or silt deposits, this facies is expanded to include massive, fine grained deposits a few to tens of centimetres in thickness, as proposed by Miall and Gibling (in press) and Rust (this volume).

[1]Geological Survey of Canada, 3303 - 33rd St. N.W., Calgary, Alberta, Canada, T2L 2A7.

Table 1. Lithofacies and sedimentary structures of modern and ancient braided stream deposits (modified from Miall, 1977, Table III).

Facies Code	Lithofacies	Sedimentary structures	Interpretation
Gms	massive, matrix supported gravel	none	debris flow deposits
Gm	massive or crudely bedded gravel	horizontal bedding, imbrication	longitudinal bars, lag deposits, sieve deposits
Gt	gravel, stratified	trough crossbeds	minor channel fills
Gp	gravel, stratified	planar crossbeds	linguoid bars or deltaic growths from older bar remnants
St	sand, medium to v. coarse, may be pebbly	solitary (theta) or grouped (pi) trough crossbeds	dunes (lower flow regime)
Sp	sand, medium to v. coarse, may be pebbly	solitary (alpha) or grouped (omikron) planar crossbeds	linguoid, transverse bars, sand waves (lower flow regime)
Sr	sand, very fine to coarse	ripple marks of all types	ripples (lower flow regime)
Sh	sand, very fine to very coarse, may be pebbly	horizontal lamination, parting or streaming lineation	planar bed flow (l. and u. flow regime)
Sl	sand, fine	low angle (<10°) crossbeds	scour fills, crevasse splays, antidunes
Se	erosional scours with intraclasts	crude crossbedding	scour fills
Ss	sand, fine to coarse, may be pebbly	broad, shallow scours including eta cross-stratification	scour fills
Sse, She, Spe	sand	analogous to Ss, Sh, Sp	eolian deposits
Fl	sand, silt, mud	fine lamination, very small ripples	overbank or waning flood deposits
Fsc	silt, mud	laminated to massive	backswamp deposits
Fcf	mud	massive, with freshwater molluscs	backswamp pond deposits
Fm	mud, silt	massive, desiccation cracks	overbank or drape deposits
Fr	silt, mud	rootlets	seatearth
C	coal, carbonaceous mud	plants, mud films	swamp deposits
P	carbonate	pedogenic features	soil

The code *Fl* was proposed by Miall (1977) for laminated sand, silt and mud formed in overbank environments. This lithofacies may be of minor importance in many braided river deposits and, as such, a term which groups together a variety of fine grained lithofacies may be satisfactory. However, many fluvial deposits contain thick and varied floodplain sequences, and a subdivision of such deposits may be desirable. McLean and Jerzykiewicz (this volume) propose the following facies types:

Fsc: siltstone, silty claystone or claystone, horizontally laminated to massive.

Fcf: claystone with freshwater molluscs.

These two facies are distinguished mainly by fossil content, the presence of molluscs indicating the existence of temporary backswamp ponds.

In addition to the above, the terms *P* for pedogenic carbonate, *Fr* for root beds (seatearth) and *C* for coal or carbonaceous mudstone, may be useful.

Care must be taken in using these lithofacies codes for descriptive and interpretive purposes, because they are not a universal panacea for sorting out the complexity of fluvial deposits. It is possible for many of the facies types to be found in more than one environment within a river; for example facies *Sp* could represent mid-channel linguoid or transverse bars or sand waves migrating across a sand flat; Boothroyd and Nummedal (this volume) describe an eolian variety. Facies *Sl* could represent crevasse splay deposits, or the fill of low relief scours, or antidunes. Each example of each lithofacies must be examined with care from the point of view of its scale, grain size, internal structures, orientation and facies associations, to ensure that facies of dissimilar origin are not grouped together under one descriptive code. However, bearing these reservations in

Table 2. The six principal facies assemblages in gravel- and sand-dominated braided river deposits.

Name	Environmental setting	Main facies	Minor facies
Trollheim type (G$_I$)	proximal rivers (predominantly alluvial fans) subject to debris flows	Gms, Gm	St, Sp, Fl, Fm
Scott type (G$_{II}$)	proximal rivers (including alluvial fans) with stream flows	Gm	Gp, Gt, Sp, St, Sr, Fl, Fm
Donjek type (G$_{III}$)	distal gravelly rivers (cyclic deposits)	Gm, Gt, St	Gp, Sh, Sr, Sp, Fl, Fm
South Saskatchewan type (S$_{II}$)	sandy braided rivers (cyclic deposits)	St	Sp, Se, Sr, Sh, Ss, Sl, Gm, Fl, Fm
Platte type (S$_{II}$)	sandy braided rivers (virtually non cyclic)	St, Sp	Sh, Sr, Ss, Gm, Fl, Fm
Bijou Creek type (S$_I$)	Ephemeral or perennial rivers subject to flash floods	Sh, Sl	Sp, Sr

Modified from Miall (1977) and Rust (this volume).

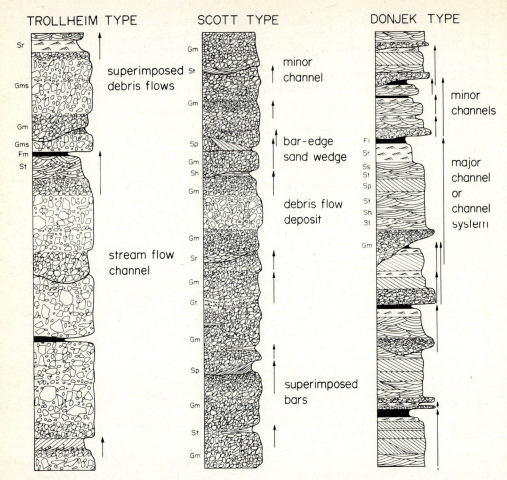

TROLLHEIM TYPE   SCOTT TYPE   DONJEK TYPE

Fig. 1. Vertical profile models for braided stream deposits. Facies codes to left of each column are given in Table 1. Arrows show small-scale cyclic sequences. Conglomerate clasts are not shown to scale.

mind, it is felt that the use of the code system should aid in standardizing lithologic descriptions and will facilitate comparisons between different fluvial sequences.

VERTICAL PROFILE MODELS

Four vertical profile models for braided rivers were erected by Miall (1977) on the basis of a survey of available information on modern sedimentary processes and ancient deposits. These models were intended to encompass all the variability found in braided rivers, but further work has shown that greater clarity could be achieved by subdividing two of the profile models, so that a total of six are now offered as a basis for interpreting the ancient record (Table 2, Fig. 1). These modifications are discussed briefly below.

The Scott model was erected for proximal braided stream deposits, including those occurring on alluvial fans, where gravel is the predominant facies, particularly facies *Gm*, with rare units of *Gp* and *Gt*, and some interbedded sandy channel fill deposits. Rust (this volume) refers to this as his G_{II} facies assemblage. A distinctly different gravelly facies

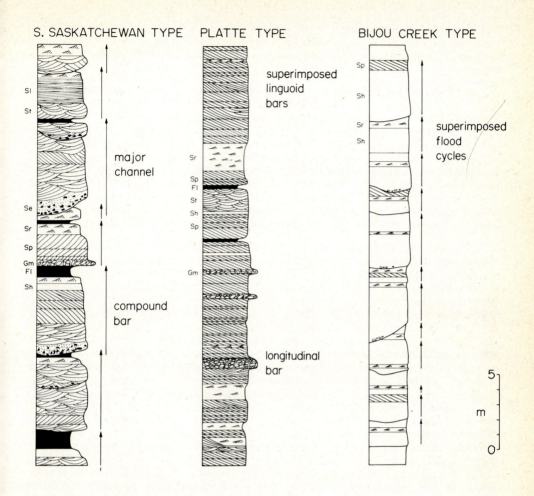

assemblage, termed G_I by Rust (this volume), is characterized by debris flow deposits (*Gms*), and by great lithological variability. Subaerial debris flows require a steep slope, an abundance of clastic debris and a high discharge for their initiation; these conditions are commonly, but not exclusively, met in arid or semi-arid environments where long, dry periods, during which abundant clastic detritus is generated by mechanical weathering, are punctuated by flash floods, with little or no vegetation to inhibit run-off. Alluvial fans with abundant debris flow deposits are particularly common in the desert areas of California and Nevada (Blackwelder, 1928; Blissenbach, 1954; Beaty, 1963, 1970; Bull, 1963, 1964; Bluck, 1964; Hooke, 1967, 1968) and the Trollheim Fan, California (Hooke, 1967) may be chosen as a modern analogue for this facies assemblage. Wasson (1977) describes some New Zealand examples.

Characteristics of the "Trollheim type" of vertical profile include the following (based mainly on Hooke, 1967; Wasson, 1977; Rust, this volume): 1. An abundance of poorly sorted, matrix-supported gravel of facies *Gms*. 2. Debris flow units may reach 3m in thickness, although superimposed flow deposits may not be readily distinguishable in outcrop. 3. Debris flow units commonly have flat (not channelled), generally abrupt bases and a lobate geometry, except where they infill stream flow channels. 4. Interbedded units of *Gm* generally are finer grained and may occupy prominent scours, reflecting the fact that stream flows require a lower slope than debris flows, and therefore commonly cause fan incision. Stream flow sheet flood deposits may also be present. 5. Minor units of *St*, *Sp*, *Fl*, and *Fm* may be present in crude fining-upward cycles (*see* Miall, 1977, for discussion of the mechanisms causing cyclicity in braided stream deposits).

Debris flows do not travel far from their source, so that the presence of Trollheim and Scott type sequences in the same braided river deposit may reflect within-fan proximal-distal variations, as suggested by Rust (this volume). Wasson (1977) records a similar down-fan change. However, tectonic, climatic or geomorphic effects may also influence fan composition (Hooke, 1967; Heward, this volume). Hooke emphasized the importance of sediment source types in generating debris flows, and demonstrated that different fans in the same tectonic and climatic setting may be characterized by different proportions of debris flow (*Gms*) and stream flow (*Gm*) gravels, depending on the availability in the source area of readily weathered detritus. Variations in source area relief or rainfall may also be a factor, and changes with time, such as scarp recession and downward erosion will also influence depositional processes in the alluvial basin.

The Donjek model was erected by Miall (1977) to encompass most types of cyclic braided river deposit. It is a common misconception that the deposits of braided rivers are disordered, whereas Miall (1977) listed a variety of mechanisms that can give rise to cyclic sequences. The Donjek River is one of the few modern braided rivers for which cycles have been clearly documented, particularly in the middle reaches studied by Williams and Rust (1969) and Rust (1972). However, most of the information available for the Donjek concerns gravel-dominated deposits. Recent work by Cant (this volume), Cant and Walker (1976; in press), Rust (this volume), Minter (this volume), Hobday (this volume) and Miall and Gibling (in press) shows that sand-dominated cycles are equally common in the ancient record. The only modern analogue for these deposits that has received sufficient sedimentological study to be used as a model is the South Saskatchewan River, Saskatchewan (Cant, Cant and Walker, *op. cit.*), and this is the basis for proposing another profile model, the "South Saskatchewan type" (Fig. 1), corresponding to facies assemblage S_{II} of Rust (this volume). In most cases facies *St* is the dominant component, with a varying proportion of *Sp, Sr, Sh, Se, Gm, Fl* and *Fm* arranged in a thinning- and fining-upward cyclic sequence. Markov chain analysis is particularly useful in studying this type of fluvial sequence (Miall, 1973; Cant and Walker, 1976; Miall and Gibling, in press). The reader is referred to Cant and to Cant and Walker (*op. cit.*) for details of the depositional processes and cyclic mechanisms prevailing in South Saskatchewan-type rivers.

The Scott, Donjek and South Saskatchewan profile types may form a gradational proximal-distal sequence in some ancient braided river deposits, reflecting a downstream decrease in gravel/sand ratio. It is proposed that the following numerical limits of gravel content be used to distinguish the three types: Scott >90%, Donjek 10 - 90%, South Saskatchewan <10%, where the total cumulative gravel thickness in a vertical section is expressed as a proportion of total section thickness.

The position of the Platte and Bijou Creek models (Table 2, Fig. 1) in this spectrum is at present unclear. Rust (this volume) did not include an equivalent of the Platte type in his discussion except as a variant of his assemblage S_{II}, but there is no doubting its existence as a discrete type (*see* examples quoted by Miall, 1977). It may represent a variety of the South Saskatchewan type, in which large bars and sand waves, rather than dunes, are the dominant depositional mode, but whether the difference relates to variations in channel topography or depth, flow velocity, discharge variations, or other causes is not clear. Further work on bedform hydraulics may throw some light on this problem.

The Bijou Creek type (Miall, 1977) was equated by Rust (this volume) with his S_I facies assemblage. It is interpreted as a proximal sandy braided stream deposit, occurring in areas that lack a gravel supply. The most characteristic feature of the Bijou Creek model is the evidence it contains of high energy flow conditions, in particular, the abundance of facies *Sh*, representing an environment dominated by flash floods, possibly ephemeral in nature, and contrasting with the perennial, less variable flow of the Platte, South Saskatchewan and other sandy braided rivers. Rust (this volume) discusses a variant of the S_I assemblage containing an abundance of facies *Se, Sl* and *St*.

Discussion

Undoubtedly our ideas about braided river deposits will be modified by future work, and the lithofacies code system and the facies models may require further expansion or modification. Both are designed in a flexible way to accommodate such improvements, and it is to be hoped that in future research important observations will not be glossed over in an attempt to force-fit every braided fluvial sequence into the published mould.

What is particularly needed now is more information on lateral variability in individual braided river deposits. A problem with Markov chain analysis is that it focusses attention on vertical profiles, whereas information regarding lateral variability may be critical in arriving at correct interpretations. Jackson (this volume) and Nijman and Puigdefabregas (this volume) point out that sand or gravel dominated point bar deposits in some single-channel, high-sinuosity rivers are very similar to South Saskatchewan or Donjek type braided river profiles, and could be identified as such in the absence of outcrop-scale information regarding facies geometry, lateral accretion surfaces and channel dimensions. Detailed paleocurrent analysis may also be of assistance in identifying channel morphology.

It should also be pointed out that the profiles tend to emphasize channel processes, it being assumed that overbank deposits are of little importance in braided river sediments. This may not always be correct. McLean and Jerzykiewicz (this volume) and Friend (this volume) describe examples of fluvial sequences with channel fills similar to those described in this paper, yet containing thick overbank deposits. A mechanism of lateral channel restriction on the floodplain, coupled with rapid subsidence, may be the cause.

On a broader scale, how do the six profile types relate to one another? Can some super-assemblages be erected to encompass proximal-distal variability in environments with different climates, sediment calibre and discharge characteristics? Most of the examples of braided river deposits described in the literature include only one or two of the facies models described in this paper, so that information on gross lateral facies variability is sparse.

Acknowledgments

Thanks are due to J. R. McLean, D. Long and N. D. Smith for stimulating discussions of the braided fluvial environment. M. Cecile, D. J. Cant and B. R. Rust read the manuscript and provided many useful comments.

References

Beaty, C. B., 1963, Origin of alluvial fans, White Mountains, California and Nevada: Ann. Assoc. Am. Geogr., v. 53, p. 516-535.

————, 1970, Age and estimated rate of accumulation of an alluvial fan, White Mountains, California: Am. J. Sci., v. 268, p. 50-77.

Blackwelder, E., 1928, Mudflow as a geologic agent in semiarid mountains: Geol. Soc. Am. Bull., v. 39, p. 465-480.

Blissenbach, E., 1954, Geology of alluvial fans in semiarid regions: Geol. Soc. Am. Bull., v. 65, p. 175-190.

Bluck, B. J., 1964, Sedimentation of an alluvial fan in southern Nevada: J. Sediment. Petrol., v. 34, p. 395-400.

Bull, W. B., 1963, Alluvial fan deposits in western Fresno County, California: J. Geol., v. 71, p. 243-251.

——, 1964, Alluvial fans and near surface subsidence in western Fresno County, California: U.S. Geol. Surv. Prof. Paper 437-A.

Cant, D. J., and Walker, R. G., 1976, Development of a braided-fluvial facies model for the Devonian Battery Point sandstone, Quebec: Can. J. Earth Sci., v. 13, p. 102-119.

—— and ——, in press, Fluvial processes and facies sequences in the sandy, braided South Saskatchewan River, Canada: Sedimentology.

Hooke, R. LeB., 1967, Processes on arid-region alluvial fans: J. Geol., v. 75, p. 438-460.

——, 1968, Steady-state relationships on arid-region alluvial fans in closed basins: Am. J. Sci., v. 266, p. 609-629.

Miall, A. D., 1973, Markov chain analysis applied to an ancient alluvial plain succession: Sedimentology, v. 20, p. 347-364.

——, 1977, A review of the braided river depositional environment: Earth Sci. Revs., v. 13, p. 1-62.

——, and Gibling, M. R., in press, The Siluro-Devonian clastic wedge of Somerset Island, Arctic Canada, and some regional paleogeographic implications: Sediment. Geol.

Rust, B. R., 1972, Structure and process in a braided river: Sedimentology, v. 18, p. 221-246.

Wasson, R. J., 1977, Last-glacial alluvial fan sedimentation in the Lower Derwent Valley, Tasmania: Sedimentology, v. 24, p. 781-800.

Williams, P. F., and Rust, B. R., 1969, The sedimentology of a braided river: J. Sediment. Petrol., v. 39, p. 649-679.

DEPOSITIONAL MODELS FOR BRAIDED ALLUVIUM

BRIAN R. RUST[1]

ABSTRACT

Braided alluvial deposits form in rivers and on alluvial fans and plains, and comprise three lithotypes: gravel-, sand- and silt-dominant. The gravel lithotype is characterised by framework-supported gravel, which in proximal deposits is mainly horizontally bedded and imbricate, occurring together with matrix-supported (debris flow) gravel in alluvial fan deposits. Horizontally bedded gravel accumulates on longitudinal bars, which are thought to originate as primary bedforms of high stage flow. Distal gravel deposits are characterised by cycles in which trough cross-bedded framework-supported gravel fines upwards to sand (commonly trough cross-stratified) and massive to laminated mud. The coarser components of the cycle are interpreted as deposits of an active braided tract; grain size decreases as the tract aggrades and becomes inactive.

Proximal sandy braided deposits typically lack mud in primary layers, but it is abundant as intraclasts on erosion surfaces. Horizontally stratified, low-angle (<10°) stratified, and trough cross-stratified sand are also abundant, in vertically and laterally variable successions. Distal sandy braided deposits commonly have fining-upward cycles, lateral continuity, and a significant primary mud content. They are transitional to deposits of meandering systems.

Silty braided deposits are typically traction- rather than suspension-laid, and lack association with coarser channel sediments. They are rare today, because most fine-grained alluvium accumulates on floodplains of meandering systems, for which vegetation is an important stabilising component. Terrestrial vegetation was essentially absent before Late Paleozoic times, when braided alluvium, including the silt-dominant type was widespread, particularly on alluvial plains of humid regions.

INTRODUCTION

Compared with meandering fluvial systems, the processes and products of braided rivers are poorly understood, partly because of their greater variability, and partly from earlier emphasis on meandering rivers. However, our knowledge of braided alluvial systems has now advanced to the stage where sedimentary models can be proposed (Walker, 1976; Miall, 1977, this volume). The present paper approaches the model concept by distinguishing three lithological types of braided alluvium, which are subdivided on the basis of facies assemblages, in relation to major environmental controls. It is based primarily on the author's studies of modern fluvioglacial systems and Paleozoic alluvial deposits, of which the latter will be described more fully elsewhere. This approach combines the advantages of modern-system studies (documented relationships between sediments and their environmental controls) with those of extensive vertical sections in ancient successions.

The term "braided" is used here for multi-channel water courses of low sinuosity, a definition discussed in more detail elsewhere (Rust, this volume). The term "alluvial" includes well-defined rivers, as well as the more dispersed channel systems of alluvial fans and plains. Deltaic systems are excluded here because they are commonly distributary rather than braided, and are generally composed of high-sinuosity (anastomosing) channels.

---

[1]Department of Geology, University of Ottawa, Ottawa K1N 6N5, Canada.

605

BRAIDED ALLUVIAL ENVIRONMENTS

Modern Environments

Braided alluvium accumulates in three intergradational environments: braided rivers, alluvial fans and alluvial plains. As there are many descriptions of modern braided alluvial environments in the literature, it will suffice here to discuss the major differences between the three types, and their influence on the sedimentary record.

Braided rivers are confined between valley walls, and cover most of the valley floor during floods. Fans and plains spread in two dimensions, but even in flood only a small part of the surface is inundated. In terms of geological time, however, all parts of the fan or plain surface become active within a relatively short period. Fans differ from plains in being localised features, with an arcuate morphology centered on an apex, anchored where a drainage system emerges from a mountain front. They occur today mostly along tectonic fronts in semiarid areas (Bull, 1964), or bordering recently deglaciated valleys (Church and Ryder, 1972; Rust, this volume, Fig. 5). Alluvial plains slope in one general direction, and result from drainage of extensive glacial margins (Nummedal *et al.*, 1974), or from distal coalescence of alluvial fans or valley-confined braided rivers.

The different environments discussed above give rise to sediment bodies of differing size, shape and internal composition. Braided river deposits are essentially linear, whereas alluvial fan and plain deposits are two dimensional, the latter being much more extensive. Fan deposits differ from those of rivers and plains in that they represent a more immediate response to adjacent high relief, and therefore show more abrupt proximal-distal facies changes, and commonly contain debris flow deposits (discussed later). In view of these distinctive features, it is unfortunate that several types of alluvial landform have been classed as fans. For example, the Scott River, Alaska was termed an outwash fan by Boothroyd and Ashley (1975), but it has a glacial source, which migrates as the ice margin shifts. The Scott thus lacks a fixed apex, and its morphology and sediment types are more like those of braided rivers.

The sediments of modern braided rivers and alluvial plains are very similar, and they are discussed together in this paper, although descriptions of modern deposits are mostly from rivers. One distinction is that alluvial plain deposits are much more extensive normal to the mean transport direction, and are more likely to be encountered in the ancient record. They also tend to be transitional upslope to alluvial fan deposits, whereas if fans are present in braided river systems they occur as lateral tributaries (Rust, 1972a).

Ancient Environments

There are significant differences between pre-Late Paleozoic and younger braided alluvial deposits, mainly due to the development of terrestrial vegetation during the Late Paleozoic (Schumm, 1968; Cotter, this volume). Today, braided alluvium is largely restricted to drainage systems with highly varied discharge and reduced plant activity, but it was much more widespread in Devonian and earlier times. The pre-Late Paleozoic record is probably biased towards braided alluvial deposits of humid regions, because of increased sediment yield under conditions of high and consistent rainfall, without vegetative protection. Floods were deeper and more frequent than in today's relatively ephemeral braided streams, the large volume of detritus building extensive braided alluvial plains, which are comparatively rare features of the modern landscape. In contrast with younger deposits, we can therefore expect that pre-Late Paleozoic braided alluvium was more abundant, and was deposited mostly on alluvial plains, in response to deeper and more frequent flood flows.

The above conclusions also apply to alluvial fan deposits, but the relative abundance of debris flow deposits in pre-Late Paleozoic versus younger fans is uncertain. Debris flows occurred in pre-vegetation times, for they have been recognised in Precambrian fan deposits by Selley (1965, p. 375) and by Maycock (cited by Bull, 1972, p. 79). According to Bull (1964, p. A22) debris flows are favoured by the presence of sufficient clay to make unconsolidated material slippery when wet, steep slopes, short periods of abundant water, and insufficient vegetative protection. McGowen and Groat (1971) suggested that alluvial fan deposits in the Ordovician or older Van Horn Sandstone lacked debris flows because of low clay mineral production in the source area, due to the absence of vegetation and to lithological factors. However, relative shale abundances of Phanerozoic and Precambrian strata (Garrels and MacKenzie, 1971, Table 9.4) show no evidence for significantly lower clay production before terrestrial plants evolved. As postulated above, the pre-Carboniferous alluvial fan record is probably biased towards deposits in regions with high and consistent rainfall. These conditions favour steady erosional processes rather than mass movement, but on the other hand the absence of vegetation should enhance debris flows. This is not certain, however, for partial plant cover aids retention of soil moisture, and in some cases may actually promote debris flow production. It is concluded that alluvial fan deposits were more abundant in pre-vegetation times, and contained debris flow layers, unless source lithologies were such as to prevent the production of fine sediment. Further work is needed before we can determine whether the incidence of debris flows has increased or decreased with the development of terrestrial vegetation.

Major Lithotypes of Braided Alluvium

Braided alluvial deposits can be divided into three types on the basis of their dominant lithology: gravel, sand and silt. Within these lithotypes further subdivisions (facies) can be recognised, chiefly on a combination of lithology and sedimentary structures (Table 1). In many cases there is a downstream transition from one lithotype to another, for example from gravel- to sand-dominant reaches in the Platte Rivers (Smith, 1970, 1971). The Slims River, Yukon shows transitional changes within its 22 km length from gravel- to sand- to silt-dominant reaches (Fahnestock, 1969). However, in most cases (fans excepted) each lithotype is extensive enough to be described separately, with its particular assemblage of facies, some of which may be shared with other lithotypes. For convenience, alluvial fans are treated as a single entity in this paper, and are discussed as a gravel lithotype.

The distinction made here between gravel- and sand-dominant systems is that the former are characterised by framework-supported gravel (framework gravel for short), with no upper size limit. Gravel-sized particles may be present in sandy deposits, but they are dispersed, unless reworked as lag concentrates. In effect, this criterion distinguishes between rivers which in flood transport gravel as bedload and sand in suspension, from those which have a mixed sand/pebble bedload. In the former case gravel is deposited first during falling stage, and is infiltrated later by sand, or occasionally is preserved in an openwork state. Harms et al. (1975, Fig. 7-3) presented a graph relating the size of gravel rolling on a bed to that of sand suspended by the same flow. A misfit between the two components in a gravel indicates a distinct decrease in flow strength before sand infiltration occurs; this is commonly the case in alluvial framework gravels. In contrast, pebbly sand results from essentially simultaneous deposition of both fractions, perhaps with minor current fluctuation. In this case the coarser clasts are rarely larger than medium pebble size (about 15 mm).

Some meandering rivers are gravelly, for example the Endrick River, Scotland, which deposits pebbly sand at high stage, partly reworking it at lower stages into isolated layers of framework gravel (pers. comm., B. J. Bluck, 1972). Fining-upward cycles are common, and contain one to two thirds overbank silt and clay (Bluck, 1971, Figs. 21-23). Thus the Endrick deposits conform to the meandering fluvial model (Allen, 1965; Walker, 1976), except for a thicker gravelly base to the channel unit, and scattered pebbles higher in each cycle. Pebbly sand sequences reported by Jackson (1976) from the meandering lower

Table 1. Lithotypes and facies assemblages of braided alluvium. Facies are arranged in order of approximately decreasing frequency from left to right, symbols explained in text.

Lithotype	Facies	Special characteristics
Gravel Fac. assemblage G$_I$: (alluvial fans)	*Gms, Gm, St, Sp, Fl/Fm*	Great lithological variation. Crude cycles may be present.
Fac. assemblage G$_{II}$: (Proximal braided rivers and alluv- ial plains)	*Gm, Gp, Sp, Sh*	*Gm* dominant; usually with well-developed imbrication. *Gp* more abundant in pre- U. Paleozoic deposits.
Fac. assemblage G$_{III}$: (Distal braided rivers and all- uvial plains)	*Gt, St, Gm, Sh, Fl*	Fining-upward cycles.
Sand S$_I$: Fac. assemblages: (Proximal braided rivers and allu- vial plains)	*Sh, Sp, Sr* (Bijou Ck. type) *Se, Sl, St (Ss)* (Malbaie type)	*Sh* dominant. Vertical and lateral variability.
Fac. assemblage S$_{II}$: (Distal braided rivers and all- uvial plains)	*Se, St (Ss), Sp, Sr, Fl/Fm*	Fining-upward cycles. Transitional to deposits of meandering rivers.
Silt (Distal braided rivers and allu- uvial plains)	*Fl, Fm*	Lack of association with sandy channel deposits. Eolian reworking and thix- otropic deformation common.

Wabash River are basically similar to those of the Endrick. Also similar are bed sediments in the meandering Amite and Colorado Rivers, which mostly have mean sizes in the medium to coarse sand range (McGowen and Garner, 1970, Figs. 22 and 23), although the authors' descriptions emphasised gravelly material (Rust, in press). The Amite deposits are essentially sand and pebbly sand, locally concentrated into lag veneers, but they are unusual in their lack of fining-upward sequences. This is due to the introduction of coarse material into the upper parts of point bar units by chute bar migration during flood (McGowen and Garner, 1970). The Endrick, Wabash and Amite sediments are unusually gravelly for meandering rivers, but they can be distinguished from the deposits of coarse braided systems by the presence of framework gravel only as localised lag concentrates.

The distinction made here between sand- and silt-dominant systems is that the principal lithology of the former is sand with very little interstitial silt or clay, the fine fractions accumulating elsewhere as separate minor bodies. In silty systems the main lithology is an intimate mixture of very fine sand and silt, with mean size commonly close to the sand-silt boundary (4ϕ, 0.0625 mm).

FACIES

Miall (1977) devised a letter code for facies nomenclature which is more logical than previous proposals, and is flexible enough to be adapted to various schemes. It is adopted here with the modifications and additions described below (see also Miall, this volume).

Gravel Facies (G)

Gm: Massive or horizontally bedded framework gravel, commonly imbricate.

Gms: Massive, matrix-supported gravel. This facies is commonly coarse, with angular megaclasts and a poorly sorted matrix ranging in size from pebbles to mud. In addition to matrix support, facies *Gms* commonly is distinguished from facies *Gm* by its lack of bedding and imbrication.

Other symbols for gravel facies are used here with the same meaning as that of Miall (1977), namely:

Gp: Planar cross-bedded framework gravel and
Gt: Trough cross-bedded framework gravel.

Sand Facies (S)

Two additions to the sand facies defined by Miall (1977) are proposed here:

Sl: Low angle (<10°) stratified sand. This is the same as facies G of Cant and Walker (1976, p. 110), and was briefly mentioned by Miall (1977, p. 30). Extensive cliff exposures in the Malbaie Formation (described below) show that dipping strata of facies *Sl* are laterally transitional to horizontally stratified sandstone. Both dipping and horizontal strata commonly bear primary current lineation, which may be parallel, perpendicular or oblique to the strike of dipping strata. Facies *Sl* is interpreted as the result of shallow, high velocity flow into low-relief scours.

Se: Erosional scours in sandstone successions. Identical to facies SS of Cant and Walker (1976, p. 104), this feature is included as a sand facies because of its abundance in sandstone successions. The erosion surface is commonly overlain by a layer of mudstone intraclasts varying from one intraclast to several meters thick. The layers commonly have a matrix of massive coarse sand, but in some cases they are entirely composed of intraclasts.

Other symbols for sand facies are used here essentially as defined by Miall (1977):

St: Trough cross-stratified sand.
Sp: Planar cross-stratified sand.
Sr: Ripple cross-laminated sand.
Sh: Horizontally stratified sand.
Ss: Scour-fill sand.

Fine-Grained Facies (F)

Redefinition is proposed here for one of the fine-grained facies of Miall (1977, p. 30):

Fm: Massive fine sandy mud or massive mud. Miall (1977, p. 30) used *Fm* for mud or silt drape, but in most cases drapes can be accomodated as thin units of facies *Fm* (as defined here) or *Fl*, which is used here with the same meaning as Miall, namely:

Fl: Laminated or cross-laminated very fine sand, silt or mud.

GRAVEL-DOMINANT BRAIDED SYSTEMS

Facies Assemblage G$_I$ Alluvial Fans

The most distinctive features of alluvial fan deposits are the highly varied facies assemblages, and the presence of debris flow deposits (facies *Gms*), as discussed previously. Variability is apparent on a local (within outcrop), and on a fan-wide scale. Fans commonly show, within a few kilometers, downslope facies trends that require tens or hundreds of kilometers in braided rivers or alluvial plains. The major trends are decreased mean and maximum size, and increased sorting and clast roundness (Blissenbach, 1954).

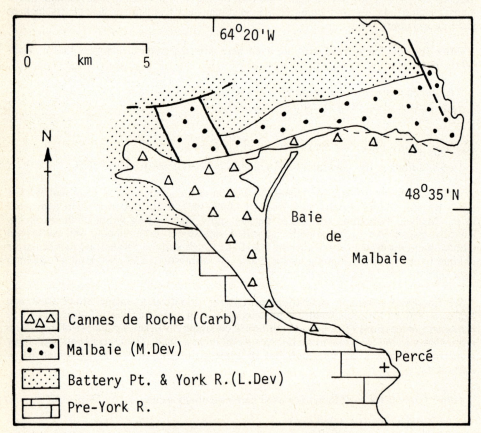

Fig. 1. Simplified geological map of the Baie de Malbaie region, Gaspé, Quebec. Small outliers of the Malbaie and Cannes de Roche Formations to the west of the main outcrop areas have been omitted.

Examples of ancient alluvial fan deposits are present in the Cannes de Roche Formation of Gaspé, Quebec, Canada (Fig. 1), a Mississippian succession with three distinct members (Alcock, 1935; McGerrigle, 1950). The Lower Member is composed chiefly of sharpstone conglomerates, which are locally very coarse, and are interpreted as debris flow deposits. Bedding is absent, except where revealed by impersistent lenses of silty sandstone (Fig. 2). The clasts are sharply angular, range up to 1 m in maximum dimension, and are commonly supported by a matrix of muddy sandstone, although matrix proportion rarely exceeds 50%. Imbrication has not been observed; clast

Fig. 2. Matrix-supported gravel (facies *Gms*) in the Lower Member, Cannes de Roche Formation, interpreted as a debris flow deposit. Scale 20 cm.

orientation is commonly random, but in some of the coarser deposits tabular particles are subhorizontal (Fig. 2). In a few of the finer, more distal deposits flat megaclasts make high angles with the bedding, being randomly distributed about a subvertical axis. Bull (1972, p. 70-71) interpreted horizontal as opposed to vertical clast orientation as a function of relatively low and high flow viscosity respectively. According to Johnson (1970, p. 433-434), the characteristically steep terminations of debris flows indicate that they have finite strength, as well as viscous properties. This suggests that the subvertical clast orientation in some finer Cannes de Roche debris flow deposits is due to increased viscosity and strength in distal lobes, which were close to stopping as a result of water loss by infiltration.

The distal debris flow deposits are interbedded with and dominated by water-laid sediments arranged in crudely-developed cycles. The cycles comprise fine-grained framework conglomerate, passing up gradationally into sandstone and silty sandstone, which at the top commonly has a concretionary limestone, interpreted as pedogenic calcrete (Fig. 3). The conglomerates are mainly horizontally stratified and imbricate, while planar and trough cross-stratification characterise the sandstones. In several cases the conglomerate is present in channels with sharply erosional bases; most channels are wide and shallow, but a few have steep lateral margins. The channeled conglomerates and cross-stratified sandstones respectively resemble what Bull (1964) termed stream-channel and braided-stream deposits on modern alluvial fans in California. However, most of the Cannes de Roche conglomerates are not channeled, and their gradational association with sandstones in cyclic successions implies a common origin in response to waning flood cycles. In other words, gravel is not necessarily limited to channels, but can spread out as a braided deposit on the proximal fan surface if stream power is high enough, as is the case on some modern coarse-grained alluvial fans, such as Spring Fan, Yukon (Koster, 1977).

Fig. 3.   Water-laid sediments in relatively distal deposits of the Lower Member, Cannes de Roche Formation. *P*: Concretionary limestone, interpreted as pedogenic calcrete; *Gm* Horizontally stratified conglomerate; *Fl*: Mudstone; *St*: Trough cross-stratified sandstone; *Sh*: Horizontally stratified sandstone. Scale 50 cm.

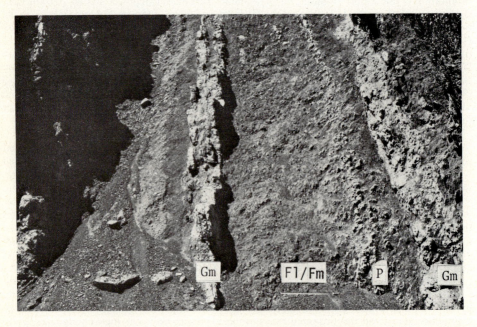

Fig. 4.   Typical succession in Middle Member, Cannes de Roche Formation. *Fl/Fm*: Laminated and unlaminated mudstone. Scale 50 cm.

The Middle Member of the Cannes de Roche Formation is interpreted as a distal fan/valley flat deposit. It is characterised by faintly laminated to apparently massive red mudstone and siltstone (facies *Fl* and *Fm*), with minor interbeds of fine conglomerate and horizons of concretionary limestone (Fig. 4). The conglomerates resemble the water-laid conglomerates of the Lower Member, but are finer-grained. They are almost entirely composed of thin (about 5 cm) horizontal strata, and are interpreted as the distal equivalents of the more extensive stream flow deposits of the Lower Member (Rust, in press).

Facies Assemblage G_{II}: Proximal Braided Rivers
and Alluvial Plains

The proximal outwash of the Donjek River (Rust, 1972a, 1975) resembles the water-laid deposits of alluvial fans, with *Gm* the dominant facies: horizontally bedded framework gravel (Fig. 5), commonly imbricate (Fig. 6), with *ab* dipping upstream and *a* transverse to flow (Rust, 1972b). It occurs in longitudinal bars, whose origin is basic to the processes of sediment accumulation and the formation of stratification. The commonly accepted model of bar genesis is that of Leopold and Wolman (1957), based on observations in a flume and of individual braid bars in small gravelly streams. They proposed that longitudinal bars are initiated during falling stage by deposition of the coarser fractions of the bedload in mid-channel. Subsequent growth takes place by addition of finer sediment on top of and downstream of the bar nucleus, which may rise above water, and become vegetated. However, it is not certain that this model is applicable to large gravelly braided systems, for which direct observation is impossible during flood stage, when major bar changes are to be expected.

Smith (1974) and Hein and Walker (1977) attempted to overcome this problem by studying the effects of diurnal rise of discharge on bar formation and migration in the Kicking Horse River, British Columbia. Smith termed active bars with predominantly

Fig. 5. Horizontally stratified framework gravel in a proximal reach of the Donjek River, Yukon (Area 1, Rust, 1972a); scale indicated by shotgun on outcrop (arrowed; about 1 m long).

depositional morphology 'unit bars', and observed their growth downstream and laterally. Hein and Walker (1977) postulated an initial stage of bar formation from 'diffuse gravel sheets', which develop into longitudinal or diagonal bars with horizontal stratification, or transverse bars with cross-strata. Again, unfortunately, it is not certain that these observations are a suitable model for processes occurring during major floods, which probably form the greater part of deposits preserved in ancient successions. As direct observation during floods is impossible, our only recourse is to deduction based on the features preserved in braided alluvium.

As already described, facies *Gm* dominates proximal outwash gravels, almost to the exclusion of cross-bedded gravel (McDonald and Banerjee, 1971, p. 1297; Church and Gilbert, 1975, p. 61 and Fig. 34; Rust, 1975, p. 240-1). The internal fabric of both horizontally stratified and apparently massive *Gm* gravel is consistent with deposition on subhorizontal surfaces (Rust, 1975, p. 245). The same combination of structure and fabric is prevalent in other types of coarse braided alluvium (Ore, 1964, p. 9; Smith, 1970, p. 2999). However, if Leopold and Wolman's (1957) model were valid for flood stage processes in rivers of this sort, one would expect cross-bedded gravel to be common, whether formed as low-angle cross-strata by migration of riffles, or as high-angle sets by slip face migration. Both processes were observed by Smith (1974, p. 219) in diurnally-induced small scale bar movements.

Another implication of Leopold and Wolman's model is that if longitudinal gravel bars form by coarse-fraction deposition during falling stage, the bars should be in a state of disequilibrium at stages high enough to transport all available material, and a flat bed would result if this stage were maintained long enough. That this is not so is indicated by the giant longitudinal bars of the Melon Gravel (Malde, 1968) and other deposits formed by catastrophic floods (Bretz *et al.*, 1956). These bars are so large (1.5 - 2.5 km long, 15 - 45 m high, Malde, p. 18) that they must have been primary structures, only superficially modified during falling stage. It is reasonable to assume that a similar condition holds for proximal braided alluvium during flood stage, that is, the longitudinal gravel bars are

Fig. 6.   Coarse imbricate gravel (facies *Gm*) in a cut bank, Donjek River (same reach as Fig. 5). Notebook is 19 cm long. Tributary alluvial fans can be seen in the background.

primary bedforms that equilibrate with the flow when all fractions of the bed material are in motion. Thus the braid bars are maintained during flood, and probably migrate upstream or downstream, or accrete vertically, depending on the flow conditions. The resulting deposit is characterised by poorly defined subhorizontal stratification because the low ratio of water depth to mean particle diameter suppresses the development of slip faces and the generation of cross-strata. This was not so with the Melon Gravel, for which Malde (1968) estimated water depths of up to about 100 m; hence the giant longitudinal bars developed slip faces, and generated giant sets of cross-beds.

An ancient equivalent of the proximal Donjek gravels is found in conglomerate units of the Devonian Malbaie Formation (McGerrigle, 1950; McGregor, 1973). The formation is located in eastern Gaspé (Fig. 1), and consists of abruptly alternated units of sandstone and conglomerate 10 to 200 m thick (Rust, 1976, 1977a). The latter units are mainly medium to coarse framework conglomerate (maximum clast size 10 - 40 cm), of which 80% is horizontally bedded, and 20% planar cross-bedded, with mean set thickness 1 m (Fig. 7). Sandstone lenses (facies *Sp* and *Sh*) form about 3% of the conglomerate successions.

Clast imbrication is essentially ubiquitous in the horizontally bedded conglomerate, with *ab* planes dipping upstream, and *a* transverse to paleoflow directions. Paleocurrents determined from this fabric are very consistent over the whole area (Fig. 8), with high vector magnitudes. Relatively low directional variance of gravel fabric was also noted in recent outwash by Bluck (1974), and is due to formation at high stage, when gravel is transported over bars and in channels with minimum deflection from the downslope direction. Paleocurrents derived from the Malbaie conglomerate cross-beds are more variable, but yield a similar overall mean vector. The cross-beds could be formed by migration of independent transverse bars, but this is considered unlikely. The cross-bed sets are relatively small (mean volume of 15 sets: 125 m³), commonly show lateral

Fig. 7. Horizontally bedded (facies *Gm*) and planar cross-bedded (facies *Gp*) framework conglomerate in the Malbaie Formation. Apparent curvature of cross-beds is largely an artifact of exposure. Notebook (arrowed) indicates scale: 19 cm long.

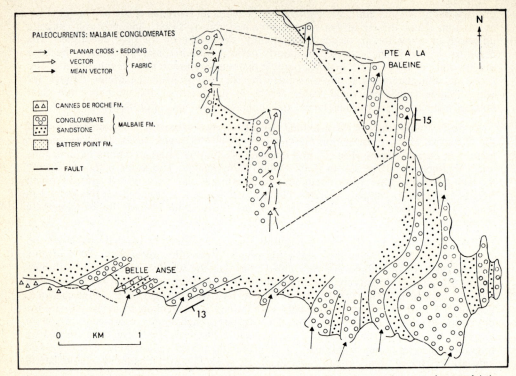

Fig. 8. Paleocurrent directions for Malbaie conglomerates. Note the consistency of mean fabric vectors over the whole area.

transition to horizontally bedded conglomerate, and make up only 20% of the conglomerate succession. It is therefore concluded that, with the possible exception of the largest cross-bed sets, they form as lateral modifications of longitudinal bars during falling stage, when flow diverges away from bar axes.

The Malbaie conglomerates closely resemble the proximal gravels of the Donjek and other similar rivers. The main difference is that the Malbaie conglomerate contains nearly 20% cross-beds, in sets averaging 1 m deep, whereas cross-beds are virtually absent from the modern gravels. This is interpreted as indicating that the longitudinal bars of the Malbaie had higher relief than those of modern systems; probably their other dimensions were also larger. This implies that the Malbaie bars formed in significantly deeper flood flows, which can be explained best by postulating a humid climate, in which floods were not only deeper, but also more frequent and of longer duration. An important effect of deeper and more prolonged flood flow is that the maintenance of longitudinal bars as primary bedforms was a much more significant stratification-forming process than it is for modern proximal braided gravels. In the modern situation, wide and rapid stage fluctuations cause considerable modification of the primary bars, for example by lateral migration of channels to form erosional banks (Rust, 1972a, p. 234).

The generalised cycle of the Scott type (Miall, 1977) is in most respects similar to the proximal Donjek gravels* and the Malbaie conglomerates. It comprises alternations of facies *Gm* and various sand facies, interpreted as formed under progressively decreasing

---

*To avoid confusion, it should be noted that Miall's Donjek model was named after the middle reaches of the river, described by Williams and Rust (1969). See Miall (this volume) for further discussion.

energy levels during flood cycles. Sand beds of similar type were observed on the proximal Donjek outwash (Rust, 1972a), but seldom surive the following flood, as indicated by their rarity (less than 1%) in vertical sections. Evidently most flood cycle deposits are interrupted, so that gravel units commonly occur in erosional contact with the underlying gravel unit. The Scott model of Miall (1977, Fig. 12) contains a debris flow deposit, but although they occasionally reach proximal braided rivers, debris flows are much more characteristic of alluvial fans (facies assemblage G_I of this paper).

Facies Assemblage G_{III}: Distal Braided Rivers and Alluvial Plains

Gravel persists as the dominant lithotype into distal reaches of many braided rivers. For example, Area 2 of the Donjek (Rust, 1972a) is 50 km from the river's glacial source, but has an active tract dominated by framework gravel, and inactive areas on which finer sediment accumulates and supports vegetation (Levels 1 to 4, Williams and Rust, 1969). Compared with proximal reaches, the finer gravel load and the relatively deep active channels probably produce bedforms with slip faces during high stage. Under similar conditions in the Knik River, Alaska, echo sounding by Fahnestock and Bradley (1973, p. 241) revealed dunes of fine gravel. Their three dimensional form was not recorded, but there were probably like the gravel bedforms with crescentic slip faces observed in the North Saskatchewan River by Galay and Neill (1967). Migration of such bedforms would generate sets of trough cross-strata (facies Gt), although they might be identified as planar (facies Gp) in small exposures of ancient rocks.

The Upper Member of the Cannes de Roche Formation shows many of the features described above (Rust, 1977b), with Gt the most abundant facies (Fig. 9). Facies Gt commonly occurs in multiple sets, with a sharply erosional base to the coset, and fines upwards to smaller sets of trough cross-stratified sandstone and pebbly sandstone (facies St). Minor sets of horizontally stratified fine framework conglomerate and sandstone are also present, usually above their respective troughed units, and in turn pass up into silty mudstone with minor troughed sandstone (facies Fl). Plant fossils are present as logs up to 5 m long in the conglomerate, and as abundant fine fragments in the sandstone and siltstone. Although the coarser units are well exposed, the finer sediments are commonly represented by covered intervals, and there are too few completely exposed cyclothems for statistical analysis of the succession.

On average, the multiple conglomerate/sandstone units make up 85-90% of the Cannes de Roche Upper Member, the rest comprising fine sediment (silty mudstone and minor sandstone). The succession is interpreted as a distal braided deposit like those of Areas 2 and 3 of the Donjek (Rust, 1972a). By analogy with the modern example, the conglomerate/sandstone units are interpreted as deposits of the active tract, while the mudstone and plant fragments accumulated on inactive areas. The vertical section at any one location represents gradual filling of an active channel complex, with dunes generating units of facies Gt. Continued aggradation in the channel and/or migration of a bar over the channel resulted in deposition of facies Gm, St or Sh. Eventually the active tract migrated elsewhere on the floodplain; the area became inactive, and started to accumulate mud and support vegetation, although minor channels still transported sand during flood.

The cyclic nature of the succession described above resembles the Donjek type model of Miall (1977, Fig. 13), except that framework conglomerate is much more abundant in the Cannes de Roche, as is framework gravel in the middle reaches of the Donjek River (Williams and Rust, 1969). Presumably this is due to the model being based mainly on ancient examples (Miall, 1977, p. 44 and Table V), most of which contain little conglomerate, but the method of model derivation was not described. The higher proportion of fine-grained deposits in the Cannes de Roche relative to the Donjek River is

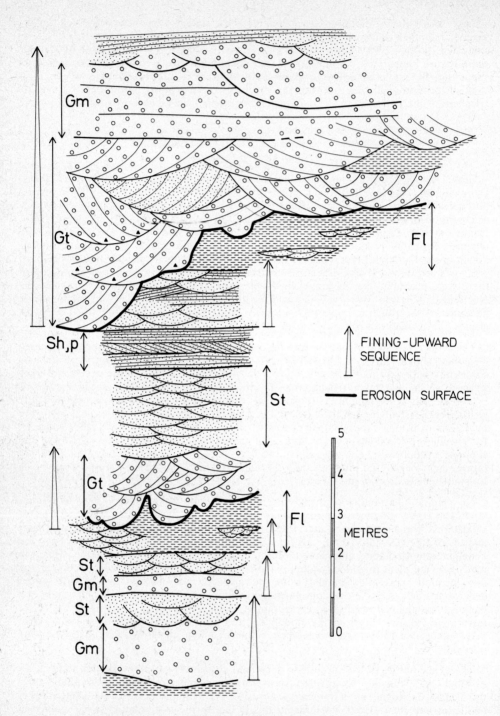

Fig. 9. Stratigraphic section of part of the Upper Member, Cannes de Roche Formation. *Gt*: trough cross-bedded conglomerate; *Sp*: planar cross-stratified sandstone; other symbols in earlier captions.

to be expected. The Donjek is confined within its glacial valley, whereas most ancient braided systems had space for extensive inactive areas to develop, and the relatively thick fine units and fining-upward channel sequences should not be considered unusual. Both these features are commonly regarded as typical of meandering fluvial deposits (Allen, 1965), but the presence of framework gravel as the dominant facies is regarded as evidence that the system was braided.

SAND-DOMINANT BRAIDED SYSTEMS

S_i: Facies Assemblages of Proximal Braided Rivers and Alluvial Plains

Sandy braided alluvium forms in a proximal environment only if gravel is not available. Since considerable relief and surges of runoff are implied, these conditions are rarely met. A modern example is an ephemeral flood in Bijou Creek, Colorado, which was competent to transport bridge girders, but deposited medium sand, of which 90-95% was horizontally stratified (facies *Sh*), with minor units of facies *Sp* and *Sr* (McKee *et al.*, 1967). Facies *Sh* is formed by shallow upper regime flow, and is analogous to the horizontal strata formed on longitudinal bars in proximal gravel systems.

A somewhat different assemblage in sandstone units of the Malbaie Formation comprises mainly facies *Se, Sl* and *St* (including *Ss*) (Fig. 10), and is also present in the Peel Sound Formation of Somerset Island, N.W.T. (pers. comm., M. R. Gibling, 1977). Intraclasts in facies *Se* range up to 40 cm in the Malbaie, giving an indication of the high competence of its streams. Grouped sets of trough cross-strata with 30 cm mean trough thickness (facies *St*), and similar isolate structures up to 1.5 m deep (facies *Ss*) are present, but in some cases the two types could not be distinguished on the geometric grounds

Fig. 10. Malbaie Formation sandstone. *Se*: Mudstone intraclasts on a scoured surface. *Sl*: Low angle stratified sandstone. The layers of mudstone intraclasts at the base of the upper *Sl* unit indicate a break in sedimentation, with intraclasts derived from an erosional scour upstream from this location. Scale 30 cm.

Fig. 11. Cliff section of Malbaie Formation sandstone showing typical lateral and vertical variability, mainly due to abundant scoured surfaces (facies *Se*, commonly only one intraclast thick). Scale 50 cm.

proposed by Cant and Walker (1976, p. 109), and they were therefore grouped as facies *St*. Much rarer, in order of decreasing abundance, are facies *Sp, Sr* and *Fl*. The most characteristic feature of the Malbaie sandstone succession is its vertical and lateral variability (Fig. 11), which, with only three major facies, makes it unsuitable for statistical analysis.

The Bijou Creek and Malbaie facies assemblages both reflect the large ranges in flow conditions of ephemeral streams. Facies *Sh* and *Sl* both result from upper regime flow, but the latter is thought to form in shallow scours. This feature, together with the abundance of scours and trough cross-beds in the Malbaie succession indicates that it was subject to deeper flow, implying a somewhat less ephemeral system than that which flooded Bijou Creek. The extreme conditions of the Bijou Creek flood are probably rare; Miall (1977, Table V) cited two ancient examples, but neither are characterised by the abundance of facies *Sh* found in the Bijou Creek deposit. It is concluded that the Malbaie type of proximal braided sandstone succession is likely to be more common in the stratigraphic record.

Facies Assemblage S$_{II}$: Distal Braided Rivers and Alluvial Plains

Miall (1977) used the Platte River (Smith, 1970, 1971) as a depositional model, but a closer integration of modern and ancient deposits of this type was achieved by Walker (1976) and Cant (this volume) for the South Saskatchewan River and the Devonian Battery Point Formation. Cant and Walker (1976) identified repetitive sequences in this formation, which fine upwards, starting with intraclasts on a scoured surface, interpreted as a channel-floor lag. This is overlain by trough cross-strata, analogous to channel-bed dunes in the river. They are in turn overlain by solitary sets of planar cross-strata with divergent paleocurrent directions, interpreted as deposits of sandy foreset bars migrating at high angles to the channel trend (Walker, 1976, Fig. 10). The sequence ends with rippled sandstone and mudstone (see Miall, this volume, for additional discussion).

Studies of other modern systems indicate some of the variety to be expected in ancient deposits due to variation in bedform shape and scale, in turn related to river size. Sandy foreset bars in the Lower Platte River have sinuous to lobate depositional fronts (Smith, 1971), which give rise to predominantly planar cross-strata (Smith, 1972). The Brahmaputra is notable for the magnitude of its bedforms, the largest having heights from 7.5 to 15 m and lengths of 180 to 900 m (Coleman, 1969). Disequilibrium between bedforms and flow during falling stage leads to erosion of crests and deposition in troughs. However, the larger the bedforms, the longer they persist (Allen, 1973), and the flood-stage structures are likely to be preserved, at least partly. Conaghan and Jones (1975) interpreted large cross-beds in the Triassic Hawkesbury Sandstone of New South Wales as partly preserved flood-stage structures of a braided river comparable in size to the Brahmaputra, although other interpretations have been made.

There are no absolute criteria for distinguishing the deposits of distal braided and meandering fluvial systems. In some cases it may be impossible (and relatively unimportant) to determine reliably the nature of the paleochannel pattern. A Devonian example of this type is present in the Peel Sound Formation at Cape Anne, Somerset Island, N.W.T. (Gibling and Rust, 1977). Markov analysis shows a repetition of the sequence: *Se, St* (large-scale followed by small-scale) *Sh* or *Fl* (Fig. 12). The latter two facies make up about 15% of the succession, and lateral continuity is good. The succession shows many aspects of a meandering sequence, but the presence of numerous minor scours and vertical transition into a section with a facies assemblage very like that of the Malbaie Formation sandstones (interpreted as proximal braided), implies an environment intermediate between distal braided and meandering.

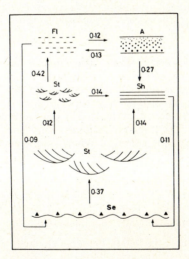

Fig. 12. Summary Markov diagram for 75 transitions in the Peel Sound Formation at Cape Anne, Somerset Island, N.W.T. (Gibling and Rust, 1977). Numbers indicate transition probabilities (Miall, 1973). A: Massive graded sandstone, rare in other successions, and not given a formal facies designation.

Silt-Dominant Systems

The Yellow River, China is a silt-dominant braided system (Chien, 1961), as is the lower reach of the Slims River, Yukon (Fahnestock, 1969), but neither author provided much information on sediment types. The sedimentological model proposed here, based mainly on a brief investigation of the lower Slims River, is therefore tentative.

Fig. 13. Aerial view of distal silty reach of the Slims River, Yukon, June 1967. The main channel flows away from the observer to the Alaska Highway bridge (directly below aircraft wingtip). Beyond the highway is the Slims delta and Kluane Lake. To the right of the main channel are minor channels and emergent silty bars; in right foreground is the distal portion of a tributary alluvial fan, Vulcan Creek.

In its silty reach the Slims has a single main channel, but this may be partly due to confinement by the abutments of the Alaska Highway bridge (Fig. 13). Many reaches of the Lower Yellow River are also dominated by a single main channel, and likewise are influenced to some extent by engineering works (Chien, 1961). Away from the main channel, the Slims is characterised by low relief bars and channels, which show little variation in sediment type. The principal facies are laminated and ripple cross-laminated sandy silt (*Fl*), and massive sandy silt (*Fm*). When water-saturated, the sediment is prone to thixotropic disturbance, like similar deposits in the Donjek River (Williams and Rust, 1969, p. 664), and it is possible that in many cases primary lamination is lost by secondary deformation. It is concluded that, despite its fine texture, much of the primary sediment is deposited by traction currents. Eolian reworking of dry bar surfaces is also an important phenomenon on the lower Slims. Isolated logs and accumulations of plant debris are present, and plants are colonising higher, marginal areas.

It was not possible to sample the bed of the main channel, which may transport coarser sediment than that observed on the bars. However, Fahnestock (1969, p. 166) reported that the bed as well as the banks are silty 1.8 km above the Alaska Highway bridge, and there is no evidence of coarser abandoned channel sediments on higher bar complexes. It is therefore suggested that dominance by silty facies, almost to the exclusion of coarser channel sediment, may be characteristic of silty braided systems. Abundant eolian reworking and soft sediment deformation are other typical features.

The existence of silty braided alluvium in the Yellow River appears to be largely due to provenance, for a vast area of the Yellow River basin is covered with loess (Chien, 1961). However, the silty braided reach of the Slims River cannot be explained in this way, for the Kaskawulsh Glacier provides material ranging from boulders to clay-sized to the Slims headwaters. Hence the downstream change to a silt lithotype is presumably a function of decreasing competence, largely related to the approach to base level at

Kluane Lake (Fig. 13). The Slims and Yellow Rivers are unusual, however, for silty braided alluvium is a rare feature of the modern landscape. Today, most fine sediment in alluvial environments is deposited from suspension on the floodplains of meandering rivers, or to a lesser extent in sandy or gravelly braided rivers. In all cases vegetation has an important function in stabilising the alluvial surface and trapping sediment. It is probable that in pre-Late Paleozoic times, in the absence of terrestrial vegetation, large amounts of silt were deposited on distal reaches of braided alluvial plains. Deposits of this type have not been widely recognised in the ancient record, probably because the likelihood of their existence has not been realised previously.

CONCLUSIONS

1. Alluvial fans form as a response to high contrast in relief, and a drainage system that is stationary long enough to produce a localised fan-shaped accumulation. The internal facies assemblage is characteristic; alluvial fans should therefore be distinguished from braided rivers and alluvial plains, which deposit different facies suites. The latter two differ only in extent: braided river deposits are one-dimensional, whereas braided alluvial plain deposits are sheet like.

2. Before the development of terrestrial vegetation, essentially in Late Paleozoic times, virtually all alluvial deposits were braided. Braided alluvium was particularly abundant in humid regions, especially on alluvial plains, which included silty braided deposits in distal reaches.

3. The three lithotypes of braided alluvium are distinguished from equivalents in meandering systems as follows: (i) gravel, by dominance of framework gravel; (ii) sand, by the presence of suspension-deposited sediment mainly as intraclasts rather than primary strata; and (iii) silt, by the lack of association with sandy channel deposits.

4. Facies assemblages within the lithotypes, and their depositional environments, are recognised as follows (Table 1):

Gravel. G_I: Alluvial fans; G_{II}: Proximal braided rivers and alluvial plains; G_{III}: Distal braided rivers and alluvial plains.

Sand. S_I: Proximal braided rivers and alluvial plains. S_{II}: Distal braided rivers and alluvial plains, with transitions to meandering rivers.

Silt. Distal braided rivers and alluvial plains.

ACKNOWLEDGEMENTS

I am grateful to Prof. R. W. R. Rutland for the provision of facilities at the University of Adelaide, and to V. A. Gostin, A. D. Miall, N. D. Smith and R. G. Walker for comments on the manuscript.

REFERENCES

Alcock, F. J., 1935, Geology of Chaleur Bay region: Geol. Surv. Can. Mem. 183.

Allen, J. R. L., 1965, A review of the origin and characteristics of recent alluvial sediments: Sedimentology, v. 5, p. 89-191.

————, 1973, Phase differences between bed configuration and flow in natural environments, and their geological relevance: Sedimentology, v. 20, p. 323-329.

Blissenbach, E., 1954, Geology of alluvial fans in semiarid regions: Geol. Soc. Am. Bull., v. 65, p. 175-190.

Bluck, B. J., 1971, Sedimentation in the meandering River Endrick: Scot. J. Geol., v. 7, p. 93-138.

————, 1974, Structure and directional properties of some valley sandur deposits in southern Iceland: Sedimentology, v. 21, p. 533-554.

Boothroyd, J. C. and Ashley, G. M., 1975, Processes, bar morphology, and sedimentary structures on braided outwash fans, northeastern Gulf of Alaska: *in* A. V. Jopling and B. C. McDonald, *eds.*, Glaciofluvial and glaciolacustrine sedimentation, Soc. Econ. Paleont. Mineral. Spec. Pub. 23, p. 193-222.

Bretz, J. H., Smith, H. T. U., and Neff, G. E., 1956, Channeled scabland of Washington: new data and interpretations: Geol. Soc. Am. Bull., v. 67, p. 957-1049.

Bull, W. B., 1964, Alluvial fans and near-surface subsidence in western Fresno County, California: U.S. Geol. Survey, Prof. Paper 437-A.

————, 1972, Recognition of alluvial-fan deposits in the stratigraphic record: *in* J. K. Rigby and W. K. Hamblin, *eds.*, Recognition of ancient sedimentary environments, Soc. Econ. Paleont. Mineral. Spec. Pub. 16, p. 63-83.

Cant, D. J. and Walker, R. G., 1976, Development of a braided-fluvial facies model for the Devonian Battery Point Sandstone, Quebec: Can. J. Earth Sci., v. 13, p. 102-119.

Chien, N., 1961, The braided stream of the lower Yellow River: Scientia Sinica, v. 10, p. 734-754.

Church, M. and Gilbert, R., 1975, Proglacial fluvial and lacustrine environments: *in* A. V. Jopling and B. C. McDonald, *eds.*, Glaciofluvial and glaciolacustrine sedimentation, Soc. Econ. Paleont. Mineral. Spec. Pub. 23, p. 22-100.

———— and Ryder, J. M., 1972, Paraglacial sedimentation: a consideration of fluvial processes conditioned by glaciation: Geol. Soc. Am. Bull., v. 83, p. 3059-3072.

Coleman, J. M., 1969, Brahmaputra River: channel processes and sedimentation: Sediment. Geol., v. 3, p. 129-239.

Conaghan, P. J. and Jones, J. G., 1975, The Hawkesbury Sandstone and the Brahmaputra: a depositional model for continental sheet sandstones: J. Geol. Soc. Australia, v. 22, p. 275-283.

Fahnestock, R. K., 1969, Morphology of the Slims River: Icefield Ranges Research Project, Sci. Results, v. 1, p. 161-172.

———— and Bradley, W. C., 1973, Knik and Matanuska Rivers, Alaska: a contrast in braiding, *in* M. Morisawa, *ed.*, Fluvial geomorphology, State Univ. New York, Binghampton, Proc. 4th Geomorph. Symp., p. 220-250.

Galay, V. J. and Neill, C. R., 1967, Discussion of "Nomenclature for bed forms in alluvial channels": J. Hydraulics Div., Am. Soc. Civil Eng., v. 93, p. 130-133.

Garrels, R. M. and MacKenzie, F. T., 1971, Evolution of sedimentary rocks: New York, Norton.

Gibling, M. R. and Rust, B. R., 1977, Proximal and distal sandy braided alluvium in Devonian successions of the Arctic and Gaspé: Geol. Assoc. Canada, Progr. Abstr., v. 2, p. 20.

Harms, J. C., Southard, J. B., Spearing, D. R. and Walker, R. G., 1975, Depositional environments as interpreted from primary sedimentary structures and stratification sequences: Soc. Econ. Paleont. Mineral. Short Course 2, Dallas.

Hein, F. J. and Walker, R. G., 1977, Bar evolution and development of stratification in the gravelly, braided, Kicking Horse River, British Columbia: Can. J. Earth Sci., v. 14, p. 562-570.

Jackson, R. G., 1976, Depositional model of point bars in the lower Wabash River: J. Sediment. Petrol., v. 46, p. 579-594.

Johnson, A. M., 1970, Physical processes in geology: San Francisco, Freeman.

Koster, E. H., 1977, Experimental studies of coarse-grained sedimentation: Unpub. Ph.D. thesis, Dept. of Geology, Univ. of Ottawa.

Leopold, L. B. and Wolman, M. G., 1957, River channel patterns: braided, meandering, and straight: U.S. Geol. Survey, Prof. Paper 282-B, p. 39-85.

Malde, H. E., 1968, The catastrophic Late Pleistocene Bonneville Flood in the Snake River Plain, Idaho: U.S. Geol. Survey, Prof. Paper 596.

McDonald, B. C. and Banerjee, I., 1971, Sediments and bed forms on a braided outwash plain: Can. J. Earth Sci., v. 8, p. 1282-1301.

McGerrigle, H. W., 1950, The geology of Eastern Gaspé: Quebec Dept. Mines, Geol. Rpt. 35.

McGowen, J. H. and Garner, L. E., 1970, Physiographic features and stratification types of coarse-grained point bars: modern and ancient examples: Sedimentology, v. 14, p. 77-112.

——— and Groat, C. G., 1971, Van Horn Sandstone, West Texas: an alluvial fan model for mineral exploration: Bur. Econ. Geol., Univ. Texas, Rpt. Invest. 72.

McGregor, D. C., 1973, Lower and Middle Devonian spores of Eastern Gaspé, Canada. I. Systematics: Palaeontographica, v. 142, p. 1-77.

McKee, E. D., Crosby, E. J. and Berryhill, H. L., 1967, Flood deposits, Bijou Creek, Colorado: J. Sediment. Petrol., v. 37, p. 829-851.

Miall, A. D., 1973, Markov chain analysis applied to an ancient alluvial plain succession: Sedimentology, v. 20, p. 347-364.

———, 1977, A review of the braided-river depositional environment: Earth Sci. Revs., v. 13, p. 1-62.

Nummedal, D., Hine, A. C., Ward, L. G., Hayes, M. O. Boothroyd, J. C., Stephen, M. F. and Hubbard, D. K., 1974, Recent migrations of the Skeidarasandur shoreline, southeast Iceland: Coastal Res. Div., Dept. of Geology, Univ. of S. Carolina, Rpt. N60921-73-C-0258.

Ore, H. T., 1964, Some criteria for recognition of braided stream deposits: Univ. Wyoming Contr. Geol., v. 3, p. 1-14.

Rust, B. R., 1972a, Structure and process in a braided river: Sedimentology, v. 18, p. 221-245.

———, 1972b, Pebble orientation in fluvial sediments: J. Sediment. Petrol., v. 42, p. 384-388.

———, 1975, Fabric and structure in glaciofluvial gravels: in A. V. Jopling and B. C. McDonald, eds., Glaciofluvial and glaciolacustrine sedimentation, Soc. Econ. Paleont. Mineral. Spec. Pub. 23, p. 238-248.

———, 1976, Stratigraphic relationships of the Malbaie Formation (Devonian), Gaspé, Quebec: Can. J. Earth Sci., v. 13, p. 1556-1559.

———, 1977a, The Malbaie Formation: sandy and conglomeratic proximal braided alluvium from the Middle Devonian of Gaspé, Quebec: Geol. Soc. Am., Abs. with Prog., v. 9, p. 313-314.

———, 1977b, The Cannes de Roche Formation: Carboniferous alluvial fan and floodplain deposits in eastern Gaspé: Geol. Assoc. Can., Prog. with Abs., v. 2, p. 46.

———, in press, The interpretation of ancient alluvial successions in the light of modern investigations: 5th. Geomorph. Symp., Univ. of Guelph, Ontario, Canada.

Schumm, S. A., 1968, Speculations concerning paleohydrologic controls of terrestrial sedimentation: Geol. Soc. Am. Bull., v. 79, p. 1573-1588.

Selley, R. C., 1965, Diagnostic characters of fluviatile sediments of the Torridonian Formation (Precambrian) of northwest Scotland: J. Sediment. Petrol., v. 35, p. 366-380.

Smith, N. D., 1970, The braided stream depositional environment: comparison of the Platte River with some Silurian clastic rocks, north-central Appalachians: Geol. Soc. Am. Bull., v. 81, p. 2993-3014.

———, 1971, Transverse bars and braiding in the lower Platte River, Nebraska: Geol. Soc. Am. Bull., v. 82, p. 3407-3420.

———, 1972, Some sedimentological aspects of planar cross-stratification in a sandy braided river: J. Sediment. Petrol., v. 42, p. 624-634.

———, 1974, Sedimentology and bar formation in the upper Kicking Horse River, a braided outwash stream: J. Geol., v. 82, p. 205-223.

Walker, R. G., 1976, Facies models 3: Sandy fluvial systems: Geoscience Canada, v. 3, p. 101-109.

Williams, P. F. and Rust, B. R., 1969, The sedimentology of a braided river: J. Sediment. Petrol., v. 39, p. 649-679.

DEVELOPMENT OF A FACIES MODEL FOR SANDY BRAIDED RIVER SEDIMENTATION: COMPARISON OF THE SOUTH SASKATCHEWAN RIVER AND THE BATTERY POINT FORMATION

DOUGLAS J. CANT[1]

ABSTRACT

Detailed comparisons of individual facies and independently constructed facies sequences in the Devonian Battery Point Formation (BP) of Quebec and the South Saskatchewan River (SS) reveal both similarities and differences.

At the bottom of the summary facies sequence (BP) which was derived by Markov analysis, trough crossbeds correspond to the deposits of dunes in the deeper channels (SS). Solitary planar crossbed sets (BP) with paleocurrent directions which diverge from those of the trough crossbeds are analogous to deposits of oblique slipface-bounded bars in the channels (SS). Cosets of up to 4 sets of planar crossbeds (BP) resemble the stratification of complex sand flats (SS). Small planar crossbed sets near the top of the sequence (BP) correspond to deposits of sand waves, common in shallow areas of the river (SS). Interbedded mudstone and rippled sandstone near the top of the sequence (BP) resemble the fine-grained cohesive floodplain deposits (SS). Some asymmetric scour fillings and low-angle cross-stratified deposits (BP) are unlike any sediment in the river (SS).

The vertical arrangement of channels, sand flats, and floodplains in the river suggests facies sequences which are similar to the summary sequence (BP). Some observed sequences (BP) are more complex, reflecting rapid aggradation in the channels of the Battery Point river which promoted lateral migration of shallow channels over bars and sand flats.

A facies model for this type of sedimentation consists mainly of channel, bar, and sand flat deposits interfingering laterally and succeeding one another vertically. Minor amounts of cohesive floodplain sediments are also present. Lateral migration of the river back and forth while aggrading causes deposition of a sheet sand elongated parallel to the paleoflow.

FACIES MODEL - GENERAL

A facies model is a summary of a sedimentary environment and its deposits. The purposes of a facies model (Walker, 1976) include acting as a norm or standard against which specific examples can be compared, as a predictor, and as a guide for observation. Most facies models are based on data from both modern environments and ancient sediments; for example the meandering stream model (Sundborg, 1956; Allen, 1970; Jackson, 1976). However, the relative amount of input from modern or ancient sediments depends on a number of factors, such as difficulty in collecting data from certain modern environments, or economic incentives for detailed work on one specific example, whether modern or ancient.

Facies models based solely on modern sediments commonly lack the dimension of time because in most cases it is impossible to study medium to long term processes. In specific instances this problem has been overcome, as in the Mississippi Delta where subsidence and preservation of delta lobes are well understood (Coleman and Gagliano, 1964). However, in most sedimentary environments, this type of integration of study of surficial processes with studies of long term processes has not been possible.

Facies models based solely on ancient sediments commonly lack detailed information about processes of sediment transport and deposition and rely very heavily on interpretation. In the study of sedimentary rocks, it is extremely difficult to investigate the controls on sedimentation such as climate or tectonism. Ancient sediments, however,

[1]McMaster University, Hamilton, Ontario; present address: Union Research Center, P.O. Box 76, Brea, California, 92621, U. S. A.

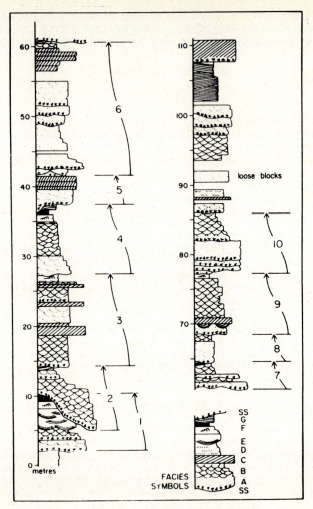

Fig. 1. Measured section through the Battery Point Formation. The facies designations are explained in the text. The sketched projection of the beds to the right is an approximate measure of grain size — maximum projection indicates coarse sandstone, about 1 mm. Facies sequences are numbered on the right. After Cant and Walker (1976).

may provide data about modern depositional environments in which observation and data collection are difficult; for example, deep marine environments.

Facies models based on data from both modern and ancient sediments are therefore more likely to be generally applicable. This paper is an attempt to integrate work on the sandy braided South Saskatchewan River (Cant and Walker, in press) with a study of Devonian braided river deposits (Cant and Walker, 1976).

GENERAL DESCRIPTIONS OF THE ROCK UNIT AND THE RIVER

The Battery Point Formation

A detailed section measured through part of this unit (Fig. 1) is dominated by trough (facies A, B) and planar crossbeds (facies C, D) developed in medium to coarse

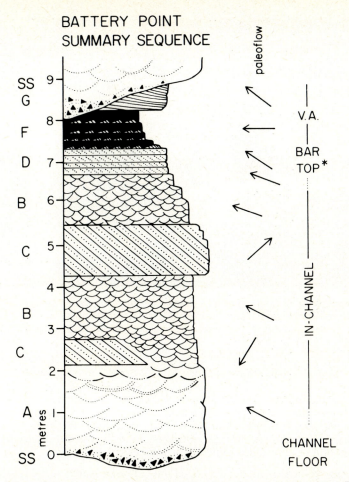

Fig. 2. The Battery Point summary sequence compiled from the measured section. The average paleocurrent direction of facies A and B is 290 degrees, but directions measured on planar crossbed sets of facies C may deviate over 90 degrees north or south of this. After Cant and Walker (1976).

sandstone. Lesser amounts of scour-filling sandstone (facies E) interbedded rippled sandstone and mudstone (facies F) and low angle to horizontally laminated sandstone (facies G) are also present. Intraclast-strewn scoured surfaces or channels up to 5 m deep (facies SS) are common in the section. More detailed descriptions of each of these types will be given in the facies by facies comparison below. The channels or scoured surfaces were used to define the limits of facies sequences, ten of which are indicated on Figure 1. Because the order of the facies within the facies sequences proved to be extremely variable, Markov analysis (Miall, 1973) was used to analyse the entire measured section. By choosing facies transitions which occurred more frequently than random, a facies relationship diagram was obtained, and a summary facies sequence (Fig. 2) constructed from it.

This sequence has a scoured, intraclast-strewn base (facies SS) above which large, poorly defined trough crossbeds are developed in coarse sandstone (facies A). These are succeeded by smaller, better defined trough crossbeds developed in slightly finer sandstone (facies B). Intercalated within these two facies are solitary sets of planar

Fig. 3. An oblique air photo of the South Saskatchewan River near Outlook, Saskatchewan. A floodplain (bottom), vegetated island (middle), channels and sand flats (top) are shown. A large, oblique bar front is present in the middle right of the photo. Flow is to the left. The river is approximately 600 m wide here.

crossbeds (facies C) which have paleocurrent directions at high angles to those of the trough crossbeds above and below them. These three facies comprise approximately 80 percent of the facies sequence. Above these, there are small planar crossbed sets (facies D), interbedded rippled sandstone and mudstone (facies F), and low angle to horizontally stratified sandstone (facies G). The scour filling sandstone (facies E) has been omitted from the diagram because it occurs so infrequently in the measured section that its most common location of occurrence could not be established.

This Battery Point sequence summarized the entire measured section, and comprises a local model for this deposit. It will be compared and contrasted with other sequences later in this paper.

The South Saskatchewan River

This river is the major stream draining the southern prairies of western Canada. Near Outlook, Saskatchewan in the studied reach, the river flows in a valley about 0.6 km wide and is straight to irregularly curving. It flows on an average slope of .0003 and transports fine to medium sand of average diameter 0.3 mm. Mapping on different scales, study of air photos, and aerial reconnaissance have shown that the river system consists of the following types of geomorphologic elements; channels, slipface-bounded bars, complex braid bars termed sand flats, and vegetated islands and floodplains (Fig. 3). At most times the sand flats are emergent, but at discharges greater than 280 m³/sec (achieved for several days during June floods most years) they are submerged. At discharges greater than 1240 m³/sec (achieved every 2.2 years) the islands and floodplains are inundated.

The major channels in the river average about 150 m wide and 3 m deep below nearby sand flats. They generally flow parallel to the local river direction, averaging 13 degrees deviation from this, but curve in places, crossing the system diagonally. Echo sounding and direct observation of the beds of the major channels have shown that they are covered by dunes at all river stages. During flood stages, the sizes of the dunes and the

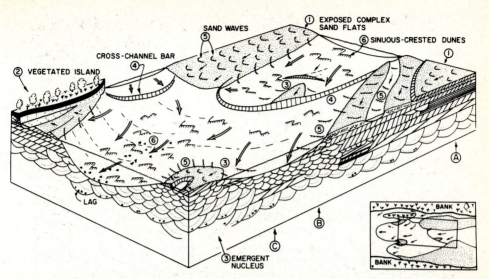

Fig. 4. A block diagram of the river showing the active environment and the preserved deposits. The inset shows a vertical view of the same area. The circled letters A, B, and C show the locations of formation of the Sand Flat, Mixed Influence, and Channel sequences respectively in Fig. 5. From Cant and Walker (in press).

depths of their scour troughs increase, probably depositing trough crossbeds up to 1 m thick. The topographically higher minor channels are floored by smaller dunes or the assemblage of sand waves and ripples. Trough and planar crossbeds less than 30 cm thick are deposited in these channels.

Where the major channels curve, they deposit slipface-bounded bars which may be up to 3 m high and several hundred meters long. The larger ones extend between major geomorphic elements, and are termed cross-channel bars. In most cases, the direction of bar advance is oblique to the local river direction, averaging 69° from this (Cant and Walker, in press). Trenching and boxcoring shows that these bars deposit sets of planar crossbeds. Where a small area of the bar top becomes topographically higher than the remainder of the bar, a small sand flat may form from this "nucleus" (Cant and Walker, in press).

The sand flats in the river range in length from 50 m to 2 km. Smaller sand flats have morphologies which are indicative of their origins from large bars, but larger sand flats are more complex because they have been modified during many episodes of erosion and accretion. Sand flats accrete downstream, laterally, and to a lesser extent upstream by bars forming around their margins and migrating toward them. At high stages, bars form on their top surfaces causing them to aggrade vertically. Because the initiation and growth of sand flats result from bar migration, the stratification of sand flats is believed to consist of stacked planar crossbed sets. At lower stages, the sand flats are covered by small dunes, sand waves, and ripples which deposit a veneer of small crossbed sets and ripple laminations on them.

The islands and floodplains in the river stand 1 to 2 m above the level of the sand flats. They are composed of alluvial sand overlain by as much as 1 m of vertical accretion muds.

The relationships between the active environment and the potentially preserved deposits are shown in Figure 4. By assuming that the topographic relationships between the different geomorphic elements would be preserved, vertical facies sequence for

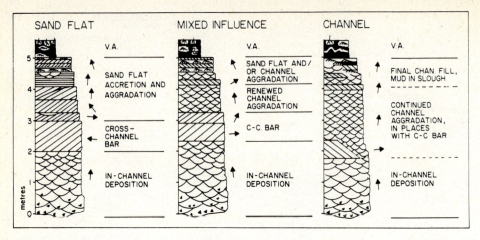

Fig. 5. The range of facies sequences which could be produced by deposition in the river. The arrows represent only one of the possible paleocurrent patterns, with the average channel direction being represented by a vertical arrow. Muddy vertical accretion deposits with lenses of sand containing ripple and convolute laminations cap each sequence. From Cant and Walker (in press).

different parts of the system were built up (Fig. 5). Large trough crossbeds deposited in deep channels are present at the base of each sequence. In the "Channel" sequence, one planar crossbed set deposited by an oblique bar is followed by the deposits of a topographically higher, minor channel. In the "Sand Flat" sequence, the stacked planar crossbeds comprising the sand flat are present above the channel deposits. The "Mixed Influence" sequence is an attempt to show partial development of a sand flat, with the re-establishment of a minor channel above. Each sequence has minor crossbeds near the top which were generated by small bedforms in shallow parts of the river. Each is capped by vertical accretion floodplain deposits.

It should be emphasized that the South Saskatchewan sequences were independently constructed from data available from the river, with no reference made to the Battery Point sequence. However, it is possible that the synthesis of the sequences was influenced by the author's previous Battery Point study.

Detailed Comparison of Facies

The base of the Battery Point facies sequence (Fig. 2) is the scoured surface facies (SS), consisting of an erosion surface overlain by as much as .25 m of coarse massive sandstone with intermixed mudstone intraclasts (Cant and Walker, 1976, Fig. 3). The base of a newly established channel in the South Saskatchewan has never been observed, probably because the channels are established or deepened at high stage, with deposition on the channel bed taking place on the falling stage. However, it is known that at high stages, banks of islands and floodplains are undercut locally by the flow, causing bank caving. This creates many cohesive mud intraclasts of highly variable size (maximum observed length, 3 m) which are scattered throughout the system. Many of these are deposited on the bases of the deepest channels.

Above the scoured surface in the Battery Point sequence, the poorly defined trough crossbedded facies (A) is present (Fig. 2). This facies is composed of coarse sandstone with scattered pebbles in which are developed large trough crossbeds. These troughs are up to .6 m deep and 3 m wide. The internal stratification of the troughs is poorly defined because the sediment is not well sorted, and little fine material is present to outline laminations. The channels in the South Saskatchewan are floored by sinuous-crested dunes (Cant, in prep.) which deposit sets of trough crossbeds (Harms and Fahnestock, 1965). During flood stages, depths and flow velocities rise (maximum observed mean flow velocity, 1.75 m/sec), and slightly coarser sediment is transported in the channels than at low stages. Repeated echo sounding has shown that the heights of the dunes (maximum of 2 m) and the depths of their scour troughs increase at high stage. Because dunes created during falling and low stage flow have shallower scour troughs, they cannot erode the deposits of the flood stage bedforms. The large trough crossbeds deposited by the large dunes at flood stage will be selectively preserved. The deep channel deposits therefore consist of trough crossbedded coarse sand, similar to facies A of the Battery Point Formation.

Above facies A in the Battery Point sequence (Fig. 2) is the well defined trough crossbedded facies (B) which is composed of medium sand with smaller trough crossbeds, averaging .15 — .20 m in depth (Cant and Walker, 1976, Fig. 5). The stratification of the troughs is better defined, being outlined by finer sediment. In the South Saskatchewan, the shallower, topographically higher channels also have dunes on their beds. However, these dunes never build up to the heights achieved by those in the deeper channels (Cant, in prep.). Boxcoring in many minor channels has confirmed that their deposits consist mainly of small trough crossbeds. A strong resemblance exists between the trough crossbedded sands of these channels and facies B of the Battery Point.

Intercalated within facies A and B (Fig. 2) are large solitary planar crossbed sets (C). These sets have paleocurrent directions which are at high angles to the directions of the trough crossbeds above and below them (Cant and Walker, 1976, Fig. 7). Smith (1970, 1971) and Collinson (1970) have shown that planar crossbed sets are deposited by transverse or linguoid bars with slipfaces forming their downstream margins. In the channels of the South Saskatchewan, long slipface-bounded cross-channel bars are present (Cant and Walker, in press, Fig. 4). These bars are commonly oblique to the main channel trend and the direction of dune migration. Boxcoring and trenching have confirmed that these bars deposit sets of planar crossbeds. The similarity of structure and paleocurrent direction between these oblique bars and the solitary sets of Facies C of the Battery Point is striking.

In some occurrences of facies C, cosets of up to 4 sets of planar crossbeds are present. These sets are smaller than most of the solitary sets, rarely exceeding .40 m in thickness. In the South Saskatchewan, the stratification of large complex sand flats is believed to consist of sets of planar crossbeds stacked one on top of another (Fig. 5) because bar migration is the major process of formation and accretion of sand flats (Cant and Walker, in press). Trenching to the water table on sand flats has revealed planar crossbeds with a veneer of small crossbed sets and ripple accretion. Some parallel laminated sand is also present. Exposed banks of the river which are composed of older sand flat material exhibit the same pattern. It is believed, therefore that the stacked planar crossbeds of facies C correspond to the deposits of sand flats in the South Saskatchewan.

The small planar crossbeds of facies D (Cant and Walker, 1976, Fig. 8) occur high in the facies sequence (Fig. 2). These closely resemble the stratification of sand waves (Southard, *in* Harms *et al.,* 1975) which are found along the margins of channels and on the tops of the sand flats in the South Saskatchewan (Cant, in prep.).

Facies F of the Battery Point consists of fine grained rippled sandstone or interbedded rippled sandstone and mudstone (Cant and Walker, 1976, Fig. 10, 11). This facies occurs at or near the top of the facies sequence (Fig. 2). In the river, the uppermost deposits of sand flats consist of rippled sand. In the banks, this rippled sand is overlain by muddy vertical accretion deposits of the floodplain (Cant and Walker, in press, Fig. 14). The floodplain muds of the South Saskatchewan are thicker and contain proportionally fewer interbeds of sand than the facies F muds. However, the two types of deposits are comparable.

In the Battery Point, there are a restricted number of occurrences of the asymmetric scour facies (E) (Cant and Walker, 1976, Fig. 9). These scours are filled by sediment with laminations which parallel the bottom surfaces of the scours. Nothing resembling this facies has been seen in the South Saskatchewan.

At the extreme top of the Battery Point sequence (Fig. 2), the low angle stratified facies (G) is present (Cant and Walker, 1976, Fig. 12)m This facies consists of fine sandstone with low angle to flat laminations which show parting lineations on their upper surfaces. This parting lineation is parallel to the channel direction established from facies A and B. Facies G is unlike anything present in the South Saskatchewan.

Although it is impossible to know exactly the reasons why these facies are present in the Battery Point, but not in the South Saskatchewan, speculations can be made. It is believed that facies E represents scours cut into channel deposits, but these scours were not filled by dune avalanching into them. It is possible that the scours were cut at a high river stage and filled at a lower one. This facies may therefore represent rapid discharge variations.

Facies G occurs above the floodplain deposits at the top of the facies sequence, and because of the parting lineation, is believed to reflect upper flow regime conditions. This facies is therefore interpreted as the deposits of high, swift floods which carried sand out of the channels and deposited it on the floodplain. Facies G resembles to some extent the flood deposits of an ephemeral stream reported by McKee *et al.* (1967). This facies may represent the deposits of flood stages which were much higher than any observed in the South Saskatchewan.

The interpretations of Facies E and G suggest that the Battery Point deposits were laid down by a river which had greater discharge variation than the South Saskatchewan. The Battery Point was laid down over a relatively long period, during which climatic conditions may have varied. In addition, land plants were restricted in extent during the Lower Devonian with the result that discharge variations were likely to have been more extreme than at the present day (Schumm, 1968).

COMPARISON OF FACIES SEQUENCES

The Battery Point summary sequence (Fig. 2) and the South Saskatchewan sequences (Fig. 5) are comparable in that: (1) each is dominantly sandy and is composed of trough and variable amounts of planar crossbed sets (2) the planar crossbed sets may be present at different stratigraphic levels within the sand bodies, and may have highly divergent paleocurrent directions compared to the trough crossbeds above and below them (3) there is little vertical trend in grain size in the sand bodies (4) each has a relatively thin muddy unit at or near its top.

Individual sequences in the Battery Point (as opposed to the summary sequence) show a great deal of variability. For example, sequences 2, 4, 7, 8, and 10 (Fig. 1) are composed almost entirely of trough crossbeds and can be related to an end member type of "Channel" sequence of the South Saskatchewan (Fig. 5). These sequences represent aggrading channels where no cross-channel bars formed, and deposition took place only from dunes. Aggradation of one part of a braided system and diversion of flow away to another part of the system has been discussed by Chien (1961), and has been observed in the South Saskatchewan (Cant and Walker, in press). Sequences 7 and 10 show multiple scours near their bases, suggesting that channel aggradation may have been intermittent, with a few periods of erosion also occurring. Sequence 9 is also interpretable in terms of the "Channel" sequence, but shows the development of a cross-channel bar near its base.

Sequences 5 and 6 have cosets of up to 5 sets of planar crossbeds, and can be interpreted in terms of the "Sand Flat" sequence (Fig. 5). Sequence 6 may represent two superimposed channel sequences, but this is not certain because of poor outcrop.

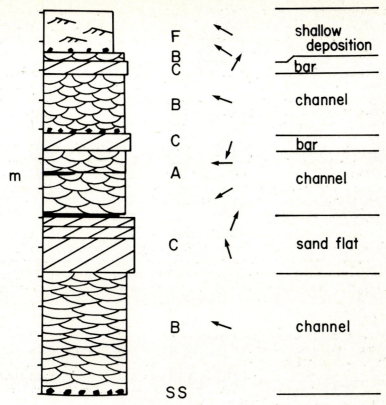

Fig. 6. Sequence 3 (Fig. 1) from the Battery Point. Note the unimodal paleocurrent pattern of the trough crossbedded units. The directions of the planar crossbed sets diverge markedly both left and right from the average direction of the troughs.

Sequence 3, shown in more detail in Fig. 6, is an example of a "Mixed Influence" sequence (Fig. 5). It shows channel, sand flat, and bar deposits interbedded with one another. Lateral to the basal trough crossbedded facies, other bar and sand flat deposits are also developed. Therefore, the order in which the three different types of facies are found is a function of the random location of the outcrop section through the deposit. Sequence 3 is considerably more complex than the South Saskatchewan sequences. This complexity may result because of large amounts of aggradation which must have occurred in the Battery Point river. Aggradation within the channels would cause these to be laterally unstable, and to migrate over sand flats and bars. These longer term processes, therefore introduce added complexity to the resulting sequences.

COMPARISON WITH MEANDERING RIVER SEQUENCES

Some meandering rivers have scroll bars which migrate onto the point bars (Sundborg, 1956; Jackson, 1976). These bars deposit sets of planar crossbeds which are at high angles to the trough crossbeds developed under them. This type of river lays down facies sequences (Fig. 7) which resemble the braided river sequences somewhat. In braided sequences however, planar crossbed sets may be developed at any level within the sandy member of the sequence. Planar sets deposited by scroll bars in meandering rivers are topographically above the channel deposits. In some cases, these bars are directly overlain by floodplain muds (Jackson, 1976).

Fully Developed   Transitional   Intermediate

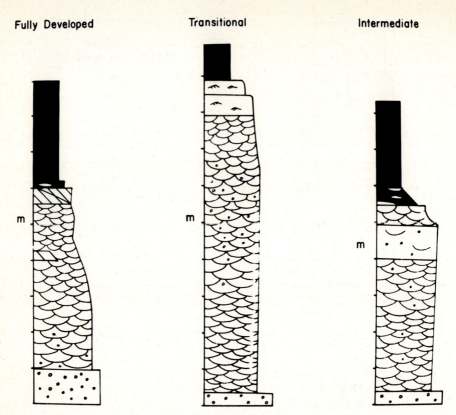

Fig. 7.   The three types of facies sequences deposited by the meandering Wabash River. Each sequence is composed mainly of trough cross-beds, but the "Fully Developed" sequence may have one or more planar crossbed sets immediately under the muddy floodplain deposits. The paleocurrent directions of these sets may diverge from the directions of the trough crossbeds under them by 90 degrees, reflecting the orientations of the scroll bars in the river. Modified from Jackson (1976).

Within one meandering facies sequence, the paleocurrent directions of the planar crossbeds may deviate in only one direction (left or right) from the trough crossbeds of the channel deposits. In a braided facies sequence, the planar crossbeds may deviate in either direction, or less likely, be parallel to the direction of the trough crossbeds. The paleocurrent directions of the Battery Point facies sequence in Fig. 6 illustrate this. The trough crossbedded facies generally have directions heading WNW, while planar sets may head north or south. In the South Saskatchewan, the orientations of the long oblique bars in channels may diverge in either direction from the river trend.

Detailed knowledge of the organization of facies and their paleocurrent directions is therefore necessary to decide which sedimentary model is appropriate in dealing with an ancient deposit. This is particularly true in cases where the fine grained floodplain deposit has been entirely removed by erosion. This conceivably could happen in either meandering or braided stream environments. However, where many facies sequences are stacked upon one another, as in most ancient fluvial deposits, the relatively thick fine grained members of meandering stream sequences are likely to be preserved in many cases. It is improbable that erosion would completely remove these from all the sequences. In stacked braided river sequences, however, the fine members are likely to be absent from more of the sequences.

DEVELOPMENT OF A FACIES MODEL

Integration of the Battery Point and South Saskatchewan studies with others reported in the literature leads to a facies model for one type of braided river sedimentation.

A small to moderate sized sandy braided river like the South Saskatchewan lays down a series of laterally interfingering channel, bar, and sand flat deposits. These types of deposits may be stacked upon one another by local aggradation and channel migration (Cant and Walker, in press). The channel deposits are mainly medium to coarse sand in the form of trough crossbeds generated by dunes on the beds (Smith, 1970; Cant and Walker, in press). The bar deposits consist of single sets of planar crossbeds (Collinson, 1970; Smith, 1970; Cant and Walker, in press). The sand flat deposits comprise several planar crossbed sets along with minor amounts of parallel lamination and a thin veneer of smaller crossbeds and ripple cross-laminations (Cant and Walker, in press). Fine grained cohesive vertical accretion deposits occur topographically above these other types (Cant and Walker, in press).

In the preserved deposits of this type of river, bar, sand flat, and channel deposits are laid down upon one another without much order (Cant and Walker, 1976, Fig. 14). These types of deposits comprise from 80 to 100 percent of each complete facies sequence in the Battery Point, and about 80 percent of the constructed facies sequences in the South Saskatchewan. The remainder, in all cases, is composed of small sandy crossbeds formed by small dunes and sand waves in shallow parts of the river, and by vertical accretion floodplain deposits.

Campbell (1976) has shown that sandy braided river deposits of the Westwater Canyon Member of the Jurassic Morisson Formation are composed of a series of infilled channels of two different scales. The smaller channel fills (180 m in width, 4 m in thickness) occur within the larger channel fills (11 km in width, 15 m thickness). The smaller channel fills suggest the migration of channels within the river system, while the larger channel fills may be the result of migration of the entire river system. Because of continuous

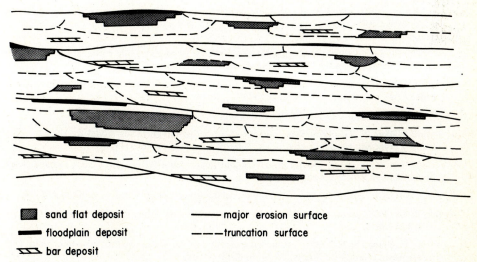

■ sand flat deposit   ——— major erosion surface

■ floodplain deposit   — — —truncation surface

⊞ bar deposit

Fig. 8. A hypothetical picture of stacking of the fluvial facies sequences. Two scales of channels are present, the smaller due to channel migration and the larger to migration of the entire river system. The diagram represents a vertical thickness of about 50 m, and a width exceeding 1 km. This diagram should be compared with the similar diagrams of meandering stream deposits of Allen (1974).

aggradation during back and forth migration of the river, the facies sequence (Fig. 2, 5) contained within the larger scale channel cuts are deposited upon older ones (Fig. 8). As Campbell (1976) has shown, this leads to a sheet sandstone body elongated in the direction of paleoflow.

This model of sandy braided rivers does not extend to very large rivers. For example, the Brahmaputra River (Coleman, 1969) shows migratory bedforms in the channels up to 15 m in amplitude. Plane bed also forms in parts of the channels at very high stage. The exposed bars in the river which may correspond to sand flats are much larger, up to 11 km in length (Coleman, 1969, Fig. 21). These and other differences between the Brahmaputra and the smaller rivers may be reflected not only in the scales of the deposits, but also in their facies types.

The model presented in this paper is based on only a few examples and is therefore, not a general model. However, it may be useful for comparative purposes. Comparison with other braided rivers or braided river deposits is necessary before the range of variability of this type of fluvial system or its deposits can be understood.

ACKNOWLEDGEMENTS

This paper is based on two theses supervised by Roger Walker at McMaster University. Funds for both studies were provided by the National Research Council of Canada. The manuscript was written while the author was the recipient of an Isaac Walton Killam Postdoctoral Fellowship at Dalhousie University. Drs. R. G. Walker and P. F. Friend improved the manuscript with their critical comments.

REFERENCES

Allen, J. R. L., 1970, Studies in fluviatile sedimentation: A comparison of fining upwards cyclothems, with special reference to coarse-member composition and interpretation: J. Sediment Petrol., v. 40, p. 298-323.

———, 1974, Studies in fluviatile sedimentation: implications of pedogenic carbonate units, Lower Old Red Sandstone, Anglo-Welsh outcrop: Geol. J., v. 9, p. 181-208.

Campbell, C. V., 1976, Reservoir geometry of a fluvial sheet sandstone: Am. Assoc. Petrol. Geol. Bull., v. 60, p. 1009-1020.

Cant, D. J., in prep., Bedforms and bar types in the South Saskatchewan River.

———, and Walker, R. G., 1976, Development of a braided fluvial facies model for the Devonian Battery Point Sandstone, Quebec: Can. J. Earth Sci., v. 13, p. 102-119.

———, and ———, in press, Fluvial processes and facies sequences in the sandy, braided South Saskatchewan River: Sedimentology.

Chien, N., 1961, The braided stream of the lower Yellow River: Scientia Sinica, v. 10, p. 734-754.

Coleman, J. M., 1969, Brahmaputra River: channel processes and sedimentation: Sediment. Geol., v. 3, p. 129-239.

———, and Gagliano, S. M., 1964, Cyclic sedimentation in the Mississippi River deltaic plain: Gulf Coast Assoc. Geol. Soc. Trans., v. 14, p. 67-80.

Collinson, J. D., 1970, Bedforms of the Tana River: Geogr. Ann., v. 52A, p. 31-56.

Harms, J. C., and Fahnestock, R. K., 1965, Stratification, bedforms, and flow phenomena (with an example from the Rio Grande), in G. V. Middleton, ed., Primary sedimentary structures and their hydrodynamic interpretation: Soc. Econ. Paleont. Mineral. Spec. Pub. 12, p. 84-115.

————, Southard, J. B., Spearing, D. B., and Walker, R. G., 1975, Depositional environments as interpreted from primary sedimentary structures and stratification sequences: Soc. Econ. Paleont. Mineral. Short Course 2, 161p.

Jackson, R. G., 1976, Depositional model of point bars in the lower Wabash River: J. Sediment. Petrol., v. 46, p. 579-595.

McKee, E. D., Crosby, E. J., and Berryhill, H. L., 1967, Flood deposits, Bijou Creek, Colorado, June 1965: J. Sediment. Petrol., v. 37, p. 829-851.

Miall, A. D., 1973. Markov chain analysis applied to an ancient alluvial plain succession: Sedimentology, v. 20, p. 347-364.

Schumm, S. A., 1968, Speculations concerning paleohydrologic controls of terrestrial sedimentation: Geol. Soc. Am. Bull., v. 79, p. 1573-1588.

Smith, N. D., 1970, The braided stream depositional environment: comparison of the Platte River with some Silurian clastic rocks, north-central Appalachians: Geol. Soc. Am. Bull., v. 81, p. 2993-3014.

————, 1971, Transverse bars and braiding in the Lower Platte River, Nebraska: Geol. Soc. Am. Bull., v. 82, p. 3407-3420.

Sundborg, A., 1956, The river Klarälven; a study of fluvial processes: Geogr. Ann., v. 38, p. 125-316.

Walker, R. G., 1976, Facies models 1, General introduction: Geoscience Canada, v. 3, p. 21-24.

PROGLACIAL BRAIDED OUTWASH:
A MODEL FOR HUMID ALLUVIAL-FAN DEPOSITS

JON C. BOOTHROYD[1] and D. NUMMEDAL[2]

ABSTRACT

The recent proglacial outwash fans of the Northeast Gulf of Alaska and the southern coast of Iceland exhibit similar depositional patterns and suites of sedimentary structures along the paths of individual braided-stream systems. Each system has a regional gradient ranging from 6 - 17 m/km in the proximal zone to 2 - 3 m/km at the distal margin. Maximum clast size (long-axis) decreases regularly downfan from 50 cm (proximal) to sand (distal). Proximal bars are longitudinal; distal bars are linguoid.

Individual fans have different types of distal margins, and may also be different laterally, in the areas between active streams. Sandy distal facies are best developed on Icelandic fans and are absent on Alaskan fans less than 8 - 10 km in length. A meandering-stream facies is sometimes present at the distal margin of Alaskan fans. Inactive stream areas support extensive dune fields and wind-tidal flats in Iceland, but are marsh-and-swamp covered in Alaska.

Fan lengths vary from 3 - 30 km. A standard proximal to distal (0 - 15 km) succession of bar and channel sedimentary structures exists for all fans studied and allows the identification of the individual or "core" braided-stream system. The addition of: 1) suites of sedimentary structures of the various distal margins (15 - 30 km), and 2) the lateral, "overbank and floodplain" facies assemblages, then leads to the construction of a multivariate model that characterizes the deposits of these humid alluvial fans.

INTRODUCTION

There is a developing body of geologic literature on the facies and depositional subenvironments of small to medium-sized, Recent, proglacial braided rivers (Boothroyd, 1972; Boothroyd and Ashley, 1975; Fahnestock, 1969; Gustavson, 1974; Williams and Rust, 1969; Rust, 1972; Nummedal *et al.*, 1974). Where the rivers are not constrained by valley walls they form, and flow on, alluvial fans. These fans may coalesce to form broad glacial-outwash plains and, if adjacent to bodies of water, glacial outwash-plain shorelines (Hayes *et al.*, 1972). Landsat images of two glacial outwash-plain shorelines, the Skeidarásandur of southern Iceland and the Malaspina Foreland on the northeast Gulf of Alaska, are illustrated in Figure 1; locations are given by Figure 2.

The purpose of this paper is to compare and contrast the facies and depositional subenvironments of the Alaskan and Icelandic fans and to offer proglacial braided-outwash fans as a model for alluvial-fan deposition in a humid environment.

RATIONALE FOR THE MODEL

Recent reviews of alluvial-fan deposition (Bull, 1972; LeBlanc, 1972; Klein, 1975) have stressed the development of models based on arid to semi-arid fans of the southwestern United States. In addition, LeBlanc (1972) and Brown *et al.* (1973) suggest alluvial-fan and braided-stream deposits usually are separate in space at any given time. That is, braided streams are a down-the-paleoslope depositional environment distinct from alluvial fans.

---

[1]Department of Geology, University of Rhode Island, Kingston, Rhode Island 02881.
[2]Department of Geology, University of South Carolina, Columbia, South Carolina 29208.

Fig. 1.   Landsat band 5 images of glacial-outwash plain shorelines. Image scale is 1:1,000,000. A. Malaspina Glacier and Foreland, Northeast Gulf of Alaska. The three active stream systems, west to east, are: Yana, Fountain, and Alder (Scene 1708 20035, 1 July 1974). B. Skeidarársandur shoreline, southern Iceland. The three active stream systems, west to east, are: Sulá, Gĭja, and Skeidará (Scene 1372 12080, 30 July 1973).

We suggest that braided streams and alluvial fans are not separate in space, but that braided-stream deposits are a major component of semi-arid alluvial-fan deposition (although not the subject of this paper), and that glacial-outwash fans may serve as a perfectly good example of braided-stream deposition leading to a model of humid alluvial-fan deposits.

The outwash fans used as examples in this paper (Figs. 1, 3) are being built by braided streams issuing from active glaciers with a maximum summer discharge of the streams of 100 - 400 ms⁻¹ (Table 1). Precipitation over the drainage basin area ranges from 1000 - 3000 mm yr⁻¹ (Boothroyd, 1972; Gustavson, 1974; Nummedal *et al.*, 1974). The nature of the flood hydrograph is such that discharge increases rapidly to the yearly maximum with the advent of the summer melt season, then declines slowly to minimum discharge values by early fall (Boothroyd, 1972; Nummedal *et al.*, 1974). We believe that the nature of the hydrograph allows the development of the predictable pattern of grainsize decrease downstream, discussed later, that is absent on semi-arid fans (Bull, 1972; Denny, 1965; Hooke, 1967). Also, no debris flows exist on the fans examined by us or by Gustavson (1974).

Relative size of the various fans is illustrated in Figure 4. The Icelandic fans are the largest. Compare the size of the proglacial fans with the two Death Valley fans we studied (Nummedal and Boothroyd, 1976).

The applicability of the glacial-outwash fan facies succession (Fig. 5) to ancient rock sequences has been pointed out to us by a number of authors including McGowen and Groat (1971), Van Horne Sandstone, West Texas; Vos (1975), Rand Basin, South Africa; and George Pedlow (personal communication), anthracite coal basins, eastern Pennsylvania. We would like to suggest that vertical and lateral facies successions of the glacial-outwash fans discussed below can function as a model for braided-stream, alluvial-fan deposits in a humid environment. We, of course, do not imply that glacial conditions need to exist, or are necessary for application of the model.

DEVELOPMENT OF THE CORE FACIES

Introduction

The proximal to distal-braided facies succession for glacial-outwash fans presented in Figure 5 is essentially that developed for the Scott and Yana fans of the Northeast Gulf of Alaska by Boothroyd and Ashley (1975). We have added information of Gustavson (1974) and Nummedal *et al.* (1974) as well as inspecting other fans on the Northeast Gulf of Alaska (Fig. 2a). These new data confirmed that the Scott-Yana facies succession applies to the other outwash fans and can be used as a general model. The following discussion of the core facies reviews some information on gradient and clast size, adds some new data, and then presents a photographic atlas of bar and channel morphology and internal sedimentary structures.

Gradient and Clast Size

The gradient and clast size of selected Alaskan and Icelandic glacial-outwash fans are shown in Figure 6. The fans include the Scott and Yana (Boothroyd and Ashley, 1975), the Sheridan (Boothroyd, 1976), Fountain and Alder (Gustavson, 1974), and Gija and Skeidará (Nummedal *et al.*, 1974). The first five are located on the Northeast Gulf of Alaska and the last two on the southern coast of Iceland (Fig. 2). Regional gradient was measured by theodolite down the active stream system of each fan, with a station every 200 - 300 m, located either by: method 1) on the high points of active bar surfaces; or by method 2) water surface in riffle areas. Clast size was obtained by measuring the 10 average largest clasts on the apex of active longitudinal bars. Long axis measurements are presented because due to the shape of the clasts it is thought the L-axis better represents the competence of the stream at any given point (Boothroyd and Ashley, 1975).

TABLE 1. Some characteristics of selected braided rivers.

RIVER	DISCHARGE (Q) m³/sec				GRADIENT m/km		SEDIMENT SIZE (MAX)	OTHER FEATURES	ENVIRONMENT	REFERENCES
	MAX	MIN	AVG	Extreme Flood	Prox	Distal				
Ganges-Brahmaputra, Bangladesh	140000	5700	31500			0.7	Sand	sand waves, large linear features, transverse bars(?)	Tropical, humid	Coleman (1969)
Durance (Rhone) France	7500-5200	40-25	188			2.6	Pebble, cobble		Warm-temperate, humid	Doeglas (1962)
Tana, Norway	3644	158				0.2	Sand	350 km long, 600 m – 2 km wide	Cold-temperate, humid.	Collinson (1970)
Donjek, Yukon	1400				6.0	0.6-.85	Cobble		Glacial, humid	Williams and Rust (1969), Rust (1972)
Platte, Nebraska	665 / 3172	2 / 3.6	109		2.0	1.0	Pebble, sand	450-600 m wide	Warm-temperate, semi-arid	Smith (1972, 1971, 1970)
Slims, Yukon	570						Cobble, boulder		Glacial, humid	Fahnestock (1969)
Skeidará, Iceland	400			1500-50000 glacier burst	6.0	3.0	Boulder		Glacial, humid	Nummedal et al. (1974)
Kicking Horse, British Columbia	125				7.2	3.4	Pebble		Glacial, humid	Smith (1974)
Yana, Alaska	125				6.0	3.0	Boulder		Glacial, humid	Boothroyd and Ashley (1975)
Lake Eyre Basin, Australia				4000-8000			Sand		Arid	Williams (1971)

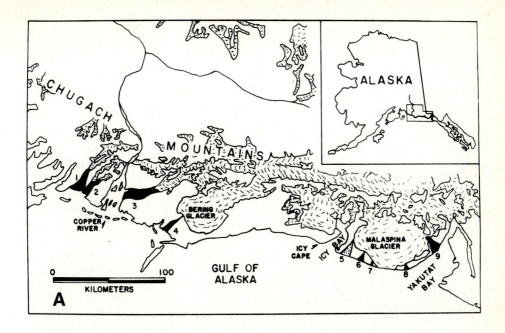

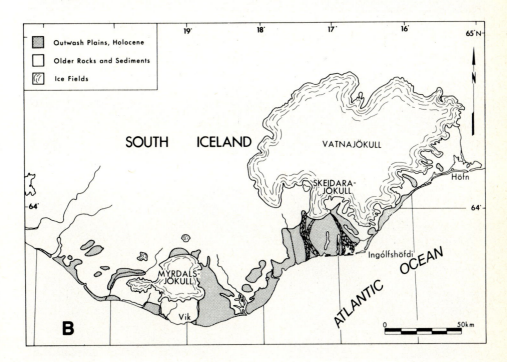

Fig. 2. Location maps. A. Northeast Gulf of Alaska. The glacial outwash fans studied or inspected (studied marked with an asterisk) are: 1) Scott*; 2) Sheridan*; 3) Martin River; 4) Edwardes; 5) Yahtse; 6) Yana*; 7) Fountain*; 8) Alder*; and 9) Kwik. B. Southern Iceland.

Fig. 3.   A. Aerial view of Yana outwash fan, West Malaspina Foreland, Northeast Gulf of Alaska, an example of a small glacial fan. B. Aerial view of the Gíja and Sulá stream systems, Skeidarársandur, southern Iceland.

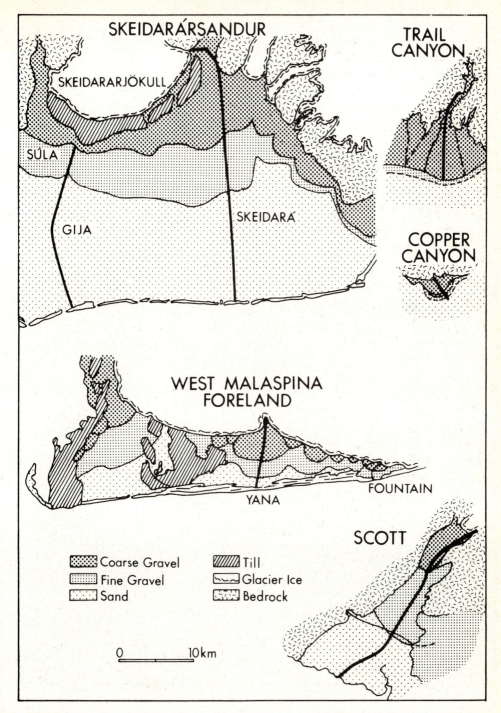

Fig. 4. Selected glacial-outwash fans plotted to the same scale to compare relative sizes and area of facies among the fans. Two Death Valley arid fans (Trail Canyon and Copper Canyon) are also plotted for comparison (from Nummedal and Boothroyd, 1976).

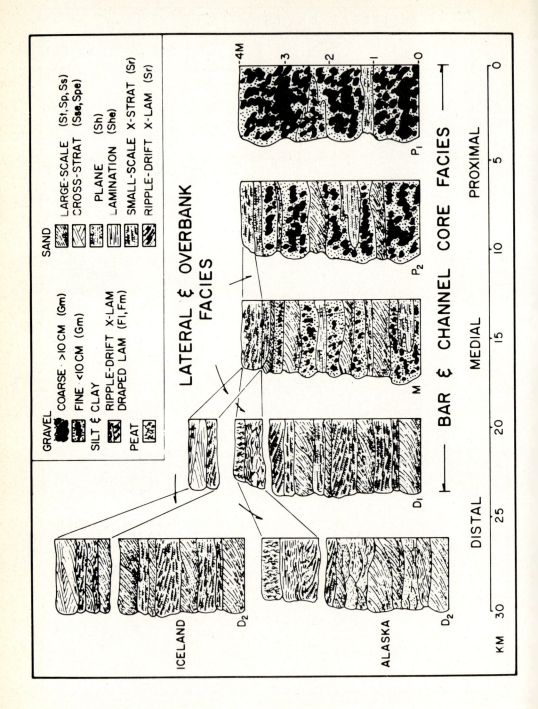

Fig. 5. Downstream facies succession for glacial-outwash fans. Sequences P_1, P_2, M, and D_1 comprise the core bar and channel. The most distal facies are given by sequences D_2 (Alaska and Iceland). Lateral and overbank facies for each area are shown above the bar and channel facies.

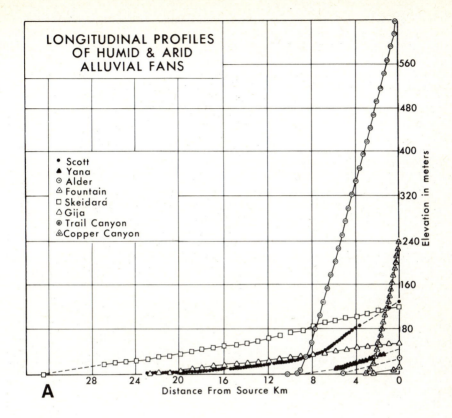

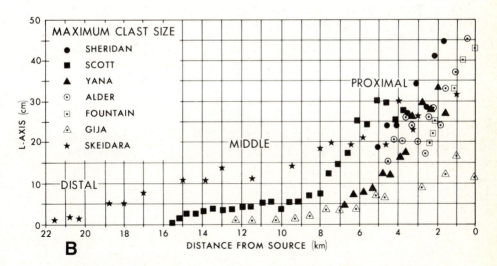

Fig. 6. A. Longitudinal profiles of glacial-outwash fans with two Death Valley fans (Trail and Copper Canyon) plotted for comparison. Gradients range from 50 m/km^{-1} (proximal, Fountain) to 2 m/km^{-1} (distal, Gija and Scott). The arid fan gradients are up to 100 m/km^{-1}. B. Maximum clast size (L-axis) measured on the apices of bars along the active stream systems.

Surface gradient (Fig. 6a) decreases from a maximum of 50 m/km [1] (Fountain) in the proximal area to approximately 2 m/km [1] at the distal margin (Scott, Gija). Gradients of the other fans fall within these proximal-distal extremes. Maximum clast size (L-axis) along the longitudinal profile lines (Fig. 6b) ranges from 45 cm (Alder, Sheridan) in the proximal area, to sand-sized sediment at the maximum of 21.5 km (Skeidará) downstream in the distal portion of the fan.

A plot of gradient versus clast size (gravel-sized material) that combines the data illustrated in Figure 6a and 6b is shown in Figure 7. It is readily apparent that gradient and clast size of these glacial-outwash fans is interdependent, and even more important, clast size and gradient decrease in a somewhat predictable manner from proximal to distal environments. Our very limited information from two arid fans in Death Valley (data plotted on Fig. 6a and 7) indicate that gradient/clast size behavior is very different. That is, gradient and clast size do not show an orderly decrease downfan. The point is that supposed alluvial fans in the ancient record need to be examined with the above ideas in mind.

Sediment Size Distribution and Channel Pattern

Segments of individual fans are grouped according to clast size thus: 1) proximal average clast size greater than 10 cm (L-axis); 2) medial clast size less than 10 cm but greater than 2 mm; and 3) distal sand-sized material. Figures 4 and 8 illustrate this sediment-size distribution downfan.

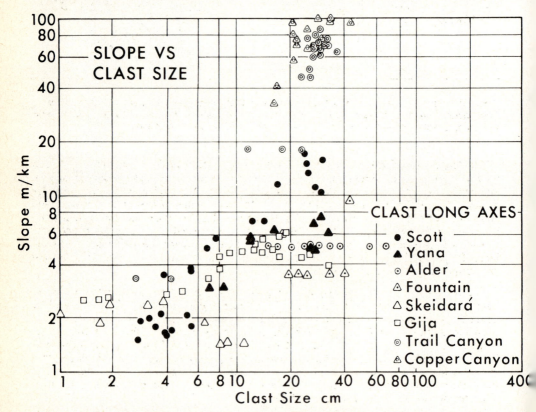

Fig. 7. Gradient (slope) versus clast size for glacial-outwash fans. There is a downstream decrease of both clast size and gradient. Compare this trend with that of the arid fans (Trail and Copper Canyon).

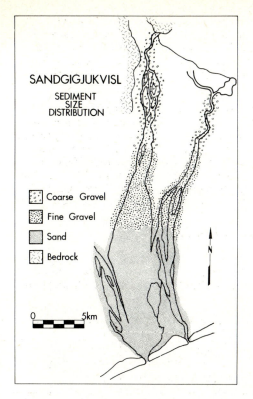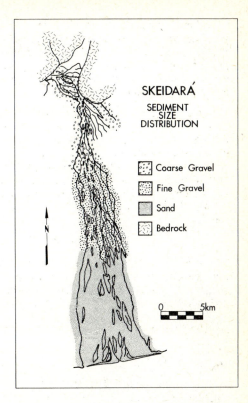

Fig. 8. Sediment-size distribution and channel pattern in two active stream systems, Iceland. The channel pattern changes from incised, to 'coarse-braided', to 'fine-braided' in a downfan direction, reflecting coarse gravel, fine gravel, and sand lithofacies, respectively.

The length of individual fans varied from a maximum of 30.5 km (Skeidará) to a minimum of 2.4 km (Fountain). It is interesting to note that downstream change in gradient, clast size and other parameters on each fan occurred over generally similar distances in a downfan direction (Figs. 6, 7). That is, smaller fans such as Fountain and Alder did not show a downstream fining to sand as did the longer fans, particularly the Icelandic examples (Gíja and Skeidará). Thus, the smaller fans exhibit proximal characteristics over their entire length, in contrast to proximal-distal changes on the larger fans. Relative fan size should also be taken into account when comparing ancient rock sequences to this Recent model.

The channel pattern on actively accreting glacial outwash fans is distinctive in that it consists of three distinct segments related to gradient and clast size (Figs. 1, 3, 8). The upper fan usually has a single, to two or three, channels incised in coarse gravel. These channels extend some 2 - 5 km from the fountain source then bifurcate into a larger number of channels that bisect around individual and multiple longitudinal bars. The nature of this channel bifurcation leads to a channel pattern termed 'coarse-braided', that is readily identifiable on aerial photographs and satellite images (Figs. 3, 8). The fact that the pattern appears to be limited to the gravel facies means it can be used as an indirect indicator of clast size on glacial-outwash fans.

The channel pattern on the sandy distal fan results from the bifurcation around, and dissection of, emergent linguoid bars. This pattern, termed 'fine-braided', also is readily identifiable on aerial photographs and satellite images, and types the area where it occurs

Fig. 9. Photographs of proximal bar and channel morphology and internal sedimentary structures. A. Aerial view of coarse-grained gravel longitudinal bars on the upper Scott fan. Downstream is to the upper left; bar lengths range from 30 - 50 m. B. Ground view looking downstream in the same general vicinity as the aerial view (9a). The vehicle is sitting on a thin, sheet-like longitudinal bar. C. Trench in longitudinal bars of the upper Skeidará fan (facies *Gm*, Miall, 1977). Scale is 100 cm long. D. Longitudinal bar slipface consisting of gravelly planar cross-stratification (facies *Gp*, Miall, 1977) exposed in wall of a kettlehole on the upper Skeidará fan, Iceland. E. Longitudinal trench in a sand-wedge slipface developed around a gravel longitudinal bar on the upper Scott, Alaska (facies *Sp*, Miall, 1977). Scale is 30 cm long.

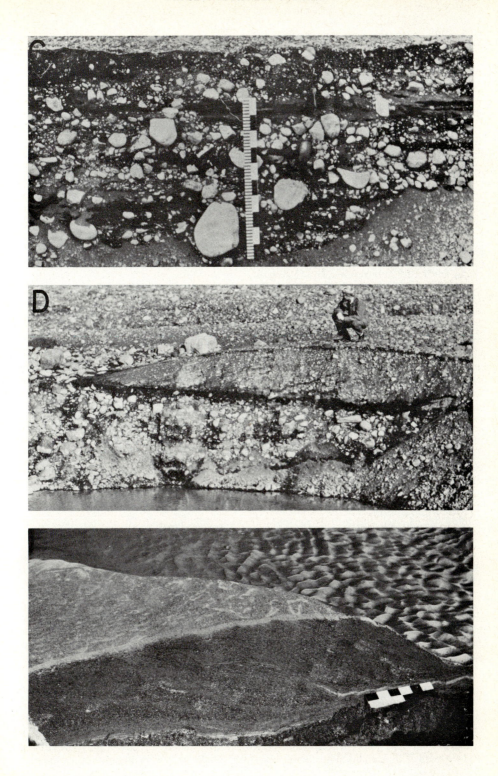

as a sand facies (Figs. 1, 3). During flood stage linguoid bars of the distal fan are submerged and the active channel area is characterized by a wide sheet of shallow water (Fig. 3b), as described by Fahnestock (1969).

Photographic Atlas of the Core Facies

Introduction

As stated above, the facies succession that is the result of active bar and channel processes (Fig. 5) was originally suggested by Boothroyd and Ashley (1975). Development of most of the facies were discussed in that paper and will not be duplicated here. This section will illustrate the facies succession with a series of photographs of bar and bedform morphology and accompanying lithofacies and internal sedimentary structures. We have included many Icelandic examples intended to complement and expand the coverage of Boothroyd and Ashley (1975). The lithofacies classification scheme of Miall (1977) has been used so our photographs may be compared to his analysis. The explanation for Figure 5 coordinates our terminology for sedimentary structures with Miall's lithofacies scheme. It is our intent that workers dealing with ancient fluvial sequences use these photographs as a guide for, and explanation of, the glacial-outwash fan facies succession (Fig. 5).

Proximal and Medial Facies

Figure 9a is an aerial photograph of longitudinal bars (following the terminology of Boothroyd and Ashley, 1975; Nummedal *et al.*, 1974; Gustavson, 1974; Rust, 1972; Smith, 1970, 1974; Williams and Rust, 1969; Hein and Walker, 1977; and Miall, 1977); Figure 9b is a ground view of the same general locality on the Scott fan. The bars exist as thin gravel sheets, but are probably not equivalent to the unit bars of Smith (1974) or diffuse gravel sheets of Hein and Walker (1977). Figure 9c is a view of facies *Gm* (Miall, 1977) in a trench on the proximal Skeidará fan, Iceland.

Two types of cross-stratification are associated with the processes of longitudinal bar formation: 1) planar cross-stratification in gravel (facies *Gp*, Miall, 1977) (Fig. 9d); and 2) planar cross-stratification in sand (facies *Sp*, Miall, 1977) (Fig. 9e). Gravel foresets, here developed on the downstream margin of a longitudinal bar in Skeidará, are rare on glacial-outwash fans that we have studied. We speculate that the relatively small maximum discharges do not allow sufficient depth of channels so that flow separation processes can develop over the gravel bars. Gravel foresets are best displayed in the Skeidará river which does have the greatest maximum discharge of those streams we studied, but is also subject to glacier-burst flooding (jökulhlaups). The gravel cross-strata may be the result of that process.

Sandy foresets (Fig. 9e) are common and develop during declining flow stages around the margins of gravel longitudinal bars. They are the sand-wedge slipfaces of Rust (1972). This facies becomes more prevalent in the fine-gravel medial areas of the fans.

Megaripples (dunes), bedforms with a crest-to-crest spacing of 60 cm - 6 m develop during declining flow stages and migrate in shallow channels on bar surfaces (Fig. 10a, Skeidará) and on the floors of larger channels at low flow stages. They are ubiquitous in the medial and distal areas of the fans studied. A trench cut into megaripples on the Yana fan is shown in Figure 10b (facies *St*, Miall, 1977).

Distal Braided Facies

Linguoid bars, sandy spoon-shaped forms with a fan-shaped slipface (following the terminology of Collinson, 1970; Boothroyd and Ashley, 1975; Nummedal *et al.*, 1974; Miall, 1977; transverse bars of Smith, 1970, 1971, 1972) are the dominant bar form of the distal braided portions of the glacial-outwash fans studied. They are particularly well-developed in Iceland (Figs. 1, 3, 11a, 11c). Those shown in Figure 11a developed

Fig. 10. Megaripple (dune) morphology and internal structure. A. Megaripples developed during waning flow stage, in a shallow channel on a longitudinal fan surface, on the medial Skeidará, Iceland. Flow was from right to left; bedform spacing is about 2 m. B. Longitudinal and horizontal cut in megaripples on the medial Yana fan, Alaska (facies *St*, Miall, 1977). Two sets are shown; scale is 30 cm long.

Fig. 11. Linguoid-bar morphology and internal sedimentary structures. A. Aerial view of high-stage linguoid bars during a minor glacier-burst flood in the Sulá river system. Flow is left to right and the bars become successively emergent and complex toward the river margin (bottom of photograph). Scale across bottom of photo is about 150 m. B. Longitudinal cut in a high-stage linguoid bar on the distal Scott fan (facies *Sp, Sh, Sr,* Miall, 1977). Flow was from right to left; scale is 30 cm long. C. Aerial view of low-stage linguoid bars on the distal Skeidará fan, Iceland.

Emergence and dissection of bars leads to a complex lobate pattern of slipfaces. Flow is toward the bottom of the photograph. Scale across the bottom of photo is about 200 m. D. Longitudinal (with scale) and transverse trench in a low-stage linguoid bar. Complex sedimentary structures are the result of bar slipface migration, and vertical accretion (ripple-drift cross-lamination) on the bar surface (facies *Sp, Sr,* Miall, 1977). Flow was from lower left to upper right. Scale increments are 10 cm long.

during a minor glacier-burst flood in the Sulá River, Iceland. These rhombic-shaped forms are termed 'high-stage' linguoid bars. A trench cut into what is thought to be a high-stage bar is shown in Figure 11b (facies *Sp, Sh*, Miall, 1977).

As the flow stage drops, the bars emerge and are dissected, with the development of numerous lobate slipfaces (Fig. 11c, Skeidará) such as those well documented by Smith (1971, 1972). These complicated forms are called 'low-stage' linguoid bars. In addition to the planar cross-stratification developed by migrating bar slipfaces, we believe that when sediment supply is adequate, vertical accretion occurs on the tops of the bars as shown in Figure 11d. This leads to a complex arrangement of lithofacies (facies *Ss, Sp, Sr,* Miall, 1977).

Summary

The Scott braided-river depositional profile of Miall (1977) is based on the proximal facies of the Scott outwash fan, but fits the other fans studied just as well. This profile is depicted by sequences P1 and P2 of Figure 5. The distal glacial-outwash fan facies is similar to the Platte-type depositional profile (Miall, 1977 and is depicted by sequence D$_1$ of Figure 5. The medial or middle fans are a mix of the two profiles (sequence M, Fig. 5).

LATERAL AND DISTAL FACIES

Introduction

The core facies of the Alaskan and Icelandic fans are flanked by a number of lateral and distal lithofacies, each with a distinct suite of internal sedimentary structures. The complete facies assemblages for both areas are illustrated in Figure 12. The basic difference between Alaska and Iceland is that Alaskan fans have an extensive vegetation cover in the inactive stream areas that is absent in Iceland. These inactive stream areas support extensive wind-tidal flats in Iceland, but are marsh-and-swamp-covered in Alaska. In addition, vegetation cover in the distal areas of some Alaskan fans may result in the development of a meandering-stream facies (Fig. 12).

The reason for the variation in vegetation growth is related to climatic conditions, specifically rainfall and wind velocity and duration. The Skeidarársandur area of southern Iceland is subject to 1000 mm of precipitation per year and is buffeted by strong winds from the east and west that are over 10 ms^1 thirty percent of the time (Nummedal *et al.*, 1974). In contrast, the Northeast Gulf of Alaska coast receives from 3600 to 5500 mm yr^{-1} precipitation and, even though this coast lies along the North Pacific storm track, the area is not subjected to constant winds of high velocity (Hayes and Kana, 1976). The frequency of glacial-burst floods in Iceland is not the reason for the sparse vegetation there, for old records indicate that sections of Skeidarársandur have not been subject to flooding for several hundred years (Nummedal *et al.*, 1974), whereas sequential aerial photographs document the growth of heavy vegetation in 30 years on inactive stream areas of the West Malaspina Foreland (Fig. 1).

The following discussion will present photographic evidence of the lateral and distal variations and incorporate each into the facies succession framework (Fig. 5).

Alaska

Since these facies were discussed in Boothroyd and Ashley (1975), only a brief review will be presented here.

The intensity of coarse-gravel sediment transport and the lack of suitable fine-grained material evidently keeps the proximal fans mostly vegetation-free (Fig. 3) unless they are totally abandoned by active streams. However, the fine-grained medial and distal areas support vegetation between active streams (grasses, alders and cottonwoods). The baffle

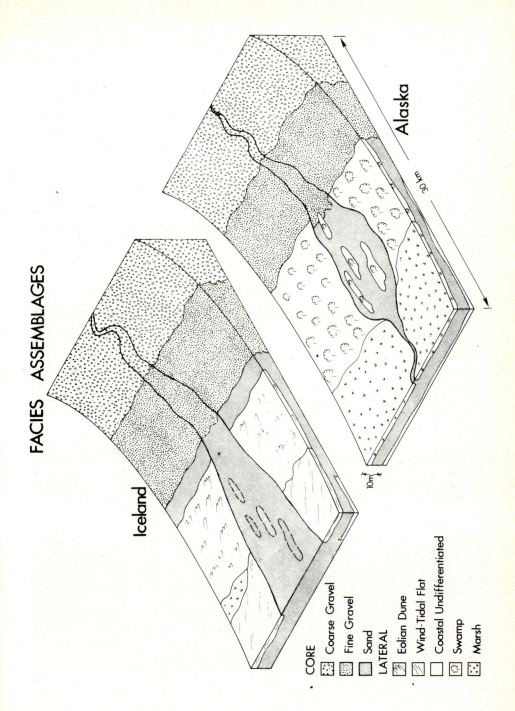

Fig. 12. Core and lateral facies assemblages for glacial-outwash fans in Alaska and Iceland. Note the differences in lateral facies, and in the most distal facies, between the two localities. This model is based on the study of 8 fans and the inspection of many more.

Fig. 13. Alaskan overbank facies. A. Aerial view of a silty crevasse-splay-like lobe on the lower to medial Scott fan. Length of lobe is about 50 - 75 m and is deposited over grass and alders on an elevated island..B. Longitudinal cut in overbank levee deposits of the lower Scott (facies *Sr, Fl,* Miall, 1977). Flow was from left to right; scale increments are 5 cm.

Fig. 14. Distal-meandering facies, Alaska. A. Aerial view of meandering stream and point bars during a waning flood stage on the lower Scott fan. Active and inactive levees are adjacent to the stream; fresh and saltwater marsh occupies the interstream areas. Point bars are about 50 m in length. B. Trench cut in point bar in the direction of bar slipface migration. The complex internal structure is the result of point bar-slipface migration, bar-top megaripple migration, and low-stage channel fill (facies *St, Sp, Sr, Fl,* Miall, 1977). Scale is 30 cm long.

Fig. 15. Lateral eolian facies, Iceland. A. Aerial view of streamlined, elongate dunes aligned parallel to dominant wind direction (left to right). Distance across photograph is about 400 m. B. Ground view of dunes shown in Figure 15a. Dunes are vegetated, longitudinal mounds with unvegetated interdune flats subject to ground water discharge. Dune in foreground is 2 m high. C. Unvegetated transverse dune with slipface that resulted from sustained winds from one direction. D. Natural cut normal to transverse dune-migration direction. Most cross-stratification is the result of stoss and lee vertical accretion (facies *Sse, Spe, She*). Scale is 100 cm long.

effect of the vegetation allows sand and silt-sized material to be deposited during overbank flooding (Fig. 13a). A trench cut in the overbank lithofacies is shown in Figure 13b (facies *Sr, Fl, Fm,* Miall, 1977). This sequence may be interbedded with, or capped by, up to a metre of fresh or saltwater marsh peat.

Along the distal margin of some fans, most notably the Scott, a combination of vegetation and fine-grained overbank sediment causes the stream systems to be forced into a meandering mode (Fig. 14a), with the development of point and lateral bars (Boothroyd and Ashley, 1975). A trench cut in a lower Scott point bar is shown in Figure 14b (facies *St, Sp, Sr, Fl,* Miall, 1977).

Iceland

Eolian Facies

The inactive stream areas between the Skeidará and Gíja stream systems support either active eolian dune fields or extensive wind-tidal flats (Figs. 1, 3, 12) (Nummedal *et al.,* 1974). The dune fields occupy the upper distal areas, and the wind-tidal flats the lower distal margin. An aerial view of a dune field is shown in Figure 15a with an accompanying ground view of the same area (Fig. 15b). The dunes shown are generally elongate parallel to dominant wind direction ranging from 10 - 20 m long, 5 - 10 m wide, and 2 - 5 m high. They may or may not be grass covered. Inspection of aerial photographs indicates that the dominant long-axis orientation is ENE-WSW. The interdune flats are largely unvegetated and may be subject to shallow flooding and runoff due to groundwater discharge during maximum summer discharge in the stream systems.

Many of the unvegetated dunes are not longitudinal in form but exists as low mounds or as small transverse dunes with or without slipfaces. They range from 10 - 60 m spacing between crest lines and from 10 m to 2 m in height (Hine and Boothroyd, in press). Winds of sufficient duration from one direction allow the development of slipfaces (Fig. 15c), although this condition is not common to the dune fields as a whole. A cut through an unvegetated dune area is shown in Figure 15d. The internal sedimentary structures consist of long (3 - 10 m), shallow (50 cm to 1 m) trough cross-stratification sets, and abundant, delicate horizontal lamination. Planar cross-stratification sets up to 2 m in height do exist, but are rare.

The eolian dune cross-stratification is not considered in Miall's (1977) facies classification scheme but, following the general framework of that scheme, we shall designate the trough sets facies *Sse;* the horizontal lamination, facies *She;* and the planar foresets, facies *Spe.* The interdune runoff areas consist of facies *Sr* with limited *Sp* and *Sh* (Miall, 1977).

Wind-Tidal Flat Facies

The most distal areas adjacent to the Skeidará and Gíja stream systems, particular to the west of Skeidará, are occupied by extensive wind-tidal flats (Fig. 1, 3b). A ground view of a portion of the western Skeidará flat is shown in Figure 16a. This flat, measuring roughly 12 by 12 km, is subject to flooding by waters blown out of the Skeidará stream system and also by overbank flooding during maximum summer discharge (Nummedal *et al.,* 1974). An aerial view of overbank flooding is shown in Figure 16b. A complicated arrangement of straight-crested bars and small spoon-shaped (linguoid ?) forms is developing in the wake of isolated eolian dunes. The interference of wakes creates the beautiful rhombic pattern.

Unfortunately, the high water table and the extreme thixotropic nature of the sediments precluded any trenching. Thus, the nature of the internal structures is based on interpretation of bedforms and surfaces. The internal structure is thought to be a combination of planar and trough cross-stratification, representing migration of megaripples and small bars (facies *St, Sp,* Miall, 1977), interbedded with some small-scale

Fig. 16. Wind-tidal flat lateral facies, Iceland. A. Almost featureless flat covered by a thin sheet (5 cm) of ground water discharge. B. Aerial view of wake interference pattern caused by overbank flow around isolated eolian dunes. Flow is from upper left to lower right. Linear to spoon-shaped bedforms are thought to give rise to facies *St, Sp, Sr* (Miall, 1977).

trough cross-stratification and horizontal lamination (facies *Sr* and *Sh*, Miall, 1977). This sequence may be capped by finer-grained material deposited during waning flooding of the flats (facies *Fl*, Miall, 1977).

Concluding Statement

Two different sets of facies successions, the bar and channel core facies and the lateral and overbank facies, have been developed to expand the scheme developed by Boothroyd and Ashley (1975). These are presented as alternate sequences and additions in Figure 5. The Alaska sequences, including the core facies P_1, P_2, M, D_1, and the most distal (D_2, Alaska), are those of Boothroyd and Ashley (1975). The Icelandic lateral and distal sequences are new. Essentially, the bar and channel portion of sequence D_2 (Iceland) is a repetition of sequence D.

The concept of core facies with lateral and distal variation may be useful for interpretation of ancient fluvial sequences. The Alaska example could be applied to areas where coal is present, whereas the Icelandic example may be a better analog for pre-Devonian fluvial sequences.

CONCLUSIONS

1. A model for buried alluvial-fan deposits has been developed based on Alaskan and Icelandic glacial-outwash fan examples.

2. A core facies is present that is repeated in a logical order downstream in all of the fans studied.

3. A difference in vegetation cover gives rise to a difference in lateral and distal facies between Alaska and Iceland. Interstream areas in Alaska are heavily vegetated; those in Iceland are not.

4. The lateral and distal variations allow the model to be applied to a variety of ancient fluvial sequences.

5. The glacial-outwash fan model offers an alternative example to the alluvial-fan model based on arid southwestern U.S. examples. Glacial conditions are not implied.

ACKNOWLEDGMENTS

Funds for the Alaskan research were provided by the Office of Naval Research, Geography Branch (NOO-14-67-A-0230-001, Miles O. Hayes, principal investigator); National Oceanographic and Atmospheric Administration (03-5-002-82, Miles O. Hayes and Jon C. Boothroyd, co-principal investigators); and the Department of Geology and Geography, University of Massachusetts. Funds for the Icelandic research were provided by the Naval Ordinance Laboratory (N60921-73-6-0258, Miles O. Hayes, principal investigator; Dag Nummedal, co-investigator). The Department of Geology, University of South Carolina, aided research in both locales.

This work has benefitted from many discussions, particularly with Thomas C. Gustavson, Norman D. Smith, Frances J. Hein, and Gail M. Ashley. Gail M. Ashley and Albert C. Hine reviewed the manuscript.

REFERENCES

Boothroyd, J. C., 1972, Coarse-grained sedimentation on a braided outwash fan, northwest Gulf of Alaska: Coastal Research Division, Dept. Geology, Univ. South Carolina, Columbia, S.C., Technical Report No. 6-CRD, 127p.

——, 1976, Sandur plains, northeast Gulf of Alaska: a model for alluvial fan - fan delta sedimentation in cold-temperate environments: *in* T. P. Miller, *ed.*, Recent and ancient sedimentary environments in Alaska; Alaska Geol. Soc., p. N1-N13.

——, and Ashely, G. M., 1975, Process, bar morphology, and sedimentary structures on braided outwash fans, northeastern Gulf of Alaska; *in* A. V. Jopling and B. C. McDonald, *eds.*, Glaciofluvial and Glaciolacustrine Sedimentation: Soc. Econ. Paleont. Mineral. Spec. Pub. 23, p. 193-222.

Brown, L. F., Cleaver, A. W., and Erxleben, A. W., 1973, Pennsylvanian depositional systems in north-central Texas: Geol. Soc. Am. Guidebook No. 14, Annual Meeting, 1973, 122p.

Bull, W. B., 1972, Recognition of alluvial fan deposits in the stratigraphic record: *in* J. K. Rigby and W. K. Hamblin, *eds.*, Recognition of ancient sedimentary environments; Soc. Econ. Paleont. Mineral. Spec. Pub. 16, p. 63-83.

Collinson, J. D., 1970, Bedforms of the Tana River, Norway: Geogr. Ann., v. 52A, p. 31-55.

Denny, C. S., 1965, Alluvial fans in the Death Valley region, California and Nevada: U.S. Geol. Survey Prof. Paper 466, 62p.

Fahnestock, R. K., 1969, Morphology of the Slims River: Icefield Ranges Res. Proj., Sci. Res., v. 1, p. 161-172.

Gustavson, T. C., 1974, Sedimentation on gravel outwash fans, Malaspina Glacier Foreland, Alaska: J. Sediment. Petrol., v. 44, p. 374-389.

Hayes, M. O., and Kana, T. W., *eds.*, 1976, Terrigeneous Clastic Depositional Environments: Coastal Research Division, Dept. Geology, Univ. North Carolina, Tech. Report No. 11-CRD, II, 171p.

——, Hobbs, C. H., III, Raffaldi, F. J., and Hague, P. R., 1972, Coastal sedimentation in a tectonically active geosynclinal basin: the glacial-outwash plain shoreline of the northeastern Gulf of Alaska (abs.): Am. Assoc. Petrol. Geol. Bull., v. 56, p. 624.

Hein, F. J., and Walker, R. G., 1977, Bar evolution and development of stratification in the gravelly, braided, Kicking Horse River, B.C.: Can. J. Earth Sci., v. 14, p. 562-570.

Hine, A. C., and Boothroyd, J. C., 1978, Morphology, processes, and recent sedimentary history of a glacial-outwash plain shoreline, South Iceland: J. Sediment. Petrol., in press.

Hooke, R. L., 1967, Processes on arid-region alluvial fans: J. Geol., v. 75, p. 438-460.

Klein, G. de V., 1975, Sandstone depositional models for exploration for fossil fuels: Cont. Education Pub. Co., Champaign, Ill., 108p.

LeBlanc, R. J., 1972, Geometry of sandstone reservoir bodies, *in* T. D. Cook, *ed.*, Underground waste management and environmental implications; Am. Assoc. Petrol. Geol. Mem. 18, p. 133-190.

McGowen, J. H., and Groat, C. G., 1971, Van Horne Sandstone, West Texas, an alluvial fan model for mineral exploration: Bur. Econ. Geology, Univ. Texas, Rept. Invest. No. 72, 57p.

Miall, A. D., 1977, A review of the braided-river depositional environment: Earth Sci. Revs., v. 13, p. 1-62.

Nummedal, D., *et al.*, 1974, Recent migration of the Skeidarársandur shoreline, Southeast Iceland: Final Report for Contract No. NG0921-73-6-0258, Naval Ordinance Laboratory, Dept. Geology, Univ. South Carolina, 183p.

——, and Boothroyd, J. C., 1976, Morphology and hydrodynamic characteristics of terrestrial fan environments: Coastal Research Division, Dept. Geology, Univ. South Carolina, Tech. Report No. 10-CRD, 61p.

Rust, B. R., 1972, Structure and process in a braided river: Sedimentology, v. 18, p. 221-245.

Smith, N. D., 1970, The braided stream depositional environment, comparison of the Platte River with some Silurian clastic rocks, north-central Appalachians: Geol. Soc. Am. Bull., v. 81, p. 2993-3014.

———, 1971, Transverse bars and braiding in the lower Platte River, Nebraska: Geol. Soc. Am. Bull., v. 82, p. 3407-3420.

———, 1972, Some sedimentological aspects of planar cross-stratification in a sandy braided river: J. Sediment. Petrol., v. 42, p. 624-634.

———, 1974, Sedimentology and bar formation in the upper Kicking Horse River, a braided outwash stream: J. Geol., v. 82, p. 205-224.

Vos, R. G., 1975, An alluvial plain and lacustrine model for the Precambrian Witwatersrand deposits of South Africa: J. Sediment. Petrol., v. 45, p. 480-493.

Williams, P. F., and Rust, B. R., 1969, The sedimentology of a braided river: J. Sediment. Petrol., v. 39, p. 649-679.

ALLUVIAL FAN SEQUENCE AND MEGASEQUENCE MODELS:
with examples from Westphalian D — Stephanian B coalfields, Northern Spain

ALAN P. HEWARD[1]

ABSTRACT

Sequences, megasequences and basin-fill sequences of progressively changing character appear common to many alluvial fan accumulations. Their study can provide evidence of depositional processes (sequences), short term fan behaviour (sequences), longer term fan behaviour (megasequences) and the depositional basin setting (basin-fill sequences). The occurrence and identification of these sequences appear dependent on down-fan trends in sediment character and depositional process, the localisation, switching and migration of the region of active fan sedimentation, a regularity of depositional event magnitude, and an absence of reworking by subsequent floods.

"The prime requisite of fan formation is the setting of highland and lowland side by side", Denny (1967). Where alluvial fans are dependent solely on initial fault scarp or erosional topography, geographically limited, relatively thin accumulations of decreasing grain size result. Fan sediments of greater geographical extent and thickness, repeated vertical stacking of sequences and megasequences, and the occurrence of alluvial fan basin-fill sequences result from continued movement along fault lines.

INTRODUCTION

"An alluvial fan is a body of detrital sediments built up by a mountain stream at the base of a mountain front" and commonly having the shape of a segment of a cone (Blissenbach, 1954, p. 176). Sedimentation is characterised by an accumulation of debris within a drainage basin and its sporadic transference to the mountain front. Beaty (1970), for example, estimated that one 'typical' debris flow every 350 years was sufficient to explain the growth of a White Mountain fan over the past 700,000 years. Deposition results from deceleration due to the increase in flow width and decrease in flow depth as floods emerge from the confines of the feeder canyon or fanhead channel (Bull, 1964a, 1968; Denny, 1965).

This paper attempts to add to the understanding of ancient alluvial fan deposits through the discussion of vertical sequences of related beds. Such sequences of progressively changing character are of fundamental importance for, as Walker (1970) emphasised in a discussion of turbidite sequences "a trend of any type forces the geologist to think in terms of long period control of . . . sedimentation". In this analysis previous descriptions are utilised[2] of both ancient alluvial fan and submarine fan vertical sequences and of modern and ancient alluvial fan deposits, and an attempt is made to marry the development of vertical sequences with models of alluvial fan behaviour. Some of the concepts derived are then applied to examples of alluvial fan vertical sequences from Westphalian D — Stephanian B coalfields, northern Spain.[3]

---

[1]Department of Geology and Mineralogy, Parks Road, Oxford, OX1 3PR, England; Present address: Department of Geological Sciences, University of Durham, South Road, Durham, DH1 3LE, England.

[2]My usage of submarine fan sequence data stems from their more extensive documentation, and from a belief that in important respects these point source accumulations of gravity induced sediment flows are similar.

[3]Since this analysis was completed relevant new data has been published by Bull (1977) and Wasson (1977a, b) on alluvial fan sediments, and by Martini and Sagri (1977) on submarine fan sequences.

Table 1 : Ancient Alluvial Fan Sequences

Author and Author's Terminology	Scale	Style	Suggested Causes of Sequences.
BLUCK (1967) Upper Old Red Sandstone, Clyde area, western Scotland.			
Sequences	eg. 140 m	Fining (and thinning) upward, mudflow – braided stream deposits.	Recession of source area, or gradual reduction in slope (relief) of source area.
WESSEL (1969; not seen: data from Klein, 1975) Upper Triassic, north-central Massachusetts, U.S.A.			
Sequences.	3–6 m	Coarsening upward single event. Coarsening upward multiple event.	Fan progradation.
WILLIAMS (1969) Torridonian (late Precambrian), northwest Scotland.			
		Upward fining.	Continuous retreat of source area.
		(Thickening and coarsening).	Distal-proximal fan relationship (his fig.14).
MIALL (1970) Devonian, Prince of Wales Island, Arctic Canada.			
	eg. > 40 m	Upward coarsening associated with an 'inverted stratigraphy' in clast types.	Rejuvenation and downcutting in source area.
DEEGAN (1973) Lower Carboniferous, Kirkcudbrightshire, Scotland.			
Megacycles or large-scale cycles.	10's – 100's m	Fining upward.	Basin subsidence and fan aggradation. Progressive decrease in depositional gradient causing finer detritus to be supplied.
		Coarsening upward.	Relatively gradual basin subsidence.
SCHLUGER (1973) Devonian, Perry Formation, New Brunswick, Canada, and Maine, U.S.A.			
Sequences.	10's – 100's m	Upward fining, conglomerate-sandstone-mudstone.	Gradual reduction of highland and lateral shift, alluvial fan-marginal lacustrine sedimentation.
STEEL (1974) New Red Sandstone, Hebridean Province, Scotland.			
Sequences.	10's m	Upward fining, upward increase in % of sandstone beds, upward decrease in extra-formational v. intra-formational clasts, mudflow-streamflood-braided stream.	Initial fault movement followed by lowering of relief and recession of source area.
		Vertically stacked upward fining etc. sequences as above.	Tectonic movements causing drainage basins to be periodically rejuvenated.
BRYHNI and SKJERLIE (1975) Middle Devonian, Old Red Sandstone, Kvamshesten district, western Norway.			
Rhythms or sequences.	10–40 m	Upward coarsening, siltstone-sandstone.	Formed at foot of fans, each rhythm indicates deposition in increasingly higher flow regimes, caused by progressive downcurrent spreading of braid bar sediments resulting from tectonic or climatic changes in source region.
STEEL and WILSON (1975) ?Permo-Triassic, Stornaway Formation, Lewis, northwest Scotland.			
Sequences.	100's m	Fining upward, mudflow – streamflood – braided stream.	Basin margin faulting of gradually decreasing intensity accompanied by sourceward migration of locus of sedimentation, probably implying a lowering of relief within drainage basin.
		Coarsening upward, braided stream – streamflood – mudflow.	Increasing rate of uplift in source area, probably resulting in fan progradation.
STEEL (1976) Devonian, western Norway			
Sequences, cycles, or cyclothems.	50 – > 200 m	Coarsening upward and decrease in amount of interbedded sandstone.	Major period of fan-building in response to fault movement, probably basin subsidence.
Units.	10 m	Coarsening upward, sometimes capped by slight fining upward.	Prograding fan-building lobes, may be in response to fault movement in fan head region.
STEEL ET AL. (1977) Devonian, Hornelen Basin, western Norway.			
Cycles, cyclothems or sequences.	50–80 m	Coarsening and thickening upward, decreasing sandstone percentage, strongly asymmetric.	Fan progradation and increasingly greater volumes of debris, marked asymmetry due to abrupt lateral tectonic shifting of the locus of subsidence and sedimentation.
	100 – 200 m	Coarsening and thickening upward, decreasing sandstone percentage, less asymmetric.	Fan progradation and increasingly greater volumes of debris, lesser asymmetry.
Subcycles, subunits, subcyclothems or sequences.	10–25 m	Coarsening and thickening upward.	Progradation of main dispersal system in response to discrete episodes of fault movement, or an autocyclic mechanism.

PREVIOUS DESCRIPTIONS OF ALLUVIAL FAN AND SUBMARINE FAN SEQUENCES

Tables 1 and 2 list many of the published descriptions of alluvial fan and submarine fan vertical sequences, *defined on progressive changes in grain size, bed thickness, interpreted depositional processes and clast types*. As can be seen, a confusing terminology exists of cycles, rhythms, sequences, megasequences, megacycles, large scale cycles, cyclothems etc. Throughout this paper the term sequence is used in a general sense, and includes: sequences in a specific sense, m-10's m thick, consisting of a single bed or a series of related beds; megasequences, 10's - 100's m thick consisting of arrangements of related beds and sequences; and basin-fill sequences, 100's - 1000's m thick, consisting of arrangements of sequences and megasequences.

Two types of *alluvial fan sequences and megasequences* have been described, fining upward and coarsening upward (Table 1). Fining upward sequences and megasequences, often accompanied by increasingly distal depositional processes, have generally been considered to result from the reduction of source area relief and/or scarp retreat. Steel and Wilson (1975) suggested that very thick fining upward megasequences might result from basin margin faulting of gradually decreasing intensity. Coarsening upward sequences and megasequences, paralleled by increasing bed thicknesses and by increasingly proximal processes, have been attributed to the progradation of fan building lobes or of the fan itself. Very thick coarsening upward megasequences were interpreted by Steel and Wilson (1975) to result from faulting with a history of progressively greater movements.

Considerably greater attention has been focussed on *submarine fan sequences and megasequences* (Table 2). Early workers (e.g. Kimura, 1966; Sestini, 1970; Sagri, 1972) proposed numerous factors which might control their development, including variations in the rate of subsidence, rhythmic tectonic movements, changes in sediment supply through variations in depositional or erosional rates in source areas, bottom topography, changes in slope angles, combination of turbidity currents from different sources, changes in the spatial arrangement of successive turbidites, and varying distances of the point of deposition from the source of the turbidity current. More recently, however, the specific ideas of Mutti, Ricci-Lucchi and Walker have received fairly general acceptance (Table 2). Fining and thinning upward turbidite and deep water conglomerate sequences have been considered to result from gradual channel abandonment (e.g. Mutti and Ricci-Lucchi, 1972; Mutti, 1974, 1977; Ricci-Lucchi, 1975), although Walker (1975a, 1977) noted that a gradual diminution of sediment supply from the source was equally likely. Van Vliet (in press) emphasised the incompatability of fining and thinning upward sequences occurring in gradually abandoned channels, and progradational coarsening and thickening upward depositional lobe sequences forming below channel mouths. He followed Ghibaudo and Mutti (1973) and Mutti (1974) in suggesting that fining and thinning upward sequences might form by lateral accretion on very low angle point bars. Walker (1975b), in contrast, presented evidence that gradual channel filling may have occurred subsequent to abandonment, a conclusion reached also by Nelson *et al.* (1977).

Coarsening and thickening upward sequences have been considered to result from the progradation of fan depositional lobes, their asymmetry, or symmetry reflecting the abrupt or gradual nature of fan lobe abandonment (Table 2). Coarsening upward megasequences have likewise been attributed to phases of fan progradation. Mutti (1977) and van Vliet (in press) also described refined versions of the basic coarsening and thickening upward sequence considered to be indicative of specific locations on submarine fans (e.g. the thickening upward mouth bar cycles of Mutti, or the abruptly based middle fan megacycles, the gradationally based outer fan megacycles and the diluted fan fringe megacycles of van Vliet).

With the exception of Kimura (1966), Bluck (1967), Sestini (1970), Sagri (1972), Steel (1974), Walker (1975a, 1977) and Steel *et al.* (1977) who also consider gradual changes in

Table 2 : Ancient Submarine Fan Sequences

(sand grade turbidite deposits except where indicated)

Author and Author's Terminology	Scale	Style	Suggested Cause of Sequences
KIMURA (1966) Permo-Triassic, Sambosan Group, and Jurassic-Cretaceous, Shimanto Group, Japan			
Major cycles.	100's m	'Increasing type', coarsening and thickening upward. 'Decreasing type', thinning and fining upward.	Regressions allowing sand into depositional basins. Transgressions excluding sand from depositional basins.
Minor cycles.	10's-100's m	'Increasing type', coarsening and thickening upward, 'decreasing type', thinning and fining upward. Tendency for 'increasing type' to occur during regressions and 'decreasing type' to accompany transgressions.	Series of suggestions: 1. Angle of slope may change intensity of turbidity currents. 2. Change in rate of sedimentation in turbidity current source area. When rate is high, turbidite may transport more sediment. 3. A stronger trigger may produce stronger turbidity currents. 4. Distance of point of deposition from source of turbidity current.
SESTINI (1970) Upper Cretaceous and Tertiary, Italy (northern Apennines), Greece and Turkey, Palaeozoic of Great Britain			
Megarhythms or rhythms.	20-80 m	Periodic changes in strato-facies.	1. Changes in sediment supply. 2. Changes in tectonic activity. 3. Interbedding of turbidites from differing sources.
Cycles or sub-cycles.	m-10's m	Thinning or thickening upward	1. Varying sediment supply. 2. Spatial arrangement of successive turbidites such that progressively thinner or thicker beds accumulate.
MUTTI and RICCI-LUCCHI (1972) Tertiary, northern Apennines, Italy			
Large scale fan sequence (their fig. 14).	100's m		Progradation of fan system.
Megasequence.	10's m	Positive; grain size and bed thickness decrease upwards, channellised.	Gradual abandonment of fan channel.
		Negative; grain size and bed thickness increase upwards, minor channels near top.	Prograding depositional lobes.
SAGRI (1972) Upper Cretaceous and Tertiary, northern Apennines, Italy			
Rhythms or megarhythms.	30-60 m	Thick turbidites at base, upper parts argillaceous with thin beds of fine grained sediments.	Various factors: 1. Rhythmic alternation of tectonic movements and tectonic quiesence. 2. Combination of turbidity currents from different sources. 3. Reduction in clastic inflow due to the increasing stability of continental shelf. 4. Variation in bottom topography.
WALKER and MUTTI (1973) Turbidite facies and facies associations in general			
Large scale fan sequence (their fig. 12).			Progradation of fan system over basin plain.
Slope channel, inner fan channel, and middle fan channel sequences.		Thinning and fining upward.	Gradual abandonment of fan distributary channel.
Middle fan sequence.		Thickening and coarsening upward.	Prograding and aggrading fan depositional lobes.
MUTTI (1974) Tertiary, Apennines, Italy, and Island of Rhodes			
Major fan cycles.	100's m	Thickening and (or) coarsening upward.	Fan progradation.
Middle fan cycles.	10's m	Thinning and (or) fining upward, channellised.	Gradual channel abandonment, vertical (or possibly lateral) accretion.
Outer fan cycles.	10's m	Thickening and coarsening upward, minor channels near top.	Prograding fan lobes.

Table 2 : continued

RICCI-LUCCHI (1975) Tertiary, Apennines, Italy

1st order cycles, bounded vertically by slope deposits.	3000 m	Marnoso-arenacea, thickening and coarsening upward.	Fan progradation, basin plain- outer fan - middle and inner fan.
	3000 m	Laga, thinning and fining upward.	Fan recession, inner and middle fan - outer fan and minor basin plain deposits.
2nd order cycles.	2-70 m	Asymmetric - positive, thinning and fining upward, mostly channellised.	Gradual channel abandonment.
		Asymmetric - negative, thickening and coarsening upward, mostly non-channellised, minor channels near top.	Progradation and rapid abandonment of fan lobes.
		Symmetric, thickening and coarsening, and thinning and fining. Composite.	Progradation and gradual abandonment of fan lobes.

WALKER (1975a) Upper Cretaceous, Wheeler Gorge, California, U.S.A.

Sequences.	25-45 m	Fining and thinning upward, conglomeratic at base.	Questions Italian workers gradually abandoned channel hypothesis and suggests a gradual diminution of supply in hinterland, giving rise to progressively finer grained and smaller flows, is equally likely.

WALKER (1975b) Miocene, Capistrano Formation, San Clemente, California, U.S.A.

Sequences.	m-10's m	Channellised fining and thinning upward.	Gradual channel abandonment.
		Channellised thickening and coarsening upward.	Progradation of turbidite lobe down a channel.

(Walker's examples occur in channels where mudstone/siltstone drapes provide evidence that channels were cut and abandoned prior to infill)

MUTTI (1977) Eocene, Hecho Group, south-central Pyrenees, Spain

Channel-fill sequences.		Fining and thinning upward.	Gradual channel abandonment.
(channels cut prior to infilling and are characteristically straight rather than meandering)			
Channel-mouth bar cycles.		Thickening upward cycle of distinctive thin bedded sediments.	Downcurrent progradation of mouth bar.
Outer fan cycles.		Thickening upward symmetric.	Lobe and lobe fringe progradation and sudden shifting of related feeder system.
		Thickening upward asymmetric.	Lobe and lobe fringe progradation and gradual shifting of related feeder system.

NELSON, MUTTI and RICCI-LUCCHI (1977) Upper Cretaceous, Wheeler Gorge, California, U.S.A.

Upper and middle fan sequences.		Fining upward.	Following channel abandonment, result from infill of channel by overbank flows from adjacent channel.

WALKER (1977) Upper Mesozoic, southwestern Oregon, U.S.A.

Large scale sequences.	50-100 m	Thinning and fining upward.	
Inner-fan? sequences.	4-85 m	Thinning and fining upward, conglomeratic at base.	Progressive channel abandonment or long-term changes in the source area leading to smaller and finer grained flows.

RUPKE (in press) Upper Carboniferous, Cantabrian Mountains, northern Spain

1st order basin fill sequences.	1000's m		
2nd order sequences.	100's m	Thickening and coarsening upward facies triplet, conglomeratic at top.	Complete cycle of fan lobe progradation and avulsion.
3rd order sequences.	m-10's m	Thickening upward.	Progradation of outer fan lobes.
		Symmetrical thickening and thinning upward.	Progradation and lateral shifting of middle fan depositional lobes.
		Thinning upward.	Construction of inner fan valley levees.

VAN VLIET (in press) Lower Tertiary, Guipuzcoa, northern Spain

Inner fan megacycles (not present in this area).		Channellised fining and thinning upward.	Channel lateral accretion.
Middle fan megacycles.	20-70 m	Abruptly based, coarsening and thickening upward, channelling near top.	Avulsion into interlobe depression followed by progradation of depositional lobe.
Outer fan megacycles.	10-50 m	Gradationally based, coarsening and thickening upward.	Progradation of depositional lobe, when strongly asymmetrical indicating rapid abandonment.
Fan fringe megacycles.		Gradationally based, coarsening and thickening upward, diluted with thin distal turbidites particularly near base.	Progradation of depositional lobe, dilution by thin turbidites from adjacent lobe(s) until depositional relief causes their exclusion.

the magnitude of causative or depositional events, most authors have related alluvial fan or submarine fan vertical sequences to the accumulation of increasingly proximal or distal deposits through the progradation, lateral migration, gradual abandonment or recession of the location of active sedimentation. Comparing the background of the interpretations for alluvial fan and submarine fan vertical sequences, alluvial fan interpretations draw on geomorphological 'cycle of erosion' concepts, whilst the majority of those of submarine fans arise from limited modern morphological data (e.g. Normark, 1970) and from analogies with deltaic mouth bars, and fluvial and distributary channels. Whilst the sequences are essentially similar in character and similar criteria are used in their definition, the resulting interpretations show considerable variance, particularly for fining and thinning upward sequences.

CHARACTERISTICS OF ALLUVIAL FANS IMPORTANT TO VERTICAL SEQUENCE DEVELOPMENT

Down-fan and down-flow trends in grain size, bed thickness, sorting, clast shape, number of channels, and processes (Table 3, Fig. 1), abundantly recorded from modern alluvial fans and from ancient fan deposits, are pre-requisites of vertical sequences, in that gradual or abrupt progradation or recession of all or part of the fan, leads to the super-position of deposits of gradually or abruptly changing character. More rapid down-fan trends may, perhaps, preferentially favour the development of sequences on smaller fans (Fig. 1b).

Table 3 : Down-Fan and Down-Flow Trends

Modern Alluvial Fans	Ancient Alluvial Fan Deposits
i. DECREASE IN GRAIN SIZE DOWN-FAN OR DOWN-FLOW (Fig.1) Eckis (1928), Chawner (1935), Sharp and Nobles (1953), Bull (1963,1968,1972), Beaty (1963), Bluck (1964), Ruhe (1964), Denny (1965,1967), Lustig (1965), Hunt and Mabey (1966), Scott and Gravlee (1968), Williams (1970), McPherson and Hirst (1972), Meckel (1975), Tanner (1976).	Bluck (1965,1967), Laming (1966), Meckel (1967), Nilsen (1968, 1969), Williams (1969), Miall (1970), Wilson (1970), McGowen and Groat (1971), Groat (1972), Schluger (1973), Steel (1974, 1976), Steel and Wilson (1975), Steel et al. (1977).
ii. DECREASE IN BED THICKNESS DOWN-FAN OR DOWN-FLOW (Fig.1c) Sharp and Nobles (1953), Beaty (1963,1974), Bull (1964a,b,1972).	Bluck (1965,1967), Williams (1969), Steel (1974).
iii. INCREASE IN SORTING DOWN-FAN OR DOWN-FLOW Blissenbach (1954), Scott and Gravlee (1968), Bull (1972), McPherson and Hirst (1972).	
iv. CHANGES IN CLAST SHAPE DOWN-FAN OR DOWN-FLOW (Trends ambiguous and variable due to original clast shapes, lithologies, distance of transport, weathering and abrasion on abandoned fan areas, and selective transport of different shaped clasts) Blissenbach (1954), Bluck (1964), Lustig (1965), Scott and Gravlee (1968), McPherson and Hirst (1972).	Bluck (1965), Laming (1966), Nilsen (1968,1969), Miall (1970), Schluger (1973). N.B. Meckel (1967), and McGowen and Groat (1971) noted no trends in clast shape.
v. DOWN-FAN DISTRIBUTION OF CHANNELS a. Decrease in number and depth apex to toe: Beaty (1963), Bull (1964a,b,1972). b. More abundant near mountain front - in mid-fan downstream from washes heading in pavement - lower part of fan: Denny (1965). c. More abundant in down-fan regions where infiltrated water emerges: Hooke (1967).	
vi. DOWN-FAN CHANGES IN TRANSPORTING AND DEPOSITIONAL PROCESSES Beaty (1963,1974), Denny (1965,1967), Hooke (1967).	Bluck (1967), Steel (1974), Steel and Wilson (1975).

a. *Decrease in Grain Size Down-fan : lithological variations on Bucaramanga Fan, Colombia* (after Tanner, 1976)

	near fan apex	near fan toe	decline in grain size
Gneiss	1 m	10-20 cm	30 cm/km
Schist	1 m	North part of fan South part of fan	20 cm/km 10 cm/km
Vein Quartz	20-45 cm	10-13 cm	8 cm/km

b. *Decrease in Grain Size Down-fan : variation due to fan size* (after Denny, 1965)

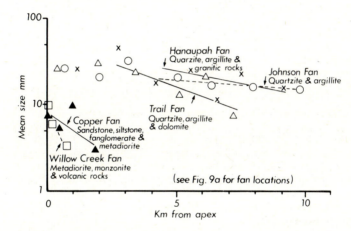

c. *Decrease in Grain Size & Bed Thickness Down-flow* (after Sharp and Nobles, 1953)

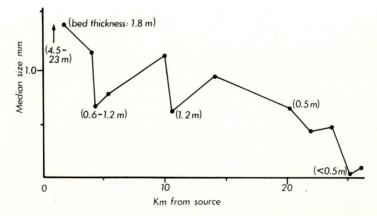

Fig. 1. Examples of down-fan decrease in grain size, and down-fan decrease in grain size and bed thickness from recent alluvial fan deposits.

The permanence of the locus of deposition on a fan surface, important to the development and magnitude of fan sequences, depends on the presence of entrenched channels. The latter typify two regions of a fan (Eckis, 1928), the fan apex where the fanhead trench controls the site of active fan sedimentation, and the fan toe where entrenched channels determine the location of secondary fans (Fig. 7). Over a considerable period of time the cutting, infilling, migration and abandonment of channels results in fairly even, overall, fan sedimentation (Beaty, 1963, 1970, 1974; Denny, 1965, 1967; Lustig, 1965; Hooke, 1967, 1968; Bull, 1968, 1972). In the short and medium term, however, the presence or absence of a fanhead trench is one of the most important features controlling the site of sedimentation (e.g. Buwalda, 1951; Bull, 1968). Numerous factors have been considered as causes of *fanhead entrenchment*. These factors, the estimated duration of their influence, and their possibly characteristic depositional products are summarised in Table 4, and discussed in relation to vertical sequences in a later section.

At any time much of the fan surface is inactive, being the site of weathering, pedogenesis and erosion (Bluck, 1964; Denny, 1965, 1967; Lustig, 1965; Hooke, 1967, 1968, 1972). The resultant non-depositional horizons may occur within, or form the boundaries between, fan sequences.

THE INDIVIDUALITY OF THE ALLUVIAL FAN FLOOD EVENT — FACT OR FICTION

Many students of modern alluvial fans have noted the variability of fan flood events, in terms of amount of precipitation required to initiate an event, the character of the flood discharge and the volume of sediment deposited. Thus Beaty (1963, p. 526) remarked that "debris flowage is undoubtedly a highly irregular process, dependent upon the chance concentration of a cell of heavy rain in a drainage basin the truck canyon of which is floored with unconsolidated materials". Likewise Bull (1964a, p.A4) commented that "floods are controlled by the areal distribution, intensity, and duration of rainfall, and by vegetable cover, lithology, and slopes of the drainage basin. The resulting flow may range from clear water to viscous mud".

Observations, such as these, perhaps prompted Bull (1972, p. 66) to state that "each bed of a fan represents a single depositional event that has resulted from one of a wide spectrum of precipitation and erosion events within the source area. The runoff that is supplied to the main stream channel leading to the fan may be the result of rainfall over the entire basin, or snowmelt runoff from all or part of the basin. Thus, differences in runoff characteristics, source and amount of sediment load, mode of transport, and other factors vary greatly and are reflected in the individual beds preserved in a fan".

Despite the apparent individual characteristics of modern fan flood events, vertical sequences of progressively changing grain size, bed thickness and depositional process characterise many ancient alluvial fan deposits (and submarine fan deposits; Tables 1 and 2; Figs. 12-14). Such sequences are generally interpreted to represent the gradual progradation, avulsion, abandonment, or lateral migration of depositional elements; rather than fractionation of sediment within a depositional element, or progressive changes in the sediment supplied. If the former are the case, the occurrence and identification of sequences require a regularity of depositional event magnitude, for with greater variation sequences would be obscured by the variable size, grain size and properties of the deposits. This possible discrepancy between modern observations and ancient fan sequences can be reconciled in that modern observations refer to flood events of very variable depositional importance and cover only a short time span. In contrast the volumetrically important event of a fan sequence probably occur once/100's or 1000's of years. However, the regularity in magnitude (volume of deposits), within certain limits, of these larger depositional events requires further explanation.

Curry (1966), Statham (1976) and Renwick (1977), when describing modern debris flow events, noted that the exceptional rainfall which preceded flows was no more exceptional than precipitation in previous years which had not resulted in debris flowage. Beaty (1963), Johnson and Rahn (1970) and Scott (1971) documented the accumulation of debris within alluvial fan feeder canyons by landsliding, slumping and rain and rill wash, and Statham (1976) and Beaty (1963, p. 526) noted that "a buildup of fresh debris on their floors must take place before they can again yield large amounts of rubble". These observations when combined with Schumm's (1973) concepts of geomorphic thresholds, can explain why exceptional precipitation frequently does not initiate large-scale fan depositional events (e.g. debris flows) and also why major depositional events for a single fan system are likely to be of similar magnitude.

Consider Fig. 2a (based on Schumm, 1973) where increasing sediment instability is plotted against time. As time progresses colluvial debris accumulates within the feeder canyon, the whole mass becoming increasingly unstable (line 1). At the threshold of instability (line 2) failure will occur and liberate the accumulated volume of sediment (X).

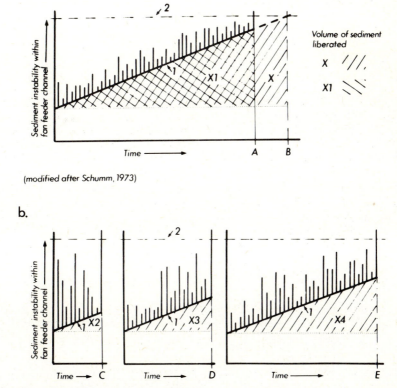

(modified after Schumm, 1973)

Fig. 2. (a) Hypothetical relationship between instability of the sediment accumulating within an alluvial fan feeder canyon, and time. Superimposed on line 1 are vertical lines representing extra instability induced by major precipitational events. When the ascending line of sediment instability (line 1), intersects line 2, the instability threshold, failure will occur (time B). However, failure occurs at time A as the result of a precipitational event. Diagonally hatched areas X and X1 represent the volume of accumulated sediments liberated at the time of failure. (b) As above, except extra instability induced by precipitational events is more variable and extreme, relative to the instability threshold (line 2). Hence, the considerable variability in volume of accumulated sediments liberated (X2, X3, X4) at the times of failure (C, D, E).

Superimposed on line 1 are vertical lines representing the extra instability induced by exceptional precipitation. At time A the threshold of instability is exceeded and flow occurs (failure would anyway have occurred at time B). The amount of extra instability induced by flood events in Fig. 2a is minor relative to the level of the instability threshold (line 2), with the result that when failure occurs similar volumes of accumulated sediment are liberated (e.g. X or X1). When the system lies below the instability threshold exceptional precipitation cannot induce flowage, so explaining the observations of Curry (1966) and Renwick (1977). Scott (1971), when describing the 1969 debris flows near Glendora, California, commented that the most striking aspect of these events was the degree to which channels in all parts of the watershed had been scoured to bedrock. He noted that six months after the storms, debris was again accumulating by gravitational processes.

In Fig. 2b the extra instability of each flood event is larger relative to the instability threshold (line 2) and very variable volumes of sediment could be liberated (e.g. X2, X3, X4). Sequences of progressively changing character representing fan segment progradation, migration, abandonment etc., however, imply an inherent regularity in event magnitude. That such sequences commonly occur in alluvial fan and submarine fan deposits suggests that drainage basin settings of the type illustrated by Fig. 2a may be

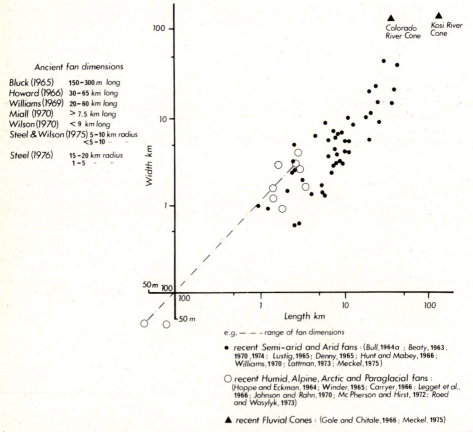

Fig. 3. Dimensions of recent and ancient alluvial fans. (Note log scales; n = 58). Anstey (1965) provides numerous additional dimensions of recent semi-arid and arid fans from the western United States and West Pakistan.

more normal in nature. For flood deposits which have suffered no subsequent reworking, the concepts of geomorphic thresholds and regularity in depositional event magnitude can help explain the maximum particle size/bed thickness relations (flood competence) described from ancient alluvial fan (Bluck, 1967; Steel, 1974) and submarine fan sediments (Walker, 1977).

TECTONIC SETTINGS OF ALLUVIAL FANS

"The prime requisite of fan formation is the setting of highland and lowland side by side", Denny (1967, p. 83). Alluvial fans normally extend from mountain fronts into valleys or bordering lowland areas for only 5-20 km (Fig. 3). Most modern fans are associated with fault scarps (Davis, 1925; Eckis, 1928; Sharp and Nobles, 1953; Blissenbach, 1954; Hunt and Mabey, 1966; Rahn, 1967; Beaty, 1970; Scott, 1971, 1973; Hooke, 1972; Wasson, 1974; Meckel, 1975; Tanner, 1976). Faults provide the initial relief required for fan formation and continued movement can lead to further accumulation and preservation of fan sediments.

There are three contrasting basin margin/alluvial fan settings (Fig. 4) which govern the thickness and extent of alluvial fan successions, and also the character and arrangement of internal vertical sequences (discussed later).

Adjacent to a *relatively permanent basin margin fault* alluvial fan sediments are stacked up along a prominent fault or fault zone (Fig. 4a). This type of basin margin/alluvial fan setting commonly occurs on the downthrown side of major strike-slip faults and may persist through several geological epochs. Crowell (1973, 1974) described the alluvial fan Violin Breccia from this type of setting in Ridge Basin, California. The Violin Breccia is 10,000 m thick, 1500 m along strike, extends only a km or so into Ridge Basin, and is now displaced laterally by strike-slip movement 28 km from its distinctive source area. Crowell (1974) suggested that the strewing out of coarse debris across a strike-slip fault, with a dip-slip component, may occur commonly in such basins. Tanner (1976), similarly documented the Bucaramanga fault, a major strike-slip fault in Colombia. Six conglomeratic alluvial fan formations, Permo-Triassic to recent in age, and 1000's m thick occur on the downthrown side of this fault. Steel (1976), considered that Devonian alluvial fan deposits from the Hornelen Basin, Norway, accumulated adjacent to a strike-slip fault. He noted that the fans commonly extended 1 km from the basin margin and that internal vertical sequences were comparatively thick.

Limited back-faulting of the basin margin (Fig. 4b) results in linear, moderate thickness, alluvial fan accumulations along prominent basin margin faults, particularly those bounding grabens or half grabens (Sharp, 1948; Howard, 1966; Belt, 1968; Miall, 1970; Groat, 1972; Schluger, 1973; Hubert et al., 1976). Continued uplift and dissection of the source area is commonly indicated by an 'inverted stratigraphy' of conglomerate clast types (Sharp, 1948; Miall, 1970; Wilson, 1970). Back-faulting results in limited directional younging of alluvial fan formations. Sharp (1948) and Miall (1970) have respectively described Eocene (48 km long, 5 km wide and 1200 m thick) and Devonian (93 km long, 16 km wide and 300 m thick) alluvial fan accumulations of this type.

Repeated back-faulting of the basin margin (Fig. 4c) can result in geographically extensive coarse grained alluvial fan successions which young in the direction of back-faulting. Although the total stratigraphic thickness of such accumulations may be very considerable, the thickness deposited and preserved on any single downthrown block is notably less. Steel and Wilson (1975) first described Permo-Triassic alluvial fan deposits in terms of this setting and commented that under such conditions of progressive back-faulting, simple 'inverted stratigraphies' of conglomerate clast types were less likely to occur. More recently Steel (1976) and Steel et al. (1977) have interpreted the Devonian 5 km thick conglomerate fill of the Solund Basin, and the majority of the 25 km thick sandstone fill of the Hornelen Basin, Norway, in terms of this model.

a. Relatively Permanent Basin Margin Fault or Fault Zone
(after Crowell, 1973)

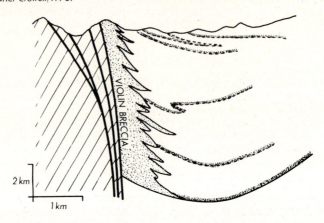

b. Limited Back-faulting of Basin Margin
(after Belt, 1968)

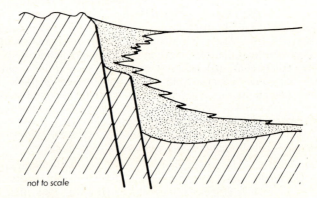

not to scale

c. Repeated Back-faulting of Basin Margin
(after Steel and Wilson, 1975)

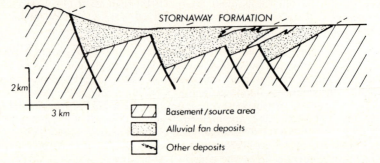

Fig. 4.   Alluvial fan / basin margin settings.

a. *Response to Initial Topography*

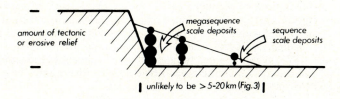

b. *Short-Moderate Duration Fanhead Entrenchment* : resulting from intrinsic or climatic factors (Table 4)

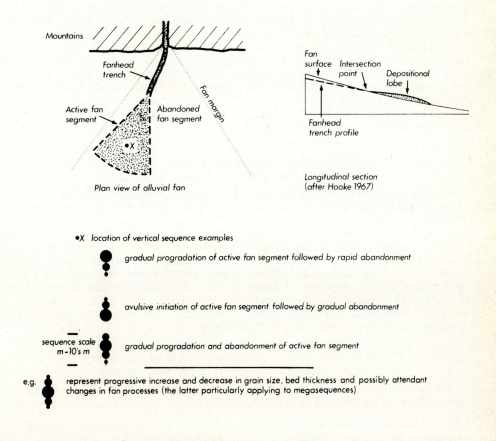

Fig. 5. Hypothetical alluvial fan behavioural models. (a) Response to initial topography. (b) Short-moderate duration fanhead entrenchment.

HYPOTHETICAL ALLUVIAL FAN BEHAVIOURAL MODELS

This section attempts to marry some of the interpretations of ancient fan sequences (Tables 1 and 2) with characteristics of modern alluvial fans and their possible behavioural patterns. In such a comparison one is faced with the problem of determining the relative importance of tectonic, climatic and sedimentologic events. Schumm (1976) predicted that five scales of fining upward sequence would be associated with erosional evolution (cycles in his description); 1st order sequences reflecting the initiating tectonic uplift, 2nd order sequences (occurring within the 1st order sequences etc.) resulting from isostatic rebound or climatic change, 3rd order sequences representing the influence of intrinsic thresholds (Schumm, 1973, 1976), and the 4th and 5th order sequences reflecting complex response, seasonal and flood effects. The models below follow the relative order of magnitude of Schumm's (1976) and Bull's (1968, p. 102-103) events.

Response to initial topography (Fig. 5a)

From a starting point of tectonic or erosional topography, alluvial fans may form where streams leave highland and enter lowland regions. As fans build upward and outward, older distal fan deposits are progressively overlain by younger mid-fan and proximal fan sediments. The products of original down-fan trends (Table 3, Fig. 1) become arranged in megasequences and sequences of increased proximality (coarsening, thickening and increasingly proximal processes upward). In the absence of uplift, as source area relief is reduced, lesser amounts of finer grained debris become available, resulting in a partially symmetrical megasequence (Fig. 5a). Close to the mountains a relatively constant, coarser grained megasequence may eventually fine upward. The ultimate thickness of the fan accumulation is the extent of the original topography, and fan deposits formed under such tectonically non-reactivated settings have been reported 10's - 100's and occasionally 1000's m in thickness (e.g. Allen, 1965; Selley, 1965; Laming, 1966). Megasequences near the mountain front are equivalent to sequences near the fan toe. An 'inverted stratigraphy' of conglomerate clast types may occur as successive rock types are eroded in the source area (e.g. Laming, 1966).

Secondary fans resulting from localised drainage of abandoned areas of primary fan

Secondary fans occurring at the toes of primary accumulations can be of diverse character and origin (Fig. 7). In their simplest form they result from drainage established on abandoned areas of the primary fan and debris is entirely reworked from the primary fan (Fig. 7a.2; Blissenbach, 1954; Ruhe, 1964; Denny, 1967; Meckel, 1975). Such accumulations of weathered and reworked fan material, which may include multi-cycle conglomerate clasts (Tanner, 1976), are better sorted and finer grained than the primary fan (e.g. Ruhe, 1964, primary fan 20 mm median clast diameter, secondary fan 1.5 mm median clast size at comparable distance from source). Secondary fans of this type are unlikely to be of great volumetric significance and may be typified by asymmetrical — partially symmetrical coarsening upward sequences[1], the nature of the sequence termination reflecting the character of abandonment.

Short-moderate duration fanhead entrenchment (Fig. 5b)
— resulting from intrinsic or climatic factors

For varying periods of time deposition on alluvial fans may be localised through the entrenchment of fan channels, particularly the fan feeder channel. Fanhead entrenchment

---

[1]Throughout this section, the terms coarsening or fining upward also imply parallel trends in bed thickness and depositional processes accompanying the increased proximality or distality (Table 3, Fig. 1).

Table 4 : Fanhead Entrenchment

Factors Causing Entrenchment	Estimated Duration of Influence *1	Depositional Results *1
INTRINSIC FACTORS (those inherent within the system)		
i. Variable nature of storm events (Denny, 1967), extensive scour and fill can occur during a storm (Sharp and Nobles, 1953; Beaty, 1970,1974; Scott, 1971,1973), or at the declining floodwater stage (Blackwelder, 1928; Beaty, 1963, 1974).	Short-lived	Accumulation of small volume of similar grade sediment below the intersection point (intersection point is where fanhead trench merges with the alluvial fan surface, Hooke, 1967).
ii. Natural result of an alternation of debris flow and stream flow processes (Bluck, 1964; Hooke, 1967).	Short-lived	Accumulation of small volume of similar grade sediment below the intersection point.
iii. When the locus of deposition shifts to a topographically low area that has not received sediment for some time (Hooke, 1967).	Short-lived	Accumulation of small volume of similar grade sediment below the intersection point.
iv. Build up of slope of fan apex until it exceeds the stability threshold. Trenching then flushes sediment down-fan and reduces slope near apex (Schumm, 1973; Weaver and Schumm, 1974).	Short-lived	Accumulation of small volume of similar grade sediment below the intersection point.
v. Capture of fan feeder channel by an adjacent channel or a minor channel heading in abandoned area of fan (Denny, 1965,1967; Goreau and Burke, 1966; Hooke, 1967,1968; Troxel , 1974).	Short - prolonged	Accumulation of small-considerable volume of similar or coarser grade sediment, perhaps transported by a differing process. Channel capture may be indicated by abrupt differences in clast type (Hunt and Mabey, 1966).
EXTRINSIC FACTORS (those external to the system)		
CLIMATE CHANGES vi An increase in the volume of storm floodwater (Eckis, 1928), due to a temporary increase in precipitation (Bull, 1964b,c).	Short - moderate duration	Accumulation of small-moderate volume of similar or coarser grade sediment, perhaps transported by a differing process.
vii A decrease in the volume of storm floodwater due to a climatic change towards increased aridity (Antevs, 1952; Lustig, 1965; Williams, 1970).	Short - moderate duration	Accumulation of small-moderate volume of similar or differing grade sediment, perhaps transported by a differing process. Deposition may be confined to channels (Lustig 1965).
MAN viii As a consequence of overgrazing and human activities (eg. Antevs, 1952; Schumm and Hadley, 1957).	Prolonged	Fortunately, irrelevant to geological past.
DECREASING DEBRIS SUPPLY ix Resulting from a decrease in debris supply (paraglacial fans in particular, Carryer, 1966; Ryder, 1971), perhaps due to a lowering of relief or to the exposure of more resistant rock types in the drainage basin (Eckis, 1928).	Prolonged	The decreasing amounts of debris supplied to fan and reworking of weathered former fan deposits will probably result in the accumulation of small-moderate volume of increasingly fine grained debris.
x As the product of downcutting during the cycle of erosion. As highland areas are reduced in altitude the slopes of stream channels are also reduced causing entrenchment of the fanhead (Eckis, 1928; Carryer,1966).	Prolonged	The decreasing amounts of debris supplied to fan and reworking of weathered former fan deposits will probably result in the accumulation of small-moderate volume of increasingly fine grained debris.
LOWERING OF BASE LEVEL xi Resulting from lowering of base level of the fan or of the depositional basin, by the truncation and dissection of a fan by a major river and its tributaries (Drew, 1873; Eckis, 1928; Blissenbach 1954; Ryder, 1971), or by a change in basin character from internal to external drainage (Denny, 1967; Groat, 1972).	Prolonged	Moderate-large accumulation of debris derived from source area and from reworked fan. Extent of eventual preservation questionable.
TECTONIC CHANGES xii Following continued tectonic relative uplift of source area (Bull, 1964b; Hunt and Mabey, 1966; Denny, 1967; Williams, 1970), or tilting (Hooke, 1972).	Prolonged	Large accumulation of debris (probably relatively coarse grained) transferred from actively dissected drainage basin onto the fan. Gradual changes in clast type may reflect progressive dissection of source area (eg. Miall, 1970).

*1 predictions of author generally based on previous writer's comments.

results in an active fan segment, whilst adjacent temporarily abandoned fan surfaces are subject to weathering, pedogenesis and erosion (Fig. 5b).

Various factors which can lead to short-moderate duration fanhead entrenchment are listed in Table 4. These are mainly intrinsic factors, the variable nature of storm events, the alternation of differing transportational processes, and the development of depositional topography, the products of which are probably indistinguishable. However, entrenchment resulting from the capture of the fan feeder channel may be indicated by changes in clast type and clast size within the accumulated sediments (Hunt and Mabey, 1966). Such entrenchment may be prolonged and result in complete abandonment of the primary fan (Goreau and Burke, 1966). Entrenchment following climatic changes may be apparent through differing transporting processes being operative. Whilst adjacent fans will probably have differing geomorphic thresholds and hence response periods following climatic changes, such changes should occur regionally (Hooke, 1967; Schumm, 1973, 1976).

Short-moderate duration fanhead entrenchment probably results in the accumulation of small-moderate volumes of sediment below the intersection point (Hooke, 1967; Fig. 5b). Such accumulations have proximal-distal characteristics as in Table 3, and internal vertical sequences will reflect the gradual or abrupt initiation and termination of sedimentation (Fig. 5b). Such m-10's m thick alluvial fan accumulations appear comparable to the suprafan lobes of some modern submarine fans (e.g. Normark, 1970), whose progradation, migration, and gradual or abrupt abandonment are considered to result in submarine fan vertical sequences of similar scale and style (Table 2; Mutti and Ricci-Lucchi, 1972; Walker and Mutti, 1973; Mutti, 1974, 1977; Ricci-Lucchi, 1975; Rupke, 1977, van Vliet, in press).

Secondary fans resulting from stream capture

The feeder streams of secondary fans which head in and rework abandoned areas of the fan may eventually capture the feeder channel of the primary fan and receive sediment directly from the source area. Secondary fans of this type may be identifiable due to the abrupt change from finer grained reworked fan sediments to coarser debris derived directly from the source area, perhaps by differing processes (Fig. 7a.3).

Prolonged fanhead entrenchment (Fig. 6a)
— resulting from decreasing sediment supply or lowering of base level (Table 4).

Prolonged fanhead entrenchment results in the active fan segment occurring at the fan toe in the form of a secondary fan. Secondary fans due to entrenchment following a decrease in sediment supply are predominantly of debris reworked from primary accumulations (Carryer, 1966). Those resulting from a lowering of base level may initially be of reworked debris, but with their gradient advantage they are likely, eventually, to capture the supply from the source area (Fig. 7b). Asymmetric coarsening upward sequences, or symmetrical coarsening and fining upward sequences may typify such accumulations due to gradual progradation and abandonment. However, on those fans characterised by a decline in amount and grade of source area debris, fining upward sequences may predominate. Fan accumulations of these types are unlikely to be of great volumetric importance and in the case of those initiated by lowering of base level, the eventual preservation of such deposits is questionable (Fig. 7b; Groat, 1972).

Scarp retreat and lowering of relief (Fig. 6b)

Scarp retreat and the lowering of relief results in the reworking of the proximal parts of successive fan accumulations as a broad alluvial fan apron develops. At the site of the original progradational fan accumulation a remnant coarsening upward sequence may occur. However, fining upward sequences predominate as localities become more distant

a. _Prolonged Fanhead Entrenchment_ : resulting from decreasing sediment supply, the latter perhaps due to the advanced state in the cycle of erosion; or due to lowering of base level (Table 4)

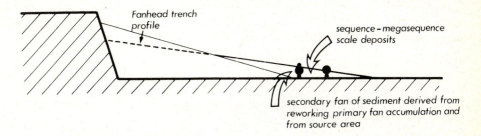

b. _Scarp Retreat and Lowering of Relief_ :

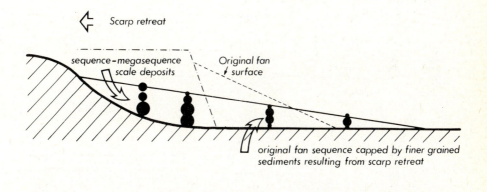

e.g. represent progressive increase and decrease in grain size, bed thickness and possibly attendant changes in fan processes (the latter particularly applying to megasequences)

Fig. 6. Hypothetical alluvial fan behavioural models (continued). (a) Prolonged fanhead entrenchment. (b) Scarp retreat and lowering of relief.

from a source area of reduced relief (Fig. 6b; Bluck, 1967; Williams, 1969; Deegan, 1973; Steel, 1974; Steel and Wilson, 1975). Deposits 10's m (e.g. Steel, 1974, Fig. 14) to 100's m thick may result (Bluck, 1967, Fig. 13). Changes in conglomerate clast types may occur as differing lithologies are eroded.

Response to tectonic uplift (Fig. 8)

Recent alluvial fans respond to tectonic uplift in two ways, (1) where relative uplift exceeds streams dissection, active fan segments occur immediately adjacent to the fan apex (Fig. 8a), (2) where the rate of stream dissection exceeds relative uplift, prolonged entrenchment occurs (Table 4) and the active fan segment occurs at the fan toe (Fig 8b). In each case progradation of the active fan segment results in an asymmetrical coarsening upward sequence or megasequence which may be terminated by a further phase of uplift. Progressive offlap of fan segments as in Fig. 8a, or an eventual decline in the grade of supplied debris prior to the next phase of uplift may result in sequences and

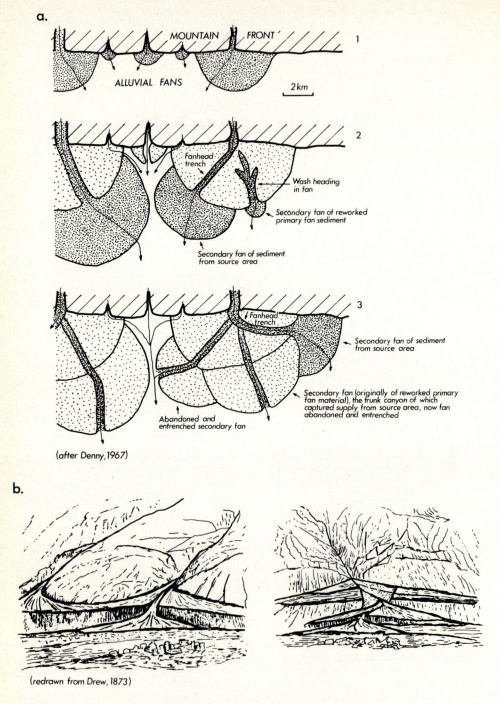

Fig. 7. Secondary fans. (a) Development of alluvial fans, after Denny (1967), illustrating secondary fans of diverse origins. (b) Secondary and triple fans, Upper-Indus Basin, resulting from the lowering of base level by the erosion of fan toes by degrading trunk river.

a. <u>*Response to Tectonic Uplift*</u> : *relative uplift exceeds rate of stream dissection*

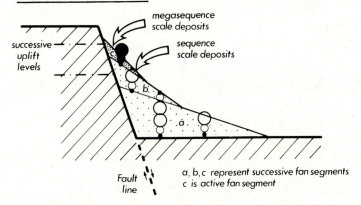

b. <u>*Response to Tectonic Uplift*</u> : *prolonged entrenchment (Table 4), as stream dissection exceeds rate of relative uplift*

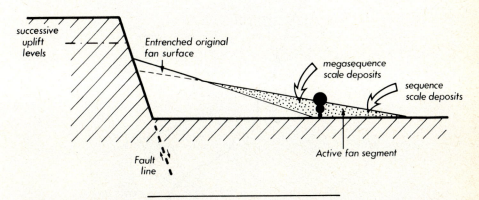

Fig. 8. Hypothetical alluvial fan behavioural models (continued). (a) Response to uplift, where relative uplift exceeds rate of stream dissection, (b) Response to uplift, where stream dissection exceeds rate of relative uplift.

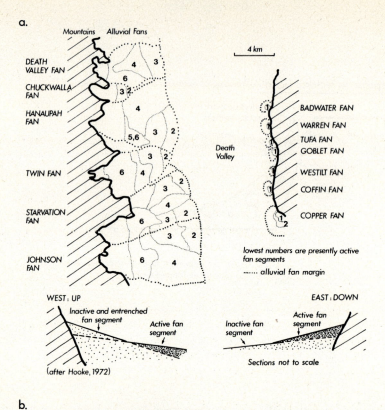

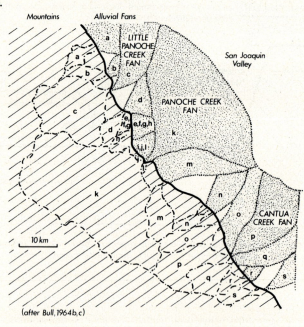

Fig. 9.   Examples of the responses of alluvial fans to relative uplift (see text for details).

Table 5 : *Summary of Alluvial Fan Behavioural Models*

Response to Initial Topography (Fig.5a):

  Asymmetrical - partially symmetrical coarsening[*1] upward megasequence near fan
  apex - sequence near fan toe.

  Immediately adjacent to mountain front relatively constant coarser grained megasequence may eventually fine upward.

  Secondary fans resulting from streams reworking abandoned areas of primary fan.

Short-Moderate Duration Fanhead Entrenchment (Fig.5b):

  resulting from - variable nature of storm events, alternation of differing trans-
                portational processes, development of fan topography, capture of
                fan feeder channel, or climatic changes.

  Asymmetrical or symmetrical coarsening or fining upward sequence reflecting gradual
  or abrupt initiation or termination of sediment supply.

  Secondary fans resulting from the capture of fan feeder channel and deriving
  sediment directly from the source area.

Prolonged Fanhead Entrenchment (Fig.6a):

  resulting from - decrease in sediment supply, or lowering of base level.

  Asymmetrical coarsening upward sequence, or symmetrical coarsening and fining
  upward sequence. Fining upward sequences may predominate or fans characterised
  by a decline in amount and grade of source area debris.

Scarp Retreat and Lowering of Relief (Fig.6b):

  Asymmetrical fining upward sequence - megasequence.

Response to Tectonic Uplift (Fig.8):

  resulting in - deposition immediately adjacent to fan apex, or prolonged
                entrenchment and deposition concentrated at fan toe.

  Asymmetrical - partially symmetrical coarsening upward megasequence (proximal
  locations) - sequence (distal locations).

*1 In addition to grain size changes, parallel trends in bed thickness and depositional
   processes are also implied, accompanying the increased proximality or distality
   (Table 3, Fig.1).

---

megasequences of a more symmetrical character. Large accumulations of debris, perhaps including 'inverted stratigraphies' of conglomerate clast types, probably result from active drainage basin dissection following repeated uplift.

These responses of recent alluvial fans to uplift are derived from the excellent descriptions of Bull (1964b) and Hooke (1972). Bull (1964b) described fans in Fresno County, California, fed predominantly by ephemeral streams (Fig. 9b). Little Panoche Creek fan, Panoche Creek fan and Cantua Creek fan have drainage basins extending further into the mountains and are fed by intermittent streams. With the exception of Little Panoche Creek fan and Wildcat Canyon fan (b and Fig. 9b), all the other fans have responded to relative uplift of the mountains by developing active fan segments, with progressivley steeper slopes, adjacent to the fan apex. On Little Panoche Creek fan and Wildcat Canyon fan stream dissection has exceeded the rate of uplift such that the fans are strongly entrenched and the active fan segment occurs near the fan toe.

Similarly, Hooke (1972) documented fans on the east and west sides of Death Valley, California (Fig. 9a). Progressive tilting of Death Valley (west side up, east side down) causes fans on the west side to be large, with active fan segments developed at the fan toes. In contrast fans on the east side are smaller and have active segments adjacent to the apex.

Table 5 summarises the fan behavioural models described in this section.

VERTICAL ARRANGEMENT OF ALLUVIAL FAN SEQUENCES AND MEGASEQUENCES

If alluvial fans are dependent solely on initial erosive or tectonic topography only relatively thin successions of decreasing grain size accumulate (e.g. Allen, 1965; Selley, 1965; Laming, 1966; Bluck, 1967; Williams, 1969). Greater thicknesses of fan sediments and more complex vertical arrangements of fan sequences result from continued uplift along fault lines (e.g. Belt, 1968; Nilsen, 1969; Steel and Wilson, 1975; Steel, 1976; Steel *et al.*, 1977). "The presence of a thick sequence of fanglomerate in the geologic column indicates that deformation and sedimentation were concurrent processes", Denny (1965, p. 58). The minor faulting and slight angular unconformities recorded commonly from alluvial fan successions provide evidence of such deformation (Eckis, 1928; Sharp, 1948; Hunt and Mabey, 1966; Bluck, 1967; Miall, 1970; McGowen and Groat, 1971; Bryhni and Skjerlie, 1975; Steel and Wilson, 1975, Riba, 1976; Tanner, 1976; Steel *et al.*, 1977).

Except for the suggestion by Steel (1974, 1976) and Steel *et al.*, (1977), that stacked arrangements of sequences result from periodic drainage basin rejuvenation caused by tectonic movements, there has been little discussion of the vertical arrangement of fan sequences. Several distinct types of arrangement can be envisaged reflecting the basin margin/alluvial fan setting and hence type of depositional basin (Fig. 4). In Fig. 4a, whilst the effects of repeated fault movement may be evident in sediments occurring at some distance out into the depositional basin, fan vertical sequences or megasequences will persistently be stacked adjacent to the fault line. Strike-slip faulting may result in some lateral displacement of successive fan accumulations. In Fig. 4b, limited back-faulting of the basin margin may result in a series of sequences stacked up against a single fault line until the succeeding fault takes over as the basin margin. At this time the sediments occurring in a proximal position adjacent to the first fault may lie at a very distal locality relative to the second. Abrupt changes from proximal to distal fan deposits, or from fan to laterally equivalent environments (e.g. fluvial, deltaic, lacustrine etc.) may indicate this type of setting. Figure 4c, repeated back-faulting of basin margin, illustrates a more extreme version of the previous case in which abrupt changes may be more common. The geographical extent of alluvial fan sediments (cf. Fig. 3), complexity of successive fan palaeocurrents (e.g. Steel and Wilson, 1975) and progressive basin marginward displacement of fan accumulations (Steel *et al.*, 1977) may be further indicators of this latter setting.

EXAMPLES OF ALLUVIAL FAN SEQUENCES, MEGASEQUENCES AND BASIN-FILL SEQUENCES FROM WESTPHALIAN D — STEPHANIAN B COALFIELDS, NORTHERN SPAIN

The Upper Carboniferous succession in the Cantabrian Mountains, northern Spain, excedes 15,000 m in thickness and becomes increasingly continental upward. Alluvial fan and lacustrine sediments predominate during the Stephanian. Sequences, megasequences and basin-fill sequences have been recorded from the La Magdalena, Matallana (Ciñera-Matallana) and Sabero coalfields (Heward, in press) and the Tejerina syncline (work in progress; Fig. 10).

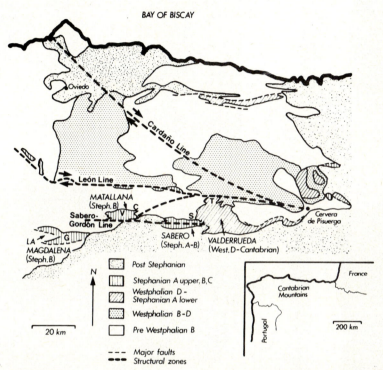

Fig. 10.   Geological map of the Cantabrian Mountains, northern Spain. Stratigraphic ages (based on floras and faunas) for the coalfields from Wagner (1970) and Knight (1975). C = Correcillas, G = Garaño, S = Saelices, T = Tejerina, V = Vegacervera, refer to locations mentioned in text or figure captions.

Sequences

Three types of sequence-scale accumulations occur, consisting of single beds or series of related beds. They are m - 10's m thick, have progressive grain size trends and other characteristics, and are separated by non-depositional or slow depositional horizons, such as coals, prominent rootlet beds, or lacustrine shales.

Single beds attaining sequence scale and having internal grain size and morphological trends are generally conglomeratic. In the examples illustrated (Fig. 11), the well rounded conglomerates are poorly sorted and unstratified. They are composed of relatively homogeneous clast supported conglomerate but contain occasional cross-bedded sandstone lenses. These conglomerates are inversely graded or inversely coarse tail graded (Fig. 11) and have an imbricate fabric in which long axes parallel the flow direction determined from adjacent cross-bedding. The sharp non-erosive bases, lateral extent, absence of stratification, lack of sorting, presence of inverse grading, and nature of the fabric, are all in accord with the transportation and deposition of these conglomerates by a debris flow mechanism (Johnson, 1970; Fisher, 1971; Middleton and Hampton, 1973; Walker, 1975c; Enos, 1977). *These single event vertical sequences provide evidence of processes operative on these Stephanian alluvial fans.*

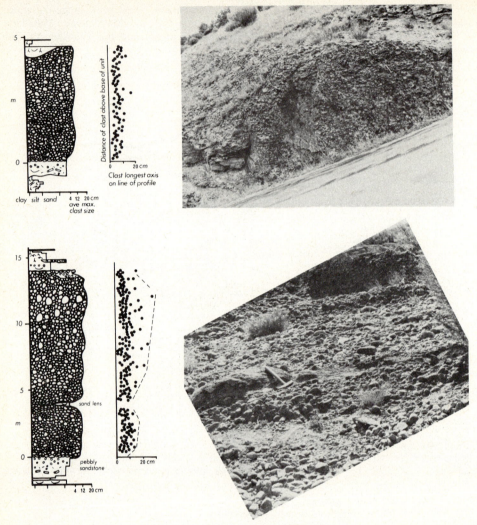

Fig. 11. Examples of poorly sorted, inversely graded, clast supported unstratified conglomerate sequences from the La Magdalena coalfield. The average maximum clast size in these conglomerate sequences was measured from the long axes of 25 largest clasts, 1-5 m either side of the section line, at intervals normally not greater than 1 m. Both conglomerate sequences young to top right; hammer handle 30 cm long. La Magdalena — Mora de Luna road, north of Garaño. (Legend on Fig. 12).

15 - 65 m thick sequences of conglomerate beds comprise the second type of vertical sequence recognised (Fig. 13a). Laterally extensive conglomerate beds are crudely organised into sequences in which maximum clast size, overall clast size and bed thickness decline upward, and matrix support becomes more prominent upwards (fining and thinning upward sequences). Conglomerate beds are poorly sorted, lack stratification and again have characteristics of debris flow deposits. The sequences may represent gradually abandoned, relatively proximal, depositional lobes (following short-moderate duration fanhead entrenchment, Table 5). Their character, scale and stacked nature appear to argue against origins resulting from decreasing sediment supply, lowering of

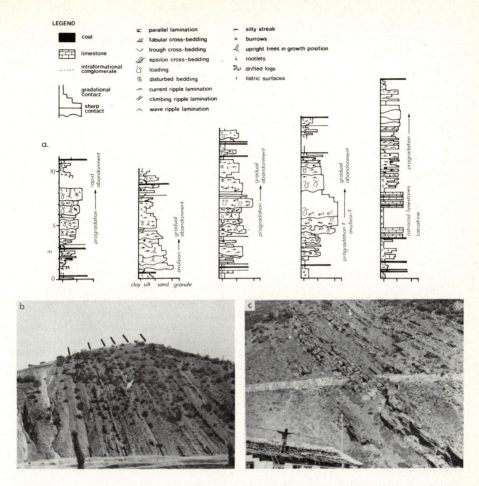

Fig. 12. Examples of distal fan sheetflood sandstone coarsening and thickening, and fining and thinning upward sequences. (a) With the exception of the last sequence from near La Magdalena, La Magdalena — Mora de Luna road. Last sequence from Vegacervera, Matallana coalfield. (b) Asymmetrical coarsening and thickening upward sheetflood sandstone sequences, except sequence containing channel (arrowed, 1.5 m deep) which is partially symmetrical. Succession youngs to top right. North of Saelices, Sabero coalfield. (c) Above prominent lacustrine coarsening upward sequence (10 m thick), series of asymmetrical lacustrine / distal fan coarsening and thickening upward sequences. Succession youngs to top right; aqueduct traverses centre of photograph. Vegacervera, Matallana coalfield.

base level, or from scarp retreat. These conglomeratic alluvial fan fining and thinning upward sequences appear closely comparable to submarine fan sequences of similar style, generally considered to result from gradual channel abandonment (Table 2). The conglomerate beds of this example are traceable for 100's m - km and show no evidence of internal channelling, even though they occur within a palaeovalley eroded in pre-Stephanian sediments (2 km wide and 300 m deep). The crude organisation of the described alluvial fan sequences (Fig. 13a) and their irregular thicknesses may result from a drainage basin setting of the type illustrated by Fig. 2b (rather than Fig. 2a). Alternatively, it could be argued that the sequences are too poorly defined to really warrant distinction. Closer analysis is clearly required.

The third type of sequence observed are 1.5 - 15 m thick and consist of a series of laterally extensive sheet sandstones separated by siltstones, mudstones, rootlet beds and coals (Fig. 12). The 10 cm - 1.5 m thick sheet sandstones are sharply based, occasionally solemarked (grooves) and individual beds fine upward. Palaeocurrents, from current ripple cross-lamination within and grooves on the bases of these sandstones, parallel the general direction of sediment dispersal. Well preserved plants occur within the sandstones and upright (in growth position) and tilted fossil tree trunks abound (Fig. 12a). These features, in combination with abundant syn-sedimentary loading and water escape structures, indicate rapid accumulation. The sandstone beds are arranged in sequences separated by coals, lacustrine shales, or prominent rootlet horizons. Sequences of beds may be asymmetrical, and coarsen and thicken, or fine and thin upward; or may be symmetrical, and consist of the superimposition of the two (Fig. 12). Some of the coarsening and thickening upward sequences may directly overlie lacustrine shales (Fig. 12a, c).

The rapidly accumulated sheet sandstones are interpreted as sheetflood deposits (Heward, in press). Their fine grain size and interbedding with lacustrine deposits, suggest a distal fan position. Such distal fan sediments could be derived directly from the source area or might be reworked from proximal fan areas (Denny, 1965; Rahn, 1967). However, the organisation of these sheetflood sandstones into sequences suggest that they are distal representatives of depositional lobes undergoing progradation or abandonment, rather than a succession of variable storm events reworking older proximal fan deposits. The sequences probably result from short-moderate duration fanhead entrenchment (Table 5) and closer observation may allow distinction of possible causes of such entrenchment (e.g. of sediment type, depositional process, character and thickness of sequence compared with surrounding sequences etc.). The fine grain size of these sequences (compared with associated deposits), similar sequence thickness and repetitive stacked nature (Fig. 12b, c) would appear to preclude origins from more prolonged fan head entrenchment resulting from decreasing sediment supply, lowering of a base level, or a response to tectonic uplift (Table 5). As has been noted with submarine fan sequences of similar type, minor channels (as in Fig. 12b) occur near the top of some coarsening and thickening upward sequences (Table II; Mutti and Ricci-Lucchi, 1972; Mutti, 1974, 1977; Ricci-Lucchi, 1975; van Vliet, in press).

The above multiple event vertical sequences appear to provide evidence of short term fan behaviour; more careful study of their character and sediment type may allow elucidation of its causes. In other fan deposits, multiple event sequences may also reflect fractionation of sediment within a fan environment and hence, indicate depositional processes (e.g. lateral migration of a meandering fan channel, gradual abandonment of a fan channel, levee progradation etc.).

Megasequences

Four types of alluvial fan megasequences can be recognised within the described coalfield deposits. In each case they are asymmetrical and are characterised by progressive changes in grain size, bed thickness, interpreted depositional processes and/or depositional environments (Figs. 13, 14). *Megasequences appear to provide some insight into longer term alluvial fan behaviour.*

Figure 13a illustrates the first type, conglomeratic megasequences, 110 - 120 m thick in which maximum clast size, overall clast size and bed thickness decrease upwards, and matrix supported conglomerates become increasingly abundant, compared with the predominant clast supported conglomerates. These megasequences probably were deposited at progressively more distal locations, relative to the source area, caused by scarp retreat and/or a lowering of relief (Table 5). Their thickness and stacked nature argue against gradual fan segment abandonment following short-moderate duration fanhead entrenchment, or from prolonged entrenchment due to decreasing sediment supply or lowering of base level.

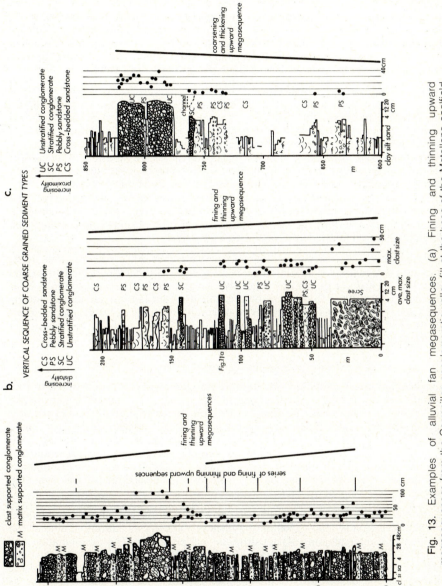

Fig. 13. Examples of alluvial fan megasequences. (a) Fining and thinning upward megasequences from the Corecillas conglomeratic valley-fill at the base of the Matallana coalfield succession. (b and c) Fining and thinning, and coarsening and thickening upward megasequences from the La Magdalena coalfield succession (Fig. 14c). The average maximum clast size in these conglomeratic megasequences was measured from the long axes of the 25 largest clasts, 1–5 m either side of the section line, at intervals normally not greater than 1 m. Maximum clast size represents the largest long axis encountered whilst measuring the above.

A fining upward megasequence of different character occurs near the base of the La Magdalena coalfield succession north of Garaño (Figs. 10, 13b). In this megasequence, fining and decreasing bed thickness are associated with a trend from scree deposits — predominantly unstratified conglomerates — stratified conglomerates — pebbly and cross-bedded sandstones. These deposits are interpreted as resulting from progressively down-flow processes (Heward, in press), and hence indicate increasing distality. Mudstones and siltstones with abundant well preserved plant remains and occasionally thin coal seams occur interbedded throughout. This fining and thinning upward megasequence is interpreted in a similar fashion to the previous description.

Immediately overlying this fining upward megasequence are two coarsening upward megasequences, 210 and 220 m thick (Fig. 14c). Figure 13c illustrates the second of these, where an increase in maximum clast size, general conglomerate clast size and bed thickness is paralleled by a transition from pebbly and cross-bedded sandstones, through stratified conglomerates, and into unstratified conglomerates (increasing proximality). These coarsening and thickening upward megasequences are considered to reflect progradation in response to tectonic uplift, deposition being concentrated by prolonged entrenchment at the fan toe (Table 5). Their scale, consistency of the very maximum clast size (40-50 cm), absence of multi-cycle clasts, stacked nature of the megasequences and their position within the coalfield succession would appear to suggest that a secondary fan origin resulting from reworking the primary fan accumulation, or an origin resulting from short-moderate duration fanhead entrenchment, was unlikely.

The fourth type of megasequence, coarsens upward over 300 - 650 m, from thick, laterally extensive lacustrine shales into coal bearing distal fan sheetflood sandstones and occasional fine grained conglomerates (Fig. 14c; Knight, 1975, and pers. comm.). Four megasequences of this type, from the Sabero coalfield record fan progradation following rapid basin deepening and probably result from relative tectonic uplift (Table 5). Their scale and stacked nature suggest they are not the distal fan products of short-moderate duration, or prolonged fanhead entrenchment.

Basin-fill sequences

The depositional model proposed for these Stephanian coalfields is illustrated in Fig. 14a (Heward, in press). Slight modifications to this model are required to explain the Westphalian D — Cantabrian, Tejerina succession, where quartzitic conglomerates interfinger with limestone rich conglomerates, stratified conglomerates are more abundant, and distal fan sediments interfinger with marine shales. In the following discussion of basin-fill sequences, composite vertical successions are presented for Matallana and Sabero, whilst those of Tejerina and La Magdalena represent continuous vertical sequences (Fig. 14). In none of the coalfields has detailed analysis of single fan morphologies been attempted, to date.

Three general aspects of the basin-fill sequences of Fig. 14 appear worthy of note. Firstly, each succession is 1500 - 2500 m in thickness and consists entirely of alluvial fan deposits, with thin intercalations of lacustrine or marine shales. The accumulation and preservation of such thicknesses of fan sediment indicate repeated fault movements. Overall, these basin-fill sequences fine upward, a feature common to thick alluvial fan accumulations as the depositional basin becomes distant from the source area, or as source area relief becomes subdued (Howard, 1966; Nilsen, 1969; McGowen and Groat, 1971). Secondly, each succession has a marked horizon where conglomerates and coarse grained proximal fan deposits are abruptly overlain by very fine grained distal fan, lacustrine or marine shales. This abrupt change occurs at varying levels within each succession and may reflect back-faulting (Fig. 4b, 4c), thus, deposits initially close to a fault line suddenly lie distant to the succeeding fault line. Thirdly, each succession comprises a series of stacked sequences and megasequences (Fig. 14). The stacking of alluvial fan sequences and megasequences, whatever their internal characteristics,

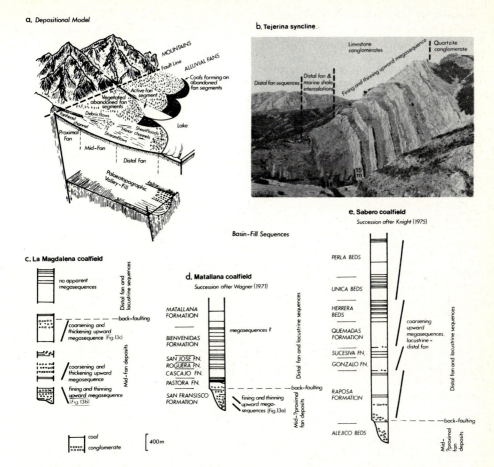

Fig. 14. Depositional model and basin-fill sequences (see text for details). Photograph (b) illustrates part of the Tejerina syncline alluvial fan succession and youngs to bottom left.

probably indicates repeated basin margin faulting and source area rejuvenation (Steel, 1974, 1976; Steel *et. al.*, 1977).

CONCLUSIONS

The functions of vertical sequence or facies models were summarised by Walker (1975d). If properly applied they can be powerful tools in the development of sedimentological understanding. However, all too commonly we are content to re-document an existing model for strata of a different age, in a different geographical area, and the comparative power of the model and even the processes and controls upon which the model depends are forgotten. The intention of this analysis has been the collation of some of the processes and controls which might affect alluvial fan vertical sequences, and to illustrate the necessity for detailed analysis of individual fan sequences rather than the establishment of a 'general or local sequence model' through the distillation of many.

Sequences, megasequences and basin-fill sequences of progressively changing character appear common to many alluvial fan accumulations (Table 1). Their study can provide evidence of depositional processess (sequences), short term fan behaviour (sequences), longer term fan behaviour (megasequences) and the depositional basin setting (basin-fill sequences). The occurrence and identification of these sequences appear dependent on downfan trends in sediment character and depositional process, the localisation, switching and migration of the region of active fan sedimentation, a regularity of depositional event magnitude, and an absence of reworking by subsequent floods. At the present day, alluvial fans occur most commonly under semi-arid climatic conditions (Blissenbach, 1954; Bull, 1972). In the geological past one can only speculate on the climatic setting most favourable to alluvial fan formation (Schumm, 1968). However, under conditions of increased rainfall and more permanent streams, flood debris may be subject to more intense stream reworking and primary sequences of fan flood events be modified or obscured.

ACKNOWLEDGMENTS

This study represents an extension of ideas presented in the author's D. Phil. thesis. Harold Reading supervised this thesis and Shell International Petroleum Company Ltd. provided a post-graduate grant. The paper was written whilst the author was employed at Koninklijke/Shell Exploratie en Produktie Laboratorium, Rijswijk, The Netherlands. Howard Johnson, John Knight, Mike Leeder, Bruce Levell, Harold Reading and Ron Steel kindly criticised the manuscript. Some of the ideas expressed were alien to certain of the above and the author is solely responsible for misconceptions.

REFERENCES

Allen, J. R. L., 1965, The sedimentation and palaeogeography of the Old Red Sandstone of Anglesey, North Wales: Yorks. Geol. Soc. Proc., v. 35, p. 139-185.

Anstey, R. L., 1965, Physical characteristics of alluvial fans: U.S. Army Natic Laboratories, Tech. Rept. ES-20, 109p.

Antevs, E., 1952, Arroyo — cutting and filling: J. Geol., v. 60, p. 375-385.

Beaty, C. B., 1963, Origin of alluvial fans, White Mountains, California and Nevada: Ann. Assoc. Am. Geogr., v. 53, p. 516-535.

———, 1970, Age and estimated rate of accumulation of an alluvial fan, White Mountains, California: Am. J. Sci., v. 268, p. 50-77.

———, 1974, Debris flows, alluvial fans and a revitalised catastrophism: Z. Geomorph. Suppl. Bd. 21, p. 39-51.

Belt, E. S., 1968, Carboniferous continental sedimentation, Atlantic Provinces, Canada: in G. deV. Klein, ed., Late Paleozoic and Mesozoic Continental Sedimentation, Northeastern North America; Geol. Soc. Am. Spec. Paper 106, p. 127-176.

Blackwelder, E., 1928, Mudflow as a geological agent in semi-arid mountains: Geol. Soc. Am. Bull., v. 39, 465-484.

Blissenbach, E., 1954, Geology of alluvial fans in semi-arid regions: Geol. Soc. Am. Bull., v. 65, p. 175-190.

Bluck, B. J., 1964, Sedimentation of an alluvial fan in southern Nevada: J. Sediment Petrol. v. 34, p. 395-400.

———, 1965, The sedimentary history of some Triassic conglomerates in the Vale of Glamorgan, South Wales: Sedimentology, v. 4, p. 225-245.

———, 1967, Deposition of some Upper Old Red Sandstone conglomerates in the Clyde area: A study of the significance of bedding: Scott. J. Geol., v. 3, p. 139-167.

Bryhni, I., and Skjerlie, F. J., 1975, Syndepositional tectonism in the Kvamshesten district (Old Red Sandstone), western Norway: Geol. Mag., v. 112, p. 593-600.

Bull, W. B., 1963, Alluvial fan deposits in western Fresno County, California: J. Geol., v. 71, p. 243-251.

———, 1964a, Alluvial fans and near surface subsidence in western Fresno County, California: U.S. Geol. Survey Prof. Paper 437-A, 70p.

————, 1964b, Geomorphology of segmented alluvial fans in western Fresno County, California: U.S. Geol. Survey Prof. Paper 352-E, p. 89-129.

————, 1964c, History and causes of channel trenching in western Fresno County, California: Am. J. Sci., v. 262, p. 249-258.

————, 1968, Alluvial Fans: J. Geol. Ed., v. 16, p. 101-106.

————, 1972, Recognition of alluvial fan deposits in the stratigraphic record: *in* J. K. Rigby and W. K. Hamblin, *eds.* Recognition of Ancient Sedimentary Environments; Soc. Econ. Palaeont. Mineral. Spec. Pub. 16, p. 63-83.

————, 1977, The alluvial fan environment: Progress in Phys. Geogr., v. 1, p. 222-270.

Buwalda, J. P., 1951, Transport of coarse material on alluvial fans: Geol. Soc. Am. Bull., v. 62, p. 1497.

Carryer, S. J., 1966, A note on the formation of alluvial fans: New Zealand Jour. Geol. Geophys., v. 9, p. 91-94.

Chawner, W. D., 1935, Alluvial fan flooding, the Montrose, California flood of 1934: Geogr. Rev., v. 25, p. 77-88.

Crowell, J. C., 1973, Ridge Basin southern California, Sedimentary Facies Change in Tertiary Rocks California Transverse and Southern Coast Ranges: Soc. Econ. Palaeont. Mineral. field trip guide, p. 1-7.

————, 1974, Sedimentation along the San Andreas Fault, California: *in* R. H. Dott, Jr., and R. H. Shaver, *eds.,* Modern and Ancient Geosynclinal Sedimentation; Soc. Econ. Palaeont. Mineral. Spec. Pub. 19, p. 292-303.

Curry, R. R., 1966, Observations of alpine mudflows in the Tenmile Range, central Colorado: Geol. Soc. Am. Bull., v. 77, p. 771-776.

Davis, W. M., 1925, The basin range problem: Nat. Acad. Sci. Proc., v. 11, p. 387-392.

Deegan, C. E., 1973, Tectonic control of sedimentation at the margin of a Carboniferous depositional basin in Kirkudbrightshire: Scott. J. Geol., v. 9, p. 1-28.

Denny, C. S., 1965, Alluvial fans in the Death Valley region, California and Nevada: U. S. Geol. Survey Prof. Paper 466, 62p.

————, 1967, Fans and pediments: Am. J. Sci., v. 265, p. 81-105.

Drew, 1873, Alluvial and lacustrine deposits and glacial records of the Upper Indus basin: Quart. J. Geol. Soc. London, v. 29, p. 441-471.

Eckis, R., 1928, Alluvial fans in the Cucamonga district, southern California: J. Geol., v. 36, p. 111-141.

Enos, P., 1977, Flow regimes in debris flow: Sedimentology, v. 24, p. 133-142.

Fisher, R. V., 1971, Features of coarse-grained, high-concentration fluids and their deposits: J. Sediment. Petrol., v. 41, p. 916-927.

Ghibaudo, G., and Mutti, E., 1973, Facies ed interpretazione paleoambientale delle Arenarie di Ranzano nei dintorni di Specchio (Val Pessola, Apennino Parmense): Mem. Soc. Geol. Italia, v. 12, p. 251-265.

Gole, C. V., and Chitale, S. V., 1966, Inland delta building activity of the Kosi River: J. Hydraulics Div., Am. Soc. Civil Eng., v. 92 (HY2), p. 111-126.

Goreau, T., and Burke, K., 1966, Pleistocene and Holocene geology of the island shelf near Kingston, Jamaica: Marine Geology, v. 4, p. 207-225.

Groat, C. G., 1972, Presidio Bolson, Trans-Pecos Texas and adjacent Mexico: Geology of a desert basin aquifer system: Bur. Econ. Geol. Univ. Texas, Rept. Invest., No. 76, 46p.

Heward, A. P., in press, Alluvial fan and lacustrine sediments from the Stephanian A and B (La Magdalena, Ciñera-Matallana and Sabero) coalfields, northern Spain: Sedimentology.

Hooke, R. Le B., 1967, Processes on arid-region alluvial fans: J. Geol., v. 75, p. 438-460.

————, 1968, Steady-state relationships on arid-region alluvial fans in closed basins: Am. J. Sci., v. 266, p. 609-629.

————, 1972, Geomorphic evidence for Late Wisconsin and Holocene tectonic deformation, Death Valley, California: Geol. Soc. Am. Bull., v. 83, p. 2073-2098.

Hoppe, G., and Ekman, S. R., 1964, A note on the alluvial fans of Ladtjovagge, Swedish Lapland: Geogr. Ann., v. 46, p. 338-342.

Howard, J. D., 1966, Patterns of sediment dispersal in the Fountain Formation of Colorado: Mountain Geologist, v. 3, p. 147-153.

Hubert, J. F., Reed, A. A., and Carey, P. J., 1976, Paleogeography of the East Berlin Formation, Newark Group, Connecticut Valley: Am. J. Sci., v. 276, p. 1183-1207.

Hunt, C. B., and Mabey, D. R., 1966, Stratigraphy and structure, Death Valley, California: U. S. Geol. Survey Prof. Paper 494-A, 162p.

Johnson, A. M., 1970, Physical processes in geology: Freeman, Cooper and Co., San Fransisco, 577p.

———, and Rahn, P. H., 1970, Mobilisation of debris flows: in New Contributions to Slope Evolution; Z. Geomorph. Suppl. Bd. 9, p. 168-186.

Kimura, T., 1966, Thickness distribution of sandstone beds and cyclic sedimentations in the turbidite sequences at two localities in Japan: Bull. Earthquake Res. Inst. Tokyo, v. 44, p. 561-607.

Klein, G. de V., 1975, Sandstone depositional models for exploration for fossil fuels: Continuing Education Publishing Company, Champaign, Illinois, 109p.

Knight, J. A., 1975, The systematics and stratigraphic aspects of the Stephanian flora of the Sabero coalfield, Part 1: The stratigraphy and general geology of the Sabero coalfield: Unpub. Ph.D. Thesis, Univ. Sheffield.

Laming, D. J. C., 1966, Imbrication, palaeocurrents and other sedimentary features in the Lower New Red Sandstone, Devonshire, England: J. Sediment. Petrol., v. 36, p. 940-959.

Lattman, L. H., 1973, Calcium carbonate cementation of alluvial fans in southern Nevada: Geol. Soc. Am. Bull., v. 84, p. 3013-3028.

Legget, R. F., Brown, R. J. E., and Johson, G. H., 1966, Alluvial fan formation near Aklavik, North West Territories, Canada: Geol. Soc. Am. Bull., v. 77, p. 15-30.

Lustig, L. K., 1965, Clastic sedimentation in Deep Springs Valley, California: U.S. Geol. Survey Prof. Paper 352-F, p. 131-192.

Martini, I. P., and Sagri, M., 1977, Sedimentary fillings of ancient deep-sea channels: Two examples from the Northern Apennines (Italy): J. Sediment. Petrol., v. 47, p. 1542-1553.

McGowen, J. H., and Groat, C. G., 1971, Van Horn Sandstone, west Texas: An alluvial fan model for mineral exploration: Bur. Econ. Geol. Univ. Texas, Rept. Invest. No. 72, 57p.

McPherson, H. J., and Hirst, F., 1972, Sediment changes on two alluvial fans in the Canadian Rocky Mountains: in H. O. Slaymaker and H. J. McPherson, eds., Mountain Geomorphology; Geomorphological Processes in the Canadian Cordillera; British Columbia, Geographical Series, 14, p. 161-175.

Meckel, L. D., 1967, Origin of Pottsville conglomerates (Pennsylvanian) in the Central Appalachians: Geol. Soc. Am. Bull., v. 78, p. 223-258.

———, 1975, Holocene sand bodies in the Colorado Delta area, northern Gulf of California: in M. L. Broussard, ed., Deltas Models for Exploration; Houston Geol. Soc., p. 239-265.

Miall, A. D., 1970, Devonian alluvial fans, Prince of Wales Island, Arctic Canada: J. Sediment. Petrol., v. 40, p. 556-571.

Middleton, G. V., and Hampton, M. A., 1973, Sediment gravity flows: Mechanics of flows and deposition: in G. V. Middleton, and A. H. Bouma, co-chairmen, Turbidites and Deep-Water Sedimentation; Soc. Econ. Paleont. Mineral. Short Course 1, Anaheim, p. 1-38.

Mutti, E., 1974, Examples of ancient deep-sea fan deposits from circum-Mediterranean geosynclines: in R. H. Dott, Jr., and R. H. Shaver, eds., Modern and Ancient Geosynclinal Sedimentation; Soc. Econ. Palaeont. Mineral. Spec. Pub. 19, p. 92-105.

———, 1977, Distinctive thin-bedded turbidite facies and related depositional environments in the Eocene Hecho Group (south-central Pyrenees, Spain): Sedimentology, v. 24, p. 107-131.

———, and Ricci-Lucchi, F., 1972, Le torbiditi dell'Appennino settentrionale: Introduzione all'analisi di facies: Soc. Geol. Italiana Mem., v. 11. p. 161-199.

Nelson, H., Mutti, E., and Ricci-Lucchi, F., 1977, Discussion: Upper Cretaceous resedimented conglomerates at Wheeler Gorge, California: Description and field guide: J. Sediment. Petrol., v. 47, p. 926-934.

Nilsen, T. H., 1968, The relationship of sedimentation to tectonics in the Solund district of southwestern Norway: Universitetsforlaget, Oslo Norges Geologiske Undersokelse No. 359, 108p.

———, 1969, Old Red sedimentation in the Buelandet-Vaerlandet Devonian district, western Norway: Sediment. Geology., v. 3, p. 35-57.

Normark, W. R., 1970, Growth patterns of deep-sea fans: Am. Assoc. Petrol. Geol. Bull., v. 54, p. 2170-2195.

Rahn, P. H., 1967, Sheetfloods, streamfloods and the formation of pediments: Ann. Assoc. Am. Geogr., v. 57, p. 593-604.

Renwick, W. H., 1977, Erosion caused by intense rainfall in a small catchment in New York State: Geology, v. 5, p. 361-364.

Riba, O., 1976, Syntectonic unconformities of the Alto Cardener, Spanish Pyrenees: A genetic interpretation: Sediment. Geol., v. 15, p. 213-233.

Ricci-Lucchi, F., 1975, Depositional cycles in two turbidite formations of northern Appenines (Italy): J. Sediment. Petrol., v. 45, p. 3-43.

Roed, M. A., and Wasylyk, D. G., 1973, Age of inactive alluvial fans — Bow River Valley, Alberta: Can. J. Earth Sci., v. 10, 1834-1840.

Ruhe, R. V., 1964, Landscape morphology and alluvial deposits in southern New Mexico: Ann. Assoc. Am. Geogr., v. 54, p. 147-159.

Rupke, N. A., 1977, Growth of an ancient deep-sea fan: J. Geol., v. 85, p. 725-744.

Ryder, J. M., 1971, The stratigraphy and morphology of paraglacial alluvial fans in south-central British Columbia: Can. J. Earth Sci., v. 8, p. 279-298.

Sagri, M., 1972, Rhythmic sedimentation in the turbidite sequences of the Northern Apennines (Italy): 24th Intl. Geol. Congr. Proc. Sect. 6, p. 82-88.

Schluger, P. R., 1973, Stratigraphy and sedimentary environments of the Devonian Perry Formation, New Brunswick, Canada, and Maine, U. S. A.: Geol. Soc. Am. Bull., v. 84, p. 2533-2548.

Schumm, S. A., 1968, Speculations concerning palaeohydrologic controls of terrestrial sedimentation: Geol. Soc. Am. Bull., v. 79, p. 1573-1588.

———, 1973, Geomorphic thresholds and complex response of drainage systems: *in* M. Morisawa, *ed.,* Fluvial Geomorphology; Pub. in Geomorphology, SUNY, Binghampton, N.Y., p. 299-310.

———, 1976, Episodic erosion: A modification of the geomorphic cycle: *in* W. N. Melhorn, and R. C. Flemal, *eds.* Theories of Landform Development; Pub. in Geomorphology, SUNY, Binghampton, N. Y., p. 69-85.

———, and Hadley, R. F., 1957, Arroyos and the semi-arid cycle of erosion: Am. J. Sci., v. 255, p. 161-174.

Scott, K. M., 1971, Origin and sedimentology of 1969 debris flows near Glendora, California: U. S. Geol. Survey Prof. Paper 750-C, p. C242-C247.

———, 1973, Scour and fill in Tujunga Wash — A fanhead valley in urban California: U. S. Geol. Survey Prof. Paper 732-B, 29p.

———, and Gravlee, G. C., Jr., 1968, Flood surge of the Rubicon River, California — hydrology, hydraulics and boulder transport: U. S. Geol. Survey Prof. Paper 422-M, 40p.

Selley, R. C., 1965, Diagnostic characteristics of fluviatile sediments of the Torridonian Formation (Precambrian) of northwest Scotland: J. Sediment. Petrol., v. 35, p. 366-380.

Sestini, G., 1970, Vertical variations in flysch and turbidite sequences: a review: J. Earth Sci., Leeds, v. 8, p. 15-30.

Sharp, R. P., 1948, Early Tertiary Fanglomerate, Big Horn Mountains, Wyoming: J. Geol., v. 56, p. 1-15.

———, and Nobles, L. H., 1953, Mudflow of 1941 at Wrightwood, southern California: Geol. Soc. Am. Bull., v. 64, 547-560.

Statham, I., 1976, Debris flows on vegetated screes in the Black Mountain, Carmarthenshire: Earth Surf. Proc., v. 1, p. 173-180.

Steel, R. J., 1974, New Red Sandstone floodplain and piedmont sedimentation in the Hebridean province, Scotland: J. Sediment. Petrol., v. 44, p. 336-357.

————, 1976, Devonian basins of western Norway — sedimentary response to tectonism and to varying tectonic context: Tectonophysics, v. 36, p. 207-224.

————, and Wilson, A. C., 1975, Sedimentation and tectonism (?Permo-Triassic) on the margin of the North Minch Basin, Lewis: J. Geol. Soc. London, v. 131, p. 183-202.

————, Maehle, S., Nilsen, H., Roe, S. L., Spinnangr, A., 1977, Coarsening-upward cycles in the alluvium of Hornelen Basin (Devonian) Norway: Sedimentary response to tectonic events: Geol. Soc. Am. Bull., v. 88, p. 1124-1134.

Tanner, W. F., 1976, Tectonically significant pebble types: sheared, pocked, and second-cycle examples: Sediment. Geol., v. 16, p. 69-83.

Troxel, B. W., 1974, Man-made diversion of Furnace Creek Wash, Zabriske Point, Death Valley, California: California Geol., Oct. 1974, p. 219-223.

Vliet, A. van, in press, The early Tertiary deep-water fans of Guipuzcoa: *in* D. J. Stanley and G. Kelling, *eds.,* Submarine Canyon and Fan Sedimentation, Dowden, Hutchison and Ross.

Wagner, R. H., 1970, An outline of the Carboniferous stratigraphy of northwest Spain: Congr. Coll. Univ. Liège, v. 55, p. 429-463.

————, 1971, The stratigraphy and structure of the Ciñera-Matallana coalfield (prov. León, N.W. Spain): Trabajos de Geologia Fac. Ci. Univ. Oviedo, v. 4, p. 385-429.

Walker, R. G., 1970, Review of the geometry and facies organisation of turbidites and turbidite bearing basins: *in* J. Lajoie, *ed.,* Flysch Sedimentology in North America: Geol. Assoc. Canada Spec. Paper 7, p. 219-251.

————, 1975a, Upper Cretaceous resedimented conglomerates at Wheeler Gorge, California: Description and field guide: J. Sediment. Petrol., v. 45, p. 105-112.

————, 1975b, Nested submarine-fan channels in the Capistrano Formation, San Clemente, California: Geol. Soc. Am. Bull., v. 86, p. 915-924.

————, 1975c, Conglomerate: Sedimentary structures and facies models: *in* Depositional Environments as Interpreted from Primary Sedimentary Structures and Stratification Sequences; Soc. Econ. Palaeont. Mineral. Short Course 2, p. 133-161.

————, 1975d, From sedimentary structures to facies models: example from fluvial environments: *in* Depositional Environments as Interpreted from Primary Sedimentary Structures and Stratification Sequences; Soc. Econ. Palaeont. Mineral. Short Course 2, p. 63-79.

————, 1977, Deposition of upper Mesozoic resedimented conglomerates and associated turbidites in southwestern Oregon: Geol. Soc. Am. Bull., v. 88, p. 273-285.

————, and Mutti, E., 1973, Turbidite facies and facies associations, *in* G. V. Middleton, and A. H Bouma, co-chairmen, Turbidites and Deep Water Sedimentation: Soc. Econ. Palaeont Mineral. Short Course 1, Anaheim, p. 110-157.

Wasson, R. J., 1974, Intersection point deposition on alluvial fans: An Australian example: Geoogr. Ann., v. 56, p. 83-92.

————, 1977a, Catchment processes and the evolution of alluvial fans in the lower Derwent Valley, Tasmania: Z. Geomorph., Bd. 21, p. 147-168.

————, 1977b, Late-glacial alluvial fan sedimentation in the Lower Derwent Valley, Tasmania: Sedimentology, v. 24, p. 781-799.

Weaver, W., and Schumm, S. A., 1974, Fan-head trenching: An example of a geomorphic threshold: Geol. Soc. Am., Abs. with Program, v. 6, p. 481.

Williams, G. E., 1969, Characteristics and origin of a Precambrian pediment: J. Geol., v. 77, p. 183-207.

————, 1970, Piedmont sedimentation and late Quaternary chronology in the Biskra region of the northern Sahara: Z. Geomorph., v. 10, p. 40-60.

Wilson, M. D., 1970, Upper Cretaceous-Palaeocene synorogenic conglomerates of south-western Montana: Am. Assoc. Petrol. Geol. Bull., v. 54, p. 1843-1867.

Winder, C. G., 1965, Alluvial cone construction by alpine mudflow in a humid temperate region: Can. J. Earth Sci., v. 2, p. 270-277.

PALEOHYDRAULICS

RECONSTRUCTING PALEOCHANNEL MORPHOLOGIC AND FLOW CHARACTERISTICS: METHODOLOGY, LIMITATIONS, AND ASSESSMENT

Frank G. Ethridge[1]
AND
Stanley A. Schumm[1]

Abstract

Empirical relations among hydrologic and sediment characteristics of alluvial channels and channel morphology have been developed by geomorphologists and hydrologists. Recently sedimentologists have begun to use these equations in order to estimate the morphologic and hydrologic characteristics of paleochannels from data obtained at paleochannel cross sections.

The estimates obtained in this way must be viewed with caution because of the limited data base from which the original equations were developed and the difficulty in obtaining data on paleochannel widths, depths, and sediment character. Because of the inherent error in such estimates the most reasonable approach is to use the simplest and most direct method involving only use of paleochannel dimensions. The additional effort required to determine the silt-clay content of sediment forming the channel perimeter may not be warranted.

Introduction

In studies that attempt to determine the character of ancient fluvial deposits, the objectives are usually to determine answers for three important and interrelated questions as follows:

1) What were the specific environments of deposition of the deposits?

2) What is and was the size, shape, and extent of the deposits?

3) What was the nature and magnitude of the processes responsible for the transport and deposition of the deposits?

Answers to the last two questions are particularly pertinent to geologists involved in the search for and the evaluation of potential economic deposits of minerals and fuels in fluvial sandstones. The cost of exploration and exploitation of such deposits could be considerably reduced if the size and shape of ancient fluvial systems could be predicted from information in a few exceptional outcrops.

Assessment of the morphology and flow characteristics of ancient fluvial systems by the application of empirical relationships derived from Holocene rivers is a relatively new means of providing answers to questions 2 and 3 above. This methodology provides quantitative data for the reconstruction of paleochannel morphology. Use of empirical equations such as those developed by geomorphologists and hydraulic engineers, in addition, provides a methodology that may permit development of more efficient sampling programs. Of course it was always possible to reconstruct the geometry of existing sedimentary deposits assuming that the time and money required for extensive surface reconnaissance and drilling programs was available.

It seems appropriate at this time to review these efforts and to evaluate the potential of the approach. Therefore, the objectives of this paper are as follows:

1) to document the historical development of methodologies designed to reconstruct the morphologic and flow characteristics of ancient fluvial systems,

[1]Department of Earth Resources, Colorado State University, Fort Collins, Colorado 80523, USA

2) to evaluate critically the Holocene data base,

3) to review the methodologies involved and the problems inherent in their implementation,

4) to summarize and evaluate the results of an analysis of the morphologic and flow characteristics of ancient river systems including new data on several Carboniferous channels, and

5) to suggest some future directions for research.

BACKGROUND AND NATURE OF HOLOCENE DATA

Early attempts to reconstruct flow characteristics of paleochannels were restricted for the most part to the evaluation of flow conditions responsible for the development of sedimentary structures. Jopling's (1966) calculations of the paleoflow regime of a small Pleistocene Gilbert-type delta is a case in point. In his very detailed analysis Jopling estimated values for the rate of movement of the bed material, rate of advance of the delta front, and the time required for the deposition of one lamina.

Such a detailed analysis may seldom be required to answer the practical problems involved in the search for economic deposits. However, his methodology did suggest that it is possible to obtain reasonable estimates of paleoflow regimes using relationships developed by hydraulic engineers and information contained in the rock record. More recently, hydraulic relations have been applied to the study of the paleoflow characteristics of Pleistocene terrace gravels and outwash (Baker, 1974; Clague, 1975; Cheetham, 1976).

Beginning in the late 1950's and continuing to the present, geomorphologists have been studying the relations between channel morphology and hydrology, thereby providing a means for predicting channel morphology and flow parameters based on empirical relationships developed from measurements taken on modern rivers (Leopold and Wolman, 1957, 1960; Schumm, 1960a, b, 1963, 1968, 1969; Leopold et al., 1964; Carlston, 1965; Dury, 1964). In addition, hydrologists have been attempting to predict discharge characteristics of small ungaged streams from cross-sectional data (Dury, 1976; Riggs, 1976; Hedman and Kastner, 1977).

Sedimentologists have used these quantitative relations to develop a methodology for predicting paleochannel characteristics. This methodology relies heavily on the relationships developed by Schumm (1960a, b; 1968; 1972) which were developed from data collected on some 33 to 69 stable alluvial stream channels in the semiarid to subhumid Great Plains Region of the U.S. The stream channels range in elevation from one-half to one mile above sea level and are far removed from the coast. The range of values for the important variables used in developing the empirical relationships are given in Table 1. Note that values for Sb (% silt-clay in channel bank) are significantly lower than corresponding values for all paleostreams studied to date. Other variables such as Qm (mean annual discharge) have a limited range.

Data from two additional rivers, the Murrumbidgee (Schumm, 1968) and the Guadalupe (Morton and Donaldson, this volume) appear to conform to the relationships developed for the Great Plain streams (Table 1) and they provide data for coastal plain rivers of moderate size, which are more like channels preserved in the rock record than are the Great Plains streams.

In summary, the data used to develop empirical relations among channel morphology, discharge, and sediment type are very limited and for the most part were obtained from rivers that are quite different from those near sea level that are preserved in the ancient rocks.

Table 1.. Holocene data base for some of the empirical relations used in assessing morphologic and flow characteristics of ancient river systems.

Variable [a]	Schumm[b] (1960a)	Schumm[c] (1968)	Morton and Donaldson[d] (this volume and 1978)
W	8-194 m	50-83 m	57 m
D	0.7-5.5 m	5.-9.5 m	7 m
W/D	4-75	7-14	8.1
Sc	0.5-88% Ave. = 9.9%	0.07-40% Ave. = 17.1%	6-54%
Sb	15-97% Ave. = 61.9%	40-87% Ave. = 66.2%	55-82%
P	1.05-2.5	1.6-2.3	2.24
Qm	0.6-146 cm	68-129 cm	46 cm

[a]Symbols: W = bankfull width; D = bankfull depth; W/D = width to depth ratio; Sc = silt-clay percent in channel bed; Sb = silt-clay percent in channel bank; P = sinuosity; Qm = mean annual discharge.

[b]33-69 alluvial stream channels from the Great Plains of the U. S.

[c]Murrumbidgee River in southeastern Australia.

[d]Guadalupe River in Texas.

METHODOLOGY AND PROBLEMS OF IMPLEMENTATION

Basic Data and Relationships

The basic data necessary for the calculation of morphologic and flow characteristics of ancient river systems are:

1) W   (bankfull stream width);

2) D   (bankfull stream depth);

3) Sc   (percent silt-clay in channel bed); and

4) Sb   (percent silt-clay in channel bank [band]).

Using these data, equations and tables as presented by Schumm (1960a, b; 1963; 1968; 1972), equations presented by Leopold and Wolman (1960), and graphs and equations presented by Hack (1957) and Dury (1965), a number of morphologic and flow characteristics can be estimated. The characteristics and the relationships necessary for their estimation are given in Tables 2 and 3. Many of the variables can be estimated from more than one relationship and thus are presented as Methods I and II (Table 2) which use sediment data or simply information on channel width and depth, respectively.

Problems Associated with Estimates of Input Variables

One of the most serious problems is associated with the estimation of paleochannel width and depth. Bankfull width and depth measurements for the Great Plains streams were made in straight reaches where the channel cross-section is symmetrical. In paleochannels, estimates of both width and depth are usually obtained from point bar deposits which are formed in bends. Moody-Stuart (1966) states that a common characteristic of ancient meander belt deposits is large-scale sigmoidal cross-stratification

Table 2. Stream characteristics and equations used in reconstructing the morphologic and flow characteristics of paleochannels.

Stream Characteristic	Symbol	Units[a]	Equation Number[b]	Method I	Method II	Other Techniques and Equations	Se[c]	r[d]	Source
Silt-clay in channel perimeter	M	%	1	$M = \dfrac{(Sc \times W) + (Sb \times 2D)}{W + 2D}$		Analysis of composite sample of channel perimeter	—	—	Schumm (1960a)
							—	—	Schumm (1968)
Type of sediment load	—	—				Obtain from Table 3	—	—	Schumm (1968)
Width-depth ratio	F	—	2	$F = 255 M^{-1.08}$			0.20	0.91	Schumm (1960a)
			3		$F = W/D$				
Sinuosity	P	—	4	$P = 0.94 M^{+0.25}$			0.06	0.91	Schumm (1963)
			5		$P = 3.5 F^{-0.27}$		0.06	0.89	Schumm (1963)
Mean annual discharge	Qm	cfs	6	$Qm^{0.38} = WM^{0.39}/37$			0.14	0.93	Schumm (1968)
			7		$Qm = W^{2.43}/18 F^{1.13}$		0.12	0.89	Schumm (1968)
			8			$Qm^{0.29} = D/0.6 M^{0.34}$	0.20	0.90	Schumm (1972)
Mean annual flood	Qma	cfs	9	$Qma^{0.58} = WM^{0.37}/2.3$			0.13	0.94	Schumm (1968)
			10		$Qma = 16(W^{1.56}/F^{0.66})$		0.13	0.86	Schumm (1968)
			11			$Qma^{0.42} = D/0.09 M^{0.35}$	0.20	0.90	Schumm (1972)
Channel slope	S	ft/mi	12	$S = 60 M^{-0.38}Qm^{-0.32}$			0.15	0.84	Schumm (1968)
			13		$S = 30(F^{0.95}/W^{0.98})$		0.16	0.84	Schumm (1972)
			14			$S = \eta V/1.49R^{2/3}$	—	—	Manning Equation (Cotter, 1971)
			15			$S = (\tau V)\eta/1.49\gamma D^{2/3}$	—	—	Manning Equation (Cotter, 1971)

			[b]		[c]	[d]	
Velocity	V	ft/sec	16	$V = 1.49\eta R^{2/3} S^{1/2}$	—	—	Manning Equation (Cotter, 1971)
Meander wavelength	L	ft	17	$L = 1890\, Qm^{0.34}/M^{0.74}$	0.16	0.96	Schumm (1968)
			18	$L = 234\, Qma^{0.48}/M^{0.74}$	0.19	0.92	Schumm (1968)
			19	$L = 18(F^{0.53} W^{0.69})$	0.21	0.91	Schumm (1972)
			20	$L = 10.9\, W^{1.01}$	0.3[e]	?	Leopold and Wolman (1960)
			21	$L = 106\, Qm^{0.46}$	0.05	?	Carlston (1965)
			22	$L = 30\, Qm^{0.5}$	?	?	Dury (1964)
Drainage area	Ad	sq mi	23	Estimated from graph	?	?	Dury (1965, p. C-7, Fig. 6)
Stream or channel length	Ls	mi	24	$LS = 1.4\, Ad^{0.6}$	?	?	Hack (1957)
Curvature radius	Rm	ft	25	$Rm = 2.7\, W^{1.1}$	?	?	Leopold and Wolman (1960)
Percent of total load as bedload	—	%	26	55/M	—	—	Schumm (1968)

[a] English units were used in the development of these equations, but conversion to metric units is made in the computer program.
[b] Numbers are assigned to each equation for easy reference.
[c] Standard Error of the estimate in log units.
[d] Correlation coefficient.
[e] Approximate value after Leeder (1973).

Table 3.. Classification of stable alluvial channels (after Schumm, 1968, p. 1579, Table 1).

Sediment Load	Channel Sediment (Percentage of Silt and Clay in Channel Perimeter)	Bedload (Percent of Total Load)	Type of River
Suspended load and dissolved load	>20	<3	Suspended-load channel; width-depth ratio <10; sinuosity >2.0; gradient relatively gentle.
Mixed load	5 to 20	3 to 11	Mixed-load channel; width-depth ratio >10, <40; sinuosity <2.0, >1.3; gradient moderate; can be braided.
Bed load	<5	>11	Bedload channel; width-depth ratio >40; sinuosity <1.3; gradient relatively steep; can be braided.

units which were classified as epsilon cross-strata by Allen (1965). Each epsilon cross-stratified unit represents a point bar deposit (Allen, 1965; Fig. 1). Most investigators (Moody-Stuart, 1966; Donaldson, 1969; Cotter, 1971; Leeder, 1973; Elliott, 1976; Morton and Donaldson, this volume) follow the two-thirds rule in estimating bankfull width from point bar deposits. This rule is adopted from Allen (1965) who states that point bars extend two-thirds of the distance across the channel. Bankfull width is thus calculated by determining the average horizontal width of all epsilon cross-stratification units in an exposure and multiplying this value by 1.5 (Fig. 1).

Some investigators define bankfull depth as the average thickness of the epsilon cross-section unit (Moody-Stuart, 1966; Elliott, 1976; Fig. 1). Others apply a correction factor to compensate for the fact that bankfull depth in straight reaches is less than in meander bends and a second correction to compensate for compaction. Donaldson (1969) adjusted his measured values of depth by converting the asymmetric cross-section of a meander bend to a rectangular cross-section characteristic of a straight reach by holding width and cross-section area constant. He then multiplied the corrected depth by 1.2 to compensate for compaction. Cotter (1971), using only one correction, reduced the thickness by 1/16 (an arbitrary factor) to obtain a more realistic depth measure.

A cursory examination of the literature reveals a dearth of tabulated information on the ratio between bankfull depth in a straight reach versus bankfull depth in meander bends. Data on this ratio are, however, available from experimental studies of meandering thalweg channels (Khan, 1971). These data (Table 4) show an average ratio of 0.585 with no consistent trend related to channel slope or cross-sectional area. Field measurements along a meandering reach of the Arkansas River (T. J. Jackson and C. T. Nadler, personal communications) appear to substantiate the experimental results given above.

The problem of correcting for compaction of sands is considerably more complex, and no simple answer is readily available, as demonstrated by the extensive literature on the subject recently summarized by Chilingarian and Wolf (1975, 1976). The problem is complicated by the fact that most studies to date are concerned with porosity losses, and data are presented in the form of graphs that relate depth of burial to total porosity. All of these graphs show a more or less curvilinear relationship with decreasing loss of porosity with increasing depth. A major problem, of course, is that reduction in porosity can result from cementation as well as from compaction. The relative importance of each of these

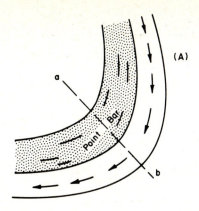

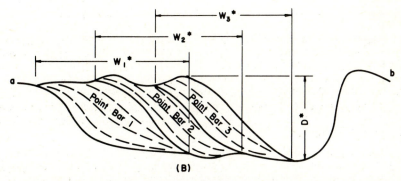

Figure 1. (A) Plan view map of hypothetical meander bend showing location of cross-section in (B).

(B) Cross-section of hypothetical lateral accretion deposits illustrating large-scale sigmoidal cross-stratified units (point bars) and depth and width measures.

Formulas used to convert width and depth measures to bankfull width and depth are as follows:

Bankfull channel width (W) = W* x 1.5

Bankfull channel depth (D) = D* x 0.585/0.9

Note: For details of internal stratification and texture within each epsilon cross-stratification unit, *see* Allen (1965, Fig. 2).

factors is still a matter of some debate. Furthermore, the degree to which unconsolidated sands can be compacted is a function of: (a) original packing and/or bridging, (b) original void ratio, (c) shape of the grains, (d) roundness and sphericity of the grains, (e) composition of the sand, and (f) size grading (Allen and Chilingarian, 1975, p. 69). Well-sorted, rounded, well-packed clean quartz sands do not compact readily, whereas poorly sorted sands are more susceptible to compaction. Finally, the nature of the stratigraphic section must be taken into consideration. For example, a section composed of sandstone alone would have a very different compactional history than a section of sandstones interbedded with shales and/or siltstones (Wolf and Chilingarian, 1976).

A tentative solution to this problem is based on data presented by Ingles and Grant (1975, p. 305, Table 6-III) and the formulas for pseudobulk compressibility (Chilingarian *et*

Table 4. Ratio of bankfull depth in straight reaches (crossing) to bankfull depth in meander bends (pool) as related to slope and cross-sectional area. Data are from 19 experimental meandering-thalweg channels (Khan, 1971).

	Slope (m/m)	0.0043	0.0059	0.0085	0.0026 to 0.13	Average
1.	Max. Depth Crossing (m)	0.0219	0.0204	0.0159	0.0155	—
2.	Maximum Depth Pool (m)	0.0377	0.0334	0.0278	0.0266	—
3.	Ratio of 2 to 3 (%)	58	61	57	58	58.5
4.	Area Crossing (sq. m)	0.0192	0.0157	0.0153	0.0141	—
5.	Area Pool (sq m)	0.0169	0.0158	0.0134	0.0121	—
6.	Ratio of 5 to 6 (%)	114	99.8	114	117	111

al., 1975, p. 39; after Fatt, 1958). Typical values of porosity of natural coarse-grained sediments together with the approximate apparent porosities to which these sediments can be compacted with modern heavy mechanical equipment by vibration shock suggest that the maximum decrease in porosity for poorly graded sand is 10% (Ingles and Grant, 1975). Ingles and Grant also suggest that the compressibility of sand is directly related to porosity (or void ratio). This statement is supported by Chilingarian et al., (1975) who state that changes in sample thickness can be substituted for changes in void ratio in formulas for pseudo-bulk compressibility of coarse-grained sediment. Using these data, a 10% reduction in thickness for the conversion of sand to sandstone as a result of compaction appears conservative. Thickness of epsilon cross-stratification units are, therefore, divided by 0.9 to convert these measures to original bankfull depth (Fig. 1).

A persistent problem in the determination of the morphology and paleohydrology of ancient river systems is the lack of well-exposed point bar sequences (epsilon cross-stratification units) in the rock record and hence an inability to estimate bankfull width. Because of this difficulty, Leeder (1973) explored an alternative method of estimating bankfull width. A plot of width versus depth for streams with sinuosities from 1.0 to 2.5 shows no direct relationship between width and depth. However, because of the interest in relatively sinuous channels, Leeder (1973) constructed a revised plot for 57 highly sinuous streams (P > 1.7). A reduced major axis regression line computed on values of width and depth for these highly sinuous streams gives the following relationship.

$$\log W = 1.54 \log D + 0.83$$

or

$$W = 6.8D^{1.54}$$

St. Dev. = 0.35 log units

r = 0.91

Using this relationship, bankfull width for high sinuosity streams can be estimated from bankfull depth, estimates of which are more readily available in the rock record.

Some problems are also involved with the determination of the final two input variables, Sc and Sb (percent silt-clay in the bed and bank). Estimates of silt-clay percent for modern streams are determined on the basis of the percentage of sediment passing through the 200 mesh sieve (sediment less than 0.074 mm in diameter). Percent silt-clay, therefore, includes some sediment classified as very fine sand in most grain size classifications used by sedimentologists. The justification for this boundary is discussed by Schumm (1960a, p. 18). The major problem involved in the determination of Sc and Sb for ancient sedimentary rocks is the recognition of diagenetic clays. Detailed petrographic and geochemical studies in recent years have revealed that much of the clay size fraction of ancient sandstones is authigenic. Percent silt-clay in paleostream deposits should, therefore, be determined by direct measurement and counting techniques using thin sections, which allow for the differentiation of diagenetic and detrital matrix. Conversion of thin-section size distributions to equivalent sieve-size distributions is not possible at present although certain statistical parameters such as mean, standard deviation, etc., can be converted using equations developed by Adams (1977). The problem of conversion is, however, not considered serious because only the percentage of sediment below a certain prescribed size limit is needed; therefore, knowledge of the entire grain-size distribution is not required.

Methods of Paleochannel Analysis

Once the measurements and samples of the rocks in a paleochannel have been made there are two approaches to the determination of paleochannel variables of interest. These are outlined in Table 2.

Method I involves the use of the percentage of silt-clay in the channel perimeter (M) which is a reflection of the type of load transported through the paleochannel. This method utilizes equations 1, 2, 4, 6, 9, 12, 17 and 26 (Table 2).

Although M can be determined by sampling of the bed and bank material of modern rivers, the difficulty of obtaining information on the variable in paleochannels led Schumm (1972) to develop a series of equations that are based on the relation between M and the width-depth ratio (equation 2, Table 2). Width-depth ratio is substituted for M and width is substituted for discharge in equations 5, 8, 11, 13 and 19, which provide a simpler way of estimating paleochannel characteristics. When equation 3 is added, Method II is developed.

Problems and Limitations of the Relationships

Two methods are developed for calculating various morphologic and hydraulic characteristics of ancient streams (Table 2). The first method involves the use of the percent silt-clay exposed in the channel perimeter as well as estimates of bankfull width and depth (W and D). The basic hypothesis underlying these relations is that the percent silt-clay in the perimeter of an alluvial stream channel reflects the nature of the sediment load moved by the stream (Table 3) which is of primary importance in determining the width-depth ratio of alluvial channels (Schumm, 1960a, b).

The methodology used by Schumm to test this hypothesis has been questioned (Melton, 1961). According to Melton, the regression relation of channel width-depth ratio (F) to weighted mean percent silt-clay (M) (equation 2, Table 2) does not support Schumm's hypothesis because the variable M contains the variable F in disguised form. In a reply to Melton, Schumm (1961) admitted that width and depth are present in both sides of the relationship. However, he proceeded to justify the use of a weighted mean percent silt-clay and with the use of multiple regression demonstrated that the width-depth ratio was, in part, controlled by the percentage of silt-clay in the bed and bank of the channel (Sc and Sb). Later work (Schumm, 1968) showed that the weighted mean percent silt-clay accurately reflects the percentage of silt-clay in the channel perimeter, as determined by analysis of composite samples taken from the perimeter of alluvial channels.

The effect of climate on the relations is difficult to evaluate, but comparison of Carlston's (1965) and Schumm's (1969, 1972) relations between meander wavelength and discharge reveal that the equations developed for the semiarid-subhumid Great Plains data overestimate discharge, as compared to the estimates obtained using Carlston's humid region equations (Saucier and Fleetwood, 1970a, b).

A computer program developed by the senior author allows calculation of paleochannel variables by either method and thus provides a check on the estimates of the variables provided accurate measures of channel form and sediment type can be obtained. If accurate measures of M cannot be obtained, the program provides the option of using only the second method. Estimates of drainage area, channel length, and radius of curvature can be obtained using relationships after Dury (1965), Hack (1957) and Leopold and Wolman (1960) (Equations 23, 24 and 25; Table 2) with both methods described above. The computer program is available on request.

Some of the equations in Table 2 are not included in either method. Equations 14, 15 and 16 are eliminated because of the difficulty in obtaining reliable estimates of the roughness coefficient η. This coefficient is a function of bed material size, regularity of banks, sinuosity, vegetation influences, variation in channel cross-section, and type of bed form (Barnes, 1967). In addition, it is usually determined by visual estimation of a stream rather than by a precise formulation. Elimination of equations 14 and 15 have no effect on the ability to estimate channel slope as equations 12 and 13 (Table 2) provide reasonable estimates of this variable. Elimination of equation 16, however, makes it impossible to obtain an estimate of paleovelocity. Equations 4, 10, 18, 20, 21 and 22 are not included because the equations selected for use in Methods I and II have higher correlation coefficients and/or lower standard errors.

Equations 23 and 24 must be used with caution because they were developed from data on river systems in humid regions such as western Europe and the eastern United States, and because drainage area (Ad) is estimated graphically from Dury's figure. Under arid conditions these equations may not be valid.

As Leeder (1973, p. 270) emphasizes, there is considerable scatter about many of the regression lines represented by the equations in Table 2. It is, therefore, more appropriate to present ranges of values for a stream variable rather than a single value which gives no information about the possible error involved. The most appropriate method for calculating the upper and lower limits of a given variable would be to calculate confidence limits for that variable using the Student's t distribution. This methodology, however, requires information from the original data matrix used to determine the relations in Table 2. Since this information is not readily available, a less accurate but more readily obtainable estimate of possible ranges of values has been suggested by Leeder (1973) and outlined in more detail by Riggs (1968). Graphical 95% confidence limits are estimated by using 1.96 times the standard error of the estimate given in Table 2. Because of the log scales used, these graphical confidence limits will not be symmetrical about the point estimate of a variable. Confidence intervals and point estimates are determined for most stream characteristics.

Another possible source of error involved in the application of equations in Table 2 is that resulting from estimation of variables that are then used in subsequent equations. For example, equation 12 in Method I involves Qm which is estimated using equation 6. Mean annual flood (eq. 9), meander wavelength (eq. 17, 18, 21, 22) and stream length (eq. 24) also involve the use of previously estimated variables (Table 2). We have attempted to utilize only equations that do not involve previously estimated variables; however, this is not always possible. Therefore, errors may be compounded by the use of certain equations.

In summary, there are numerous problems involved in the estimation of the morphology and hydrology of paleochannels. Hence, the estimates of paleochannel morphology and hydrology obtained by Methods I and II must be used with caution and with an awareness of the potential for serious error. Therefore, statement of the range of values associated with each estimated variable constitutes a conservative and realistic approach to estimation of paleochannel characteristics.

MORPHOLOGY AND FLOW CHARACTERISTICS OF ANCIENT RIVER SYSTEMS

Review of Previous Literature

During the twelve-year period between 1966 and 1977 morphologic and flow characteristics have been estimated for some 15 different ancient river systems ranging in age from Devonian to Quaternary (Table 5). Comparison and evaluation of the data as reproduced in Table 5 is difficult because of the lack of uniformity in collecting data and estimating paleochannel characteristics.

First there is no consistent method for determining bankfull depth. As previously discussed, some investigators use the height of the epsilon cross-stratification as a measure of bankfull channel depth (Elliott, 1976) while others (Donaldson, 1969; Cotter, 1971; Morton and Donaldson, this volume) make various types of corrections to take into account the fact that bankfull depths for modern streams are determined in straight reaches. Donaldson (1969) is the only previous investigator to correct for the effects of compaction. Miall (1976), in a study of Cretaceous braided stream deposits, determined bankfull depth by using a relationship between mean ripple height and mean water depth (Allen, 1968).

Secondly, relationships and methods used in estimating stream characteristics vary greatly. Most previous investigators use the relationship developed by Schumm (1960a) to calculate the percent silt-clay in the channel perimeter (M) (equation 1, Table 2). Donaldson (1969), however, modifies this relationship as follows:

$$M = \frac{(Sc \times W) + (Sb \times D)}{W + D}$$

He suggests that bankfull depth rather than 2D should be used because in meander bends only the outside bank is actively eroding. No data are given to justify this modification. Some investigators use the relationships as listed under Method I (Table 2) (Donaldson, 1969; Cotter, 1971; Padgett and Ehrlich, 1976; Morton and Donaldson, this volume). Others use relations as listed under Method II (Leeder, 1973, his Method B; Derr, 1974), while a few use equations 20 and 21 (Table 2) (Leeder, 1973, his Method B; Elliott, 1976).

Perhaps the most circuitous approach was that of Eicher (1969) who, in an analysis of Lytle Front Range streams in Colorado, began by determining the distance from a hypothetical source terrane. Using this distance, he determined stream order and drainage area. From stream order he determined channel width and discharge from relations given by Leopold and Miller (1956). He then estimated channel slope by comparison with modern Front Range streams. Baker (1974) employed engineering hydraulic calculations to estimate flow depths, gradients, and flood discharges for Quaternary terrace deposits near Golden, Colorado.

Finally, recognition of exposed arcuate point bar ridge and swale deposits in sedimentary rock sequences ranging in age from Carboniferous to Miocene has allowed direct measurement of meander wavelength and radius of curvature in addition to the height and width of epsilon cross-stratification units (Puigdefabregas, 1973; Nami, 1976; Padgett and Ehrlich, 1976).

Other difficulties exist in comparing previous results. Some investigators estimate only a few stream variables (Moody-Stuart, 1966; Berg, 1968; Baker, 1973; Leeder, 1973), whereas others have estimated numerous variables (Donaldson, 1969; Cotter, 1971; BeMent, 1976; Gardner, 1977), and Puigdefabregas (1973) presents only variables that can be measured directly. The method of reporting estimates varies considerably. In some papers only point estimates are given and in others a range of possible values is reported. Only Elliott (1976) gives both point and interval estimates of stream characteristics.

Table 5.   Morphologic and flow characteristics of paleochannels. (Symbol notations given in Table 2.)

Stream Character	Units of Measure	Dev. Spitsbergen Moody-Stuart (1966)	L. Cret. Wyo. Berg (1968)	Penn.-W.V. Donaldson (1969)	Cret.-Colo. Eicher (1969)	Cret.-Utah Cotter (1971)	Upper O.R.S. England Leeder (1973)	Quaternary Colo. Baker (1974)
Sc	%			10-25		15		
Sb	%			100		100		
W	m	8-126	457	97.5	183-549	91.4	7(315)[a]	
D	m	0.5-6.25		6.7		7.62	1(4.25)[a]	2-3
M	%			20		27		
Type sediment load	—					suspended load		
% total load as bedload	%			2.75				
F	—			16		8-12		
P	—	1.4-1.9		1.99		1.8-2.1		
Qm	cms	100-10000		153-394	566	170-198	0.01-3 (24300)[a]	
Qma	cms			605		623		1400
S	m/km			0.197		0.189		8-10
V	m/sec			0.29-1.14	1.52	0.914		
L	m		7924-12192	1273-1932		762-1250	20-530 (14470)[a]	
Ad	sq. km			103600	13727-58275	15540-20720		
Ls	km			805-966	644	322		
Rm	m		1372-2134	465				

Table 5. Continued.

Stream Character	Units of Measure	Miocene, Spain Puigdefabregas (1973)	Jurassic Utah Deer (1974)	Jurassic England Nami (1976)	Carboniferous Morocco Padgett Ehrlich (1976)	Carboniferous England Elliott (1976)	Cretaceous Arctic Canada Miall (1976)	Carboniferous Ky. BeMent (1976)	Pennsylvanian Ky. Gardner (1977)
Sc	%				5				
Sb	%				100				
W	m	(3-5)[b]	100-393	21 (17-120)[a]	113-268	120		40-110	130-140
D	m	(1-2)[b]	10-22	4	5(?)	7.5	4-5	5-7	10 ± 0.5
M	%		6.6-10.3						
Type sediment load	—								
% total load as bedload	%								
F	—		10-18.2	4.3-30(5.25)[b]				13-18	12-15
P	—					1.66	1.03-1.38	1.7-1.8	1.5-2.1
Qm	cms		8.5-122	0.5-10.3	130-763	4-2500 (av. = 100)	250-640	60-330	400 ± 180
Qma	cms		130-765				1000-1600	700-1500	500-1800
S	m/km		0.25-0.55		0.15-0.19			0.3-0.6	0.1-0.2
V	m/sec				0.19-0.466			0.5	0.4-0.6
L	m		457-1554	110-436 (130-180)[b]	1987-3139 (3960-4785)[b]	345-5460 (av. = 1370)			1000-3000
Ad	sq. km						500-30000	6000-180000	5000-100000
Ls	km							500	100-600
Rm	m	(200)[b]		77.5[b]					

[a] Maximum values for fluvial cycles without epsilon cross-strata units; Method A, Leeder (1973).

[b] Measured directly as a result of excellent exposures of arcuate point bar ridges.

Methods used to determine the basic input data are not clear in many instances and thus estimates of stream variables cannot be compared with those from other studies.

For the above reasons, morphologic and flow characteristics of selected ancient streams have been recalculated using the methodology suggested earlier to achieve uniformity of estimates. The paleochannels selected include those for which basic data (height and width of epsilon cross-stratification units plus Sc and Sb or M) are available. Unfortunately, only three examples out of the 15 listed in Table 6 contain sufficient data for this type of reconstruction. The characteristics of these three paleochannel systems can then be compared with the characteristics of three additional paleochannels from Pennsylvanian-age rocks in West Virginia (described elsewhere in this volume, Morton and Donaldson), Pennsylvania (Beutner et al., 1967), and southern Illinois (Ethridge et al., 1973).

Pennsylvanian Channels

Morton and Donaldson (this volume) investigated point bar deposits in the middle part of the Conemaugh Group (Pennsylvanian) of northern West Virginia. Using their basic data, estimates of paleochannel characteristics are obtained by the method described here (Table 6).

Beutner et al., (1967) described a cross-bedded sandstone of the Pennsylvanian Kittanning Formation exposed in a strip mine near Philipsburg, Pennsylvania. Such features as inclined sigmoidal sedimentation units and the internal arrangement and scale of sedimentary structures led these investigators to propose a point bar origin for this sandstone body. Six epsilon cross-stratification units from this exposure have been measured and sampled. Average height of the epsilon cross-stratification units is 6.6 meters and width is 24.4 meters. Estimates of Sc and Sb from this section analysis yield values of 13.8 and 100%, respectively. These measurements form the basic input data for the calculations of stream characteristics in Table 6.

The upper Caseyville Pounds Sandstone of Lower Pennsylvanian age is exposed in a roadcut on Interstate 24 north of Vienna in southern Illinois. The sequence of sandstones, siltstones, and claystones described by Ethridge et al., (1973) is interpreted as a progradational delta sequence. The upper-most unit in the section contains trough cross-bedded sandstones interbedded with siltstones which fill channel shaped bodies. This unit has been interpreted as a series of distributary channel environments. Channel development was apparently contemporaneous, in part, with laterally equivalent overbank and crevasse splay deposits. Lateral migration of some channels is indicated by local development of point bar deposits as evidenced by epsilon cross-stratification units. A single point bar surface was traced for a lateral distance of 35.4 meteres before it terminated against an abandoned channel fill deposit. This point bar deposit has a thickness of 3.9 meters. Values for Sc and Sb determined from thin section analysis of bed and bank samples are 21 and 100%, respectively. Again, these measures form the basic input data for estimation of the stream given in Table 6.

Inspection of Table 6 reveals several interesting facts concerning the nature of the paleochannels as well as the methodology involved in making estimates of paleochannel characteristics. The largest of the six ancient streams (Table 6) is the Stephanian River of Late Carboniferous age from southern Morocco. It can best be described as a broad, shallow, moderately sinuous, mixed-load channel and was comparable in terms of discharge to the Santee River, South Carolina (Padgett and Ehrlich, 1976). Based on regional paleographic studies by V. V. Cavaroc (Padgett and Ehrlich, 1976), the estimated distance from the source uplands to the basin of deposition was about 100 km. Our estimate of stream length based on equation 24 (Table 2) suggests a much larger stream length. Extrapolation beyond the limits of the data presented by Dury (1965) may account for this difference.

Table 6. Morphologic and flow characteristics of six ancient channels and the modern Murrumbidgee River in Australia. Characteristics of ancient channels are determined by the methods outlined in this paper. Ranges are not reported for Ad, Ls, and Rm because of lack of information on standard error. Symbol notations are given in Table 2.

Stream Character	Units of Measure	Sewickly River Penn.-W.V. Donaldson (1969)		Ferron River Cret.-Utah Cotter (1971)		Stephanian River Carboniferous-Morocco Padgett & Ehrlich (1976)		Grafton River Penn.-W.V. Morton & Donaldson (this vol.)		Kittanning River Penn.-Penn.		Pounds River Penn.-Illinois		Murrumbidgee River, Australia Schumm (1968)
		Method I	Method II	Method I	Method II	Method I	Method II	Method I	Method II	Method I	Method II	Method I	Method II	
Sc	%	15.		15.		5.				13.8		21.		
Sb	%	100.		100.		100.				100.		100.		
W	m	97.53		91.44		192.8		75.		36.7		53.08		67.
D	m	5.45		5.94		3.25(?)		5.85		4.3		2.52		6.4
M	%	23.5		24.78		8.10		15.		30.17		27.84		25.
Type of sed. load	—	Suspended-load ch.		Suspended-load ch.		Mixed-load channel		Mixed load channel		Suspended-load ch.		Suspended-load ch.		Suspended-load ch.
% of total load as bed load	%	2.3		2.22		6.79		3.67		1.82		1.98		10.
F (Av.)	—	8.41	17.9	7.96	15.38	26.6	59.3	13.7	12.8	6.43	8.53	7.02	21.10	
(Range)		3.41-20.75		3.23-19.63		10.8-65.7		5.6-33.8		2.61-15.87		2.85-17.31		
P (Av.)	—	2.07	1.61	2.1	1.67	1.59	1.16	1.85	1.76	2.20	1.96	2.16	1.54	2.
(Range)		1.58-2.71	1.22-2.11	1.6-2.75	1.28-2.19	1.21-2.08	0.89-1.52	1.41-2.42	1.34-2.3	1.68-2.89	1.50-2.57	1.65-2.83	1.17-2.01	
Qm (Av.)	cm	212.	74.	188.	75.	425.	100.	67.	57.	21.	15.8	50.7	14.0	104.
(Range)		113.-398.	30.-182.	100.-354.	30.-185.	226.-800.	40.5-246.	35.-126.	23.-140.	11.-39.	6.5-39.1	27.-95.	5.7-34.5	
Qma (Av.)	cm	1054.	546.	974.	546.	1727.	717.	503.	452.	229.	194.	411.	190.	312.
(Range)		586.-1895.	221.-1347.	542.-1751.	221.-1346.	961.-3105.	291.-1768.	279.-904.	183.-1114	127.-411.	79.-478.	228.-739.	77.-468.	
S (Av.)	m/km	.197	.309	.201	.285	.236	.494	.338	.291	.37	.398	.292	.655	.133
(Range)		.1-.388	.15-.636	.102-.395	.138-.586	.12-.465	.24-1.018	.172-.666	.141-.599	.191-.741	.193-.82	.148-.575	.318-1.349	
L (Av.)	m	1154.	1355.	1068.	1196.	3224.	4092.	1088.	947.	437.	466.	627.	971.	853.
(Range)		561.-2376.	525.-3495.	519.-2199.	464.-3085.	1566.-6637.	1586.-10556.	529.-2241.	367.-2443.	212.-899.	180.-1201.	305.-1291.	376.-2506.	
Ad*	sq. km	23310.	38850.	20720.	25900.	518000.	777000.	20720.	15540.	2330.	2850.	5700.	12950.	
Ls	km	531.	722.	495.	566.	3415.	4356.	495.	417.	133.	150.	228.	373.	
Rm	m	263.	469.	247.	437.	521.	992.	203.	351.	99.	160.	143.	240.	

*Estimated from Dury's (1965) graph and rounded to nearest 10 sq. km.

Stream characteristics of the ancient Ferron River of Cretaceous age in Utah and the Sewickly River of Pennsylvanian age in West Virginia suggest that these two inferred river systems are comparable and are intermediate in size between the Stephanian River and the other rivers analyzed (Table 6). These two ancient rivers were relatively deep, narrow, highly sinuous, suspended-load streams that are comparable in terms of average discharge to the modern Sabine River, Texas (data on the Sabine River from Padgett and Ehrlich, 1976, Fig. 6).

The three remaining Pennsylvanian-age streams were similar in terms of size and discharge to the modern Guadalupe River, Texas. Some differences exist, however, among these remaining channels. The Pounds distributary and the Kittanning are suspended-load channels whereas the Grafton is a mixed-load channel. The Kittanning and Pounds channels were similar to the modern Murrumbidgee channel in Australia in most respects with the exception of average discharge.

A review of the data in Table 6 reveals a considerable amount of agreement as well as disagreement between stream characteristics estimated by Method I and the same characteristics estimated by Method II. In general, Method I tends to underestimate width-depth ratio (F), channel gradient (S), and meander wavelength (L), and to overestimate sinuosity (P), mean annual discharge (Qm), and mean annual flood (Qma) with respect to Method II. Estimates made by the two methods show close agreement for the Kittanning and Grafton rivers and poor agreement for the Pounds distributary channel. For the Ferron and Sewickly channels there is excellent agreement between the characteristics estimated by the two methods for S, L, and Ls, and a lack of agreement for F, P, Qm, Qma, and Rm. This lack of agreement between the results obtained by the two methods may be due to errors in the determination of the input data or more likely to basic inadequacies in the data base used to determine the equations in Table 2. The differences in estimates of width-depth ratio and sinuosity obtained by the two methods also results in a change in the classification of the channel (Table 3). For example, the Sewickly, Ferron and Pounds Rivers become mixed-load streams based on the results obtained by the use of Method II and the Stephanian River becomes a bedload river.

SUMMARY AND SUGGESTIONS

Recent attempts to reconstruct the morphologic and flow charateristics of ancient rivers rely heavily on the use of several groups of empirical relationships developed for modern streams. The majority of these relations were derived from studies of stable alluvial channels in the western United States. With two exceptions these relations have not been tested on modern coastal plain streams which have a greater likelihood or preservation in the rock record. The range of values for the morphologic and flow characteristics of the modern streams is restricted and does not include data from large rivers. Because of these limitations, existing empirical relationships must be tested on modern coastal plain rivers. In addition, studies need to be conducted on the changing morphology and sediment character of aggrading channels, and further work is needed on the effect of sediment load on channel morphology. Finally, the application of this methodology to the study of ancient rivers has suffered because of a lack of consistency in determining the basic input variables and the choice of relations.

Despite these problems, geomorphologists have provided sedimentologists with a potentially powerful method for obtaining reasonably accurate estimates of the characteristics of ancient river systems. Therefore, we have presented two methods of obtaining estimates of morphology and hydrology of ancient channels. The first method requires data on width, depth and percent silt-clay from inferred paleochannels, whereas the second method only requires width and depth measurements. Both methods find their principal application in the study of ancient point bar deposits. The methods are not readily applicable to the study of braided stream deposits because of the difficulty of obtaining estimates of channel width. An alternate approach to the study of these deposits has recently been suggested by Miall (1976).

Application of the approach presented here to six previously described ancient channels show, as expected, considerable variability in the estimated results. We believe that the methodology can provide useful information about paleochannel morphology and hydrology, but great care must be exercised in the application of the results. In view of the results of the analysis, Method II utilizing channel dimensions is recommended. It is the easiest method to use and is one that, in most instances, will provide a reasonable estimate of paleochannel morphology and hydrology.

ACKNOWLEDGMENTS

We wish the thank David Houseknecht who provided measurements of epsilon cross-stratification units and samples of inferred bed and bank material from the Kittanning Formation in Pennsylvania. Robert Morton kindly provided a copy of his unpublished manuscript from which data on the ancient Grafton River was obtained.

REFERENCES

Adams, John, 1977, Sieve size statistics from grain measurements: J. Geol., v. 85, p. 209-227.

Allen, D. R., and Chilingarian, G. V., 1975, Mechanics of sand compaction: *in* G. V. Chilingarian and K. H. Wolf, Compaction of coarse-grained sediments, I; Elsevier, N.Y., p. 43-77.

Allen, J. R. L., 1965, The sedimentology and paleogeography of the Old Red Sandstone of Anglesey, North Wales: Yorks. Geol. Soc. Proc., v. 35, p. 139-185.

————, 1968, Current ripples, their relation to patterns of water and sediment motion: North Holland Pub. Co., Amsterdam, 433p.

Baker, V. R., 1974, Paleohydraulic interpretation of Quaternary alluvium near Golden, Colorado: Quaternary Research, v. 41, p. 94-112.

Barnes, H. H., 1967, Roughness characteristics of natural channels: U. S. Geol. Survey Water Supply Paper 1849, 213p.

BeMent, W. O., 1976, Sedimentological aspects of middle Carboniferous sandstones on the Cumberland overthrust Sheet: Unpubl. Ph.D. thesis, Univ. of Cincinnati.

Berg, R. R., 1968, Point-bar origin of Fall River Sandstone reservoirs, northeastern Wyoming: Am. Asso. Petrol. Geol. Bull., v. 52, p. 2116-2122.

Beutner, E. C., Flueckinger, L. A., and Gard, T. M., 1967, Bedding geometry in a Pennsylvanian channel sandstone: Geol. Soc. Am. Bull., v. 78, p. 911-916.

Carlston, C. W., 1965, The relation of free meander geometry to stream discharge and its geomorphic implications: Am. J. Sci., v. 263, p. 864-885.

Cheetham, G. H., 1976, Palaeohydrological investigations of river terrace gravels: *in* D. A. Davidson and M. L. Shackley, *eds.,* Geoarchaeology, Earth Science and the Past; Westview Press Inc., Boulder, CO, p. 335-344.

Chilingarian, G. V., and Wolf, K. H., 1975, Compaction of coarse-grained sediments, I: Elsevier, N.Y., 552p.

————, and ————, 1976, Compaction of coarse-grained sediments, II: Elsevier, N.Y., 808p.

————, and ————, and Allen, D. R., 1975, Introduction: *in* G. V. Chilingarian, and K. H. Wolf, Compaction of coarse-grained sediments, I; Elsevier, N.Y., p. 1-42

Clague, J. J., 1975, Sedimentology and paleohydrology of Late Wisconsinan outwash, Rocky Mt. Trench, Southeastern British Columbia: *in* A. V. Jopling, and B. C. McDonald, *eds.,* Glacio-fluvial and glaciolacustrine sedimentation; Soc. Econ. Paleont. and Mineral. Spec. Pub. 23, p. 223-237.

Cotter, Edward, 1971, Paleoflow characteristics of a Late Cretaceous river in Utah from analysis of sedimentary structures in the Ferron Sandstone: J. Sediment. Petrol., v. 41, p. 129-138.

720   *FRANK G. ETHRIDGE and STANLEY A. SCHUMM*

Derr, M. E., 1974, Sedimentary structure and depositional environment of paleochannels in the Jurassic Morrison Formation near Green River, Utah: Brigham Young University Geology Studies, p. 1-39.

Donaldson, A. C., 1969, Paleostream analysis: *in* A. C. Donaldson, *ed.,* Some Appalachian coals and carbonates: Models of ancient shallow-water deposition: Pre-convention guidebook; Geol. Soc. Am., Geol. Dept. West Virginia Univ. and West Virginia Geol. and Economic Survey, Morgantown, Pa., p. 265-277.

Dury, G. H., 1964, Principles of underfit streams: U. S. Geol. Survey Prof. Paper 452-A, 67p.

———, 1965, Theoretical implications of underfit streams: U.S. Geol. Survey Prof. Paper 452-C, 43p.

———, 1976, Discharge prediction, present and former, from channel dimensions: J. Hydrology, v. 20, p. 219-245.

Eicher, D. L., 1969, Paleobathymetry of cretaceous Greenhorn sea in eastern Colorado: Am. Assoc. Petrol. Geol. Bull., v. 53, p. 1075-1090.

Elliott, T., 1976, The morphology magnitude and regime of a Carboniferous fluvial-distributary channel: J. Sediment. Petrol., v. 46, p. 70-76.

Ethridge, F. G., Fraunfelter, George and Utgaard, John, 1973, Depositional environments of selected Lower Pennsylvanian and Upper Mississippian: Field Conference, Southern Ill. Univ. 158p.

Fatt, I., 1958, Compressibility of sandstones at low to moderate pressure: Am. Assoc. Pet. Geol. Bull., v. 42, p. 1924-1957.

Gardner, T. W., 1977, Paleohydrology, paleomorphology, and depositional environments of fluvial sandstones of Pennsylvanian age in eastern Kentucky Unpub. Ph. D. thesis, Univ. of Cincinnatti.

Hack, J. T., 1957, Studies of longitudinal stream profiles in Virginia and Maryland: U. S. Geol. Survey Prof. Paper 294-B, p. 45-97.

Hedman, E. R., and Kastner, W. M., 1977, Streamflow characteristics related to channel geometry in the Missouri River basin: U. S. Geol. Survey J. Res., v. 5, p. 285-300.

Ingles, O. H., and Grant, K., 1975, The effect of compaction on various properties of coarse-grained sediments: *in* G. V. Chilingarian and K. H. Wolf, Compaction of coarse-grained sediments, I; Elsevier N.Y., p. 293-348.

Jopling, A. V., 1966, Some principles and techniques used in reconstructing the hydraulic parameters of a paleo-flow regime: J. Sediment. Petrol., v. 36, p. 5-49.

Khan, H. R., 1971, Laboratory study of alluvial river morphology: Unpub. Ph. D. thesis, Colorado State Univ., 189p.

Leeder, M. R., 1973, Fluviatile fining upward cycles and the magnitude of paleochannels: Geol. Mag., v. 110, p. 265-276.

Leopold, L. B., and Miller, J. P., 1956, Ephemeral streams — hydraulic factors and their relation to the drainage net: U. S. Geol. Survey Prof. Paper 282-A, p. 1-37.

———, and Wolman, M. G., 1957, River channel patterns: braided, meandering, and straight: U. S. Geol. Survey Prof. Paper No. 282-B, p. 39-85.

———, and ———, 1960, River meanders: Geol. Soc. Am. Bull., v. 71, p. 769-794.

———, and ———, and Miller, J. P., 1964, Fluvial processes in geomorphology: W. H. Freeman and Co., San Francisco and London, 522p.

Melton, M. A., 1961, Discussion: The effect of sediment type on the shape and stratification of some modern fluvial deposits: Am. J. Sci., v. 259, p. 231-233.

Miall, A. D., 1976, Paleocurrent and paleohydrologic analysis of some vertical profiles through a Cretaceous braided stream deposit, Banks Island, Arctic Canada: Sedimentology, v. 23, p. 459-483.

Moody-Stuart, M., 1966, High— and low— sinuosity stream deposits, with examples from the Devonian of Spitsbergen: J. Sediment. Petrol., v. 36, p. 1102-1117.

Morton, R. A., and Donaldson, A. C., 1978, Hydrology, morphology, and sedimentology of the Guadalupe fluvial-deltaic system: Geol. Soc. Am. Bull. (in press).

Nami, M., 1976, An exhumed meander belt from Yorkshire, England: Geol. Mag., v. 113, p. 47-52.

Padgett, G. V., and Ehrlich, Robert, 1976, Paleohydrologic analysis of a late Carboniferous fluvial system, southern Morocco: Geol. Soc. Am. Bull., v. 87, p. 1101-1104.

Puigdefabregas, Cayo, 1973, Miocene point bar deposits in the Ebro Basin, northern Spain: Sedimentology, v. 20, p. 133-144.

Riggs, H. C., 1968, Techniques of water resources investigations of the United States Geological Survey, Book 4, Chapter A1: Some statistical tools in hydrology: U. S. Geol. Survey, 39p.

———, 1976, A simplified slope-area method for estimating fluvial discharge in natural channels: U. S. Geol. Survey J. Res., v. 4, p. 285-291.

Saucier, R. T., and Fleetwood, A. R., 1970a, Origin and chronologic significance of late Quaternary terraces, Ouachita River, Arkansas and Louisiana: Geol. Soc. Am. Bull., v. 81, p. 869-890.

———, and ———, 1970b, Origin and chronologic significance of late Quaternary terraces, Ouachita River, Arkansas and Louisana: empirical equation correction: Geol. Soc. Am. Bull., v. 81, p. 3155-3156.

Schumm, S. A., 1960a, The shape of alluvial channels in relation to sediment type: U. S. Geol. Survey Prof. Paper 352-B, 30p.

———, 1960b, The effect of sediment type on the shape and stratification of some modern fluvial deposits: Am. J. Sci., v. 258, p. 177-184.

———, 1961, A Reply: The effect of sediment type on the shape and stratification of modern fluvial deposits: Am. J. Sci., v. 259, p. 234-239.

———, 1963, A tentative classification of alluvial river channels: U. S. Geo. Survey Circ. 477, 10p.

———, 1968, River adjustment to altered hydrologic regimen — Murrumbidgee River and paleochannels, Australia: U. S. Geol. Survey Prof. Paper 598, 65p.

———, 1969, River metamorphosis: J. Hydraulics Div., Am. Soc. Civil Eng., v. 95 (HY1) Proc. Paper 6352, p. 255-273.

———, 1972, Fluvial paleochannels: in J. K. Rigby and Wm. K. Hamblin, eds., Recognition of ancient sedimentary environments; Soc. Econ. Paleont. Mineral., Spec. Pub. 16, p. 98-107.

Wolf, K. H., and Chilingarian, G. V., 1975, Diagenesis of sandstones and compaction: in G. V. Chilingarian and K. H. Wolf, Compaction of coarse-grained sediments, II; Elsevier, N.Y., p. 69-444.

PALAEOHYDRAULIC INTERPRETATION USING MATHEMATICAL MODELS OF CONTEMPORARY FLOW AND SEDIMENTATION IN MEANDERING CHANNELS

JOHN S. BRIDGE[1]

ABSTRACT

A mathematical model to calculate steady, nonuniform flow, bed topography, bed sediment size and internal structure in open channel bends is briefly described and evaluated. The model is used to predict the three-dimensional facies distributions of single point bar deposits arising from lateral deposition at bankfull stage. The results show the expected diversity of lateral deposits (as found in ancient sequences) formed under different physiographic conditions, and also the variability in thickness, grainsize and internal structure of lateral deposits from a single point bar. It follows that single vertical sections through 'fining-upward' coarse members cannot necessarily be considered representative of a whole point bar deposit, or of deposition in a channel reach containing several point bars. Guidelines are given for obtaining quantitative estimates of the geometry and hydraulics of ancient curved channels using the model, and other published techniques of palaeohydraulic reconstruction are critically appraised in the light of the method presented here.

INTRODUCTION

The well known fluviatile fining-upward model has been widely used to explain ancient sedimentary sequences in terms of meandering river deposition. Allen (1970a, b) showed theoretically, from a consideration of the interaction between (fully developed, uniform and steady) flow and sediment transport over idealised point bar surfaces, that mean grain size should decrease from the talweg to the inner bank of curved channels, and that equilibrium bed configuration (hence internal sedimentary structure) should also vary systematically. Hence the different varieties of vertical patterning in coarse (sand) members of fining upwards sequences, produced by lateral channel migration at bankfull stage, could be explained by generation in differing hydraulic and physiographic settings. Unfortunately the power of Allen's model in reconstruction of palaeochannel geometry and hydraulics, by mathematically simulating observed fining-upward sequences, has not been generally realised (but *see* Bridge, 1974, 1975).

One assumption of the qualitative and quantitative fining-upward models for meandering channels is that the pattern of flow, sediment texture and structure in a given river cross-section is constant around the length of the bend. It has been realised for many years, however, that flow in bends is nonuniform, and recent empirical sedimentation studies have confirmed the expected nonuniformity of sediment properties in the downstream direction (*see* Bridge, 1977b for references). Complete vertical sequences through a single lateral deposit are thus expected to vary with streamwise position.

To give a generalised physical basis to the limited field observations to date it is necessary to account theoretically for the interaction between nonuniform flow, sedimentation and bed topography in curved alluvial channels. This has been made possible recently using a theoretical approximation to flow and sediment transport·in channels of idealised planform (Engelund, 1974a, b). This theory has been incorporated in a model to predict equilibrium grain size and bed configuration (hence internal sedimentary structure) for steady (bankfull) flow in regular noncircular bends, hence the three-dimensional facies distribution of single lateral deposits. The model is briefly described here (full details in Bridge, 1977a, b) and its use in simulating (hence interpreting) ancient fluviatile coarse-members is examined.

---

[1]Geology Department, Queen's University, Belfast, N. Ireland, BT7 1NN

Model of Flow and Sedimentation in Curved Channels

The model begins in the usual way by considering the representative motion of single sedimentary particles at the bed. Due to helical flow in bends the bed shear stress vector deviates by an angle δ from the mean flow direction (Fig. 1), thus tending to move particles inwards. With steady flow the net rate of transverse sediment transport must be zero or a stationary bed cannot be attained; the channel cross-profile actually migrates laterally but this is small compared with the speed of bed load particles, and the assumption of a stationary bed is justified. Thus to counterbalance the inward transport due to helical flow a transverse bed slope develops (a point bar), and a sediment particle will, on average, travel parallel to the mean flow direction.

By consideration of the balance of fluid, frictional and gravity forces acting on bed load particles in the transverse (z) and mean flow (x or longitudinal) directions, and taking the planform of the curved channel as a sine-generated curve (Langbein and Leopold, 1966), an expression for bed topography is given as

$$\frac{y}{y_c} = \left(1 + \frac{z}{R}\cos\frac{2\pi x}{M}\right)^{11\tan\phi} \tag{1}$$

The system of orthogonal, curvilinear coordinates used in equation (1) is shown in Fig. 1: y is local flow depth; y_c is centreline depth (lying midway between talweg and inner bank, and approximately equal to mean depth, y_m); x is distance along channel centreline in the flow direction; z is perpendicular to x, lying in the horizontal plane, measured positively to the left of the flow direction; M is centreline channel length in one wavelength; R is the minimum value of the centreline bend radius, at the bend axis; $\tan\phi$ is the dynamic friction coefficient due to collisions with other grains.

Assumptions made in the development of equation (1) are; (i) longitudinal bed slopes are negligible; (ii) helical flow develops quickly from the straight 'crossover' section, such that δ can be described by $\tan\delta = 11y/r$, where r is local radius of curvature; (iii) minor alteration to bed topography, arising from erosion and deposition due to changes in local flow velocity in the longitudinal direction, can be neglected; (iv) suspended sediment transport is not appreciable (but *see* Engelund, 1976 for a brief examination of the effects of appreciable suspended load); (v) areas close to banks, where vertical flow components are appreciable, are not dealt with; (vi) centrifugal force on bed load grains is negligible. Despite these approximating assumptions, equation (1) has shown good agreement with bed topography in natural and laboratory channels compared so far (Bridge, 1977b; Engelund, 1974a).

By considering the balance of forces acting on bed load particles moving over a bed now given by equation (1), using an approach similar to Allen (1970a, b), an expression for equilibrium (spherical) grain diameter, D, is

$$D = \frac{3\tau_x}{2(\sigma-\rho)g(\cos\alpha\tan\phi\cos\gamma-\sin\gamma)} \tag{2}$$

where τ_x is effective longitudinal bed shear acting on bed-load grains, σ and ρ are sediment and fluid densities respectively, and g is gravitational acceleration. The local transverse and longitudinal bed slopes, α and γ, are obtained by differentiating equation (1) with respect to z and x respectively (Bridge, 1977a, b). Equation (2) assumes availability of all sizes in the bed sediment, and that the lift force on particles is negligible.

To evaluate equation (2) it is necessary to calculate effective local shear acting on bed-load grains, and prediction of local equilibrium bed configuration requires also knowledge of the mean velocity across the width and along the length of the channel.

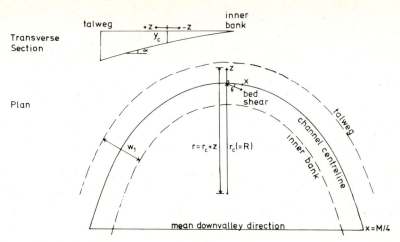

Fig. 1   Definition diagram for geometry and flow in open-channel bend.

Following Engelund (1974a) a first order approximation to mean longitudinal velocity for a vertical is

$$V = V_m(1 + u) \tag{3}$$

where V_m is mean velocity for a cross-section, occurring at the position of mean depth, and u is a small amount by which local mean velocity, V, differs from V_m, given by

$$u = z(a \; \sin\frac{2\pi x}{M} \; + \; b \; \cos\frac{2\pi x}{M}) \tag{4}$$

where a and b are functions of tan ϕ, y_m R, M and the Darcy-Weisbach friction coefficient, f.

By virtue of the relationship between bed shear stress and the square of mean velocity, a first order approximation to local bed shear can be written

$$\tau_x = \tau_{xm}(1 + 2u) \tag{5}$$

where τ_{xm} is the mean value for a cross section, occurring at the position of mean depth. Mean cross-sectional velocity and bed shear can be calculated using

$$V_m = \sqrt{\frac{8\tau_{xm}}{\rho f}} \; = \; \sqrt{\frac{8gy_m S_m}{f}} \tag{6}$$

where S_m is longitudinal water-surface slope along the locus of mean depth ($\approx$ channel centreline). A constant average value of S_m for a bend can only be assumed if flows are not strongly nonuniform (e.g. Yen and Yen, 1971, p. 317). It is important that values of S_m and f in equation (6) are those which account only for effective shear acting on bed load grains themselves if defining bed shear for use in equation (2). If *actual* water surface slope is used in equation (6), additional head losses due to various types of secondary flows would be accounted for (e.g. in the case of rippled and duned beds) and 'total' shear would be calculated instead of 'effective' shear (*see* Bridge, 1977b, p. 414-415, for further discussion of this point). Comparison of results from equations (3) and (5) with natural and laboratory data available so far have shown encouraging agreement, predicting the fundamental shift of mean velocity and bed shear maxima from the inner to the outer side of the channel with distance around a bend.

It should now be possible to predict local equilibrium bed configuration (hence sedimentary structure) throughout a bend using one of the hydraulic stability schemes proposed. Such prediction should also afford a valuable check on the validity of f. Two easily used and acceptably based schemes are the depth-velocity-grainsize diagrams of Southard (1971, 1975) and the stream power - grain size diagram of Allen (1970c) with modifications to account for upper regime bed configurations (Allen, 1970a, b). Southard's attractive scheme is at present limited to depths less than about half a metre. Although extrapolation of Southard's bed-state boundaries to greater depths is generally supported by the field data of Boothroyd and Hubbard (1974, 1975), Jackson (1976a) and Rubin and McCulloch (1976), many more field studies are required before agreement is reached on location of bed state boundaries (*see* Fig. 2), or before a comprehensive range of grain sizes is covered. For the present, therefore, Allen's (1970a, b) scheme will be mainly used, as this has support (at least for lower regime states) from field data (*see* Bridge and Jarvis, 1976; Smith, 1971b) as well as laboratory data.

The lower part of the dune existence field in the stream power-size diagram has been subdivided and designated 'diminished dunes' (Bridge and Jarvis, 1976; Smith, 1971b); these forms have well defined avalanche faces, amplitudes less than about 5 cm in the rivers in which they were studied, and wavelength height ratios greater than about 20. From a geometrical and hydraulic viewpoint they appear to be the same bed state as the straight crested 'sandwaves' of Southard (1975) and Boothroyd and Hubbard (1974, 1975), the 'bars' of Costello (1974), the 'intermediate flattened dunes' of Pratt (1973) and Pratt and Smith (1972), the 'sand bars' of Yalin (1972, p. 247), and some of the 'flat dunes' of Znamenskaya (1963). Such features have also been recorded in the flume experiments of Guy *et al.* (1966), Southard and Boguchwal (1973), and Williams (1967). Costello (1974) and Jackson (1976a) suggest also that the large 'transverse bars' in rivers also fall into this category. The wavelength/height ratios of dunes show maxima at the lowest and highest values of dimensionless bed shear covering the dune state (Fredsoe, 1975; Yalin, 1972, p. 248), leading Allen (1976) to the conclusion that diminished dunes (and washed-out dunes) form a continuum with 'normal' dunes but are relatively flat because they are generated near one or other limit of the dune state. A noteworthy problem when attempting to construct bed state stability diagrams from field data (apart from disequilibrium effects) is the recognition and differentiation of ripples, lower regime plane beds, and the smaller relatively flat dunes using standard sounding equipment (Jackson, 1976a, p. 607, 613, 616).

It is generally accepted that lower-regime plane beds, ripples, dunes and upper-regime plane beds give rise, respectively, to 'horizontal' stratification, smallscale crossstratification, largescale cross-stratification, and 'horizontal' stratification (Harms, 1975). One of the reasons for distinguishing the lower part of the dune field as a separate unit is that these straight crested forms appear to give rise to planar cross-strata (Harms, 1975; Harms and Fahnestock, 1965; Jackson, 1976a; Smith, 1970, 1972) as opposed to the trough-cross-strata of the more three-dimensional dunes. This simple relationship is complicated by the formation of horizontal stratification by migration of relatively flat bedwaves (ripples as well as dunes) whilst close to a transition to one of the plane bed states (Bridge and Jarvis, 1976; Coleman, 1969; Einstein and Chien, 1953; Jopling, 1964; McBride et. al. 1975; Smith, 1971a). Bearing this in mind, and also that there is no clear cut distinction between straight-crested and more three-dimensional dunes, bed configurations from the lower part of the dune field will be assumed to form largescale planar cross-stratification in this model.

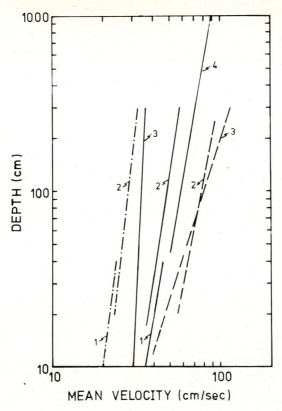

Fig. 2 Velocity-depth diagram showing boundaries between equilibrium bed states, as defined by various authors. 1 = Southard (1975), 0.45 - 0.54 mm sand; 2 = Boothroyd and Hubbard (1975), 0.35 - 0.4 mm sand; 3 = Costello (1974), 0.49 - 0.51 mm sand; 4 = Jackson (1976a), 0.4 - 0.53 mm sand.— . — . — . No sediment movement/ripple; ——— ripple/diminished dunes (and dunes); — — diminished dunes/dunes.

The model proposed herein considers equilibrium grainsize and bed configuration under steady flow conditions with all sediment sizes in motion over a stationary bed. As most sediment transport activity occurs at or around bankfull stage it is natural to use the model to predict sedimentation at bankfull flow conditions. In reality, of course, it is necessary to consider the possibility of bed configurations in disequilibrium with unsteady flows. For instance, the superimposition of smallscale active bed configurations on the surface of larger bed configurations migrating in disequilibrium can have a marked effect on the nature of (preserved?) internal structures (e.g. Bluck, 1971; Collinson, 1970; Jackson, 1976a; Jones, 1977; McCabe and Jones, 1977; Smith, 1972).

The occurrence of scroll bars (Jackson, 1976a) or zones of flow separation (e.g. Leeder and Bridges, 1975) at the downstream inner bank parts of curved channel cannot be simulated in this model. The occasional prediction of negative bed shear stresses in this region is probably due to the first order approximation used in obtaining equation (5) from equation (3).

Bearing in mind the assumptions made in this model, and that much more research is required before it can be tested rigorously and improved (Bridge, 1977b) its performance will be illustrated by simulating sedimentation under a set of different input conditions.

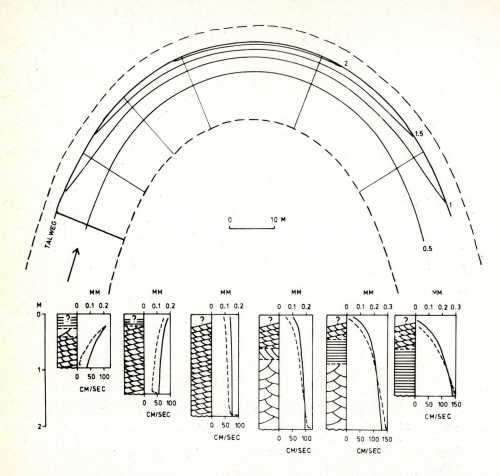

Fig. 3a   Predicted vertical variation of sedimentary structure, mean grain size (dashed lines) and mean flow velocity (solid lines) from various positions along curved channels. Vertical sections correspond to channel cross sections on map; from left to right, sections are progressively further downstream. On the map, approximate positions of inner and cut bank are shown (dashed lines). Bed topography contours are in metres below bankfull level. Input parameters to experiments are shown in table 1. Experiment a is shown above.

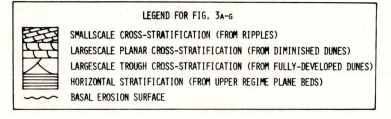

LEGEND FOR FIG. 3A-G

SMALLSCALE CROSS-STRATIFICATION (FROM RIPPLES)
LARGESCALE PLANAR CROSS-STRATIFICATION (FROM DIMINISHED DUNES)
LARGESCALE TROUGH CROSS-STRATIFICATION (FROM FULLY-DEVELOPED DUNES)
HORIZONTAL STRATIFICATION (FROM UPPER REGIME PLANE BEDS)
BASAL EROSION SURFACE

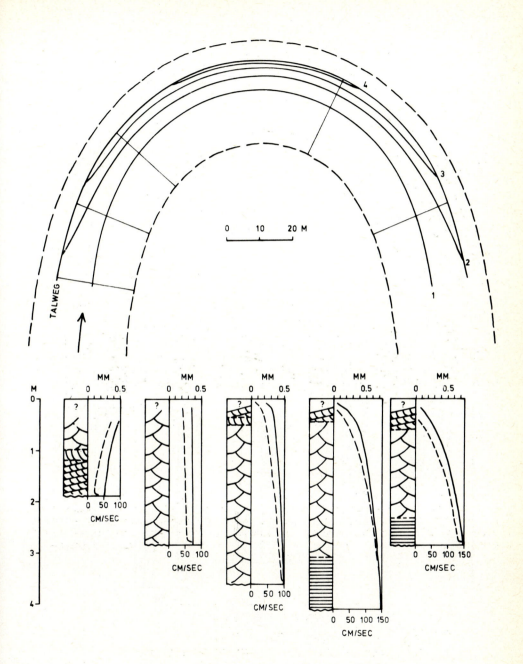

Fig. 3b Experiment b.

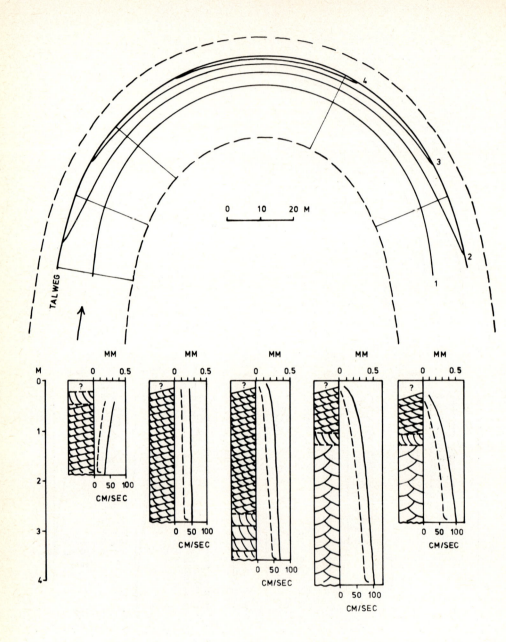

Fig. 3c Experiment c.

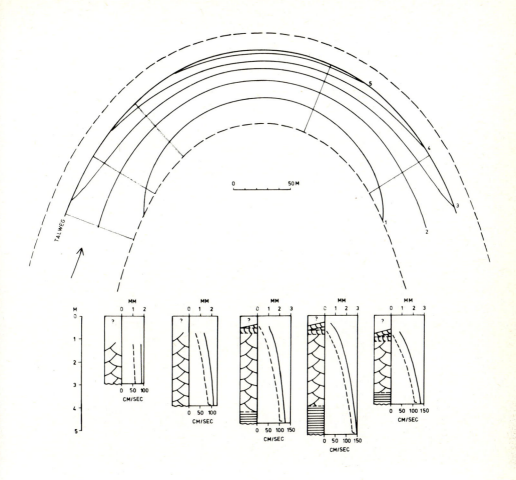

Fig. 3d   Experiment d.

SEDIMENTARY SEQUENCES GENERATED BY THE MODEL

A full examination of the behaviour of the model using ranges of different values of inputvariables in all possible combinations is not feasible. Furthermore, due to the complex interaction of the variables concerned, it is not generally possible to systematically study the effect of a single variable whilst all others are held constant. Instead, representative input conditions were chosen such that the combination of hydraulic and geometric variables for a given channel bend are consistent with those found in nature (see next section). Table 1 shows full details of the input conditions used, and Fig. 3 shows planforms of the simulated bends, and vertical profiles of grainsize, internal structure and mean flow velocity for various positions on each bend. It has been assumed that lateral migration (combined with some form of channel abandonment) can

TABLE 1 — Input used in the experiments

Experiment:	a	b	c	d	e	f	g
Mean depth (m)	0.5	1.0	1.0	2.0	2.0	2.0	5.4 - 4.0[1]
Bankfull channel width measured from talweg to inner bank (m)	16.0	24.0	24.0	64.0	64.0	120.0	200.0
Mean centreline 'water surface' slope	0.0002	0.0002	0.0001	0.0003	0.0001	0.0001	0.0001
Centreline channel length in one wavelength (m)	250.0	400.0	400.0	1005.0	1005.0	1885.0	3200.0
Minimum centreline radius of curvature (m)	31.3	43.4	43.4	125.9	125.9	334.0	401.0
Centreline sinuosity	1.5	1.8	1.8	1.5	1.5	1.2	1.5
Dynamic friction coefficient	0.85	0.8	0.8	0.5	0.65	0.65	0.55
Mean Darcy-Weisbach 'friction' coefficient	0.015	0.03	0.03	0.06	0.04	0.04	0.04

In all experiments gravitational acceleration is 981 cm/sec^2, fluid density is 1 gm/cm^3, sediment density is 2.65 gm/cm^3.
[1]Mean depth decreases from 5.4 m at the crossovers to 4.0 m at the bend axis.

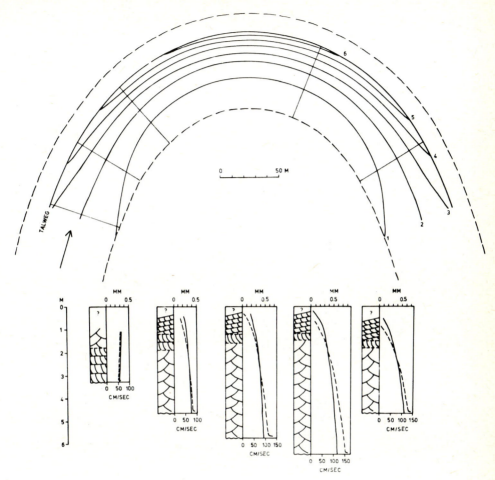

Fig. 3e Experiment e.

preserve all of the profiles shown, albeit in varying proportions (*see*, for example, Bluck, 1971; Jackson, 1976b).

It is well known that models of this kind can simulate the various different types of fining-upward coarse-members reported from ancient fluviatile sediments in different geographic locations (Allen, 1970a, b). These results, however, point (quite predictably) to the variability in thickness, grain size and sedimentary structure patterns expected within the lateral deposit of a *single* bend. Bed shear stress (hence velocity, as f is constant) and texture patterns predictably follow each other closely, giving rise to 'transitional' depositional facies at bend entrances (and sometimes at exits) grading downchannel into 'fully developed' depositional facies (*see* Bridge and Jarvis, 1976; Jackson, 1976b). It *may* seem that the 'intermediate' facies of Jackson (1976b) is defined because the measured flow pattern is not that responsible for measured grainsize pattern. In reality, although maximum mean velocity may be over the talweg (i.e. fully developed), maximum effective bed shear stress (hence grainsize) may not be in this position as well, i.e. if f is not constant as asumed here. Variation of effective f may result

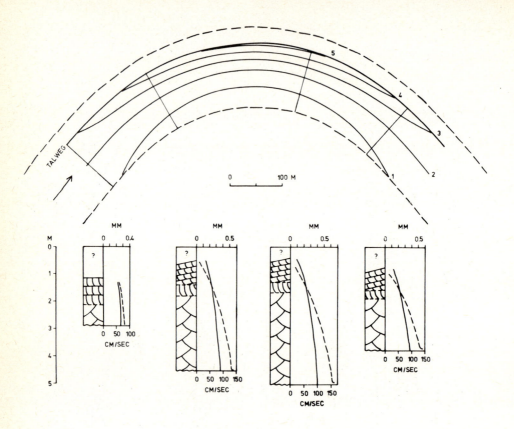

Fig. 3f Experiment f.

from bed configuration variation in a given section; however, it cannot be stated at present which is the most likely reason for the absence of an intermediate facies.

More specifically the model indicates that as S_m and y_m increases, and tan ϕ decreases, mean grainsize for a cross section increases (compare Figs. 3a or 3c with 3d). Also the *range* of grainsize across a fully developed flow section increases with decreasing r_c/w_1 (r_c is local centreline radius of curvature; w_1 is width between talweg and inner bank) but with only a weak dependence on tan ϕ (for small $w_1/2r_c$). Compare, for example, Figs. 3e and 3f. Therefore variability of vertical grainsize profiles between different sections across one point bar depends considerably on minimum r_c/w_1 in the fully developed zone because at bend entrances there are always vertical profiles that have constant (and coarsening upward) grain size. Longer transitional zones will naturally give rise to more prominent coarsening upward sequences. The relative length of the transitional zones appear to be increased by decreasing f and increasing y_m for a given bend, however an analysis of *all* variables involved has not been attempted.

Generalisations about patterns of equilibrium bed configuration around bends are less easy to make, due to the complex interaction of bed configuration, depth, slope, friction factor and grainsize. However it is expected that the average bed configuration for a cross

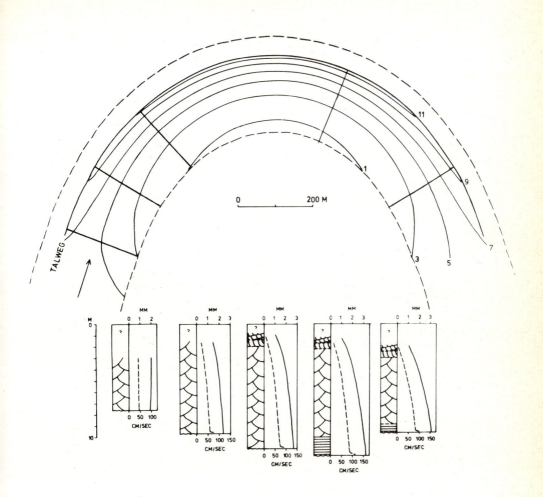

Fig. 3g Experiment g. Note that the results of this experiment are an approximation to the flow and sedimentation in the Lower Wabash river (e.g. Jackson, 1976b).

section will gradually change from upper to progressively lower regime types as y_m and/or *actual* S_m decreases. Upper regime bedforms may form at some of the lower values of y_m and/or S_m and higher $\tan\phi$, partly due to the accompanying decrease in grainsize. Where lower flow regime configurations are dominant in a bend, higher powered configurations are found in the deeper parts of the developed flow region and near the inner bank at the bend entrance, whilst lower-powered configurations are found along the inner, shallow side of the developed flow zone and near the bend entrance just downstream of the inflection point (e.g. Fig. 3c, e). The model does not predict an increase in upper-regime plane beds as r_c increases as suggested by Allen (1970a, b).

A major outcome of the experiments is that single vertical sections of coarse members attributable to meandering river deposition cannot necessarily be thought of as representative of deposition in a single point bar, in substantial support of field investigations by Bluck (1971), Bridge and Jarvis (1976), Jackson (1976b). Furthermore, as geometry (e.g. sinuosity) of individual bends in a given reach is expected to vary, a degree of variability in vertical profiles is expected between the transitional and fully developed facies of identical stratigraphic position but from different point bars. To properly assess the variability of lateral deposits three-dimensional exposure must be available, or at least a number of stratigraphically equivalent coarse members in close juxtaposition.

GUIDELINES FOR PALAEOHYDRAULIC/GEOMETRIC RECONSTRUCTION USING THE MODEL

If the input to the model can be chosen such that simulated lateral deposits compare with suspected ancient lateral deposits, then a quantitative estimate of the geometry and hydraulics of ancient curved channels can be obtained. Because of the number of input parameters required (see Table 1 and Bridge, 1977a) it may seem that there is no single unequivocal set of geometric and hydraulic conditions that can simulate one lateral deposit (see Allen, 1970b, p. 319). It is therefore important to define as much input information as possible from fluvial coarse-members themselves.

At first the thickest part of a single complete coarse member should be found; according to the model this should show fining-upward of grainsize and corresponds to a deposit from near the bend axis where talweg depth (= coarse member thickness) is at a maximum for the bend. Mean depth, y_m, may be taken as half to one third of the talweg depth initially, and may also be assumed approximately constant around the bend.

Bankfull width may be estimated quite accurately if epsilon cross-stratification (representing transverse profiles of point bars) is present. The horizontal width of these strata equals the distance between inner bank and talweg, w_1, as long as they are measured *perpendicular to local flow direction*. Full bankfull width w, may then be estimated from w = kw_1 where k varies from 0.75 to 0.9 according to Allen (1970a). In the absence of epsilon cross-stratification bankfull width may be estimated using empirical data cited in Leeder (1973).

Values of M and sinuosity, sn (hence R) can sometimes be obtained directly from aerial photographs (e.g. Nami, 1976; Padgett and Ehrlich, 1976; Schumm, 1968) but in most cases they must be estimated. M may readily be obtained from M $\approx 4\pi$w (Hey, 1976), and sn can be estimated from palaeocurrent azimuths as long as the reasons for their variability are understood (*see* Bluck, 1971, 1976; Ferguson, 1977; Miall, 1976). Alternatively sn may be estimated from width/(max.)depth ratio (Schumm, 1972), keeping in mind the drawbacks of this method (see later section).

Gravitational acceleration, g, is 981 cm/sec^2, and fluid and sediment densities are normally taken, respectively, as 1 gm/cm^3 and 2.65 gm/cm^3. The (constant) value of dynamic friction coefficient, tan ϕ, is obtained from Bagnold (1966, fig. 4), taking mean grain diameter as that occuring about mid-way up the coarse member. (*see* Bridge, 1976). In the (unlikely) event that horizontal lamination (from upper regime plane beds) dominates the coarse member an initial estimate of tan ϕ is 0.75, which can be checked when local values of dimensionless bed shear stress and grainsize are calculated. Note that the effects of appreciable suspended sediment transport (in the case of upper regime flows) can be accounted for by multiplying the chosen tan ϕ by a positive factor greater than unity (Engelund, 1976).

The mean friction coefficient, f, defines the relationship between square of mean velocity and effective bed shear stress acting on grains. If the coarse member is dominated by horizontal lamination (from upper regime plane beds), f $\approx$ 0.02. If large or small-scale cross stratification is dominant estimation of f is more difficult (but *see* Yalin, 1972, p. 276). If largescale trough cross-stratification (from fully developed dunes) dominates the coarse member an initial estimate of f is about 0.04 - 0.06 (e.g. Bridge and

Jarvis, 1977). If tabular cross-stratification (from diminished dunes) or smallscale cross-stratification (from ripples) are dominant f can be estimated as 0.03.

It is finally required to define either S_m or V_m (i.e. at bend axis centreline; see equation (6)). These parameters are chosen by trial and error such that the model predicts the thickness, grainsize and sedimentary structure variation in *all* vertical sections through the coarse member. Adjustments may be required to some of the other input variables but these are not expected to be fundamental. It is essential at this stage to recognise features of falling stage (e.g. mud drapes, desiccation cracks) in the coarse member, which, of course, the model cannot simulate. Bankfull discharge may then be determined from the product of mean velocity and cross-sectional area if a realistic geometry is chosen for the flow section between talweg and outer bank.

COMPARISON WITH OTHER METHODS OF PALAEOHYDRAULIC/ GEOMETRIC RECONSTRUCTION

With the exception of Allen (1970a, b) and Bridge (1975) other attempts to quantify interpretation of fluviatile sedimentary deposits have been restricted to general geometrical and hydraulic properties of ancient channels based on observed geometry, sedimentary structures and grainsizes of channel sediments, and empirical or theoretical relationships from modern channels. These are discussed below.

Hydraulics

A very general hydraulic interpretation can be given in terms of flow regime (for discussion see Blatt *et al.*, 1972, p. 119-124; Walker, 1975, p. 76-77), however more specific information can be obtained by considering the velocity-depth-size or stream power-size diagrams already discussed. By associating sedimentary structures in a given sediment size with particular bed configurations the *range* of velocity-depth or stream power can be defined in which the sediment was deposited (e.g. Cotter, 1971; Friend and Moody-Stuart, 1972; Miall, 1976). The assumption embodied herein, that original bed configurations were in equilibrium with the flow, probably matters little when dealing with such large ranges of hydraulic variables.

Where preserved, bed wave dimensions (especially antidunes) may yield an estimate of local velocity and depth of flow using one of the schemes relating (equilibrium) height and length to the controlling hydraulic variables (e.g. Allen, 1970c, 1977; Fredsoe, 1975; Jopling, 1967; Shaw and Kellerhals, 1977; Yalin, 1972). Unfortunately the preserved thickness of cross-stratified sets can only indicate a minimum height for the generating bed wave because of erosional tops to foresets; their interpretative value is thus limited. Bankfull depth in the talweg of curved channels is normally taken as the thickness of the coarse (i.e. sand) -members of 'fining-upward' cycles.

Some workers have estimated the minimum (dimensionless) bed shear stress or velocity required to move sediment of a given size as bed load, or to cause its suspension (Baker, 1974; Friend and Moody-Stuart, 1972; Jopling, 1966; Miall, 1976). Such determinations seem of limited value as deposition probably occurred at bed shears or velocities much higher than those critical for entrainment (or suspension) as evidenced by the frequent occurrence of stratification indicative of appreciable sediment transport rates.

Given velocity and depth, and by assuming suitable (empirical) values for resistance coefficient, mean water surface slope has been estimated from the well known Chezy, Darcy-Weisbach, or Manning (uniform) flow equations. (Cotter, 1971; Friend and Moody-Stuart, 1972; Jopling, 1966; Padgett and Ehrlich, 1976).

In performing these operations it is essential to be aware *exactly* what parameter is being determined (and how) from the field evidence. Are velocities and depths local values, average values, or maximum values for a given channel reach or cross section?

How relevant are calculated hydraulic parameters to the channel forming (flood) flows? Are the hydraulic equations applicable? From the literature available to date it appears that these questions often remain unanswered. Certainly none of the methods mentioned can give the comprehensive three-dimensional information afforded by the model presented here, or to the two-dimensional versions for fully developed bankfull flow in bends (Allen, 1970a, b; Bridge, 1976).

Channel Geometry and Discharge

As seen in a previous section, some geometrical properties of ancient channels can be obtained directly from coarse member geometry. In the absence of direct outcrop evidence various authors have made use of Schumm's (1971, 1972) empirical regression equations which relate 'dependent' morphologic parameters (width, depth, meander wavelength, sinuosity and channel slope) to each other and to the 'independent' variables, discharge and type of sediment load (the latter supposed related to weighted percentage silt and clay in channel bed and banks).

Sinuosity, sn, and width/maximum depth ratio, w/y_{max}, were both related to weighted silt-clay percent, sc, and thus to each other. Some authors have guessed sc based on general sediment characteristics (Friend and Moody-Stuart, 1972; Miall, 1976) whereas others have attempted to define it from outcrop (Cotter, 1971; Padgett and Ehrlich, 1976; Schumm, 1968). To do the latter it is necessary to have a former cut bank exposed, and a lateral deposit with recognisable original silt-clay content (Leeder, 1973). The sn, w/y_{max} relationship was used by Elliott (1976), Miall (1976) and Moody-Stuart (1966). Some problems with Schumm's sn, w/y_{max}, sc relationships have been identified as (i) w/y_{max} of high sinuosity (>1.7) streams increases with stream size, thus the w/y_{max}, sn relation does not hold for larger streams (Leeder, 1973), (ii) the w/y_{max}, sc relation only appears to hold for streams with sandy, mobile beds which are neither aggrading or degrading (Riley, 1975), (iii) As sc is weighted it is a function of w/y_{max}, and thus their correlation is spurious; the correlation between unweighted percentage of silt and clay in bed and banks and w/y_{max} is not as good between sc and w/y_{max} (Melton, 1961; Riley, 1975).

Other fairly well established empirical relationships are $L \approx 11w$ (L is meander wavelength) and $r_c \approx 2.5w$ (Dury, 1976; Leopold *et al*., 1964; Richards, 1976) although Hey (1976) points out that these relationships are not general; they have been used by Leeder (1973) and Padgett and Ehrlich (1976). Leeder (1973) defined and used an empirical relation between w and y_{max} (for sn >1.7); except for his analysis it is often difficult to know what part of a curved channel w and y_{max} refer to, each varying with location on a river bend.

Schumm (1971, 1972) also presented empirical relations between one of the variables depth (maximum), width, slope, wavelength and the two 'independent' variables, sc and discharge. Thus, given sc, a geometric parameter or a discharge measure can be estimated if the other is known (e.g. Cotter, 1971; Friend and Moody-Stuart, 1972; Leeder, 1973; Miall, 1976; Padgett and Ehrlich, 1976; Schumm, 1968). Problems involved with use of sc (or w/y_{max}) are discussed above. In this respect, empirical relationships between morphological parameters and (dominant) discharge alone cannot be general as other independent variables (e.g. sediment discharge, nature of bank materials, valley slope) exert some control, even if not the major one (see Bridge, 1973, for discussion). For this reason it is doubtful whether a single hydraulic geometry equation containing only one (or even two) independent variables can represent a single stream never mind a regional group of streams. Such equations are, however, used by Dury (1976), Leeder (1973) and Moody-Stuart (1966). Discharge can always be determined as the product of velocity and cross-sectional area.

In summary, empirical equations like those of Schumm can give an estimate of generalised channel morphology and discharge only if used within their limits of definition, and it is always useful to specify confidence limits on any estimates made

(Leeder, 1973). It must again be emphasised that compatible values of parameters are used in the various empirical equations.

CONCLUDING REMARKS

The model presented here has not been tested rigorously and a great deal of field and laboratory research is required before the various assumptions made, and hence its accuracy, can be assessed, hopefully leading to improvement of the method.

Comparison of the model with natural channel data so far are encouraging, and the experiments have shown some interesting features of coarse member variability within single bends. The model is a generalised and useable means of obtaining quantitative estimates of the geometry and hydraulics of ancient lateral deposits, but must be looked at critically if techniques of palaeoenvironmental reconstruction are to be refined..

ACKNOWLEDGEMENTS

I am most grateful to R. G. Jackson, M. R. Leeder and J. R. L. Allen who kindly criticised the initial manuscript. I am also indebted to the Fluvial Conference sponsors, The Royal Society, and the Queen's University of Belfast whose generous financial assistance made my visit to Calgary possible.

REFERENCES

Allen, J. R. L., 1970a, A quantitative model of grain size and sedimentary structure in lateral deposits: Geol. J., v. 7, p. 129-146.

———, 1970b, Studies in fluviatile sedimentation: A comparison of fining upwards cyclothems, with special reference to coarse member composition and interpretation: J. Sediment. Petrol., v. 40, p. 298-323.

———, 1970c, Physical processes of sedimentation: George Allen and Unwin, 248p.

———, 1976, Bed forms and unsteady processes: some concepts of classification and response illustrated by common one-way types: Earth Surf. Proc., v. 1, p. 361-374.

———, 1977, The plan shape of current ripples in relation to flow conditions: Sedimentology, v. 24, p. 53-62.

Bagnold, R. A., 1966, An approach to the sediment transport problem from general physics: U.S. Geol. Survey Prof. Paper 422-I, p. 1-37.

Baker, V. R., 1974, Palaeohydraulic interpretation of Quaternary alluvium near Golden, Colorado: Quaternary Research, v. 4, p. 94-112.

Blatt, H., Middleton, G. V., and Murray, R., 1972, Origin of sedimentary rocks: Prentice Hall, 576p.

Bluck, B. J., 1971, Sedimentation in the meandering River Endrick: Scott. J. Geol., v. 7, p. 93-138.

———, 1976, Sedimentation in some Scottish rivers of low sinuosity: Trans. Roy. Soc. Edinburgh, v. 69, p. 425-456.

Boothroyd, J. C. and Hubbard, D. K., 1974, Bedform development and distribution pattern, Parker and Essex Estuaries, Massachusetts: U.S. Army Corps of Engineers, Coastal Engineering Research Center, Miscellaneous Paper 1-74.

———, and ———, 1975, Genesis of bedforms in mesotidal estuaries: in L. E. Cronin, ed., Estuarine Research, Academic Press, v. 2, p. 217-234.

Bridge, J. S., 1973, Computer simulation of sedimentation in meandering streams: Unpub. Ph.D. Thesis, University of St. Andrews.

———, 1974, Mathematical model and FORTRAN IV program for computer simulation of sedimentation in meandering streams: Computer Applications, Geogr. Dept., Univ. Nottingham, v. 2, p. 217-266.

———, 1975, Computer simulation of sedimentation in meandering streams: Sedimentology, v. 22, p. 3-43.

————, 1976, Bed topography and grain size in open channel bends: Sedimentology, v. 23, p. 407-414.

————, 1977a, Mathematical model and FORTRAN IV program to predict flow, bed topography and grain size in open channel bends: Computers and Geosciences, v. 2, p. 407-416.

————, 1977b, Flow, bed topography, grain size and sedimentary structure in open channel bends: a three-dimensional model: Earth Surf. Proc., v. 2, p. 401-416.

————, and Jarvis, J., 1976, Flow and sedimentary processes in the meandering River South Esk, Glen Clova, Scotland: Earth Surf. Proc., v. 1, p. 303-336.

————, and ————, 1977, Velocity profiles and bed shear stress over various bed configurations in a river bend: Earth Surf. Proc., v. 2, p. 281-294.

Coleman, J. M., 1969, Brahmaputra River: channel processes and sedimentation: Sediment. Geol., v. 3, p. 129-239.

Collinson, J. D., 1970, Bedforms of the Tana River, Norway: Geogr. Ann., v. 52A, p. 31-56.

Costello, W. R., 1974, Development of bed configurations in coarse sands: Rept 74-1. Department of Earth and Planetary Sciences, Massachusetts Institute of Technology, Cambridge, Massachusetts.

Cotter, E., 1971, Paleoflow characteristics of a late Cretaceous river in Utah from analysis of sedimentary structures in the Ferron Sandstone: J. Sediment. Petrol., v. 41, p. 129-138.

Dury, G. H., 1976, Discharge prediction, present and former, from channel dimensions: J. Hydrology, v. 30, p. 219-245.

Einstein, H. A., and Chien, N., 1953, Transport of sediment mixtures with large ranges of grain sizes: U.S. Corps of Engineers, Missouri River Division, Sediment Series No. 2, 49p.

Elliott, T., 1976, The morphology, magnitude and regime of a Carboniferous fluvial-distributory channel: J. Sediment. Petrol., v. 46, p. 70-76.

Engelund, F., 1974a, Flow and bed topography in channel bends: J. Hydraulics Div., Am. Soc. Civil Eng., v. 100, p. 1631-1648.

————, 1974b, Experiments in curved alluvial channel: Progress Report No. 34, p. 31-36. Inst. of Hydrodynamics and Hydraulic Engineering, Technical University, Denmark.

————, 1976, Experiments in curved alluvial channel, part 2: Progress Report No. 38, p. 13-14. Inst. of Hydrodynamics and Hydraulic Engineering, Technical University, Denmark.

Ferguson, R. I., 1977, Meander sinuosity and direction variance: Geol. Soc. Am. Bull., v. 88, p. 212-214.

Fredsoe, J., 1975, The friction factor and height-length relations in flow over a dune-covered bed: Progress Report 37, p. 31-36. Inst. of Hydrodynamics and Hydraulic Engineering, Technical University, Denmark.

Friend, P. F., and Moody-Stuart, M., 1972, Sedimentation of the Wood Bay Formation (Devonian) of Spitsbergen: regional analysis of a late orogenic basin: Norsk Polarinstitutt Skrifter Nr. 157, p. 1-77.

Guy, H. P., Simons, D. B. and Richardson, E. V., 1966, Summary of alluvial channel data from flume experiments, 1956-1961: U.S. Geol. Surv. Prof. Paper 462-I, 96p.

Harms, J. C., 1975, Stratification produced by migrating bedforms: *in*: Depositional environments as interpreted from primary sedimentary structures and stratification sequences, Soc. Econ. Paleont. Mineral., Short Course No. 2, p. 45-61.

————, and Fahnestock, R. K., 1965, Stratification, bedforms and flow phenomena (with an example from the Rio Grande): *in* G. V. Middleton, *ed.*, Primary sedimentary structures and their hydrodynamic interpretation, Soc. Econ. Paleont. Mineral. Spec. Pub. 12, p. 84-115.

Hey, R. D., 1976, Geometry of river meanders: Nature, v. 262, p. 482-484.

Jackson, R. G., 1976a, Largescale ripples of the Lower Wabash River: Sedimentology, v. 23, p. 593-623.

————, 1976b, Depositional model of point bars in the Lower Wabash River: J. Sediment. Petrol., v. 46, p. 579-594.

Jones, C. M., 1977, Effects of varying discharge regimes on bed-form sedimentary structures in modern rivers: Geology, v. 5, p. 567-570.

Jopling, A. V., 1964, Interpreting the concept of the sedimentation unit: J. Sediment. Petrol., v. 34, p. 165-172.

——, 1966, Some principles and techniques used in reconstructing the hydraulic parameters of a paleo-flow regime: J. Sediment. Petrol., v. 36, p. 5-49.

——, 1967, Some techniques used in the hydraulic interpretation of fluvial and fluvio-glacial deposits: Geomorph. Symp., Univ. Guelph, Ontario, Canada, p. 93-116.

Langbein, W. B. and Leopold, L. B., 1966, River meanders — theory of minimum variance: U.S. Geol. Survey Prof. Paper 422-H, 15pp.

Leeder, M. R., 1973, Fluviatile fining-upwards cycles and the magnitude of palaeochannels: Geol. Mag., v. 110, p. 265-576.

——, and Bridges, P. H., 1975, Flow separation in meander bends: Nature, v. 253, p. 338-339.

Leopold, L. B., Wolman, M. G., and Miller, J. P., 1964, Fluvial processes in Geomorphology: Freeman, 522p.

McBride, E. F., Shepherd, R. G. and Crawley, R. A., 1975, Origin of parallel near-horizontal laminae by migration of bed forms in a small flume: J. Sediment. Petrol., v. 45, p. 132-139.

McCabe, P. J., and Jones, C. M., 1977, Formation of reactivation surfaces within superimposed deltas and bedforms: J. Sediment. Petrol., v. 47, p. 707-715.

Melton, M. A., 1961, Discussion on 'The effect of sediment on the shape and stratification of some modern fluvial deposits': Am. J. Sci., v. 259, p. 231-233.

Miall, A. D., 1976, Palaeocurrent and palaeohydrologic analysis of some vertical profiles through a Cretaceous braided stream deposit, Banks Island, Arctic Canada: Sedimentology, v. 23, p. 459-483.

Moody-Stuart, M., 1966, High- and low-sinuosity stream deposits, with examples from the Devonian of Spitsbergen: J. Sediment. Petrol., v. 36, p. 1102-1117.

Nami, M., 1976, An exhumed Jurassic meander belt from Yorkshire, England: Geol. Mag., v. 113, p. 47-52.

Padgett, G. V. and Ehrlich, R., 1976, Palaeohydrologic analysis of a late Carboniferous fluvial system, southern Morocco: Geol. Soc. Am. Bull., v. 87, p. 1101-1104.

Pratt, C. J., 1973, Bagnold approach and bed-form development: J. Hydraulics Div., Am. Soc. Civil Eng., v. 99, p. 121-137.

——, and Smith, K. V. H., 1972, Ripple and dune phases in a narrowly graded sand: J. Hydraulics Div., Am. Soc. Civil Eng., v. 98, p. 859-874.

Richards, K. S., 1976, The morphology of riffle-pool sequences: Earth Surf. Proc., v. 1, p. 71-88.

Riley, S. J., 1975, The channel shape-grain size relation in eastern Australia and some palaeohydrologic implications: Sediment. Geol., v. 14, p. 253-258.

Rubin, D. M. and McCulloch, D. S., 1976, Bedform dynamics in San Francisco Bay, California: Geol. Soc. Am. Abstracts with Programs, v. 8, p. 1079.

Schumm, S. A., 1968, River adjustment to altered hydrologic regimen — Murrumbidgee River and Paleochannels, Australia: U.S. Geol. Survey Prof. Paper 598.

——, 1971, Fluvial geomorphology: the historical perspective. in H. W. Shen, ed., River Mechanics, Fort Collins, Colorado, Water Resources Pubs., v. 1, ch. 4, p. 1-30.

——, 1972, Fluvial palaeochannels: in J. K. Rigby and W. K. Hamblin, eds., Recognition of ancient sedimentary environments, Soc. Econ. Paleont. Mineral., Spec. Pub. 16, p. 98-107.

Shaw, J. and Kellerhals, R., 1977, Paleohydraulic interpretation of antidune bedforms with applications to antidunes in gravel: J. Sediment. Petrol., v. 47, p. 257-266.

Smith, N. D., 1970, The braided stream depositional environment: comparison of the Platte River with some Silurian clastic rocks, North-Central Appalachians: Geol. Soc. Am. Bull., v. 81, p. 2993-3014.

——, 1971a, Pseudo-planar stratification produced by very low amplitude sand waves: J. Sediment. Petrol., v. 41, p. 69-73.

——, 1971b, Transverse bars and braiding in the Lower Platte River, Nebraska: Geol. Soc. Am. Bull., v. 82, p. 3407-3420.

————, 1972, Some sedimentological aspects of planar cross-stratification in a sandy braided river: J. Sediment. Petrol., v. 32, p. 624-634.

Southard, J. B., 1971, Representation of bed configurations in depth-velocity-size diagrams: J. Sediment. Petrol., v. 41, p. 903-915.

————, 1975, Bed configurations, *in*: Depositional Environments as interpreted from primary sedimentary structures and stratification sequences, Soc. Econ. Paleont. Mineral., Short Course No. 2, p. 5-43.

————, and Boguchwal, L. A., 1973, Flume experiments on the transition from ripples to lower flat bed with increasing sand size: J. Sediment. Petrol., v. 43, p. 1114-1121.

Walker, R. G., 1975, From sedimentary structures to facies models: example from fluvial environments: *in*: Depositional environments as interpreted from primary sedimentary structures and stratification sequences, Soc. Econ. Paleont. Mineral. Short Course No. 2, p. 63-79.

Williams, G. P., 1967, Flume experiments on the transport of a coarse sand: U.S. Geol. Survey Prof. Paper 562-B, p. 1-31.

Yalin, M. S., 1972, Mechanics of sediment transport: Pergamon. 290p.

Yen, C-L and Yen, B. C., 1971, Water surface configuration in channel bends: J. Hydraulics Div., Am. Soc. Civil Eng., v. 97, p. 303-321.

Znamenskaya, N. S., 1963, Experimental study of the dune movement of sediment: Soviet Hydrology, 1963, p. 253-275.

PALAEOHYDROLOGICAL RECONSTRUCTIONS FROM A HOLOCENE VALLEY FILL

MICHAEL CHURCH[1]

ABSTRACT

At the head of Ekalugad Fjord, eastern Baffin Island, a sequence of outwash deposits preserves a 6,000 year history of Holocene fluvial activity. Four depositional phases occurred:

1) deposits of sublateral drainage under glacial ice prior to 6,000 years BP (referred to as T1);

2) major outwash deposits in front of retreating ice before 4,300 years BP (T2) followed by an erosional interval;

3) aggradation correlated with Neoglacial events at 2,500 years ± BP (T3) followed by an erosional interval;

4) aggradation associated with recent Neoglacial activity (300 to 100 years BP) (T4).

Equivalent effects are found in many other east coast valleys in Baffin Island.

The T2 sediments are mainly foreset deltaic beds, deposited into the sea. Beds vary from cobbles with little matrix to fine, laminated sands. About 2,000 beds, comprising 30% of the total thickness, were sampled to determine grain size. These results form the basis for palaeoflow calculations using tractive force theory. The assumptions underlying the calculations are critically examined and several alternative calculations are made. The relative frequency distribution of the high flows is presented: discharges were about 10x larger than present-day ones, with a maximum of about 300 m^3s^{-1}, against an observed maximum of 200 m^3s^{-1}. It is concluded that present knowledge of sediment transport mechanics constrains palaeohydrological reconstructions from sediment textural information to be at best an order-of-magnitude exercise.

INTRODUCTION

A major goal of sedimentology is to be able to interpret from morphological evidence of deposits the conditions under which they were made. An important method has been to classify the materials and structures of deposits and to deduce, by comparison with presently forming sediment bodies, the environment of their occurrence (cf. reviews in Middleton, 1965). Interpretation of a sequence of beds yields qualitative information of fluctuating palaeoenvironmental conditions. Some workers have attempted to deduce the hydraulic conditions accompanying sediment transport and deposition more or less exactly from measurements of sedimentary textures or structures and use of hydraulic formulae.

In this study, a late-glacial and postglacial alluvial valley fill of Holocene age in eastern Baffin Island is examined. The fill consists of a gravel outwash deposit which has undergone several episodes of aggradation and dissection during the past 6,000 years. It is located at the head of Sarvalik Arm of Ekalugad Fjord (60°50'N., 69°20'W.: see Fig. 1) in eastern Baffin Island, and is typical of the valley alluvial fills that are the dominant postglacial depositional landscape feature in the mountainous eastern Canadian arctic.

In the first part of the paper the deposit is described, and the history of its development is deduced from morphological and stratigraphical evidence. The results are compared with regional knowledge of the Holocene environment to indicate the quality of information that may be gained from such studies. In the second part of the paper,

[1]Department of Geography, The University of British Columbia, Vancouver, British Columbian V6T 1W5.

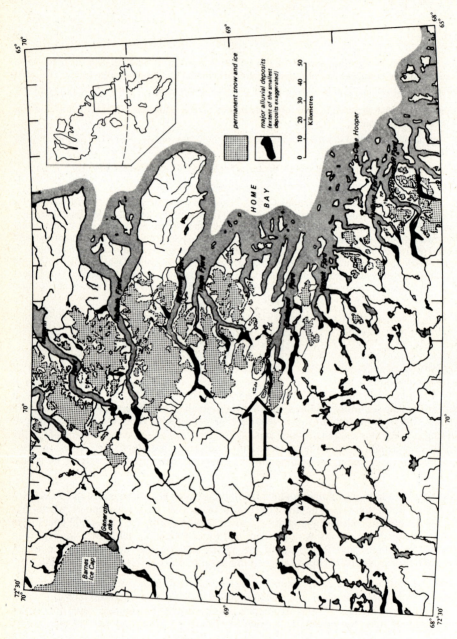

Fig. 1. Location map: Ekalugad Fjord, on the Home Bay Coast of Baffin Island, showing the major Holocene alluvial deposits in the region.

Fig. 2. View of Ekalugad Valley, showing the major units of the alluvial valley fill and the location of materials that have been absolutely dated (Table 1). Symbol "T" designates an alluvial surface; "Mw" indicates moraine (water washed).

textural information is used to make palaeoflow calculations (based on tractive force theory) for a portion of the major late-glacial outwash. The results are appraised by comparison with data of the contemporary hydrological regimen in the valley. A critique of such palaeohydrological reconstructions is then offered.

HISTORY OF THE FJORD HEAD

Deglaciation

Ekalugad Fjord is the largest of a group of fjords that runs directly ESE to WNW on the Home Bay coast of eastern Baffin Island. The mountains at Home Bay are lower than anywhere else on the east coast, and the fjord valleys are relatively broad and open. Hence, inland ice found a main line of efflux here during Pleistocene times. Ekalugad Fjord carried one of the largest of the outlet glaciers.

Andrews *et al.* (1970) described the history of deglaciation in the Home Bay area. In general, the glaciers had retreated to the fjord heads before about 7,000 years ago, but at Ekalugad Fjord the ice stood at the large moraine at the end of Tasiujaq Cove in Sarvalik Arm about 6,100 years ago (Fig. 2) (cf. Table 1 for a list of relevant radiocarbon age determinations. "Radiocarbon years" is understood in reference to all such citations). The moraine is sandy and was probably laid down underwater. On the distal side the sea stood at 54 m ± asl. By 5,700 years BP the glacier had retreated and broken into three valley tongues that stood at moraine loops at the head of Ekalugad Valley. The average rate of retreat over the 6.8 km was 17 m per annum, mainly by calving at the icefront.

Andrews (1968, 1969) worked out the regional and local pattern of isostatic rebound consequent upon the waning and disappearance of the Pleistocene ice. Immediately after removal of the ice the rate of uplift was about 2.25 m/century whereas today it has slowed down to only about 0.3 m/century.

Of some significance for studies of sedimentation is the fact that, due to the difference in the time of removal of the ice as one moves inland, postglacial uplift began later and has carried on at a consistently higher rate at sites farther inland. This implies that a net tilt up inland must occur for any projected former sea level (i.e., plane surface). Such deformation of the plane at the head of Ekalugad Fjord appears to be proceeding at the rate of 1.1 cm km^{-1}/century (Andrews, 1969). The strike of the plane is along 140° 8 320° True.

TABLE 1

RADIOCARBON DATES FROM THE HEAD OF EKALUGAD FJORD[1]

Site No.	Radiocarbon age determination[2] (years B.P.)	"Most probable date"[3]	Associated marine limit (m asl.)	Laboratory Number
1	5900 ± 260	6100	54	I-2412
2	5840 ± 300	5700	43	I-3062
3	4990 ± 350			I-2442
4	4430 ± 220	4350	23	I-2584
5	3170 ± 200			Gif-3956

1. Material for all dates except No. 5 consisted of marine mollusc shells, preserved in deltaic bedding. Details of site and collection are given in Andrews (1967). Site 5 consisted of peat: the date is not previously published.

2. Based on 2 standard error range about the determined age.

3. After J. T. Andrews (1969). Note that the dates quoted here refer to the date of formation of the feature in which the materials were found. At sites 3 and 4 the shells must have been deposited more or less synchronously with the feature they are intended to date. However, sites 1 and 2 refer to subjacent or nearby moraines, so that the radiocarbon age determinations represent minimum age estimates only. Any error that would result is probably well within the range of the date, however.

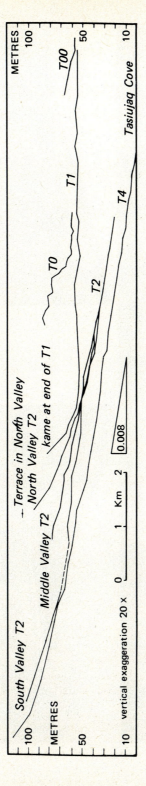

Fig. 3. Long profiles of glacial-fluvial and fluvial surfaces in Ekalugad Valley.

Fig. 4. (a) View along the T0 and T1 surfaces (upvalley).
(b) T2 surface: view upvalley. The pebble surface is largely stabilized by lichen growth and *Cassiope* heath. Major channel and bar features can still be defined on the surface.

(c) Exposure of peat associated with the T3 surface in South Valley. Material from this site was dated at 3170 ± 200 yrs BP (Table 1: entry 5).

Glacial-fluvial and Fluvial Deposits

The major sedimentary units in the valley are illustrated in Figure 2. On top of the Tasiujaq moraine on both sides of the fjord lie kame delta deposits, built by lateral meltwater which flowed into the sea. The north side delta is considerably larger, as one would expect, since there was probably much more water on this, south facing, side of the valley. This delta is the terminus of a short, steep lateral terrace (T00) (Figs. 2, 3) and of a large, subglacially formed terrace (T1) that runs for about 7 km to the mouth of North Valley. The front of terrace T1 is an ice-contact face upon which crevasse fillings have been deposited. The terrace was probably already forming 6,000 years ago under the aegis of meltwaters which flowed down the north valley side and under the glacier edge (Fig. 4). On the south side several terrace fragments persist that might have been part of a much more restricted lateral drainage system. At some time during the retreat of the ice to the valleyhead moraines, drainage was initiated from a pond in North Valley across the shoulder of the valley and down a gully. This water mainly drained sublaterally onto the T1 terrace, but a true lateral terrace can be detected along the valleyside (TO: Figs. 2, 3, 4a).

When the ice stood at the valleyhead moraines, formation of the major relict alluvial deposit began at the fjordhead, the T2 surface (Figs. 2, 3, 4b). Material was initially deposited into a sealevel at about 43 m asl, and sedimentation continued until about 4,300 years ago, by which time sea level had fallen to about 20 m. The continuity of the surface indicates that it continuously regraded itself during this time. Toward the end of its development a good deal of material probably was derived from reworking of earlier deposits. The period of active development probably coincided with the time between the formation of the valleyhead moraines and the retreat of the ice over the watershed boundary. This 1400 year period of relatively minor retreat spans the cool period of 4500-5500 years BP locally called the Flint phase (Andrews *et al.*, 1970: *see* also Table 2).

After sediment ceased to be supplied to the T2 deposit, the rivers began to entrench themselves in order to accommodate themselves to continued uplift. A sequence of meander terraces at various elevations along North and Middle Rivers (Fig. 2) indicates the period. *In situ* peat near present river level in South Valley, dated at 3170 ± 200 years BP indicates a stable surface near the end of this period (Fig. 4c). That peat is buried by 1 or 2 m of fluvial gravels which are inferred to be associated with aggradation about 2500 years BP following an early neoglacial climax (*see* Andrews *et al.*, 1970). This surface is designated T3. Renewed degradation has virtually destroyed this level. The total degradation between T3 and T4 is not known, but was probably not much below the level of the present surface.

The present T4 surface developed after the climax of the recent neoglacial period in the eighteenth century. Table 2 summarizes the correspondence between periods of significant fluvial activity and the Holocene history of Baffin Island. The correlation of events confirms the glacially conditioned rhythm of significant runoff and sedimentation in Home Bay. Andrews *et al.*, (1970) identified a sequence of "strandlines" in the Home Bay region, dateable shorelines characterized by prominent accumulations of deltaic and littoral sediments. These should correspond to times of high sediment flux, presumably associated with periodic glacial activity. These strandlines are given in Table 2, and also confirm the pattern of glacially conditioned clastic sedimentation.

Interestingly, the Home Bay pattern departs from that of Cumberland Peninsula, 150 km farther south, where the presence of an independent centre of ice dispersal in Penny Highland and greater exposure to North Atlantic weather have apparently produced a somewhat different history of minor climatic fluctuations (*see* Miller, 1973). Whilst some discrepancies may be the result of imprecision in dating, the three-phase Neoglacial history of Home Bay after 6,000 years BP is matched by an apparent two-phase history in

TABLE 2

HOLOCENE CLIMATIC FLUCTUATIONS IN EASTERN BAFFIN ISLAND

Date (years BP)	This Study (inferred from alluvial sequence)	Strandline Sequence (Andrews et al. 1970)	Central Baffin and Home Bay (after J.T.Andrews)	Northern Cumberland Peninsula (Miller, 1973)
10000			Cockburn stade; cool, damp	late glacial maximum
9000		I	warmer, drier	
8000		II Ekalugad Phase	Isortoq phase; cool, damp	
7000	warm, major glacier retreat, runoff, and outwash development	III / IV Isortoq Phase		
6000		V	warm, dry	warm, dry (Hypsithermal)
5000	cooler phase; pause in alluvial sedimentation	VI Flint Phase	Flint phase; cool, damp, moraines ca. 4800 BP	?
4000	warm; outwash development, later river incision	VII(?)	warm, dry	cold Neoglacial moraines ca. 3 200 BP, >1650 BP
3000	cool, damp; peat formation	VIII Early Neoglacial	cool, damp; Neoglacial moraines ca. 2800 BP	
2000	warmer; outwash growth then incision	IX	warmer	warmer
1000	cool	X[1] Late Neoglacial	cool, damp; moraines ca. 200 BP and later	cooler, damper; moraines 780 BP, 350 BP, 65 BP
PRESENT	warmer; outwash		warmer	

1 – Added by writer from evidence in Andrews, et al.

Cumberland Peninsula, with out-of-phase climatic change near 4,500 years BP. In absence of more precise evidence, it appears that alluvial sediments may be useful for discriminating such regional variations in pattern.

THE T2 DEPOSITS: STRATIGRAPHY AND STRUCTURE

The T2 deposits were chiefly built by sedimentation from Middle Valley. Though the surface features are well attenuated now (Figs. 2, 4b), they still permit lines of sediment movement to be inferred for the final period of activity. Middle Valley, having the most direct and open access to the interior, was the major route for the effluent valley glacier whose meltwaters gave rise to T2. North River built a large, low-angle alluvial fan out toward the "Middle Valley sandur," but there appears to have been relatively little input from South Valley — in contrast with the present day situation.

Under the top cover of sandur deposits, which varies from 1 to about 5 m in thickness and represents channel and bar materials, the bulk of the visible deposits is foreset beds (Fig. 5a) deposited in the same manner as the delta currently developing at the distal end of the T4 deposit (Church, 1972). Texture varies from coarse cobble beds with little matrix, to fine, laminated, sometimes rippled sands. The sediments are frozen except where erosion has exposed them. Ground ice is not uncommon (Fig. 5b), though it does not occupy a significant proportion of the total sedimentary volume. Though the characteristic depth of the active layer may have fluctuated somewhat, it is supposed that the bulk of the sediments has remained frozen since formation.

The T2 deltaic materials are well exposed along Middle River from the distal end to within about 0.5 km of the moraine in Middle Valley. Detailed observations were carried out on some 2,000 foreset beds (Fig. 6). The chief interest in studying the beds was to see whether they could be used as palaeodischarge indicators and so the comments that follow will concentrate on characteristics relevant to this purpose.

Beds were observed directly in the exposed section, so that apparent dips were measured. In calculating true thickness, it was assumed that the beds had initially been deposited at the angle of repose. Soundings at the present delta front indicate that this value is approximately 35°, a datum which is approached or reached in several dip sections along the exposure. Since the beds were examined near the contact with the sandur topsets, it is assumed that they were originally deposited near the top of the delta, so that the dip assumption seems reasonable. It also appears, from studies in Tasiujaq Cove, that slumping occurs on the delta front, so that the exposed sequence may not represent in all details the actual depositional sequence. Furthermore, coarse materials may be abnormally small as the result of slip face sorting and hence the sampled sequence may not be completely representative.

The thickest bed measured was 34 m, and it was one of the coarsest gravels encountered. There was no internal structure and it is not, evidently, a true deltaic bed. Though apparently conformable foreset contacts occur at both ends of the bed, it truncates a foreset, grey sand below for part of its length. On the T2 surface above the exposure a major channel scar is found, and so it appears that this is a deep channel deposit. The largest true foreset bed is about 9 m thick, and it is one of the few graded beds. The vast majority of beds are well under a metre thick, many fine beds being less than 10 cm thick.

The bedding is remarkably distinct; there are only rarely gradational contacts or changes within a bed. This suggests that each bed represents a discrete sedimentation event. It is probable that each bed represents a different flow peak — daily melt floods at times, possibly irregular storm events at others. It is certain, as well, that the exposed transect of the beds possesses variability due to varying distance from the channel mouths as they moved back and forth at the delta front, so that not all the changes necessarily

Fig. 5. (a) View of deltaic foreset beds under the T2 surface.
        (b) Foreset beds and sandur topsets just outside the valleyhead moraine. Lenses of
             clear ice (arrows) are conformable with the bedding and clear ice also fills the
             interstices amongst clasts in some coarse gravel beds.

represent hydraulic differences. Random turning point tests (see Bradley, 1968) applied to
the sequences for grain size and bed thickness (Fig. 6) confirm that they are both random
at the $\alpha = 0.05$ level, suggesting that they are in fact derived from sequences of unrelated
events.

The coarse materials vary in their occurrence from beds of cobbles completely devoid
of fine materials to pebbles dispersed sparsely in beds of otherwise uniform sand. Most
commonly, however, poorly sorted gravels are found with abundant larger cobbles.
Normal and reverse (coarsening upward) grading occurs occasionally; most often this is
not effected by a change in the coarsest material, but by a change in the proportion of
fines, so that the net result is a change in mean grain size through the bed. Reverse
grading may be the result of jökulhlaup floods. Notable oxidation has occurred in some
beds.

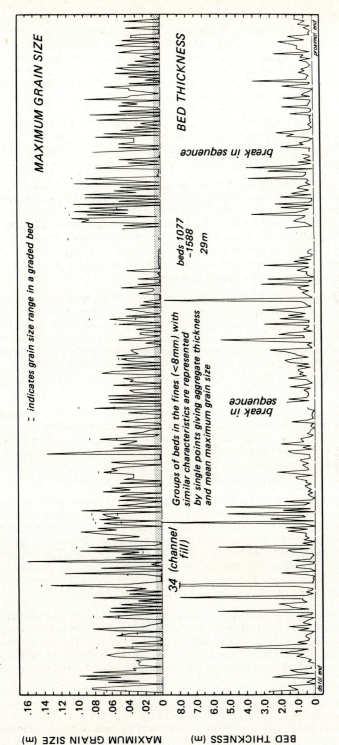

Fig. 6. Characteristics of the measured sequence in the T2 deposits. Abscissa is an arithmetic scale of bed number in the measured sequence, proximal end to right.

The fines commonly consist of medium to coarse sands. Often, divisions finer than those designated as "beds" occur in the fines. Heavy minerals in discrete bands, or micas, sometimes define very thin laminae. In one instance 19 laminae of order 1.2 mm thickness each were counted within a single coarse sand bed. In other places the microfabric of the bed betrays the presence of laminae even when there is no mineralogical segregation or obvious change in grain size. Each lamina undoubtedly represents a fluctuation in sedimentation related to hydraulic conditions. It is most probable, particularly where mineralogical separation is involved, that the fluctuations are very short term, probably diurnal or semidiurnal (conditioned by stream flow or tides). In such cases, the bed that contains the laminations represents a discrete "event" only insofar as it represents a period of relatively uniform conditions which may have persisted for days, weeks, or through an entire season.

The fine beds occasionally exhibit ripple marks. Some beds are contorted internally, while in other instances groups of beds are deformed. Whilst the former occurence is probably related to slumping on the delta face, the latter effect is most likely a consequence of freezing. In the fine beds, variations in strike are much more apparent than in gravels. Wedges of beds often truncate each other (always from above) indicating that variation in current direction commonly occurred, and that erosion sometimes occurred on the delta face. This situation confirms that examination of a linear transect of beds cannot be expected to yield a complete sequence.

The measured length of the T2 deposit represents 30% of the total distance (along a line normal to the strike of sedimentation) between the valleyhead moraines and the distal end of the T2 deposit. If the bedding sequence is typical, there are some 6,500 beds in the complete sequence. This implies, in turn, that there were, on average, 4 to 5 significant sedimentation events during each season of the 1,400 year history of deposition. This seems reasonable. Furthermore, it seems possible that individual laminae in the fines, which are probably about 10 times as numerous as the beds, may represent diurnal events.

Marine mollusc shells have been recovered from six sites within the T2 beds. All the samples represented a community dominated by *Mya truncata* and *Hiatella arctica*. An assemblage was recovered from marine silt under T2 deposits on the south side of Ekalugad Valley, as follows (identifications by J. T. Andrews): *Hiatella arctica* (Linné); *M. pseudoarenaria* Schlesch; *Macoma calcarea; Clinocardium ciliatum* (Fabricius); *Nucula tenuis; Nuculuna,* cf. *buccata* or *minuta; Portlandia arctica*; Gastropoda; *Oenopota* sp.

The assemblage is characteristic of a muddy bottomed environment in cold water. The age of this deposit is about 5,000 years (cf. Table 1; site 3), which is near the middle of the T2 sedimentation period. That none of the other samples appears to differ radically from this one suggests that inshore marine environmental conditions did not vary greatly during the entire period; indeed they do not appear to have been greatly different from what is experienced today. Significantly, none of the species mentioned by Andrews (1972) as indicating warmer conditions, and found by him in deposits both somewhat older and somewhat younger than 5,000 BP on the east Baffin coast, is present.

A thin detrital peat layer was found in the deltaic beds about half way along the section under a surface at 31 m. If this is approximately the emergence that has occurred since its deposition, it also appears to have been deposited ca. 5,000 BP. This occurs in the midst of a long sequence of fine beds (516 beds with an aggregate thickness of 360 m) which may be associated with the cold period that occurred about 4,500-5,000 years BP (cf. Table 2).

PALAEOFLOW ESTIMATION

In a recent paper, Baker (1974) outlined a scheme of palaeohydraulic computations for coarse-grained (gravel-cobble) fluviatile deposits, some aspects of which had been presented earlier (Baker, 1973). Jopling (1966) has given some considerations for deducing hydraulic parameters of depositional sequences in finer (sand) fluvial sediments, but a variety of unknown or indeterminate conditions, largely associated with the occurrence of bedforms, made that analysis difficult. Several workers have made isolated estimates of streamflow velocity or depth required to move certain clastic materials — chiefly boulders associated with extraordinary floods (for one example, *see* Birkeland, 1968; references to earlier studies are given therein). Most have utilized the Manning formula, with an estimated flow resistance coefficient, to arrive at the results.

The utility for Quaternary studies of a workable scheme for palaeohydraulic computations is obvious. Extensive sequences of glaciofluvial outwash deposits occur in formerly glaciated regions; extraordinary riverine deposits occur beyond the limits of Pleistocene ice as well as in regions that received glacial runoff or experienced enhanced fluvial runoff. Many of these deposits have not been significantly altered since initial deposition, and some contain evidence of former channel form and slope that is sufficient to permit complete palaeohydrologic reconstructions to be attempted. Such is the case in the present study. The computations made for the T2 sequence of cobble-gravel beds at Ekalugad Fjord will be preceded by a discussion of the basis and assumptions for them.

Sediment Transport Mechanics: Some Problems of Formulating a Palaeoflow Criterion

The usual means by which information about streamflow is derived from coarse sedimentary materials is to assume that the largest grains present in the deposit represent the largest material that the stream was competent to carry. Provided that an acceptable competence relationship is available, data about the flow may be deduced. The entrainment function of Shields has generally been applied for this purpose. It is given by

$$\theta = \frac{\tau_c}{(\gamma_s - \gamma_f)D} \quad \propto \quad \frac{v_{*c}D}{\nu} = R_* \tag{1}$$

where θ is a dimensionless representation of the shear stress necessary to entrain a particle and R_* is a particle Reynolds Number (*see* Appendix — Notation for definition of the other terms). Gessler (1971) has reviewed Shields' work recently, and Yalin (1972) has incorporated it into a formal theory for sediment entrainment.

For small values of R_*, the value of θ varies. Stresses in this range entrain very small particles which subsist within a viscosity-dominated laminar flow very close to the boundary. However, for $R_* > 10^2$ particles protrude into the turbulent flow, viscous effects are no longer important, hence R_* no longer affects the result. θ then has a constant value of approximately 0.05. Since this result has been found for all reasonable values of $(\gamma_s - \gamma_f)$, and since v_* may be expressed as $(\tau/\rho_f)^{1/2}$, the only free parameters remaining are tractive force and sediment size (provided ρ_f is accepted as being sensibly constant: i.e., the concentration of suspended material in the flow is not very high). The direct relation between these two parameters is presented in Figure 7, and provides the immediate basis for palaeohydraulic computations. The range of validity of the simplification is $D > -3\phi$ or 8 mm approximately.

Almost all laboratory confirmations of the Shields relation have been carried out with materials in the size range $D < 2$ mm: the 8.50 mm material of Meyer-Peter and Müller (1948) appears to be the largest material used in trials explicitly designed to investigate threshold conditions of sediment motion. Field data plotted on figure 7 for $D > 8$ mm have been derived from a variety of studies, none of them designed to critically test Shields' relationship. There is considerable scatter, possible reasons for which merit some attention.

Some question has been raised about the completeness of Shields' criterion, particularly in respect of the exclusion of flow depth. The "relative roughness," d/D,

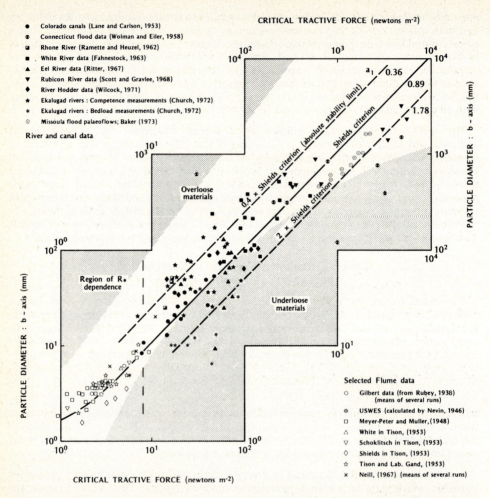

Fig. 7. Critical tractive force versus particle diameter, based on data for rivers and canals and selected flume data (D > 1.0 mm), and comparison with the range of empirical scour relations proposed from engineering studies (non-stippled region). The bases for the data vary widely in a manner that influences the plot position of the results:

Rubicon River data: identified large boulders in river channel after catastrophic failure of rock-fill dam; hydraulic parameters based on slope-area reconstruction. The very large boulders, which travelled least far downstream, may have been moved in an earth-water slurry.

White River data: individual cobbles trapped on screens during transport: hydraulic data measured.

Connecticut flood data: large boulders identified as having moved during floods on several rivers. Source of boulder sometimes known. Hydraulic parameters based mainly on slope-area reconstructions.

River Hodder data: marked cobbles set on streambed and distances of movement measured for various flows: present values based on extrapolation to zero distance datum of results grouped into narrowly size-graded ranges. Hydraulic results based on measurements or interpolation from observed hydraulic geometry relationships.

Ekalugad rivers: competence measurements based on removal of marked cobbles set on streambed; hydraulic parameters derived from crest-stage data interpreted via observed hydraulic geometry relationships.

where d indicates flow depth, is an important scaling factor for flow/boundary interaction. For "well-behaved" flows (ones in which the bed materials remain fully submerged and in which a well-defined profile of mean velocity persists), the variable v_* appears sufficient to define flow conditions near the bed (and remote from the surface). This does not remain true for very shallow wake or wave dominated flows, including some that may occur over gravel deposits in real rivers (cf. Neill, 1967, for a formulation that includes relative roughness).

A more refractory problem underlying the determination of Shields' relation is the question of what constitutes the "threshold condition" for entrainment. Movement of material begins as sporadic displacements of isolated grains which become more and more numerous as flow increases. Shields (1936) determined his threshold condition by observing sediment transport at various rates of flow and then extrapolating back to the conditions indicated for zero transport. Though indirect, the procedure has the merit of being reasonably consistent. The actual entrainment of individual particles will no doubt be strongly influenced by the recent history of deposition and scour. This implies that, interpreted in terms of Shields' criterion, the largest particles subsequently found in an alluvial deposit may over-represent *or* under-represent the shear forces present in the flow that moved them.

Most important, entrainment at the bed depends not on mean velocity/pressure conditions, but on instantaneous values. In turbulent flow, velocity fluctuations permit entrainment of far larger material than would be expected from Shields' criterion based on mean values. On a natural streambed, the variability of bed materials would impose an effective spatial variability on the resistance to displacement of individual particles as well. Furthermore, variations in the flow and bed geometry may interact in a stochastic way. Gessler (1971) has presented an empirical study of entrainment probabilities, and Yalin (1972) has made theoretical considerations of such circumstances. This effect undoubtedly contributes to the scatter of results observed in figure 7.

A variety of alternative entrainment or "scour" criteria based on tractive force have been proposed, mainly from experimental results. They have most recently been

---

Bedload measurements based on trap samples; hydraulic parameters measured at trap site and calculated as $\tau = \rho_f v_*^2$

Eel River data: removal of marked natural streambed material (confirmed by photography: hydraulic parameters from stream gauge data).

Rhone River data: removal of marked cobbles set on bed: hydraulic parameters from stream gauge data and reach survey. Plotted data represent mean values interpolated into the field results of behaviour of three size-graded samples.

Colorado canals: Grain size data represent D_{75} of material in which unlined canal was dug: hydraulic parameters based on conditions at maximum flow.

Missoula breakout flood data: palaeoflow data based on slope and maximum flow depth inferred from deposits or channel morphology, and D based on the largest material recovered. Included for comparison.

Meyer-Peter and Müller data: direct observation of commencement of particle motion. Grain size data are D_{50} of narrowly graded mixtures: 1 m wide flume.

USWES data: basis unchecked.

White data: basis unchecked.

Neill data: direct observation of commencement of particle motion. Unigranular material. Plotted points are means of several runs: 0.9 m flume.

Gilbert data: Data notated by Gilbert as "few or very few" or "occasional" grain displacements. Plotted points are means of several runs: 0.6 m flume.

Schoklitsch data: basis unchecked.

Shields data: bed-load transport measured at several flows and conditions extrapolated to zero transport.

summarized by Graf (1971), and their collective range is shown in Figure 7. The range mainly reflects problems of defining incipient motion and the effects of varying bed material geometry.

Many analyses have been carried out which seek a direct relation between flow velocity and the grain size of material which the flow is competent to move (cf. the review *in* Graf, 1971). Novak (1973) has recently reappraised the data available for coarse particle behaviour, with particular attention paid to field results from rivers. The major problem of specifying velocity criteria for stream competence lies in deciding what is the appropriate velocity in the vicinity of the streambed, and then in establishing a reliable field procedure for gaining measurements. In the past, a variety of arbitrary but consistent procedures for determining a reference velocity has been applied, and the empirical association with maximum particle size moved is maintained because of the existence of a well-defined velocity profile. Unfortunately, in absence of prior information about flow depth, the results cannot be applied in reverse in order to make palaeohydraulic deductions. Estimation of "palaeovelocity" from grain size information alone is not, in principle, possible.

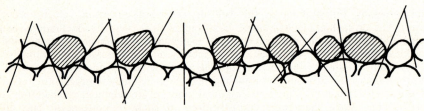

A : Normally loose boundary

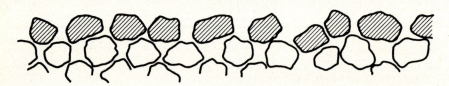

B : Overloose boundary

C : Underloose boundary (imbricated)

Fig. 8. Sketch of noncohesive bed states: lines are contact tangents between adjacent clasts. See text for further description.

Ultimately, velocity criteria cannot avoid the most serious conceptual problem noted in respect of tractive force criteria, that entrainment in fact is a probabilistic phenomenon depending upon the frequency and intensity of flow excursions accompanying eddy impingement onto a spatially varied bed. It does not depend functionally upon mean conditions.

Application of a Palaeoflow Criterion

Natural Grain Size Distributions

To this point, concern has centred upon what might constitute an appropriate formulation of sediment entrainment conditions. Beyond that, there remains a series of problems associated wih the application of any criterion in a real river.

All of the foregoing considerations tacitly assume that the character of the sediment plays no direct part in the determination. Yalin (1972) argued that the grain size distribution of the bed materials must influence the result, because grain size, D, is a surrogate measure for equivalent particle roughness. Two different grain size distributions will yield the same Shields relationship only if D is chosen so as to represent their attributive roughness. Normally, however, one chooses a constant characteristic of the size distribution (say, D_{50}) rather than a constant characteristic of the hydrodynamic effect of the mixture. In interpreting the conditions at entrainment for sedimentary deposits, one faces the possibility for some error due to the failure to account for the particular grain size distribution of the sediments.

The readiness with which an individual particle may be entrained depends upon the balance of forces to which it is subject. In addition to the hydrodynamic forces of lift and drag, and the opposing effects of particle weight and friction between particles, an individual may be restrained by neighbouring grains. Hence, the entire geometrical environment of a particle — not merely its size — will influence whether or not it is moved. This simple argument can be extended to encompass the entire state of a streambed. A noncohesive bed generally exists in one of three states (*see* also Fig. 8):

> *Normal boundary:* Materials are resting in a non-dilated state, without imbrication, and with usual packing (this implies a random arrangement of grains, such as one might expect to find in a pile of sand dumped from a hopper). This state corresponds to Yalin's (1972; p. 77) "settled mobile bed." More or less half the grains in the uppermost layer are "available" for movement.

> *Overloose boundary:* Materials are resting in a dilated state, normally because of the presence of a large volume of water within the sediment and flow toward the surface exerting a dispersive stress; also included are materials with completely open packing. Such "quick" sediments are common in river channels, and have been observed to occur in materials up to medium gravels.

> *Underloose boundary:* Materials are resting in a state of close packing or of imbrication. This is the common condition in gravel or cobble floored streams.

Most experimental work has been done with normally loose boundaries: observations in nature apply mainly to one of the other two states. This can be seen in Figure 7.

The principal conclusion to be drawn from the foregoing considerations is that, in nature, a range of conditions occurs that could introduce nearly an *order of magnitude* range in the values of Shields' dimensionless shear parameter (or any other deterministic parameter) for entrainment of coarse materials, depending on the initial condition of the materials to be entrained. It appears that the variety of bed conditions exhibited in natural rivers might be entirely subsumed only by the introduction of probability statements or of empirical coefficients.

There are some further problems introduced by the nature of the data used for palaeohydraulic computations. Cobbles are examined at their site of deposition, *not* that of entrainment. Gross error could be introduced if one by chance measured anomalously large particles that had been introduced into the channel by bank erosion, by mudflow, or (in glaciofluvial context) perhaps by a dirty avalanche or by melting out of glacial ice (see also Fahnestock, 1963; Baker and Ritter, 1975). Beyond that, however, there remains the question of whether the hydraulic conditions at the site of deposition need be the same as at entrainment. There has been little direct study of the competence of a stream to maintain materials in motion once entrained. Some of the data in Figure 7 indicate that proportionally larger material may be moved than might be entrained. A recent experimental study by Graf and Pazis (1977) indicates that sands are deposited at values of θ about 5 per cent lower than those for entrainment: the discrepancy will be greater for larger materials since they may initially be underloose.

There has been little critical study of fluvial deposition (see Bagnold, 1968, for comments). Size grading is, however, the usual adjunct of deposition in stream channels. Such evidence as is available does not suggest that the process works in simple reverse fashion to that of entrainment. Finally, deposition often occurs in zones of local flow divergence where conditions are highly nonuniform. Nearly all the available laboratory and field data discussed above pertain to uniform or quasi-uniform flows.

Ancillary Computations

The shear stress term in Shields' dimensionless shear ratio describes conditions at the bed at the point at which entrainment occurs. To make further progress, it is necessary to assume that this value is equal to the average shear stress in the vicinity, which is given by $\tau = \gamma_f RS$, where $R = A/P$ is the hydraulic radius of the flow, A is cross-sectional area of flow and P is the wetted perimeter. Baker and Ritter (1975) discuss the basis for use of this relation in stream competence studies. For wide, shallow channels, such as are typical of gravel rivers, $R \approx \bar{d}$, the mean flow depth, and $P \approx w$, the width of the flow. If the energy gradient is known, then flow depth can be estimated. The energy gradient can often be determined, to a satisfactory approximation, either by surveying the slope of the surface of the sedimentary body, or by measuring the slope on bedding planes within the sedimentary deposit.

By use of an appropriate flow resistance equation, some additional information about the flow may be obtained, provided that an appropriate resistance coefficient may be determined. The best known total flow resistance formula is the Manning formula, long used in engineering practice. The Manning resistance coefficient, n, is a relatively conservative parameter and may be estimated for present purposes in one of two ways. A direct estimate based on experience in other rivers may be made. Alternately, it may be assumed that the channel is reasonably uniform and that substantially all flow resistance derives from particle resistance at the bed (low sinuosity gravel streams typical of proximal outwash conform reasonably well with this assumption for a given reach). Then an equation for estimating the resistance coefficient of a particulate boundary may be applied, such as that due to Limerinos (1970):

$$n = \frac{0.113 \, R^{1/6}}{1.16 + 2.0 \log (R/D_{84})} \qquad (2)$$

(The equation is not dimensionless. The numerator coefficient is given for SI units. The British units coefficient is 0.0926. Limerinos reviews a variety of alternative equations in his paper.)

The denominator introduces the effect of relative roughness. The equation is empirical, and was derived for gravel rivers. However it conforms almost exactly with an equation derived from the logarithmic velocity law using Nikuradse's classical experimental data. A variety of alternate power-law or logarithmic-law flow resistance formulae could also be chosen (*see* Day, 1977, for discussion of resistance laws in shallow channels).

If some basis exists for deciding whether the velocity value derived from the resistance relation is representative of the entire channel, and if some information on channel cross-sectional form is available (as may occur from the morphology or stratigraphy of the sedimentary body) then an estimate of total discharge of the channel might be made.

T2 PALAEOFLOWS AT EKALUGAD FJORD

Computational Formulae

Here, it is proposed to compute specific discharge for a sequence of beds so that the frequency distribution of the flows may be considered. The initial result will depend only upon the measurement of grain size and certain fixed parameters, so that it represents nothing more than a transformed measurement of particle size. The transformation is justified only if (1) there exists a sufficiently well-defined relationship between stream competence and tractive force, and (2) particle size sufficiently determines flow resistance. The basis for these assertions has been examined above.

The simplified form of Shields' relation is used, vis:

$$\tau_c = a_1 D \tag{1a}$$

It is apparent in figure 7, however, that the usual constant, $a_1 = 0.89$, is not at all universal. In the present-day Ekalugad rivers the constant apparently lies closer to 1.8 because of characteristically underloose conditions. Results are presented for several values. Values of τ_c are used in the formula for average shear stress to obtain flow depth as

$$d \sim \tau_c / \gamma_f S \tag{3}$$

The hydraulic gradient, S, was approximated by the mean slope of the T2 surface outside the valleyhead moraines, corrected for the isostatic deleveling since 4,300 years BP (Fig. 3) and found to be 0.008. The value is very close to the mean gradient of the present T4 surface and is thought to be reasonably accurate. Hence, we have

$$d \sim a_2 D \tag{3a}$$

In order to obtain $\bar{v}$, it is necessary to adopt a resistance equation. Because equation (2) can conveniently be used for computing n, Manning's equation will be adopted. Making appropriate substitutions from (2) and (3a) in the usual form of Manning's formula (for metric units: coefficient for British units is 1.49),

$$\tag{4}$$

gives
$$\bar{v} = R^{2/3} S^{1/2} / n$$

$$\bar{v} \sim a_3 D^{1/2} \tag{4a}$$

n may not be very well estimated, but since it is conservative, this should not lead to serious errors.

Combining (3a) and (4a) yields specific discharge as

$$q \sim a_4 D^{3/2} \tag{5}$$

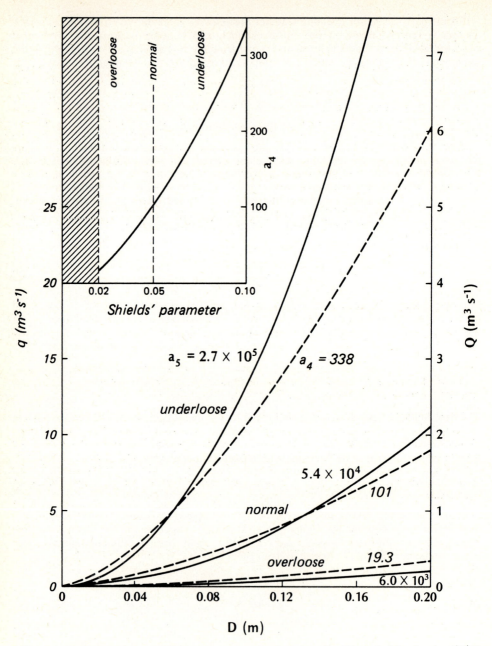

Fig. 9. Transformations between grain size and discharge for T2 materials. Inset: relation between Shields' parameter and the discharge coefficients.

The range of a_4 is shown in Figure 9 with respect to the range of Shields' constant and several functions are displayed. It is apparent that the result is rather sensitive to the choice of Shields' criterion.

Values of D were determined either by sampling 25 large cobbles from each bed and then obtaining the mean of the *5 largest*, or by taking D_{85} value in representative sieve analyses (beds with $D_{max} < 8$ mm).

The computations may be extended to derive estimates of total discharge if some means of assessing stream width can be established. Two approaches may be considered here. First, it is possible to measure the width of extant channel traces on the T2 surface. The mean of 85 measurements taken from air photographs at the scale of 1:10,000 (which provided more representative sampling than was available on the ground, though less precision) was 16.3 m with a standard deviation of ± 6.2 m. The value is very small by the standard of contemporary T4 channels and must be interpreted as indicating either that the last rivers on the surface carried far smaller discharges than the contemporary channels, or that active braiding persisted to the end and that several "16 metre channels" were always active at once.

A second approach is to suppose that the scaling of width with discharge in T2 channels was similar to that found in the present-day channels. The relation $w_s \approx 35Q^{1/4}$ is appropriate for the contemporary streams (Church, 1972). Then we may deduce that

$$w_s^{3/4} \sim 35 \, q^{1/4}, \text{ or}$$

$$w_s \sim 115 \, q^{1/3} \tag{6}$$

Combining equations (5) and (6) yields a relation

$$Q \sim a_5 D^2 \tag{7}$$

which is also graphed in figure 9. This relation was used in the computations. Table 3 summarizes values of constants a_1 to a_5. It should be noted that constants a_2 to a_5 are empirical and must be derived for each new problem.

For the largest cobble size, D = 0.16 m, Q may be variously estimated, as in Table 3. It is probable that in the proglacial environment that characterized the development of most of the T2 deposit, materials available for entrainment usually exhibited a condition closer to "normal looseness" than do the contemporary alluvial channel deposits. Compared with the highest measured normal discharge of 194 m^3s^{-1} in the combined contemporary Ekalugad Rivers, and recalling that the T2 river probably drained an area of order 10x larger size, the maximum palaeoflow estimates seem reasonable.

TABLE 3
CONSTANTS IN X.D RELATIONS AND MAXIMUM COMPUTED DISCHARGE

Equation	1	1a	3a	4a	5	7	
Parameter		τ	d	v	q	Q	Q_{max}
Constant	θ	a_1	a_2	a_3	a_4	a_5	(D = 0.16 m)
Overloose	0.02	0.36	4.6	4.2	19.3	6.0×10^3	154 $m^3 \, s^{-1}$
Normal	0.05	0.9	11.5	8.8	101	5.4×10^4	1382
Underloose	0.07	1.3	16.1	11.3	183	1.2×10^5	3072
	0.10	1.8	23.0	14.7	338	2.7×10^5	6912

Results

The frequency distributions of grain size and thickness of the sampled beds (Fig. 6) are displayed in Figure 10. Equations (5) and (7) were used to transform the grain size data into flow data, except that for D < 8 mm the coefficients have been adjusted to account for the nonlinearity in Shields' relation. Results for normal and two underloose bed specifications are shown in Figure 11a along with contemporary, observed flow distributions in South River (49 per cent glacial cover) for comparison. For D < 8 mm only "normal" boundary conditions are used, since underloose conditions are not expected to apply to small, compact grains. The effect of grain sheltering cannot reasonably be assessed. Several other assumptions and limitations in the data must be borne in mind in assessing these results.

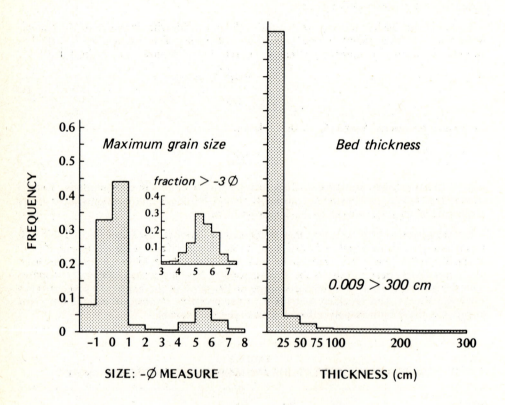

Fig. 10. Frequency distributions of maximum grain size and bed thickness for the sampled T2 beds at Ekalugad Fjord.

1) The results are simply transformations of grain size data and will reflect any biases present in them. The data exhibit a prominent gap in the size range 2 - 16 mm. This is typical of alluvial sediments, although here the gap reveals itself in a larger size range than is normal. The shift is perhaps a consequence of having measured maximum grain sizes rather than means. Nevertheless, it is probable that the results are biased by this effect of selective deposition, and that this is responsible for the levelling of the plots at about 12.5% "exceedance."

2) The frequency distribution of flows is based on the distribution of *number of occurrences* of grain size only. There is no information included on the duration of individual flows. This might be derived from bed thickness information. There would not be a simple proportionality between thickness and flow duration, however, because of the effect of varying grain size between beds. It would be possible to compute a set of "standardized thicknesses" as bed thickness/grain size. Since little is known about packing effects, the proportion of large grains transported (or deposited), or slumping, this did not seem to be a reasonable extension of the calculations. The distribution of bed thicknesses is highly skewed in any case (Fig. 10), with more than 85% being less than 250 mm thick — that is, of order 1 to 10 grain layers thick for grain sizes greater than 16 mm, which yield virtually the entire plot of Figure 11.

3) For reasons discussed previously, it is not clear that grain sizes less than 8 mm, and certainly those less than 2 mm, reflect the flows that transported them. They may represent deposition relatively far from an active channel mouth, hence only the finer portion of the load transported. The finest materials certainly reflect flows that transported material only in suspension, the physics of which is radically different from that considered here. The extent of these effects cannot be separated from the effect of real flow variability. One way to circumvent this problem is to delete the fines from the sequence. A decision could be taken to reject data at 8 mm (on the behaviour of Shields' criterion for entrainment) or 16 mm (on the basis of the apparent deficiency of smaller sizes).

Since the full significance of the distribution below 16 mm is not known, rejection was made at 8 mm. As the original event sequence consisted of independent events, this procedure should not further bias the new, "contracted" sequence. This sequence contains 241 beds — about 12% of the entire sequence. If the sequence is typical of the entire deposit, there are about 800 such beds, gaining a return period for the partial duration series of flows that created them of about 1.75 years. This seems to be high: since the distal 30% of the beds was examined, the assumption of representativeness for the occurrence of these coarsest beds may not be reasonable. Figure 11b presents flow frequencies for the new sequence, and compares it with contemporary flows that were in excess of the *observed* threshold for bedload transport in South River of 15 m^3s^{-1}.

The exceedance plots are approximately parallel for the larger flows (the upper 5% of the flows), lending some confidence to an assertion that the frequency distribution of this portion of the palaeoflow range has been reasonably recovered. This corresponds to the upper 30% of competent flows (Fig. 11b). The writer has no means to select one palaeoflow plot in order that absolute flows may be specified unequivocally: the plots for $a_1 = 1.3$ are preferred on experience of bed condition in active, proglacial streams. All the results imply palaeoflows an order of magnitude larger than contemporary flows.

Gross Sediment Yield

One inexact check may be made on the order of magnitude results presented above. The total volume of sediments in the valley has been estimated using data derived from vertical electrical resistivity profiles. The measurements were carried out using the Schlumberger configuration. Electrode spreads of up to 450 m were used, giving a theoretical vertical range of 225 m: in practice the probable limit of vertical resolution was 100 m±. The stratigraphical situation under the sandur may be very complex. The existence of foreset bedding, irregular variations in grain size, and possibility for massive ground ice to occur would further complicate the picture. Nevertheless, most of the unconsolidated material is coarse, frozen alluvium, electrically distinct from the expected granite-gneisses below, and so it was hoped that a clear break in electrical properties

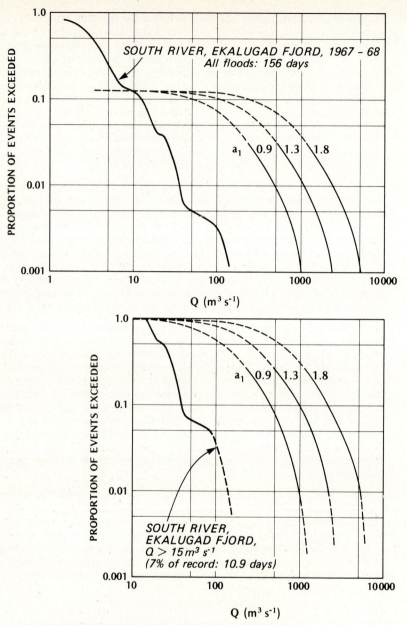

Fig. 11. (a) "Exceedance" plot of T2 palaeoflows using all data, for several boundary conditions, and comparison with observed flows in contemporary Ekalugad Rivers. The logarithmic plot emphasizes the distribution of high flows. For the palaeoflows the proportional "exceedance" is based on numbers of events (beds) only and there is no true duration criterion (see text for further discussion). The South River flows form true exceedance plots. Dashed portions of the curves are not reasonably known.

    (b) "Exceedance" plot of T2 palaeoflows using only beds with D > 8 mm, and comparison with observed flows in contemporary rivers truncated at the bed material entrainment discharge.

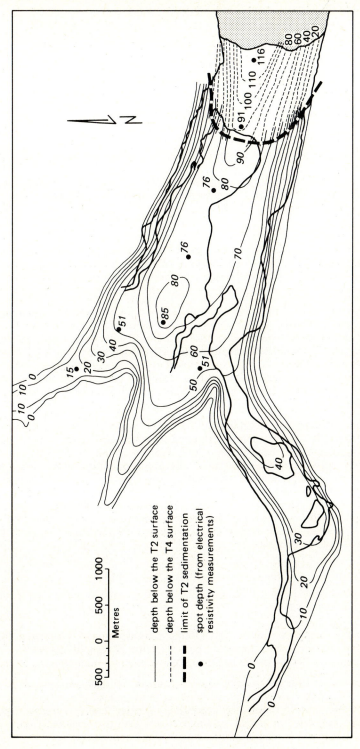

Fig. 12. Inferred depths of unconsolidated materials in Ekalugad Valley. The data base is very sparse and the interpretation should be regarded as conjectural.

MICHAEL CHURCH

TABLE 4
MEAN ANNUAL SEDIMENT YIELD RATE TO T2 AND POST-T2 SEDIMENTS[1]

	Sediment Volume[2]	Period of Actual Growth	Time of Growth	Mean Annual Yield
T2	4×10^8 m^3	5700 - 4300 yrs. BP	1400 yrs.	29×10^4 m^3
post-T2	0.7×10^8 m^3	4300 yrs. BP - present	4300 yrs.	1.6×10^4 m^3

1. Comparison assumes that the bulk density of all sediments is the same. Ice in T2 sediments may introduce some error.
2. Coarse alluvium only, transported mainly as bedload and deposited as deltaic forests or floodplain sediments. Inclusion of all unconsolidated materials does not significantly affect the comparison.

would occur. This was in fact detected in most cases in empirical interpretation of the resistivity profiles, and provided a series of data on probable bedrock depth which are of the size expected from projected valleyside profiles and a projection of the fjord bottom topography.

The resulting map of bedrock contours (Fig. 12) is drawn on the basis of sparse data and must be considered to be mainly subjective. Nevertheless, it can be used to indicate the magnitude of alluvial sediments. The total amount of sedimentary material deposited over bedrock under the completely reconstructed T2 surface was about 5×10^8 m^3. If it is assumed that the upper 50 m of depth is composed of coarse alluvial material (by analogy with the present situation at the delta front in Tasiujaq Cove), then about 4×10^8 m^3 of alluvium occur. The balance would be marine silt derived from suspended load of the river, and/or glacial till.

Post T2 sediments comprise an additional 1.1×10^8 m^3, of which 0.7×10^8 m^3 are in the upper 50 m of depth (it can be reliably estimated from the surface topographic map that 0.46×10^8 m^3 of material has been derived from erosion of the T2 deposits, so that the volume of newly derived coarse material is only about one-third of the total). Table 4 indicates the mean annual sediment volumes delivered to the alluvium during T2 and post-T2 time. The result indicates a delivery rate for coarse alluvial material in T2 time about 20 times greater than in post-T2 time. This would be consistent with order of magnitude larger runoff, and the usual, moderately nonlinear relation between discharge and sediment transport. In one season's observations (1967) bedload sediment delivery was calculated as follows (Church, 1972):

for South River, 1.1×10^4 m^3, and a peak discharge of 157 m^{3}s^{-1}
for the combined Ekalugad Rivers, 7.3×10^4 m^3 and a combined peak discharge of 194 m^{3}s^{-1}.

The flows are hourly mean values. The sediment transport was calculated from the Meyer-Peter and Müller (1948) formula, but with the threshold discharge adjusted according to direct observation of the beginning of bed material motion in the rivers.

These results are considered to satisfactorily confirm the order of magnitude greater discharges of T2 time. In view of the probable order of magnitude greater drainage area, it is not expected that specific runoff was appreciably greater than it is today. Occasional jökulhlaup floods may have produced very high flows, just as occasionally occurs today in South River. The peak contemporary discharge quoted above is for such an event.

DISCUSSION

In this paper, palaeohydrological information has been sought from a Holocene alluvial valley fill in eastern Baffin Island, using two approaches. Morphological and stratigraphical evidence allows reconstruction of a sequence of depositional and erosional episodes, and some absolute dates allow the sequence to be compared with regional evidence for Holocene environmental changes. If we assume that rapid alluvial deposition is associated in this arctic region with glacier retreat, accompanied by high runoff and abundant sediment transport away from the ice front, whereas stable or degradational phases are associated with periods of climatic stability or deteriorating climate, a view of Holocene climatic fluctuations may be gained from alluvial sequences. Alluvium constitutes the most extensive postglacial terrestrial sedimentary material in Baffin Island: in absence of more precise indicators, it can be used to infer and compare local postglacial environmental sequences. By themselves, however, morphological-stratigraphical interpretations remain entirely qualitative in nature.

It appears that, in the Home Bay region, there have been three periods of notably cool climate during the past 6,000 years. The first is reflected in an apparent pause in deposition of the main late glacial outwash. Without fortuitous dateable material, the sequence could not be absolutely fixed. With such data, the alluvial record compares well with the record of Holocene climatic fluctuations for central Baffin Island (cf. Table 2), although it is difficult to delimit periods precisely. Interestingly, the sequence is at variance in places with the record from Cumberland Peninsula, 200 km farther south. The detailed record there may be expected to be different because of greater exposure to the oceanic influence of the North Atlantic, and especially because of local climatic effects accompanying glaciation around Penny Highland.

By the study of bedding texture, information on the magnitude of palaeoflows was sought. Whilst quantitative results can be achieved, two sets of considerations compromise their accuracy. First, the selection of a criterion by which to transform grain size into flow parameters presents difficulties associated with the variability of sediment entrainment conditions. In this study, Shields' tractive force criterion was used, but its implementation was made difficult by:

i) lack of a reasonable means to account for the probabilistic character of sediment entrainment;

ii) the effect of variable streambed structure on sediment entrainment.

Secondly, the observed sediment textures are the result of sorting processes associated with transport and deposition that are not well understood. Here, two aspects of this problem are important:

i) the possibility that the largest materials sampled in the beds may not represent the competency of the flow, as the result of local unsystematic variability in the bed (the effect of slumping, cross-channel gradients of material size, etc.);

ii) the well known loss of granules and fine gravels from alluvial materials may bias calculations for materials in the size range 2 mm - 16 mm, or more.

Computations were made for a sample of some 2,000 beds, comprising about 30% of the main, T2 outwash deposit at Ekalugad Fjord. The former set of problems makes it difficult to assign an absolute magnitude to the flows; the second set clearly biases the exceedance plot for moderate flows. Nevertheless, the finite range of observations about Shields' criterion allows an order of magnitude to be assigned to flows, and it appears that the upper range of flows can be reconstructed in frequency with some confidence.

It appears that flows to the head of Ekalugad Fjord during the final ablation of the Pleistocene ice cap were about 10x greater than at present. This appears to be the most reasonable sort of statement that can be made in the present state of the art of palaeohydrological reconstruction from the evidence of the sediments.

ACKNOWLEDGEMENTS

Data for this study were collected during field investigations for the Terrain Sciences Division, Geological Survey of Canada. Alan Graves and John Knight helped in the data collection. The radiocarbon date Gif-3956 was kindly provided by Mme. G. Delibrias. The paper has been reviewed by John T. Andrews of the University of Colorado and Victor R. Baker of the University of Texas at Austin.

REFERENCES

Andrews, J. T., 1967, Radiocarbon dates obtained through Geographical Branch field observations: Geogr. Bull., v. 9, p. 115-162.

————, 1968, Pattern and cause of variability of postglacial uplift and rates of uplift in arctic Canada: J. Geol., v. 76, p. 404-425.

————, 1969, The importance of the radiocarbon standard deviation in the determination of relative sea levels and glacial chronologies: an example from East Baffin Island, N.W.T., Canada: Arctic, v. 22, p. 13-24.

————, 1972, Recent and fossil growth rates of marine bivalves, Canadian Arctic, and late-Quaternary Arctic marine environments: Palaeogeog., Palaeoclim., Palaeoec., v. 11, p. 157-176.

————, Buckley, J. and England, J. H., 1970, Late-glacial chronology and glacio-isostatic recovery, Home Bay, East Baffin Island, Canada: Geol. Soc. Am. Bull., v. 81, p. 1123-1148.

Bagnold, R. A., 1968, Deposition in the process of hydraulic transport: Sedimentology, v. 10, p. 45-56.

Baker, V. R., 1973, Paleohydrology and sedimentology of Lake Missoula flooding in eastern Washington. Geol. Soc. Am., Spec. Paper 144, 79p.

————, 1974, Paleohydraulic interpretation of Quaternary alluvium near Golden, Colorado: Quaternary Research, v. 4, p. 94-112.

————, and Ritter, D. F., 1975, Competence of rivers to transport coarse bedload material: Geol. Soc. Am. Bull., v. 86, p. 975-978.

Birkeland, P. W., 1968, Mean velocities and boulder transport during Tahoe-age floods of the Truckee River, California-Nevada: Geol. Soc. Am. Bull., v. 79, p. 137-141.

Bradley, J. V., 1968, Distribution-free statistical tests: Prentice-Hall, Englewood Cliffs, N.J., 388p.

Church, M., 1972, Baffin Island sandurs: Geol. Surv. Can. Bull. 216, 208p.

Day, T. J., 1977, Resistance equation for alluvial-channel flow: discussion: J. Hydraulics Div., Am. Soc. Civil Eng., v. 103, p. 582-585.

Fahnestock, R. K., 1963, Morphology and hydrology of a glacial stream — White River, Mount Rainier, Washington: U.S. Geol. Survey Prof. Paper 422-A, 70p.

Gessler, J., 1971, Beginning and ceasing of sediment motion: in H. W. Shen, ed., River Mechanics; Water Resources Publications, Fort Collins, Colorado, v. 1, Ch. 7, 22p.

Graf, W. M., 1971, Hydraulics of sediment transport: McGraw-Hill, New York, 573p.

Graf, W. L. and Pazis, G. C., 1977, Les phénomènes de déposition et d'érosion dans un canal alluvionnaire: J. Hydraulic Res., v. 15, p. 151-166.

Jopling, A. V., 1966, Some principles and techniques used in reconstructing the hydraulic parameters of a paleoflow regime: J. Sediment. Petrol., v. 36, p. 5-49.

Lane, E. W., and Carlson, E. J., 1953, Some factors affecting the stability of canals constructed in coarse granular materials: Internat. Assoc. Hydraulic Res. and Hydraulics Div., Am. Soc. Civil Eng., Minnesota International Hydraulics Conference, University of Minnesota, Minneapolis, Sept. 1-4, 1953, Proc., p. 37-48.

Limerinos, J. T., 1970, Determination of the Manning coefficient from measured bed roughness in natural channels: U.S. Geol. Survey Water-Supply Paper 1898-B, 47p.

Meyer-Peter, E., and Müller, R., 1948, Formulas for bed-load transport: Internat. Assoc. Hydraulic Res., Meeting of Stockholm, 1948, Proc., Appendix 2, p. 1-26.

Middleton, G. V., *ed.*, 1965, Primary sedimentary structures and their hydrodynamic interpretation: Soc. Econ. Paleont. Mineral. Spec. Pub. 12, 265p.

Miller, G. H., 1973, Late Quaternary glacial and climatic history of northern Cumberland peninsula, Baffin Island, N.W.T., Canada: Quaternary Research, v. 3, p. 561-583.

Neill, C. R., 1967, Mean velocity criterion for scour of coarse uniform bed-material: Internat. Assoc. Hydraulic Res., 12th Congr., Colorado State University, Fort Collins, September 11-14, 1967, Proc., v. 3, p. 46-54.

Nevin, C., 1946, Competency of moving water to transport debris: Geol. Soc. Am. Bull., v. 57, p. 651-674.

Novak, I. D., 1973, Predicting coarse sediment transport: the Hjulström curve revisited: *in* M. Morisawa, *ed.*, Fluvial Geomorphology; 4th Annual Geomorphology Symposium, Binghampton-SUNY, September 27-28, 1973. Proc., p. 13-25.

Ramette and Heuzel, 1962, Le Rhône à Lyon. Étude de l'entrainment des galets à l'aide de traceurs radioactifs: La Houille Blanche 17, p. 389-399.

Ritter, J. R., 1967, Bed-material movement, Middle Fork Eel River, California: U.S. Geol. Survey Prof. Paper 575-C, p. C219-C221.

Rubey, W. W., 1938, The force required to move particles on a stream bed: U.S. Geol. Survey Prof. Paper 189-E, p. 121-141.

Scott, K. M., and Gravlee, G. C., 1968, Flood surge on the Rubicon River, California — hydrology, hydraulics and boulder transport: U.S. Geol. Survey Professional Paper 422-M, 40p.

Shields, A., 1936, Anwendung der Ähnlichkeitsmechanik und der Turbulenzforschung auf die Geschiebebewegung: Preussische Versuchsanstalt für Wasserbau und Schiffbau, Berlin, Mitteilungen, Nr. 26, 26p. Translation by W. P. Ott and J. C. van Uchelen, Application of similarity principles and turbulence research to bedload movement: U.S. Dept. Agr., Soil Conservation Service, Cooperative Laboratory, California Institute of Technology, 70p.

Tison, L. J., 1953, Recherche sur la tension limite d'entrainement des matériaux constitutifs du lit: Internat. Assoc. Hydraulic Res. and Hydraulics Div., Am. Soc. Civil Eng., Minnesota International Hydraulics Conference, University of Minnesota, Minneapolis, Sept. 1-4, 1953. Proc., p. 21-35.

Wilcock, D. N., 1971, Investigation into the relations between bedload transport and channel shape: Geol. Soc. Am. Bull., v. 82, p. 2159-2176.

Wolman, M. G., and Eiler, J. P., 1958, Reconnaissance study of erosion and deposition produced by the flood of August, 1955, in Connecticut: Am. Geophys. Union Trans., v. 39, p. 1-14.

Yalin, M. S., 1972, Mechanics of sediment transport: Pergamon, Oxford, 290p.

APPENDIX: NOTATION

A	Cross-sectional area of flow	(L^2)
a_i	Constants in equations (1a) - (7)	
D	Grain size of particles (b-axis diameter or equivalent measure)	(L)
D_x	Size than which x percent of the particles is smaller	
d	Depth of flow	(L)
n	Manning total flow resistance coefficient	$(L^{-1/3} T)$
P	Wetted perimeter of flow	(L)
Q	Total discharge of water	$L^3T^{-1})$
q	Specific discharge of water (discharge per unit width)	$(L^3T^{-1}L^{-1})$
R	Hydraulic radius of flow, $= A/P$	(L)
R_*	Particle Reynolds number, $= Dv_*/\nu$	(0)
S	River slope (energy gradient: approximately equivalent to water surface slope)	(0)

v	Flow velocity	(LT^{-1})
v_*, v_{*c}	Friction or shear velocity, $= (\tau/\rho_f)^{1/2}$; critical shear velocity for sediment entrainment	(LT^{-1})
w, w_s	Width of flow; surface width of flow	(L)
γ_f, γ_s	Specific gravity of water; of sediment	(MLT^{-2})
θ	Dimensionless shear stress $= \tau/(\gamma_s - \gamma_f) D$	(0)
ν	Kinematic viscosity of water	(L^2 T^{-1})
ρ_f	Mass density of water	(ML^{-3})
τ, τ_c	Shear stress at the streambed; critical shear stress for sediment entrainment	(ML^{-1} T^{-2})
ϕ	Notation for phi measure	

THE GUADALUPE RIVER AND DELTA OF TEXAS — A MODERN ANALOGUE FOR SOME ANCIENT FLUVIAL-DELTAIC SYSTEMS[1]

Robert A. Morton[2] and Alan C. Donaldson[3]

Abstract

Fluvial-deltaic systems within the Conemaugh Group (Pennsylvanian) of northern West Virginia are similar to the Guadalupe River and delta of Texas. Similarities include gross lithology, facies distribution, and sandstone geometry. Fluvial environments and closely associated marine environments recognized in the Guadalupe and Conemaugh Group sediments are (1) alluvial channels, (2) distributary channels, (3) distributary-mouth bars, and (4) destructive marine facies of abandoned deltas. The fluvial environments represent small suspended-load and mixed-load rivers that are characterized by meandering alluvial channels with numerous point-bars, which give way downstream to relatively straight delta distributaries. Fluvial processes dominated both the modern Guadalupe and Conemaugh Group deltas.

Reconstructions of paleohydrology, three dimensional channel characteristics and sediment load for a river of Pennsylvanian age (Grafton alluvial channel) are comparable to hydrological, morphological, and sedimentological data for the Guadalupe River. This comparison and previously published paleoflow data indicate that the Guadalupe fluvial-deltaic system is analogous to some ancient coastal-plain rivers and shallow-water deltas that prograded into and filled relatively stable cratonic basins.

Introduction

Recent descriptions of modern rivers and deltas have resulted in a better understanding of the myriad of extant fluvial-deltaic systems. Most studies of modern deltas deal with facies analysis of depositional environments and the physical processes responsible for sediment dispersal but the attributes of associated fluvial systems are commonly accorded secondary importance. Notable exceptions, for example, are studies of the Mississippi River (Fisk; 1944, 1947, 1955) and Brahmaputra River (Coleman, 1969) that document the hydrological and sedimentological characteristics of modern rivers associated with large deltas.

As a result of these modern studies numerous ancient fluvial-deltaic systems have been recognized in the stratigraphic record and efforts are being made to quantify some ancient fluvial systems by reconstructions of river morphology and hydrology. These paleoflow studies, however, rarely provide data that allow close comparisons with modern analogues.

The present study emphasizes application of the Guadalupe River and delta of Texas as an analogue of some ancient coastal-plain rivers and shallow-water deltas. The anatomy of the Guadalupe delta and its depositional history were described by Donaldson et al. (1970) from cores, textural analyses, biological assemblages, and facies maps. These data, in turn, were the basis for developing a model of fluvial-dominant shallow-water deltas (Guadalupe type) (Donaldson, 1974). The model contained fluvial components but documentation of the physical characteristics of the Guadalupe River was limited. Consequently, Morton and Donaldson (1978) described hydrological, sedimentological, and three dimensional channel characteristics of the suspended-load and mixed-load

[1]Publication authorized by the Director, Bureau of Economic Geology, The University of Texas at Austin.
[2]Bureau of Economic Geology, The University of Texas at Austin, Austin, Texas 78712
[3]Department of Geology and Geography, West Virginia University, Morgantown, West Virginia 26506

streams that together comprise the Guadalupe fluvial system. Morton and Donaldson also demonstrated that the empirical relationships established for modern rivers by Schumm (1960a, 1960b, 1963a, 1963b, 1968a) are applicable to small coastal-plain rivers in subhumid climates. This extension of Schumm's concepts lends additional support to the validity of quantitative paleostream analysis of ancient fluvial-deltaic systems.

The present study utilizes the previous studies to interpret ancient fluvial-deltaic systems within the middle part of the Conemaugh Group (Pennsylvanian) in northern West Virginia. Specific results include (1) descriptions of alluvial channel and distributary channel facies of Guadalupe and Conemaugh Group sediments (2) paleochannel reconstruction of the Grafton sandstone according to relationships presented by Schumm (1968a, 1972), and (3) comparison of channel morphology and hydrology of the ancient Grafton and modern Guadalupe Rivers.

SHALLOW-WATER DELTA MODEL

The Guadalupe River and delta are situated within a stable depositional basin where fluvial processes are dominant over marine processes and where sediment supply is greater than compactional subsidence. This physical setting gives rise to other distinguishing features of shallow-water fluvial-deltaic systems (Guadalupe type) such as relatively thin prodelta facies, relatively thick delta-plain facies, and the cannibalization of previously deposited deltaic sediments, which are replaced by active and abandoned alluvial channel and distributary channel deposits during delta progradation (Donaldson, 1974). Channel meandering and active point bars are restricted to the alluvial plain and upper delta plain whereas delta distributaries are single, straight symmetrical channels. These fluvial facies represent the greatest total sand preserved in the stratigraphic record. Because sandstones are important for establishing the three dimensional framework of the delta, correct interpretation and understanding of the sandstones through their sedimentary structures, geometry, stratigraphic position, and trend (Donaldson and Morton, 1971) are vital to the overall interpretation of deltaic sediments.

Dominantly sand facies, Guadalupe River and delta

Alluvial plain

Point-bar and flood-basin deposits of the Guadalupe River are composed predominantly of coarse to fine sand that grades vertically into flood-basin muds. Gravel bars in the active alluvial channel and gravel beds in the Holocene valley fill are reworked late Tertiary and Quaternary gravels that form an extensive cover over Coastal Plain sediments in the Guadalupe drainage basin.

Alluvial valley fill consists principally of meander belt sand and silty sand that averages between 6.5 and 10 km wide with thicknesses ranging from 7.5 to 11 m. The surface of the entrenched valley fill exhibits numerous courses of previously active fluvial channels in various stages of abandoment; small oxbow lakes are also common. Although most of the valley fill is sand size or greater, the surface is fine grained, usually mud veneered, and lacks accretionary depositional features such as ridge and swale topography (McGowen and others, 1976).

Distributary channels

Channel fill in abandoned distributaries of the Guadalupe delta consists of fine sand and silty sand which grades upward into clay and silty clay denoting the change from active to abandoned conditions (Donaldson et al., 1970). The sands and silty sands exhibit large- and small-scale trough crossbedding; logs and plant fragments are abundant but fauna is rare. An erosional contact separates these sediments from underlying deposits.

Distributary channels are limited in width to several hundred feet, suggesting lateral stability both spatially and temporally. Maximum cored thickness of distributary channel fill was 6.7 m; however, most channel-fill deposits were between 3 and 4.5 m thick.

Distributary-mouth bar

Distributary-mouth bar deposits are fine sand, sandy silt, and silty sand interbedded with thin beds and laminae of silty clay (Donaldson et al., 1970). Silty clay increases and clay beds thicken seaward from the distributary mouth. Sedimentary structures are primarily accretion ripples that grade into starved ripples and burrow-mottled structures from the bar crest toward the distal limits of the bar.

Distributary-mouth bars extend laterally from 1.2 to 1.8 km and thicknesses range from 1 to 2 m and average about 1.2 m; however, local accumulations up to 2.5 m thick in alternate positions along the Guadalupe River suggest frequent abandonment of distributaries contemporaneously with delta growth.

Beach ridge

The strandplain adjacent to the Guadalupe delta represents reworked sediment locally derived from the bay and nearby Pleistocene escarpment. The abandoned beach ridge, the most prominent feature on the plain, is dominantly sand but contains caliche nodules, pebbles, and shell fragments. Grain size increases upward in the vertical sequence and the coarsest material is concentrated at the surface as a storm-lag deposit. The caliche nodules and much of the sand are reworked Pleistocene sediments. The beach ridge ranges in thickness from 0.6 to 2.6 m and extends laterally about 1.5 km. Primary sedimentary structures and bedding are not commonly preserved within these beach ridge deposits because of extensive bioturbation.

Other Characteristics of the Guadalupe River

Three dimensional channel characteristics, bed forms, sediment load, and discharge for the lower reaches of the Guadalupe River were presented by Morton and Donaldson (1978). In general, width-depth ratio, channel sinuosity, channel gradient, valley gradient, mean grain size of channel sediment, and percent sand in channel sediment and bank sediment decrease downstream. The greatest changes in these parameters commonly occur between the alluvial plain and the delta plain.

The Guadalupe River is a suspended-load stream according to the classification of Schumm (1968b) which is based on channel width-depth ratio, gradient, and sinuosity. The downstream decrease in width-depth ratio and sedimentological parameters appears to be a common characteristic of suspended-load fluvial-deltaic systems. The downstream decrease in sinuosity, however, is best explained by the decrease in channel gradient (Morton and Donaldson, 1978).

Discharge for the Guadalupe River and its major downstream tributaries is highly variable and depends largely on meteorological conditions. Lowest discharges are recorded during droughts whereas extreme flooding is commonly associated with aftermath rainfall from hurricanes. Average discharge in the vicinity of delta is about 65 m³/sec (Morton and Donaldson, 1978).

FLUVIAL-DELTAIC SANDSTONE FACIES OF CONEMAUGH GROUP

Fluvial-deltaic environments within the middle part of the Conemaugh Group (Pennsylvanian) of northern West Virginia were examined in detail by Morton (1972). Rock properties, including grain-size distribution, sedimentary structures, composition, sandstone geometry, and crosscutting relationships as well as the three dimensional facies distribution were used to determine depositional environments. The stratigraphic framework was based on measurements and descriptions of nearly 100 measured sections from outcrop and core data.

Characteristics of Sandstone Facies

Alluvial channel

Alluvial channel deposits are characterized by coarse- to fine-grained, moderately well-sorted sandstones. Transverse cross sections of sandstone bodies are broadly convex downward; cross stratification types are predominantly point bar (epsilon bedding of Allen, 1963), tabular, and festoon or large-scale troughs. Grain size decreases from very coarse sand and granules at the base to silty sand and sandy silt at the top. Basal lag deposits consist of shale clasts and ironstone nodules (1.25 to 15 cm), coal fragments, and carbonized plant remains. Alluvial channels generally range from 6 to 12 m thick except where two channels have coalesced vertically to produce a composite sandstone up to 20 m thick.

Lower point-bar, chute-fill, and chute bar deposits similar to those of the Colorado River described by McGowen and Garner (1970) are preserved in the Grafton sandstone (Fig. 1). Sedimentary structures and grain-size data suggest that the Grafton point-bar sequence falls somewhere between the fine-grained point bar of the Brazos River (Bernard and Major, 1963; Bernard and others, 1970) and the coarse-grained point bar of the Colorado River (McGowen and Garner, 1970). Abandoned chute fill of the postulated mixed-load stream is composed of shale rather than fine sand with thin mud drapes, as was reported for the Colorado River; furthermore, trough cross-stratification is the dominant sedimentary structure in upper point-bar deposits (chute bar) of the Grafton sandstone rather than more planar foreset cross-stratification. Lower point-bar deposits of the Grafton sandstone are characterized by bed sets, bounded by scour surfaces, that exhibit graded bedding from basal conglomerate to sandstone and siltstone.

Distributary channel

Aggradational distributary channels are preserved as sandstones ranging in thickness from 4.5 to 10 m. The geometry of sandstone units normal to flow direction is convex downward. Other distinguishing features are erosional basal contacts and a vertical decrease in grain size from sand to sandy silt. Bedding is more massive at the base but decreases in thickness towards the channel top. Large- and small-scale trough cross bedding are the most common sedimentary structures.

Distributary channels that were abandoned periodically are recognized by their overall upward decrease in total sand and by erosional basal and lateral contacts. Furthermore, asymmetrical channel infilling is dominated by shale interbedded with shaly siltstone, and alternating thin beds of fine sandstone and siltstone representing channel reactivation. Examples of these active and abandoned channel facies are preserved in an outcrop of the Buffalo sandstone (Fig. 2), interpreted as a longitudinal section of a lower delta plain distributary channel. This outcrop shows sand waves with an amplitude of 1.3 m and a wave length of 45 m overlain by siltstone deposited during channel abandonment.

Distributary-mouth bar

Distributary-mouth bars, 1.5 to 7.5 feet thick, are characterized by interbedded sandstone, siltstone, and shale. Thin bedded, fine-grained sandstones with shale clasts are common sediments of proximal bars (near the river mouth) whereas more distal bars are usually interbedded siltstone and shale up to 6 m thick. Both lower and upper contacts are usually gradational with underlying and overlying shale although some outcrops show distinct lower contacts of proximal bars. Horizontal bedding is common in most distributary-mouth bar sandstones, but some exhibit small-scale ripples with amplitudes

alluvial channel
(Grafton sandstone)

shelf
(Ames shale)

Fig. 1. Alluvial channel facies of Grafton sandstone exhibiting lower point-bar, abandoned chute fill, and upper point-bar deposits. Outcrop located on Interstate 79 approximately 11.3 km south of Morgantown, West Virginia.

destructive marine
(Pine Creek limestone)
delta plain

distributary
channel
(Buffalo sandstone)

Fig. 2.   Outcrop of Buffalo sandstone interpreted as a longitudinal section of a distributary channel. Overlying the sandstone and siltstone is a fossiliferous, bioturbated destructive marine unit. Outcrop is on Interstate 79 approximately 7.2 km south of Morgantown, West Virginia.

ranging from 0.3 to 2.5 cm; burrows are common where thin (less than 1 m) sandstones form the upper part of the bar. X-ray radiographs show that proximal bars are relatively homogeneous whereas distal bars contain microripples that occur as slightly undulose alternations of coarser and finer laminae.

Destructive marine units

The destructive marine units, or sediments representing delta abandonment and reworking are usually characterized by a thin (15 to 45 cm) sandy limestone or calcareous sandstone bed (Fig. 2). Within the bed, crinoid stems, brachiopods, and bryozoa are abundant and burrow structures are numerous. Some destructive marine units probably represent carbonate shoal areas of the subaqueous abandoned lower delta plain whereas others are probably remnants of thin beach deposits.

Typical Stratigraphic Sequences

An outcrop containing the Grafton sandstone (Fig. 3) also includes a progradational sequence from normal marine to alluvial channel. An underlying sequence forming the lower 13 m of the outcrop, consists of shale and calcareous shale with limestone and ironstone nodules interbedded with thin beds of siltstone and limestone that probably represent marsh and lake deposits of an older deltaic cycle. The overlying Harlem coal (15 cm) represents swamp deposits and marks the top of the previous cycle. Overlying the coal are 7.5 m of interbedded fossiliferous limestone and calcareous shale which are the record of a widespread marine transgression. This unit, the Ames Shale, grades vertically into 3 m of nonfossiliferous shale or regressive prodelta deposits. In turn, the thin prodelta grades upward into 1 m (total exposed) of siltstone and sandstone representing the distributary-mouth bar. The prodelta and distributary-mouth bar were partially removed by a prograding distributary channel that was later abandoned, as indicated by the asymmetrical fill of shale and silty shale. The upper surface of the lower delta plain is truncated by the overlying alluvial channel facies of the Grafton sandstone. Overall, the distribution of fluvial-deltaic sediments within this outcrop is remarkably similar to that of the Guadalupe delta model (Donaldson *et al.*, 1970).

COMPARISON OF GUADALUPE AND GRAFTON FLUVIAL-DELTAIC SYSTEMS

Facies characteristics and distribution

Collation of specific facies demonstrates the similarities and differences of the Grafton and Guadalupe systems. For example, geometry of fluvial counterparts and vertical and lateral facies relationships within the two systems are analogous. The compound birdfoot-lobate Guadalupe delta (2.7-3.7 m thick) overlies bay sediments. Cycles within the Conemaugh Group (12-20 m thick) appear to be relatively thin, elongate, high constructive deltas that prograded over previously deposited deltaic or offshore marine facies.

The predominance of suspended sediment transported by the Guadalupe River gives rise to dominantly mud deltas. Similarly, the volume of mudstone and the greater proportion of mudstone versus sandstone in the Conemaugh Group sediments suggest that the Pennsylvanian fluvial systems were predominanty suspended-load streams. Grain size of bed load for both the Guadalupe and Grafton Rivers decreases downstream (Fig. 4) but sediment transported by the Grafton fluvial system was slightly coarser and included a greater size range than sediment presently transported by the Guadalupe River. Distributary-mouth bars of Conemaugh Group deltas are thin (3 to 7.5 m) and wide (1.2 to 2.1 km) lenticular bodies having a geometry similar to distributary-mouth bars of the Guadalupe delta, which are not as thick (1 to 2 m) and extend laterally for approximately 1.2 to 1.8 km. The difference in thickness between delta front facies of the

alluvial channel
(Grafton sandstone)

distributary-mouth bar

prodelta

shelf
(Ames shale)

swamp (Harlem coal)

interdistributary bay
and lake
(Ewing limestone)

Fig. 3. Outcrop of Ames shale and Grafton sandstone showing vertical distribution of fluvial-deltaic facies representative of depositional environments ranging from normal marine (shelf) to alluvial channel. The Harlem coal is partially covered by talus on the bench. Outcrop is located on Interstate 79, approximately 10.5 km south of Morgantown, West Virginia.

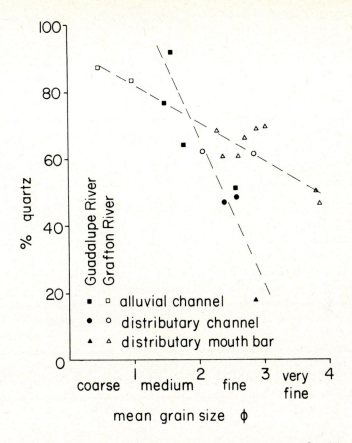

Fig. 4. Superimposed plots of mean grain size and per cent quartz, Guadalupe River and Grafton River fluvial systems. Data were averaged for within-segment variation.

Guadalupe and Grafton deltas is probably related to water depth of the basin. The Guadalupe River debouches into a bay less than 2 m deep; however, the bay was about 4 m deep during initial progradation. Stratigraphic relationships indicate that the ancient Grafton delta likewise prograded into shallow water probably less than 9 m deep. The entire Grafton delta sequence is approximately 9 to 12 m thick; however, this is minimum thickness because of post-depositional compaction.

Major differences in the two systems are related to scale, duration, and possibly total suspended sediment transported by the fluvial systems. Individual facies such as prodelta, delta front, and delta plain are thicker within the Conemaugh Group deltas than counterpart facies of the Guadalupe delta. Furthermore, Conemaugh Group deltas prograded much farther than the Guadalupe delta but this may be due, in part, to differences in time and greater suspended sediment of the Pennsylvanian rivers.

Paleohydrology of Grafton River

Similarities between the Guadalupe fluvial-deltaic system and Conemaugh Group sediments leads to reconstruction of the Grafton River paleochannel and comparison of paleochannel characteristics and paleohydrology with channel morphology and hydrology of the Guadalupe River. Analytical methods used to estimate paleochannel characteristics

Table 1. Interrelationships of channel morphology, streamload, and flow characteristics[*]

Variable	Symbol	No.	Equation	Reference
WIDTH/DEPTH RATIO	F	(1)	$F = 255M^{-1.08}$	Schumm, 1960b
SINUOSITY	P	(2)	$P = 0.94M^{0.25}$	Schumm, 1963b
		(3)	$P = 3.5\,F^{-0.27}$	Schumm, 1963a
MEAN-ANNUAL FLOOD	Qma	(4)	$\log Qma = 0.268 + 0.469 \log M + 1.378 \log W$	Cotter, 1971 (after Schumm)
MEAN-ANNUAL DISCHARGE	Qm	(5)	$Qm^{0.38} = \dfrac{WM^{0.37}}{37}$	Schumm, 1968a
GRADIENT	Sc	(6)	$S_c = 60M^{-0.38}\,Qm^{-0.32}$	Schumm, 1968a
BED-LOAD %	S	(7)	$S = \dfrac{55}{M}$	Schumm, 1968a
WIDTH/DEPTH RATIO	F	(8)	$F = \dfrac{21\,Qma^{0.18}}{M^{0.74}}$	Schumm, 1968a
WIDTH	W	(9)	$W = \dfrac{2.3\,Qma^{0.58}}{M^{0.37}}$	Schumm, 1968a
DEPTH	D	(10)	$D = 0.6M^{0.34}Qm^{0.29}$	Schumm, 1968a

[*]Empirical relationships based on feet (length) and feet per second (discharge). Values were converted to metric units for text and tables.

have developed in the following stages: (1) determination of quantitative relationships of modern streams of the Great Plains by Schumm (1960a, 1960b, 1963a, 1963b), (2) reconstruction of the paleohydrology of ancient streams (Schumm, 1968a; Donaldson, 1969; Cotter, 1971; Derr, 1974; Elliot, 1976; Miall, 1976; Padgett and Ehrlich, 1976), (3) comparison of quantitative expressions of Great Plains streams with coastal plain streams by Morton and Donaldson (1978) and finally, (4) the comparison of ancient and modern streams with similar geometry and flow characteristics (this paper). During the past decade, Schumm developed empirical equations (Table 1) which relate three dimensional channel characteristics to stream discharge and to the type of stream load (Schumm,

1972) on the basis of river studies in the Western United States (Great Plains) and Australia. Basic to these equations is the relationship of width-depth ratio (F in Table 1) to percent silt-clay (less than 0.074 mm) in the channel perimeter (M in Table 1).

Equations listed in Table 1 were established for rivers in a semiarid to subhumid climate with modern vegetation. Climate during the Pennsylvanian Period was probably warm and humid, perhaps with seasonal rainfall (Schopf, 1975) that promoted growth of abundant tropical vegetation; but grasses had not evolved. Therefore, because the climate and vegetation, which affect flood peak, runoff, and bank stability, were not the same for the Great Plains and the ancient Grafton River, direct application of these equations could be in error.

Evaluation of paleochannel reconstruction

Two general methods have been accepted for obtaining paleochannel dimensions from outcrop measurements. The first method pertains to outcrops with epsilon cross stratification (Allen, 1963) and depends on the following assumptions: (1) that height from toe to crest of epsilon cross stratification is equivalent to bankfull depth, and (2) that horizontal extent of the same surface measured perpendicular to the channel axis is two-thirds bankfull width.

The second method assumes that the coarse channel fill, usually sandstone, is equal in thickness to bankfull depth (Allen, 1965; Leeder, 1973); the depth thus obtained can be used to derive bankfull width according to the equation proposed by Leeder (1973). This assumption appears to be valid judging from the stratigraphic cross sections presented by Fisk (1947), Frazier and Osanik (1961), and Bernard and others (1970). Those studies show that coarse-member thickness is a good approximation of bankfull channel depth except where vertical aggradation has occurred.

Both methods for paleochannel reconstruction have inherent limitations and sources of inaccuracy, but they appear to be basically sound for estimating the desired parameters. The overriding factor that has not been fully evaluated, however, is the equivalency of width-depth ratios determined in meander bends and straight reaches of the same stream. This uncertainty effects both methods of reconstruction since the first method deals strictly with point-bar surfaces and the second method is also predicated on a depth that is representative of conditions in a meander bend. One might intuitively expect different width-depth ratios in straight reaches and in meander bends, knowing that both channel width and depth are greater in meander bends than in straight reaches (Fisk, 1947; Sundborg, 1956). Data for the Guadalupe River indicate that there is considerable natural variance of width-depth ratios along a stream segment but, when paired for comparison, adjacent values of width-depth ratio in meander-bend and straight reaches are remarkably similar.

Reconstruction of Grafton River

In light of preceding evidence and in order to remain consistent with previous paleochannel reconstructions (Donaldson, 1969; Cotter, 1971; Leeder, 1973; Elliott, 1976), the general methods described for outcrops containing epsilon cross stratification were applied with one notable exception. The principle that the width of the point-bar surface is two-thirds the actual width of the river, suggested by Moody-Stuart (1966) and subsequently accepted by other workers (Donaldson, 1969; Cotter, 1971; Leeder, 1973; Elliott, 1976), is limited to fine-grained point bars and is not applicable to two-tiered point bars separated by a chute. Suspended-load streams such as the Guadalupe River rarely have chutes, but mixed-load streams like Coleto Creek, a tributary of the Guadalupe (Morton and Donaldson, 1978) have upper and lower point bars. McGowen and Garner's (1970) example from the Colorado River, as well as air photo and field data for Coleto Creek, indicate that the best estimate of straight-reach width can be made by applying the two-thirds correction to the width of the lower point-bar surface.

A point-bar surface of the Grafton sandstone (Fig. 1) was traced laterally for about 50 m normal to the direction of flow before it terminated against chute-bar deposits. A value of approximately 75 m was obtained by using 50 m as two-thirds the river width. If scour by the chute removed some of the lower point bar, this estimate of width is slightly low. Bankfull depth was also adjusted because fathometer profiles of the Guadalupe River as well as channel cross sections for the Mississippi River (Fisk, 1947) clearly show that bankfull depth in a straight reach is approximately two-thirds bankfull depth in a meander bend. The meandering channel facies of the Grafton sandstone is generally 9 m thick. Applying the two-thirds correction gives a paleochannel bankfull depth 6 m. The percent silt-clay in the channel perimeter (M) was estimated from point-counts of thin sections. Using estimates of width = 75 m, depth = 6 m, and M = 15%, additional properties of the ancient river were derived using the equations in Table 1. Substituting M = 15 in equation (1) gives a width-depth value of 13.7, which compares favorably with the field estimate of 12.5.

Sinuosity was calculated from equations (2) and (3) using M = 15 and F = 12.5 and 13.7. Sinuosity for equation (2) is 1.85 whereas the values calculated by equation (3) are 2.11 and 2.07 respectively. Mean annual flood estimated from equation (4) was 364 m^3/sec and mean annual discharge calculated from equation (5) was 58.8 m^3/sec. Calculations from equation (6) indicate that the gradient of the river was low, approximately 0.35 m/km. The percent of total load transported as bed load, 3.7 % calculated by equation (7), as well as estimates for sinuosity and gradient, suggest that the Grafton fluvial system was a mixed-load stream according to Schumm's classification of alluvial channels (Schumm, 1968b). Width, depth, and width-depth ratio were checked by equations (8), (9), and (10). Substituting the previously estimated value for mean annual flood (364 m^3/sec) in equation (8) yields a width-depth ratio of 15.6. The width calculated from equation (9) is 62.5 m and the depth from equation (10) is 4.3 m. These values are reasonably close to the values obtained from the field estimates and other calculations.

Although a single value was calculated for each parameter of the Grafton River, a range of values for each parameter can be obtained by varying the correction factors and assumptions used in the calculations or by obtaining values for the confidence limits of the regression line (Leeder, 1973). The sources of error in the computations ranked from lowest to highest in relative importance, are: (1) errors in measurements for both outcrops and thin section data, (2) compounding error as a result of substituting erroneous values from previous calculations, (3) sampling error, (4) validity of assumptions in estimating correction of width and depth as well as lack of correction for compaction, and (5) validity of application of empirical equations to a system which has not been defined empirically. All sources of error except the validity of application affect only the precision of the computations.

COMPARISON WITH GUADALUPE RIVER AND OTHER PALEOCHANNELS

Hydrology and channel morphology estimated for the Grafton fluvial system and comparable data compiled for the Guadalupe River Victoria gaging station (U.S. Geological Survey, 1974 and U.S. Congress, 1954) are presented in Table 2. Even though the calculations for the Grafton River are only gross estimations, if the relationships of Schumm are applicable to conditions during Pennsylvanian time, then it is evident that the Grafton fluvial system closely resembled the Guadalupe and other small coastal plain rivers. The data, however, should not be construed as representing indentical fluvial systems, but rather as representative data from relatively similar systems. The assumptions involved in estimating discharge for paleochannels obviously do not permit precise determinations, but, order of magnitude estimates of hydrologic parameters can be inferred. The size of the two systems and their load are similar; therefore, it is not surprising that the calculations result in similar hydrologic and three dimensional characteristics.

Table 2.. Comparison of three dimensional characteristics and paleohydrology for the Grafton fluvial system with measurements for the Guadalupe River (Victoria gaging station) 50.7 miles above the mouth of the Guadalupe River. Hydrologic data for the Guadalupe River were compiled and converted from surface water records (U.S. Geological Survey, 1974) and House Document 344 (U.S. Congress, 1954).

	Modern Guadalupe River	Ancient Grafton River
Width	57 m	75 m
Depth	7 m	6 m
Width/Depth	8.1	12.5
Sinuosity	2.24	2.0
Gradient	0.33 m/km	0.35 m/km
Mean-annual flood	389 m³/sec	364 m³/sec
Mean-annual discharge	45.8 m³/sec	58.8 m³/sec
Mean grain size (channel)	0.45 mm (sieve)	0.62mm (thin section)

In a review of paleochannel reconstruction, Schumm (1972) emphasized the paucity of ancient paleochannels with dimensions comparable to modern rivers with drainage basins of continental proportions. Schumm also noted the distinct differences between sandstones associated with single paleochannels and alluvial valley fills comprising multiple channels. Previously published reconstructions of single paleochannels suggest that small coastal plain rivers or perhaps fluvial systems associated with subdeltas of larger delta systems are important components of the rock record. Ancient rivers studied by Cotter (1971), Derr (1974), Donaldson (1969), Leeder (1973) and Elliott (1976) have three dimensional characteristics and calculated paleohydrology that are the same order of magnitude as the Guadalupe River and other modern coastal plain rivers.

ACKNOWLEDGMENTS

Appreciation is expressed to S. A. Schumm, E. Cotter, V. R. Baker, L. F. Brown, Jr., and J. H. McGowen, who critically read the manuscript and made suggestions that substantially improved its clarity and content.

REFERENCES

Allen, J. R. L., 1963, The classification of cross-stratified units, with notes on their origin: Sedimentology, v. 2, p. 93-114.

———, 1965, A review of the origin and characteristics of recent alluvial sediments: Sedimentology, v. 5, p. 89-191.

Bernard, H. A., and Major, C. F., Jr., 1963, Recent meander deposits of the Brazos River: an alluvial "sand" model: (abs.) Am. Assoc. Petrol. Geol. Bull., v. 47, p. 350.

———, Major, C. F., Jr., Parrott, B. S., and LeBlanc, R. J., Sr., 1970, Recent sediments of southeast Texas — a field guide to the Brazos alluvial and deltaic plains and the Galveston barrier island complex: Univ. Texas, Austin, Bur. Econ. Geology, guidebook 11, 83 p.

Coleman, J. M., 1969, Brahmaputra River: channel process and sedimentation: Sediment. Geol., v. 3, p. 129-239.

Cotter, E., 1971, Paleoflow characteristics of a late Cretaceous river in Utah from analysis of sedimentary structures in the Ferron sandstone: J. Sediment, Petrol., v. 41, no. 1, p. 129-138.

Derr, M. E., 1974, Sedimentary structure and depositional environment of paleochannels in the Jurrasic Morrison Formation near Green River Utah: Brigham Young Univ. Geol. Studies, v. 21, pt. 3, p. 3-40.

Donaldson, A. C., 1969, Paleostream analysis, *in* Some Appalachian coals and carbonates: models of ancient shallow-water deposition: West Virginia Geol. and Econ. Survey, p. 265-277.

————, 1974, Pennsylvanian sedimentation of Central Appalachians: Geol. Soc. Am. Spec. Paper 148, p. 47-78.

————, Martin, R. H., and Kanes, W. H., 1970, Holocene Guadalupe Delta of Texas Gulf Coast, *in* J. P. Morgan, *ed.*, Deltaic sedimentation modern and ancient, Soc. Econ. Paleont. Mineral. Spec. Pub 15, p. 107-137.

————, and Morton, R. A., 1971, Paleodrainage analysis of some Pennsylvanian sandstones in Central Appalachians: Geol. Soc. Am. Bull., abstracts with program, v. 3, no. 7, p. 547-548.

Elliot, T., 1976, The morphology, magnitude and regime of a carboniferous fluvial-distributary channel: J. Sediment Petrol., v. 46, p. 70-76.

Fisk, H. N., 1944, Geological investigations of the alluvial valley of the lower Mississippi River: Mississippi River Comm., Vicksburg, 78 p.

————, 1947, Fine grained alluvial deposits and their effect on Mississippi River activity: U.S. Army Corps Engineers, Waterways Experiment Station, Vicksburg, Mississippi, 82 p.

————, 1955, Sand facies of Recent Mississippi delta deposits: World Petroleum Cong., 4th, Proc. Sec. 1, p. 377-398.

Frazier, D. E., and Osanik, A., 1961, Point-bar deposits, Old River locksite, Louisiana: Gulf Coast Assoc. Geol. Soc. Trans., v. 11, p. 121-137.

Leeder, M. R., 1973, Fluviatile fining-upwards cycles and the magnitude of paleochannels: Geol. Mag., v. 110, p. 265-276.

McGowen, J. H., and Garner, L. E., 1970, Physiographic features and stratification types of coarse-grained point bars: modern and ancient examples: Sedimentology, v. 14, p. 77-111.

————, Proctor, C. V., Jr., Brown, L. F., Jr., Evans, T. J., Fisher, W. L. and Groat, C. G., 1976, Environmental geologic atlas of the Texas Coastal Zone — Port Lavaca area: Univ. Texas, Austin, Bur. Econ. Geology, 107 p.

Miall, A. D., 1976, Paleocurrent and paleohydrologic analysis of some vertical profiles through a Cretaceous braided stream deposit, Banks Island, Arctic Canada: Sedimentology, v. 23, p. 459-483.

Moody-Stuart, M., 1966, High- and low-sinuosity stream deposits, with examples from the Devonian of Spitsbergen: J. Sediment. Petrol., v. 36. p. 1102-1117.

Morton, R. A., 1972, A study of some ancient, modern, and experimental deltaic sediments: Unpub. Ph. D. thesis, West Virginia Univ. Morgantown, West Virginia, 171 p.

————, and Donaldson, A. C., 1978, Hydrology, morphology and sedimentology of the Guadalupe fluvial-deltaic system: Geol. Soc. Am. Bull., (in press).

Padgett, G. V., and Ehrlich, R., 1976, Paleohydrologic analysis of a late Carboniferous fluvial system, southern Morocco: Geol. Soc. Am. Bull., v. 87, p. 1101-1104.

Schopf, J. M., 1975, Pennsylvanian climate in the United States, *in* McKee, E. D., Crosby, E. J., and others, Paleotectonic investigations of the Pennsylvanian System in the United States: U.S. Geol. Survey Prof. Paper 853, pt. II, p. 23-31.

Schumm, S. A., 1960a, The effect of sediment type on the shape and stratification of some modern fluvial deposits: Am. J. Sci., v. 258, p. 177-184.

————, 1960b, The shape of alluvial channels in relation to sediment type: U.S. Geol. Survey Prof. Paper 352-B, p. 17-30.

————, 1963a, A tentative classification of alluvial river channels: U.S. Geol. Survey Circ. 477, 10 p.

————, 1963b, Sinuosity of alluvial rivers on the Great Plains: Geol. Soc. Am. Bull., v. 74, p. 1089-1100.

————, 1968a, River adjustment to altered hydrologic regimen-Murrumbidgee River and paleochannels, Australia: U.S. Geol. Survey Prof. Paper 598, 65 p.

————, 1968b, Speculations concerning paleohydrologic controls of terrestrial sedimentation: Geol. Soc. Am. Bull., v. 79, p. 1573-1588.

————, 1972, Fluvial paleochannels, *in* J. K. Rigby and W. K. Hamblin, *eds.,* Recognition of ancient sedimentary environments: Soc. Econ. Paleont. Mineral. Spec. Pub. 16, p. 98-107.

Sundborg, A., 1956, The River Klaralven: a study of fluvial processes: Geog. Ann., v. 38, p. 127-316.

U.S. Congress, 1954, Guadalupe and San Antonio River, Tex: House of Representatives, 83d Congress, 2nd Session, House Doc. 344, 60 p.

U.S. Geological Survey, 1974, Water resources data for Texas, Part 1, Surface water records: Austin, Texas, Water Res. Div., 609 p.

AN ANALYSIS OF TWO TECTONICALLY CONTROLLED INTEGRATED DRAINAGE NETS OF MID-CARBONIFEROUS AGE IN SOUTHERN WEST VIRGINIA

Guy Padgett[1] and Robert Ehrlich[2]

Abstract

Simple inspection of the orientation and pattern of a drainage net provides direct information concerning regional slope and degree of structural control in the net as well as, indirectly, hydrologic and climatic information. All of these factors are of interest in interpreting environments of deposition of ancient sedimentary sequences. Inspection of integrated paleodrainage nets can thus provide significant paleogeographic insight. Although rarely exposed at the surface, paleodrainage nets are common in the subsurface and are frequently delineated on maps of mature coal mines.

As an example, two Mid-Carboniferous drainage nets from West Virginia are described and analyzed. Net geometries show a transport direction from southeast to northwest and a slope of about six feet per mile. The orientation of the nets and the trellis arrangement of the channel lengths indicate structural control by the same tectonic framework that uplifted and deformed the resultant strata (Alleghenian deformation). This observation is substantiated by cross sections and isopachous maps for the strata. Discharge, drainage density, drainage area and other hydrologic parameters are calculated from the nets.

Introduction

Many coals were deposited as peats in regions subjected to non-uniform rates of sedimentation controlled, for the most part, by the interaction of migrating depositional environments. These migrations directly reflect rates of sediment influx, discharge of the associated rivers, and nature of pre-existing substrata (e.g. sand, clay, bedrock, etc.). The controlling factors are themselves manifestations of underlying tectonic and climatic factors.

Much of this can be deduced from examination of paleodrainage nets. Such factors as the direction of paleoslope can be determined directly; whereas other factors, such as discharge, can be estimated from net geometry using empirical relations determined from Recent systems (Schumm, 1968, 1972; Leopold & Wolman, 1960; Leopold et al., 1964; Ethridge and Schumm, this volume). In addition, net geometry may reflect the timing and nature of tectonic and climatic events.

However, there are few descriptions of exhumed segments of paleodrainages (Padgett and Ehrlich, 1976; Baker, 1973). These have demonstrated the potential usefulness of paleodrainage analysis but have also underscored the rarity of preserved field examples, the study of which requires their preservation and burial for tens or hundreds of millions of years followed by gentle exhumation of near-horizontal sequences.

Portions of paleodrainage nets are much more numerous in the subsurface. Essentially two-dimensional vertical slices through such nets are commonly observed on outcrop or from well data. The subject of this report concerns the fact that fairly large, continuous portions of subsurface paleodrainage nets are preserved on mine maps from mature coal mines. These nets are abundant enough to enable analysis of paleodrainages to become a significant tool in interpreting ancient continental sequences. These "paleonets" emerge on mine maps of mature coal mines because only coal exceeding a certain minimum thickness is mined underground; hence, the pattern of mined versus unmined areas can faithfully reflect paleodrainage. This can occur in two ways.

---

[1]Central Exploration, Consolidation Coal Company, 6451 Pleasant St., Library, PA 15129.
[2]Dept. Geology, University of South Carolina, Columbia, S.C. 29208.

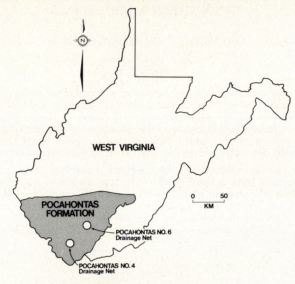

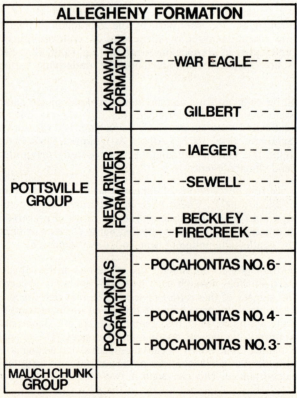

Fig. 1. Location Map and table of stratigraphy.

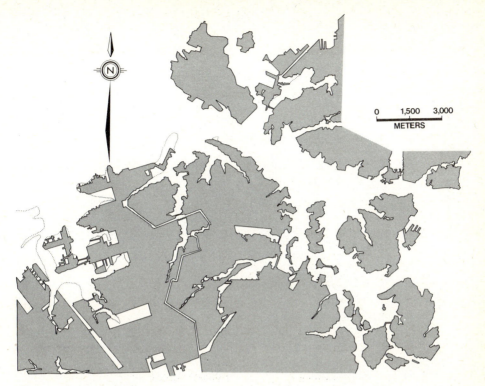

Fig. 2. Composite Mine Map — Pocahontas Number Four Coal. The shading represents mined areas within the coal seam. Unshaded are the local unconformities where coal is absent or thin due to channel scouring from the overlying sandstone. The mine location is shown in Figure 1.

In the first case, a drainage is established on alluvial silts and shales overlying an older buried peat. This drainage system incises the thin detrital cover (amounting to a few feet) and impinges on and partly erodes the underlying peat. The erosion, verified by observation of the cross-cutting relationships in the mines, is clearly delineated on mine maps routinely compiled by coal company personnel.

In the second case, an existing drainage system is abandoned, either by piracy or by structural movements. The abandoned channelways then become the sites for accumulation of coal. With time, the coal environment overtops the channel system and the entire region is converted to a coal swamp. Again, mine map patterns portray the network, and inspection of mine workings is necessary to establish the unconformable relationship between the base of the coal and underlying strata.

The Pocahontas Formation

We will discuss two nets from the Mid-Carboniferous Pocahontas Formation of southern West Virginia (Fig. 1). The Pocahontas Number Six seam is younger and occurs about 18 m above the other, the Pocahontas Number Four seam, in the same general geographic area (Fig. 1). In the older system, involving the Number Four seam, the drainage is manifested by incision of a younger channel system into a peat; whereas, the upper net is outlined by filling an abandoned channel network with peat (now the Pocahontas Number Six seam).

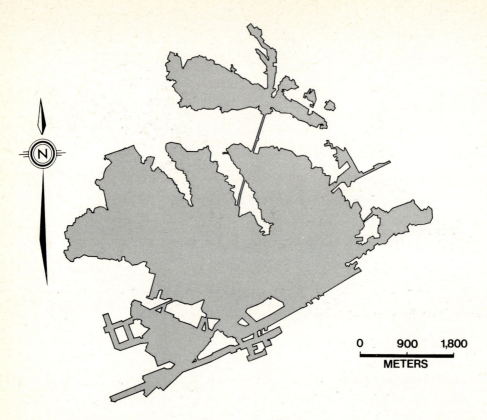

Fig. 3. Mine Map — Pocahontas Number Four Coal. The shading represents mined areas within the coal seam. Unshaded are the local unconformities where coal is absent or thin due to channel scouring from the overlying sandstone. This mine is located southwest of the composite mine illustrated in Fig. 1 (*See* Fig. 7).

About 52 km² of a drainage net were incised into the Pocahontas Number Four coal (Fig. 2). Net geometry simply and clearly indicates that the drainage direction was from the southeast to the northwest. This is supported by facies relationships, crossbed directional data and mineral facies (Davis and Ehrlich, 1974; Englund, 1974; Galloway, 1972; Padgett, 1972). The coal is not as extensively eroded to the southeast. In that direction, the incising channels can be seen in outcrop scouring the overlying silt and shale, but not cutting the coal. We have calculated the gradient from this point to the area where the coal is scoured to be about .00114. A second drainage net, located to the southwest, is illustrated in Figure 3.

The overlying drainage net in the Number Six coal cover covers about 8 km² (Fig. 4). In this case the mined area represents thick coal occupying abandoned channelways. The unmined areas are interfluves. The flow direction is from the northeast to the southwest. This is in contrast to the northwesterly flowing Pocahontas Number Four net. As will be seen, this change in flow direction can be tied to an underlying structural control.

In both cases the pattern of the drainage net is a modified dendritic type. Inasmuch as the substrate is homogeneous, this suggested to us that the oriented pattern is structurally controlled. In order to test this, a series of cross sections and isopachous maps were

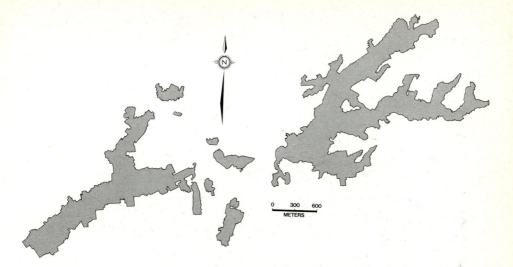

Fig. 4. Mine Map — Pocahontas Number Six Coal. The shading represents mined areas within the coal seam. Unshaded are thin coal areas. The thicker (mined) coal accumulated in abandoned channelways. The mine is located in Figure 1.

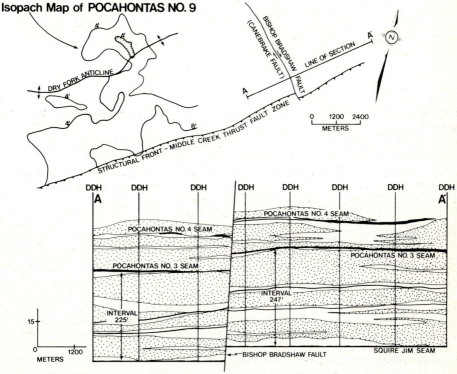

Fig. 5. Isopachous Map — Pocahontas Number Nine Seam — Cross Section Through Bishop/Bradshaw Fault. The interval between the Squire Jim and Pocahontas Number Three seam increases abruptly across the Bishop/Bradshaw fault. The Pocahontas Number Nine seam thins over the Dry Fork anticlinal axis. Both observations support the contention that a single long-lived structural framework controlled both sediment distribution and deformation of the resultant strata.

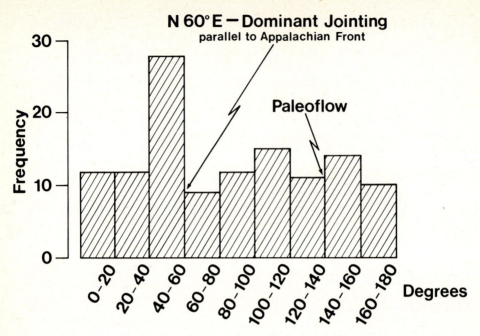

Fig. 6. Histogram of Channel Segment Orientations for the Pocahontas Number Four Drainage. The orientation of each discreet channel segment was measured and the results displayed as a histogram. The dominant range (N 40 E to N 60 E) coincides with the direction of major jointing (N 60 E).

constructed (Fig. 5). Isopach maps of coal seams reveal a general thinning trend over the present anticlinal axis, while cross sections through the same area exhibit abrupt sedimentary thickening across faults that are normal to the Appalachian front such as the Bishop/Bradshaw fault. The Bishop/Bradshaw fault has been interpreted as a right lateral strike slip fault with about 210 - 300 m of lateral movement (Elder *et al.*, 1974). The vertical offset of the strata (about 12 m) is attributed to the gentle folding that existed prior to the strike slip movement. Our data would indicate a sequence of events involving vertical movement along the fault zone occurring during sedimentation (hence, the abrupt stratigraphic thickening as one crosses the fault), followed by strike slip movement along the same fault zone after the strata were lithified and gently folded. This suggests a single, long-lived structural framework that both controlled sediment distribution and deformed the resultant strata.

The orientation of the trellis paleodrainages and the orientation of the major jointing for the area coincide (Fig. 6). The dominant jointing (North 60° East) has been attributed to the same Alleghenian deformation that generated the faults and gentle folds of the plateau (Elder *et al.*, 1974). The trellis channel lengths of the Pocahontas Number Four parallel this jointing, suggesting that the tectonism that produced the deformation controlled the ancient drainage pattern. Another expression of the link between the paleonets and structure rests on the fact that the trend of folds in the Appalachian Plateau changes from the area of the Pocahontas Number Six paleonet to the area of the Pocahontas Number Four paleonet, corresponding to the change in dominant flow directions between the two nets (Fig. 7). This implies that Appalachian folding and faulting was forming at least as early as the Middle Carboniferous. This effect of growing structure on the Carboniferous sequence in Pennsylvania and W. Virginia has been reported by others, including Arkle (1959) and Williams and Bragonier (1974).

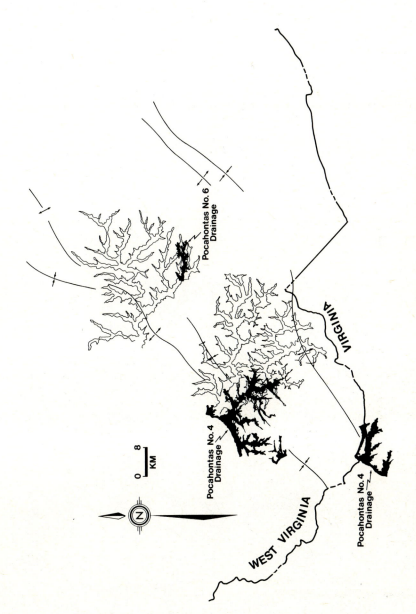

Fig. 7. Ancient and Recent Drainage — Southern West Virginia. The ancient drainage patterns from the Pocahontas Number Four and Number Six seams (shaded) are superimposed over the existing structurally controlled drainage nets in Southern West Virginia (unshaded outline). Major fold axes are as shown.

The present day Appalachian region is a classic area for showing the control of drainage by structure. Thus, the orientation of modern drainage shows a change in direction similar to what we observe in the ancient. Since we say that the structural controls on recent drainage are the same as those of the ancient nets, it is instructive to superimpose and so directly compare the two drainages separated in time by more than 200 m.y. (Fig. 7). The similarities in size and orientation are striking.

Not all Appalachian nets are trellis or dendritic; nor do those that begin as dendritic or trellis remain so. For example, in the case of the Pocahontas Number Four net discussed earlier, small scale structures and grain size distribution within the overlying sandstone deposit and its gross morphology would suggest it was deposited from a meandering channel system. Patterns preserved in other coal seams, such as the Pittsburgh seam in northern West Virginia, are neither dendritic nor trellis and would suggest quite a different geologic history than the examples cited in this paper.

Some statements can be made about the paleohydrology of the Pocahontas nets. Reynolds (1972) and Sharp (1974) have determined that discharge, drainage area, and link magnitude are interdependent in drainage nets of the coastal plain of South Carolina — an area that can be considered analogous to the Pocahontas nets. Reynolds (1972) presented a relationship between link magnitude and area. By substituting the link magnitude from our basins into his equations we can calculate areas. These calculated areas agree closely with areas measured directly from the mine maps (Fig. 8, Table 1). Encouraged by these

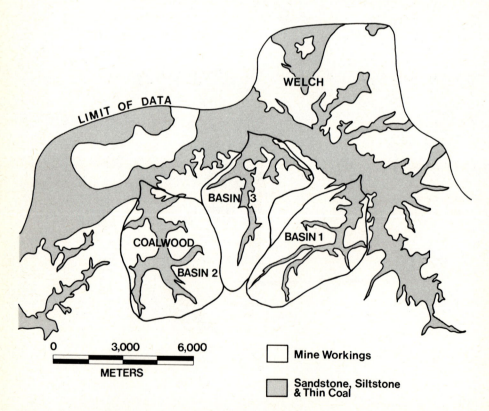

Fig. 8. Interpretation of Composite Mine Map — Pocahontas Number Four Seam. Drainage channels are shaded (as determined from the mine workings map). Hydrologic parameters for the three basins outlined on the map are listed in Table 1.

Table 1. Hydrologic Parameters. The hydrology of the Pocahontas Number Four (Fig. 8) and Pocahontas Number Six (Fig. 4) are as summarized.

HYDROLOGIC PARAMETERS

	Channel Length Measured (KM)	Channel Length Calculated[1] (KM)	Co-efficient Calculated[2]	Basin Drainage Area Measured (KM²)	Basin Drainage Area Calculated[1] (KM²)	Sum of Channel Lengths (KM)	Drainage Density[3]
Pocahontas #4							
Basin #1	7.06	6.52	1.52	15.23	17.40	19.68	2.08
Basin #2	7.77	6.47	1.68	15.05	20.41	24.51	2.62
Basin #3	6.41	6.05	1.48	13.47	14.76	18.51	2.21
Pocahontas #6	3.54	3.62	1.3	6.19	5.49	14.21	3.69

HORTON STATISTICS

	No. of Streams			Bifurcation Ratios[4] (1st to 2nd Orders)
	1st Order	2nd Order	3rd Order	
Pocahontas #4				
Basin #1	14	5	1	2.8
Basin #2	18	6	1	3.0
Basin #3	15	4	1	3.7
Pocahontas #6	16	6	1	2.7

LINK VS AREA (KM²)

Basin	Link	Area	Calculated[5]
#1	14	15.23	14.37
#2	18	15.05	18.59
#3	15	13.47	15.41
Composite	47	43.75	48.37

(1) $L = 1.4A_d^{.6}$; 1.4 av. co-efficient (range 1-2.5) (L in miles, A_d in miles²)

(2) $L = aA_d^{.6}$ where a = co-efficient; L (in miles) & A_d (in miles²) measured

(3) Drainage Density $= \dfrac{\Sigma L}{Ad}$ where L is length of contributing channel (in miles); A_d measured area (in miles²)

(4) Average for U.S. rivers 3.5

(5) Area = .4 Link − .05 (r² = .82) from Reynolds (1972) (miles²)

results, we estimated mean annual discharge by Reynolds regression equation to be between 2.8 and 3.4 m³/sec. This would correlate to 100 to 127 cm of rainfall a year. Table 1 summarizes the various hydrologic parameters calculated.

ECONOMIC CONSEQUENCES

The location and the extent of economically important coals in the Pocahontas Formation have been well defined since the turn of the century. Consequently, facies relationships and other stratigraphic analyses offer little information of value for regional coal exploration. However, lessons learned during mining, especially insight into the local distribution of the economic reserve, contribute much when applied to new mine projects in the same strata.

Development drilling (closely spaced pre-mine drilling) in the Pocahontas measures often result in an array of holes with thick and thin coal heights, seemingly chaotically spaced across the prospective mine block. If these borings are interpreted without regard to lessons learned in adjacent coal mines, the resulting thickness isopachs might bear little relation to the actual pattern of coal thickness. If, however, the data are interpreted using adjacent mined areas as models, the "thin" and "thick" zones naturally fall into easily defined patterns. Additional test drilling will verify the accuracy of such interpretations. If test drilling verifies the interpretation, isopach patterns can be validly constructed using paleodrainage net models for interpretation between control points. In such cases, the refinement can pay dividends in mine planning, especially in terms of the orientation of production panels (longwall or continuous mine sections) and main haulageways; as in one of the examples used in this paper, the Pocahontas Number Six Seam. To develop the coal, mains were directed up the channelways (thick coal locales) with sub-mains angling off individual dendrites (Fig. 4). The mine plan is a logical consequence of the distribution of economic reserves; any other mine plan for this seam, not tailored to the coal distribution, would have had disasterous economic consequences.

SUMMARY AND CONCLUSIONS

Portions of paleodrainage nets are commonly delineated on mine maps of mature coal mines. Analysis of these paleonets can provide useful paleoenvironmental information concerning climate, hydrology, local and regional slopes, as well as detecting the influence of contemporary tectonism. Such paleonets are probably not restricted to the Carboniferous of the Appalachians, but should be present wherever coal is mined from similar facies. Therefore, it is to be expected that nets of many ages in many areas should become available for analysis as a byproduct of coal mining.

REFERENCES

Arkle, T., Jr., 1959, Monongahela Series, Pennsylvania System, and Washington and Greene Series, Permian System, of the Appalachian Basin: Guidebook series, Geol. Soc. Am., Pittsburgh Meeting, 1959, p. 117-138.

Baker, V. R., 1973, Paleohydrology and Sedimentology of Lake Missoula Flooding in eastern Washington: Geol. Soc. Am. Spec. Paper 144, 75p.

Davis, M. W. and Ehrlich, R., 1974, Late Paleozoic Crustal Composition and Dynamics in the Southeastern United States: *in* G. Briggs, *ed.*, Carboniferous of the Southeastern United States; Geol. Soc. Am. Spec. Paper 148, p. 171-185.

Elder, C. H., Jeran, P. W., and Keck, D. A., 1974, Geologic Structure Analysis of the Coal Mining Area of Buchanan Co., Va.: Bur. Mines, U.S. Dept. Int., Rept. Invest. 7869, 29p.

Englund, K. J., 1974, Sandstone Distribution Patterns in the Pocahontas Formation of Southwest Virginia and Southern West Virginia: *in* G. Briggs, *ed.*, Carboniferous of the Southeastern United States; Geol. Soc. Am. Spec. Paper 148, p. 31-45.

Galloway, Malcolm, 1972, Carboniferous deltaic sedimentation in Fayette and Raleigh Counties, Southeastern West Virginia: Unpub. Ph.D. thesis, Univ. South Carolina, Columbia, South Carolina, 108p.

Leopold, L. D., and Wolman, M. G., 1960, River meanders: Geol. Soc. Am. Bull., v. 71, p. 769-794.

——, ——, and Miller, J. P., 1964, Fluvial processes in geomorphology: W. H. Freeman, San Francisco, 522p.

Padgett, G. V., 1972, Carboniferous sedimentation in McDowell and Wyoming counties, southeastern West Virginia: Unpub. M.S. thesis, Univ. South Carolina, Columbia, South Carolina, 58p.

——, and Ehrlich, R., 1976, Paleohydrologic Analysis of a Late Carboniferous Fluvial System, Southern Morocco: Geol. Soc. Am. Bull., v. 87, p. 1101-1104.

Reynolds, D. E., 1972, The Discharge of South Carolina streams as it relates to Link Magnitude: Unpub. M.S. thesis, Univ. South Carolina, Columbia, South Carolina, 59p.

Schumm, S. A., 1968, River adjustment to altered hydrologic regimen — Murrumbidgee River and paleochannels, Australia: U.S. Geol. Survey Prof. Paper 598, 65p.

——, 1972, Fluvial paleochannels: in J. K. Rigby and W. K. Hamblin, eds., Recognition of ancient sedimentary environments; Soc. Econ. Paleont. Mineral. Spec. Pub. 16, p. 98-107.

Sharp, W. E., 1974, The Dilution Capacity of Small Streams in South Carolina: Water Resources Research Inst., Clemson University, Report 48, 70p.

Williams, E. G., and Bragonier, W. A., 1974, Controls of early Pennsylvanian Sedimentation in Western Pennsylvania: in G. Briggs, ed., Carboniferous of the Southeastern United States; Geol. Soc. Am. Spec. Paper 148, p. 135-152.

ECONOMIC   APPLICATIONS
OF   FLUVIAL SEDIMENTOLOGY

A SEDIMENTOLOGICAL SYNTHESIS OF PLACER GOLD, URANIUM AND PYRITE CONCENTRATIONS IN PROTEROZOIC WITWATERSRAND SEDIMENTS

W. E. L. Minter[1]

Abstract

Placer ore bodies were developed during deposition of the Proterozoic Upper Witwatersrand Group in braided river environments on fan deltas around the margin of the Witwatersrand Basin, which was created by gentle synclinal warping of the Kaap-Vaal Craton. Tectonic instability around the margins of the Witwatersrand Basin, during the final stages of filling and during deposition of the overlying Ventersdorp Group, created marginal fault scarps to the basin in places, and yoked basins, into which humid alluvial fans, partly composed of Witwatersrand material, were deposited.

These various environments of deposition are illustrated with examples from four placers that are being exploited at present. Stratigraphically, the economically important placers occur as: regressive deposits on unconformities at the base of sedimentary units; transgressive deposits on angular unconformities at the base of sedimentary units; or as terminal deposits on disconformities at the top of sedimentary units.

Heavy detrital particles of gold, and hydraulically equivalent-sized particles of uraninite and pyrite were hydraulically concentrated and preserved in pebble-supported conglomerate and typically trough-crossbedded pebbly quartz arenite facies. These economic placer minerals, which were derived from the granitic-greenstone basement, probably survived exposure in subaerial and shallow water environments because of a low oxygen pressure in the atmosphere.

Mapped details of the internal and external geometry of the placers, various scalar parameters, and paleocurrent analysis are used to define the facies and complicated distribution patterns of placer concentrates for exploration and mining activities.

Introduction

In 1886, eighteen years after the discovery of alluvial placer and quartz lode gold deposits in South Africa, gold-bearing conglomerates of the Upper Witwatersrand Group were discovered cropping out in the Transvaal near Johannesburg. A recent chronological review by Pretorius (1975) described the progress in the fields of exploration, mining and geological endeavour that followed the first discovery of the Witwatersrand deposits. These activities eventually outlined six major goldfields in the Witwatersrand depository, thus placing South Africa in the forefront of the major gold producing countries of the world.

Up to 1960 a considerable amount of literature concerning mining, metallurgy and general geology, with an emphasis on structure and stratigraphic correlations, had accumulated, but very little sedimentological information was available. Indeed, little credit for the main discoveries can be attributed to any sedimentological analysis.

In 1957 the gold mining industry established a research unit, which was designed, under the Directorship of D. A. Pretorius, to investigate the Witwatersrand deposits. It was through this unit that the first systematic sedimentological study was made (Steyn, 1964).

Two papers by Brock and Pretorius (1964a, b) are a landmark in the geological assessment of these deposits and finally laid the hydrothermal versus placer controversy to rest. They defined clearly the structurally closed nature of the Witwatersrand Basin with stratigraphic borehole data. They also presented a conceptual model relating the structural processes to sedimentary responses through which the erosion products of gold-bearing Archean rocks were concentrated into rich fluvial placers and deposited around the margin of the Witwatersrand Basin.

[1]Geological Department, Anglo American Corporation, P.O. Box 20, Welkom 9460, South Africa.

More detailed aspects of this concept, which were developed and published by
Pretorius (1966, 1974a) are substantiated by the results of systematic sedimentological
studies presented in theses by: Armstrong (1966) on parts of the Kimberley placer east of
Johannesburg; Knowles (1967) on parts of the Ventersdorp Contact placer near
Carletonville; Minter (1972) on the entire Vaal placer near Klerksdorp; and Hodgson
(1967), Sims (1969), Pienaar (1969) and Hutchison (1971) on parts of the Basal placer near
Welkom. Although very little of this work has been published (Knowles, 1966; Minter,
1976) it has been summarized in a comprehensive synopsis of the nature of the
Witwatersrand deposits by Pretorius (1974b). Since 1964, concurrent with this progress,
detailed sedimentological studies have been undertaken on many of the gold mines for the
purpose of selective mining and for further exploration.

The purpose of this paper is not to review previous work but to illustrate, with
heretofore unpublished data, the features that characterize a typical Witwatersrand
placer. Because the entire spectrum of sedimentary features has not been analysed or is
not exhibited or preserved in any single placer, supplementary evidence is drawn from
four separate deposits. These particular deposits, which are located in the Welkom,
Klerksdorp and Carletonville Goldfields (Fig. 1) are all mined at depths of approximately
2000 m below surface, and are completely covered by younger rocks. They have all been
discovered since 1934 by means of boreholes, which were sited largely as a result of
gravimetric and magnetometer surveys. Collectively they account for a total production
of 7 million kilograms of gold and over 40 million kilograms of uranium.

GEOLOGICAL SETTING

Location

The sediments of the Upper Witwatersrand Group occupy an oval-shaped basin whose
long axis extends northeastwards through Welkom and Johannesburg for 160 km and
whose shorter northwestern axis is 80 kilometres wide (Fig. 1). The preserved area covers

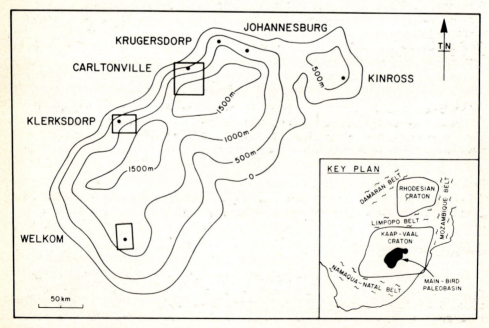

Fig. 1. Isopach plan of the Main-Bird paleobasin showing its position in the Kaap-Vaal craton
and the location of study areas. (Modified after Brock and Pretorius, 1964a.)

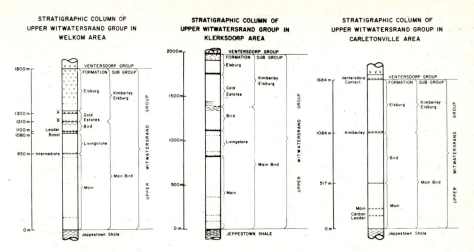

Fig. 2. Stratigraphic columns of the Upper Witwatersrand Group in the Welkom, Klerksdorp and Carletonville Goldfields, showing the position of important placers. The Khaki Shale and Leader Quartzite strata which lie between the Basal and Leader placers ought to be grouped separately. However, they are left unnamed in this column because the entire Upper Witwatersrand nomenclature is under review. In the Klerksdorp Goldfield the Vaal placer lies at the base of the Bird formation.

9750 km² and is superimposed on the more extensive Lower Witwatersrand Basin in the centre of the Archean Kaap Vaal Craton.

The four deposits that have been chosen to represent Proterozoic placers in the Witwatersrand Basin are situated in three goldfields which are located near the towns of Welkom, Klerksdorp and Carletonville respectively (Fig. 1). These placers, referred to in mining jargon as reefs, are known as the Basal Reef (Welkom Goldfield), the B Reef (Welkom Goldfield), the Vaal Reef (Klerksdorp Goldfield) and the Ventersdorp Contact Reef (Carletonville Goldfield).

Stratigraphic setting

The Basal, B and Vaal placers occur in the Livingstone, Gold Estates and Bird Formations, as shown in the stratigraphic columns (Fig. 2), at positions between 1000 and 1200 m above the base of the Upper Witwatersrand Group. The Ventersdorp Contact placer occurs at the base of the Ventersdorp Group.

The Basal and Steyn placers

The Basal Reef is the major gold and uranium bearing ore body in the Welkom Goldfield. It is composed of at least two laterally coalescing placers, deposited on practically the same paleosurface. They will be referred to as the Basal and Steyn placers. These placers are generally less than a metre thick and each covers an area of approximately 200 km².

The sediments in the Livingstone Formation that underlie the Basal and Steyn placers are composed of khaki to greenish-grey trough-crossbedded coarse-grained arenites arranged in cosets that are between one and two metres thick. The cosets are separated by argillaceous veneers, or thin mud drapes, and by scour surfaces. Solitary tabular planar crossbeds up to one metre thick, occur interspersed throughout this sequence of sand which was transported towards the east (Sims, 1969). Exposures through the total

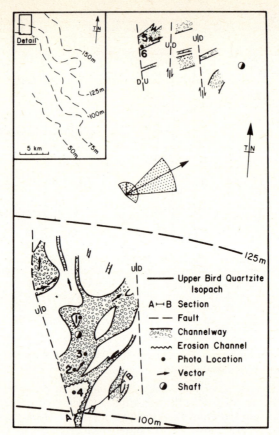

Fig. 3.   Details of an area in the northern part of the Welkom Goldfield showing paleochannels of the B placer superimposed on the Bird Formation isopachs.

sequence are sparse but measurements of its thickness indicate a decrease eastwards down the paleoslope at a rate of 10 m/km.

The Basal and Steyn placers rest disconformably on a scour surface at the top of the quartzitic Livingstone Formation of which they represent a regressive terminal phase. They were subsequently covered by 20 metres of muds and muddy sands, known as the Khaki Shale and Leader Quartzite respectively, which prograded from the south, into standing water, and covered the entire goldfield.

The B placer

The B placer of the Welkom Goldfield is economically of secondary importance compared with the Basal and Steyn placers because, although it is distributed over an area of 400 km², it is confined to shallow channelways, up to 2 metres deep, which cover less than 35% of the paleosurface and occupy a very short paleoslope (Fig. 3).

This placer lies at the base of the Gold Estates Formation (Fig. 2) and represents a regression over the underlying Bird Formation which in itself is a transgressive fining upward sequence from a basal conglomerate unit to a very fine-grained argillaceous quartzite and shale unit. Deposition of the B placer was followed by the deposition of 90 m of coarse-grained fluvial sediments representing the remainder of the Gold Estates Formation. During this period of subsequent deposition, a number of coarse gravel units

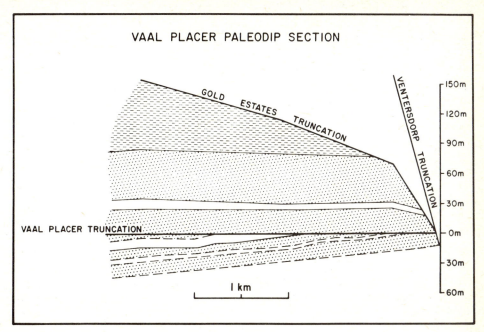

Fig. 4. A section drawn down the paleoslope of the Vaal placer to illustrate prominent angular unconformities.

were deposited. Their occurrence may represent internal responses on the fan caused by particular floods or to the fan's geomorphic threshold having been exceeded.

The Vaal placer

The stratigraphic setting of the Vaal placer, which is the major gold and uranium bearing orebody in the Klerksdorp Goldfield and represents the base of the Bird Formation in that area (Fig. 2), is different from the stratigraphic settings of the previously mentioned placers in that it rests upon a prominent angular unconformity (Fig. 4). The placer, which has an average thickness of 30 cm, covers an area of 260 km². It transgressively truncated an underlying fluvial sandy fan delta at a gradient which changes from 1:120 to 1:350 down a paleoslope distance of 10 km. This paleoslope is very regular and bears the drainage etches of the Vaal placer (Fig. 5a). The placer was buried by coarse sands that were transported by a longshore current which flowed parallel to the truncation surface isopachs (Fig. 5b).

The Ventersdorp Contact placer

All the well mineralized Ventersdorp Contact placers in the Witwatersrand are associated with prominent angular unconformities that cut across lithified Witwatersrand strata. The Ventersdorp Contact placer in the Carletonville Goldfield truncates at an average angle of 5 degrees through over 1500 m of Upper Witwatersrand strata from the Elsburg down to the Main Formation (Fig. 2) with increasing hiatus to the northeast (Fig. 6).

Although the Upper Witwatersrand strata were deposited down a south-southeast paleoslope, as indicated by the strike of isopachs of the strata and minor angular unconformities between them, a major fault along the eastern part of the goldfield, accompanied by a downward tilt to the southwest, created a yoked basin on the margin of

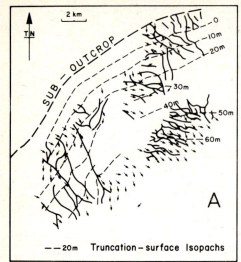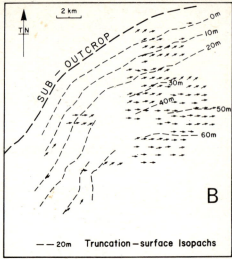

Fig. 5.   A, The paleosurface upon which the Vaal placer accumulated. Isopachs to an underlying marker from its suboutcrop position define the gradient of the planar truncation and reflect secondary subsidence along a centrally placed southeast axis. Shallow braided channels etched onto the paleosurface reflect the placer distribution pattern which is substantiated by moving-average vectoral means based on 440 trough-crossbedded foreset measurements. B, Moving-average vectoral means based on 500 measurements of the axes of trough-crossbedded sets in the transgressing argillaceous quartzites which buried the Vaal placer. Parallelism to the strike of the angular unconformity may indicate longshore drift.

the Upper Witwatersrand Basin (Fig. 6). A south-southeast tilt again prevailed in this area during Transvaal times so that truncations of the Ventersdorp and Upper Witwatersrand Groups strike northeast.

Development of the Ventersdorp Contact placer at Carletonville was terminated and buried by thick lava flows. Geomorphic features which would not have survived a long period of erosion were consequently preserved.

Age

The Archean granites of the Craton have been dated at 3100 ± 100 m.y. by the U/Pb method and the Upper Witwatersrand sediments have been dated at between 2800 ± 60 and 2620 ± 50 m.y. from the ages of underlying Jeppestown lavas and overlying Ventersdorp lavas, respectively (Fig. 2).

Uraninite grains extracted from Upper Witwatersrand placers have been dated at 3040 ± 100 m.y. This age supports the concept that detrital gold and uraninite in the placers was originally derived from Archean greenstone belts in the craton.

Sedimentary Nature

Lithology

The lithology of Witwatersrand placer deposits is distinguished from their bounding lithologies by their maturity. The placers are composed of coarse-grained, light-gray quartz arenite and an oligomictic pebble assemblage. For instance, the Vaal placer has the following composition:

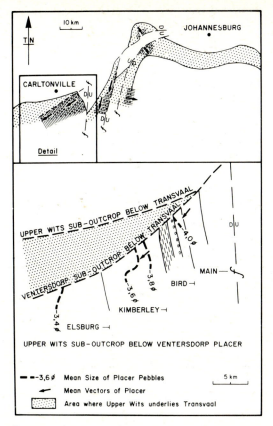

Fig. 6. Upper map illustrates the structurally-yoked Ventersdorp basins around the northwestern margin of the Upper Witwatersrand Basin. Lower map illustrates how the Carletonville Ventersdorp Contact placer (detail area) has truncated through the entire Upper Witwatersrand sequence which was temporarily tilted down towards the southwest before the southeast Witwatersrand Basin tilt was reasserted during Transvaal times. Mean pebble size decrease and vectoral means substantiate the paleoslope determined from the angle of unconformity.

The arenite is composed of: quartz ± 95%, rock fragments 0 - 1%, sericite 2 - 5%, chlorite 0 - 3%.

The pebble assemblage comprises: vein quartz 85.0%, chert 12.0%, quartzite 2.6%, quartz-prophyry 0.2%, yellow, silicified, shaly fragments, black carbonaceous mud-balls, and dark chloritic schist 0.2%.

The underlying and overlying sediments are not as mature as the Vaal placer. They contain less than 70% quartz in the sand-sized fraction and more than 30% polymictic components in the pebbly fraction.

In the Welkom Goldfield, the Basal placer and the oldest channel sediments of the B placer, contain gray oligomictic conglomerates, but younger placer sediments of the B placer are polymictic: vein quartz 60%, chert 9%, quartzite 1%, quartz-porphyry 4%, yellow shale 26%. The Steyn placer is also polymictic, a feature that distinguishes it from the Basal placer which occupies an adjacent position on the same paleosurface.

The Ventersdorp Contact placer at Carletonville, however, having eroded and reworked a great part of the previously lithified Upper Witwatersrand stratigraphy, contains conglomerate deposits with up to 50% quartzite clasts although the assemblage is generally 98% vein quartz.

Facies

The facies recognised in Witwatersrand placers are essentially: a scour surface; a pebble lag; horizontally layered conglomerate; planar crossbedded pebbly quartzite; trough-crossbedded quartzite; planar crossbedded quartzite; horizontally bedded quartzite; and thin mud drapes.

The vertical sequence of these facies in proximal locations is usually a scour surface followed by a conglomerate lag which grades normally into a trough-crossbedded quartzite. In places lags develop into thick conglomerates either filling depressions or forming longitudinal bars. The other facies may occur above these in various sequences and multiple scour surfaces are common. The internal geometry is therefore lenselike. A change in the pattern of the vertical sequence of facies is evident in distal placer environments. It is marked by a decrease in pebbles, a predominance of trough-crossbedded quartzite, rare units of tabular planar crossbedded quartzite, horizontally bedded quartzite and prominent mud drapes and mud clasts.

The Steyn placer at present provides the longest, most extensively accessible, paleoslope exposure available in the Witwatersrand Basin and therefore affords an ideal opportunity to examine facies changes of various features that are related to an hydraulic energy decrease down the paleoslope.

Along the paleostrike marked by the 40 mm maximum pebble-size isopleth (Fig. 7a), many major channel scours, like the one featured in Figures 8a and 8b, are preserved. Multiple scour surfaces with associated lag gravels are evident in Figure 8a with the last scour overlapping on to similar but older adjacent channels. The mean size of the quartz-pebble population in this part of the depositary is up to -4.5ϕ.

Twenty kilometres further down the paleoslope, the Steyn placer is essentially a quartz arenite unit which was built by shallow sandbars and dunes (Figs. 9a, b). Thin yellow mud veneers less than 1 mm thick occur on bedform surfaces and mark periods of fine sediment settling between periods of sand transport. Small pebbles lie at the base of trough sets and also form thin lag veneers in places on scour surfaces and in particular on the scalloped scoured base of the placer unit. the mean size of the quartz-pebble population in this more distal part of the Steyn placer is -2.5ϕ.

Younger channelling is evident in this facies of the placer and the multiple depositional surfaces within the placer sediment are marked by yellow mud drapes, up to five centimetres thick, with mud balls and flakes in the succeeding trough-crossbedded sets of quartzite (Fig. 10a). These features indicate that the extensive thin sheet of Steyn placer sediment was produced in a fluvial environment by the unconfined flow of continuously shifting shallow braided streams.

Fig. 7. A, Paleocurrent plan of Steyn and Basal placers showing: their entry positions; boundary of lateral coalescence (marked with T symbols); mean pebble-size isopleths of ten largest pebbles; main trends of best mineralization based on detailed moving-average contours; vectoral means; and the location of heavy mineral sample sites (1 to 5). B, The location and distribution pattern of the high gold content facies in the Basal and Steyn placer deposits. Upslope contour re-entrants are associated with channelways. C, The location and distribution pattern of the high uranium content facies in the Basal and Steyn placer deposits. The fit between gold distribution and paleocurrent data indicates hydraulic control. The position of the uranium facies, downslope from the gold facies, may be related to an apparent lack of coarse-grained uraninite. D, A mineralogical ratio change down the paleoslope, expressed by the ratio of uranium to gold, is a useful indicator of the paleoslope gradient and the placer facies. The high ratio area marks the beginning of the heavy mineral distal facies.

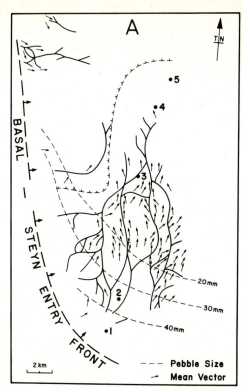

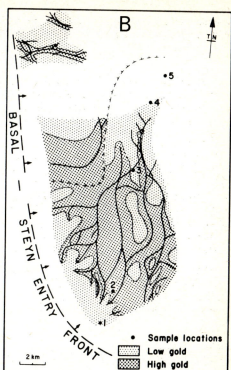

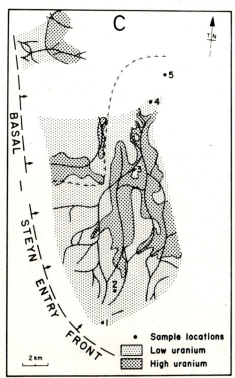

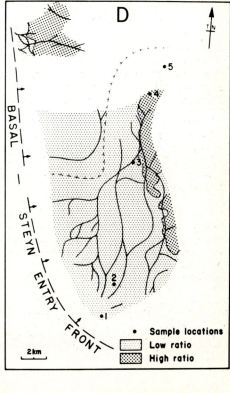

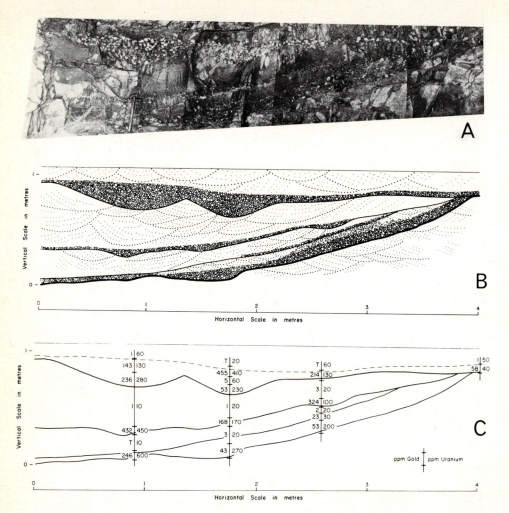

Fig. 8. A, Composite underground photograph of the edge of a Steyn placer channelway in a proximal facies. B, Line drawing of same location to illustrate multiple scour surfaces within the channel with associated lag gravels and intercalated trough-crossbedded quartz arenite. C, Gold and uranium concentrations related to the channel sediments. (Mapped by G. Oram.)

The B placer is composed of similar facies the most prominent of which are scour surfaces overlain by thin pebble lags, low relief pebble bars, pebble filled depressions, and tabular planar crossbedded quartzites (Figs. 11a, b, c, d). Facies changes down the paleoslope of the B placer are rapid and the mean quartz-pebble size decreases from $-3.7\,\phi$ to $-2.5\,\phi$ over a distance of 4 km. A similar change in pebble size in the Steyn placer occurred over approximately 15 km.

The most proximal placer facies being mined is evident in the Ventersdorp Contact placer at Carletonville where approximately 25 km of paleoslope have been exposed and the hydraulic energy gradient is reflected by a pebble size decrease from $-4.0\,\phi$ to $-3.4\,\phi$ and improvement in sorting reflected by a decrease in the standard deviation of pebble sizes from $1.42\,\phi$ to $0.56\,\phi$.

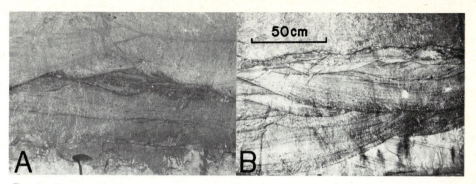

Fig. 9. Transverse views of trough-crossbedded distal facies of the Steyn placer. Dark heavy mineral accumulations are concentrated on foresets and particularly on the scoured bases of sets with small pebbles associated in places. A, The removal of previously deposited, reworked, heavy mineral accumulations by the large set on the left is well illustrated. B, The lowest trough set marks the base of the mature placer unit. A mud veneer followed by apparently homogeneous sericitic Leader Quartzite overlies the placer sediment.

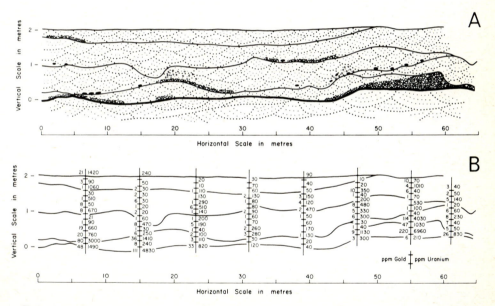

Fig. 10. A, Very distal Steyn placer sediment composed of trough-crossbedded quartz arenite resting on a well defined disconformity. Younger channelling, which has partly removed and reworked the placer, is distinguished by scour surfaces, yellow mud drapes and mud clasts. B, Scour surfaces, and coarser sediment accumulations are associated with higher gold and uranium concentrations. Samples with trace amounts of gold have been left blank. (Mapped by S. Buck.)

A gross facies change is evident within 6 km of the fault scarp where sheet gravels up to 4m thick and composed of well rounded cobbles up to 40 cm in diameter, occur (Fig. 12a). These are the coarsest clasts ever observed in the Witwatersrand and are interpreted as mid-fan deposits. Internal clean lithic arenite lenses mark deposition during waning floods and in many places separate individual conglomerate layers. Assemblage and textural changes seem to indicate that there are at least two sheet-gravel mid-fan

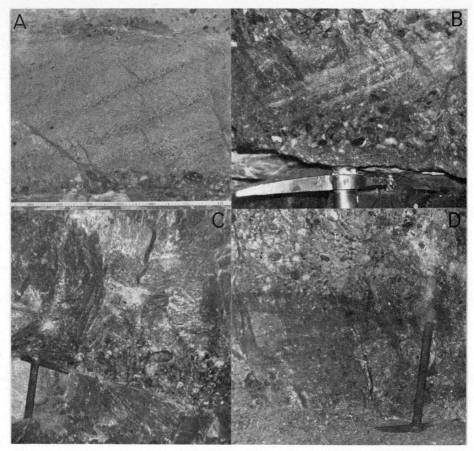

Fig. 11. Primary structures in the B placer. A, Planar cross-bedding formed by a solitary transverse sandbar of coarse argillaceous quartzite containing small pebbles arranged on foreset planes, overlying a thin lag deposit. Centimetre tape for scale. Taken near location 5 (Fig. 3). B, Heavy mineral concentrations with abundant nodular pyrite in foresets on the lee end of a thin gravel bar. C, The edge of a 2-metre deep conglomerate-filled channel that was scoured into the underlying fine-grained argillaceous quartzite of the Bird Formation. Taken at location 1 (Fig. 3). D, Avalanche foresets of mature pebbly quartz arenite representing a transverse bar channel sediment in the B placer. Coarse nodular pyrite is concentrated on the foresets and in a winnowed layer on top of the bar. The hammer head rests on the scoured base of the channel. Overlying coarse gravel with mature mineralized quartz arenite matrix filled the remainder of the channel. Taken at location 2 (Fig. 3).

deposits which are entrenched by, and also overlap, the more widespread channel deposits (B. Krapez, personal communication) (Fig. 12b).

The Vaal placer appears to contain only the scour surface, thin pebble lag, and trough crossbedded quartzite facies. These may be repeated in the vertical sequence within deeper channels, grading normally upwards. In addition, an inverse grading is sometimes evident at the top of the placer unit where reworking of the surface has produced a winnowed pebbly concentrate. The pebble supported conglomerate facies is not common; the average pebble packing density is 23%.

Besides a lack of facies changes down the 10 km of paleoslope that is exposed, the mean quartz-pebble size is $-3.19 \, \phi$ throughout with a standard deviation of $0.51 \, \phi$

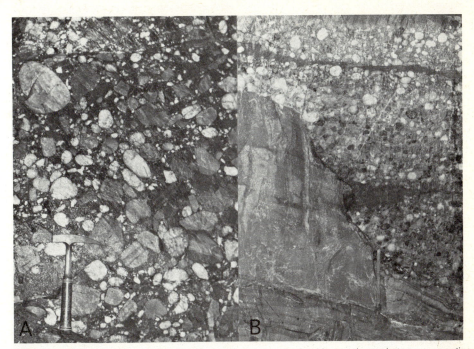

Fig. 12. A, Cobble and large pebble conglomerates in the Ventersdorp placer representing superimposed sheets of gravel preserved on a 30-metre high terrace. Half the clasts are composed of Lower Witwatersrand quartzites. B, A channel deposit of the Ventersdorp placer at a location which is 10 kilometres further down the paleoslope. The channel, which eroded into previously lithified siliceous Kimberley-Elsburg quartzites, was symmetrically filled by three layers of gravel interbedded with muddy sediment and then preserved by lava.

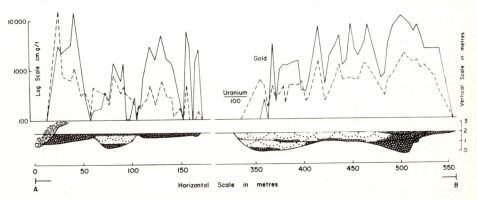

Fig. 13. A section drawn along line A - B, shown on figure 3, to illustrate that channel sediment of the B placer and associated gold and uranium mineralization is confined to channelways. A younger erosion channel has removed placer sediments to the left of the section. (Mapped by R. Metcalfe.)

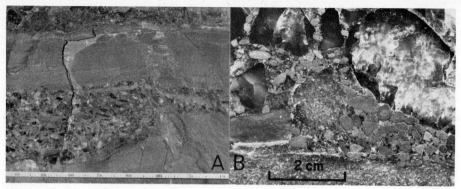

Fig. 14. A, A shallow gravel-filled channel in the B placer which has scoured into the fine-grained argillaceous quartzite of the Bird Formation. The channel placer deposit is buried beneath medium-grained argillaceous quartzite which is covered with a second placer lag deposit. Coarse-grained nodular pyrite is abundant in both layers which contain gold concentrations of approximately 500 ppm. This vertical sequence is rare. Centimetre tape for scale. Taken near location 5 (Fig. 3). B, A polished slab of well mineralized lag gravel from the B placer. Coarse pyrite nodules occupy the pebble matrix with scattered granules of carbonaceous material and fine grains of gold near the basal contact.

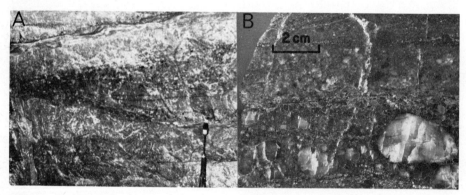

Fig. 15. A, The edge of a shallow channel containing the Vaal placer. B, A polished slab of a 5-centimetre thick layer of Vaal placer. The second thin lag concentrate is composed of small oligomictic pebbles scattered in a gritty mature quartz arenite. Thin black carbonaceous seams representing fossil algal mats occur in situ with disrupted granules of carbon along the surface of the angular unconformity, at the base, and on the top of the lag accumulation. Fine to medium-grained pyrite nodules are abundant in the lag and along foreset planes higher up.

These features are consistent with each other and indicate that the depositional site of the Vaal placer moved transgressively up the paleoslope as truncation took place, thereby obscuring scalar indications of the paleoslope.

To the west of the Vaal placer, on the same paleosurface, an abrupt change in pebble assemblage and increase in pebble size mark the lateral coalescence of a second placer which has been called the Witkop facies (Minter, 1976). Amongst the pebble population of the Vaal placer there are numerous ventifacts which are considered to have been windblasted while lying on the exposed surface of the underlying fan before truncation occurred. Aeolian sediments have not been recognized in the rock record.

Channel morphology

Channels up to 4m deep and over 500m wide have been recorded in the Steyn placer. They evidently branch and join in a braided pattern and have been filled in a generally symmetrical manner by mature placer sediment which has eventually overlapped between the deepest channels to form an extensive thin sheet of sediment. The main distribution pattern has been interpreted from an integration of detailed moving-average trends of thickness and gold and uranium content with paleocurrent vectors (Fig. 7a). Paleosurface topographic highs very rarely project as islands through the sheet of placer sediment.

By contrast, the B placer is confined to discrete interconnected channels (Fig. 3). The channel profiles are seldom symmetrical (Fig. 13), but generally show one steeply-inclined side and an opposite low-angled side (Fig. 11c). The channels were not filled sideways, as by pointbar migration, but symmetrically. Transverse bars of avalanche-bedded pebbly sand (Fig. 11d) and longitudinal bars of gravel grew down the channels and spread laterally. In some instances gravel accumulations appear to hug one side of the channel. The placer channelways range in width from 1 to 200 m and are up to 2 m deep. They are shallower in distal locations (Fig. 14a). Mapping of their edges indicates a sinuosity that meanders within an arc of 110° in a braided pattern that bifurcates and joins.

Channels in the Vaal placer are up to 1m deep but average 30 cm (Fig. 15a) and have a width/depth ratio of between 200 and 1000. The placer sediment eventually overflowed the channels to produce a dendroidal sheet of sediment, only a few centimetres thick in places (Fig. 15b). In rare instances, topographic elevations protrude up to one metre above the surface of placer accumulation in the form of islands. The mapped pattern of shallow channelways, which were filled by the Vaal placer, is sinuous and braided and represents the final drainage etch on the paleosurface of truncation. However, at this stage the pattern is still incomplete in areas that have not been mined (Fig. 5a).

The distribution of the Ventersdorp Contact placer has been strongly dominated by the topography of the paleosurface (personal communication G. F. Wagener; C. Engelbrecht and A. Turner). It is evident in areas midway down the paleoslope that conglomeratic placer deposits occur on top of, and as terraces on some flanks of topographic elevations of the paleosurface that are up to 7m high. Information available at present is inadequate to define the overall shapes, sizes and frequency of these features. Towards the eastern fault scarp, in the proximal part of the deposit, topographic elevations up to 30 m high and 500 m across have been recorded. Where channels are well defined the steep nature of their sides is evident and indicates that the bedrock was lithified (Fig. 12b). Channel edges in Witwatersrand placers have slopes that are usually less than 15° which implies that the channel beds were poorly consolidated sands. The very fine grained argillaceous quartzite and shale beneath the B placer was apparently a sufficiently cohesive sediment to form steeper, stable channel sides.

Paleocurrents

Paleocurrent analyses of Witwatersrand placers have confirmed unimodal transport and braided distribution patterns. Vectoral means calculated from over a thousand foreset measurements of trough-crossbedded quartzite in the Steyn and Basal placers indicate that the pebbly Basal placer sediment regressed slightly over the sandy facies of the Livingstone fan from a western source. The sediment supply evidently switched to a southwestern location during deposition of the Steyn placer. The Steyn placer regressed even more than the Basal placer, which it overlaps in places along their approximate line of coalescence. The vectoral mean azimuths of the respective dispersal patterns are unimodal and show a variance of less than 3000.

In the B placer, paleocurrent vectors measured from: avalanche crossbedded transverse bars of arenite (Fig. 11d); trough-crossbedding in intercalated quartz arenite within placer deposits; and from avalanche crossbedded transverse bars immediately above thin lag deposits (Fig. 11a), indicate a northeasterly vectoral mean with a variance of 1850. This direction conforms well with the general direction of channelways that travelled down the northeast paleoslope indicated by isopachs of the Bird Formation (Fig. 3).

In the Vaal placer, paleocurrent analysis was conducted by measuring ten foresets, each from a separate, trough-crossbedded set, within 500 m cells of a superimposed grid. Moving-average vectoral means of 440 foreset azimuths indicate a unimodal southeast paleocurrent direction and substantiate the detailed sinuosity of the etched drainage-channel pattern (Fig. 5a). This pattern reflects the control of paleosurface shape on the placer transportation and indicates that placer deposition was penecontemporaneous with, and followed, channel scouring. The variance of the paleocurrent data is much higher (4721) than in the underlying fan (2484). This variance probably reflects the comparison between partly confined transport of Vaal placer sediments in the incised braided pattern and the unconfined transport of the sediment in the underlying fan.

In the Ventersdorp Contact placer the detailed paleocurrent distribution is more complex because of the multiplicity of the deposit and its incised paleosurface topography. The vectoral means calculated from trough-crossbedded foreset azimuths indicate transport down a southwest paleoslope. This slope is also implied by isopachs and suboutcrop strikes of Upper Witwatersrand strata that underlie the Ventersdorp Contact placer.

A number of paleocurrent analyses of Witwatersrand placers, that were summarized into rose-diagrams, showed bimodal distributions which were interpreted as indicating reworking by a longshore current. However, it is evident that by measuring trough-crossbedded foresets in transverse sections, the same type of bimodality is produced. This error may be further enhanced by mixing data from diverging braided channels. Therefore, these interpretations may be discounted. But in the case of the Vaal placer, where the axes of trough-crossbedded sets in quartzite immediately overlying the placer indicate transport parallel to the strike of sub-outcropping strata beneath the placer (Fig. 5b), there seems to be little doubt that a longshore current was responsible. The variance of these azimuths is only 279 and moving-average vectors follow the curvature of the sub-outcrop strike.

Detrital placer minerals

Numerous minerals have been identified in abraded detrital forms in Witwatersrand placers (Ramdohr, 1959). The most abundant is pyrite which occurs in an average concentration of 3% and comprises 90% of the sulphide minerals present. The pyrite occurs predominantly as rounded nodules (Saager and Esselaar, 1969). Other detrital sulphides identified are cobaltite, linnaeite and arsenopyrite. After pyrite, the most abundant detrital minerals are chromite, zircon and leucoxene, with uraninite and gold occurring in relatively minor amounts. Mineralogical studies have indicated that many of these minerals have been remobilised during metamorphism.

A spectroscopic scan of samples from the sediments below, within and above the B placer, illustrates the relative concentrations of a selection of elements (Table 1). Conspicuous geochemical anomalies associated with placer concentrates do not appear to extend along the disconformity away from the placer channel edge.

Mineral concentration

In the proximal parts of the Steyn placer, near location 2 (Fig. 7a) where deep channels containing large pebble conglomerates occur, gold and uranium is concentrated in the

Table 1.   Spectrographic analysis of the B placer and its bounding lithologies.

Element	Below B placer ppm	Within B placer ppm	Above B placer ppm
Ag	<0.1	36	<0.1
As	<100	237	<100
Au	0.2	358	—
Ba	650	200	300
Be	3	0.5	1
Bi	<1.0	45	<1.0
Ca	2125	300	475
Co	<10	312	22
Cr	192	209	139
Cu	6	1325	57
Fe	>3000	>3000	>3000
La	100	100	100
Mg	1750	>3000	1250
Na	>3000	2375	1750
Ni	57	925	50
Pb	12	650	27
Sb	<50	57	<50
Sn	12	8	11
U	—	260	—
Y	70	60	50
Zn	12	132	32
Zr	75	300	87

pebble supported gravels and in pebble lags on scour surfaces (Figs. 8a, b, c). The winnowed tops of gravel bars are also sites of heavy mineral concentration.

On average, the trough-crossbedded mature quartz arenite that is intercalated between gravel layers, and with which the detrital mineral grains are considered to have been in hydraulic equilibrium, contains approximately 11% of the total gold content. Concentrations occur on foreset planes and particularly along the scoured bases of trough-crossbedded sets.

In the more distal parts of the Steyn placer, channelled bedforms are not prominent, the disconformity is subtly scalloped, and the placer sediment is essentially a sandbody (Fig. 9b). It is in this facies of the fan environment that primary structures illustrating the process of heavy-mineral concentration are preserved. It appears that the hydraulic process which produces trough-crossbedding is directly responsible because heavy minerals are generally concentrated on the tangential toes of foresets and merge along the scoured base of each set. The scour and fill mechanism appears to have reworked the sediment repeatedly as the train of sand dunes migrated downstream. When the migration of sediment dominated and preservation was low, only the bottom well-mineralized part of each set was preserved. The operation of this process over a long period produced a layer of apparently small scale trough remnants which was very mature and contained a high mineral concentrate at its base. These concentrations were generally held there by small pebble accumulations and by algal mats and for this reason they have been referred to in the literature as 'carbon seam reefs'.

Where multiple channelling is evident in distal placer deposits, the scour surfaces marking channel bottoms were also often well mineralized by the migration of overlying trough-crossbedded dunes (Figs. 10a, b).

Gold and uranium distribution

The Steyn placer has been extensively mined. Approximately 360,000 routinely sampled sections of the placer unit, taken at two-metre intervals along lines 200 m apart, have been combined from 100 m square cells and, by using a simple moving-average procedure, have been used to produce trend surfaces of gold and uranium content.

The detailed trend surface pattern of gold content in the Steyn placer described by flow lines conforms with the paleocurrent data (Figs. 7a, b). An outline of the high gold content area, which has an average concentration of from 15 to 60 ppm, illustrates that a mineralogical gold facies area, probably hydraulically related to the mode of the particle size — frequency supplied exists (Fig. 7b).

The beginning of the high gold content facies is marked by contour re-entrants up the paleoslope, along a curved paleostrike which lies between locations 1 and 2, parallel to the entry front. This facies terminates about 10 km down the paleoslope in the form of high gold content tongues. The position of this gold facies in the distal parts of a fluvial fan placer is consistent with the model conceived by Pretorius (1966) and with the alluvial fan exploration model derived from a study of the Van Horn Sandstone by MacGowen and Groat (1971).

The distribution of uranium content, obtained by the same process used for gold, indicates a similar hydraulically controlled pattern of distribution with contour re-entrants upslope and terminal tongues further down the paleoslope (Fig. 7c). However, it is interesting to observe that although gold and uranium distributions are sympathetic, the commencement of the high uranium content facies area, with average uranium concentrations of between 150 and 500 ppm, is displaced 2 km downslope from commencement of the gold facies area. These data lead to good statistical correlation on a local scale but poor correlation on a regional scale.

The relative displacement of gold and uranium facies is further illustrated by the ratio of uranium to gold content shown in Figure 7d. Contour levels of this ratio change from less than 5 in proximal areas to more than 40 in the very distal areas and provide a useful paleoslope indicator.

The heavy detrital minerals in the B placer are associated with the host sediments in the same way as they are in the Steyn placer. Unlike those in the Steyn placer, the channel sediments of the B placer have not spread sufficiently to produce a sheet of sediment but are confined to channelways. As a result of this confinement the gold distribution follows a simple drainage pattern. Average concentrations are found on the basal contact in thin lags (Fig. 14b), in pebble-supported conglomerates, and on sedimentary partings within the placer. The gold particles are between 0.5 and 0.005 mm in size. Scanning electron photomicrographs of gold particles show folded and scored shapes which indicate transportation.

The uranium content in the B placer along section A — B (Fig. 3) is low but is approximately four to ten times more abundant than the gold content and therefore belongs to a facies resembling the proximal part of the Steyn placer model. The uranium - gold ratio changes from 4 to 42 down a paleoslope distance of 4 km.

The same association between trough-crossbedding and detrital mineral concentration is evident in the sediment of the Vaal placer as was described for the Steyn placer. However, pebble-supported conglomerates are rare in the Vaal placer. Concentrations containing 72% of the total gold content simply occur on the basal contact of the placer. Because preservation was therefore dependant on protection by burial, subsequent scouring by the longshore current discussed earlier interfered in places with concentrates at the base of thin deposits of the Vaal placer.

A moving-average trend surface of the thickness of the Vaal placer has been generalized in Figure 16a to illustrate the relationship between thicker placer and channels. The synclinal southeast axis of subsidence down the centre of the depository,

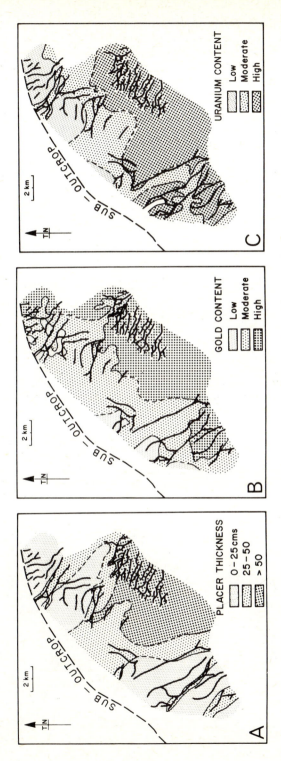

Fig. 16. Generalized moving-average distribution patterns of thickness (A), gold content (B), and uranium content (C), in the Vaal placer related to the paleoslope and channel patterns.

which is indicated by an embayment of the truncation surface isopachs (Fig. 5a), is also indicated by a thicker layer of placer. Part of this anomaly might represent data contamination caused by including overlying sediments, which are not part of the placer, in the sample. However, the discontinuation of this broad thicker area up the paleoslope may indicate incompleteness of the transgressive accumulation as it approached a hinge point in the basin which is marked by a convergence of later truncation (Fig. 4). The thinner layer of placer upslope from this area has been vulnerable to later scouring.

The average concentration of gold in the Vaal placer is 15 ppm. The generalized moving-average pattern of gold content in the Vaal placer illustrated in Figure 16b, indicates broad correlations with the placer isopach plan. Detailed trend-surface deviation plans demonstrate strong parallelism between channelling and payshoot directions (Minter, 1976).

The average concentration of uranium in the Vaal placer is 500 ppm. The generalized uranium moving-average distribution (Fig. 17) also resembles the gold and isopach plans. Detailed contours illustrate a close correlation between channels of thick placer sediment and total uranium content (Minter, 1976). This correlation is the result of only 20% of the uranium being concentrated at the base of the placer and the remainder being distributed on sediment concentration surfaces throughout the placer. This vertical distribution of uranium may be an effect of the uraninite being finer grained than the size that would be the theoretical hydraulic equivalent of the quartz grains (Coetzee, 1965). No systematic change in the uranium/gold ratio occurs down the paleoslope of the Vaal placer. This fact and the lack of other paleoslope changes support a transgressive concept. It also implies that gold and uranium contents may continue to remain constant further down the paleoslope.

The average concentration of pyrite in the Vaal placer is 3%. Like uranium, it is distributed throughout the placer sediment and is concentrated on foresets, trough bases and other sediment concentration surfaces. Consequently, pyrite content is related to channel depth. Only a small area of the Vaal placer has been systematically analysed for sulphur content which is proportional to the pyrite content. Moving-average contours of the uranium and pyrite contents in this area are shown in Figure 17. There is good correlation between their main trends which also correspond to isopachs and paleocurrent channels.

The distribution of gold within the Ventersdorp Contact placer is sporadic but associated with pebble-supported conglomerates. The areal distribution of gold is difficult to assess realistically because there are so many different gravel populations. In simply

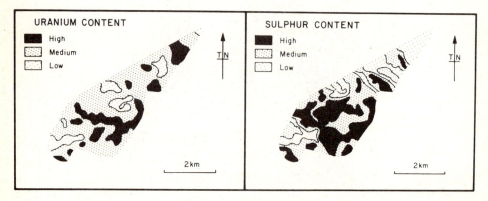

Fig. 17. Detailed moving-average of uranium and sulphur content in a part of the Vaal placer deposit to illustrate the paleocurrent control on uraninite and pyrite distribution.

channelled areas, moving-average trend surfaces can be successfully interpreted. The average gold concentration mined is approximately 25 ppm.

The average uranium content in the Ventersdorp Contact placer is very low. The uranium - gold ratio is generally less than 5 but increases of up to 34 have been record 20 km down the paleoslope.

Detrital mineral sizes

Pyrite nodules are the most abundant of the detrital placer minerals and are ubiquitous in all Witwatersrand placers. A number of varieties are evident. The most abundant is one with a shiny compact form. From trace element and lead isotope analysis, Saager (1976) has established a correlation between this variety and the crystalline pyrite that occurs in Archean greenstone belts, resembling the Barberton Mountain Land, the type area from where Witwatersrand sediments are though to have been derived. Other less abundant varieties, originally described by Ramdohr (1959), include oolitic, porous, layered and replacement types. Utter (1977) has suggested that the porous varieties were iron sulphide muds or gels.

It is clear from the sizes and spatial arrangement of this genetically mixed population of pyrite nodules that occur together on foreset planes, trough bottoms and winnowed surfaces, that they were all heavy detrital particles in hydraulic equilibrium with each

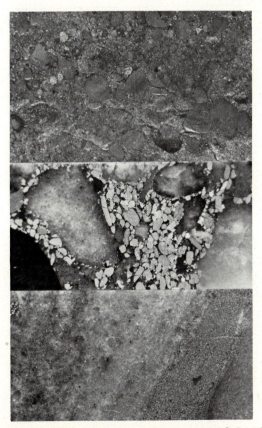

Fig. 18. Polished slabs taken from well mineralized layers in the Steyn placer at intervals along twenty kilometres of paleoslope to illustrate the progressive decrease in pyrite nodule size. Actual size.

other when they were deposited. In the Steyn placer these nodules display an obvious size decrease down the paleoslope as illustrated in Figure 18.

Relative changes in their sizes on an areal scale have been gauged from contouring the mean sizes of pyrite nodules. These were calculated by moments from measurements of the maximum exposed axes of all compact pyrite nodules in 56 smoothed slabs. The slabs were cut across the bedding of samples which were well mineralized with gold. Contours of the results (Fig. 19c), illustrate that the mean pyrite nodule size decreases down a paleoslope which was independently interpreted from the vectors of paleocurrent measurements, pebble sizes and mineralization trends. The contours also confirm the broad entry front of the Steyn placer (2.2 ϕ contour) and the merging of the very distal facies with a similar Basal placer facies.

A paleodip section constructed from measurements at sample locations 1 to 5 illustrates the gradient in mean pyrite nodule size from coarser than 1.6 ϕ to finer than 3.0 ϕ along 20 km of slope (Fig. 19a). The size frequency distribution in Sample 1 appears to be bimodal as a result of very coarse pyrite nodules in the population. The standard deviation of sizes at this location is 1.48 ϕ and decreases with the improvement in sorting downslope to 0.67 ϕ in Sample 5 at location 5.

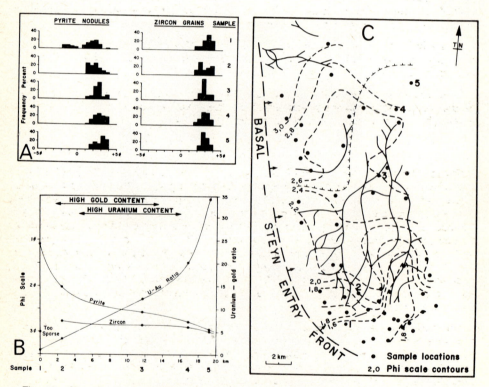

Fig. 19. Size distribution of detrital heavy minerals in the Steyn placer. A, Size frequency distribution of pyrite nodules and zircon grains at five locations distributed down 20 kilometres of paleoslope. B, The size relationship between pyrite and zircon is the inverse of that expected and reflects a lack of coarse zircons. This discrepancy decreases downslope. Similar reasoning could account for the changing ratio between uranium and gold contents. C, Contours of the mean size of compact pyrite nodules reveal the paleoslope dispersal of the Steyn placer and the more distal nature of the Basal placer.

By contrast, the mean zircon size decreases only slightly over the same distance. Although the mean decreases from 2.77 ϕ to 2.98 ϕ, the sorting improves from a standard deviation 0.70 ϕ to 0.44 ϕ (Fig. 19a). Because pyrite nodules have a greater specific gravity than zircon grains, their mean size should be less than that of the zircons so as to be in hydraulic equilibrium with them. However, the pyrite nodules are in fact coarser grained, as has been previously observed (Coetzee, 1965). It is evident from the comparison of relative mean sizes down the paleoslope (Fig. 19b), that the degree of discrepancy between mean pyrite and mean zircon sizes decreases progressively downslope.

From the sparse number and small size of zircons, in samples from the proximal part of the Steyn placer (Location 1), it appears that the upper size limit of the zircon grain population originally supplied from the provenance area was too fine-grained to be concentrated in the relatively higher hydraulic energy environment. Zircon grains are very abundant in more distal samples from Location 5. Because zircon and uraninite grains are considered to have been supplied from the same granitic provenance, the displaced position of the uranium facies area relative to the gold facies area, which is illustrated by the uranium - gold ratio curve in Figure 19b, may be due to an original uraninite grain size population that was finer grained than one which would have been hydraulically equivalent to the gold particle size population.

Pyrite nodules in the B placer display the same change in grain size over 4 km as the nodules in the Steyn placer do over 20 km.

An analysis of the mean pyrite nodule sizes in the Vaal placer (Minter, 1972) indicates that there is no significant decrease down the paleoslope. This again reflects the transgressive nature of the Vaal placer.

Coarse-grained pyrite nodules up to 2 cm in diameter occur in the very coarse deposits of the Ventersdorp contact placer which are also well mineralized with gold. Pyrite nodules become finer grained further down the paleoslope.

Plant life

During the Proterozoic, plant forms were primitive and the only fossil remains preserved occur as carbonaceous material in the form of granules and thin seams that are rarely up to 5 cm thick. Examination of the material using scanning electron microscope techniques (Hallbauer, 1975) and mass spectrometer analysis after vacuum pyrolysis (Zumberge, 1976) indicate a plant origin. The material occurs generally in patches on the bottom scour surface of the Steyn and Basal placers where it appears to have grown in situ. Periods during which little sediment accumulated, due either to continuous dune migration or to lack of sediment supply, probably favored growth and in special circumstances of location resulted in thick carbonaceous seams. Figure 20a illustrates a thick seam from the base of the Steyn placer that has entrapped sand grains, pebbles and placer mineral concentrates.

The carbonaceous material also occurs on scour surfaces within the placer unit; sometimes on a thin yellow mud veneer (Fig. 10a); on the scoured base of trough-crossbedded sets; on the toes of foresets (Fig. 20b); wrapped around pebbles in oncolite fashion (Fig. 20c); as detrital granular accumulations occurring on the stoss side of shallow ripples (Fig. 20d); and on reactivation surfaces in transverse placer sandbars. The spatial arrangements of the fossil plant material on fluvial bedforms, and within primary structures, leads one to favor an algal plant form.

Fragments of this carbonaceous material have been observed in the coarser gravel facies of placers but seldom appear to have been preserved there as mats. This is also true of the proximal facies of the B placer. In distal locations, the B placer contains well-developed seams of carbon. Abundant visible gold, which is associated with the

Fig. 20. Fossilized black carbonaceous material from the Steyn placer representing primitive Proterozoic plant life — possibly algal. A, A thick algal mat enclosing pebbles, sand and detrital minerals. B, Thin algal mats on foresets and bottomsets. C, Thin algal skins attached to pebbles. D, Granular algal material banked behind ripples one centimetre high.

carbonaceous seams in these distal deposits, may have originally been trapped in the algal mats.

Seams or granules of carbonaceous material are always present at the base of the Vaal placer, directly on the truncated paleosurface, and also on other sedimentary surfaces as in the Steyn placer. The carbonaceous material obviously accumulated during or after the major truncation with which the Vaal placer was penecontemporaneous.

Although high gold and uranium concentrations are associated with the carbonaceous seams, the areal distribution of gold and uranium concentration patterns do not coincide with the abundance of carbonaceous material. Therefore, their association appears to be indirect and is probably related to the original spatial coincidence of algal mats and detrital accumulations on surfaces of concentration that were not rapidly buried.

Black carbonaceous material in seams up to 2 cm thick has been observed in the Ventersdorp Contact placer in the Klerksdorp Goldfield indicating that the algal plant it may represent was still active at this date. Seams have also been recorded in the Carletonville Ventersdorp Contact placer.

SYNTHESIS

The most striking and important feature of these Proterozoic placers is their lithological and mineralogical maturity. They are composed of distinctively light to dark gray, coarse-grained quartz arenites and oligomictic conglomerates. The polymictic assemblage of grains and pebbles in the Steyn and B placers of the Welkom Goldfield is an exception to the rule. By contrast, the bounding lithologies of placer sediments are more argillaceous and polymictic.

The Upper Witwatersrand placers occupy shallow, low sinuosity channels, which in proximal areas are up to 4 m deep, but average between 100 and 50 cm deep further down their paleoslopes. Their channel width/depth indices are between 200 and 1000. In some instances, the placers are confined to separate channels that are interconnected in a braided pattern, but generally the placers being mined are thin irregular sheets comprising multiple layers of braided channel sediment. The placer channel sediments generally show a horizontal top which may reflect reworking.

Paleocurrent measures obtained from trough-crossbedded foresets substantiate a braided pattern of channel flow and indicate unidirectional fluvial transport down paleoslopes indicated by isopachs of the Upper Witwatersrand Group or by truncation-surface isopachs measured from underlying strata. Foreset azimuth variance calculated from data collected over entire placer deposits ranges from 1850 to 4721. The existence of longshore currents in the Witwatersrand Basin has been established in at least one instance where the transgressing Vaal Reef placer, deposited on an angular unconformity, was buried beneath sediments that were transported parallel to the paleostrike as the placer became progressively submerged.

The coarse-grained sediment facies observed in the placers comprise: crudely bedded conglomerate forming sheets, longitudinal bars and filling depressions; trough-crossbedded conglomerates in the form of channel-lined lags; and rare tabular sets of planar-crossbedded conglomerates and pebbly quartzites, representing transverse bars (Smith, 1974). Conglomerate-filled depressions, lags and bars are pebble supported and planar-crossbedded units are usually matrix-supported.

The sandy facies include: trough-crossbedded quartzite; horizontal bedded quartzite, both coarse- and fine-grained; and rare solitary sets of planar-crossbedded quartzite. A very fine-grained facies is represented by shaly muddrapes and clasts.

The vertical sequence of these placer sediment facies always begins with a scour surface and is generally followed by a pebble lag which grades normally into trough-crossbedded quartzite. Proximal locations may be distinguished from distal locations by means of: the size of clasts and heavy minerals; changes in the vertical sequence (Smith, 1970); and by mineral ratio changes.

Many abraded detrital heavy minerals comprising noble metals, sulphides and oxides have been identified as inherent sedimentary concentrates in the mature placer sediments. Pyrite is the most abundant and gold and uraninite are economically the most important. Their accumulation and preservation lend support to the hypothesis that the Proterozoic atmosphere had a low oxygen pressure. Concentrations of these minerals are spatially linked to primary sedimentary structures within the placers, particularly with lag accumulations on scour surfaces. These heavy minerals are generally hydraulically equivalent to the sand-sized fraction of the placers and are found concentrated on the toes of foresets and on the basal scour surfaces of sets. Therefore the flow-regime producing trough-crossbedding is considered to be an important element in placer mineral concentration.

Abundance ratios of placer minerals appear to change from proximal to distal environments depending on the relative particle sizes originally supplied from each specific provenance area. The most useful indicator is the ratio between uranium and gold which becomes progressively larger down many placer paleoslopes.

Fossilized black carbonaceous material has been preserved as seams and granules on: channel scour surfaces; transverse bar reactivation surfaces; foreset planes; ripple surfaces; and around pebbles in both the Witwatersrand and Ventersdorp Contact placers. The morphology of this material, examined by means of the scanning electron microscope, and analysis of its organic chemistry, favours a plant origin which, from its spatial arrangement in the sediment, is interpreted as algal. In many instances the surfaces occupied by this carbonaceous material coincide with heavy mineral concentration surfaces and as a result, due to original trapping and digestion and to later adsorption, mineralization has become associated. The carbonaceous material appears to be barren of gold mineralization when it is not associated with a heavy mineral concentration surface.

Upper Witwatersrand placers fit a sequence of braided river models from the Scott through the Donjek and South Saskatchewan to the Platte types (Miall, 1977, this volume). The Scott type is rarely preserved because of basin edge truncation and the Platte type is rarely explored because of its patchy low-grade mineral content. Therefore, most of the placers being exploited best fit the Donjek model (Table 2). There is no evidence for a gradation from braided to meandering systems. The placers appear to have been deposited by unconfined flow in braided channels on fan-deltas, being built into the Witwatersrand depository.

The Upper Witwatersrand strata have been intensely faulted. The major structures are block faults which strike parallel to the Basin edge and fortuitously displace the placers up to elevations that are accessible to mining. There are no paleocurrent or sedimentary features in the Upper Witwatersrand placers discussed which indicate even the incipient presence and influence of these major faults. They are therefore considered to be of late Witwatersrand and Ventersdorp age. Tectonic influence during the Upper Witwatersrand appears to be largely related to cratonic warping. Primary folding along a northeast axis created the major basin axis and inward tilt while secondary crossfolds induced local entry points which drained erosional products from Archean Greenstone Belts in the surrounding craton.

Periods of warping probably caused the rejuvenation which is marked by prograding placers at the bases of many Upper Witwatersrand formations. The transgressive Vaal placer may have been caused by the rate of subsidence exceeding sedimentation, and internal responses related to geomorphic gradients on the fans probably produced repeated but less extensive placer deposits within, and at the top of, formations (Basal and Steyn placers).

By contrast, most of the Ventersdorp Contact placers are clearly yoked deposits. They occur along the shallow rim of the Upper Witwatersrand, can be clearly related to major fault scarps, and contain clasts derived from erosion of lithified Witwatersrand sediments. Although the maturity of the Ventersdorp Contact placers, their sedimentary facies, vertical sequences and proximal-distal indicators resemble those in the Upper

Table 2.. Features relating Witwatersrand placers to braided river models.[1]

Placer	Major Facies	Minor Facies	Mean Pebble Size	Uranium/ Gold Ratio	Model
Steyn (proximal)	Gt, Gm	Gp, St	−4.5φ	<5	Donjek
Steyn (distal)	St, Sp	Gt, Sh, Fm	−2.5φ	40	Platte
B (proximal)	Gt, Gm	St, Sp	−4.7φ	4 - 10	Donjek
B (distal)	St	Gt, Sp	−2.5φ	>25	Platte
Vaal	St	Gt	−3.2φ	10 - 30	Platte
Ventersdorp Contact (proximal)	Gm, Gt	St, Fm	Max 40 cm	<5	Scott
Ventersdorp Contact (distal)	Gt	St	−3.4φ	10 - 34	Donjek

[1]Nomenclature and abbreviations after Miall, 1977.

Gm – massive gravel, Gt – trough-crossbedded gravel, Gp – Planar-crossbedded gravel, St – trough-crossbedded sand, Sp – planar-crossbedded sand, Sh – horizontally-bedded sand, Fm – mud drape.

Witwatersrand placers, their drainage pattern, incised into lithified Witwatersrand strata and complicated by geomorphic factors, is more complex.

EXPLORATION APPLICATION

None of the present major gold and uranium producing orebodies, amongst the Upper Witwatersrand placers, crop out at the surface. They have all been discovered beneath thousands of metres of younger strata by diamond drilling.

Magnetometer surveys have been useful in identifying the suboutcrop location of magnetic shales in the Lower Witwatersrand Group and gravimetric surveys have been used to indicate the position where Vendersdorp lava, overlying the Upper Witwatersrand Group, is thinnest, thus marking shallow borehole sites.

Underground exposures have subsequently provided windows into the Upper Witwatersrand through which structural and sedimentological information has been obtained from which to understand the deposits and interpret borehole probes beyond.

Exploration within mining leases has been tackled with short underground boreholes, drilled from access tunnels, and by tunnels in the plane of a placer, across its paleocurrent trend.

Routine sampling data on many of the gold mines have been stored on a computer data base from which trend surfaces are automatically contoured. Computer programs are used for grain-size and paleocurrent analysis. Ore reserves in sub-economic areas are selected on the basis of detailed mapping of channel scours, channel sediments and bedforms. Ore-reserve blocks are oriented on the basis of paleocurrent measurements.

An appreciation of the lognormal distribution of the placer mineral content and of the variance caused by the association of heavy minerals with the heterogeneous geometry of the fluvial sediments has enabled one to make better estimates of the mean grades of mineralization in the placers. Applied fluvial sedimentology has therefore become a valuable new tool in the Witwatersrand gold and uranium mining industry.

ACKNOWLEDGEMENTS

I gratefully acknowledge the permission and support granted by the Management of Anglo American Corporation to publish this material. I also wish to thank D. A. Pretorius, E. S. A. Antrobus, P. M. Strydom and J. P. McMagh for reviewing the paper and to G. de Vries Klein, N. D. Smith and A. D. Miall for critically reading the manuscript. I am indebted to R. Swift for drafting the drawings, J. Tatalias for preparing the photographs, M. Young and M. Schiffer for typing the manuscript and a number of colleagues who mapped the details of suitable underground exposures.

This paper was presented at the 1977 SEPM Research Symposium on "Predictive Sedimentary Models for Earth Resources" on June 14, 1977 in Washington, D.C.

REFERENCES

Armstrong, G. C., 1966, A sedimentological study of the U.K.9 Kimberley reefs in part of the East Rand: Unpub. M.Sc. thesis, Univ. of Witwatersrand, Johannesburg, 65p.

Brock, B. B., and Pretorius, D. A., 1964a, An introduction to the stratigraphy and structure of the Rand Goldfield: *in* S. H. Haughton, *ed.*, The geology of some ore deposits of Southern Africa, v. 1, Geol. Soc. S. Afr., p. 25-62.

—— and ——, 1964b, Rand Basin sedimentation and tectonics: *in* S. H. Haughton, *ed.*, The geology of some ore deposits of Southern Africa, v. 1, Geol. Soc. S. Afr., p. 549-600.

Coetzee, F., 1965, Distribution and grain-size of gold, uraninite, pyrite and certain other heavy materials in gold-bearing reefs of the Witwatersrand Basin: Trans. Geol. Soc. S. Afr., v. 68, p. 61-88.

Hallbauer, D. K., 1975, Plant origin of the Witwatersrand carbon: Minerals Sci. Engng., v. 7, p. 111-131.

Hodgson, F. D. I., 1967, The relationship between the sedimentology and the gold distribution of the Basal reef zone, Harmony Gold Mine: Unpubl. M.Sc. thesis, Univ. Orange Free State, Bloemfontein, 97p.

Hutchison, R. J., 1971, The geology of the Western Holdings mine, with special reference to the sedimentology of the Basal Reef zone: Unpubl. M.Sc. thesis, Univ. of Orange Free State, Bloemfontein, 131p.

Knowles, A. G., 1966, A geological approach to determine the pattern of gold distribution in a reef: *in* symposium on mathematical statistics and computer applications in ore valuation, S. Afr. Inst. Min. Metall., p. 157-168.

––––––, 1967, A paleocurrent study of the Ventersdorp Contact Reef at Western Deep Levels Limited on the Far West Rand: Unpubl. M.Sc. thesis, Univ. of Witwatersrand, Johannesburg, 122p.

McGowen, J. H., and Groat, C. G., 1971, Van Horn Sandstone, West Texas: An alluvial fan model for mineral exploration: Bur. Econ. Geol., Univ. Texas at Austin, 57p.

Miall, A. D., 1977, A review of the braided-river depositional environment: Earth Sci. Revs., v. 13, p. 1-62.

Minter, W. E. L., 1972, The sedimentology of the Vaal Reef in the Klerksdorp area: Unpub. Ph.D. thesis, Univ. of Witwatersrand, Johannesburg, 170p.

––––––, 1976, Detrital gold, uranium, and pyrite concentrations related to sedimentology in the Precambrian Vaal Reef placer, Witwatersrand, S. Afr.: Econ. Geol., v. 76, p. 157-176.

Pienaar, J. P., 1969, Geological aspects of the Basal Reef in the Free State Saaiplaas gold mining area: Unpub. M.Sc. thesis, Univ. of Orange Free State, Bloemfontein, 177p.

Pretorius, D. A., 1966, Conceptual geological models in the exploration of gold mineralization in the Witwatersrand Basin: *in* symposium on mathematical statistics and computer applications in ore valuation, special pub., J. S. Afr. Inst. Min. Metall., p. 225-266.

––––––, 1974a, Gold in the Proterozoic sediments of South Africa: systems paradigms, and models: Inf. Circ. No. 87, Econ. Geol. Res. Unit, Univ. of Witwatersrand, Johannesburg, 22p.

––––––, 1974b, The nature of Witwatersrand gold-uranium deposits: Inf. Circ. No. 86, Econ. Geol. Res. Unit, Univ. of Witwatersrand, Johannesburg, 50p.

––––––, 1975, The depositional environment of the Witwatersrand goldfield: a chronological review of speculations and observations: Minerals Sci. Engng., v. 7, no. 1, p. 18-47.

Ramdohr, P., 1959, New observations on the ores of the Witwatersrand in South Africa: Trans. Geol. Soc. S. Afr., annex to v. 61, 50p.

Saager, R., 1976, Geochemical aspects of the origin of detrital pyrite in Witwatersrand conglomerates: Econ. Geol. Res. Unit, Univ. of Witwatersrand, Johannesburg, Inf. Circ. No. 105, 21p.

––––––, and Esselaar, P. A., 1969, Factor analysis of geochemical data from the Basal Reef, Orange Free State Goldfield, South Africa: Econ. Geol., v. 64, p. 445-451.

Sims, J. F. M., 1969, The stratigraphy and paleocurrent history of the upper division of the Witwatersrand System on President Steyn mine and adjacent areas in the Orange Free State goldfield with specific reference to the origin of auriferous reefs: Unpub. Ph.D. thesis, Univ. of Witwatersrand, Johannesburg, 181p.

Smith, N. D., 1970, The braided stream depositional environment: comparison of the Platte river with some Silurian clastic rocks, north-central Appalachians: Geol. Soc. Am. Bull., v. 81, p. 2993-3014.

––––––, 1974, Sedimentology and bar formation in the upper Kicking Horse river; a braided outwash stream: J. Geol., v. 82, p. 205-223.

Steyn, L. S., 1964, The sedimentology and gold distribution pattern of the Livingstone reefs on the West Rand: Unpub. M.Sc. thesis, Univ. of Witwatersrand, Johannesburg, 132p.

Utter, T., 1977, The morphology of different types of pyrite from the Upper Witwatersrand system of the Klerksdorp goldfield (Abs.): Geokongress 77, Rand Afrikaans Univ., Johannesburg, p. 120-122.

Zumberge, J. E., 1976, The organic analyses and the development of the Vaal reef carbon seams of the Witwatersrand gold deposit: Unpub. Ph.D. dissertation, Depart. Geo-sciences, Univ. Arizona.

SEDIMENTARY PATTERNS OF URANIUM MINERALISATION IN THE BEAUFORT GROUP OF THE SOUTHERN KAROO (GONDWANA) BASIN, SOUTH AFRICA[1]

BRIAN R. TURNER[2]

ABSTRACT

Uraniferous Upper Permian sediments of the Beaufort Group, Southern Karoo (Gondwana) Basin, South Africa, cover an area of about 25,000 km[2], attain a maximum thickness of 2,700 m and consist predominantly of sandstones and mudstones. A systematic upward decrease in the ratio of these two lithologies, together with changes in their sedimentary properties enable the succession to be divided into low sinuosity channel, high sinuosity channel and floodbasin facies associations. This overall fining in the succession may be fundamentally linked to decreasing sediment supply and channel gradients in response to denudation of the source area and the migration (or the changing locations) of the depositional system. The facies associations probably represent the proximal, intermediate and distal components of an extensive fluvial depositional system which built out from a distant source to the south and west.

Significant uranium mineralisation only occurs in high sinuosity channel sandstones, probably because of more numerous permeability barriers. The distribution of mineralised zones suggest that local permeability differences and carbonaceous debris were the major controls of mineralisation. Host sandstones are arkosic wackes containing carbonate cement, sulphides and volcanic fragments, interbedded with occasional cherts of tuffaceous origin. The uranium was probably derived by the leaching of volcanic material and transported by mildly reducing alkaline ground water as uranyl carbonate complexes. Channel sands provide permeable pathways for these solutions and on encountering a strong reductant uranium was precipitated. Remobilisation and further concentration has played no part in the genesis of these deposits, which probably formed *in situ*, shortly after deposition and before compaction and diagenesis had proceeded to completion.

INTRODUCTION

Although uranium occurs in a variety of favourable stratigraphic settings it shows a distinct preference for continental sandstones, a relationship that has long been appreciated (95% of all United States reserves are in continental sandstones). In recognising that depositional environment plays a key role, many attempts have been made to establish the nature of the host sandstone and develop a regional depositional model that can assist exploration and help establish the controls of uranium mineralisation (intrinsic or extrinsic). However, only a very small fraction of suitable host sediments generally contain economic quantities of uranium and commonly as uranium concentration decreases so the volume of mineralised rock increases (Gabelman, 1971). The uraniferous Upper Permian sediments of the Beaufort Group in the southern Karoo (Gondwana) Basin, South Africa (Fig. 1), are a classic case in point, covering an area of about 25,000 km[2] and attaining a maximum thickness of 2,700 m. These terrigenous clastic deposits contain only small localised concentrations of uranium, none of which have proved economically exploitable despite more than 5 years intensive exploration.

The purpose of the present study is to describe and interpret regional facies patterns and to develop a depositional model. The model may have economic significance for future exploration in the basin as well as genetic significance in other sedimentary basins, especially in southern hemisphere continents where the Karoo is just one of many similar and related Gondwana sequences (Turner, 1975).

---

[1]The South African Committee for Stratigraphy recommends Karroo be spelt Karoo, but both spellings are found in the literature.

[2]Bernard Price Institute for Palaeontological Research, University of the Witwatersrand, Johannesburg 2001, South Africa.

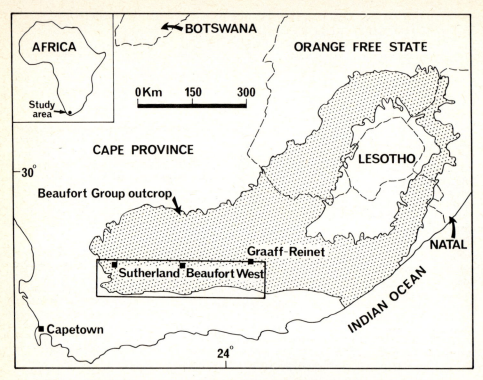

Fig. 1.   Beaufort Group outcrop in the main Karoo Basin and location of study area.

GEOLOGICAL SETTING

The Beaufort sediments were deposited in an intracratonic basin receiving detritus from a volcanically active source located to the south and west. The sediments are mainly sandstones, siltstones and mudstones with a few thin lenticular limestones and abundant calcareous concretions; the latter are particularly common in the argillaceous strata. Underlying these deposits is the Ecca Group which can be divided into a lower part, deposited under conditions of rapid subsidence mainly by turbidity currents (not necessarily marine), a middle part which records a transition from a deep water open shelf facies to a landward and progressively shallower prodelta facies (the lack of invertebrates and biogenic structures indicating an enclosed body of water where conditions were generally inhospitable to animal life) and an upper part forming up to three progradational (coarsening-up) delta front sequences. Together the Ecca and Beaufort, as defined here, represent net progradation, with the Ecca Basin withdrawing northwards prior to progradation of the Beaufort sediments in the same general direction. Thus argillaceous sediments assigned to the Ecca Group in the northern part of the basin may be time equivalent of the Beaufort Group in the south. Along the southern margins of the basin the Beaufort sediments have been caught up in the Cape Orogeny and extensively folded; northwards the folding decreases and dies out completely approaching the northern boundary of the study area (Fig. 1) where the sediments are flat-lying.

The succession can be divided according to the changing ratio of sandstone to mudstone. This shows a systematic upward decrease which, together with changes in sedimentary properties enable the succession to be divided into low sinuosity channel, high sinuosity channel and floodbasin facies associations (Fig. 2). Uranium mineralisation

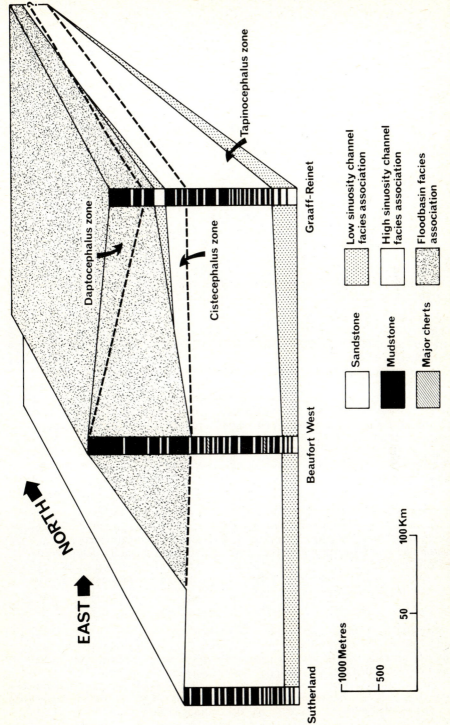

Fig. 2. Block diagram showing distribution of facies associations and biozones in the Beaufort Group of the southern Karoo Basin. The northern limit of the block diagram is the northern outcrop boundary of the Beaufort Group north of the study area (Fig. 1).

is present in all three facies associations but is only significantly developed in high sinuosity channel sand bodies (Turner, 1975; Kübler, 1977). The host rocks are weakly metamorphosed (laumontite facies) and composed of fine to very fine and locally coarse, moderately sorted, arkosic wackes containing carbonate cement and volcanic fragments of predominantly acid character. These are interbedded with minor cherts of tuffaceous origin (Martini, 1974), ranging from a few centimetres to 2 m thick.

Fossil plant material occurs in the sandstones and a few of the mudstones. Vertebrate remains are common throughout the succession and form the basis of a biostratigraphic zonation. Five zones are currently recognised (Kitching, 1973), only three of which are represented along the southern margin of the basin; these are from west to east, the *Tapinocephalus* zone, the *Cistecephalus* zone and the *Daptocephalus* zone (Fig. 2). In addition a few fish (*Atherstonia*), fresh water molluscs (*Palaeomutela, Palaeonodonta*) and the trace fossil *Planolites?* occur.

Sedimentary Facies Associations

Low Sinuosity Channel Facies Association

Description

This facies association is characterised by the presence of fining-upwards sequences comprising thick (3-23 m), laterally persistent (>1 km) tabular sandstones overlain by subordinate amounts of siltstone and mudstone (Fig. 3).

Basal scoured surfaces of low relief are overlain by intraformational conglomerates containing mudstone, siltstone and carbonate clasts. The sandstones are predominantly fine-grained but locally are coarse-grained, particularly along the base or associated with intraformational conglomerates. Large calcareous concretions, plant material and bone fragments are present locally. Planar, curved and discordant internal erosion surfaces occur in the lower part of some sandstones dividing them up into several sedimentation units (Fig. 3A). Horizontal lamination with current lineation is the dominant internal structure; massive bedding is locally developed in the lower part of some sandstones. Planar cross-bedding is uncommon and trough cross-bedding rare, the planar types consisting of solitary wedge-shaped sets or less commonly cosets (Fig. 3) with tangential bottom contacts and sharp, slightly scoured uneven bases, occasionally strewn with intraformational clasts. These generally merge laterally and vertically with horizontal laminations. Palaeocurrents are unidirectional and indicate paleoflow to the north.

The slightly finer uppermost sandstones in these sequences are ripple cross-laminated or horizontally laminated. These grade sharply into siltstones or bluish-grey to maroon mudstones which are generally massive and unlaminated (Fig. 3). Locally the mudstones are interbedded with siltstones and fine sandstones, and contain mudcracks, ripple lamination and scattered calcareous concretions. Mudstones make up only a very small part of this facies association and may be absent from the lowermost sequences; however, in the upper sequences mudstones become thicker and more frequent as the ratio of sandstone to mudstone decreases sharply.

Interpretation

The fining-upwards texture and pattern of sedimentary structures are typical of fluvial deposits. The basal scoured surface and intraformational conglomerate are comparable to channel lag deposits (Allen, 1964). Upper phase horizontally laminated sandstones are common in fluvial deposits where upper flow regime conditions prevail (Harms *et al.*, 1963; Harms and Fahnestock, 1965; McKee *et al.*, 1967); the dominance of the structure suggests that these conditions persisted throughout most of the preserved depositional phase. The internal erosion surfaces imply fluctuations in flow and more than one flood event, each successive flood scouring the deposits of the previous flood (Picard and High,

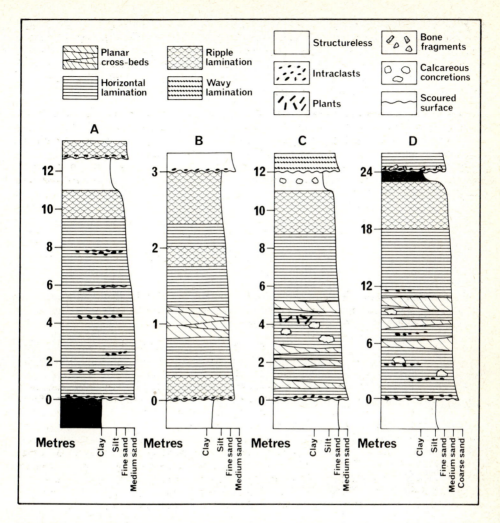

Fig. 3.   Vertical sequences of the low sinuosity channel facies association. Localities: A and B —
43 km east-southeast of Graaff-Reinet; C and D — 40 km south-southwest of Sutherland.

1973). The planar cross-bedding resembles the bar deposits of some modern braided streams (Collinson, 1970; Smith, 1970; Miall, 1977), the tangential bottom contacts and lower scoured set boundaries corresponding to the infilling of channel floor depressions under high intensity lower flow regime conditions (McKee *et al.,* 1967; Smith, 1971). Unlike most modern braid bar deposits, the planar cross-bedding is often overlain by relatively thick upper phase horizontal lamination.

The ripple cross-laminated and horizontally laminated sandstones higher in the sequence probably formed during falling water stages. The overlying siltstones and mudstones closely resemble modern overbank and floodplain deposits where exposure and pedogenesis are common features (Fisk, 1944; Picard and High, 1973). Inundation during floods and the deposition of rippled fine sandstones and siltstones is consistent with this interpretation (Shantzer, 1951; Coleman, 1969; Miall, 1977, Table III).

The prevalence of horizontal lamination and solitary planar cross-beds suggests streams of relatively large flow power (Allen, 1974) whose channels carried occasional bars (Smith, 1970; Collinson, 1970). These conditions are perhaps most commonly encountered in low sinuosity streams, an interpretation consistent with sandstone geometry, palaeocurrent trends and the high ratio of sandstone to mudstone (Allen, 1965; Schumm, 1968). Estimates of channel depth (megaripple height proportional to mean water depth) and width have been obtained from Allen (1968, Fig.).4, 3.20) and the ratio of the two values substituted in the equation for channel sinuosity derived by Schumm (1963). This yields sinuosity values of between 1.2 and 1.3, indicating straight to transitional channel patterns, according to Schumm's (1963) classification. Modern streams of this type provide only limited parallels, although the facies shows some similarity to the flood deposits of Bijou Creek, Colorado, which consist of horizontally bedded sand with subordinate amounts of planar cross-bedding and ripple cross-laminated sand (McKee *et al.*, 1967). Upper flow regime conditions persisted throughout most of the constructional phase of flood activity when discharge and velocity were exceptionally high, as in many ephemeral streams (Williams, 1971; Picard and High, 1973; Miall, 1977).

Despite bed form analogies with modern braided streams there is no *direct* evidence of bar emergence (plant colonisation, minor terraces etc.) and dissection typical of such streams during low water stage (Fisher and Brown, 1972; Smith, 1976), indicating perhaps that the bars were permanently submerged. In this respect a modern analogue may be the Cimarron River which has characteristics intermediate between braided and meandering streams. Deposits comprise fining-upwards sequences containing horizontal lamination, tabular and trough cross-bedding and ripple cross-lamination, the abundance of horizontal lamination helping to distinguish it from braided or meandering models (Shelton and Noble, 1974). Thus this facies association may have been deposited by meandering streams of low sinuosity as their similarity in grain size and sedimentary structures suggest (Fisher and McGowen, 1969). Deposition of overbank fines was probably limited by repeated avulsion and channel migration across the floodplain, promoted by weak (sandy) bank conditions and the lack of well developed confining ridges.

High Sinuosity Channel Facies Association

Description

This resembles the low sinuosity channel facies association in the presence of lower scoured surfaces with thin intraformational conglomerates overlain by sandstones from 1-24 m thick (Fig. 4). The sandstones are typically lenticular and grade upwards and intertongue laterally with mudstone. They are fine- to very fine-grained with scattered mudstone and siltstone clasts and occasional calcareous concretions. Horizontal lamination is less abundant than in the low sinuosity channel facies association and locally may be subordinate to trough and planar cross-bedding (Fig. 4). Individual cross-beds are up to 1.5 m thick and generally confined to the lower and middle parts of sandstones. Palaeocurrents throughout the facies association are bimodal and directed towards the north and east. Locally the crossbedding shows soft sediment deformation (Fig. 4C) referred to as slumped structures by Kitching (1973) and intraformational deformation by Kübler (1977). The slightly finer grained sandstones and siltstones higher in the sequence contain ripple and climbing ripple cross-lamination, horizontal lamination and a few burrows (*Planolites*?). Current ripple-marked surfaces are common, the ripples being of straight, sinuous-crested or linguoid types (Allen, 1968).

Locally the siltstones are structureless and may terminate the sequence (Fig. 4D and H), although more commonly the siltstones and fine sandstones pass gradationally into maroon mudstones. These frequently contain calcareous concretions, often concentrated into ill-defined bands or merging to form nearly continuous limestone beds up to 2 m thick. Maroon to red and green mottled horizons are found as well as thin siltstone and

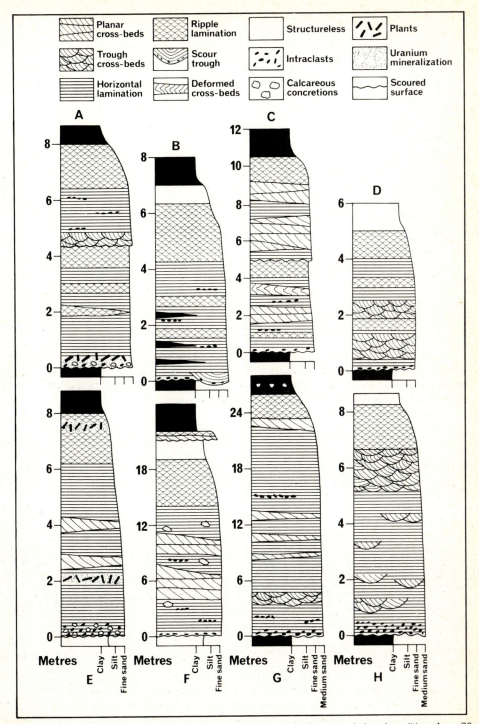

Fig. 4. Vertical sequences of the high sinuosity channel facies association. Localities: A — 72 km southeast of Graaff-Reinet; B and C — 61 km south of Graaff-Reinet; D — 77 km southeast of Graaff-Reinet; E and F — 16 km south of Sutherland; G and H — 74 km southeast of Graaff-Reinet.

sandstone layers, mudcracks and ripple cross-lamination in which the laminae are locally convoluted (Kübler, 1977). The amount of mudstone increases significantly towards the top of the facies with a corresponding increase in mottling and calcareous concretions, some of which contain vertebrate fossils (Kitching, 1973). The mudstones include local coarsening-upwards sequences (0.3 to 2.5 m thick) consisting of mudstone grading up into silty mudstone and capped by ripple cross-laminated and horizontally laminated siltstones and fine sandstones.

Interpretation

Unlike the low sinuosity channel facies association these sandstones are lenticular in character and contain a higher proportion of trough cross-bedding. The ratio of sandstone to mudstone is much lower and there is a gradual transition from coarse to fine within fining-upwards sequences. These features resemble the high sinuosity meandering stream model of Allen (1965, 1970a) and Visher (1972). Thus the scoured surface and intraformational conglomerate correspond to channel lags, with the overlying sandstone recording point bar lateral accretion (epsilon cross-bedding has been found at only one locality). The sequence from horizontally laminated and cross-bedded sandstone near the base of fining-up sequences to the ripple cross-laminated and horizontally laminated finer uppermost sandstones compares with point bar development in high sinuosity streams in response to declining bed shear stress as currents move obliquely over the point bar surface. Allen (1970a) showed theoretically that upper phase plane beds could make up substantial thicknesses of fining-upwards sequences in high sinuosity meandering rivers provided water surface slope and channel curvature are relatively large, as for example in the Red River, Louisiana (Harms *et al.*, 1963).

In plan the sandstones reproduce all the features of modern arcuate point bar ridges (Kübler, 1977, frontispiece). Meander belts were narrow compared to floodplain width, indicating the early abandonment of meander belts, probably in favour of gradient advantages offered by interfluve basins (Ferm and Cavaroc, 1968). As point bars accreted laterally their deposits became overlain by overbank fines, deposited mainly from suspension (Allen, 1965). Repeated inundation of the overbank environment is suggested by fine sandstone and siltstone interbeds, common features in the floodplain deposits of modern rivers (Fisk, 1944). Mudcracks, calcareous concretions (cornstones) and colour mottling indicate exposure, pedogenesis and a fluctuating ground water table (Allen, 1974). The development of thick ($<$ 2 m) beds of calcareous concretions suggests periods of prolonged exposure (Reeves, 1970). Local coarsening-upwards sequences probably represent deposition by channel overspill and crevasse-splays in shallow water floodplain environments (Turner, 1975).

There are a number of variations to the basic fining-upwards sequence as many of the sandstones are multistory and contain more than one point bar accumulation (Fisher and Brown, 1972). Consequently it is often more practical to recognise lower sandy and upper argillaceous members with overlapping and transitional relations between them. Kübler (1977) defined a modal cycle (Duff and Walton, 1962) by Markov Chain analysis which showed little departure from the observed trend, namely that horizontal and cross-bedded sandstones are overlain by ripple cross-laminated sandstones followed by siltstone and mudstone. Although Kübler (1977) considered the coarser member to be most variable it appears that the upper more argillaceous member shows the greater variability. This partly results from the fact that the several overbank subenvironments (Turner, 1975) were not included as separate facies states in Kübler's (1977) analysis and also because of the poor continuity of exposure in the part of the basin he studied. The implications are important since uranium mineralisation occurs in the lower coarse sandy members of fining-upwards sequences. If there is little variability then it could provide a useful predictive tool where mineralisation is preferentially associated with certain sedimentary structures or certain parts of the fining-upwards sequence.

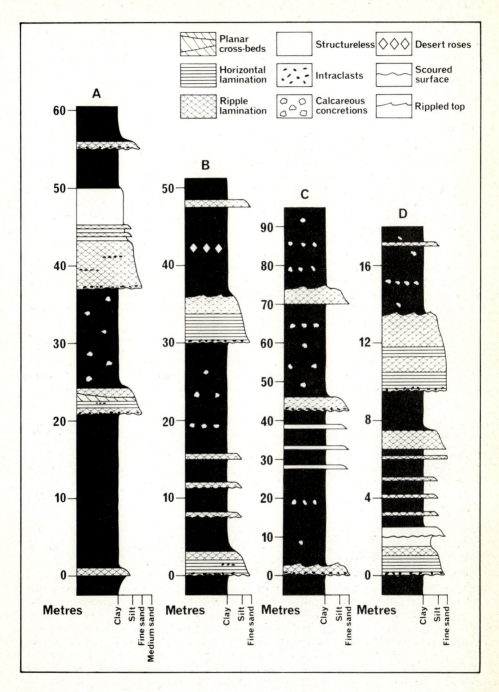

Fig. 5. Vertical sequences of the floodbasin facies association. Localities: A — 16 km west of Beaufort West; B, C and D — 43 km east-southeast of Graaff-Reinet.

Floodbasin Facies Association

Description

This facies association consists of red, generally massive mudstones with occasional interbeds of sandstones and siltstones (Fig. 5). The mudstones form thick sequences (up to 30 m) separating sandstone and siltstone beds. Desiccation cracks occur along the bases of sandstones and siltstones, and calcareous concretions are common throughout the mudstones, sometimes concentrated into concretionary layers and associated with vertebrate fossils. 'Desert roses' composed of pseudomorphs of calcite after gypsum appear to be mainly confined to the eastern part of the basin around Graaff-Reinet (Keyser, 1966).

Three basic types of sandstone interbed have been recognised. The commonest type consists of thin (<3 m), graded, fine-grained sandstones resting on scoured bases and often associated with mudstone clasts. Internally the sandstones are horizontally laminated and ripple cross-laminated with the tops of beds frequently ripple-marked (Fig. 5); most of the ripples are straight to sinuous in plan. The second type of sandstone resembles channel sand bodies in that they are generally thicker (> 3 m), laterally more persistent (up to 0.5 km) and contain rare planar cross-bedding, sometimes occurring between ripple cross-laminated sandstones at the base and horizontally and ripple cross-laminated sandstones at the top (Fig. 5A). The least common type of sandstone (< 1 m thick) has a slightly scoured base, is graded and lacks internal stratification. The siltstones are generally less than 1 m thick and reddish-grey in colour with sharp or slightly scoured and uneven bases. Most siltstones are ripple cross-laminated and grade upwards into silty mudstone.

Interpretation

This facies association closely resembles modern floodbasin sediments which are usually thick and laterally extensive and consist predominantly of overbank fines with occasional channel deposits (e.g. Hwang-Ho floodplain, Kukal, 1971). The mudstones compare with floodbasin sediments deposited from suspension in slowly moving or ponded waters remote from the main stream channels (Allen, 1965; Coleman, 1969). In dry climates floodbasin deposits are normally brown or red and lack organic debris. Evidence of exposure and pedogenesis abounds in such deposits, especially in semi-arid regions such as the Indus alluvial plain (Allen, 1970b). The prevalence of calcareous concretions implies periods of prolonged exposure consistent with aridity, and the development of 'desert roses'. Plant cover of basins under these conditions was probably sparse and generally restricted to channel margins. Since vegetation capable of colonising and stabilising interfluve hillslopes had not evolved, run-off and sediment yield were exceptionally high (Schumm, 1968), promoting thick sequences of flood sediments.

The sharp erosive bases and normal grading of many of the sandstones suggest the sudden introduction of sediment laden waters during floods, flow power decreasing during deposition (Kuenen, 1967; McKee *et al.,* 1967). The thicker, laterally persistent sandstones with local planar cross-bedding may represent former active channels or crevasse-channels which discharged flood waters onto the floodplain. Scattered mudstone clasts and upper phase plane bed deposits indicate some measure of the flow velocity. The vertical repetition of two or three such sandstones separated by floodplain fines suggests temporary stabilisation of the alluvial ridge (meander belt) and repeated avulsion. Stabilisation would be further encouraged by channel margin vegetation (Smith, 1976). The normally graded and structureless sandstones probably reflect rapid deposition as crevasse-splays from small floods (Leeder, 1973). The normally graded rippled siltstones could have originated as the distal equivalents to some of the thinner sandstones.

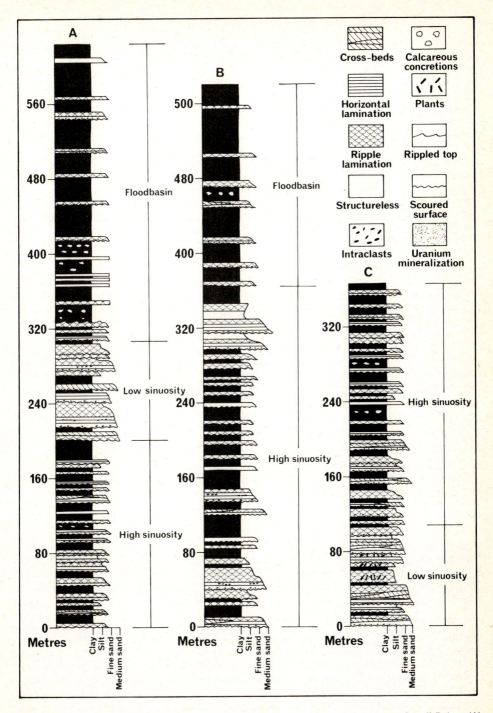

Fig. 6.   Generalised stratigraphic sections of part of the Beaufort Group in the Graaff-Reinet (A) Beaufort West (B) and Sutherland (C) areas. Note the overall fining-upwards trend with floodbasin facies dominating towards the top in the Graaff-Reinet and Beaufort West areas.

Vertical Succession of Facies and Depositional Model

Palaeocurrent data indicate that the sediments were deposited by streams flowing to the north and east. Lateral equivalents of the low sinuosity, high sinuosity and floodbasin facies associations are not exposed, but if the vertical succession of rock types and sedimentary structures reflects the overall distribution of depositional environments then they may represent the proximal, intermediate and distal components of a fluvial distributary system. The overall upwards-fining trend (Fig. 6) and inferred changes in channel morphology may be linked to decreasing sediment supply and channel gradients in response to denudation of the source area and migration of the depositional system. As channel gradients declined and the depositional system shifted sourcewards discharge and bed load decreased and increasing quantities of suspended sediment were supplied to the basin. Schumm (1960) noted that sinuosity increased as the percentage of silt and clay in the stream bed and bank rose. Consequently the river system changed from low sinuosity to high sinuosity channels. The low sinuosity channel facies association probably formed the most proximal element in the depositional system, accumulating in high gradient, upstream areas where channels may have been braided and characterised by a heavy traction load and fluctuating discharge.

Any one of the many interacting variables controlling fluvial systems could have been responsible for the local development of low sinuosity channel sediments in the east (Fig. 2 and 6). However, if the upward increase in red colouration of the mudstones and calcareous concretions, as well as the presence of 'desert roses' signifies increased aridity (Keyser, 1966; Haughton, 1969), then climate may have been the critical variable. Through reduction in vegetation aridity increases peak discharge and the amount of sand transported. Since the channels normally contain less water under such conditions the only way they can transport the increased load would be to increase gradient by converting the channel system to one of low sinuosity (Schumm, 1965). Good examples of similar changes in channel morphology have been documented for some of the rivers of the Great Plains in the United States, as well as the Mississippi which showed similar evolutionary trends in response to changes in gradient, sediment load and discharge between the Wisconsin glaciation and the present (Schumm, 1965).

The rivers may have flowed into the remnant Ecca Basin whose centre had withdrawn northwards prior to progradation of the Beaufort sediments in the same general direction (Ryan, 1967; Turner, unpublished data). As the ultimate depository the basin acted as the locus for accumulation of fine-grained sediments, and towards the close of Ecca times it became progressively shallower as indicated by the abundant current and wave ripples in the uppermost beds. In spite of this the energy of the water body appears to have been incapable of constructing beaches, barriers and other features characteristic of active coastal and shallow water marine environments. In the absence of these features, fine-grained detritus would be transported basinwards and deposited over the shallow basin floor, inhibiting the formation of carbonates and providing a base for fluvially induced progradation. The palaeogeography was probably one of a gently sloping alluvial plain passing directly into a shallow low energy water body.

Following deposition of the low sinuosity channel facies association (lower-most Beaufort Group) the Ecca Basin probably ceased to exist (Ryan, 1967) and the site of the major depository changed. Palaeocurrent data suggests that the depository now lay to the east (Turner, 1975) where there are deeper water facies equivalents of the Beaufort strata to the west (high sinuosity channel facies association and floodbasin facies association) which include lacustrine sequences (Johnson, 1976). Application of the depositional model to delineating areas of potential uranium mineralisation suggests that the most likely site can be expected where the fluvial sediments pass into deeper water facies equivalents. Some mineralisation has been reported here, although exploration has not been exhaustive.

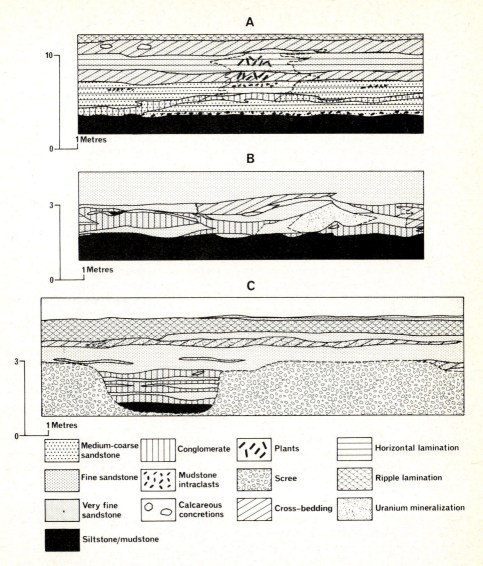

Fig. 7. Cross-sections through mineralised channel sandstones (modified after Turner, unpublished data, and Kübler, 1977). Localities: A — 43 km southeast of Sutherland; B and C — 88 km northwest of Beaufort West.

URANIUM MINERALISATION

Relationship of Mineralisation to Sedimentary Features

Because of the lack of evidence for structural control of mineralisation (Turner, 1975) sedimentary features at both regional and local level become important. Areas of known mineralisation appear to be mainly related to the high sinuosity channel sediments. Mineralisation has been noted in the low sinuosity channel and floodbasin sediments but is insignificant. This preferential association probably reflects more numerous

BRIAN R. TURNER

permeability barriers in the high sinuosity channel deposits where sandstone: mudstone ratios average about 1:4. This facies association is laterally extensive and has a thickness of up to 1,000 m. Together with the facies association above and below it appears that during Beaufort times the southern part of the Karoo Basin lay within a uranium province, receiving detritus from a volcanically active source to the south and west (Turner, 1975).

Mineralisation is generally found in the thicker parts of channel sandstone lenses, particularly in the lower half (Fig. 7) where they contain discontinuous siltstones and mudstones, or interfinger with argillaceous rocks. Most mineralised sandstones are dark grey to black and calcareous, a few are bleached white or yellowish-brown and some have associated copper minerals. The mineralised zones are separated by large areas of barren sandstone, and range from small pods and lenses a few metres in extent to larger discontinuous bodies traceable for up to 1 km (Kübler, 1977). Individual mineralised zones vary in thickness from a few centimetres up to a maximum of about 7 m, and contain both primary (uraninite and coffinite) and secondary uranium minerals (metatorbanite, uranospinite and uranophane). Within channel sandstones sedimentary structures frequently affect the distribution of mineralisation (Fig. 7). Intraformational conglomerates, penecontemporaneous deformation structures, horizontal laminations, cross-beds and ripple cross-lamination may be preferentially mineralised and invariably associated with carbonaceous debris. These features suggest two main controls on the location of uranium in channel sandstones, firstly, local permeability differences and secondly, the availability of carbonaceous debris. Grain size appears to have exerted little influence, since many of the mineralised zones are either coarser or finer than adjacent barren sandstones, although broadly similar in composition. The geometry of mineralised zones shows that they are generally elongate in the direction of sandstone thickening, and assume an arcuate trend corresponding to the shape of the channel depository. This, together with their position within the sandstones suggest that they may have formed in the deeper channel thalwegs where intraformational debris and organic matter tend to concentrate. These may enhance permeability and facilitate passage of mineralising fluids as well as providing a suitable local environment for precipitation.

Most of the best developed mineralised areas occur in the upper part of the high sinuosity channel facies association (as around Beaufort West). Unfortunately, here the ratio of sandstone to mudstone is lower and the uranium is localised in individual channels but widespread because of the distribution of host channel sandstones. The only worthwhile deposits are likely to be those in the thickest and laterally most persistent sandstones, or those occurring as successive superimposed mineralised sandstones; an unlikely situation in view of the depositional character of floodplain sediments and the general *en echelon* arrangement of channel sands (Visher, 1972, Fig. 2).

Source and Transportation

The volcanic fragments in the host sandstones and interbedded tuffaceous material (Martini, 1974) suggest a volcanic source for the uranium. The uranium was probably leached from the acid-rich volcanic source and transported by alkaline ground waters (which provide the most favourable transporting medium for the uranium ion, Gruner, 1956; Weeks and Eargle, 1963; Norton, 1970, particularly under conditions of dry climatic weathering, Eargle et al., 1975). The degree of dispersion of uranium in ground waters depends on its chemical properties. In the Beaufort it normally occurs in the insoluble +4 oxidation state (uraninite and coffinite), but it can be oxidised to the +6 oxidation state in the oxidising zone of rocks where the uranium forms soluble complexes (Adler, 1974).

Hoestetler and Garrels (1962) have shown that uranium can be transported as uranyl carbonate complexes by oxidising or mildly reducing solutions containing abundant carbon dioxide and having a neutral to alkaline pH. However, only reducing solutions can pass through the host sandstones without markedly altering them and on coming into contact with a strong reductant will precipitate uranium as insoluble U^{+4}. The solutions responsible for mineralisation of the Beaufort sediments were probably of this type, as evidenced by: (1) the close association of uranium with fossil plant material, sulphides and carbonate cement, (2) the absence of roll-type ore bodies and large areas of oxidised sandstone (Rackley et al., 1968), (3) the geometry of known mineralised bodies, (4) the presence of zeolites whose formation is favoured by alkaline ground water and (5) the secondary overgrowths on quartz grains suggesting alkaline conditions favourable to silica mobilisation but generally unfavourable to feldspar alteration (pH from 8 to 10, similar to that required for synthesis of coffinite).

Precipitation and Emplacement

The size and concentration of the deposits depends on the permeability of the host rock and the amount of mineralising solutions coming into contact with reductant. Interruption of flow is also essential to allow time for uranium to precipitate (Gabelman, 1971), for if continuous flushing occurs leaching rather than fixation may be the dominant process (Doi et al., 1975). In this respect low structural dips ($<5°$) are important (Grutt, 1975) as they help to reduce excessive flushing whilst providing adequate uraniferous groundwater throughput for the formation of deposits. Since the high sinuosity channel facies commonly dips at more than $5°$, the mineralisation probably predates deformation.

The most favourable host rocks are fluvial sandstones with discontinuous permeability zones, as in the Beaufort. These acted as permeable pathways for the mineralising groundwater, their texture, highly varied lithology and sedimentary structures providing suitable conditions for uranium transport and precipitation. Since most of the sandstones are fine-grained and poorly sorted they may have restricted the volume of mineralising ground water flowing through the rock and thus the size and concentration of the deposits (Gabelman, 1971). On encountering strong reductants in zones of differential permeability fluid flow would be interrupted or delayed and the uranyl carbonate complexes in solution reduced and precipitated mainly as uraninite and coffinite; the carbonate in solution providing a source of cement for the host sandstone (Kübler, 1977). The major reductant in the sandstones was probably hydrogen sulphide produced by anaerobic bacteria acting on detrital carbonaceous debris. Evidence of this is the close association of carbonaceous debris with uranium, its frequent replacement by ore minerals, the abundant pyrite in many of the sandstones, and the black iron monosulphides which along with carbonaceous debris give non-oxidised sandstones their dark grey colour. Mobile reductants such as humic compounds, petroleum or gases appear to have played no part in precipitation; neither is there evidence to suggest that the pseudocoals found in parts of the high sinuosity channel facies association (Rossouw and De Villiers, 1953) acted as reductants. Petrographic studies indicate that some of the interstitial uranium mineralisation may have formed by the replacement of clay matrix (Turner, 1975), carbonate cement and sulphides in the sandstones (Kübler, 1977).

Remobilisation and concentration of the uranium minerals by multiple migration and accretion (Gruner, 1956) is unlikely because of: (1) the peneconcordant nature of the deposits, (2) the general absence of large redox fronts, (3) the lack of structural control and (4) the highly compacted, impermeable sandstone hosts (Turner, 1975, and unpublished data; Kübler, 1977). This suggests that the uranium probably formed in situ shortly after deposition of the host rocks and before compaction and diagenesis had proceeded to completion, thereby preventing further transmissability by ground water or pore solutions (Garrels et al., 1957). Additional evidence along these lines is the replacement of undeformed plant cell structures by ore minerals interpreted by Kübler (1977) as an indication that mineralisation occurred before deep burial of the sediments (Fisher, 1968) from epigenetic fluids (Gabelman, 1971). Contemporary oxidation of the surficial deposits probably led to some loss of uranium in solution, with a corresponding lowering of grade and the formation of secondary minerals.

ACKNOWLEDGEMENTS

The author would like to thank Dr. A. Heward (Durham) and Dr. D. K. Hobday (Natal) for their constructive reviews of the manuscript. I am also grateful to Dr. M. R. Johnson (Geological Survey of South Africa) and Dr. A. W. Keyser (Geological Survey of South Africa) for useful discussions. I am grateful to Esso Minerals, Africa, for permission to use unpublished research data.

REFERENCES

Adler, H. H., 1974, Concepts of uranium ore formation in reducing environments in sandstone and other sediments: *in* Formation of uranium ore deposits; International Atomic Energy Agency, Vienna, p. 141-168.

Allen, J. R. L., 1964, Studies in fluviatile sedimentation: Six cyclothems from the lower Old Red Sandstone, Anglo-Welsh Basin: Sedimentology, v. 3, p. 163-198.

————, 1965, A review of the origin and characteristics of recent alluvial sediments: Sedimentology, v. 5, p. 89-191.

————, 1968, Current Ripples, their relation to patterns of water and sediment motion: North-Holland Pub. Co., Amsterdam, 433p.

————, 1970a, Studies in fluviatile sedimentation: a comparison of fining-upwards cyclothems, with special reference to coarse member composition and interpretation: J. Sediment. Petrol. v. 40, p. 298-323.

————, 1970b, Physical Processes of Sedimentation: George Allen and Unwin, London, 248p.

————, 1974, Sedimentation of the Old Red Sandstone (Siluro-Devonian) in the Clee Hills area, Shropshire, England: Sediment. Geol., v. 12, p. 73-167.

Coleman, M. J., 1969, Brahmaputra River: channel processes and sedimentation: Sediment. Geol., v. 3, 129-239.

Collision, J. D., 1970, Bedforms of the Tana River, Norway: Geogr. Ann., v. 52A, p. 31-56.

Doi, K., Hirono, S., and Sakamaki, Y., 1975, Uranium mineralisation by ground water in sedimentary rocks, Japan: Econ. Geo., v. 70, p. 628-646.

Duff, P. McL. D., and Walton, E. K., 1962, Statistical basis of cyclothems; a quantitative study of the sedimentary succession in the East Pennine Coalfield: Sedimentology, v. 1, p. 235-255.

Eargle, D. H., Dickinson, K. A., and Davies, B. O., 1975, South Texas uranium deposits: Am. Assoc. Petrol. Geol. Bull., v. 53, p. 30-54.

Ferm, J. C. and Cavaroc, V. V., 1968, A non-marine model for the Allegheny of West Virginia: *in* G. deV. Klein, *ed.,* Late Paleozoic and Mesozoic Continental Sedimentation, Northeastern North America; Geol. Soc. Am., Spec. Paper 106, p. 1-19.

Fisher, R. P., 1968, The uranium and vanadium deposits of the Colorado Plateau region: *in* J. D. Ridge, *ed.,* Ore deposits of the United States, 1933-1967, Vol. 1: Am. Inst. Mining Metall. Petrol. Eng., New York, p. 734-746.

Fisher, W. L., and McGowen, J. H., 1969, Depositional systems in Wilcox Group (Eocene) of Texas and their relation to occurrences of oil and gas: Am. Assoc. Petrol. Geol. Bull., v. 51, p. 30-54.

Fisher, W. L., and Brown, L. F., 1972, Clastic depositional systems — genetic approach to facies analyses: Bureau of Economic Geology, Univ. of Texas at Austin, 211p.

Fisk, H. N., 1944, Geological investigation of the alluvial valley of the lower Mississippi River: Mississippi River Comm., Vicksburg, Miss., 78p.

Gabelman, J. W., 1971, Sedimentology and uranium prospecting: Sediment. Geol., v. 6, p. 145-186.

Garrels, R. M., Hoestetler, P. B., Christ, C. L., and Weeks, A. D., 1957, Stability of uranium, vanadium, copper and molybdenum minerals in natural waters at low temperatures and pressures (abs.): Geol. Soc. Am. Bull., v. 68, p. 1732.

Gruner, J. W., 1956, Concentration of uranium in sediments by multiple migration-accretion: Econ. Geol., v. 51, p. 495-520.

Grutt, E. W., 1975, Search for uranium: a perspective: Mining Yearbook, p. 119-126.

Harms, J. C., Mackenzie, D. B., and McCubbin, D. G., 1963, Stratification in modern sands of the Red River, Louisiana: J. Geol., v. 71, p. 566-580.

———, and Fahnestock, R. K., 1965, Stratification, bed forms and flow phenomena (with an illustration from the Rio Grande), in G. V. Middleton, ed., Primary sedimentary structures and their hydrodynamic interpretation: Soc. Econ. Paleont. Mineral. Spec. Pub. 12, p. 84-115.

Haughton, S. H., 1969, Geological History of Southern Africa: Geol. Soc. S. Africa, 535p.

Hoestetler, P. B., and Garrels, R. M., 1962, Transportation and precipitation of uranium and vanadium at low temperatures with special reference to sandstone-type uranium deposits: Econ. Geol., v. 57, p. 137-167.

Johnson, M. R., 1976, Stratigraphy and sedimentology of the Cape and Karoo Sequences in the eastern Cape Province: Unpub. Ph. D. thesis, Rhodes University, Grahamstown, 336p.

Keyser, A. W., 1966, Some indication of an arid climate during the deposition of the Beaufort Series: Ann. Geol. Surv. Dep. Min. S. Afr., v. 5, p. 77-80.

Kitching, J. W., 1973, On the distribution of the Karroo vertebrate fauna with special reference to certain genera and the bearing of this distribution on the zoning of the Beaufort Beds: Unpub. Ph. D. thesis, University of the Witwatersrand, Johannesburg, 256p.

Kübler, M., 1977, The sedimentology and uranium mineralisation of the Beaufort Group in the Beaufort West — Fraserburg — Merweville area, Cape Province; Unpub. M.Sc. thesis, University of the Witwatersrand, Johannesburg, 106p.

Kuenen, P. 1967, Emplacement of flysch-type sand beds: Sedimentology, v. 9, p. 203-243.

Kukal, Z., 1971, Geology of Recent Sediments: Academic Press, London and New York, 490p.

Leeder, M. R., 1973, Sedimentology and palaeogeography of the Upper Old Red Sandstone in the Scottish Border Basin: Scott, J. Geol., v. 9, p. 117-144.

Martini, J. E. J., 1974, The presence of ash beds and volcanic fragments in the greywackes of the Karroo System in the Southern Cape Province: Trans. Geol. Soc. S. Afr., v. 77, p. 113-116.

McKee, E. D., Crosby, E. J., and Berryhill, H. L., 1967, Flood deposits, Bijou Creek, Colorado, June, 1965: J. Sediment. Petrol., v. 37, p. 829-851.

Miall, A. D., 1977, A review of the braided-river depositional environment: Earth Sci. Revs., v. 13, p. 1-62.

Norton, D. L., 1970, Uranium geology of the Gulf coastal area: Corpus Christi Geol. Soc. Bull., v. 10, p. 19-26.

Picard, M. D., and High, L. R., 1973, Sedimentary structures of ephemeral streams: Developments in Sedimentology 17, Elsevier, Amsterdam, 215p.

Rackley, R. I., Shockey, P. N., and Dahill, M. P., 1968, Concepts and methods or uranium exploration: Wyoming Geol. Assoc. Guidebook, Black Hills Area, 12th Annual Field Conf., p. 115-124.

Reeves, C. C., 1970, Origin, classification, and geological history of caliche on the southern high plains, Texas and eastern New Mexico: J. Geol., v. 78, p. 352-362.

Rossouw, P. J., and De Villiers, J., 1953, The geology of the Merweville area, Cape Province: Explan. Sheet Geol. Survey Dept. Mines, S. Afr., 189, 78p.

Ryan, P. J., 1967, Stratigraphic and palaeocurrent analysis of the Ecca Series and the lowermost Beaufort beds in the Karroo Basin of South Africa: Unpub. Ph.D. thesis, University of the Witwatersrand, Johannesburg, 210p.

Shantzer, E. V., 1951, Alluvium of plains Rivers in a temperate zone and its significance for understanding the laws governing the structure and formation of alluvial suites (in Russian): Trav. Inst. Sci. Geol. Akad. Nauk, v. 135, p. 1-271.

Shelton, J. W., and Noble, R. L., 1974, Depositional features of braided-meandering stream: Am. Assoc. Petrol. Geol. Bull., v. 58, p. 742-748.

Schumm, S. A., 1960, The effect of sediment type on the shape and stratification of some modern fluvial deposits: Am. J. Sci., v. 258, p. 177-184.

———, 1963, Sinuosity of alluvial rivers on the Great Plains: Geol. Soc. Am. Bull., v. 74, p. 1089-1100.

———, 1965, Quaternary paleohydrology, in H. E. Wright, and D. G. Frey, eds., Quaternary of the United States: Princeton Univ. Press, p. 783-794.

————, 1968, River adjustment to altered hydrologic regimen — Murrumbidgee River and paleochannels: U.S. Geol. Survey Prof. Paper 598, 65p.

Smith, D. G., 1976, Effect of vegetation on lateral migration of anastomosed channels of a glacial meltwater river: Geol. Soc. Am. Bull., v. 87, p. 857-860.

Smith, N. D., 1970, The braided stream depositional environment: comparison of the Platte River with some Silurian clastic rocks, north-central Appalachians: Geol. Soc. Am. Bull., v. 81, p. 2993-3014.

————, 1971, Transverse bars and braiding in the lower Platte River, Nebraska: Geol. Soc. Am. Bull., v. 82, p. 3407-3420.

Turner, B. R., 1975, Depositional environments and uranium mineralisation in the Permian Lower Beaufort beds (Tapinocephalus Zone) of the Karoo (Gondwana) system in South Africa: IX me Congress International de Sedimentologie, Nice, 91-96.

Visher, G. S., 1972, Physical characteristics of fluvial deposits, *in* J. K. Rigby, and W. K. Hamblin, *eds.,* Recognition of ancient sedimentary environments: Soc. Econ. Paleont. Mineral. Spec. Pub. 16, p. 108-145.

Weeks, A. D. and Eargle, D. H., 1963, Relation of diagenetic alteration and soil forming processes to the uranium deposits of the southeast Texas coastal plain: Clays Clay Minerals, Proc. Natl. Conf. Clays Clay Minerals, v. 10, p. 23-41.

Williams, G. E., 1971, Flood deposits of the sand-bed ephemeral streams of central Australia: Sedimentology, v. 17, p. 1-40.

SYMPOSIUM ABSTRACTS

SYMPOSIUM ABSTRACTS

KEYNOTE ADDRESS:

ACHIEVEMENTS AND PROSPECTS IN FLUVIAL SEDIMENTOLOGY

J. R. L. ALLEN,

Sedimentology Research Laboratory, University of Reading, Whiteknights, Reading, England, RG6 2AB

Our present knowledge and understanding of the sedimentology of fluviatile deposits has been obtained through the combined efforts of geologists, geomorphologists, and hydraulic engineers. Useful future advances will depend on the continued close collaboration between these workers.

Channel pattern and process. Channel pattern and interconnectedness vary widely in contemporary streams but can be regarded as functions of stream power, bank erodibility, and time. The pattern of a river reach may change with time, and different reaches on the same river may differ strongly in pattern at a given time. If a satisfactory basis for palaeohydraulic interpretation is to be obtained, the attack on the question of the dependence of channel geometry on hydraulic and drainage-basin controls must be renewed in a wider and more vigorous manner than hitherto. There is a fair understanding of meandering stream processes, obtained through a combination of theoretical, experimental and field studies of contemporary rivers, but an inadequate knowledge of the processes operating in streams of low sinuosity, where secondary flow may not be important. Our inadequate knowledge of the time scales appropriate to river processes, particularly of channel-shifting and bar-building, is holding back understanding of the full implications of channel facies encountered in the older geological record. Knowledge of the processes affecting the overbank is particularly inadequate and retards appreciation of the corresponding facies.

Bed forms. In the analysis of fluvial channel facies, bed forms and their dependent internal structures afford potential information about transport directions, hydraulic conditions and regime, and channel pattern and connectedness. Extensive data have been obtained experimentally on the characteristics and hydraulic limits of the common one-way bed forms in sand-sized sediments. There is little corresponding data on bed forms in gravel, and our knowledge of bed forms in silts also cannot be described as satisfactory. Efforts are, however, being made to remedy these defects. These experiments, almost without exception, were carried out under conditions of steady-state equilibrium (flow quasi-uniform and quasi-steady). Flows in real rivers are non-uniform and unsteady, and field studies show that many bed forms display lag effects in relation to the patterns of hydraulic conditions. Are bed forms which differ geometrically necessarily different hydrodynamically, or can operational differences be explained by lag effects? How might lag affect our judgement of regime from the deposits which chance to be preserved? What are the bed forms of gravel-bed streams?

Facies and facies models. A combination of studies of contemporary fluvial deposits with the older geological record has led to the presentation of general facies models for the deposits of meandering and a variety of low-sinuosity streams. In refining and applying these models in the future, their limitations should not be overlooked. Firstly, the models are of local relevance, for what may be true of a stream or fluvial sand body at one restricted site may not necessarily be true either overall or even at sites close by. A lateral change of facies may be expected in point bars, for example, because of the influence of local channel curvature on local processes. Secondly, these models emphasize the facies originating in channels. Just as overbank processes are ignored in comparison with channel processes, so the relatively fine-grained sediments of the overbank are neglected at the expense of the more immediately attractive channel sands and gravels. Overbank facies are the more important volumetrically in many alluvial suites and are the "matrix" in which the channel sand-bodies lie. A better understanding of overbank facies, particularly those dependent on pedogenic and other superficial processes, would greatly assist our understanding of how alluvial suites are put together by the flows in and emanating from channels, under the influence of subsidence, base-level change, and climatic variation. A future emphasis on this aspect of fluvial deposits would go far towards bridging the present gap between the traditional stratigrapher's treatment of alluvial suites on the macroscale and the sedimentologist's perhaps undue concern to date with the microscale aspects of channel facies.

PRESERVATION POTENTIAL AND BEDDING FEATURES IN VARIABLE DISCHARGE STREAMS

ROBERT H. BLODGETT,

200 Main Street, Dunedin, Florida 33528

K. O. STANLEY,

Department of Geology and Mineralogy, Ohio State University, Columbus, Ohio 43210

Construction and application of actualistic sedimentological models for coarse-grained fluvial sediments must address the preservation potential of features observed in modern streams. Effective evaluation of preservation potential requires a detailed comparison of modern fluvial bedforms, barforms, and hydraulic parameters with the sedimentary structures, clast size, and vertical/lateral variability of ancient fluvial sediments.

Comparative study of the Platte River system, Great Plains, U.S.A., with fluvial sediments of the Oligocene White River Group has shown the utility of discharge and preservation potential in interpreting gravel and sand-size fluvial sediments. Results of this study include a revision of the classic Platte model for braided stream sedimentation based on discharge fluctuation data, aerial photography at high discharge, and a historical analysis of the effects of human modification on the Platte.

Application of the Platte model to Oligocene bed configuration reconstruction indicates that some widely held concepts of longitudinal and transverse bars need revision to adequately describe modern barforms and provide a workable model for recognition of ancient barforms. Vertical and lateral sequences of White River Group sedimentary structures indicate high discharge bed configuration may have a great preservation potential in variable discharge streams.

VERTICAL ACCRETION DURING HIGH-GRADIENT PROGRADATION OF FLUVIAL SYSTEMS

G. H. EISBACHER,

Geological Survey of Canada, 100 West Pender Street, Vancouver, British Columbia, V6B 1R8

Vertical accretion of fluvial conglomerate and sandstone units is common in tectonically active and rapidly subsiding basins. In such basins it is common to find coal or wind blown tuff interlayered with coarse conglomeratic deposits. Simple vertical sequences of grain size and sedimentary structures are difficult to establish. Upward-coarsening and upward-fining cycles coexist. The most characteristic features of these progradational fluvial successions are sole marks similar to flute casts which seem to have formed when pulses of coarse material scoured into finer deposits of lowland swamps. Examples of this type of environment are known in the Eocene and Oligocene of the northwestern Canadian Cordillera. Recent examples should be expected in areas where swampy lowlands are close to tectonically active mountain fronts and where alluvial fans cannot develop due to rapid subsidence.

UPPER DEVONIAN OF SOUTH-EAST IRELAND — ANATOMY OF A CONSTRICTED ALLUVIAL FAN

P. R. R. GARDINER,

Geological Survey of Ireland, 14 Hume Street, Dublin 2, Ireland

Three dimensional cross-sections permit detailed evaluation of this localised sand-deficient alluvial fan sequence. In bulk terms three valley fill episodes are recognised. Initially an 8 km wide NE-trending feeder canyon, some 150 m deep, was partially plugged by proximal gravelly material deposited by debris flows and sheet floods. This resulted in a broader, shallower canyon profile. Plugging was completed by conglomerates dominantly deposited by sheet and streamflood action.

Overstep onto lateral divides resulted in the third phase valley fill on a larger scale (> 20 km width) in the context of a stable distal fan - flood plain setting. Fine grained members, in part possibly wind transported, comprise some 50% of this 200+ m thick fill, with immature caliche profiles locally developed. Interleaved gravel deposits occur as thick braid bars, sheetflood crevasse splay units, and low sinuosity channel fills from ephemeral rivers. Low palaeocurrent variance and facies associations are considered suggestive of continued lateral constriction. Localised ponds with associated vegetation are interpreted to have been present in the central parts of the fan fringe.

The geometry and facies of this sequence is contrasted with other known alluvial fan models. The constricted model recognised here may prove to be relatively common in post-orogenic alluvial retreat situations.

BED FORMS AND STRATIFICATION TYPES OF MODERN GRAVEL MEANDER LOBES, NUECES RIVER, TEXAS

THOMAS C. GUSTAVSON,

Bureau of Economic Geology, The University of Texas at Austin , Texas 78712

All major streams draining the southwestern flank of the Edwards Plateau in south-central Texas transport large volumes of gravel and sandy, muddy gravel and are developing meander lobe sequences consisting predominantly of coarse gravel. The largest of these streams, the Nueces River, has a sinuosity index of 1.3 and a water surface slope of 1.8 m/km in the study area. Stream discharge is variable and has ranged from no flow to more than 17,000 m³/s.

Mean clast b-axis length for the 10 largest clasts at 13 sample sites ranged from 2.5 to 10.8 cm. Velocities of 2.7 to 4.4 m/s 1 meter above the streambed are required to transport these clasts. Stream velocities of these magnitudes occur about once in 8 years when discharge of the Nueces River exceeds 3,000 m³/s. Mean grain size of Nueces River alluvium ranges from 1.2 to 3.4 cm. At a flow depth of 1 m sediment of this size has a critical erosion velocity of 1.8 to 3 m/s. Velocities of this magnitude occur about once in 2 years when discharge exceeds 340 m³/s. Under these conditions flow is subcritical, with critical shear stresses on depositional surfaces ranging from 6.4 to 12.7 kg/m².

Gravel clasts are imbricated and channel bed forms are predominantly transverse gravel bars with slip faces ranging up to 2 m high and wavelengths in excess of 100 m. Stratification includes graded planar crossbeds and horizontal beds. Lower lateral accretion face sediments are also predominantly transverse bars; upper lateral accretion face deposits occur as longitudinal gravel ridges deposited in the lee of vegetation and, less commonly, as chute bars. Near the upper limit of meander lobes where vegetation is heavy, mud and muddy sand occur as overbank deposits; in these deposits sedimentary structures other than desiccation cracks are rare.

Sedimentary sequences in gravel meander lobe systems deposited by low sinuosity streams are graded or nongraded horizontal beds and planar crossbeds overlain by mud and muddy sand interbedded with horizontally bedded gravels. These deposits in turn are overlain by overbank deposits of mud and muddy sand. Similar sedimentary sequences occur in the extensive Quaternary terraces that parallel the Nueces River.

FLUVIAL MODELS IN COAL AND HYDROCARBON EXPLORATION

JOHN C. HORNE,

Department of Geology, University of South Carolina, Columbia, South Carolina 29208

RAM S. SAXENA,

The Superior Oil Company, 700 Saratoga Building, New Orleans, Louisiana 70112

Channel and levee deposits are important components of fluvial-deltaic sequences and are economic factors in the exploration and exploitation of oil and gas in the Tertiary deposits of the Gulf Coast and of coal in the Carboniferous of the Appalachians.

In the transitional zone between lower and upper delta plain sequences of the Tertiary deposits, infrequent, laterally migrating, meandering channels developed. Fine-grained point bar deposits accumulated in single-storied sequences within these channels. Associated with these deposits are brackish to marine shales and siltstones with good source quality organic matter, and because of this, these channels become reservoir targets.

However, producing hydrocarbons from fine-grained reservoirs can be difficult. Within the point bar sequences individual sandstone beds may be separated and sealed by thin shale beds. Selected multiple perforations must be made for the individual sandstone lenses or much of the hydrocarbons in the reservoir will be left untapped.

In the Carboniferous, channels that occurred directly over a coal often scoured downard onto the coal reducing its thickness and, in some instances cut it out entirely. The sandstones of channel deposits that rest directly on coals can provide excellent roof rock in mines, but when slump blocks that accumulate most abundantly on the "cut bank" side of channels are encountered, severe roof problems develop.

Levee deposits can cause splits in coal seams that reduce coal thickness below minable minimums and increase the percent rejects above tolerable limits. Where extensively rooted levees occur directly over a minable coal, severe roof problems are encountered.

Because of their effect on economic parameters of hydrocarbons, channel and levee deposits must be considered during the exploration phase with respect to drill hole spacing in order to delineate these features. In addition, these deposits must be taken into account during the exploitation phase to provide the maximum development of reservoir potential for oil and gas and during the mine planning and development for coal because of potential roof problems they may cause.

GENESIS OF FLUVIAL BEDFORMS

ROSCOE G. JACKSON II,

Department of Geological Sciences, Northwestern University, Evanston, Illinois 60201

Bedform genesis is doubly germane to fluvial sedimentology. Not only do streams display a complex variety of bedforms, despite simpler gross-flow conditions than in other environments, but bedforms are also an important component of existing facies models. This paper presents some recent findings on the origin of fluvial bedforms.

Three classes of bedforms are defined on the basis of formative factors and physical scales. The class of largest bedforms, macroforms, includes those bedforms which do not depend upon local flow conditions but which instead respond to long-term hydrological and geomorphological factors. Fluvial macroforms include alternate bars, point bars, scroll bars, and pool-and-riffle sequences. Their wavelength scales with stream width. Mesoforms include those bedforms that respond directly to flow conditions of the outer zone of the local boundary layer of the flow system. Their wavelength scales with boundary-layer thickness (= flow depth in streams), and their lifetimes are commensurate with the duration of individual hydrological events. Fluvial mesoforms include dunes, Coleman's large-scale lineation, and antidunes. The smallest bedforms, microforms, respond to flow conditions of the inner zone of the turbulent boundary layer and thus scale with wall variables. Fluvial microforms include current lineation and small scale ripples.

The conceptual simplicity of this classification is realized incompletely in fluvial streams, owing to the poorly understood bedforms variously referred to as braid bars, linguoid bars, spool bars, unit bars, etc. Theoretical models of the braiding mechanism suggest that these "braid" bars are macroforms, but observations of their behavior in natural streams show mesoformal and macroformal traits.

A new approach to bedform genesis involves the concept of flow structures. Recent laboratory experiments indicate that turbulent shear flows contain one or more inherent structures, each consisting of a more-or-less deterministic secondary flow superimposed upon the prevailing unidirectional mean flow. The main microscale structure is the wall streaks of the turbulent boundary layer, which govern the formation of microforms. Mesoscale structures include the bursting phenomenon, longitudinal roll vortices, in-phase waves, and possibly transverse roll vortices. Wavelength of mesoscale flow structures is proportional to boundary-layer thickness. A speculative explanation of the well documented sequence of bedforms with increasing flow regime involves the successive domination of microscale structure (producing ripples), bursting (dunes), longitudinal roll vortices (upper flat bed and large-scale lineation), and in-phase waves (antidunes).

Other flow structures, possibly width dependent, are associated with macroforms. Examples include spiral motion in bends, which maintains point bars, and obscure long-period oscillations in flow velocity, which may be associated with alternate bars and initiation of meandering.

There remains a critical need to test further these proposals.

EVOLUTION OF SOURCE-DISTAL TO SOURCE-PROXIMAL FLUVIAL REGIMES IN THE HIMALAYAN MOLASSE

GARY D. JOHNSON,

Department of Earth Sciences, Dartmouth College, Hanover, New Hampshire 03755

The record of Himalayan tectonism is preserved in a 7000 m sequence of molasse represented by the Neogene and early Quaternary Siwalik Group. Fining upwards sedimentary cycles from strata of the Upper Siwalik sub-Group generally reflect the transition from streams having high sinuosity, well-developed meander-belt structure typical of much of the Siwalik Group, to low-sinuosity and braided stream regimes during the late Pliocene and early Pleistocene (ca. 2.5 m.y. to 0.5 m.y. B.P.). In significant fashion, this transition in sedimentary style of the Himalayan molasse is not an isochronous event but rather a variable response to the morphogenesis of the adjacent schuppenstruktur of the Outer Himalaya, syndepositional deformation of the Himalayan foredeep margin and proximity to major antecedent stream courses which flow from these regions into the foredeep basin.

Constraints on the chronology of these events is provided by a magnetic polarity reversal stratigraphy which is established for the Upper Siwalik sub-Group. This is coupled with radiometric age determination of several volcanic ashes. The events discussed cover the entire Pliocene and Pleistocene record of the evolution of this fluvial molasse in a region of northern Pakistan.

A FLUME STUDY OF FLUVIAL GRAVEL FABRIC

EMLYN H. KOSTER,

Department of Geological Sciences, University of Saskatchewan, Saskatoon, Saskatchewan S7N 0W0

Gravel-sized sediment has not shared in the rapid expansion of experimental sedimentology since 1960. In terms of the sedimentary structure hierarchy, it is considered that 'rank 6', and to a lesser extent 'rank 5', structures embrace the possible range of conglomerate features amenable to a laboratory study. Gravel fabric appears to be a more profitable line of inquiry in view of its common usage in paleocurrent studies, and because of the low preservation potential and/or difficulty in recognising small-scale bedforms.

The laboratory flows were steady-state and spanned the lower- and upper-flow regimes. Clasts isolated on a sandbed undergo combined rotation and translation in response to current-crescent scour. In general, motion is dependent both on the degree of form roughness associated with migrating sand bedforms as well as on the geometric properties of the clasts. Imbrication to moderate angles and current-normal orientation are the dominant responses. Optimum stability of imbricated clasts is attained at the regime transition where flows best approach a steady and two-dimensional character, and is described by a simple power function of projection area versus nominal diameter. The ellipsoidal tendency of the clasts was utilised in the calculation of projection areas.

Since the sandbed co-existing with gravel deposition would likely be in the transition or upper-flow regime, where a stable clast configuration is rapidly attained in response to vigorous current-crescent scour, it follows that the 'equilibrium area' concept is a potentially useful tool for the investigation of matrix-supported gravels.

DEPOSITS OF FINE GRAINED MEANDERING RIVERS, WITH LARGE DISCHARGE VARIATIONS, IN THE CARBONIFEROUS OF THE MARITIME PROVINCES, CANADA

PETER J. McCABE,

Department of Geology, University of Nebraska, Lincoln, Nebraska 68588

A series of fining-upward sequences, up to 11 m thick and interpreted as point bar deposits, occur in the lower part of the Shepody Formation (mid-Carboniferous) in the Chignecto Bay area of New Brunswick and Nova Scotia. The sandstones are predominantly of fine to very fine sand grade but have a maximum grain size of medium sand grade. Plane parallel lamination and ripple lamination are the dominant sedimentary structures in the sandstones. The sandstones pass upward into siltstone with climbing ripple or parallel lamination and clay mudrock. Small, steep-sided channels, infilled with rippled sandstones and siltstones, cut into the upper parts of the fining-upward sequences. These are interpreted as chute-fill deposits. Desiccation features occur in the upper third of the fining-upward sequences and in the chute channels. The presence of chute channels in such a fine-grained sequence and the desiccation structures suggest rivers with considerable stage fluctuations.

In the underlying Maringouin Formation, fining-upward sequences are similar to those in the Shepody but are thinner (up to 5 m) and finer (virtually all very fine sandstone or finer). Ripple lamination is the dominant sedimentary structure. Desiccation features are seen throughout most of the fining-upward sequences and it is probable that many of the river channels were ephemeral.

BEDFORMS FROM ABANDONED FLUVIAL CHANNELS, CARBONIFEROUS, MARITIME PROVINCES

PETER J. McCABE,

Department of Geology, University of Nebraska,Lincoln, Nebraska 68588

Fining-upward fluvial sequences in the upper Shepody and Boss Point Formations (mid-Carboniferous in the Chignecto Bay area of New Brunswick and Nova Scotia) are in the order of 25 m thick. Several sequences are incomplete with the fining-upward sequence truncated by a mudrock bed. The truncation of these sequences was probably produced by abandonment of the channel by meander neck cutoff or avulsion. In a few places the mudrock drape has been removed leaving surfaces, up to 300 m^2 in area, covered with bedforms present in the river when it was abandoned. Preservation of the bedforms is excellent and even individual avalanche lobes can be seen.

Ripples, sandwaves, dunes and longitudinal bars have been recognized. The larger bedforms show evidence of falling stage modification: small dunes are washed out; ripples in the troughs of larger bedforms indicate a low stage runoff; and the crestlines of some sandwaves and longitudinal bars have been rounded or modified by minor lobes. The preservation of the bedforms allows a far better understanding of the stratification within the fining upward sequences than would otherwise be possible.

EPSILON CROSS-STRATA IN THE ATHABASCA OIL SANDS

GRANT D. MOSSOP,

Alberta Research Council, 11315 - 87 Avenue, Edmonton, Alberta T6G 2C2

The Lower Cretaceous McMurray Formation oil sands contain solitary sets of cross-strata up to 25 metres thick, with cross-stratal dips of 8 to 12 degrees. In all essential regards, these sets conform to Allen's definition of epsilon cross-strata.

The principal characteristics of a typical set are as follows: the lower bounding surface is erosional, and on the scale of a single exposure is essentially planar and horizontal; the cross-strata, which discordantly overlie the basal surface, consist of decimetre to metre thick beds of fine sand, separated by thin partings of argillaceous silt; individual beds are remarkably continuous and uniform from the base to the top of the set and, in sections normal to depositional strike, they characteristically exhibit a straight line profile; subtle fining-upwards within the set is manifest largely by an upwards increase in the proportion of silt and clay rather than as a decrease in mean sand size; depositional strike is generally consistent in any given exposure, but adjacent outcrops commonly show wide divergence in attitude.

Single beds within an epsilon set characteristically exhibit the following sequence: on a sharp base lie essentially structureless sands, followed upwards by trough cross-bedded sands (only sporadically developed) overlain by ripple-bedded sands, grading into laminated silt and clay. Preliminary paleocurrent analysis on the trough sets and the ripples indicates unidirectional flow approximately parallel to depositional strike. Common in the silt partings, and characteristically penetrating down into the underlying sands, are tubular walled burrows up to 5 mm in diameter, exhibiting a form very similar to polychaete burrows found in the fresh water portions of some present-day estuaries. Microflora within the silts is also of fresh water origin.

Details of the depositional environment of the epsilon cross-strata have yet to be established, but it is proposed that they originated as lateral accretion deposits, laid down on the sloping inner banks or point bars of deep sinuous channels. The falling flow regime indicated in the sands suggests that sand deposition may have taken place only during waning of the flood stage, with fines accumulating during the remainder of the cycle.

COMPARISON OF LABORATORY AND NATURAL MEANDERING

GARY PARKER,

Department of Civil Engineering, University of Alberta, Edmonton, Alberta T6G 2E3

Both the processes of meander initiation in laboratory flumes, and the maintenance of bed geometry in flumes with imposed meandering banks bear a close resemblance to the field situation. However, it has not proved possible to model, for example, the process of formation, cutoff, and reformation of large loops in the laboratory. Clearly laboratory meandering provides only a partial model of the field case. A lack of awareness of the extent of correspondence between the two has led to applications of laboratory results that are in some cases unduly broad and in some cases unduly narrow. For example, a major difference between natural sand-bed rivers and laboratory meandering in self-formed sand channels is the fact that no suspension can occur in the laboratory case, ruling out vertical accretion on the floodplain and bar. This fact has important implications as regards to the inability of laboratory streams to form coherent meanders of large amplitude. On the other hand, the correspondence between natural meandering in coarse gravel and laboratory meandering in coarse sand, where suspension is precluded in both cases, is perhaps better than is generally recognized. It can be demonstrated that the concept of pseudomeandering, long used to distinguish the laboratory and field cases, must apply to field meandering in coarse gravel, making the concept essentially meaningless.

SEDIMENTOLOGY OF A PALEOVALLEY FILL: PENNSYLVANIAN KYROCK SANDSTONE IN CENTRAL KENTUCKY, U.S.A.

WAYNE A. PRYOR,

Geology Department, University of Cincinnati, Cincinnati, Ohio 45221

What are the characteristics of the fill of a basal Pennsylvanian paleovalley in central Kentucky, how homogenous is it, what was its depositional environment, and how was it connected, if at all, with the Appalachian Basin to the east?

The paleovalley is about 200 feet (66 m) deep, about 3 miles (5 km) wide and has several sharply defined benches along its sides. Its fill is almost entirely sandstone and consists of two major sandstone bodies: a lower medium to coarse grained, crossbedded body and an upper, fine to medium grained crossbedded, but less homogenous body that contains some shale. The lower sandstone is interpreted to have been deposited by a low sinuosity braided stream; the upper unit was deposited by a series of small high sinuosity meandering streams 20 to 30 feet (6 to 10 m) deep. The upper unit has more variable paleocurrents and extends somewhat beyond the limits of the erosional valley.

The internal facies of the sandstone fill suggest deposition by a stream of perhaps about the size of the Wabash River (in southwestern Indiana) on an alluvial plain in a tropical climate. The change from braided to meandering pattern is probably related to lower gradients as alluviation encroached up paleoslope into the Illinois Basin in early Pennsylvanian time. There are several different ways to connect this paleovalley to the Appalachian Basin, none of which can be well established. East of the study area the sandstone becomes totally decemented and only scattered quartz pebbles and cobbles mark its former presence.

FLUVIATILE DEPOSITS OF THE ATHABASCA FORMATION OF NORTHERN SASKATCHEWAN

PAUL RAMAEKERS,

Saskatchewan Geological Survey, 201 Dewdney Avenue East, Regina, Saskatchewan S4N 4G3

Analysis of over 3000 paleocurrent measurements in association with sedimentological data and measured sections indicate the presence of several deposystems in the eastern and central Athabasca Basin. The eastern systems derive from the Wollaston Fold Belt and form broad fans with coarse gravels at their apices that grade into sands within a short distance. Silts and clays are notably scarce. Thickness of the sedimentary cycles, variance in current direction, and the type and frequency of sedimentary structures vary down-current. The sequence is interpreted as a group of broad, coalescing alluvial fans grading into a plain traversed by shallow braided streams and subject to strong eolian influence.

PARTICLE SIZE SORTING IN SUBGLACIAL ESKERS

HOUSTON C. SAUNDERSON,

Department of Geography, Wilfrid Laurier University, Waterloo, Ontario N2L 3C5

Glaciofluvial deposits in subglacial eskers consist of two groups, those deposited during (1) free-surface flow and (2) full-pipe flow. Hydraulic sorting during free-surface flow is no different from that in open-channel eskers, and consequently structural assemblages are similar. Under full-pipe flow there are only minor differences in primary structures and size sorting in sand sizes. Thus, ripple, dune and plane bed attributes are found in both supraglacial and subglacial environments during free-surface flow and during subglacial, full-pipe flow.

Near and above the limit deposit condition, at Froude numbers ranging from about 3 to 10, particle sorting becomes distinctively different under full-pipe flow than in free-surface flow. At a velocity slightly less than the limit deposit velocity, the whole bed slides along the invert of a pipe. Inside this sliding bed, particle sorting is probably restricted to that produced by particle collisions and dilation of the sediment mix caused by strong seepage flow through the bed. At flow velocities higher than the limit deposit velocity, heterogeneous and homogeneous suspension occur, but all of the sediment in the pipe is in suspension and no bed forms can be preserved. A poorly sorted sand and gravel facies of probable sliding bed origin is abundant in several eskers near Guelph and Orangeville, Southern Ontario. Beyond the inferred exits of these eskers, more effective size sorting has produced well-defined bedding in fans and deltas.

EXPERIMENTAL STUDIES OF FLUVIAL PLACERS

S. A. SCHUMM,

Department of Earth Resources, Colorado State University, Fort Collins, Colorado 80523

Experimental studies of channel incision reveal that heavy minerals are concentrated on the bedrock floor of valleys by reworking of valleyfill sediments. Inner channels formed by scour into bedrock are especially favorable locations. These deeper parts of the valley floor are located at both the inside and outside of valley bends, depending on the nature of the sediment load. Pauses in deposition and renewed scour permit heavy mineral concentrations to form within the valley-fill.

Arid alluvial fans rarely contain economic concentrations of heavy minerals, and yet the large fluvial fans of Ghana and South Africa contain important gold concentrations. During experimental studies of the growth of a fluvial or wet fan under conditions of perennial flow, fan-head trenching occurred repeatedly, and this reworking of the fan-head deposits plus rejuvenation of the source area streams remobilized and concentrated heavy minerals in the fan head and mid-fan. This process was not influenced by external variables but rather it reflected the natural morphologic and sedimentologic development of a fluvial fan.

Valley-fill, bedrock and fluvial fan placers may in many instances reflect the behavior of a complex geomorphic system, rather than the influence of external variables.

DOWNSTREAM GRAIN-SIZE CHANGES IN ALBERTAN RIVERS

JOHN SHAW,

Department of Geography, University of Alberta, Edmonton, Alberta T6G 2E3

R. KELLERHALS,

2533 Wallace Crescent, Vancouver, British Columbia V6R 3V3

Detailed lithologic and grain-size analyses of 174 samples collected along 12 rivers in Alberta were carried out by L. B. Halferdahl. This set of data is used to illustrate downstream changes in the bed material of these rivers. Each of the major rivers can be sub-divided into three reaches on the basis of downstream change in grain-size. The Mountain reaches show highly variable grain-size characteristics with a tendency to increasing size downstream. Through the Foothills and Western Plains plots of grain-size, D, against distance, x, follow the relationship $D = D_0 e^{-ax}$. Where D_0 is the grain-size at $x = 0$ and a is a diminution coefficient. D and D_0 are given as D_{50} or D_{90} obtained from cumulative grain-size curves. Bed material is predominantly gravel in the above two reaches. At the beginning of the third reach the rivers become sand bedded and there is no significant decrease in grain-size with distance downstream.

The reach characteristics can be explained qualitatively in terms of their geomorphic history. The Mountain reaches include numerous lakes and infilled lakes of glacial origin. These act as sediment traps and cause the burial of coarse material. In addition much of the present supply is from periglacial environments in which the bedrock is heavily fractured. The Foothills and Western Plains reaches have been degrading since deglaciation and are expected to show grain-size changes according to Sternberg's relationship. The Eastern Plains reaches were affected by isostatic adjustment to the Wisconsin, Laurentide Ice Sheet. Simplifying assumptions of the isostatic effect on stream slopes leads to the conclusion that, following initial degradation, these streams should now be aggrading. Aggradation accounts for the abrupt change from gravel-bedded to sand-bedded streams.

Abrasion coefficients from laboratory experiments and diminution coefficients from rivers are compared. Laboratory experiments greatly underestimate diminution coefficients for rivers, and aggrading rivers show higher diminution coefficients than degrading rivers. The diminution coefficient, a, can be divided into three components: a_t, the component of abrasion during transport; a_v, the component of *in situ* abrasion; and B, the component of differential transport. The value of the diminution coefficient for limestone is obtained using the bed material samples, and also using the diminution coefficient for quartzite and an experimentally derived ratio of the abrasion coefficients for quartzite and limestone. The second derivation of the diminution coefficient for limestone includes an assumption that $B = 0$. As the two coefficients are similar this assumption is accepted, and the ratio of abrasion in transport to abrasion *in situ* is calculated for quartzite as

$$\frac{a_t}{a_v} = 0.1$$

This result explains why laboratory experiments, which do not include a_v, greatly underestimate abrasion coefficients in rivers.

SCALE MODELING OF BED FORMS

JOHN B. SOUTHARD AND LAWRENCE A. BOGUCHWAL

Department of Earth and Planetary Sciences, Massachusetts Institute of Technology, Cambridge, Massachusetts 02139

Experiments were made in a small flume to test the validity of scale-model parameters describing transport of loose sediment. Using seven variables to characterize flow and sediment transport (mean flow depth d, mean flow velocity U, fluid density ρ and viscosity μ, sediment size D and density ρ_s, and gravity g), dimensional analysis provides four model parameters: Reynolds number $\rho U d / \mu$, Froude number $U/(gd)^{0.5}$, size ratio d/D, and density ratio ρ_s/ρ. Two scaled runs were made in the ripple regime at a scale ratio of about 1.7 to 1 using sand and water with appropriate grain sizes and viscosities. Frequency distributions of ripple height, spacing, and migration rate were determined for 100 ripples; the test for scaling is whether the two curves for each of these variables coincide when the appropriate scaling factor is applied. In each case the curves are closely coincident, indicating that scale modeling of sediment transport by variation of grain size and water temperature is feasible. This technique should be valuable in studying bed forms and transport rates in small to medium-sized rivers by means of scaled hot-water runs in laboratory flumes.

A pilot study was made in an insulated flume 0.9 m wide and 11 m long with water temperatures up to 80°C (providing a scale model ratio of up to 2.3) with sand sizes ranging from effectively 0.11 mm to 0.30 mm and a flow depth of up to effectively 0.5 m. Sand waves are present at flow velocities intermediate between those for ripples and for dunes, with heights comparable to those of ripples but with significantly longer spacings. Dune spacing increases with increasing flow velocity; dune height increases with increasing flow velocity, then decreases as velocities for upper flat bed are approached. The velocity interval for stable sand waves and dunes decreases with decreasing sand size, and disappears at about 0.11 mm.

DENSITY SORTING OF SAND-SIZE SPHERES DURING TRACTION TRANSPORT: AN EXPERIMENTAL STUDY

JAMES R. STEIDTMANN,

Department of Geology, University of Wyoming, Laramie, Wyoming 82071

It is well known that sediment sorting according to size, shape and density takes place but it is not known just how it takes place. To assess the effects of size and density sorting, sand-size spheres of different density were transported under controlled flume conditions. Observations on the motion of discrete grains show that grains smaller than the roughness elements move continuously and have the same velocities regardless of density. For grains near and slightly larger than the roughness, movement is intermittent and, for a given size, heavy particles move more slowly than lights. For grains much larger than the roughness, movement is once again continuous over the rough surface and light and heavy grains move at nearly the same velocity.

Bulk samples containing spheres of a range of sizes and two densities were also studied. Analyses of grain size for the two densities show that for planebed transport, the size of heavies decreased with respect to that of lights with distance transported. For ripple-bed transport however, the size relations between associated light and heavy grains remained essentially unchanged for the duration of transport. These results indicate that the specific process of sediment sorting is a function of the exact mechanism of sediment motion.

ANATOMY OF AN ARKANSAS RIVER SAND BAR

RICHARD STEINMETZ,

Amoco Production Company, P.O. Box 591, Tulsa, Oklahoma 74102 U.S.A.

Detailed relationships between cross-bed measurements and shifting channel patterns will be demonstrated, and grain size analyses showing a wide variety of distributions will be presented.

The sand bar is located in the Arkansas River valley approximately 10 miles upstream from Tulsa, Oklahoma. From aerial photo sequences, plus discharge and river stage records, it was determined that the entire sand bar (600,000 cubic yards) was deposited in 156 hours. Deposition occurred during two floods, May 19 to 22, 1957 (60 hours) and October 3 through 6, 1959 (96 hours).

The sand bar was studied in detail along a natural cubank (500 feet long) parallel to the valley axis and in a trench (700 feet long by 15 feet deep) dug perpendicular to the cutbank. Sand peels and box cores were taken, and cross-bed types were recorded. Closely spaced cross-bed measurements (N = 210) recorded at 12 vertical sections, and grain size analyses were performed on 210 samples from the same sections.

Results show that the highly variable patterns of cross-bed dips match the erratic and changing flow directions prevalent during flood stages. In some of the vertical sections, cross-bed dip directions are at all angles to the overall east-west orientation of the Arkansas River valley. Mean grain size ranges from 0.07ϕ (0.95 mm) to 4.44ϕ (0.05 mm), and standard deviation ranges from 0.26ϕ to 1.48ϕ. At most of the vertical sections, grain size distributions show no systematic change from bottom to top.

THE FORMULATION OF FLUVIAL FACIES MODELS: POSSIBLE "END MEMBERS" AND THE SPECTRUM OF TYPES BETWEEN END MEMBERS

ROGER G. WALKER,

Department of Geology, McMaster University, Hamilton, Ontario L8S 4M1

The wide range of different river types (meandering, straight, braided; coarse, fine) has spawned a wide range of possible facies models. The best known end member — the meandering fining-upward sequence — is itself a composite model involving facies sequences for active lateral accretion, neck cut-off, and chute cut-off. The philosophy and ultimate usefulness of "splitting" existing models will be examined in the lecture. Models for braided systems, especially the emerging models of sandy and gravelly systems, will be developed, compared, and contrasted. The possible spectrum of types from meandering to braided sandy systems will be emphasized, within the context of what a good model should do, namely 1) act as a *norm* for purposes of comparison, 2) act as a *guide* for future observations, 3) act as a *predictor* in new situations, and 4) act as a *basis* for an overall interpretation.